Bacterial Adhesion

MECHANISMS AND PHYSIOLOGICAL SIGNIFICANCE

Bacterial Adhesion

MECHANISMS AND PHYSIOLOGICAL SIGNIFICANCE

Edited by
DWAYNE C. SAVAGE
University of Illinois
Urbana, Illinois

and

MADILYN FLETCHER
University of Warwick
Coventry, England

PLENUM PRESS • NEW YORK AND LONDON

Library of Congress Cataloging in Publication Data

Main entry under title:

Bacterial adhesion.

Includes bibliographies and index.
1. Bacteria—Physiology. 2. Cell adhesion. I. Savage, Dwayne C. II. Fletcher, Madilyn M.
QR84.B15 1985   589.9′01   85-3434
ISBN 0-306-41941-6

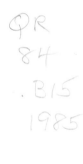

Contributors

Edwin H. Beachey, Veterans Administration Medical Center and Departments of Medicine and of Microbiology and Immunology, University of Tennessee Center for the Health Sciences, Memphis, Tennessee 38104

K.-J. Cheng, Department of Animal Science, Agriculture Canada Research Station, Lethbridge, Alberta T1J 4B1, Canada

Gordon D. Christensen, Veterans Administration Medical Center and Department of Medicine, University of Tennessee Center for the Health Sciences, Memphis, Tennessee 38104

J. William Costerton, Department of Biology, University of Calgary, Calgary, Alberta T2N 1N4, Canada

Madilyn Fletcher, Department of Environmental Sciences, University of Warwick, Coventry CV4 7AL, England

J. Gould, Department of Biochemistry, University of Cambridge, Cambridge CB2 1QW, England

Richard E. Isaacson, Department of Epidemiology, School of Public Health, University of Michigan, Ann Arbor, Michigan 48109. *Present address:* Pfizer Central Research, Groton, Connecticut 06340

S. Kjelleberg, Department of Marine Microbiology, University of Göteborg, S-413 19 Göteborg, Sweden

Daniel A. Kluepfel, Department of Plant Pathology, University of Florida, Gainesville, Florida 32611. *Present address:* Laboratory for Microbiology, Agricultural University, Wageningen, The Netherlands.

Halina Lis, Department of Biophysics, The Weizmann Institute of Science, Rehovot, Israel

George I. Loeb, Naval Research Laboratory, Washington, D.C. 20375. *Present address:* David Taylor Naval Ship R & D Center, Annapolis, Maryland 21402

Thomas J. Marrie, Department of Microbiology, Dalhousie University, Halifax, Nova Scotia B3H 1V8, Canada

Kevin C. Marshall, School of Microbiology, University of New South Wales, Kensington, New South Wales 2033, Australia

Ann G. Matthysse, Department of Biology, University of North Carolina, Chapel Hill, North Carolina 27514

D. H. Northcote, Department of Biochemistry, University of Cambridge, Cambridge CB2 1QW, England

Itzhak Ofek, Department of Human Microbiology, Sackler School of Medicine, Tel Aviv University, Tel Aviv, Israel

Hans W. Paerl, Institute of Marine Sciences, University of North Carolina, Morehead City, North Carolina 28557

Steven G. Pueppke, Department of Plant Pathology, University of Florida, Gainesville, Florida 32611. *Present address:* Department of Plant Pathology, University of Missouri, Columbia, Missouri 65211

Dwayne C. Savage, Department of Microbiology, University of Illinois, Urbana, Illinois 61801

Nathan Sharon, Department of Biophysics, The Weizmann Institute of Science, Rehovot, Israel

W. Andrew Simpson, Department of Microbiology and Immunology, University of Tennessee Center for the Health Sciences, Memphis, Tennessee 38104

G. Stotzky, Laboratory of Microbial Ecology, Department of Biology, New York University, New York, New York 10003

Anthony J. Wicken, School of Microbiology, University of New South Wales, Kensington, New South Wales 2033, Australia

Preface

Study of the phenomena of bacterial adhesion to surfaces has accelerated considerably over the past 10 to 15 years. During this period, microbiologists have become increasingly aware that attachment to a substratum influences considerably the activities and structures of microbial cells. Moreover, in many cases attached communities of cells have important effects on their substratum and the surrounding environment. Such phenomena are now known to be important in plant and animal hosts, water and soil ecosystems, and man-made structures and industrial processes.

Much work on microbial adhesion in the early 1970s was descriptive. Those studies were important for detecting and describing the phenomena of bacterial adhesion to substrata in various environments; the findings have been presented in numerous recently published, excellent books and reviews. In some studies, attempts were made to elucidate some fundamental principles controlling adhesion processes in different environments containing a variety of microorganisms. Common threads have been observed occasionally in different studies. Taken as a whole, however, the information has revealed that many disparate factors are involved in adhesion processes. Whether a particular microorganism can adhere to a certain substratum depends on the properties of the microbial strain itself and on characteristics of the substratum and of the environment. For example, the capacity of a bacterial strain to adhere to a substratum may depend on whether the cell is fimbriated, encapsulated, or a rough mutant; whether the substratum is a nonbiological or biological surface; and on the pH, temperature and ionic strength of the environment. Such conditions may affect adhesion directly, by influencing the physicochemistry of the process, or indirectly, by modifying the physiological processes of the bacteria. Moreover, when biological surfaces are involved, many additional, complex physiological factors can come into play.

When bacterial cells adhere to a given substratum, significant changes can take place in those cells and in the substratum. If the surface is a nonbiological one, then the attached cells may be deformed physically and undergo numerous poorly defined physiological changes. Concomitantly, products of the metabolism of the microorganisms may begin to alter the substratum, such as by dissolving pits in it. Such phenomena may be particularly prominent when the substratum to which the organism is attached contains, or is, its nutritional substrate. When the substratum is a biological one, such as a plant or animal surface, then the attached microbial cells may not only undergo physiological changes themselves, but may secrete substances or be involved in processes, such as those involving genetic or invasive activities, that may have profound consequences for the host. Thus, understanding of microbial adhesion processes must go beyond mere description of the processes themselves and extend to sophisticated study of their mechanisms and consequences.

Such research must be concerned with the subcellular mechanisms by which the bacterial cells adhere to the substratum and, in many cases, the genetics of the process in the adhering cells. Investigators must guard, however, against confusing mechanisms in separate adhesive events. A danger exists for such misinterpretation because of the chemical and structural heterogeneity of bacterial populations and the bacterial surface itself.

The purpose of this book is to examine in one volume what is presently known about the physicochemical and molecular bases of the processes of bacterial adhesion to surfaces, and the influences of attached bacterial communities on their substrata. In most chapters, the emphasis is on the results of experimental studies of various aspects of such processes.

The book is divided into three sections. The first, introductory section begins with a description of some of the phenomena of bacteria adhering to various substrata in different environments (Chapter 1). This is followed by descriptions of the structural and chemical properties of the surfaces involved in the adhesion processes, i.e., bacterial (Chapter 2), animal (Chapter 3), plant (Chapter 4), and nonbiological (Chapter 5) surfaces.

The second section is concerned with the mechanisms by which bacteria adhere to surfaces. The first three chapters deal with adhesive interactions at nonbiological interfaces in different environments, i.e., solid–water (Chapter 6) and air–water (Chapter 7) interfaces, and soil particle surfaces (Chapter 8). The next three chapters deal with bacterial adhesion to surfaces of higher organisms, i.e., plant–bacterium (Chapter 9) and animal–bacterium interactions. The latter topic is developed in two chapters, one on adhesive interactions involving complex macromolecular mechanisms (Chapter 10), and one on interactions known to be mediated by pili (Chapter 11).

The third section deals with the consequences of the phenomena of bacteria adhering to surfaces. The consequences for the bacterial cells themselves are discussed in two chapters. In Chapter 12, emphasis is placed on how the physiological activity of a bacterium can change when its cells are in proximity to a surface. Chapter 13 discusses how a surface can influence the properties of an attached bacterium in an extremely complex way when the surface is another organism which can act as a source of nutrients or inhibitors. The final two chapters deal with how bacteria attached to surfaces of plants (Chapter 14) and animals (Chapter 15) may influence their hosts, such as by invading tissues, inducing resistance responses and otherwise altering host physiology, sometimes to the advantage, but often to the disadvantage of the animal or plant involved.

A reading of these chapters will reveal readily that bacteria adhere to surfaces by a variety of mechanisms and that the consequences of such adhesion may be many and varied. It is seen that, at this time, it is generally not possible to elucidate fundamental processes underlying the phenomena. The chapters reveal, as well, however, that methods now exist for elucidating the mechanisms at the molecular level. The reader is encouraged to look for similarities in biochemical interactions or responses in various ecosystems, which may offer clues to underlying and dominant molecular mechanisms or the evolutionary history of interactions involving adhesion. We hope that this book will prove to be useful in the planning and execution of future experimental attempts to understand the mechanisms of bacterial adhesion and their physiological consequences. We will be especially satisfied should the book prove to be a strong stimulant for further research on these important subjects.

We wish to express our great appreciation to the authors of the chapters who labored so hard and well to produce their fine manuscripts. We also thank them for their patience during the long editorial process. We wish as well to express gratitude to Mr. Kirk Jensen, who inspired us to edit this book.

Dwayne C. Savage
Madilyn Fletcher

Contents

I. Introduction and Description of Surfaces

Chapter 1
Phenomena of Bacterial Adhesion
J. William Costerton, Thomas J. Marrie, and K.-J. Cheng

Chapter 2
Bacterial Cell Walls and Surfaces
Anthony J. Wicken

Chapter 3
Animal Cell Surface Membranes
Itzhak Ofek, Halina Lis, and Nathan Sharon

Chapter 4
Characteristics of Plant Surfaces
J. Gould and D. H. Northcote

Chapter 5
The Properties of Nonbiological Surfaces and Their Characterization
George I. Loeb

II. Mechanisms of Adhesion

Chapter 6
Mechanisms of Bacterial Adhesion at Solid–Water Interfaces
Kevin C. Marshall

Chapter 7
Mechanisms of Bacterial Adhesion at Gas–Liquid Interfaces
S. Kjelleberg

Chapter 8
Mechanisms of Adhesion to Clays, with Reference to Soil Systems
G. Stotzky

Chapter 9
Mechanisms of Bacterial Adhesion to Plant Surfaces
Ann G. Matthysse

Chapter 10
Adhesion of Bacteria to Animal Tissues: Complex Mechanisms
Gordon D. Christensen, W. Andrew Simpson, and Edwin H. Beachey

Chapter 11
Pilus Adhesins
Richard E. Isaacson

III. Consequences of Adhesion

Chapter 12
Effect of Solid Surfaces on the Activity of Attached Bacteria
Madilyn Fletcher

Chapter 13
Influence of Attachment on Microbial Metabolism and Growth in Aquatic Ecosystems
Hans W. Paerl

Chapter 14
Responses of Plant Cells to Adsorbed Bacteria
Steven G. Pueppke and Daniel A. Kluepfel

Chapter 15
Effects on Host Animals of Bacteria Adhering to Epithelial Surfaces
Dwayne C. Savage

I

Introduction and Description of Surfaces

1

Phenomena of Bacterial Adhesion

J. WILLIAM COSTERTON, THOMAS J. MARRIE, and K.-J. CHENG

1. THE SESSILE MODE OF BACTERIAL GROWTH

When we use direct microscopic techniques to examine bacteria growing in natural and pathogenic ecosystems, we are forcibly struck by the profound differences between these organisms and cells within derived *in vitro* laboratory cultures. The development of a series of new techniques for the stabilization and visualization of bacterial surface structures has intensified these differences because, as we come to understand bacterial surface structures more fully, we see that they comprise a very complex and fluid interface with the environment which the bacterium must maintain in order to survive amid the ubiquitous antibacterial agents and factors in nature.

1.1. Bacterial Glycocalyx *in Vivo* and *in Vitro*

The bacterial glycocalyx is defined (Costerton *et al.*, 1981a,b) as those polysaccharide-containing structures of bacterial origin, lying outside the integral elements of the outer membrane of gram-negative cells and the peptidoglycan of gram-positive cells. Glycocalyces are subdivided into two types:

1. S layers composed of a regular array of glycoprotein subunits at the cell surface, as described by Sleytr (1978)
2. Capsules, composed of a fibrous matrix at the cell surface, that may vary in thickness and may accurately be described by the following nonexclusive descriptors: (a) rigid—a capsule sufficiently structurally coherent to exclude particles (e.g., India ink, nigrosin); (b) flexible—a capsule sufficiently deformable that it does not exclude particles; (c) integral—a capsule that is normally intimately associated with the cell surface; (d) peripheral—a capsule that may remain associated with the cell in some circumstances and may be shed into the menstruum in others.

J. William Costerton • Department of Biology, University of Calgary, Calgary, Alberta T2N 1N4, Canada. Thomas J. Marrie • Department of Microbiology, Dalhousie University, Halifax, Nova Scotia B3H 1V8, Canada.   K.-J. Cheng • Department of Animal Science, Agriculture Canada Research Station, Lethbridge, Alberta T1J 4B1, Canada.

Because the polysaccharides of the bacterial glycocalyx do not attract the heavy metal stains routinely used in electron microscopy (EM), this structure was not effectively visualized in early EM preparations (Fig. 1). When polyanion-specific stains such as ruthenium red (Luft, 1971) and alcian blue were developed, bacterial glycocalyces were seen as con-

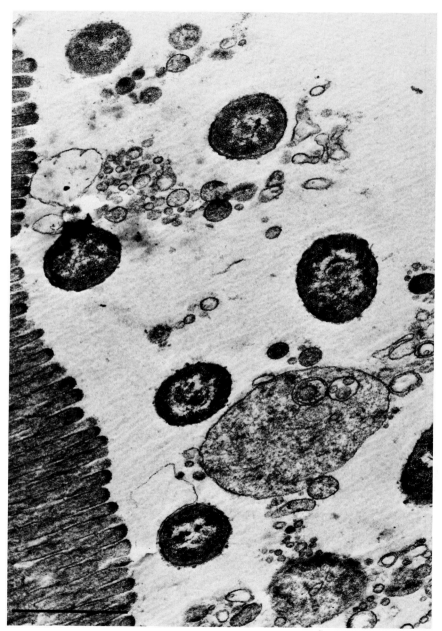

Figure 1. Transmission electron micrograph (TEM) showing cells of a fimbriated enterotoxigenic strain of *Escherichia coli* on the neonatal calf ileum. The glycocalyx surrounding these bacterial cells is not stained by conventional EM stains, and this structure is not detected. The bar in this and subsequent micrographs indicates 1.0 μm.

densed electron-dense accretions on the bacterial cell surface (Fig. 2), because the dehydration required for EM virtually destroyed these fine polysaccharide matrices, which are 99% water in their natural hydrated state (Sutherland, 1977). Modern techniques of glycocalyx stabilization using lectins (Birdsell *et al.*, 1975) and specific antibodies (Mackie *et al.*, 1979; Chan *et al.*, 1982a) preserve much of the finely fibrillar bacterial glycocalyx in its true dimensions (Fig. 3). These new methods allow us to describe the real surfaces of bacteria as

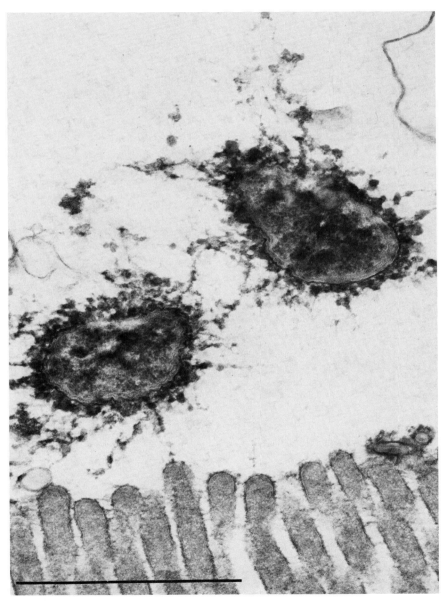

Figure 2. TEM of a ruthenium red-stained preparation of a fimbriated enterotoxigenic strain of *E. coli* on the neonatal calf ileum. This stain, which is specific for polyanions such as certain carbohydrates, has revealed the residue of the exopolysaccharide K30 glycocalyx, which is radically condensed during dehydration for EM.

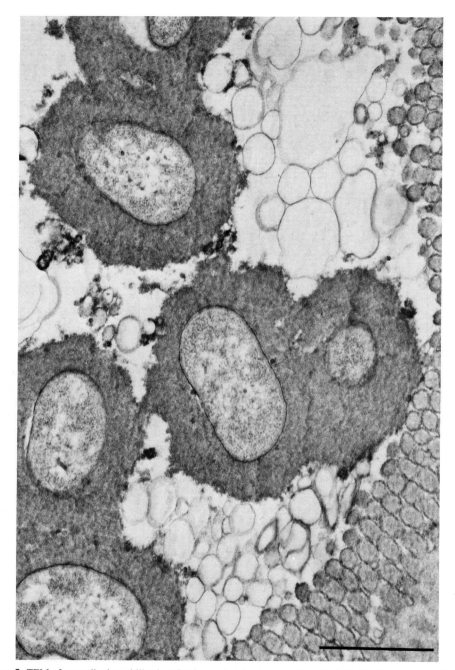

Figure 3. TEM of an antibody-stabilized, ruthenium red-stained preparation of a fimbriated enterotoxigenic strain of *E. coli* on the neonatal calf ileum. Stabilization of the K30 glycocalyx with anti-K30 antibody prevents the condensation of this hydrated structure during dehydration, and the true extent of the extensive capsular glycocalyx is revealed.

they grow in various ecosystems, and the direct examination of millions of cells in hundreds of environments has demonstrated that all bacteria are surrounded by glycocalyces as they grow in nature. *In vitro* cultures of single species do not usually contain antibacterial factors, and this suspension of selection pressure in favor of glycocalyx production often results in the overgrowth of natural glycocalyx-enclosed cells by fast-growing, mobile "swarmer" cells. This rapid drift from freshly isolated, wild, nontypeable strains towards standard, typeable laboratory cultures (Doggett *et al.*, 1964) is noted in many diagnostic laboratories, and animal passage is often used to reverse this trend and reisolate "encapsulated" strains.

The inclusion of surfactants (Govan, 1975) and/or antibiotics (Govan and Fyfe, 1978) in media has been shown to promote glycocalyx production, and selection in favor of natural adherent glycocalyx-enclosed cells is clearly seen when adhesion at an interface gives better aeration in a shaken culture (Lam *et al.*, 1980; Govan *et al.*, 1979). While standard *in vitro* bacterial cultures are very valuable in that they provide excellent replicable systems in which to study bacterial physiology or bacterial genetics, the surfaces of their component cells are profoundly altered, and they should not be used in adhesion studies or in studies of the susceptibility of bacterial cells to antibacterial factors.

1.2. Microcolony Formation by Adherent Bacteria

Because of universal bacterial glycocalyx production in natural and pathogenic ecosystems, cells of these organisms most often divide within a hydrated exopolysaccharide matrix, so that the daughter cells are trapped in a juxtaposition that results in the formation of microcolonies of morphologically identical cells (Fig. 4). The glycocalyces that mediate microcolony formation also mediate adhesion, so that most microcolonies are formed on surfaces where they burgeon and eventually coalesce to form an adherent biofilm (Fig. 5). Direct examination of bacterial growth in natural (Geesey *et al.*, 1978; Cheng *et al.*, 1977) and pathogenic (Marrie and Costerton, 1983a; Lam *et al.*, 1980) ecosystems shows that most bacterial cells grow in adherent glycocalyx-enclosed microcolonies that develop to very great (occasionally macroscopic) dimensions when nutrient conditions are favorable. This bacterial strategy is very effective because the cells are retained in the nutritionally favorable niche in a mode of growth that allows them to trap and use soluble nutrients (see Section 1.5) and provides them with a measure of protection from antibacterial agents such as chemicals (Ruseska *et al.*, 1982), surfactants (Govan, 1975), antibodies (Baltimore and Mitchell, 1980), and phagocytic amoebae and leukocytes (Schwarzmann and Boring, 1971) (see also Chapter 12). The basic bacterial strategy is, clearly, to live within protected adherent microcolonies in nutritionally favorable environments and to dispatch mobile "swarmer" cells to reconnoiter neighboring niches and to establish new adherent microcolonies in the most favorable of them.

1.3. Consortium Formation by Adherent Bacteria

Many bacterial processes in natural ecosystems are carried out by consortia of physiologically related organisms, as is the case in cellulose digestion (Costerton *et al.*, 1983) and in methane generation (Bryant *et al.*, 1967). The direct examination of bacteria growing in natural environments has provided morphological evidence that the bacterial members of these consortia are immediately juxtaposed (Fig. 6), within mixed adherent microcolonies, to facilitate their physiological cooperation (Cheng *et al.*, 1981). The formation of highly structured consortia in natural systems suggests that the bacterial members of a consortium adhere to their initial substratum, and to each other, to form a highly organized microbial

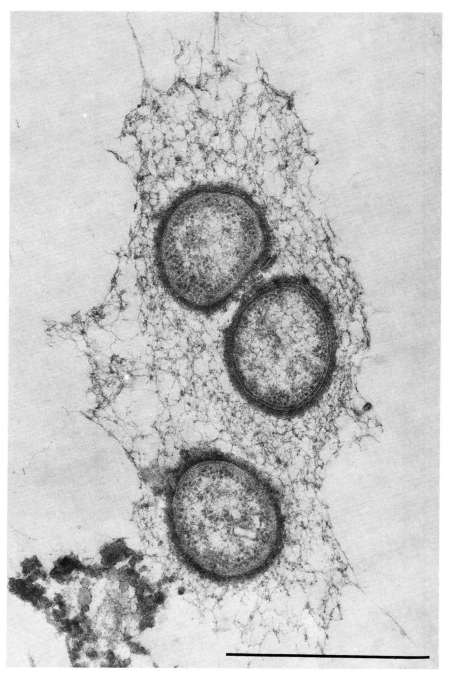

Figure 4. TEM of a ruthenium red-stained preparation of a microcolony of bacteria from an adherent biofilm on a rock surface in a subalpine stream. Note the morphological similarity of these sister cells that are held together by their finely fibrillar glycocalyces.

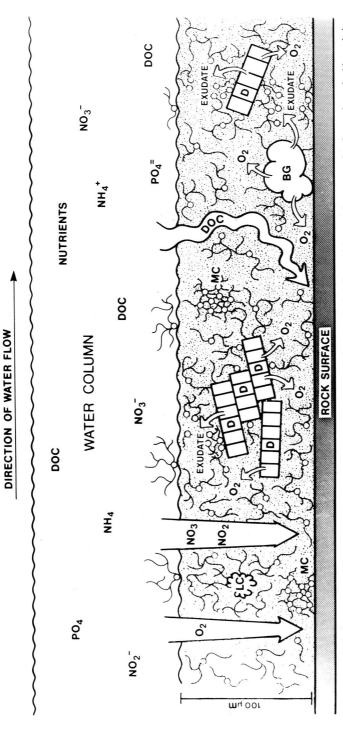

Figure 5. Diagrammatic representation of a natural adherent biofilm in which bacteria (open circles) live within a continuous matrix of exopolysaccharide made by themselves and by their algal symbionts. The diagram speculates concerning processes within this microbial biofilm, where diatoms and blue-green algae (cyanobacteria) are physiologically integrated with the adherent bacteria. BG, blue-green bacteria; D, diatoms; DOC, dissolved organic carbon; LC, lysed cyanobacteria; MC, microcolony.

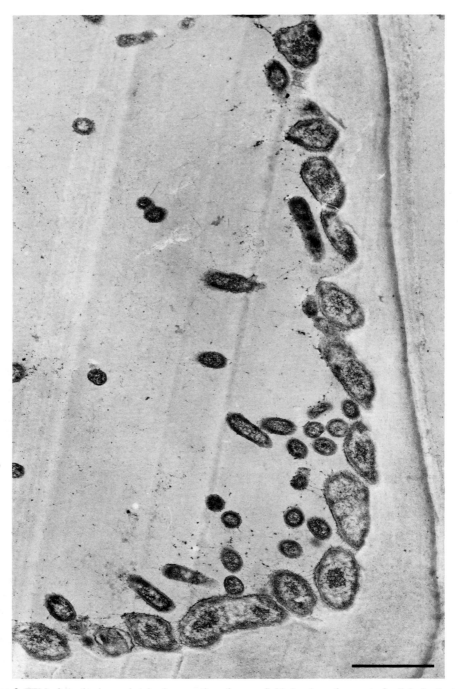

Figure 6. TEM of a ruthenium red-stained preparation of rumen fluid showing a fragment of cellulosic plant cell wall that has been colonized by a "pit-producing" cellulolytic organism that resembles *Bacteroides succinogenes*. Note that this primary colonizer has been selected from the myriad bacteria of the rumen by its affinity for cellulose, and that a slender (0.2 μm) rod-shaped organism has colonized the adjoining area to form a simple consortium.

community within which substrate transfer, and even hydrogen transfer (Mah *et al.*, 1977), is facilitated. This interbacterial affinity is clearly suggested by the observation that the butyrate-oxidizing and methane-producing members of a physiological consortium studied by Bryant's group (McInerney *et al.*, 1981) form mixed microcolonies even when they are grown in a liquid medium. The topographically focused nature of natural bacterial processes, such as cellulose digestion (Akin and Amos, 1975) and corrosion (Costerton and Lashen, 1983), typically leads to the formation of "pits" in digestible substrata (Fig. 7; Chapter 10, Section 4.1). This focused attack occurs because bacteria within an adherent microcolony find themselves in a nutritionally favorable environment and develop a "critical mass" of cells and extracellular enzymes and products so that a microniche is generated, and the substratum is eroded immediately beneath the microcolony. Thus, cellulose and metal are both pitted by bacterial activity and the bacteria responsible for this process are seen in a microcolony that sinks into the pit as it develops. Figure 7 shows a shallow pit produced in the surface of mild steel by sulfate-reducing bacteria (SRBs) that have grown as a microcolony on the metal surface and produced sufficient H_2S to initiate corrosion immediately under their developing microcolony.

1.4. Biofilm Predominance in Aquatic Systems

Direct examination and careful quantitation of the planktonic (floating) and sessile (adherent) bacterial populations of a subalpine stream (Geesey *et al.*, 1978) showed that the sessile population was clearly predominant. A 1-m "reach" of this small stream contained more than 10,000 sessile bacterial cells for each planktonic cell, and a subsequent examination of 84 ecologically varied streams and rivers throughout Alberta (Ladd *et al.*, 1979) showed the same predominance of sessile bacterial growth. This revelation of the predominance of sessile bacterial growth in aquatic ecosystems generated a concept (Lock *et al.*, 1984) of a stream as a "microbiological reactor" in which water, carrying dissolved nutrients and a few detached bacteria, flows past bacterial biofilm populations on submerged surfaces (Fig. 5). The predominance of sessile bacteria is not limited to natural aquatic ecosystems, and it has now been documented in hundreds of industrial systems from heat exchangers (Costerton and Geesey, 1979) to the water injection lines used in secondary oil recovery (Costerton and Lashen, 1983).

1.5. Physiology of Biofilm Populations

In their elegant pioneering studies of the growth of marine bacteria, both Henrici (1933) and ZoBell (1943) reported a marked tendency of these cells to grow on surfaces, and they noted that these solid–water interfaces would be expected to favor bacterial colonization because they concentrate nutrients. We now know that surfaces in aquatic environments are rapidly coated with an organic accretion (Fazio *et al.*, 1982) that contains many polysaccharides of microbial origin and that, like the saliva coat on teeth (Gibbons and van Houte, 1975), this adsorbed organic layer constitutes the surface that is actually colonized by adherent bacteria (Chapter 6, Section 4.8). Very soon after initial adsorption to a surface, bacteria are seen to produce large amounts of fibrous glycocalyx material (Fig. 8) and to "cement" themselves onto the surface so that their reversible adsorption becomes irreversible adhesion (Marshall *et al.*, 1971; Chapter 6, Sections 3 and 4). The chemical structures of many of the glycocalyces of predominant species of aquatic bacteria are now understood in considerable detail (Sutherland, 1977), and most are seen to comprise polyanionic matrices that must be expected to act as ion-exchange resins, in that they attract and bind charged ions and molecules within the matrix that surrounds the bacterial cells that initially comprise

Figure 7. Scanning electron micrograph (SEM) of the surface of a mild steel coupon subjected to corrosion by sulfate-reducing bacteria growing within an adherent biofilm. This corrosion forms pits (inset) at the bottom of which we can see metal crystals and rod-shaped bacterial cells. The bar in the inset indicates 10 μm.

adherent microcolonies and eventually comprise confluent biofilms (Fig. 5). Depending on the dissociation characteristics of the matrix–ligand association, the adsorbed molecules and ions may be available to cells within the matrix at effective concentrations very much higher than those at the surface of planktonic organisms actually immersed in the flowing aqueous phase. Because the ion-exchange matrix comprised by the glycocalyces of biofilm bacteria

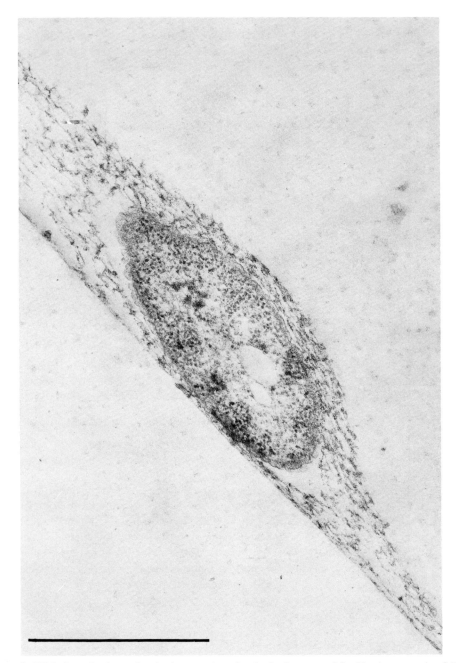

Figure 8. TEM of a ruthenium red-stained preparation of a plastic disc exposed for 30 min to a strain of *E. coli* freshly isolated from a septic hip prosthesis. This cell has adhered to the plastic surface and quickly cemented itself irreversibly to this surface, by its production of large amounts of fibrous exopolysaccharide glycocalyx material.

"loads" from its exposed surface, we must expect cells closer to this surface to have preferred access to rare ligands. However, this mode of growth promotes the access of deeper cells to common nutrients, because individual biofilms are comprised of large amounts of the fibrous matrix and deeper cells are not actually occluded by overlying cells.

We have examined the uptake of radiolabeled substrates (glutamate) by aquatic bacteria growing in thick biofilms on submerged surfaces, on an activity per cell basis, and we have compared this uptake with that of the same cells following dispersion in a liquid menstruum, and the rate of both uptake and mineralization is somewhat similar (Ladd et al., 1979). Recent work in England (Whittenbury and Dow, 1977), and in our laboratory (W. F. McCoy, B. Wright, and J. W. Costerton, unpublished data), has indicated that sessile aquatic bacteria growing in adherent biofilms may differ from their planktonic counterparts in several important respects, including growth rates, morphology, and rates of pivotal physiological processes. Taken together, these data suggest that an adherent biofilm composed of microcolonies of various aquatic bacteria may be considered to constitute a "quasitissue" that may rationally be thought of as having measurable rates of respiration and nutrient uptake (Fig. 5).

1.6. Removal of Biofilms

The natural life of a biofilm is cyclical in that its biomass is increased by cellular proliferation and exopolysaccharide production and decreased by cell death, detachment at the flowing surface, and "grazing" by benthic macroorganisms (Lock et al., 1984). Succession is clearly seen in many biofilms, as when photosynthetic primary producers die and are digested by heterotrophic saprophytes, but biofilms continue to occupy submerged surfaces in aquatic environments for very prolonged periods of time. Neither spontaneous nor biocide-induced killing of bacteria within a biofilm automatically detaches the biofilm from a colonized surface, and we have reported extensive biofilms comprised of dead bacteria within their extensive matrices (Ruseska et al., 1982). However, oxidizing biocides such as chlorine both kill biofilm bacteria and destroy the polymeric units of the biofilm matrix that anchor this structure to the colonized surface so that they are very effective in physically removing bacterial biofilms in flowing systems.

Recently, we have patented (U.S. Patent 06-365,260, April 5, 1982) a physical method of biofilm removal that uses the destructive power of ice crystal formation in the inner areas of the matrix to break anchoring polymers and to detach the bacterial biofilm. In practical terms, a cold nonfreezing solution (ethylene glycol at approximately $-12°C$) is passed through a fouled system, and this slow freezing causes the detachment of the biofilm from the metal surface to form "blisters" that break upon the reestablishment of flow (Fig. 9). Three successive freeze–thaw cycles using this method have been shown to remove bacterial biofilms completely from fouled surfaces even when these biofilms contained significant quantities of inorganic "scale." Generally, heating and acid treatment have been shown to degrade biofilms to adherent residues that may be as troublesome as the original biofilm, but bleach treatment (5% sodium hypochlorite) has been very successful in killing bacteria and dissolving biofilm matrices in studies of fouled geological cores (Shaw et al., 1984) and of operating oil wells (Clementz et al., 1982).

1.7. Control of Biofilm Formation

The rate of initial biofilm formation by bacterial adhesion to various surfaces in experimental aquatic environments has been shown to be dependent to a degree on the chemical nature of the surface (Pringle and Fletcher, 1983). However, this demonstrable influence of surface chemistry on bacterial biofilm formation is considerably modified by the fact that all such surfaces in natural environments are coated by a layer of common organic chemical constituents and by the fact that subsequent biofilm development depends on cell proliferation and cell–cell adhesion within the matrix. When different surfaces within an aquatic

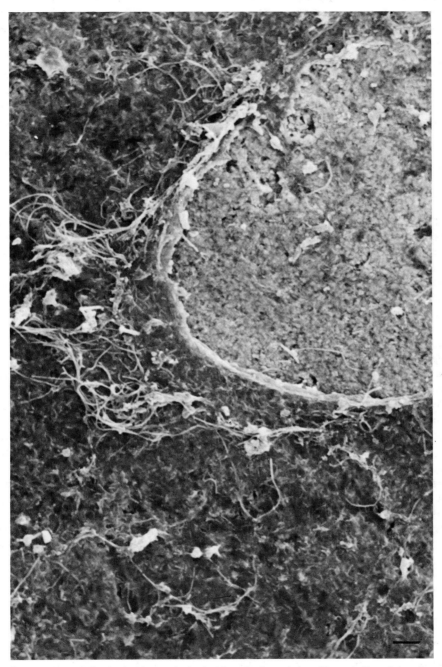

Figure 9. SEM of a steel disc that had been colonized with a mixed population of bacteria (2 weeks) until a coherent and plastic biofilm was formed. This biofilm was ''blistered'' by treatment with ethylene glycol at $-12°C$, which produces large ice crystals in the proximal biofilm, and the reestablishment of flow broke the blister (top right portion of figure) and removed a section of the biofilm to expose the underlying metal surface. Repetition of this procedure (three times) is sufficient to remove bacterial biofilms completely.

system that has developed thick mature bacterial biofilms are examined, barely perceptible differences are seen between adherent cell numbers and biofilm thickness on steel, rock, wood, and plastic. A careful statistical comparison of bacterial colonization of copper-emitting admiralty brass with that of other metals, such as mild steel and stainless steel, showed no statistical difference in the cell numbers or the thickness of the mature (3 week) biofilms that developed on these surfaces. When clean metal surfaces are submerged and exposed to the continuous presence of dead bacteria (suspensions containing 2 ppm iso-thiazalone) for as long as 30 days, no biofilm development is seen on these exposed surfaces (Ruseska *et al.*, 1982).

2. BACTERIAL ADHESION TO INERT MEDICAL PROSTHESES

Modern medical practice is necessarily invasive, and a very large number of plastic and metal prostheses are implanted into patients in both short-term and long-term applications. We have used direct EM techniques to examine the surfaces of many of these prostheses, after they have been in place for various lengths of time, and we have shown that the bacteria that colonize their surfaces grow in coherent biofilms very similar to those seen in natural ecosystems. Because host defense mechanisms are mobilized in response to the presence of these bacteria, planktonic organisms are usually very rapidly killed and are very rarely found in the early stages of these infections, but the biofilm-protected bacteria persist in spite of host responses and of antibiotic therapy and often necessitate the removal of the prosthesis. Because these colonizing organisms grow in very coherent biofilms adherent to the surface of the prosthesis, they are notoriously difficult to obtain as samples for culture and characterization, and the biofilm must be recovered by scraping and dispersed by physical means (i.e., shaking, sonication) to allow their accurate quantitation.

2.1. Biofilm Formation and Structure

The biofilms that develop on prostheses may contain a single organism that has colonized a sterilely implanted prosthesis by way of the bloodstream (hematogenous contact) (Marrie and Costerton, 1983a) leading to the formation of a monospecies biofilm (Fig. 10). Alternatively, a prosthesis may have been implanted through a normally colonized area of the body (e.g., the cervix) and acquired a variety of colonizing organisms (Marrie and Costerton, 1983b) that develop into adherent mirocolonies and finally into a coherent and continuous multispecies biofilm (Fig. 11). The structure of these prosthesis biofilms is very similar to that of biofilms in natural ecosystems in that the bacterial cells are embedded in a fibrous matrix (Peters *et al.*, 1982; Marrie and Costerton, 1983a) of their own exopolysaccharide glycocalyces (Figs. 10 and 11) that mediates cell–surface and cell–cell adhesion and constitutes an ion-exchange "resin." Bacterial microcolonies on many human prostheses develop as huge barnaclelike accretions (Fig. 12) composed of hundreds of thousands of cells, and, in areas protected from shear forces, these large microcolonies may coalesce to form a thick coherent biofilm that virtually occludes the lumen of a tubular prosthesis.

2.2. Pathogenic Consequences of Biofilm Development

The biofilm mode of growth of the bacteria that colonize human prostheses confers protection on these pathogenic organisms (see Section 2.3), but it also limits their immediate pathogenic impact on the host and many patients with heavily colonized prostheses experience few overt symptoms. All of the intrauterine contraceptive devices (IUDs) that we have

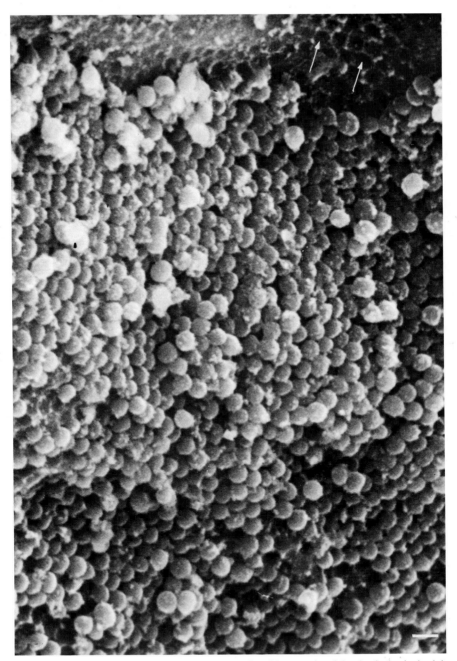

Figure 10. SEM of the biofilm formed by staphylococcal cells, of bacteremic origin, that had colonized the metal surface of a cardiac pacemaker. This thick bacterial biofilm, composed of coccoid cells and their enveloping glycocalyx material (arrows), allowed its component cells to withstand the therapeutic challenge of very high levels of antibiotics (cloxacillin and rifampicin).

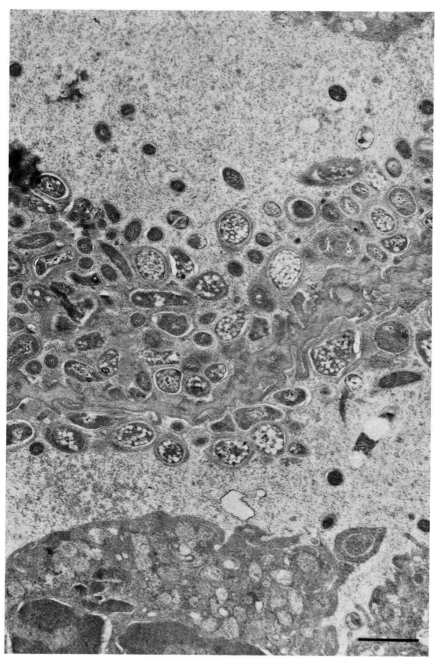

Figure 11. TEM of a ruthenium red-stained preparation of material scraped from the surface of an intrauterine contraceptive device (IUD) that had been implanted in the uterus of a healthy, entirely asymptomatic woman for several months. Note the extensive multispecies colonization of this plastic surface and the very extensive fibrous glycocalyx surrounding the small rodlike cells (top).

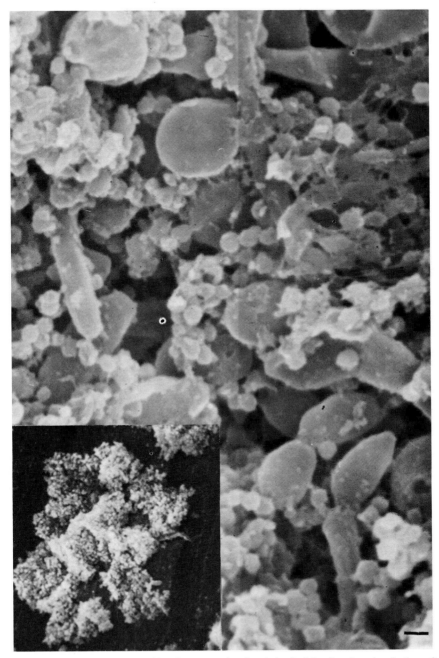

Figure 12. SEM of microorganisms within a huge accretion (inset) that had developed on the plastic surface of an intracardial Hickman catheter. This accretion was composed of coccoid bacteria and yeastlike cells. Laboratory cultures yielded *Staphylococcus epidermidis* and *Candida albicans,* and the patient developed a bacteremia from which these two organisms were also cultured. Many patients whose Hickman catheters were very heavily colonized by *S. epidermidis* did not develop any overt bacterial disease.

examined to date have been found to be coated with a thick multispecies bacterial biofilm (Fig. 11), but the patients from whom they were removed were asymptomatic (Marrie and Costerton, 1983b). We speculate that the release of endotoxin from this biofilm containing millions of gram-negative bacteria reinforces the inflammatory response of the endometrium to the presence of a bacterially colonized foreign object and may contribute to the contraceptive efficacy of these routinely colonized plastic prostheses. While this bacterial colonization of a prosthesis that lies within the uterus may enhance, or even account for, its contraceptive effect, the presence within this adherent biofilm of the pathogenic species of *Bacteroides* and *Actinomyces,* which cause pelvic inflammatory disease and tubular actinomycosis, suggests (Marrie and Costerton, 1983b) that the biofilms that develop on these prostheses may constitute a conduit between normally colonized areas (the cervix) and sensitive, normally sterile areas (the fallopian tubes) of the female reproductive tract. Similarly, urinary catheters may provide a conduit for the growth of a multispecies biofilm from the exterior toward the bladder (Marrie and Costerton, 1983c), and Hickman catheters may provide a conduit for the growth of a monospecies *Staphylococcus epidermidis* biofilm from the skin to the heart (M. Moody, S. Schimpff, and J. W. Costerton, unpublished data). Once a bacterial biofilm has developed on the surface of a plastic or metal prosthesis, by growth from a septic area or by hematogenous spread, this protected population affects the host by releasing toxic products and occasionally releasing free bacterial cells. The former activity may produce no clinical symptoms, or it may have profound effects such as the decalcification of bone adjacent to colonized orthopedic prostheses. The latter activity is usually easily countered by macrophages mobilized as a part of the inflammatory response, but the continuous release of mobile pathogens into the bloodstream or into the upper urinary and reproductive tracts is inherently dangerous (Marrie *et al.,* 1979), and clinicians usually prefer to remove prostheses known to be heavily colonized by bacteria. Thus, it is accurate to state that the threat of bacterial colonization and biofilm formation poses the most important limitation on the use and development of prostheses in human medicine.

2.3. Resistance of Biofilm Populations to Host Defense Factors and to Antibacterial Chemotherapy

Bacterial biofilms on the surfaces of plastic and metal prostheses are phenomenally persistent in spite of demonstrable activation of host defense mechanisms (i.e., antibody production, macrophage activation) and of very aggressive antibiotic therapy (Marrie and Costerton, 1983a), and we only seek to explain this well-documented and remarkable persistence. Leake and Wright (1979) have clearly shown that both peritoneal and pulmonary macrophages show unusual locomotory responses to the surfaces of the biomaterials used to manufacture prostheses, and our own studies have shown that activated macrophages are not able to phagocytose bacterial cells within biofilms. This phenomenon can be inferred in the case of IUDs because we see many macrophages in the tissue immediately adjacent to these biofilm-coated plastic prostheses, but all recovered IUDs retain their biofilm coating. *In vitro* studies have shown that even relatively small amounts of their glycocalyces protect bacteria from surfactants (Govan, 1975), antibodies (Baltimore and Mitchell, 1980), phagocytes (Schwarzmann and Boring, 1971), and antibiotics (Govan and Fyfe, 1978), and we can extrapolate this protective effect to the massively glycocalyx-enclosed cells of thick adherent biofilms. This latter protection is perhaps best illustrated in the case of the biofilm shown in Fig. 10, because this patient developed a staphylococcal bacteremia while equipped with an endocardial pacemaker. Retrospective EM studies of the pacemaker assembly showed that the pacemaker tip bore a very thick monospecies biofilm (Fig. 10), while even the subcutaneous battery pack bore developing staphylococcal microcolonies (Fig. 13). Perceiving

Figure 13. SEM of a plastic surface associated with a cardiac pacemaker that had been colonized by a bacteremic route by cells of *Staphylococcus aureus*. These coccoid bacteria can be seen within one of the microcolonies (lower), because their amorphous glycocalyx exopolymers had condensed during dehydration to reveal the enclosed bacterial cells.

that the pacemaker assembly would be colonized, but unwilling to remove the prosthesis because of a history of syncopal attacks (Marrie and Costerton, 1983a), the clinicians treated this patient with 12 g/day (intravenous) of cloxacillin for 4 weeks and subsequently with 12 g/day (intravenous) of cloxacillin and 600 mg/day (orally) of rifampicin for 6 weeks. This massive cloxacillin therapy of an infection caused by cloxacillin-sensitive *Staphylococcus aureus*, which was patently located in the circulating system and endocardium, controlled the bacteremia but failed to kill the bacteria within the biofilm, and the bacteremia recurred twice after the massive antibiotic therapy was suspended. After the pacemaker assembly was removed, intravenous cloxacillin therapy was successful in controlling the infection, which did not recur when a new extracardial pacemaker was implanted. This example is typical of many that illustrate that bacteria within biofilms are inherently resistant to antibiotic therapy and justifies the clinical consensus that colonized prostheses should be removed.

3. AUTOCHTHONOUS BACTERIAL POPULATIONS IN ANIMAL SYSTEMS

In addition to the bacteria that come into juxtaposition with animal tissues by contact, by the ingestion of food and water, and by the inhalation of air, many animal tissues attract and maintain specific bacterial species on their surfaces, and some even establish physiological cooperations with these truly autochthonous bacteria (Cheng *et al.*, 1981). Savage (1977) has defined autochthonous bacteria as those that colonize particular tissues, usually in a succession from host infancy to maturity, and maintain stable population levels in the climax communities of normal adults. The mode of growth of truly autochthonous bacterial populations is of pivotal importance because these organisms, by definition, persist for long periods of time in immediate juxtaposition to the colonized tissue surface, and they must therefore be capable of withstanding the operation of a continuously stimulated host defense system, against whose specific agents (i.e., surfactants, antibodies, phagocytes) they have no special protection. Growth within an adherent biofilm provides these autochthonous bacteria with the same inherent protection from antibacterial agents provided by this mode of growth in nature and on the surfaces of medical prostheses.

3.1. Bacterial Colonization of Digesta

Many animal digestive systems optimize the bacterial digestion of nutrient substrates as a major component of the overall digestive process (Chapter 15, Section 3.2.1), and the direct observation of insoluble nutrients shows that they are very rapidly colonized by bacteria that form initial adherent microcolonies and eventual adherent biofilms (Akin and Amos, 1975; Cheng and Costerton, 1980; Cheng *et al.*, 1981). The fluid contents of digestive systems may be thought of as a mobile reservoir of many bacterial species, and several workers have shown that insoluble nutrients (e.g., cellulose, starch) are rapidly and specifically colonized (Patterson *et al.*, 1975; Minato and Suto, 1976) by bacteria that have the enzymatic capability to digest them. Thus, pectin surfaces within forage leaves are rapidly colonized by *Lachnospira* (Cheng *et al.*, 1980), while cellulose surfaces are rapidly colonized by cellulolytic bacteria such as *Bacteroides succinogenes* (Cheng and Costerton, 1980), *Ruminococcus albus* (Patterson *et al.*, 1975), or *R. flavefaciens*. Careful morphological studies have allowed us to identify the primary colonizer of cellulosic plant cell walls (Fig. 14) and to associate these bacteria with pits formed by their digestive enzymatic activities. Once a monospecies biofilm has been formed on the surface of an insoluble

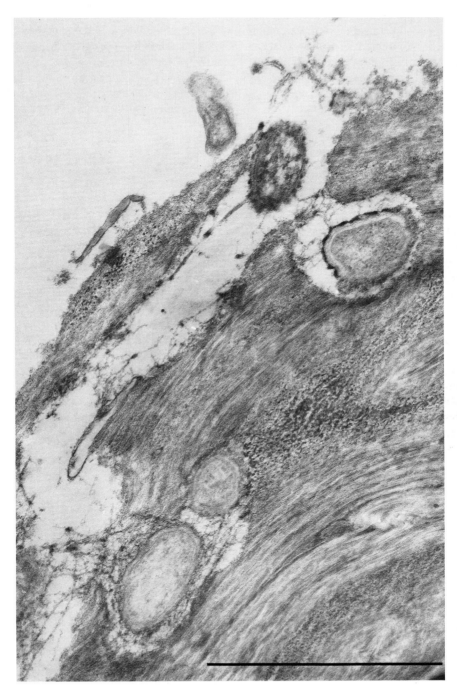

Figure 14. TEM of a ruthenium red-stained preparation of partly digested cellulosic plant cell walls from the rumen of the sheep. The bacterial cells resemble *Bacteroides succinogenes*, and they appear to have digested their specific cellulose substrate to form very extensive pits and tunnels.

nutrient, a "critical mass" of enzyme activity is achieved, and deep pits are formed even when the original colonizing cells of *B. succinogenes* have autolyzed (Fig. 15) to produce cell envelope fragments with which cellulose activity is still associated (Forsberg *et al.*, 1981). Pit formation is characteristic of bacterial digestion of insoluble nutrients and of bacterial corrosion of metals (Fig. 7), because digestive and corrosive activity is much enhanced immediately under an adherent microcolony whose matrix concentrates digestive and corrosive agents and enzymes. Direct examination of the bacterial biofilm on partially digested particulate digesta often shows the formation of consortia in which the initial layer of cells of the primary colonizing species is overlaid with a layer of cells of a morphologically distinct organism (Fig. 6). This affinity of the second organism in a consortium for the glycocalyx surface of a biofilm formed by the primary colonizer has also been noted in physiological and morphological studies of consortia formed by butyrate-oxidizing and methane-producing organisms (McInerney *et al.*, 1981).

3.2. Bacterial Colonization of Digestive Systems

Savage (1977) has reviewed the numerous clear-cut instances in which specific tissues within certain animal digestive systems are found to be colonized by monospecies bacteria biofilms. However, these monospecies colonizations tend to be on tissue surfaces (e.g., stomach) where chemicophysical conditions already limit the number of bacterial species that can survive and proliferate. Direct examinations of the non-mucous-coated tissue surfaces of such organs as the bovine rumen reveal a rich variety of bacterial morphotypes in the developed biofilm, and the isolation of these adherent organisms by the homogenization of blocks of the colonized tissue (Cheng and Costerton, 1983) showed the presence of approximately 23 bacterial species that colonize this tissue surface in the newborn calf and maintain a partly gram-positive, proteolytic, ureolytic, and partly facultatively anaerobic biofilm on the surface of this tissue throughout the life of the animal (Cheng *et al.*, 1981). The epithelial tissue of the rumen is a very special environment because very large amounts of urea diffuse through the rumen wall and must be transformed to ammonia to avoid toxic effects. The taxonomically distinct population of approximately 23 bacterial species that colonize the rumen wall includes several species that produce large amounts of urease, and this enzyme is essential to the normal physiological function of the bovine rumen (Chapter 15, Section 3.2.1). Thus, this adherent autochthonous bacterial population makes an important physiological contribution to the health of the whole animal. New methods (Rozee *et al.*, 1982) for the retention and structural preservation of mucus on the surfaces of intestinal tissues show an amazing variety of bacterial and protozoan morphotypes in a very extensive polyanionic matrix composed of mucus and bacterial exopolysaccharide (Fig. 16). These data suggest that the 400-μm "unstirred" layer on the surface of intestinal tissues (Levin, 1979) comprises a biofilm composed of bacteria, protozoans, mucus, and exopolysaccharides, and that this very thick layer constitutes a complex viscous ecosystem within which a potential pathogen must gain a measure of predominance before it can approach and colonize the actual tissue surface (Chapter 15). Direct observations of the microbial populations within this thick mucous blanket have shown the extent of population changes (Fig. 17) caused by treatment with specific lectins (J. G. Banwell, D. Cooper, and J. W. Costerton, unpublished data), and our preliminary data indicate that the autochthonous bacterial population of the intestinal mucous blanket may vary very widely in response to stress and to physiological changes.

3.3. Autochthonous "Barrier" Populations as a Protection from Disease

The presence of a thick biofilm containing bacterial species well adapted to the tissue surface ecological niche constitutes one of many defenses of animal tissues against coloniza-

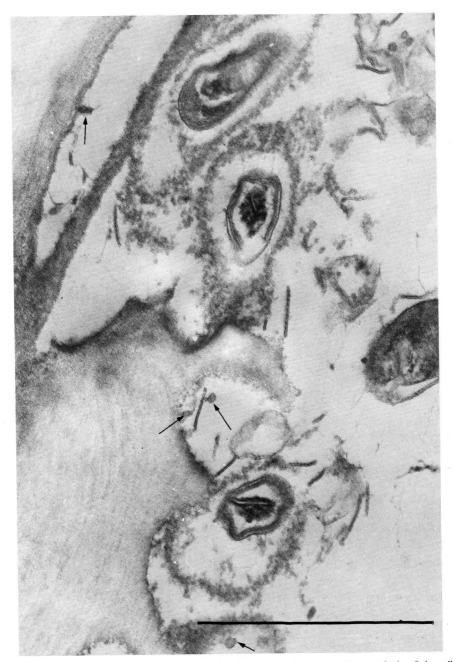

Figure 15. TEM of the same preparation seen in Fig. 14 showing the progressive autolysis of the adherent cellulolytic bacteria to form vesicular fragments (arrows) that have been shown to be capable of continued cellulolytic activity.

tion by pathogenic bacteria, in that the tissue surface is occluded by a vigorous and highly competitive autochthonous population (Chapter 15, Section 3.2.3). At the level of the mucus-covered tissues of the intestine, the development of a vigorous microbial population in the mucous blanket coincides with the time that the newborn animal is no longer suscepti-

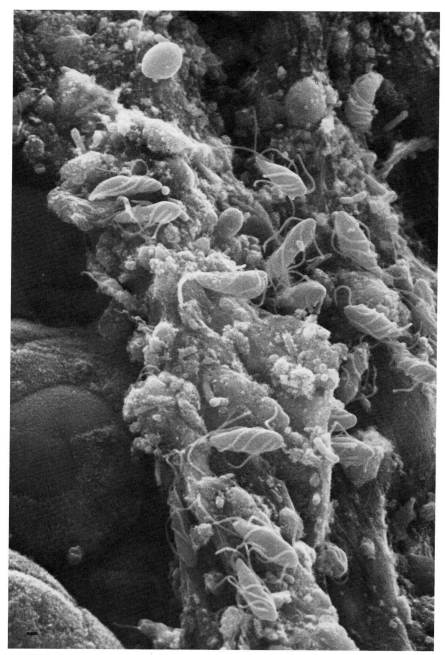

Figure 16. SEM of mouse ileum that had been treated immunologically to retain the mucous blanket. Note that there are no bacteria on the microvillar tissue surface (left portion of figure) in this young animal but that the mucous blanket is heavily colonized by bacteria and protozoans.

ble to infection by enterotoxigenic *Escherichia coli* (ETEC) (Moon, 1974). Indeed, animals ingest millions of enteric pathogens daily but most of these fail to gain predominance in the complex microbial population of the thick (ca. 400 μm) unstirred layer that comprises the mucous blanket, and thus fail to make effective contact with the intestinal epithelium. The

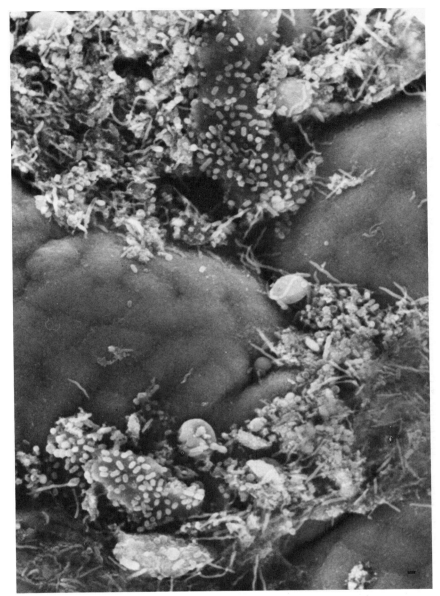

Figure 17. SEM of a preparation of the ileum from an animal whose microbiological flora had been manipulated by the feeding of a lectin. The bacterial flora is indeed profoundly altered, and we see rod-shaped bacteria in coherent microcolonies and very large numbers of spirochetes in detritus and in mucus.

extensive study of bacteria adherent to the formerly free side of sloughed uroepithelial tissue cells from the distal human female urethra has served to detect a biofilm-forming, adherent autochthonous bacterial population (Marrie *et al.*, 1980) that changes significantly with the hormonal state of the host (Marrie *et al.*, 1978). During reproductive years, the autochthonous flora of the female urethra is comprised largely of lactobacilli (Marrie *et al.*, 1980), and these gram-positive rods form an almost continuous biofilm (Fig. 18) over the free surface of the tissue cells. After menopause, the predominant population of the distal urethral epithelium changes to gram-negative rods, and this new autochthonous bacterial population

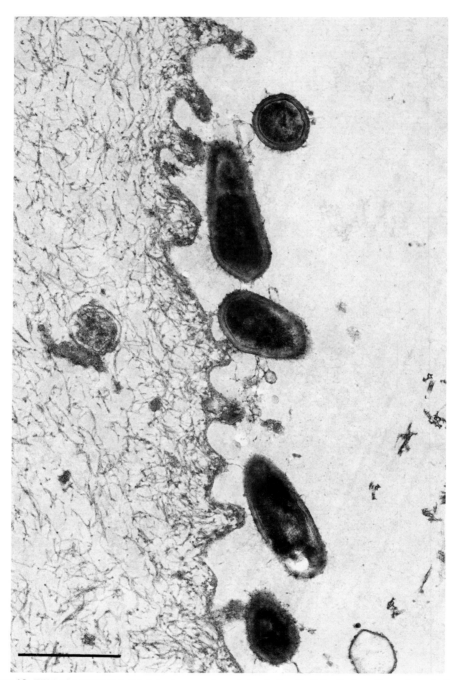

Figure 18. TEM of a ruthenium red-stained preparation of sloughed epithelial cells from the urethra of a human female. The surfaces of these keratinized epithelial cells are heavily colonized by gram-positive, rod-shaped bacteria (lactobacilli), whose glycocalyces are condensed to form fine fibers in the absence of stabilization against dehydration.

is less effective in protecting the urinary tract from invasion by fecal pathogens. Thus, the adherent biofilm populations of both the distal female urethra and the distal female reproductive tract can be thought of as ''barrier populations'' whose ecological vigor and acid tolerance allow them to produce a local environment that is unsuitable for the growth or attachment of fecal pathogens. The disturbance of this natural bacterial barrier function by manipulation, trauma, catheterization, broad-spectrum antibiotic therapy, or hormonal changes may allow an ascending infection by omnipresent fecal pathogens such as *Escherichia* and *Proteus* species. The autochthonous bacterial population of the human oropharynx is comprised largely of gram-positive cocci that adhere specifically to the fibronectin coat on these epithelial tissues (Johanson *et al.*, 1969; see also Chapter 10, Sections 2.1.6 and 3.1). Stresses such as underlying disease (e.g., cystic fibrosis) and psychological trauma (e.g., impending surgery) alter the protease content of saliva, and Woods *et al.* (1980) have shown that the fibronectin and the autochthonous bacterial population are demonstrably lost from the oropharyngeal surface in stressed individuals. This autochthonous population is rapidly replaced by a biofilm composed largely of cells of *Pseudomonas aeruginosa*, and experiments with tracheal rings have shown that these organisms grow in a glycocalyx-enclosed biofilm on the colonized tissues (Baker and Marcus, 1982). A biofilm composed of glycocalyx-enclosed microcolonies of *P. aeruginosa* also develops on the endotracheal tubes commonly used in intensive care facilities. Exposure to aspirated fragments of the bacterial biofilms from endotracheal tubes may account for the very high incidence of pneumonia caused by *P. aeruginosa* in intensive care facilities.

3.4. Disease Prevention by the Manipulation of Bacterial Barrier Populations

In systems in which a protective autochthonous bacterial population will eventually develop, pathogenic bacteria are often successful in colonizing the tissue if they make effective contact before the barrier population develops naturally. Here the strategy has been to accelerate the development of the barrier population by the infusion of the predominant natural colonizing organism as soon as the tissues are receptive for colonization. We have introduced large numbers of cells of 29 strains of beneficial digestive tract bacteria into newborn ruminants and have recorded both resistance to neonatal ETEC infection (Chapter 11, Section 1; Chapter 15, Section 2.2.1) and improved early weight gain. Barnum and his associates (Brooks *et al.*, 1983) have introduced cells of *Corynebacterium bovis* into the teat ducts of cows before freshening and have reported the effective prevention of mastitis. Sprunt *et al.* (1980) have used strains of streptococci protectively to colonize the digestive tracts of premature infants for the prevention of neonatal diarrhea. With several colleagues, we are presently seeking to develop effective bacterial inocula in order to build up autochthonous bacterial biofilms on urethras that have been inadvertently sterilized by antibiotic treatment and on urine conduits that have been constructed surgically from sterile tissues (human ileum). The reestablishment of the barrier function of the more complex autochthonous bacterial biofilms on the tissues of the mature digestive tract and of the oropharynx has not yet received much attention but the rewards of a successful manipulation of these populations would be very significant indeed.

4. PATHOGENIC BACTERIAL ADHESION IN ANIMAL SYSTEMS

Early research into the adhesion of pathogenic bacteria to target tissues has ignored many complexities in its search for simple answers. Several bona fide adhesion mechanisms have now been described in some detail, but attempts to block these mechanisms have dissolved into confusion when alternative adhesion mechanisms have been discovered or when whole animal complexities intervened. To date, the sole effective blockage of a specific adhesion mechanism (Chan *et al.*, 1982a) that has resulted in the practical control of pathogenic adhesion is the immune blockage of the pilus-mediated adhesion of ETEC in the neonatal ileum (Acres *et al.*, 1979). Many of the complexities that have bedeviled this work emanate from the complexities of the bacterial cell surface itself, in that strains that have been maintained for years in laboratory culture may have at their surface pili or outer membrane proteins that act as well-defined ligands in *in vitro* adhesion assays but are buried by the glycocalyx in the glycocalyx-encapsulated cells of truly pathogenic wild strains. Alternate bacterial adhesion mechanisms are also found to exist, and even the pilus-mediated adhesion of ETEC to the neonatal ileum (Chapter 11) is complicated because strains lacking all of the well-recognized pili can cause fatal diarrheal disease in colostrum-deprived calves (Chan *et al.*, 1982b). Other complexities arise from the presence of mucus and serum factors on the surface of the target tissue, and we now find that many autochthonous bacteria actually bind to fibronectin on the cell surface (Woods *et al.*, 1980), while many pathogenic bacteria really live in the mucus at cell surface (Chapter 15, Section 2.2.1) and react only secondarily with the target cell. The perception that arises from the extensive work of the past two decades is that *in vitro* assays of the adhesion of surface-modified cells of laboratory strains of bacteria to cultured or derived animal cells in the absence of body fluids are interesting biochemical exercises but have very little relevance to actual pathogenic processes. This is especially true when the target cells used are transformed tissue culture cells with profoundly altered surface characteristics. What is clearly required is a detailed and rational examination of the bacterial and target cell surface components, throughout the temporal course of an infection, and a gradual identification of the important bacterial and target cell ligands that mediate the initial colonization and promote the subsequent bacterial proliferation, which are the sine qua non of bacterial infection.

4.1. Bacterial Pili and Surface Proteins as Specific Ligands

The proteins of the bacterial cell surface may be located partially within the outer membrane, in an extracellular protein coat, or in protruding pili (Chapter 11). In this latter form, they protrude through the glycocalyx and are clearly available to bind to target cell ligands at a considerable distance (1 to 6 μm) from the bacterial surface. Pilus formation is notoriously dependent on environmental parameters (Morris *et al.*, 1982), and it is reassuring actually to visualize the K99 pili of ETEC infecting the neonatal ileum (Fig. 19) and to note that these structures span the space between the infecting bacterium and the target tissue. Antipilus vaccines have been successful in controlling this enteric disease in neonatal animals, and these structures are obviously the primary pathogenic ligand of the ETEC. Pili have been identified as important pathogenic ligands in several other bacterial pathogens such as *Neisseria gonorrhoeae* (Lambden *et al.*, 1982) and uropathogenic *E. coli* (Svanborg-Eden *et al.*, 1977), while nonpilus surface proteins have also been shown to be important in the pathogenicity of *N. gonorrhoeae* (Swanson, 1981).

4.2. Bacterial Glycocalyx as a Ligand and as a Bacterial Defense Mechanism

The role of the glycocalyx is much more difficult to define in studies of pathogenic adhesion because it plays a critical role in both bacterial adhesion and bacterial persistence and survival on the infected tissue. The K30 glycocalyx of ETEC forms an extensive capsule around the infecting cells (Fig. 3) on the ileal surface, and confers a major measure of

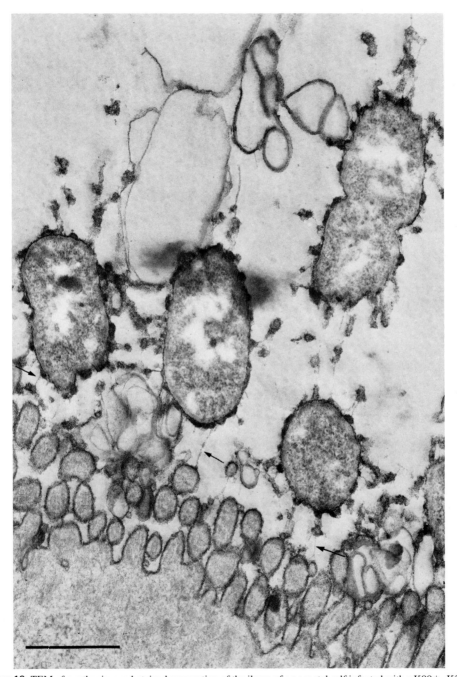

Figure 19. TEM of a ruthenium red-stained preparation of the ileum of a neonatal calf infected with a K99$^+$, K30$^+$ strain of enterotoxigenic *E. coli*. Specific treatment with monoclonal anti-K99 antibody has thickened these fimbriae so that they are resolved (arrows) and are seen to mediate attachment to the microvillar surface of the tissue. The K30 glycocalyx is condensed in the absence of stabilization against dehydration.

protection from gut surfactants (Govan, 1975), but it also comprises the effective cell surface, in the absence of pili, and is the presumptive bacterial ligand in the pathogenic pilus-minus strains (Chan *et al.*, 1982b). Even where pili such as the P pilus of uropathogenic *E. coli* are known to be the primary bacterial pathogenic ligand (Svanborg-Eden *et al.*, 1977), adherent bacteria within the infected urinary tract are seen to be embedded in very extensive masses of glycocalyx material (Fig. 20). Thus, a much more logical and less simplistic perception emerges from our simple-minded direct observations, in that primary adhesion by well-defined and specific adhesion mechanisms may be followed by glycocalyx production that both "cements" the adhesion and provides immediate protection from antibacterial antagonists, as in the colonization of inert surfaces in nature and in the human body.

The repeated observation that freshly isolated pathogenic bacteria produce large amounts of glycocalyx (Doggett *et al.*, 1964) and that this production often decreases on subsequent *in vitro* culture, but can be reestablished by animal passage, indicates that this exopolysaccharide structure plays an important role in the pathogenic process. Direct morphological examination of freshly isolated bacterial pathogens (Fig. 21) and of bacterial pathogens actually growing on infected target tissues (Figs. 3, 17, 20, and 22) has shown unequivocally that these cells are surrounded by thick glycocalyces whose outer membrane surface constitutes the effective surface of the cell as an infectious unit. Because the effective surface of the animal cell is also a glycocalyx composed of exopolysaccharides (Roseman, 1974), the pathogen–target cell interaction must be thought of essentially as an apposition of predominantly carbohydrate surfaces. This basic apposition can be reinforced by the specific binding to the target cell of pili that protrude past the bacterial glycocalyx, and polypeptide lectins clearly play an important role in mediating bacterial adhesion (Dazzo and Brill, 1977) by cross-linking the carbohydrate surfaces of bacteria with those of the target cell (Chapter 9, Section 2.3).

Several bacterial diseases are noted for their persistence in the infected host and for their resistance to clearance by antibiotic therapy. We have examined clinical materials from three of these diseases, and we have examined animal model infections by the same pathogens in considerable detail, to ascertain the mode of growth of the pathogenic bacteria and to relate this mode of growth to persistence in spite of host defense mechanisms and antibiotic therapy. Osteomyelitis is an especially persistent bacterial disease and the direct examination of infected bone has shown that the wide variety of bacterial pathogens that cause this disease grow exclusively in large glycocalyx-enclosed microcolonies. Model infections of the rabbit femur show classic symptoms of osteomyelitis, including deossification, and the pathogen is seen to produce truly phenomenal microcolonies (Fig. 23), within which the bacterial cells can be shown to be producing the fibrous glycocalyx material (Fig. 24) that comprises the matrix of these microcolonies. These rabbits were neither immunosuppressed nor deficient in cellular defense mechanisms, but it is indeed difficult to imagine how cells within this massive microcolony could be killed by antibodies or how the whole microcolony could be resolved by phagocytosis. Similar microcolonies are seen within the lungs of children with cystic fibrosis pneumonia caused by *P. aeruginosa* and within the lungs of rats infected with cystic fibrosis strains of this organism encased in agar beads (Lam *et al.*, 1980). Both clinical and model infections are chronic, and the former routinely persist for up to 15 years without resolution and without any evidence of general toxemia or bacteremia (Costerton *et al.*, 1979). The alveolar macrophages of cystic fibrosis patients have been shown to be essentially normal (Thomassen *et al.*, 1982), and both the patients and the infected rats produce a wide range of antibodies against the infecting organisms, but these intact host defense mechanisms fail to resolve the infection. Sustained high-dose antibiotic therapy, using both a

Figure 20. SEM of the bladder of a mouse 18 hr after the direct instillation of a uropathogenic strain of *E. coli*. Note that these rod-shaped cells are adherent to the tissue surface and that many of them are partly buried under an amorphous residue of polysaccharides of tissue and bacterial origin.

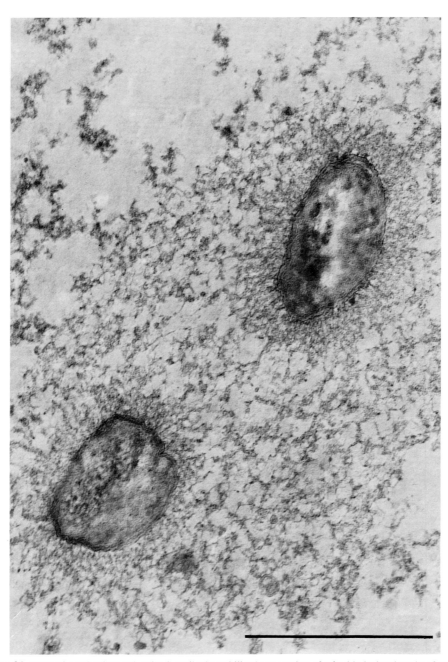

Figure 21. TEM of a ruthenium red-stained, antibody-stabilized preparation of a freshly isolated strain of *Pseudomonas aeruginosa* from a child with cystic fibrosis. Note the very extensive alginate glycocalyx that surrounds the cells of this organism and constitutes their effective outer surfaces.

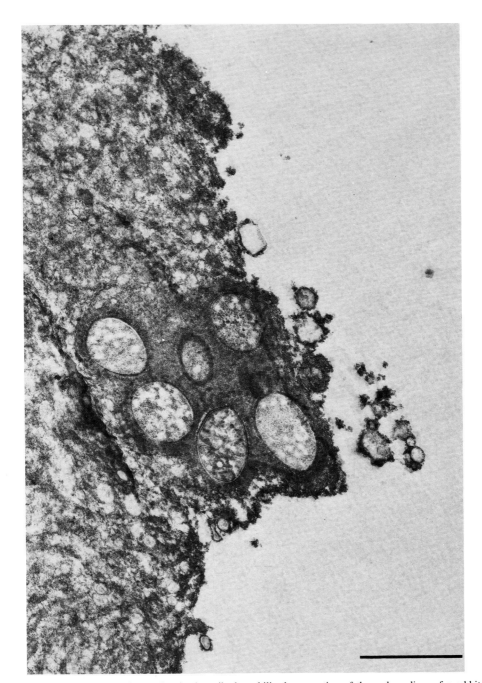

Figure 22. TEM of a ruthenium red-stained, antibody-stabilized preparation of the endocardium of a rabbit in which viridans group streptococcal endocarditis had been induced via bacteremia. Note the discrete dextran-enclosed bacterial microcolony that developed in 4 hr on the infected endocardial tissue.

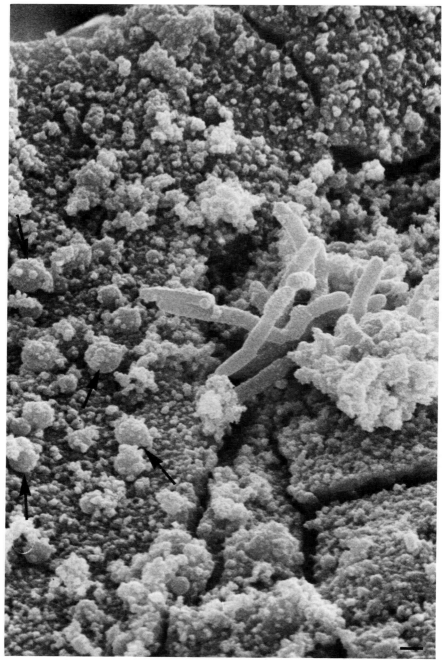

Figure 23. SEM of the microbial biofilm that developed on bone surfaces following the direct instillation of cells of *Bacteroides intermedius* and *Staphylococcus aureus* into the marrow cavity of the rabbit femur. Coccoid cells of *S. aureus* are seen (arrows) throughout this very thick accretion while the rod-shaped cells of *B. intermedius* are seen to form very discrete microcolonies within the biofilm.

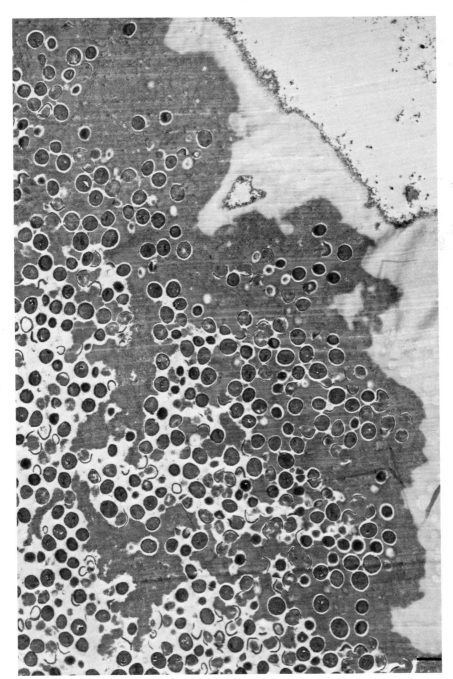

Figure 24. TEM of a ruthenium red-stained preparation of a staphylococcal biofilm that developed on the plastic catheter used to deliver the bacteria to the marrow cavity in Fig. 23. Note the phenomenal production of extracellular glycocalyx material, here stabilized with specific antibodies at the edge of the very large bacterial microcolony.

β-lactam and an aminoglycoside antibiotic, has been shown to override the protective barriers surrounding these cells and to kill the infecting organisms (Rabin *et al.*, 1980), but routine doses of antibiotics have no effect on the pathogens even when laboratory tests of dispersed cells show considerable efficacy of the antibiotic concerned (Irvin *et al.*, 1981).

While anatomical barriers have been invoked as an explanation for the failure of humoral immune factors and antibiotics to resolve bacterial infections of the bones and the lung, no such argument can be made concerning bacterial infections of the endocardium. We have examined the macroscopic "vegetations" formed on the endocardium of infected humans and have found that they consist entirely of viridans group streptococci and their glycocalyces (Chapter 10). Direct examination of the process of the formation of these adherent aggregates on the endocardium of catheterized rabbits shows that an initial glycocalyx-enclosed microcolony is formed on a fibrin clot on the endocardial surface (Fig. 22), and that this microcolony burgeons until a macroscopic bacterial colony develops, within which each streptococcal cell is surrounded by massive amounts of its dextran glycocalyx (Fig. 25). Bacterial endocarditis has a very high mortality, and antibiotics that clearly show efficacy against laboratory cultures of the causative strains are rarely therapeutically effective.

4.3. Role of Bacterial Surface Changes in the Microbial Ecology of Bacterial Infections

The phenotypic plasticity of bacteria is a central feature of their very successful adaptation to a wide variety of natural environments, and this characteristic accounts for their predominance in almost every system in spite of the ubiquity of antibacterial agents and predators. Tragically, however, infectious disease microbiologists have tended to think in terms of rigid taxonomic entities that "have a certain enzyme" or "have pili and a capsule," because these phenotypic characters happen to be expressed in an arbitrarily chosen medium at a certain stage in batch culture. Growth within the animal body is very stressful for bacteria, because of the plethora of specific and nonspecific host defense mechanisms, and yet autochthonous bacteria and a growing number of chronic pathogens accomplish very-long-term persistence—the latter in spite of intensive clinical attempts to dislodge them. Bacteria react to environmental stress by the protean reaction of rapidly altering their phenotypic characters until survival value selects a successful combination of these characters. Thus, a "captive" bacterium in a test tube is a phenotype selected for rapid growth in the total absence of antagonists, and it may bear very little resemblance to the same organism as it grows slowly, e.g., under iron limitation (Brown, 1977), or surrounded by antagonists, e.g., in the infected tissue (Costerton *et al.*, 1981b). Similarly, an initial infecting pathogenic cell may use pili to attach to the target tissue, but the polypeptide appendages may then invite phagocytic attack (Rest *et al.*, 1982), and the organism may then cease to produce them or cover them with glycocalyx. A pathogen confronted with the ruined host defenses of a burned tissue responds by forming mobile invasive cells, while the same organism forms large defensive microcolonies when confronted with the intact host defense systems of a nearly normal lung (Costerton *et al.*, 1979, 1981a,b). An ecological approach is badly needed in the study of bacterial pathogenesis, and one potent potential technique for these studies is the direct immunocytological examination of pathogenic bacteria as they grow on the surfaces of infected tissues at all stages of the disease to be examined.

5. SUMMARY

Adhesion is a basic bacterial strategem. In its simplest form, bacterial adhesion to an inert substance in an aqueous environment places the cell in a concentrated nutrient zone.

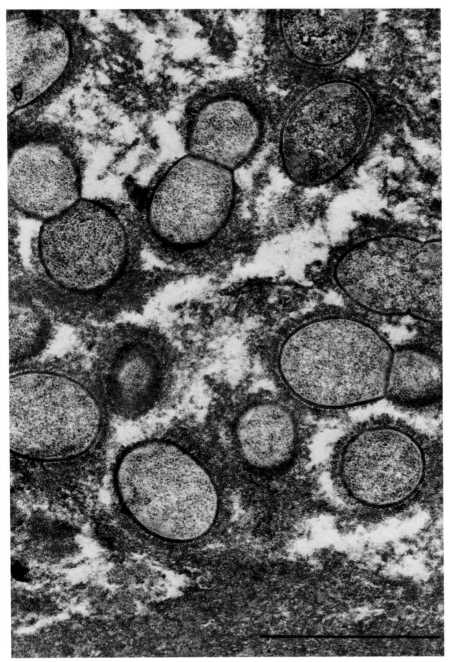

Figure 25. TEM of a ruthenium red-stained, antibody-stabilized preparation of a macroscopic bacterial ''vegetation'' that later developed on the endocardium of the rabbit in Fig. 22. Note the extent to which the viridans group of streptococcal cells that comprise this enormous microcolony are surrounded by their fibrous, dextran exopolysaccharide glycocalyces.

This nutrient concentration is reinforced when the cell produces an anionic exopolysaccharide glycocalyx that mediates adhesion and acts as an ion-exchange resin to attract and bind organic and inorganic nutrients in the area immediately surrounding the adherent cell. Cells that find themselves in a favorable nutrient environment proliferate within the glycocalyx matrix to form adherent microcolonies, and physiologically cooperative cells may be recruited by specific mutual adhesion to produce a functional consortium. These developing microcolonies condition the environment of their component cells by concentrating enzymes and metabolic products and by excluding gases such as oxygen. Because the area immediately under a developing microcolony is intensely locally affected by bacterial products, bacterial digestion and bacterial corrosion has a strong tendency to form pits in insoluble substrates. The glycocalyx-enclosed adherent microcolony is also a defensive aggregate in which the anionic matrix protects the cells from antagonists such as antibiotics, biocides, surfactants, complement, and antibodies and from antibacterial agents such as bacteriophages, free-living amoebae, and animal phagocytes. If antagonist pressure relents somewhat, bacteria within protected microcolonies are able to dispatch mobile "swarmer" cells to colonize new available surfaces. Once microcolonies become very large, or coalesce to form a coherent biofilm, their component cells can only be killed by concentrations of antagonists capable of saturating and penetrating the protective glycocalyx matrix, and sensitivity data generated using single detached cells are not predictive. Biofilms can be removed from inert surfaces by mechanical means, by the recently developed ice nucleation technique, or by digestion using bleach and other low-residue reagents.

REFERENCES

Acres, S. D., Isaacson, R. E., Babiuk, L. A., and Kapitany, R. A., 1979, Immunization of calves against enterotoxigenic colibacillosis by vaccinating dams with purified K99 antigen and whole cell bacterins, *Infect. Immun.* **25:**121–126.

Akin, D. E., and Amos, H. E., 1975, Rumen bacterial degradation of forage cell walls investigated by electron microscopy, *Appl. Microbiol.* **29:**692–701.

Baker, N. R., and Marcus, H., 1982, Adherence of clinical isolates of *Pseudomonas aeruginosa* to hamster tracheal epithelium in vitro, *Curr. Microbiol.* **7:**35–40.

Baltimore, R. S., and Mitchell, M., 1980, Immunologic investigation of mucoid strains of *Pseudomonas aeruginosa*: Comparison of susceptibility to opsonic antibody in mucoid and non-mucoid strains, *J. Infect. Dis.* **141:**238–247.

Birdsell, D. C., Doyle, R. J., and Morgenstern, M., 1975, Organization of teichoic acid in the cell wall of *Bacillus subtilis*, *J. Bacteriol.* **121:**726–734.

Brooks, B. W., Barnum, D. A., and Meek, A. H., 1983, An observational study of *Corynebacterium bovis* in Ontario dairy herds, *Can. J. Comp. Med.* **47:**73–78.

Brown, M. R. W., 1977, Nutrient depletion and antibiotic susceptibility, *J. Antimicrob. Chemother.* **3:**198–201.

Bryant, M. P., Wolin, E. A., Wolin, M. J., and Wolfe, R. S., 1967, *Methanobacillus omelianskii*, a symbiotic association of two species of bacteria, *Arch. Mikrobiol.* **59:**20–31.

Chan, R., Acres, S. D., and Costerton, J. W., 1982a, The use of specific antibody to demonstrate glycocalyx, K99 pili, and the spatial relationships of K99+ enterotoxigenic *E. coli* in the ileum of colostrum-fed calves, *Infect. Immun.* **37:**1170–1180.

Chan, R., Lian, C. J., Costerton, J. W., and Acres, S. D., 1982b, The use of specific antibodies to demonstrate the glycocalyx and spatial relationships of K99⁻,F41⁻, enterotoxigenic strain of *Escherichia coli* colonizing the ileum of colostrum-deprived calves, *Can. J. Comp. Med.* **47:**150–156.

Cheng, K.-J., and Costerton, J. W., 1980, Adherent rumen bacteria: Their role in the digestion of plant material, urea and epithelial cells, in: *Digestive Physiology and Metabolism in Ruminants* (Y. Ruckebush and P. Thivend, eds.), MTP Press, Lancaster, pp. 227–250.

Cheng, K.-J., Akin, D. E., and Costerton, J. W., 1977, Rumen bacteria: Interaction with dietary components and response to dietary variation, *Fed. Proc. Fed. Am. Soc. Exp. Biol.* **36:**193–197.

Cheng, K.-J., Fay, J. P., Howarth, R. E., and Costerton, J. W., 1980, Sequence of events in the digestion of fresh legume leaves by rumen bacteria, *Appl. Environ. Microbiol.* **40:**613–625.

Cheng, K.-J., Irvin, R. T., and Costerton, J. W., 1981, Autochthonous and pathogenic colonization of animal tissues by bacteria (a review), *Can. J. Microbiol.* **27**:461–490.

Clementz, D. M., Patterson, D. E., Aseltine, R. J., and Young, R. E., 1982, Stimulation of water injection wells in the Los Angeles Basin by using sodium hypochlorite and mineral acids, *J. Pet. Technol.* September:2087–2096.

Costerton, J. W., and Geesey, G. G., 1979, Microbial contamination of surfaces, in: *Surface Contamination* (K. L. Mittal, ed.), Plenum Press, New York, pp. 211–221.

Costerton, J. W., and Lashen, E. S., 1983, The inherent biocide resistance of corrosion-causing biofilm bacteria, Corrosion 83, Paper 246, National Association of Corrosion Engineers, pp. 246/1–246/11.

Costerton, J. W., Brown, M. R. W., and Sturgess, J. M., 1979, The cell envelope: Its role in infection, in *Pseudomonas aeruginosa*, in: *Clinical Manifestations of Infection and Current Therapy* (R. G. Doggett, ed.), Academic Press, New York, pp. 41–62.

Costerton, J. W., Irvin, R. T., and Cheng, K.-J., 1981a, The bacterial glycocalyx in nature and disease, *Annu. Rev. Microbiol.* **35**:299–324.

Costerton, J. W., Irvin, R. T., and Cheng, K.-J., 1981b, The role of bacterial surface structures in pathogenesis, *CRC Crit. Rev. Microbiol.* **8**:303–338.

Costerton, J. W., Rozee, K. R., and Cheng, K.-J., 1983, Colonization of particulates, mucous and intestinal tissue, in: *Progress in Food and Nutrition Science,* Pergamon Press, Elmsford, N.Y.

Dazzo, F. B., and Brill, W. J., 1977, Receptor site on clover and alfalfa roots for *Rhizobium, Appl. Environ. Microbiol.* **33**:132–136.

Doggett, R. G., Harrison, G. M., and Wallis, E. S., 1964, Comparison of some properties of *Pseudomonas aeruginosa* isolated from infections of persons with and without cystic fibrosis, *J. Bacteriol.* **87**:427–431.

Fazio, S. A., Uhlinger, D. J., Parker, J. H., and White, D. C., 1982, Estimations of uronic acids as quantitative measures of extracellular and cell wall polysaccharide polymers from environmental samples, *Appl. Environ. Microbiol.* **43**:1151–1159.

Forsberg, C. W., Beveridge, T. J., and Hellstrom, A., 1981, Cellulose and xanthanase release from *Bacteroides succinogenes* and its importance in the rumen environment, *Appl. Environ. Microbiol.* **42**:886–896.

Geesey, G. G., Mutch, R. J., Costerton, J. W., and Green, R. B., 1978, Sessile bacteria: An important component of the microbial population in small mountain streams, *Limnol. Oceanogr.* **23**:1214–1223.

Gibbons, R. J., and van Houte, J., 1975, Bacterial adherence in oral microbial ecology, *Annu. Rev. Microbiol.* **29**:19–44.

Govan, J. R. W., 1975, Mucoid strains of *Pseudomonas aeruginosa*: The influence of culture medium on the stability of mucus production, *J. Med. Microbiol.* **8**:513–522.

Govan, J. R. W., and Fyfe, J. A. M., 1978, Mucoid *Pseudomonas aeruginosa* and cystic fibrosis: Resistance of the mucoid form to carbenicillin, flucloxacillin, tobramycin and the isolation of mucoid variants in vitro, *J. Antimicrob. Chemother.* **4**:233–240.

Govan, J. R. W., Fyfe, J. A. M., and McMillan, G., 1979, The instability of mucoid *Pseudomonas aeruginosa*: Fluctuation test and improved stability of the mucoid form in shaken culture, *J. Gen. Microbiol.* **10**:229.

Henrici, A. T., 1933, Studies of freshwater bacteria. I. A direct microscopic technique, *J. Bacteriol.* **25**:277–286.

Irvin, R. T., Govan, J. R. W., Fyfe, J. A. M., and Costerton, J. W., 1981, Heterogeneity of antibiotic resistance in mucoid isolates of *Pseudomonas aeruginosa* obtained from cystic fibrosis patients: Role of outer membrane proteins, *Antimicrob. Agents Chemother.* **19**:1056–1063.

Johanson, W. G., Jr., Pierce, A. K., and Sanford, J. P., 1969, Changing pharyngeal bacterial flora of hospitalized patients: Emergence of gram-negative bacilli, *N. Engl. J. Med.* **281**:1137–1140.

Ladd, T. I., Costerton, J. W., and Geesey, G. G., 1979, Determination of the heterotrophic activity of epilithic microbial populations, in: *Native Aquatic Bacteria: Enumeration, Activity and Ecology* (J. W. Costerton and R. R. Colwell, eds.), American Society of Testing Materials, Philadelphia, pp. 180–195.

Lam, J., Chan, R., Lam, K., and Costerton, J. W., 1980, Production of mucoid microcolonies by *Pseudomonas aeruginosa* within infected lungs in cystic fibrosis, *Infect. Immun.* **28**:546–556.

Lambden, P. R., Heckels, J. E., and Watt, P. J., 1982, Effect of anti-pilus antibodies on survival of gonococci within guinea pig subcutaneous chambers, *Infect. Immun.* **38**:27–30.

Leake, E. S., and Wright, M. J., 1979, Variations on the form of attachment of rabbit alveolar macrophages to various substrata as observed by scanning electron microscopy, *J. Reticuloendothel. Soc.* **25**:417–423.

Levin, R. J., 1979, Fundamental concepts of the structure and function of the gastrointestinal epithelium, in: *Scientific Basis of Gastroenterology* (H. L. Duthie, ed.), Churchill Livingstone, Edinburgh, pp. 307–351.

Lock, M. A., Wallace, R. R., Costerton, J. W., Ventullo, R. M., and Charlton, S. E., 1984, River epilithon: Towards a structural–functional model, *Oikos* **42**:10–22.

Luft, J. H., 1971, Ruthenium red and ruthenium violet. I. Chemistry, purification, methods for use for electron microscopy, and mechanism of action, *Anat. Rec.* **171**:347–368.

McInerney, M. J., Bryant, M. P., Hespell, R. B., and Costerton, J. W., 1981, *Syntrophomonas wolfei* gem. nov. sp. nov., an anaerobic, syntrophic, fatty acid oxidizing bacterium, *Appl. Environ. Microbiol.* **41**:1029–1039.

Mackie, E. B., Brown, K. N., Lam, J., and Costerton, J. W., 1979, Morphological stabilization of capsules of group B streptococci, types Ia, Ib, II and III, with specific antibody, *J. Bacteriol.* **138**:609–617.

Mah, R. A., Ward, D. M., Baresi, L., and Glass, T. L., 1977, Biogenesis of methane, *Annu. Rev. Microbiol.* **31**:309–341.

Marrie, T. J., and Costerton, J. W., 1983a, A scanning electron transmission microscopic study of an infected endocardial pacemaker lead, *Circulation* **66**:1339–1343.

Marrie, T. J., and Costerton, J. W., 1983b, Bacterial colonization of intrauterine contraceptive devices, *Am. J. Obstet. Gynecol.* **146**:384–394.

Marrie, T. J., and Costerton, J. W., 1983c, A scanning electron microscopic study of the adherence of uropathogens to a plastic surface, *Appl. Environ. Microbiol.* **45**:1018–1024.

Marrie, T. J., Harding, G. K. M., and Ronald, A. R., 1976, Anaerobic and aerobic urethral flora in healthy females, *J. Clin. Microbiol.* **8**:67–72.

Marrie, T. J., Harding, G. K. M., Ronald, A. R., Dikkema, J., Lam, J., Hoban, S., and Costerton, J. W., 1979, Influence of mucoidy on antibody coating of *Pseudomonas aeruginosa*, *J. Infect. Dis.* **139**:357–361.

Marrie, T. J., Lam, J., and Costerton, J. W., 1980, Bacterial adhesion to uroepithelial cells—A morphological study, *J. Infect. Dis.* **142**:239–246.

Marshall, K. C., Stout, R., and Mitchell, R., 1971, Mechanism of the initial events in the sorption of marine bacteria to surfaces, *J. Gen. Microbiol.* **68**:337–348.

Minato, H., and Suto, T., 1976, Technique for fractionation of bacteria in rumen microbial ecosystem. I. Attachment of rumen bacteria to starch granules and elution of bacteria attached to them, *J. Gen. Appl. Microbiol.* **22**:259–276.

Moon, H. W., 1974, Pathogenesis of enteric diseases caused by *Escherichia coli*, *Adv. Vet. Sci. Comp. Med.* **18**:179–211.

Morris, J. A., Thorns, C., Scott, A. C., Sojka, W. J., and Wells, G. A., 1982, Adhesion *in vitro* and *in vivo* associated with an adhesive antigen (F41) produced by a K99 mutant of the reference strain *Escherichia coli* B41, *Infect. Immun.* **36**:1146–1153.

Patterson, H., Irvin, R., Costerton, J. W., and Cheng, K.-J., 1975, Ultrastructure and adhesion properties of *Ruminococcus albus*, *J. Bacteriol.* **112**:278–287.

Peters, G., Locci, R., and Pulverer, G., 1982, Adherence and growth of coagulase-negative staphylococci: On the surface of intravenous catheters, *J. Infect. Dis.* **146**:479–482.

Pringle, J. H., and Fletcher, M., 1983, Influence of substratum wettability on attachment of freshwater bacteria to solid surfaces, *Appl. Environ. Microbiol.* **45**:811–817.

Rabin, H. R., Harley, F. L., Bryan, L. E., and Elfring, G. L., 1980, Evaluation of a high dose tobramycin and ticarcillin treatment protocol in cystic fibrosis based on improved susceptibility criteria and antibiotic pharmacokinetics, in: *Perspectives in Cystic Fibrosis* (J. M. Sturgess, ed.), Canadian Cystic Fibrosis Foundation, Toronto, pp. 370–375.

Rest, R. F., Fischer, S. H., Ingham, Z. Z., and Jones, J. F., 1982, Interactions of *Neisseria gonorrhoeae* with human neutrophils: Effects of serum and gonococcal opacity on phagocyte killing and chemiluminescence, *Infect. Immun.* **36**:737–744.

Roseman, S., 1974, Complex carbonydrates and intercellular adhesion, in: *Biology and Chemistry of Eucaryotic Cell Surfaces* (E. Y. C. Lee and E. E. Smith, eds.), Academic Press, New York, pp. 317–347.

Rozee, K. R., Cooper, D., Lam, K., and Costerton, J. W., 1982, Microbial flora of the mouse ileum mucous layer and epithelial surface, *Appl. Environ. Microbiol.* **43**:1451–1463.

Ruseska, I., Robbins, J., Lashen, E. S., and Costerton, J. W., 1982, Biocide testing against corrosion-causing oilfield bacteria helps control plugging, *Oil Gas J.* March **8**:253–264.

Savage, D. C., 1977, Microbial ecology of the gastrointestinal tract, *Annu. Rev. Microbiol.* **31**:107–133.

Schwarzmann, S., and Boring, J. R., III, 1971, Antiphagocytic effect of slime from a mucoid strain of *Pseudomonas aeruginosa*, *Infect. Immun.* **3**:762–767.

Shaw, J. C., Bramhill, B., Wardlaw, N. C., and Costerton, J. W., 1984, Bacterial fouling in a model core system, *Appl. Env. Microbiol*, in press.

Sleytr, U. B., 1978, Regular array of macromolecules on bacterial cell walls: Structure, chemistry, assembly, and function, *Int. Rev. Cytol.* **53**:1–44.

Sprunt, K., Leidy, G., and Redman, W., 1980, Abnormal colonization of neonates in an ICU: Conversion to normal colonization by pharyngeal implantation of alpha hemolytic streptococcus strain 215, *Pediatr. Res.* **14**:308–313.

Sutherland, I. W., 1977, Bacterial polysaccharides: Their nature and production, in: *Surface Carbohydrates of the Prokaryotic Cell* (I. W. Sutherland, ed.), Academic Press, New York, pp. 27–96.

Svanborg-Eden, C., Eriksson, B., and Hansen, L. A., 1977, Adhesion of *Escherichia coli* to human uroepithelial cells "in vitro," *Infect. Immun.* **18:**767–774.

Swanson, J., 1981, Surface-exposed protein antigens of the gonococcal outer membrane, *Infect. Immun.* **34:**804–816.

Thomassen, M. J., Demke, C. A., Wood, R. E., and Sherman, J. M., 1982, Phagocytosis of *Pseudomonas aeruginosa* by polymorphonuclear leukocytes and monocytes: Effect of cystic fibrosis serum, *Infect. Immun.* **38:**802–805.

Whittenbury, R., and Dow, C. S., 1977, Morphogenesis and differentiation in *Rhodomicrobium vannielii* and other budding and prothecate bacteria, *Bacteriol. Rev.* **41:**754–808.

Woods, D. E., Boss, J. A., Johanson, W. G., Jr., and Straus, D. C., 1980, Role of adherence in the pathogenicity of *Pseudomonas aeruginosa* lung infections in cystic fibrosis patients, *Infect. Immun.* **30:**694–699.

ZoBell, C. E., 1943, The effect of solid surfaces on bacterial activity, *J. Bacteriol.* **46:**39–56.

2

Bacterial Cell Walls and Surfaces

ANTHONY J. WICKEN

1. INTRODUCTION

Three decades ago, Salton and Horne (1951) described a ballistic method for the disintegration of bacteria and isolation of a cellular fraction that retained the original shape of the bacterial cells but was relatively free of cell membrane and cytoplasmic components. We now recognize that virtually all bacterial cells are protected from osmotic disruption by the presence of a cell wall or cell envelope of high tensile strength distal to the plasma or cell membrane. Since the 1950s, enormous strides have been made in describing the chemistry, biosynthesis, and function of these chemically unique prokaryotic structures in conjunction with their topography and architecture as seen in the electron microscope. The wealth of information that is now available has formed the basis of numerous chapters, monographs, and reviews. The reader is referred to some of the more recent of these: Daneo-Moore and Shockman (1977), Duckworth (1977), Ghuysen (1977), Sargent (1978), Sleytr (1978), Tipper and Wright (1979), Tomaz (1979), Tonn and Garder (1979), Wright and Tipper (1979), Rogers *et al.* (1980), Wicken and Knox (1980), Beveridge (1981), Rogers (1981), Shockman *et al.* (1981), Ward (1981), Shockman and Barrett (1983). It should be noted that literature citations in the text of this review are generally intended as illustrative rather than exhaustive and wherever possible are directed to recent review articles.

Most species of bacteria can be divided into two groups on the basis of the gram-staining reaction. Cell wall or envelope appearance in the electron microscope is generally markedly different for the two groups. Thin sections of gram-positive bacteria usually reveal a well-defined, 15- to 30-nm-thick, rigid outer cell wall and an underlying, closely applied, cell-limiting plasma membrane. The wall represents 15 to 30% of the dry weight of the cell and its rigidity lies in its major polymeric component, peptidoglycan or murein. Generally, gram-negative bacteria are seen in thin section as having a thin peptidoglycan layer (1 to 2 nm thick) sandwiched between the inner plasma membrane and an outer membrane with a similar trilamellar appearance to the plasma membrane. The degree of separation between these components varies with the organism and preparation technique used. Freeze-fracture

Anthony J. Wicken • School of Microbiology, University of New South Wales, Kensington, New South Wales 2033, Australia.

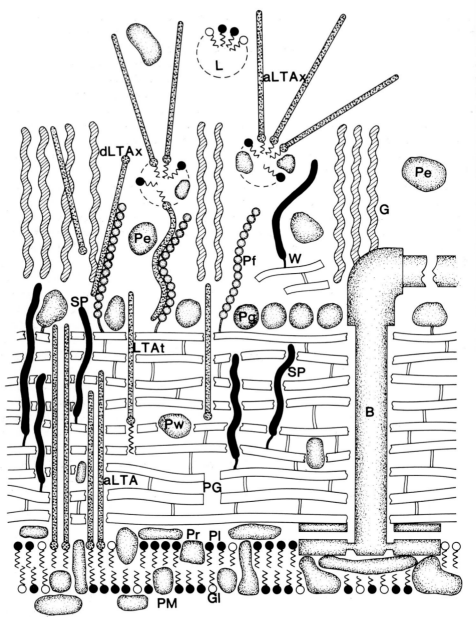

Figure 1. Schematic representation of the organization of the cell wall of gram-positive bacteria. Cross-linked peptidoglycan (PG) is shown overlaying the plasma cell membrane (PM) composed of protein (Pr), phospholipid (Pl), and glycolipid (Gl). Secondary cell wall polymers (SP) such as teichoic acids, teichuronic acids, and polysaccharides are covalently linked to peptidoglycan and extend through the peptidoglycan network with exposure, in some cases, at the surface of the rigid portion of the cell wall. Protein components of the wall include the basal bodies of flagella (B), noncovalently associated proteins (Pw) both within the peptidoglycan matrix and as part of the glycocalyx region. Covalently associated protein on the surfaces of the peptidoglycan is shown in two configurations, globular (Pg) and fibrillar (Pf). Lipoteichoic acid, in its acylated form (aLTA), extends from the upper half of the bilayer of the plasma membrane, to varying degrees depending on chain length, through the peptidoglycan net. Excreted LTA (aLTAx) may exist as mixed micellar aggregates with protein and lipid. Deacylated extracellular LTA (dLTAx) exists as monomer, and the polyglycerophosphate chains of both forms may

techniques have generally confirmed the picture built up from cross-sectional profiles for both gram-positive and gram-negative bacteria. The chemical composition of isolated cell wall/envelope fractions also differs between the two groups. In gram-positive bacteria the cell wall peptidoglycan may be variously and covalently associated with other carbohydrate-type polymers (polysaccharides, teichoic acids, teichuronic acids) and to a lesser extent with proteins. In gram-negative bacteria the outer membrane is a complex structure of lipopoly-saccharides, proteins, phospholipids, and lipoprotein with the last of these components, in part, providing a covalent linkage to the underlying peptidoglycan layer—these two regions together forming the cell wall of gram-negative bacteria. Structures intermediate between these two basic cell wall types have been described in a number of genera as indeed has variability in the gram-staining reaction.

The foregoing represents an overview of the classical picture of the polymeric composi-tions of bacterial cell walls, and the bacterial cell surface, the topic of this chapter, would similarly be classically defined as the outer reaches of the peptidoglycan–polysaccharide complex of the gram-positive bacterial cell wall and the outer surface of the outer membrane in gram-negative bacteria. However, such a definition of the bacterial cell surface is based only on a purely operational fraction—the isolated bacterial cell wall. More recently we have recognized the existence, particularly in gram-positive bacteria, of cell wall-associated com-ponents that are not covalently bonded to other cell wall components. The association may even be transient in that some of these components may be in the process of excretion into the external environment. Such cell wall-associated components are readily lost in preparing cell wall fractions but in the living cell they not only contribute to the physicochemistry of the cell surface but may also extend that surface, as a "gradient of polymers" (the bacterial glycocalyx; Chapter 1, Section 1.1), well beyond the formal boundary of the classical cell wall. Also contributing to the complexity of the bacterial cell surface are such cellular structures as flagella, pili, fimbriae (Chapter 11), etc., many of which are not cell wall components but nevertheless penetrate that structure and are expressed at the cell surface.

The major purpose of this chapter is to examine the pictures that are beginning to emerge of the real "faces" of bacterial cells and to what extent the chemistry and, in turn, the properties of these "faces" are governed by environmental conditions. Schematic repre-sentations of the various components and their locations and organization in the cell wall/membrane complexes of gram-positive and gram-negative bacteria are shown in Figs. 1 and 2. These drawings are intended as being illustrative of the concepts described and developed in the text and not as depicting bacterial cell walls in any literal sense.

2. GRAM-POSITIVE BACTERIA

The cell wall of gram-positive bacteria is conceptually and operationally easier to define than that of gram-negative bacteria. It is readily isolated as a highly purified, insoluble residue from mechanically disrupted bacteria (see Rogers *et al.*, 1980) which still retains the characteristic shape of the species. It is, however, an operational fraction and in the process

---

interact ionically with surface fibrillar proteins. LTA in the process of excretion has a transient association in either the fully acylated or the deacylated form, with the peptidoglycan net (LTAt). Similar orientations are assumed for other amphiphiles in those gram-positive bacteria that do not contain LTA. Also contributing to the glycocalyx region are excreted lipids (L), proteins (Pe), and wall turnover products (W) as well as extracellular polymeric material peculiar to the capsular or glycocalyx region of the cell (G). (The diagram is intended to be illustrative of the concepts described in the text and should not be regarded as a literal depiction or necessarily drawn to scale.)

of its preparation soluble elements, originally associated with this insoluble residue in the living cell, will have been lost. This fraction will contain peptidoglycan and one or more carbohydratelike or true-polysaccharide polymers covalently linked to the peptidoglycan. In some species a small amount of covalently linked protein may also be present. Excluded from this fraction will be capsular and wall-associated components that are not in covalent linkage to peptidoglycan. Thus, while the isolated wall is in some sense an artifact, an understanding of the chemical ultrastructure of this fraction is essential to an understanding of the nature of the cell surface, for it is the solid and rigid base to the face of the gram-positive bacterial cell.

2.1. Peptidoglycan

Peptidoglycan is a major cell wall polymer of gram-positive bacteria, accounting for 30 to 50% of the dry weight of the isolated wall fraction. It is now well recognized that it is the component primarily responsible for the shape, rigidity, and protective function of walls. Chemically, peptidoglycan consists of a network of linear glycan strands made up of repeating β1-4-linked N-acetylglucosaminyl-N-acetylmuramyl dimers which are in turn covalently linked together by cross-links between short tetrapeptides originating from muramyl carboxyl groups. This results in a three-dimensional network, covalently stabilized in all directions, which completely surrounds the cell (for reviews see Ghuysen, 1977; Tipper and Wright, 1979; Rogers et al., 1980). Interspecies variation in peptidoglycan structure is quite considerable, and while this can be useful in chemical taxonomy (Schleifer and Kandler, 1972) the effects of these structural variations on peptidoglycan function are not known. The types of variation that can occur may be summarized as:

1. The length of the glycan strands may vary from 10 disaccharide units to 200 or more both between species and within the walls of a single species. The heterogeneity of chain length may be due, at least in part, to the action of endogenous peptidoglycan hydrolases after the peptidoglycan has been synthesized (Daneo-Moore and Shockman, 1977). Apart from this there is a remarkable degree of uniformity in the structure of glycan strains between species, the only other differences recorded being the presence on some of O-acetyl substitution, phosphate residues, and the loss of some N-acetyl groups.

2. The amino acid composition of the peptides substituting muramyl carboxyl groups and the interpeptide cross-linking is the main area of interspecies peptidoglycan variation (Schleifer and Kandler, 1972). Within a species not all muramyl residues are necessarily substituted with peptides and variations can occur in the amino acid composition of cross-bridges.

3. Considerable variation occurs between species in the degree of cross-linking, being as low as 20% in *Micrococcus luteus* and as high as 75% in *Staphylococcus aureus*, in terms of the total potential cross-links. The distribution of these cross-links throughout the cell wall is also unknown (Ghuysen, 1977; Tipper and Wright, 1979).

While the chemical structure of peptidoglycan is well understood, the three-dimensional spatial arrangement of this unique and very large macromolecule still remains to be unraveled. This aspect has been well reviewed by Ghuysen (1977), Rogers et al. (1980), Beveridge (1981), and Shockman and Barrett (1983). X-ray diffraction studies, NMR, infrared spectroscopy, and other physical techniques tend, in total, to confirm the idea of a rigid and paracrystalline peptidoglycan network with NMR studies (Lapidot and Irving, 1979) indicating that most of the motional freedom lies in the peptide cross-links, the glycan chains and their peptide tails being relatively rigid. Peptidoglycan cannot, however, possess a continuous crystalline face for both exogenous and endogenous lytic enzymes can penetrate the structure, suggesting that lattice defects of considerable size must exist in the structure (Beveridge, 1981). Further, as will be discussed later, both covalently and noncovalently associated polymers penetrate this complex network. The orientation of the glycan strands has also proved difficult to ascertain but it is believed from currently available evidence that the glycan strands in rod-shaped organisms are oriented either perpendicular to or helically to the long axis of the cell. Differences in the topology of glycan strands in cross-walls (which become poles) and in cocci may relate also to differences in assembly (for a detailed discussion see Daneo-Moore and Shockman, 1977; Shockman and Barrett, 1983).

2.2. Secondary Cell Wall Polymers

Walls of gram-positive bacteria may contain one or more secondary or accessory polymers that are attached directly, or indirectly via phosphodiester bonds, to carbon 6 of N-acetylmuramyl residues of the peptidoglycan.

2.2.1. Teichoic Acids

(For reviews see Duckworth, 1977; Rogers et al., 1980; Ward, 1981.)

Teichoic acids are polymers of either ribitol phosphate or glycerophosphate joined by phosphodiester bonds to alcohol groups of the polyol residues. Secondary hydroxyl groups may be variously substituted with sugar or amino-sugar residues in glycosidic linkage and D-alanyl residues in ester linkage. The proportion of D-alanyl ester to carbohydrate residues, total substitution of the polymer, and nature of the carbohydrate substituents are variable with the species and, to some extent, with the growth conditions employed. Chain lengths of from 10 to 40 or more polyol units have been reported, again depending on the species and also the mildness of the extraction procedure used. While restricted to gram-positive bacteria, not all species contain teichoic acids and the presence of a teichoic acid within a single species is generally restricted to either the glycerol or the ribitol-polymer type. Evidence for a direct phosphodiester linkage between teichoic acid and carbon 6 of muramic acid has been obtained only in one instance (Button et al., 1966). More generally the linkage between the two polymers appears to be via a linkage unit consisting of three units of glycerophosphate attached to N-acetylglucosamine which in turn is linked through its reducing group to a phosphate ester group on carbon 6 of a muramic acid residue (see reviews listed above). Variations on the polyolphosphate polymer theme for glycerol teichoic acids are known in which glycosidically linked sugars or sugar 1-phosphate residues form part of the main chain (Duckworth, 1977). In some species glycerol teichoic acids may be 1:2-phosphodiester-linked rather than 1:3-linked.

2.2.2. Teichuronic Acids

(For reviews see Rogers et al., 1980; Ward, 1981.)

Teichuronic acids are a group of acidic polysaccharides that, in some gram-positive organisms, can replace teichoic acids when the organisms are grown under limiting phosphate conditions (see Section 2.5). In Bacillus licheniformis 6346 and B. subtilis var. niger this polymer consists of equimolar proportions of N-acetylgalactosamine and D-glucuronic acid, some 23 to 25 disaccharide units in length. Similar polymers have been found in Micrococcus luteus NCTC 2665 and strains of Staphylococcus aureus, B. megaterium, and Corynebacterium species showing more complex sugar compositions but having a common feature of a uronic or aminouronic acid constituent (Rogers et al., 1980). Linkage of teichuronic acids to carbon 6 of muramyl residues may be direct through a phosphodiester in the case of B.

licheniformis 6346, but in the case of *M. luteus*, where the polymer has a repeating unit represented by →4)-*N*-acetyl-D-mannosaminouronosyl-β(1,6)-D-glucosyl-α(1→, linkage to peptidoglycan appears to be via a linkage unit consisting of two *N*-acetylmannosaminouronic acid residues joined to an *N*-acetylglucosamine residue in phosphodiester linkage to the peptidoglycan (Rohr *et al.*, 1977).

Another type of acidic cell wall polysaccharide is represented by the phosphorylated heteropolysaccharide of *Lactobacillus salivarius* (Knox *et al.*, 1980). Phosphorylated polysaccharides containing sugar 1-phosphate residues have also been reported from *Micrococcus* sp. and *S. lactis* (see Duckworth, 1977).

2.2.3. Neutral Polysaccharides

Many gram-positive bacteria have neutral cell wall polysaccharides which are generally heteropolymers of two to four different neutral or amino sugars. The chemical complexity of the polysaccharides of gram-negative bacteria is not mirrored in gram-positive bacterial cell walls. Wall polysaccharides are particularly prevalent among the streptococci and lactobacilli and many of them are used as typing or grouping antigens (see reviews by Knox, 1970; Sharpe, 1970). Strains of *L. casei* belonging to either serological group B or C can be differentiated on the basis of group-specific cell wall polysaccharides. Group B polysaccharide contains rhamnose, glucose, *N*-acetylglucosamine, and *N*-acetylgalactosamine and sometimes galactose. α-L-Rhamnosyl residues are the major component as well as the main antigenic determinant. Group C polysaccharide lacks rhamnose, and glucose is immunodominant with the trisaccharide *O*-β-D-glucosyl-1→6-*O*-β-D-galactosyl-1→6-*N*-acetylglucosamine being a major part of the antigenic determinant of the polysaccharide (Knox, 1970). In group A streptococci, the wall polysaccharide antigen is a polymer of L-rhamnose and *N*-acetylglucosamine (in a molar ratio of 2:1), the immunodominant amino sugar being β-linked through their reducing terminals to a 1:3-linked polyrhamnose core. Group C streptococcal wall polysaccharide resembles that of Group A except the determinant amino sugar is *N*-acetylgalactosamine (McCarty and Morse, 1964).

Evidence for linkage of wall polysaccharides to muramic acid via a phosphodiester bond has been obtained in a number of cases (Liu and Gotschlich, 1967; Coyette and Ghuysen, 1970; Knox, 1970). The linkage is acid-labile and polysaccharides are readily cleaved from isolated walls by mild acid treatment. Autoclaving at neutral pH can also cleave this linkage, better yields of polysaccharide being obtained when the cell walls contain other acidic groups (Campbell *et al.*, 1978).

In *L. casei* subsp. *rhamnosus*, in addition to the group C polysaccharide, there is a type-specific rhamnose-containing polysaccharide. Both polysaccharides are linked to peptidoglycan. In a variant of the same organism, the rhamnose polysaccharide has been lost from its covalent attachment to the cell wall but is still produced as capsular material (Knox, 1970). It is possible that loss of the ability to link polysaccharide to peptidoglycan may explain the presence of polysaccharide capsules in a number of bacteria. In group B streptococci both group- and type-specific polysaccharides are found covalently linked to peptidoglycan and extracellularly as a capsular slime layer (De Cueninck *et al.*, 1982).

2.2.4. Organization of Secondary Polymers

The total amount of cell wall per cell can vary widely with growth conditions in gram-positive bacteria; however, the relation between amount of peptidoglycan and total amount of secondary polymers is maintained (Shockman and Barrett, 1983). This is not surprising given the evidence that peptidoglycan chains are synthesized with attached secondary polymers before insertion into the peptidoglycan network of the wall. It would appear, at least with acidic secondary polymers, that only one type of polymer is attached to a single glycan chain (Ward and Curtis, 1982) and molar ratio considerations would seem to demand that every glycan strand has an accompanying secondary polymer (Shockman and Barrett, 1983). NMR studies have indicated that secondary polymers are more mobile than the peptidoglycan net to which they are attached (Lapidot and Irving, 1979) and are usually depicted (Ghuysen, 1977) as extending perpendicular to the glycan strands toward the cell surface. These considerations together with what is known of the directional sequence of wall synthesis and subsequent wall turnover (Daneo-Moore and Shockman, 1977; Shockman and Barrett, 1983) would seem to obviate against any preferential layering of secondary polymers in the cell wall. Some ultrastructural and probe studies (see Beveridge, 1981, for summary) have suggested a concentration of secondary polymers at the cell surface. However, assuming a perpendicular orientation of these molecules within the wall, detection at the cell surface could be a reflection both of surface polymer as well as polymer with attachment points deeper within the wall structure.

2.3. Proteins

Walls of most gram-positive bacteria contain some protein but generally not in amounts approaching those of the secondary polymers. Further, in most cases the proteins are cell wall associated rather than covalently bound to peptidoglycan (see Section 2.4.2). However, in *S. aureus,* protein A is found as a 42,000-dalton protein bound to peptide amino groups in the peptidoglycan as well as a free excreted product in the medium (Sjöquist *et al.,* 1972; Lindmark *et al.,* 1977). In group A streptococci a surface M protein is strongly associated with the cell wall but whether or not this is due to linkage to peptidoglycan or represents a fimbrial association (see Section 4.2) is not known.

2.4. Cell Wall-Associated Polymers

Thus far we have considered the gram-positive bacterial cell wall as a complex but covalently linked heteropolymeric structure overlying and protecting the plasma membrane. Associated noncovalently with this structure are other chemical components, some of which, such as the amphiphiles, may also retain an association with the cell membrane while others, such as proteins and polysaccharides, may be variously and loosely associated with the cell wall in depth or exclusively at the cell surface as capsule or glycocalyx.

2.4.1. Amphiphiles

The term *amphiphile* has been used to describe a diverse range of polymers that share certain physicochemical properties, namely the presence of both hydrophobic and hydrophilic regions in their structure. Amphiphiles are present in all membranes, phospholipids being a simple and ubiquitous example. However, bacterial cell membranes contain a number of specialized amphiphiles that have generated considerable interest because of their wide-ranging biological activities and properties. The oldest known of these is lipopolysaccharide or endotoxin from gram-negative bacteria (see Section 3.3.1) but we now know that there are amphiphile counterparts in gram-positive bacteria that share many of the biological properties of lipopolysaccharides.

2.4.1a. Lipoteichoic Acids. *Lipoteichoic acid* is a generic term used for a group of structurally related amphiphiles. The hydrophilic portion of the molecule is a glycerol

teichoic acid, a 1→3 phosphodiester-linked polymer of 25 to 30 units of glycerophosphate variously substituted on carbon 2 with sugars or D-alanine. The phosphomonester end of the polymer is linked covalently to the hydrophobic portion of the molecule, generally either a glycolipid or phosphatidylglycolipid that are also found as free lipid constituents of the bacterial cell membrane. Lipoteichoic acids show genus and species variation in the degree of polymerization of the hydrophilic chain, in the nature and degree of glycosidic substitution, in the extent of D-alanyl ester substitution, and in the structure of the lipid moiety. Lipoteichoic acids are found in a wider range of gram-positive bacteria than the cell wall glycerol teichoic acids and their synthesis is independent of the synthesis of the wall polymer (for reviews on the structure, properties, and synthesis of lipoteichoic acids, see Knox and Wicken, 1973; Wicken and Knox, 1975; Duckworth, 1977; Wicken, 1980; Wicken and Knox, 1980, 1983; Shockman and Wicken, 1981; Ward, 1981).

A membrane lipid component present in lipoteichoic acid suggested a cell membrane location for these molecules—by intercalation of the fatty acid residues of the lipid moiety as an ''anchor'' into the upper half of the bilayer of the membrane. Serological detection of lipoteichoic acid at the cell wall surface of some bacteria and specific immunochemical labeling of thin sections of bacterial cells led to an early model of lipoteichoic acid localization in which the long hydrophilic portion of the molecule could penetrate the peptidoglycan network of the cell wall and, in some cases, be detected at the cell surface while still retaining a cell membrane association with the lipid end of the molecule (van Driel et al., 1973). Subsequently it was shown that lipoteichoic acid is more locationally mobile than this model would suggest. Many organisms excrete lipoteichoic acid into the environment during the normal process of growth and division without cell wall turnover or cell lysis and also in different molecular forms (Markham et al., 1975; Joseph and Shockman, 1975). Both cellular and extracellular lipoteichoic acid can exist as either a high-molecular-weight micellar aggregate and true amphiphile or a lower-molecular-weight deacylated monomer form which, lacking fatty acid substituents, cannot undergo hydrophobic aggregation. The relative proportions of both forms in the two locations as well as the proportion of acylated to deacylated polymer vary with the species and condition of growth (see Section 2.5). Excretion of lipoteichoic acid in either form appears to be an ongoing process throughout the bacterial growth cycle and demands passage of these molecules from the cell membrane, through the cell wall to the external environment. As a result of this, detection of lipoteichoic acid as a cell surface component may reflect either a transient situation (Wicken and Knox, 1977, 1980) or binding of excreted polymer to some other surface such as protein (Ofek et al., 1982). Lipoteichoic acids interact both ionically and hydrophobically with proteins and will also form tenacious complexes with polysaccharides. The lipid ends of lipoteichoic acids also readily allow these molecules to bind to various eukaryotic cell membranes—a property shared by all bacterial cell amphiphiles. These properties of lipoteichoic acids, and other evidence, have suggested a model for the adhesion of S. pyogenes cells to eukaryotic cells via excreted lipoteichoic acid bound ionically to a bacterial cell surface protein and hydrophobically to the eukaryotic cell membrane (Ofek et al., 1982; Chapter 10, Section 2). It would appear certain that excreted lipoteichoic acid is likely to be included in capsular or glycocalyx material whether this be protein as discussed above or polysaccharide, an example of the latter being the detection of lipoteichoic acid in oral plaque (Rølla, 1981). In S. mutans strains grown on defined media, not only is lipoteichoic acid excreted in appreciable quantities but also 15 to 30% (depending on the growth phase) of the total membrane lipid (Cabacungan and Pieringer, 1980). The extent to which lipid excretion is a general phenomenon and its relation to growth phase in gram-positive bacteria remain to be determined.

2.4.1b. Other Gram-Positive Bacterial Amphiphiles.

Not all gram-positive species of bacteria contain lipoteichoic acids. *Micrococcus* species produce instead a lipomannan which is a linear polymer of some 52 to 72 D-mannose residues, about a quarter of which are succinylated. A glycolipid is covalently linked to one end of the polymer (Owens and Salton, 1975; Powell *et al.*, 1975). A complex and incompletely characterized fatty acid-substituted heteropolysaccharide which is also negatively charged has been isolated from *Actinomyces viscosus* and detected in other actinomycetes (Wicken *et al.*, 1978). In both cases, these polymers, like lipoteichoic acid, are membrane localized, detectable at cell surfaces and also in the external environment. Some chemical variation on the common theme of a ''carbohydrate''–lipid amphipathic complex may well be a feature of all gram-positive bacteria.

2.4.2. Proteins

Reference was made earlier to wall-associated protein being more common than covalently linked protein in gram-positive bacterial cell walls. One important group of such proteins are the peptidoglycan hydrolases many of which show a strong affinity to the isolated cell wall. The role of these enzymes in cell wall synthesis, turnover, and post-synthesis remodeling has been well reviewed by Daneo-Moore and Shockman (1977) and Shockman and Barrett (1983).

Detection of proteins at the cell surface may, as in the case of streptococcal M protein, reflect the presence of ''fimbriae'' or other filamentous cellular protrusions (see Section 4). However, other surface protein antigens (T and R) have been described for group A streptococci as well as other streptococcal species. Gram-positive bacteria are also well known for their excretion of a wide range of hydrolytic enzymes—unlike gram-negative organisms which tend to retain such enzymes in the periplasmic space (for reviews see Glenn, 1976; Aaronsen, 1981). A transient association of these excreted proteins with the cell surface would seem to be inevitable if synthesis and excretion is an ongoing process during cellular growth. *Streptococcus mutans* and *S. sanguis* strains are recognized to produce a range of proteins some of which are found associated with the cell wall as well as in the extracellular environment. Some of these are enzymes involved in extracellular glucan and fructan synthesis and hydrolysis (Gibbons and van Houte, 1978). Other examples include an IgA protease (Genco *et al.*, 1975; Chapter 15, Section 2.3.2), glucan-binding proteins (Russell, 1979), and antigen B which has antigenic cross-reactivity with human heart tissue (Russell, 1980). The probable involvement of surface proteins in adhesion of oral microorganisms also seems to be generally accepted (see Doyle *et al.*, 1982, for a summary) but the status of such proteins as associated or covalent wall components remains to be determined,

Some bacteria produce regularly structured surface arrays or S-layers which are assemblies of glycoprotein subunits following a number of different geometric formats. This area has been the basis of an excellent review by Sleytr (1978) and is also discussed by Beveridge (1981). These arrays are self-assembly systems driven by minimum free energy considerations with all of the information and energy required for assembly being contained entirely within the glycoprotein subunits. Association with the cell surface is noncovalent but electrostatic and involves, in some cases, specific divalent cations as salt bridges. In some cases there is no absolute requirement for the bacterial cell, glycoprotein subunits, under physiological conditions, readily assembling as sheets, cylinders, or residues, the geometric form depending on the generic origin of the subunits.

2.4.3. Capsules, Slime Layers, and the Glycocalyx

(For reviews see Sutherland, 1977; Costerton *et al.*, 1981a,b.)

Many species of bacteria are capable of producing extracellular polymers which may take a variety of forms in terms of their physical relationship to the producing cell. As a result, descriptive terms that have been used, such as *slime layers, capsules, microcapsules,* etc., are often confusing. Costerton *et al.* (1981a,b; Chapter 1, Section 1.1) have proposed the term *bacterial glycocalyx* to define any (polysaccharide-containing) component found distal to the bacterial cell wall. The term *glycocalyx* is perhaps an unfortunate one in that it tends to suggest an essentially polysaccharide character which is somewhat of an over-simplification of a chemically complex region of the cell. This notwithstanding, the glycocalyx has been further delineated into:

1. S-layers, which are the glycoprotein arrays described in Section 2.4.2
2. Capsules, composed of fibrous material, and to which one or more of the following descriptive terms can be applied:

   a. Rigid, denoting a structural coherence that excludes particles (the classical capsule delineated by India ink staining)
   b. Flexible, denoting a deformable structure that does not exclude particles
   c. Integral, denoting a close association with the cell surface
   d. Peripheral, denoting a variable and generally loose association with the cell surface that would merge with fully extracellular and nonassociated polymeric material.

Most capsules are polysaccharide in nature although some genera, notably *Bacillus,* are capable of forming polypeptide capsules particularly under growth conditions having excess nitrogen. Polysaccharide capsules may be simple homopolymers that are generally synthesized extracellularly. The dextranlike glucans produced by oral streptococci through the action of extracellular glucosyltransferases on sucrose would be an example of such a capsule which, by the above terminology, would also be described as flexible and peripheral. Rigid or integral capsules tend to be heteropolymers whose synthesis is intracellular, involving sugar nucleotide and polyisoprenol-intermediates (see Sutherland, 1977). Many of these heteropolysaccharides may, apart from a variety of neutral sugars, contain polyols, amino sugars, uronic acids, phosphate, or pyruvate. *Streptococcus pneumoniae* strains form a wide variety of antigenically type-specific, exopolysaccharide capsules, ranging from a simple disaccharide repeating unit to more complex structures, some of which resemble cell wall teichoic acids. Mention has been made (Section 2.2.3) of a possible link between ex-opolysaccharide capsule synthesis and the loss of ability to link the polysaccharide to the peptidoglycan, a view that is supported by similarities in the routes of biosynthesis of wall and capsular heteropolymer (Sutherland, 1977). Capsules, whether integral or peripheral, will inevitably "trap" other excreted or secreted bacterial products such as amphiphiles, membrane lipids, proteins, and cell wall turnover products. They will also be penetrated to a greater or lesser extent by proteinaceous appendages of the cell such as flagella, pili, and fimbriae (see Section 4), as well as molecular extensions of the surface of the cell wall. While these "penetrations" may help to anchor the capsule, they will also increase the complexity of the chemistry of this region.

The bacterial glycocalyx is a highly hydrated structure and thus readily deformed during preparation of specimens for electron microscopy. However, special staining techniques utilizing ruthenium red or the use of lectins or specific antibodies can stabilize these structures sufficiently to demonstrate a high degree of order in the form of radially and/or concentrically arranged fibers [see Costerton *et al.* (1981a,b) and Chapter 1 for a series of excellent electron micrographs].

2.5. Turnover and Environmentally Induced Variation in Cell Wall Polymers

Turnover of cell wall polymers has been studied extensively in a number of bacteria and clearly such studies are of significance in interpretation of the nature of the cell surface. From the data presently available (Daneo-Moore and Shockman, 1977; Shockman and Barrett, 1983) it is difficult to describe or predict turnover in general terms. Some organisms exhibit turnover of peptidoglycan and secondary polymers at rates that vary from species to species while others are completely conservative in their synthesis of covalently linked cell wall polymers. In some rod-shaped organisms poles show less turnover than the cylindrical portion of the cell. Where turnover does occur it is generally from the outer surface of the cell wall while newly synthesized material is added to the inner surface adjacent to the cell membrane. Such turnover indicates that the assembled wall is not inert. Nevertheless, as wall turnover is apparently not essential for wall synthesis, what it means in terms of cell economy remains to be determined.

Environmentally induced phenotypic variation in peptidoglycan structure appears to be rare and tends to be restricted to relatively minor changes in cross-bridge peptide composition (Tipper and Wright, 1979). Changes in the growth environment can, however, cause marked changes in secondary polymers. One of the best documented is the switch in synthesis from teichoic acid to teichuronic acid in a number of organisms when grown under phosphate limitation (Ellwood and Tempest, 1972). Return to nonlimiting phosphate conditions causes teichoic acid to be synthesized once more in place of the other anionic polymer. With some organisms that exhibit cell wall turnover, the change in composition is not merely dilution of one polymer type by the other but active deletion of the polymer being replaced. The essential feature for these cells appears to be the constant presence of an anionic polymer which is presumed to be concerned with Mg^{2+} uptake (Ellwood and Tempest, 1972). In *Bacillus subtilis* W23, phage SP50 utilizes surface teichoic acid as a receptor (Archibald, 1976). Phage does not bind to phosphate-limited (teichuronic acid-containing) cultures and there is a lag of half a generation on return to phosphate-rich media before maximum phage binding occurs. This is a further illustration of the concept of incorporation of new wall at the inner face and loss of old wall at the surface with consequent lag in newly incorporated material appearing at the cell surface.

Chemostat studies on a number of oral lactobacilli and streptococci under conditions of carbohydrate limitation with different carbohydrates, generation times, and pH values (Knox and Wicken, 1984) have indicated that growth conditions can affect such "whole cell" properties as adhesion, immunogenicity of surface components, and rate of enzymatic cell lysis. With these organisms, cell wall components, as determined by analyses of the composition and relative proportions of peptidoglycan, polysaccharides, and teichoic acids, showed a high degree of phenotypic stability, while marked changes occurred in the amounts of wall-associated polymers such as capsular polysaccharide, lipoteichoic acids, and proteins as well as in the excretion of these polymers into the external environment.

In natural, particularly aquatic, environments, microbial populations are not simply collections of rods, cocci, and spirals. Prokaryotic form in these situations shows a great deal of diversity (see Dow and Whittenbury, 1980) and phenotypic response in terms of whole cell morphology to nutrient or other environmental pressures. The formation of prosthecae, polymorphic vegetative cell cycles, and the ability of a number of bacterial species to extend their cytoplasmic membrane into lamellae and tubes etc. (see Rogers *et al.*, 1980) are a few

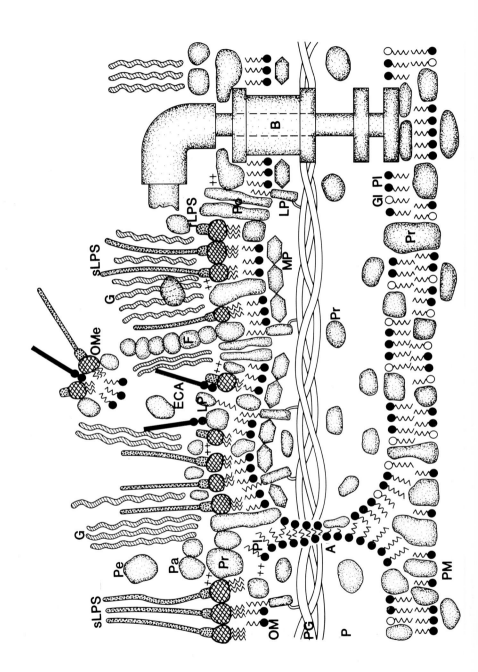

Figure 2. Schematic representation of the organization of the cell wall of gram-negative bacteria. The cell wall is shown as having three regions, namely, outer membrane (OM), peptidoglycan monolayer (PG), and periplasm (P) overlaying the inner plasma membrane (PM). The two membranes show continuity at various attachment points (A) and may also be bridged by the basal bodies (B) of flagella (pili may have a similar orientation). The plasma membrane is shown as composed of phospholipid (Pl), glycolipid (Gl), and protein (Pr); some protein is also found in the periplasm. The asymmetric outer membrane is composed of phospholipid (Pl), protein (Pr), lipopolysaccharide in smooth (sLPS) and rough (rLPS) forms, and related polysaccharide polymers such as the enterobacterial common antigens (ECA) as well as divalent metal cations (++). Peptidoglycan is closely associated with the inner face of the outer membrane, covalently through lipoprotein (LP) and noncovalently through a hexagonal array of matrix protein (MP). Other integral proteins of the outer membrane are shown as either restricted to one face of the membrane or spanning it and in the case of the porins (Po) forming narrow ionic channels across the membrane. Fimbriae (F) are suggested to arise from proteins in the outer face of the outer membrane and may extend for considerable distances beyond the outer membrane. Outer membrane-associated protein (Pa) and extracellular protein (Pe), excreted outer membrane fragments (OMe), the extended O-polysaccharide side chains of sLPS, as well as extracellular capsular polysaccharides (G) are shown as contributing to the heterogeneity of the glycocalyx region of the cell. (The diagram is intended as being illustrative of the concepts described in the text and should not be regarded as a literal depiction or one that is necessarily drawn to scale.)

examples of such diversity. Studies on the underlying biochemistry of these responses are still in their infancy, however.

3. GRAM-NEGATIVE BACTERIA

In contrast to gram-positive bacteria, the gram-negative bacterial cell wall is multi-layered and structurally and chemically more complex. In thin section a thin electron-dense peptidoglycan layer is seen underlying the outer cell membrane. Between the outer cell membrane and the plasma cell membrane is a space, the periplasmic space, the size of which depends on the osmolarity of the external milieu. The cell wall is generally regarded opera-tionally as the outer cell membrane plus the peptidoglycan layer. (For reviews see Tipper and Wright, 1979; Wright and Tipper, 1979; Rogers *et al.*, 1980; Beveridge, 1981.)

3.1. The Periplasmic Space

Unlike gram-positive bacterial cells where cell membrane is applied closely to the undersurface of the cell wall, in gram-negative bacteria there is a definite gap between cell and outer cell membranes. Hypotonic media will plasmolyse gram-negative bacteria to produce a larger periplasmic space while hypertonic media will cause it to contract. The boundaries of this space are generally defined as including peripheral proteins of the inner face of the outer cell membrane and the outer face of the plasma cell membrane. The peptidoglycan layer by this definition is therefore part of the periplasm.

Under normal conditions the periplasmic space can represent 20 to 40% of the total cell volume and it has a number of important functions:

1. It is in Donnan equilibrium with molecules of up to 1000 daltons of the external environment that readily penetrate the outer cell membrane (see Section 3.3) thus allowing, in a sense, for a "remote" sampling of the environment
2. The periplasmic space is isoosmotic with the cytoplasm of the cell which allows for a turgor pressure in the peptidoglycan–outer membrane complex that contributes to the mechanical–protective role of the cell wall
3. The periplasm contains a number of hydrolytic enzymes that allow for preprocessing of some incoming nutrients and postprocessing of some secreted or excreted material as well as binding proteins involved in active transport of nutrients across the plasma cell membrane (Beveridge, 1981; Aaronsen, 1981).

3.2. Peptidoglycan

The peptidoglycan layer in gram-negative bacteria represents 2 to 10% of the dry weight of the wall and varies in thickness from 3 to 10 nm. The structure of the polymer is similar to gram-positive bacteria but cross-linking of the muramyl peptides is generally direct between diaminopimelic acid and D-alanine residues of different muramyl peptides and does not involve interpeptide bridges. It has been suggested that the peptidoglycan forms a sheet with the glycan chains on the plasma membrane side and the peptide chains on the outer mem-brane side. In *E. coli* peptidoglycan the degree of cross-linking of chains is of the order of 30%. The observed thickness of the peptidoglycan layer would accommodate a monolayer and certainly no more than three close-packed layers. In gram-negative rods the glycan strands are probably at right angles to the long axis of the cell and wrap around the cell in a shallow helix (Beveridge, 1981).

Associated with the peptidoglycan are at least two proteins, both of which are also components of the inner face of the outer cell membrane and therefore chemically stabilize these two wall layers as a single complex. One of these, the matrix protein, of 36,500 daltons, forms a hexagonal lattice of polygonal units over the upper face of the peptidoglycan. Its mode of attachment to peptidoglycan is ionic. The other protein is a lipoprotein of approximately 7000 daltons, molecules of which are covalently attached at their COOH-terminal ends to approximately 10% of the diaminopimelic acid residues of the peptidoglycan (see Section 3.3.3) while the lipid ends are inserted in the inner half of the outer membrane bilayer. In contrast to gram-positive bacteria, there is no association of carbohydrate-type polymers with the peptidoglycan of gram-negative bacteria.

3.3. The Outer Cell Membrane

(For reviews see Di Rienzo *et al.*, 1978; Inouye, 1979; Wright and Tipper, 1979; Rogers *et al.*, 1980; Beveridge, 1981.)

The outer cell membrane of gram-negative bacteria is composed of phospholipid (20 to 25%), lipopolysaccharide and related polysaccharides (30%), and protein (45 to 50%). It is now generally accepted that the outer membrane is asymmetric with lipopolysaccharide and protein in the outer leaflet and phospholipid and protein forming the inner leaflet of the bilayer of the membrane; some of the protein, at least, spans the bilayer. The contribution of fatty acids from phospholipid and lipopolysaccharide is about the same.

3.3.1. Lipopolysaccharide and Related Polymers

Lipopolysaccharides are the oldest known examples of polymeric bacterial amphiphiles and have been studied extensively in terms of their chemistry and biological activities (for recent reviews see Lüderitz *et al.*, 1981; Rietschel *et al.*, 1982a,b). The molecule consists of three distinct regions covalently linked together, a hydrophobic lipid component (lipid A), a core polysaccharide, and *O*-antigenically specific polysaccharide side chains. Considerable diversity in the structure of lipopolysaccharides exists both between genera and species and also in the microheterogeneity within the lipopolysaccharide from a single species. In summary:

1. A wide species diversity in the structure of the *O*-antigenic-specific side-chain polysaccharides has provided the basis for detailed serological clarification of some genera (e.g., *Salmonella* and *Shigella*). All *O*-polysaccharides are polymers of repeating sequences of two to four monosaccharide units which may include a wide range of hexoses, pentoses, deoxy-, dideoxy-, amino-, and diamino-sugars. In this region of the molecule alone, lipopolysaccharides are chemically much more complex than the cell wall polysaccharides of gram-positive bacteria. Bacterial strains having lipopolysaccharides without *O*-antigenic polysaccharide side chains are known as rough or R-forms, strains possessing the complete lipopolysaccharide structure being smooth or S-forms.

2. The core polysaccharide is generally constant in composition for S-forms from a particular genus. Aldoheptose and 2-keto-3-deoxyoctonate are regarded as characteristic components of this region but there are exceptions to this general rule. Ketodeoxyoctonate, apart from contributing together with phosphate residues to the net negative charge of this amphiphile, provides the linkage, through an acid-labile bond, to the lipid A region of the molecule.

3. The lipid A regions of lipopolysaccharides share a generic similarity in that they consist of phosphorylated glucosamine residues, forming di- or higher oligosaccharides to which are attached both *O*- and *N*-fatty acyl esters; characteristically the *N*-acyl residue is a

3-hydroxyalkanoic acid. Other amino-substituents that are often present in lipid A from different genera are ethanolamine or 4-aminoarabinose.

4. Microheterogeneity can exist in lipopolysaccharide preparations in all regions of the molecule. Differences in the length and composition of the O-specific polysaccharides can exist in the same preparation. Within a species some of the normally S-form lipopolysaccharide may lack O-specific side chains and thus be R-form lipopolysaccharide. In R-form mutants the length and composition of the core polysaccharide may also vary. Isolated lipopolysaccharide which, like all amphiphiles, forms a high-molecular-weight micellar aggregate in solution, may vary considerably in the state and physical appearance of the aggregates depending on the method of extraction and nature of associated cations.

As with the amphiphiles of gram-positive bacteria, lipopolysaccharides are found to varying extents in the extracellular environment, generally in the form of lipopolysaccharide–phospholipid–protein complexes derived from the outer membrane of both growing and stationary phase cells. This "excretion" may be due to normal turnover of the outer membrane but there is evidence in *E. coli* for preferential excretion of newly synthesized material (Mug-Opstelten and Witholt, 1978). In *Xanthomonas* sp. the formation of an extracellular polysaccharide resembling a deacylated version of its lipopolysaccharide (Yadomae *et al.*, 1978) suggests a similarity to the excretion of lipoteichoic acids from gram-positive bacteria (see Section 2.4.1a).

Another polysaccharide amphiphile found in the outer membrane of Enterobacteriaceae is the enterobacterial common antigen (Makela and Mayer, 1976). The hydrophilic region appears to be a linear polymer of 1,4-linked *N*-acetyl-*O*-glucosamine and *N*-acetyl-D-mannosamine. The structure of the hydrophobic region has not been fully elucidated. This component can occur as a haptenic free form in the outer membrane or in an immunogenic form covalently linked to the core polysaccharide of the lipopolysaccharide in some R-form strains.

3.3.2. Phospholipids

The outer membrane has a lower phospholipid composition than the plasma cell membrane. Much of the phospholipid is phosphatidylethanolamine with smaller amounts of phosphatidylglycerol and cardiolipin. Exchange of phospholipids between the two membranes has been reported but this would seem to be somewhat restricted as the relative proportions of the different phospholipids in the two membranes are not the same.

3.3.3. Proteins

While more than 20 different polypeptides from outer membranes of gram-negative bacteria can be identified on SDS–polyacrylamide gels, the major outer membrane proteins, varying from four to six in different species, make up 70% or more of the protein content (see reviews by Wright and Tipper, 1977; Rogers *et al.*, 1980; Beveridge, 1981). The nomenclature used to describe outer membrane proteins is confusing in that there is little uniformity between different research groups. Two components, matrix protein(s) and lipoprotein, have already been mentioned (Section 3.2) as joining the peptidoglycan layer to the inner face of the outer cell membrane. Other major integral proteins described in a number of species in the molecular weight range 33,000 to 38,000, include the "porins" which have been demonstrated as forming hydrophilic transmembrane channels for passive diffusion of small molecules across the outer membrane.

The lipoprotein is the most abundant polypeptide of the outer membrane. One third of it is covalently linked to the peptidoglycan but all of it is strongly associated with the matrix protein. The possibility of exposure of free lipoprotein at the outer surface on rough mutants has been suggested by immunological studies.

A major peripheral protein in *E. coli* is the Tol G protein which has an outer surface location and acts as a bacteriophage receptor. Minor proteins greatly outnumber the major proteins and include receptors for a range of bacteriophages and colicins and also functional binding proteins for transport of some nutrients. Specific transport systems for vitamin B_{12}, iron, maltose, and maltodextrins as well as nucleosides have been demonstrated in *E. coli*.

3.3.4. Organization of the Outer Cell Membrane

The outer cell membrane is atypical of normal membranes in that it acts as a permeability barrier to hydrophobic substances, gram-negative bacteria being relatively insensitive to detergents and hydrophobic dyes. This property is readily explained by the known asymmetry of the outer membrane. All of the lipopolysaccharide is found in the outer half of the membrane and covers 30 to 40% of its area while the phospholipid is found in the inner half of the bilayer. The outer face in S-forms would therefore be highly hydrophilic and a barrier to the passage of hydrophobic molecules. In support of this, deep rough mutants (lacking much of the core as well as the *O*-polysaccharide portion of the lipopolysaccharide) are much more sensitive to hydrophobic drugs and detergents. In such mutants there is also a decrease in protein and a compensatory increase in phospholipid but in both halves of the bilayer. There is some evidence in *E. coli* that a surface distribution of lipopolysaccharide is not random but in domains that do not intermix freely and may be of different composition. Much of the protein of the outer membrane is integral and spans the membrane with exposure at the outer face while other protein components are peripheral to either face. Protein movement appears to be more restricted than lipopolysaccharide movement and tight packing of protein and extensive protein–protein interaction in the outer membrane has also been indicated by freeze-fracture and chemical cross-linking studies. There is also evidence for attachment between plasma and outer cell membranes, possibly in regions of peptidoglycan discontinuities. Such attachment sites are the presumed sites of export of outer membrane components synthesized in the plasma cell membrane. They may also serve as entry points for viral DNA, as phage absorption to the outer membrane in these regions has been described for a number of organisms. The nature and stability of these attachment sites remain unknown; their total number per cell has been estimated as varying from 200 to 400 (Mühlradt *et al.*, 1973) in *S. typhimurium* each measuring 25 to 50 nm in cross-section to 6% of the total surface in *E. coli* (Bayer and Starkey, 1972).

3.4. Extracellular Components and Glycocalyx

Gram-negative bacteria produce a wide variety of glycocalyces including the glycoprotein S-layers (Sleytr, 1978) and the various types of capsule as defined by Costerton *et al.* (1981a,b; Chapter 1, Section 1.1; see also Section 2.4.3). Summaries of the homo- and heteropolymeric polysaccharides that can form the glycocalyces of gram-negative bacteria are provided by the reviews of Sutherland (1977) and Beveridge (1981). Homopolysaccharides include glucans from *Acetobacter* species and some *Rhizobium* species and levans from *Xanthomonas, Aerobacter,* and *Acetobacter* species. Polymers of *N*-acetylneuraminic acid (sialic acids) have also been reported from *E. coli* K1 and *Neisseria meningitidis* and in both cases the polymers are closely associated with the cell surface. Heteropolysaccharides generally follow the gram-positive bacterial pattern in having repeating units made up of a few sugar residues together with other components such as uronic acids and pyruvate or other ketals. The K-antigens of the genera *Klebsiella* and *Escherichia* provide numerous examples of such heteropolysaccharides (Sutherland, 1977). *Azotobacter vinelandii* and *Pseudomonas aeruginosa* produce irregular, nonrepeating, heteropolymers of mannuronic and glucuronic acids that resemble alginates. These polymers are retained as flexible peripheral capsules and *P. aeruginosa* can also produce a homopolymer of 2-ketoglucuronic acid as an outer glycocalyx to the first.

In addition to the fibrous matrix of the glycocalyx, intercalation or entrapment of "excreted material" from the outer cell membrane must inevitably contribute to the chemical complexity of this region of the cell. Further, in smooth strains, the *O*-antigenic polysaccharide side chains of lipopolysaccharide located in the outer membrane will penetrate deeply into the glycocalyx region (Shands, 1966).

3.5. Turnover and Cell Wall Variation

Wall turnover, as discussed earlier (Section 2.5) for gram-positive bacteria, has been observed in a strain of *E. coli* and *Neisseria gonorrhoeae*; there are other strains and species, however, in which it does not occur. The phenomenon in gram-negative bacteria has not been investigated to the extent that it has in gram-positive bacteria (see Daneo-Moore and Shockman, 1977). Mention has been made (Section 3.3.1) of the apparent turnover of lipopolysaccharide in terms of its excretion into the external environment as a complex with protein and phospholipid. The peptidoglycan in gram-negative bacteria appears to be as phenotypically stable in its composition as its gram-positive counterpart (Daneo-Moore and Shockman, 1977). The effects of limiting growth conditions, as afforded by continuous culture, also have not been studied extensively in relation to gram-negative bacterial cell wall composition. Two studies are significant. With *Aerobacter aerogenes,* lipopolysaccharide content increased with growth rate in glycerol-limited cultures, decreased in Mg^{2+}-limited cultures, and was relatively constant in phosphate-limited cultures. However, a marked variation in the composition of the core polysaccharide region of this component (as measured by heptose/phosphate ratios) with change in growth conditions was noted (Pearson and Ellwood, 1972). Fast-growing cells were also more endotoxic than slow-growing cells. Growth limitation was also shown to affect the whole cell toxicity of *E. coli* strain MRE160 (Pearson and Ellwood, 1974), carbon limitation producing the most toxic organisms and sulfur limitation the least toxic. These observations are obviously of interest in the production of less toxic bacterial vaccines.

4. SURFACE APPENDAGES

Treatises on bacterial cell walls usually omit mention of such surface appendages as flagella, fimbriae, and pili (exceptions being the review articles of Sokatch, 1979; Beveridge, 1981) and yet it must be obvious that such appendages should be considered as part of the cell wall surface, for the appendage either rises from or passes through discontinuities in this structure. All such appendages project beyond the cell surface and into the external environment (see Chapter 11).

4.1. Flagella

Flagella are complex structures composed of at least 11 polypeptides forming a filament, a hook, and a complex basal body, the latter interacting with both cell wall and plasma cell membrane (Sokatch, 1979; Beveridge, 1981). Interaction of basal body with plasma membrane proteins is believed to be the basis of flagella rotation. How the necessary holes in the cell wall are provided to accommodate such structures is unknown. Whatever the mechanism, its control is intriguing, as flagella placement is a constant locational feature of a particular species, the equal distribution of peritrichous flagella being an extreme case. The most prominent and extracellular portion of a flagellum is the helical filament, usually composed of self-assembling subunits of a single protein, flagellin. In most bacteria the filament is naked and in *E. coli* is 13.5 nm in diameter and averages 5 μm in length. Some genera (e.g., *Vibrio*) produce sheathed flagella where the sheath is an extension of the outer cell membrane but possibly not identical in composition with this structure. Sheathed flagella bear some resemblance to the axial filaments of spiral organisms.

4.2. Sex Pili and Fimbriae

Sex pili are proteinaceous appendages which occur on donor strains of bacteria and whose formation is determined by the presence of conjugative plasmids (Sokatch, 1979). Many different types have been described on the basis of their plasmid determinants, diameter (8.5 to 9.5 nm), length (2 to 20 μm), and attachment of specific RNA and DNA phages. They are not helical structures like flagella and little is known of how they are anchored to the plasma membrane. As with flagella, however, they have to be provided with discontinuities in the cell wall.

Fimbriae are short, straight protein appendages that are not involved in movement or transfer of nucleic acid. They are variable in number and dimensions and may have either a polar or a peritrichous arrangement. They are most prominent in gram-negative bacteria and have been implicated in adhesion and bacterial aggregation (Pearce and Buchanan, 1980; see Chapter 11, Section 2). Their origin within the cell wall membrane complex is unknown but it is tempting to speculate that fimbriae are associated with outer membrane proteins. In this connection it is of interest to note that structures called spinae (Willison *et al.*, 1977) found in some marine pseudomonads are surface appendages that are attached to proteins of the outer cell membrane. *Corynebacterium* is the only gram-positive bacterial genus reported to possess fimbriae (see Pearce and Buchanan, 1980) similar to those found in gram-negative bacteria. In some gram-positive bacteria, notably streptococci, surface protein can take the form of a "fine fibrillar fuzz" that extends into the external environment (Gibbons *et al.*, 1972; Beachey and Ofek, 1976; and Chapter 10, Section 2) and be involved in adhesion of cells to surfaces. These fibers have also been referred to as fimbriae but their appearance is distinctly different from that observed with the classical fimbriae of gram-negative bacteria.

4.3. Prosthecae

Prosthecae are the distinctive polar or lateral extensions of the cell in "stalked" or prosthecate bacteria that have been variously attributed to needs for flotation or increased surface area. Prosthecae appear to be bounded by extensions of the cell envelope and membrane. Relatively little is known of possible differences in the surface chemistry of prosthecae compared to the rest of the cell.

5. PROBING THE BACTERIAL CELL SURFACE

The nature of the forces and possible mechanisms involved in bacterial adhesion are considered elsewhere in this volume (Chapters 6–11). Direct contact between bacterial cell wall and adherent surface is unlikely through energy considerations; this view has been supported by numerous direct observations, a gap of around 10 nm between bacterial cell wall and surface generally being reported. This gap represents a high-energy barrier due partly to the size of the bacterial cell and partly to the ionic nature of both surfaces. Long-range forces may well be involved in reversible adhesion but stronger short-range forces resulting in irreversible adhesion demand penetration of the energy barrier. This might be achieved by probes arising from either surface and bridging the two. As far as the bacterial cell is concerned, flagella, pili, and particularly fimbriae are ideal probes in terms of their narrow diameter in relation to their length. The radial arrangement of fibrillar glycocalyces (see Section 2.4.3) would also afford potential short-range interaction between bacteria and surfaces. In many cases glycocalyces are penetrable to varying extents by other large molecules and at one extreme may be regarded as concentrated solutions of polymers that can bridge the gap between more solid adherent surfaces. The idea of extracellular polysaccharide glues cementing bacteria to surfaces has long been held to be one mechanism of irreversible adhesion. The physicochemical factors involved in attraction and adhesion of bacteria to surfaces and interfaces would be both polar and nonpolar and both types of property can be manifested at bacterial surfaces through different molecular groupings at the surface and appendages associated with it.

5.1. Surface Charge of Bacterial Cells

Bacterial cells, like most eukaryotic cells, are generally recognized as having a net negative charge and acidic pH with isoelectric points in the region of pH 3. Electrophoretic mobility of bacteria can be observed and measured directly under the microscope (see Richmond and Fisher, 1973). While such measurements are indicative of total charge, they give no indication of surface distribution of charge. Isoelectric focusing (Langton et al., 1975) has indicated species differences in isoelectric points as well as different net negative charges within portions of a single species population. Fluorescent dyes have also been used to probe the charge and potential at the surfaces of bacteria (see Aaronson, 1981).

The origins of the bacterial surface negative charge mirror the complex chemistry of the surface. In gram-positive bacteria teichoic and teichuronic acids of the cell wall and acidic polypeptides and polysaccharides of the glycocalyx have all been cited as contributing to negative charge. Removal of the hyaluronic acid capsule from Streptococcus pyogenes or destruction of the cell wall ribitol teichoic acid with periodate in Staphylococcus aureus has been reported to reduce the negative charge in these organisms (see Rogers, 1979). With gram-negative bacteria, acidic lipopolysaccharides and proteins of the outer cell membrane in addition to extracellular polymers of the glycocalyx are all sources of negative charge. Many of the outer membrane proteins, for example, have been shown to possess ratios of acidic to basic amino acids of greater than 1 (see Aaronson, 1981).

5.2. Hydrophilic and Hydrophobic Properties of Bacterial Cells

A variety of methods have been devised to probe the relative hydrophobicity or hydrophilicity of whole bacterial cells (see also Chapter 6, Section 2.2.4; Chapter 7, Section 3.2). These include:

1. Partition between hydrophilic and hydrophobic phases in two-phase systems derived from either (a) a liquid hydrocarbon such as xylene, octane, or hexadecane and water (Rosenberg et al., 1980) or (b) aqueous polymers of dextran and polyethylene glycol linked with hydrophobic groups (Magnusson et al., 1977)
2. Hydrophobic interaction chromatography using octyl agarose gel preparations (Dahlback et al., 1981)
3. Adherence to polystyrene (Rosenberg, 1981).

No single method probably adequately measures the degree of hydrophobicity (or hydrophilicity). With populations of single cells, a good relationship generally exists between the findings obtained with the various methods. Large differences can occur, however, when the bacteria are predominantly present in chains. Hydrophilicity can further be due to charged or to neutral surface polymers. For example, R-forms of *Salmonella typhimurium* indicate both hydrophobic and charged hydrophilic properties while the S-forms indicate only extensive noncharged hydrophilicity due presumably to neutral *O*-specific side-chain polysaccharides of the lipopolysaccharide dominating the cell surface (Magnusson *et al.*, 1977).

Application of these methods to a wide variety of bacteria has shown a remarkable range of relative surface hydrophobicity or hydrophilicity. Oral streptococci have a high degree of surface hydrophobicity (Olsson and Westergren, 1982); the cooperative involvement of hydrophobic bonding in the adhesion of *Streptococcus sanguis* to hydroxylapatite has been proposed by Doyle *et al.* (1982). The constancy of this property in relation to repeated subculture, medium composition, and other growth conditions largely remains to be investigated. There are indications, however, that environmental conditions may play a major determinative role.

Surface components that may contribute to hydrophobicity include proteins and amphipathic polymers. In group A streptococci both surface M protein and lipoteichoic acid have been suggested as major contributors to both the hydrophobic properties and adherence of these bacteria to eukaryotic cells (Tylewska *et al.*, 1979; Ofek *et al.*, 1982; Miörner *et al.*, 1983). In *Salmonella typhimurium* a greater exposure of lipid A at the surface of the outer membrane in R-forms has been equated with a greater degree of hydrophobicity in R-forms as compared to S-forms (Magnusson *et al.*, 1977).

5.3. Immunoglobulins, Lectins, and Phages as Specific Surface Probes

Specific probes such as antibodies, lectins, or bacteriophages have been used extensively to define the nature of surface components. Combined with electron microscopy they also afford a means of assessing the topographical distribution of such components. Such methods have also demonstrated that the bacterial cell wall does not have the well-defined outer boundary usually seen under the microscope but rather that the cell surface is extended into the gradient or continuum of polymeric material that collectively forms the glycocalyx. Unlike the outer membrane of gram-negative bacteria or the peptidoglycan–polysaccharide complex of gram-negative bacteria, the glycocalyx is penetrable to varying extents by these probes (see Rogers, 1979; Beveridge, 1981).

Ferritin-conjugated antibodies have been used to demonstrate extension of lipopolysaccharide beyond the outer membrane of some gram-negative bacteria by as much as 150 nm (Shands, 1966) as well as the presence of membrane lipoteichoic acid external to the cell wall of some lactobacilli (van Driel *et al.*, 1973). In the latter example the pattern of labeling may indicate both lipoteichoic acid in the process of excretion as well as a wall/membrane component (Wicken and Knox, 1980). Antisera made against whole gram-negative bacteria often contain antibodies to a wide variety of "surface antigens." The concept of a surface that is a loosely organized in-depth mixture of polymers, readily penetrable by antibodies, offers an explanation for simultaneous surface expression of a variety of antigens. The alternative explanation of a patchy distribution of different antigens at the same surface level is not borne out by immunochemical labeling (Rogers, 1979). The hyaluronic acid capsules of some group A and group C streptococci similarly do not afford a barrier to antibodies reacting with cell wall components. The advent of monoclonal antibodies has opened up a new source of immunological probes with high, uniform specificities that will undoubtedly extend in the future this approach to study of the cell surface.

The immune response to whole bacterial cells also provides useful information on the cell surface because the immune system "sees" the actual surface components presented to it. Analysis of the specificities of immune sera will thus enable recognition of immunodominant surface components.

Bacteriophages are excellent topographical probes due to their high degree of specificity for receptor sites and ready visiblity in the electron microscope. In the case of gram-negative bacterial cell hosts, different phage specificities may be directed at different regions of the lipopolysaccharide molecule, outer membrane proteins, polysaccharide of the glycocalyx, pili, fimbriae, and flagella (Lindberg, 1973, 1977). Penetration of different phages through the glycocalyx region to the receptors at the surface of the outer membrane such as core lipopolysaccharide or proteins may or may not be inhibited by components of the glycocalyx. Thus, a careful selection of a "battery" of bacteriophage probes has the potential of defining the outer layers of a bacterial cell in depth. In gram-positive bacteria, phage receptors frequently involve secondary wall polymers such as teichoic acids or polysaccharides together with portions of the peptidoglycan to which they are covalently linked. Thus, secondary wall polymers must, at least in part, be exposed at the cell wall surface, a finding that has been supported in many cases by antibody and lectin-binding. As with gram-negative bacteria, the glycocalyx region may either hinder phage absorption or be "invisible" to them.

6. FUNCTIONAL ASPECTS OF BACTERIAL CELL WALLS AND SURFACES

The primary function of the bacterial cell wall is to preserve the underlying protoplast from osmotic disruption. In gram-positive bacteria the thick, multilayered peptidoglycan is solely responsible for osmotic protection and maintenance of shape. In gram-negative bacteria both the peptidoglycan layer and the outer membrane, to which it is linked covalently, contribute to the mechanical strength of the cell envelope. It is also evident that shape maintenance is a reflection not only of the peptidoglycan monolayer but also of long-range order in the arrangement of the protein components of the outer membrane.

Other cellular functions attributed to the cell wall include the ion-exchange capacities of negatively charged polymers in the cell wall and cell membrane of gram-positive bacteria and outer membrane of gram-negative bacteria in relation particularly to the uptake of Mg^{2+} ions (see reviews by Ghuysen, 1977; Tipper and Wright, 1979; Ward, 1981; Beveridge, 1981) as well as the barrier functions of the outer cell membrane and the ecological advantages of the periplasmic space in gram-negative bacteria (see reviews by Ghuysen, 1977; Wright and Tipper, 1979; Rogers et al., 1980). The negatively charged teichoic acids and lipoteichoic acids of gram-positive bacteria have also been considered to have a role in regulating endogenous autolytic enzymes (Daneo-Moore and Shockman, 1977; Shockman and Barrett, 1983).

The cell surface also has a receptor and recognition role, which may be of advantage to the cell in the case of DNA transformation or adhesion to surfaces or interfaces or disadvantageous in the cases of phage or colicin attachment.

More recent recognition of the universality of the presence of a glycocalyx surrounding bacterial cells in natural and pathogenic environments (Costerton et al., 1981a,b; and Chapter 1) has given new emphasis to a previously largely overlooked area of microbial morphology. In the past "capsules" tended to be regarded as relatively unimportant in the bacterial cell economy in that it was usually possible to grow, in laboratory culture, strains with or without their capsules. In natural situations, many bacterial species exist in two physiologically distinct forms (Whittenbury and Dow, 1977), sessile microcolonies surrounded by an extensive glycocalyx and mobile swarm cells that can colonize new surfaces. Isolation into pure culture in laboratory media favors the swarm cell by removing the competitive challenges of the natural environment and the bacterial surface changes radically and quickly through loss of the glycocalyx. This loss is generally irreversible although it may be prevented by the inclusion of antibacterial substances such as surfactants or antibiotics in the medium (Govan, 1975; Govan and Fyfe, 1978). Costerton and co-workers consider the glycocalyx to be essential to survival in natural environments and the loss of this structure *in vitro* to be of serious concern in extrapolating from laboratory studies to microbial ecology or microbial pathogenicity (Costerton *et al.*, 1981a,b; and Chapter 1, Section 1.1). Loss of the glycocalyx in many cases may mean exposure of a new "face" of the bacterial cell to the external environment which may have different adhesion and other properties from those seen in natural environments. More attention will have to be directed in the future to laboratory preservation of surface components seen in natural populations for laboratory studies aimed at describing microbial interaction with their environment to have relevance.

REFERENCES

Aaronson, S., 1981, *Chemical Communication at the Microbial Level,* Volume 1, CRC Press, Boca Raton, Fla.

Archibald, A. R., 1976, Cell wall assembly in *Bacillus subtilis:* Development of bacteriophage-binding properties as a result of the pulsed incorporation of teichoic acid, *J. Bacteriol.* **127:**956–960.

Bayer, M. E., and Starkey, T. W., 1972, The adsorption of bacteriophage ϕX174 and its interaction with *Escherichia coli*; a kinetic and morphological study, *Virology* **49:**236–256.

Beachey, E. H., and Ofek, I., 1976, Epithelial cell binding of *Group A streptococci* by lipoteichoic acid on fimbriae denuded of M protein, *J. Exp. Med.* **143:**759–771.

Beveridge, T. J., 1981, Ultrastructure, chemistry and function of the bacterial wall, *Int. Rev. Cytol.* **72:**229–317.

Button, D., Archibald, A. R., and Baddeley, J., 1966, The linkage between teichoic acid and the glycosamino peptide in the walls of a strain of *Staphylococcus lactis, Biochem. J.* **99:**11C–14C.

Cabacungan, E., and Pieringer, R. A., 1980, Excretion of extracellular lipids by *Streptococcus mutans* BHT and FA-1, *Infect. Immun.* **27:**556–562.

Campbell, L. K., Knox, K. W., and Wicken, A. J., 1978, Extractability of cell wall polysaccharide from lactobacilli and streptococci by autoclaving and dilute acid, *Infect. Immun.* **22:**842–851.

Costerton, J. W., Irvin, R. T., and Cheng, K.-J., 1981a, The role of bacterial surface structures in pathogenesis, *CRC Crit. Rev. Microbiol.* **8:**303–338.

Costerton, J. W., Irvin, R. T., and Cheng, K.-J., 1981b, Role of the bacterial glycocalyx in nature and disease, *Annu. Rev. Microbiol.* **35:**299–324.

Coyette, J., and Ghuysen, J.-M., 1970, Structure of the walls of *Lactobacillus acidophilus* strain 63 AM Gasser, *Biochemistry* **9:**2935–2943.

Dahlback, B., Hermansson, M., Kjelleberg, S., and Norkrans, B., 1981, The hydrophobicity of bacteria—an important factor in their initial adhesion at the air–water interface, *Arch. Microbiol.* **128:**267–270.

Daneo-Moore, L., and Shockman, G. D., 1977, The bacterial cell surface in growth and division, in: *Cell Surface Reviews,* Volume 4 (G. Poste and G. L. Nicolson, eds.), Elsevier/North-Holland, Amsterdam, pp. 597–715.

De Cueninck, B. J., Shockman, G. D., and Swensen, R. M., 1982, Group B, type III streptococcal cell wall: Composition and structural aspects revealed through endo-N-acetylmuramidase-catalysed hydrolysis, *Infect. Immun.* **35:**572–582.

Di Rienzo, J. M., Nakamura, K., and Inouye, M., 1978, The outer membrane proteins of gram-negative bacteria: Biosynthesis, assembly and functions, *Annu. Rev. Biochem.* **47:**481–532.

Dow, C. S., and Whittenbury, R., 1980, Prokaryotic form and function, in: *Contemporary Microbial Ecology* (D. C. Ellwood, J. N. Hedge, M. J. Latham, J. M. Lynch, and J. H. Slater, eds.), Academic Press, New York, pp. 391–417.

Doyle, R. J., Nesbitt, W. E., and Taylor, R. G., 1982, On the mechanism of adherence of *Streptococcus sanguis* to hydroxylapatite, *FEMS Microbiol. Lett.* **15:**1–5.

Duckworth, M., 1977, Teichoic acids, in: *Surface Carbohydrates of the Prokaryotic Cell* (I. W. Sutherland, ed.), Academic Press, New York, pp. 177–208.

Ellwood, D. C., and Tempest, D. W., 1972, Effects of environment on bacterial wall content and composition, *Adv. Microbiol. Physiol.* **7**:83–117.

Genco, A. J., Plant, A. G., and Moellering, R. C., 1975, Evaluation of human oral organisms and pathogenic *Streptococcus* for production of IgA protease. *J. Infect. Dis.* **131**:S17–S21.

Ghuysen, J.-M., 1977, Biosynthesis and assembly of bacterial cell walls, in: *Cell Surface Reviews,* Volume 4 (G. Poste and G. L. Nicolson, eds.), Elsevier/North-Holland, Amsterdam, pp. 463–596.

Gibbons, R. J., and van Houte, J., 1978, Bacteriology of dental caries, in: *Textbook of Oral Biology* (J. H. Shaw, E. A. Sweeney, C. C. Cappuccino, and S. M. Meller, eds.), Saunders, Philadelphia, pp. 975–991.

Gibbons, R. J., van Houte, J., and Liljemark, W. F., 1972, Parameters that affect the adherence of *Streptococcus salivarius* to oral epithelial surfaces, *J. Dent. Res.* **51**:424–435.

Glenn, A. R., 1976, Production of extracellular proteins by bacteria, *Annu. Rev. Microbiol.* **30**:41–49.

Govan, J. R. W., 1975, Mucoid strains of *Pseudomonas aeruginosa*: The influence of culture medium on the stability of mucus production, *J. Med. Microbiol.* **8**:513–522.

Govan, J. R. W., and Fyfe, J. A. M., 1978, Mucoid *Pseudomonas aeruginosa* and cystic fibrosis—Resistance of mucoid form to carbenicillan, flucloxacillin and tobramycin, and isolation of mucoid variants *in vitro, J. Antimicrob. Chemother.* **4**:233–240.

Inouye, M., (ed.), 1979, *Bacterial Outer Membranes: Biogenesis and Functions*, Wiley, New York.

Joseph, R., and Shockman, G. D., 1975, Synthesis and excretion of glycerol teichoic acid during growth of two streptococcal species, *Infect. Immun.* **12**:333–338.

Knox, K. W., 1970, Antigens of oral bacteria, *Adv. Oral Biol.* **4**:91–130.

Knox, K. W., and Wicken, A. J., 1973, Immunological properties of teichoic acids, *Bacteriol. Rev.* **37**:215–257.

Knox, K. W., and Wicken, A. J., 1984, Effect of growth conditions on the surface properties and surface components of oral bacteria, in: *Continuous Culture 8* (A. C. R. Dean, D. C. Ellwood, and C. G. J. Evans, eds.), Ellis Horwood, England, pp. 72–88.

Knox, K. W., Campbell, L. K., Evans, J. D., and Wicken, A. J., 1980, Identification of the group G antigen of lactobacilli, *J. Gen. Microbiol.* **119**:203–209.

Langton, R. W., Cole, J. S., and Quinn, P. F., 1975, Isoelectric focusing of bacteria; species location within an isoelectric focusing column by surface charge, *Arch. Oral Biol.* **20**:103–106.

Lapidot, A., and Irving, C. S., 1979, Nitrogen-15 and carbon-13 dynamic nuclear magnetic resonance study of chain segmental motion of the peptidoglycan pentaglycine chain of ^{15}N-Gly- and ^{13}C$_2$-Gly-labelled *Staphylococcus aureus* cells and isolated cell walls, *Biochemistry* **18**:1788–1796.

Lindberg, A. A., 1973, Bacteriophage receptors, *Annu. Rev. Microbiol.* **27**:205–241.

Lindberg, A. A., 1977, Bacterial surface carbohydrate and bacteriophage adsorption, in: *Surface Carbohydrate of the Prokaryotic Cell* (I. W. Sutherland, ed.), Academic Press, New York, pp. 289–356.

Lindmark, R., Movitz, J., and Sjöquist, J., 1977, Extracellular protein A from a methicillin-resistant strain of *Staphylococcus aureus, Eur. J. Biochem.* **74**:623–628.

Liu, T. Y., and Gotschlich, E. C., 1967, Muramic acid phosphate as a component of the mucopeptide of gram-positive bacteria, *J. Biol. Chem.* **242**:471–476.

Lüderitz, O., Freudenberg, M. A., Galanos, C., Lehmann, V., Rietschel, E. T., and Shaw, D. W., 1981, Lipopolysaccharides of gram-negative bacteria, in: *Microbial Membrane Lipids,* Vol. 17 (S. Razin and S. Rottem, eds.), Academic Press, New York, pp. 79–151.

McCarty, M., and Morse, S. I., 1964, Cell wall antigens of gram-positive bacteria, *Adv. Immunol.* **4**:249–286.

Magnusson, K. E., Stendahl, O., Tagesson, C., Edebo, L., and Johansson, G., 1977, The tendency of smooth and rough *Salmonella typhimurium* bacteria and lipopolysaccharide to hydrophobic and ionic interaction as studied in aqueous polymer two-phase system, *Acta Pathol. Microbial Scand. Sect. B* **85**:212–218.

Makela, P. H., and Mayer, H., 1976, Enterobacterial common antigen, *Bacteriol. Rev.* **40**:591–632.

Markham, J. L., Knox, K. W., Wicken, A. J., and Hewett, M. J., 1975, Formation of extracellular lipoteichoic acid by oral streptococci and lactobacilli, *Infect. Immun.* **12**:378–386.

Miörner, H., Johansson, G., and Kronvall, G., 1983, Lipoteichoic acid is the major cell wall component responsible for surface hydrophobicity of group A streptococci, *Infect. Immun.* **39**:336–343.

Mug-Opstelten, D., and Witholt, B., 1978, Preferential release of new outer membrane fragments by exponentially growing *Escherichia coli, Biochim. Biophys. Acta* **508**:287–295.

Mühlradt, P. F., Menzel, J., Golechi, J. R., and Speth, V., 1973, Outer membrane of *Salmonella* sites of export of newly synthesised lipopolysaccharide on the bacterial surface, *Eur. J. Biochem.* **35**:471–481.

Ofek, I., Simpson, W. A., and Beachey, E. H., 1982, Formation of molecular complexes between a structurally defined M protein and acylated or deacylated lipoteichoic acid of *Streptococcus pyogenes, J. Bacteriol.* **149**:426–433.

Olsson, J., and Westergren, G., 1982, Hydrophobic surface properties of oral streptococci, *FEMS Microbiol. Lett.* **15:**319–323.

Owens, P., and Salten, M. R. J., 1975, A succinylated mannan in the membrane system of *Micrococcus lysodeikticus, Biochem. Biophys. Res. Commun.* **63:**875–880.

Pearce, W. A., and Buchanan, T. M., 1980, Structure and cell membrane-binding properties of bacterial fimbriae, in: *Bacterial Adherence* (E. H. Beachey, ed.), Chapman & Hall, London, pp. 289–344.

Pearson, A. D., and Ellwood, D. C., 1972, The effect of growth conditions on the chemical composition and endotoxicity of walls of *Aerobacter aerogenes* NCTC 418, *Biochem. J.* **127:**72P.

Pearson, A. D., and Ellwood, D. C., 1974, Growth environment and bacterial toxicity, *J. Med. Microbiol.* **7:**391–393.

Powell, D. A., Duckworth, M., and Baddeley, J., 1975, A membrane-associated lipomannan in micrococci, *Biochem. J.* **131:**387–393.

Richmond, D. V., and Fisher, D. J., 1973, The electrophoretic mobility of microorganisms, *Adv. Microb. Physiol.* **9:**1–27.

Rietschel, E. T., Galanos, C., Lüderitz, O., and Westphal, O., 1982a, The chemistry and biology of lipopolysaccharides and their lipid A components, in: *Immunopharmacology and the Regulating Leukocyte Functions* (D. R. Webb, ed.), Dekker, New York, pp. 183–229.

Rietschel, E. T., Schade, U., Jensen, M., Wollenweber, H., Lüderitz, O., and Greisman, S. G., 1982b, Bacterial endotoxins: Chemical structure, biological activity and role in septicaemia, *Scand. J. Infect. Dis. Suppl.* **31:**8–21.

Rogers, H. J., 1979, Adhesion of microorganisms to surfaces: Some general considerations of the role of the envelope, in: *Adhesion of Microorganisms to Surfaces* (D. C. Ellwood, J. Melling, and P. Rutter, eds.), Academic Press, New York, pp. 29–55.

Rogers, H. J., 1981, Some unsolved problems concerned with the assembly of the walls of gram-positive organisms, in: β-*Lactam Antibiotics* (M. R. J. Salton and G. D. Shockman, eds.), Academic Press, New York, pp. 87–100.

Rogers, H. J., Perkins, H. R., and Ward, J. B., 1980, *Microbial Cell Walls and Membranes,* Chapman & Hall, London.

Rohr, T. E., Levy, G. N., Stark, N. J., and Anderson, J. S., 1977, Initial reactions in biosynthesis of teichuronic acid of *Micrococcus lysodeikticus* cell walls, *J. Biol. Chem.* **252:**3460–3465.

Rølla, G., 1981, Possible role of lipoteichoic acids in the pathogenicity of dental plaque, in: *Chemistry and Biological Activities of Bacterial Surface Amphiphiles* (G. D. Shockman and A. J. Wicken, eds.), Academic Press, New York, pp. 365–379.

Rosenberg, M., 1981, Bacterial adherence to polystyrene: A replica method of screening for bacterial hydrophobicity, *Appl. Environ. Microbiol.* **42:**375–377.

Rosenberg, M., Gutnick, D., and Rosenberg, E., 1980, Adherence of bacteria to hydrocarbons: A simple method for measuring cell CAP surface hydrophobicity, *FEMS Microbiol. Lett.* **9:**29–33.

Russell, R. R. B., 1979, Wall-associated protein antigens of *Streptococcus mutans, J. Gen. Microbiol.* **114:**109–115.

Russell, R. R. B., 1980, Distribution of cross-reactive antigens in *Streptococcus mutans* and other oral streptococci, *J. Gen. Microbiol.* **118:**383–388.

Salton, M. P. J., and Horne, R. W., 1951, Studies of the bacterial cell wall. I. Electron microscopical observations on heated bacteria, *Biochim. Biophys. Acta* **7:**19–42.

Sargent, M., 1978, Surface extension and the cell cycle in prokaryotes, *Adv. Microbiol. Physiol.* **18:**105–176.

Schleifer, K. H., and Kandler, O., 1972, Peptidoglycan types of bacterial cell walls and their taxonomic implications, *Bacteriol. Rev.* **36:**407–477.

Shands, J. W., 1966, Localisation of somatic antigen in gram-negative bacteria using ferritin antibody conjugates, *Ann. N.Y. Acad. Sci.* **133:**292–298.

Sharpe, A. C., 1970, Cell wall and cell membrane antigen used in the classification of lactobacilli, *Int. J. Syst. Bacteriol.* **20:**509–518.

Shockman, G. D., and Barrett, J. F., 1983, Structure, function and assembly of cell walls of gram-positive bacteria, *Annu. Rev. Microbiol.* **37:**501–527.

Shockman, G. D., and Wicken, A. J. (eds.), 1981, *Chemistry and Biological Activities of Bacterial Surface Amphiphiles,* Academic Press, New York.

Shockman, G. D., Daneo-Moore, L., McDowell, T. D., and Wong, W., 1981, Function and structure of the cell wall—Its importance in the life and death of bacteria, in: β-*Lactam Antibiotics* (M. R. J. Salton and G. D. Shockman, eds.), Academic Press, New York, pp. 31–66.

Sjöquist, J., Movitz, J., Johansson, I. B., and Helm, H., 1972, Localization of protein A in the bacteria, *Eur. J. Biochem.* **30:**190–194.

Sleytr, U. B., 1978, Regular arrays of macromolecules on bacterial cell walls: Structure, chemistry, assembly and function, *Int. Rev. Cytol.* **53**:1–64.

Sokatch, J. R., 1979, Roles of appendages and surface layers in adaptation of bacteria to their environment, in: *The Bacteria: A Treatise on Structure and Function,* Volume 7 (I. C. Gunsalus, J. R. Sokatch, and L. N. Ormston, eds.), Academic Press, New York, pp. 229–290.

Sutherland, I. W. (ed.), 1977, *Surface Carbohydrates of the Prokaryotic Cell,* Academic Press, New York.

Tipper, D. J., and Wright, A., 1979, The structure and biosynthesis of bacterial cell walls, in: *The Bacteria: A Treatise on Structure and Function,* Volume 7 (I. C. Gunsalus, J. R. Sokatch, and L. N. Ornston, eds.), Academic Press, New York, pp. 291–426.

Tomaz, A., 1979, The mechanism of the irreversible antimicrobial effects of penicillins: How the beta-lactam antibiotics kill and lyse bacteria, *Annu. Rev. Microbiol.* **33**:113–137.

Tonn, S. J., and Garder, J. E., 1979, Biosynthesis of polysaccharides by prokaryotes, *Annu. Rev. Microbiol.* **33**:169–199.

Tylewska, S. K., Hjerten, S., and Wadstrom, T., 1979, Contribution of M protein to the hydrophobic surface-properties of *Streptococcus pyogenes,* *FEMS Microbiol. Lett.* **6**:249–253.

Van Driel, D., Wicken, A. J., Dickson, M. R., and Knox, K. W., 1973, Cellular location of the lipoteichoic acids of *Lactobacillus fermentum* NCTC 6991 and *Lactobacillus casei* NCTC 6375, *J. Ultrastruct. Res.* **43**:483–497.

Ward, J. B., 1981, Teichoic and teichuronic acids: Biosynthesis, assembly and location, *Microbiol. Rev.* **45**:211–243.

Ward, J. B., and Curtis, C. A. M., 1982, The biosynthesis and linkage of teichuronic acid to peptidoglycan in *Bacillus licheniformis, Eur. J. Biochem.* **122**:125–132.

Whittenbury, R., and Dow, C. S., 1977, Morphogenesis and differentiation of *Rhodomicrobium vanielii* and other budding and prosthecate bacteria, *Bacteriol. Rev.* **41**:754–808.

Wicken, A. J., 1980, Structure and cell membrane-binding properties of bacterial lipoteichoic acids and their possible role on adhesion of streptococci to eucaryotic cells, in: *Bacterial Adherence* (E. H. Beachey, ed.), Chapman & Hall, London, pp. 139–158.

Wicken, A. J., and Knox, K. W., 1975, Lipoteichoic acids: A new class of bacterial antigen, *Science* **187**:1161–1167.

Wicken, A. J., and Knox, K. W., 1977, Biological properties of lipoteichoic acids, in: *Microbiology 1977* (D. Schlessinger, ed.), American Society for Microbiology, Washington, D.C., pp. 360–365.

Wicken, A. J., and Knox, K. W., 1980, Bacterial cell surface amphiphiles, *Biochim. Biophys. Acta* **604**:1–26.

Wicken, A. J., and Knox, K. W., 1983, Cell surface amphiphiles of gram-positive bacteria, *Toxicon* Suppl. **3**:501–507.

Wicken, A. J., Broady, K. W., Evans, J. D., and Knox, K. W., 1978, New cellular and extracellular amphipathic antigen from *Actinomyces viscosus* NY1, *Infect. Immun.* **22**:615–616.

Willison, J. H. M., Easterbrook, K. B., and Coombs, R. W., 1977, The attachment of bacterial spinae, *Can. J. Microbiol.* **23**:258–266.

Wright, A., and Tipper, D. J., 1979, The outer membrane of gram-negative bacteria, in: *The Bacteria: A Treatise on Structure and Function,* Volume 7 (I. C. Gunsalus, J. R. Sokatch, and L. N. Ornston, eds.), Academic Press, New York, pp. 291–426.

Yadomae, T., Yamada, H., and Miyazaki, T., 1978, Characterisation of an extracellular polysaccharide from a *Xanthomonas* species, *Carbohydr. Res.* **60**:128–139.

3

Animal Cell Surface Membranes

ITZHAK OFEK, HALINA LIS, and NATHAN SHARON

1. INTRODUCTION

The interior of all cells is separated from the external environment by a membrane called the plasma or surface membrane. In free-living cells the plasma membrane usually has the appearance of a structure that is homogeneous over the entire cell surface. In differentiated cells that are organized into tissues, discrete parts of the membrane may show special modifications that facilitate the performance of particular functions (Fig. 1). Common variations include: (1) tight junctions, gap junctions, and desmosomes, frequently seen at areas of intercellular contact between tissue cells; (2) outward proliferation of the plasma membrane, to promote adsorption of nutrients by increasing the cell surface in contact with the external environment (e.g., "brush borders" of intestinal epithelial cells) or to serve as motile structures (e.g., cilia of tracheal cells); and (3) inward proliferation of the plasma membrane to provide a close link between the intracellular, energy-producing metabolism and the energy-requiring activities of the plasma membrane (e.g., kidney tubule cells).

All membranes of animal cells consist of lipids and proteins, both in their nonglycosylated and glycosylated forms. The relative amounts of each of these groups of compounds vary considerably (Table I) as does the nature of the individual molecular constituents (Table II). In spite of these large variations, the organization of the membranes is remarkably similar. In this chapter we describe the chemical nature of the various membrane constituents and their mode of organization in the membrane. We then focus on those constituents which act as receptors in general and in particular for bacterial adhesion to animal cells. In order to keep the literature list within reasonable limits, references will be made, whenever possible, to recently published reviews or books. References to original literature will be limited to the subject of membrane constituents involved in bacterial adhesion.

Itzhak Ofek • Department of Human Microbiology, Sackler School of Medicine, Tel Aviv University, Tel Aviv, Israel.   Halina Lis and Nathan Sharon • Department of Biophysics, The Weizmann Institute of Science, Rehovot, Israel.

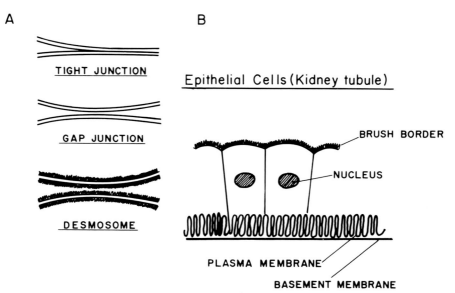

Figure 1. Specialized regions of cell membranes.

2. MOLECULAR CONSTITUENTS

2.1. Lipids

The major classes of lipid commonly found in animal cell membranes are glycerophospholipids, sphingolipids, and sterols (Table II).

Glycerophospholipids or glycerophosphatides form the most abundant and varied class of membrane lipids and are the major components responsible for the bilayer organization of the membrane. As seen in Table II, the diversity of the glycerophospholipids arises from the type of the fatty acids esterified with the hydroxyl groups at carbon 1 and 2 of the glycerol and from the type of head group attached by a diester link to the phosphate group esterified with the primary hydroxyl group at carbon 3 of the glycerol chain. More than a half-dozen fatty acids, which differ in the number of carbon atoms in the chain and in the number of double bonds (or degree of unsaturation), have been found in glycerophospholipids. Usually one of the fatty acids in the glycerol phosphate molecule is saturated, mostly that on carbon 1, and the other is unsaturated.

The structural backbone of the *sphingolipids* is sphingosine, a long-chain amino alcohol

Table I. Protein, Lipid, and Carbohydrate Content of Animal Cell Membranes[a]

Membrane	Protein (%)	Lipid (%)	Carbohydrate (%)
Human erythrocytes	49	43	8
Myelin	18	79	3
Liver cells[b]	54	36	10
Bovine retinal rod (outer segment)	60	40	—

[a]From Sim (1982) unless otherwise noted.
[b]From Lotan and Nicolson (1981).

Table II. Lipids in Biological Membranes

Glycerophospholipids $CH_3C_{x_1}H_yCOOCH_2$ $\vert$ $CH_3C_{x_2}H_yCOOCH$ $\vert$ CH_2OPO_3R	$\left.\begin{array}{l}x_1 = 14\text{--}18 \\ x_2 = 14\text{--}18\end{array}\right\}$ usually $x_1 \neq x_2$ y depends on the number of double bonds in the chain and varies from $2x - 2$ for one double bond to $2x - 12$ for six double bonds
Phosphatidic acid Phosphatidylcholine Phosphatidylserine Phosphatidylglycerol	$R = H$ $R = CH_2CH_2\overset{+}{N}(CH_3)_3$ $R = CH_2CH(NH_2)COOH$ $R = CH_2CH(OH)CH_2OH$
Sphingolipids $CH_3(CH_2)_{12}CH{=}CH{-}CHOH$ $\vert$ $CH_3C_xH_yCONH{-}CH_2$ $\vert$ CH_2OR	$x = 14\text{--}18$
Ceramide Sphingomyelin Glycosphingolipids	$R = H$ $R = PO_3CH_2CH_2\overset{+}{N}(CH_3)_3$ $R = $ saccharides
Sterols Cholesterol	$R = H$

of 18 carbon atoms with one double bond. It is *N*-acetylated by one of several fatty acids, to form ceramide. The hydroxyl group of ceramide is either esterified by a polar group consisting of phosphate and choline (as in sphingomyelin) or linked glycosidically to one of a variety of carbohydrate structures to form glycosphingolipids. The diversity of sphingolipid molecules arises from the various types of fatty acid and, perhaps mostly, from the various oligosaccharide substituents (Table III).

The most common *sterol* in membranes is cholesterol (see Table II); it is usually present in the nonesterified form.

Membrane lipids are amphiphatic molecules, i.e., they contain both a polar and a nonpolar region. When placed in a polar environment, such as found in the intra- and extracellular milieu of biological systems, they arrange themselves in bilayers with the nonpolar region inside and the polar groups facing outwards. This arrangement serves to minimize the interaction of the hydrocarbon residues with the water molecules of the environment. Thermodynamic considerations have revealed that bilayer formation is favored when the amphiphate contains one polar and two nonpolar moieties. Indeed, the bulk of the lipids of plasma cell membranes are composed of such molecules.

Under appropriate experimental conditions, mixtures of membrane lipids (e.g., phosphatidylcholine and phosphatidylserine or phosphatidylcholine and cholesterol) reassemble into closed vesicles (or liposomes), possessing the lipid bilayer structure. By providing both a membrane matrix and an internal volume separated from the outside, liposomes serve as an excellent model of a biological membrane (Bangham *et al.*, 1974). For more detailed

Table III. Structure of Major Neutral and Acidic Glycolipids in Animal Cells[a]

Neutral	
Glucosylceramide	Glc-Cer
Lactosylceramide	Galβ4Glc-Cer
Trihexosylceramide	Galα4Galβ4Glc-Cer
Globoside	GalNAcβ3Galα4Galβ4Glc-Cer
Forssman hapten	GalNAcα3GalNAcβ3Galα4Galβ4Glc-Cer

Acidic (gangliosides)

GM$_3$ Galβ4Glc-Cer
  ↑ α2,3
  NeuNAc

GM$_2$ GalNAcβ4Galβ4Glc-Cer
  ↑ α2,3
  NeuNAc

GM$_1$ Galβ3GalNAcβ4Galβ4Glc-Cer
  ↑ α2,3
  NeuNAc

GD$_{1a}$ Galβ3GalNAcβ4Galβ4Glc-Cer
  ↑ α2,3   ↑ α2,3
  NeuNAc   NeuNAc

GT$_1$ Galβ3GalNAcβ4Galβ4Glc-Cer
  ↑ α2,3   ↑ α2,3
  NeuNAc |
    NeuNAcα2,8 NeuNAc

[a]All sugars are of the D configuration.

information on the properties and organization of membrane lipids, the reader is referred to Jain and Wagner (1980) and Tanford (1980).

2.2. Proteins and Glycoproteins

Virtually all proteins of animal membranes are glycosylated, i.e., they are glycoproteins (Sharon and Lis, 1981, 1982). In many respects, membrane glycoproteins are similar to soluble glycoproteins. For example, they do not differ significantly in the overall amino acid composition and carbohydrate content and contain the same monosaccharide constituents: the hexoses D-galactose, D-mannose, and L-fucose; the N-acetylhexosamines, N-acetyl-D-glucosamine and N-acetyl-D-galactosamine; and the sialic acids. D-Glucose, found in the collagens, does not appear to occur in membrane glycoproteins; the uronic acids, D-glucuronic acid and L-iduronic acid, are confined to proteoglycans such as heparan sulfate, when these are present in membranes.

Several types of carbohydrate–peptide linkages are also common to both groups of glycoprotein—the N-glycosyl bond between N-acetyl-D-glucosamine and asparagine (N-acetyl-D-glucosaminyl-asparagine) and the O-glycosidic bond between N-acetyl-D-galactosamine and a serine or threonine of the protein polypeptide backbone.

A distinctive feature of the primary amino acid sequence of membrane glycoproteins is the presence of regions rich in hydrophobic amino acids. Such glycoproteins exhibit amphiphatic properties, which are enhanced by the presence of the highly hydrophilic carbohydrate in the nonhydrophobic regions of the glycoproteins. Hydrophilic regions in membrane glycoproteins account for their limited solubility in aqueous solution in the absence of detergents. More importantly, the hydrophobic stretches serve to anchor the proteins to the lipid bilayer (Fig. 2).

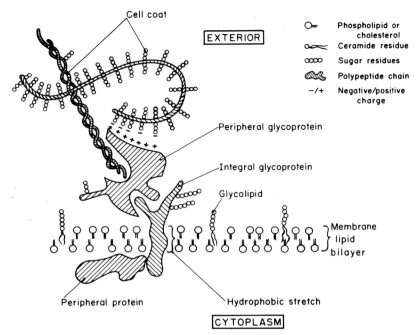

Figure 2. Organization of animal cell membrane.

3. ORGANIZATION OF CELL MEMBRANE CONSTITUENTS

All animal cell membranes share common organizational features, as described below and illustrated in Fig. 2:

1. The major membrane lipids are arranged in a planar bilayer configuration that is predominantly in a "fluid" state under physiological conditions
2. All glycolipids, and certain glycoproteins, are inserted or "intercalated" into the bilayer and thus are designated integral or intrinsic components of the membranes
3. Other glycoproteins and proteins are bound to the surface of the plasma membrane by weak ionic interactions or hydrogen bonding with the integral components and are designated peripheral or extrinsic components
4. In many animal cells there is a substantial layer of carbohydrate-containing material of variable thickness outside the plasma membrane but in close or intimate association with the membrane. This layer is known as the cell coat ("fuzz") or extracellular matrix.

The distinction between membrane constituents as being integral, peripheral, or belonging to the cell coat is based on the method required to dissociate the constituent in question from the cell membrane. The integral constituents may be released only after disruption or perturbation of the phospholipid bilayer, usually by detergents (Lichtenberg et al., 1983). Nonintegral surface constituents are commonly released by washing the cells with buffers of different pH or ionic strength, or by using chelating agents such as EDTA or EGTA. There is no general method, however, to selectively release either peripheral or extracellular matrix constituents. As a result, the distinction between these two classes of membrane constituents is sometimes difficult and very often both are referred to as "nonintegral."

One of the key features of the membrane is its asymmetry. For the nonglycosylated lipids the asymmetry is only partial, in that every phospholipid is present on both sides of the layer but in different amounts. In human erythrocytes, for example, lipids with positively charged head groups (e.g., phosphatidylethanolamine and phosphatidylserine) predominate in the internal layer, facing the cytoplasm (Marinetti and Crain, 1978). The asymmetry with respect to proteins, glycoproteins, and glycolipids is absolute: every molecule of a given membrane constituent has the same orientation across the lipid bilayer, with the carbohydrate moieties of the glycosylated compounds always exposed on the outer surface. For further information on the organization of the animal cell membrane, the reader is referred to recent reviews (Lodish *et al.,* 1981; Singer, 1981; Lotan and Nicolson, 1981; Aplin and Hughes, 1982) and books (Hay, 1981; Sim, 1982).

3.1. Integral Membrane Constituents

As mentioned above, both glycolipids and glycoproteins belong to this class of membrane constituent. With respect to their organization within the cell membrane, these compounds share several properties. Of special interest is their mobility within the plane of the membrane, made possible by the fluidity of the lipids of the membrane bilayer (Edidin, 1974; Cherry, 1979). Not all integral constituents possess the same degree of mobility. Their movement may be restricted by interactions with extrinsic proteins on the inner surface or by junctions (Nicolson, 1976; Geiger, 1983). As a result, the topography of integral glycoproteins is nonrandom and some of these constituents are confined to specific regions of the membrane. Nevertheless, the concept that the membrane lipids behave as a continuous fluid with regions of low fluidity, in which integral constituents are free to move (Singer and Nicolson, 1972; Singer, 1974), has been very useful in describing many phenomena associated with the cell surface, such as "patching" and "capping" of integral membrane macromolecules following their interaction with a multivalent extracellular agent (Karnovsky and Unanue, 1973; Raff and de Petris, 1973).

3.1.1. Glycolipids

All glycolipids of the cell membrane are constituents of the outer half of the bilayer, where they comprise 30–60% of the total lipid content. The ceramide group is responsible for anchoring the molecules to the membrane by intercalating within the lipid bilayer while the oligosaccharide chain extends to the outer surface. The glycolipids containing sialic acid (i.e., gangliosides, see Table III) contribute significantly to the net negative charge of the animal cell surface. When isolated in pure form from the cell membrane, glycolipids tend to form micelles and when mixed with animal cells they can "coat" most cells with an orientation comparable to the original one, i.e., with the oligosaccharide portion exposed to the outer surface (Sedlacek *et al.,* 1976; Callies *et al.,* 1977). This behavior is similar to that of bacterial glycolipids such as lipopolysaccharides of gram-negative bacteria and lipoteichoic acid of gram-positive bacteria (Wicken and Knox, 1981; Chapter 2, Sections 2.2 and 3.3.1). Many investigators took advantage of this property for experiments in which the type of oligosaccharide and number of molecules to be inserted in the animal cell can be controlled (Wiegandt *et al.,* 1981).

3.1.2. Glycoproteins

The hydrophobic region(s) present in all integral glycoproteins of the cell surface is responsible for anchoring the molecules to the cell membrane. Most, if not all, integral glycoproteins extend across the entire thickness of the bilayer so that they are exposed to both the internal and the external environment. Some of the glycoproteins, for example glycophorin in the human erythrocyte (Marchesi *et al.*, 1976; Tanner, 1978) and the E_2 glycoprotein of Semliki Forest virus (Garoff, 1979), span the membrane only once while others traverse it more than once [band 3, also in the human erythrocyte (Steck, 1978)] (Fig. 3). As mentioned, the oligosaccharide moiety is always exposed on the cell surface. The internal segment of a membrane-spanning intrinsic glycoprotein may be in close contact with some of the peripheral proteins on the cytoplasmic face of the membrane. The intrinsic glycoproteins may therefore play an essential role in processes that require communication between the outside of the cell and its interior, such as transport phenomena and the response of cells to external stimuli, e.g., hormones, toxins, and other cells (Nicolson, 1976, 1979; Finean *et al.*, 1978). Depending on the size and number of their hydrophobic regions, the membrane glycoproteins may display low to moderate solubility in detergent-free aqueous solutions. Unlike glycolipids, the integral glycoproteins in isolated form do not reassociate or "coat" cell membranes, but can be reassembled into liposomes (Eytan, 1982). The orientation of the reconstituted glycoproteins is, however, symmetrical in that half of the molecules insert with the glycosylated part facing outwards and half facing inwards.

3.2. Peripheral Membrane Components

The peripheral proteins and glycoproteins are anchored to the surface of the membrane by weak ionic interaction or hydrogen bonding with integral constituents of the cell membranes, e.g., glycoproteins, glycolipids, or polar head groups of phospholipids. A given peripheral component is always located on the same side of the membrane, either outside or inside. The proteins are associated mainly with the inner face. Some of them interact with elements of the cytoskeleton of the cell, thus providing a link between integral membrane constituents and the interior of the cells. Because of the difficulties mentioned earlier in

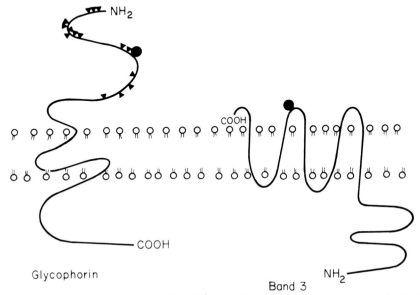

Glycophorin

Band 3

Figure 3. Organization of glycophorin and of band 3 in the human erythrocyte membrane. ●, *N*-glycosidic oligosaccharide units; ▼, *O*-glycosidic units. The drawing of band 3 is based on data in Macara and Cantley (1983).

distinguishing between peripheral glycoproteins of the cell surfaces and glycoproteins of the extracellular matrix, very few compounds have been unequivocally identified as peripheral glycoproteins. The best example of a membrane constituent of this class is fibronectin, widely distributed on the surface of normal cells. The structure of fibronectin is characterized by the presence of several distinct domains with binding sites for specific ligands, such as heparan sulfate, collagen, and hyaluronic acid (Hynes and Yamada, 1982; Ruoslahti *et al.*, 1982). By virtue of this structure, and the outermost location of fibronectin on the cell surface, this glycoprotein is eminently suitable to mediate contacts between cells and their environment. (For a discussion of fibronectin and adhesion of gram-positive bacteria to tissue surfaces, see Chapter 10, Section 2.1.6.)

3.3. Glycoproteins of the Cell Coat

Most cells are coated with a layer of glycoproteins that are bound to the cell surface by weak noncovalent bonds with either a peripheral or an integral constituent. The coat can be of variable thickness and chemical complexity. It may take the form of a matrix, as in mature cartilage where it is composed of a collagen–proteoglycan–hyaluronic acid complex which is associated with the cell membrane via fibronectin (Aplin and Hughes, 1982). Other cells, such as mucosal cells of the respiratory and urinary tract, are coated with a layer rich in highly sialylated, high-molecular-weight glycoproteins known as mucins.

The different types of coat, produced by specialized cells at specific sites in the organism, are well adapted to their specific physiological functions. Thus, the matrix of the cartilage is able to withstand stress and to bear considerable loads, while the mucins form viscous solutions that perform protective, lubricant, and transport roles in the appropriate tissues. Within each type of coat, the adaptation is further refined by the great diversity of basic structures of the various components.

4. CELL SURFACE RECEPTORS

4.1. General Concepts

The surface constituents of animal cells are important for controlling many cellular activities, such as development, differentiation, intracellular communication, and recognition or clearance of foreign particles, as well as the activation of cells and the triggering of defined biological responses. When a particular cell membrane constituent mediates one or more of such activities by specifically combining with an extracellular biological agent (or ligand), it becomes known as a receptor for that agent. Two criteria thus characterize a membrane receptor: that it should be able to interact specifically with a ligand and that this interaction should lead to a biologically relevant response (Cuatrecasas, 1974; Schulster and Levitzki, 1980). The interaction step follows general rules which are common to all receptors; it entails stereospecific and complementary binding between the receptor molecule and the ligand, and is analogous to the interaction between antibody and antigen, lectin and specific sugar, or enzyme and substrate. In contrast, the biological consequences are diverse, depending on the type of cell, the nature of the ligand, and the receptor.

The active chemical group of the receptor to which the ligand binds will be referred to as the attachment or recognition site. It is a structure consisting, as a rule, of either sugar residues or amino acid residues; occasionally, both the peptide backbone and the carbohy-

Table IV. Chemical Nature of Representative Receptors Isolated from Animal Cell Membranes

Biological agent (ligand)	Source	Receptor Chemical nature	Receptor Recognition site	Ref.[a]
Acetylcholine	Postsynaptic cells	Glycoprotein MW 200,000	A 40,000-MW subunit	6
Cholera toxin	Several types of cells	Glycolipid (GM₁)	NeuNAcGalGalNAc	2, 7
Concanavalin A	Human erythrocytes	Glycoprotein (band 3)	D-Mannose residues	3
Blood type A antibody	Human erythrocytes	Glycoprotein (band 3) Macroglycolipids	N-Acetyl-D-galactosamine	4, 5
Asialoglycoprotein (e.g., asialofetuin)	Hepatic cells	Glycoprotein MW 45,000	D-Galactose-specific binding site	1

[a]1, Ashwell and Harford (1982); 2, Critchley *et al.* (1982); 3, Findlay (1974); 4, Finne (1980); 5, Kościelak *et al.* (1976); 6, Prives (1980); 7, Yamakawa and Nagai (1978).

drate moiety of a glycoprotein may serve as the attachment site (Duk *et al.*, 1982; Tollefsen and Kornfeld, 1983). The carbohydrate attachment sites are always part of glycoprotein and/or glycolipid molecules. Table IV gives examples of receptor molecules and attachment sites for different ligands. As can be seen, the same cell membrane constituent may possess recognition sites for diverse types of agent, and therefore serve as receptor for more than one ligand. Conversely, identical attachment sites for a particular ligand may be found on different constituents.

Of the two types of attachment sites on receptor molecules, i.e., sugars and amino acids, the former are by far the more versatile. This is because of the enormous number of specific structures that can be formed from relatively few monosaccharide units, as compared with the number of peptides that can be formed from the same number of amino acids (Table V). Several reviews have appeared during the last decade on cell surface sugars as recognition sites in biological interactions (Yamakawa and Nagai, 1978; Sharon, 1979, 1981; Gahmberg, 1981; Hakomori, 1981; Fishman, 1982).

Table V. Comparison of the Number of Possible Isomeric Peptides and Oligosaccharides[a]

Monomer composition	Product	Number of isomers Peptides	Number of isomers Saccharides[b]
X₂	Dimer	1	11
X₃	Trimer	1	176
XYZ	Trimer	6	1056

[a]From Sharon (1975); calculations by Clamp (1974).
[b]Pyranose ring only.

Table VI. Methods for the Visualization of Ligand Binding to Cells

Ligand derivative	Microscopic technique
Fluorescein or rhodamine	Light microscopy
Ferritin	Electron microscopy
Colloidal gold	Electron microscopy
Horseradish peroxidase	Light microscopy; electron microscopy
Underivatized ligand followed by fluorescent antiligand antibodies	Light microscopy

4.2. Binding Studies

Numerous methods have been developed for the study of ligand binding to cell surface receptors (Cuatrecasas, 1974; Bernard, 1979; Levitzki, 1980; Schrevel et al., 1981). Binding can be easily demonstrated with the aid of suitable derivatives, obtained by coupling the ligand to compounds that can be visualized in the light or electron microscope (Table VI). Such studies serve to determine the localization and density of receptor molecules for a given ligand on the cell surface.

Quantitation of binding is usually done by measuring the amount of radioactivity bound by cells after their exposure to varying concentrations of radiolabeled ligand. The data obtained can be graphically presented in five different plots, summarized in Table VII. The various plots assume that the binding of the ligand to the cell surface receptors is saturable, (i.e., very little or no change in the amount of bound ligand occurs by increasing the concentration of the ligand beyond a certain point), that it is reversible, and that it obeys the law of mass action. Two important parameters can be derived from the plots. One is the amount of bound ligand at the saturation point, which reflects the total number of receptors available on the cell surface for the ligand; the other is the affinity of the receptor–ligand complex expressed as the molar concentration of ligand needed to achieve 50% saturation.

4.3. Expression of Receptors on Cell Surfaces

In living tissues, there is a dynamic change in the production and expression of any particular cell membrane constituent as a function of age and the physiological state of the cell. Changes are also bound to occur in cells exposed to the action of diverse drugs, some of

Table VII. Ways of Plotting Binding Data between a Ligand and Its Cell Surface Receptors[a]

Plot	Function plotted[b]
Direct	r versus c
Logarithmic	r versus $\log c$
Double reciprocal	$1/r$ versus $1/c$
Scatchard	r/c versus r
Hill	$\log r/(1 - r)$ versus $\log c$

[a]Modified from Levitzki (1980).
[b]Notation: r = amount of bound ligand (usually expressed as $\mu g/10^6$ cells); c = free concentration of ligand (usually expressed as $\mu g/ml$).

which may affect the biosynthesis and expression of cell membrane constituents. Such changes have been best documented in sugar residues of glycoproteins and glycolipids (Gahmberg, 1977; Fukuda and Fukuda, 1978; Critchley, 1979; Hakomori, 1981; Gahmberg and Andersson, 1982), largely due to the availability of specific probes for these residues [e.g., lectins (Sharon and Lis, 1975; Roth, 1980) and enzymes (Flowers and Sharon, 1979)]. Moreover, the proximity of other cell surface constituents may interfere with the ability of receptors to bind the ligand. When comparing the binding of a given ligand to cells under varying physiological conditions, it is therefore not always possible to determine whether any difference observed is due to a change in the number of receptor molecules, alterations in the recognition site, or interference by other surface constituents.

It is beyond the scope of this chapter to review the vast amount of data accumulated during the last three decades on the availability of receptors for different types of biological agents as a function of growth and differentiation. We choose to illustrate the complexity involved in this area by presenting some of the variables that affect the expression of one receptor, that for insulin, in man (Table VIII; see also Grunberger et al., 1983).

4.4. Characterization of Cell Surface Receptors

The binding studies described above (Section 4.2) are essential to demonstrate the presence of receptor molecules for a particular ligand on the cell surface, but by themselves tell only little about the identity and nature of these molecules. Identification and characterization of a cell surface constituent as a receptor for a biological agent require (1) elucidation of the nature of the attachment site(s), (2) isolation of the receptor molecule which contains the attachment site(s) from the cell membrane and its physicochemical characterization, (3) demonstration that it is available to the agent on the cell surface, and (4) proof that it

Table VIII. Variations in the Availability of the Insulin Receptor in Humans[a]

Condition	No. of receptor molecules per cell	Receptor affinity	Remarks
Normal adults, 5 hr after 100-g glucose intake	Normal	↑	
Newborn infants	↑ ↑	↑	
Hyperinsulinemic obese	↓	Normal	Chronic diet increases number of receptors
Thin, hyperinsulinemic adult diabetics	↓	Normal	Treatment with sulfonylurea increases number of receptors
Insulinoma	↓	Variable	Tumor removal normalizes all binding parameters
Acanthosis nigricans with insulin resistance			
Type A	↓ ↓	Normal	Binding parameters unchanged after 72-hr fast
Type B	Normal	↓ ↓	Circulating antireceptor antibody
Ataxia telangiectasia	↑	↓ ↓	
Hyperinsulinemic acromegalics	↓	↑	

[a]Data from Bar et al. (1979). Arrows denote increase (upward) or decrease (downward) of parameter in question.

Table IX. Procedures Commonly Used to Identify the Chemical Nature
of Recognition Sites of Cell Surface Receptors

Procedure	Materials	Information gained and comments
Hapten inhibition of the interaction of a biological agent with cells by low-molecular-weight compounds. The interaction can be measured either as binding or as biological activity of the agent (O'Brien, 1979; Schulster and Levitzki, 1980)	The inhibitors should be soluble, well-defined, and free of damaging effects on the cell	Inhibition of the interaction suggests that the structure of the attachment site is closely related to that of the inhibitor. Confirmation of this assumption can be obtained by showing that the agent binds to immobilized inhibitor
Effect of enzymatic and chemical modifications of the cell surface on the interaction of the agent with the cells (Schulster and Levitzki, 1980)	Proteolytic enzymes, glycosidases, or chemical reagents which cleave specific covalent bonds	This approach is helpful in determining whether the recognition site is a protein or a carbohydrate. The more specific the treatment for its moleculr target, the more information is gained on the structure of the recognition site
Inhibition of binding of the agent by compounds that bind to known structures on the cell surface (Schulster and Levitzki, 1980)	The inhibitors may be lectins of known specificity or antibodies (preferably monoclonals) against specific epitopes	Inhibition suggests that the molecular target site of the inhibitors may be involved in binding of the agent under study
Binding of the agent to tissue culture cells that exhibit alterations in the composition or structure of surface constituents (Pan et al., 1979)	Clones selected for resistance to lectin (Briles, 1982) or cells treated with drugs that alter specific oligomeric structures of cell surface constituents [e.g., tunicamycin (Schwarz and Datema, 1982) or swainsonine (Elbein et al., 1981; Tulsiani et al., 1982), which affect the biosynthesis of carbohydrate chains of glycoproteins]	Correlation between decreased binding of the agent and well-defined alterations of cell surface constituents indicates the involvement of the affected structures in binding of the agent

mediates the biological activity of the agent. The procedures available for tackling some of the above problems are given in Tables IX and X.

4.5. Animal Cell Surface Receptors in Bacterial Adhesion

Numerous studies on bacterial adhesion performed during the last decade and summarized in several reviews (Jones, 1977; Gibbons, 1977; Ofek and Beachey, 1980; Beachey, 1981) and books (Bitton and Marshall, 1980; Berkeley et al., 1980; Beachey, 1980; Elliott et al., 1981; Beachey et al., 1982; see Chapters 10, 11, and 15) have marshalled two major concepts. The first is that bacterial adhesion is an initial and prerequisite step in the infectious process of pathogenic bacteria of any species. The second is that the adhesion of bacteria to animal cells is a specific interaction mediated by (macro)molecules on the bacterial surface that combine with complementary structures on the animal cell surfaces. The terms *adhesins* and *receptors* have been coined to describe the corresponding molecules on the surface of

Table X. Procedures for the Identification and Purification
of Cell Membrane Receptors

Procedure	Comments
Overlay	
SDS electrophoresis or thin-layer chromatography of solubilized cell membranes, followed by overlay of the gel with a solution of radiolabeled (or fluorescent) ligand and radiography (Burridge, 1978; Furlan *et al.*, 1979; Magnani *et al.*, 1980; Momoi *et al.*, 1982)	Reveals membrane constituents capable of binding the agent
Receptor cross-linking	
Binding of the agent, derivatized with a cleavable photolabile cross-linking reagent, to cells, followed by light-sensitive cross-linking and solubilization of the receptor–agent complex. The receptor can then be released from its bond to the ligand and analyzed (Das and Fox, 1979; Jaffe *et al.*, 1979, 1980; Aplin *et al.*, 1981)	Leads to the isolation of membrane constituents available to the ligand on the surface of intact cells
Affinity chromatography	
Chromatography of solubilized radiolabeled cell membranes on columns derivatized with the agent. Elution with a suitable buffer or specific reagent, e.g., hapten inhibitor (Pricer *et al.*, 1974; Shorr *et al.*, 1981; Fujita-Yamaguchi *et al.*, 1983)	Yields membrane constituents capable of binding the agent
Reconstitution	
Incorporation of the purified receptor into artificial membranes or cells devoid of the receptor, followed by binding studies with the agent and determination of biological response (Miller and Racker, 1979; Blumenthal and Shamoo, 1979; Kelleher *et al.*, 1983)	Reconstitution of the biological response is the best proof that the isolated cell membrane constituent is indeed the receptor for the agent

Table XI. Receptor Molecules and Attachment Sites for Bacteria on Animal Cells

Bacteria	Receptor molecule	Location in membrane	Attachment site	Ref.[a]
E. coli (pyelo-nephritogenic)	Glycolipid	Integral	D-Galα4-D-Gal	4,5
Mycoplasma	Glycophorin	Integral	Sialic acid	2
Streptococci	Fibronectin	Peripheral	Fatty acid binding region	1
Shigellae	Colonic mucus	Cell coat	L-Fuc/D-Glc-specific binding site	3

[a] 1, Beachey *et al.* (1983); 2, Feldner *et al.* (1979); 3, Izhar *et al.* (1982); 4, Källenius *et al.* (1981); 5, Leffler and Svanborg-Edén (1981).

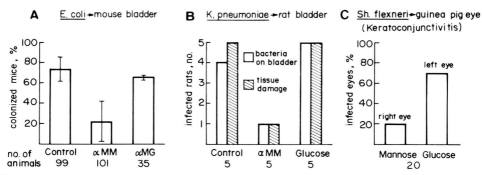

Figure 4. Prevention of infection, caused by bacteria which adhere to host tissues via mannose-specific lectins, by mannose or methyl-α-mannoside. Based on data from (A) Aronson *et al.* (1979), (B) Fader and Davis (1980), (C) Andrade (1980). αMM, methyl-α-mannoside; αMG, methyl-α-glucoside.

bacteria and animal cells, respectively. Thus, animal cell receptors of bacterial adhesins fulfill the criteria of a biological receptor, i.e., specific binding followed by a physiologically relevant response, in this case manifested as infection.

Some of the methodology described in Table IX, in particular hapten inhibition of bacterial adhesion to animal cells, was successfully utilized to identify the chemical nature of the attachment site for 12 to 14 different pathogenic microorganisms (Beachey, 1981). With the exception of the attachment sites for streptococci and shigellae, which are made up of amino acids, the remaining attachment sites are carbohydrates (see Chapter 11, Section 5). The receptor molecules that carry the various attachment sites have been identified in only a very few cases (Table XI). As can be seen, receptors for bacterial adhesion are found in all three classes of membrane constituents, namely, integral, peripheral, and cell coat. Chemically they may be either proteins, glycoproteins, or glycolipids.

Although studies on biological receptors aim at identifying both the receptor molecule and the attachment site, in bacterial adhesion knowledge of the nature and structure of the latter is of special interest. This interest derives from the concept that attachment site analogs (also called *receptor analogs*), capable of inhibiting bacterial adhesion, may prevent the development of infection. Indeed, in three separate systems it has been demonstrated that D-mannose and methyl-α-D-mannopyranoside inhibited experimental infection by bacteria that adhere to animal cells via mannose-specific lectins present on their surface in the form of type 1 fimbriae (Fig. 4). Similarly, Galα4Gal-containing sugars were shown to prevent urinary tract infection in mice (Svanborg-Edén *et al.*, 1982) and monkeys (Källenius *et al.*, 1983) by *E. coli* that adhere to urinary cells via a globoside-specific lectin (Leffler and Svanborg-Edén, 1981; Källenius *et al.*, 1981; see Chapter 11, Section 5.1).

These findings, besides providing some of the best evidence for the central role of bacterial adhesion in the infectious process, illustrate the great potential of simple receptor analogs in the prevention of bacterial infections.

REFERENCES

Andrade, J. R. C., 1980, Role of fimbrial adhesiveness in experimental guinea pig keratoconjunctivity by *Shigella flexneri*, *Rev. Microbiol.* **11**:117–125.

Aplin, J. D., and Hughes, R. C., 1982, Complex carbohydrates of the intracellular matrix: Structures, interactions and biological roles, *Biochim. Biophys. Acta* **694**:375–418.

Aplin, J. D., Hughes, R. C., Jaffe, C. L., and Sharon, N., 1981, Reversible cross-linking of cellular components of adherent fibroblasts to fibronectin and lectin-coated substrata, *Exp. Cell Res.* **134:**488–494.

Aronson, M., Medalia, O., Schori, L., Mirelman, D., Sharon, N., and Ofek, I., 1979, Prevention of colonization of the urinary tract of mice with *Escherichia coli* by blocking of bacterial adherence with methyl-α-D-mannopyranoside, *J. Infect. Dis.* **139:**329–332.

Ashwell, G., and Harford, J., 1982, Carbohydrate-specific receptors of the liver, *Annu. Rev. Biochem.* **51:**531–554.

Bangham, A. D., Hill, M. W., and Miller, N. G. A., 1974, Preparation and use of liposomes as models of biological membranes, *Methods Membr. Biol.* **1:**1–68.

Bar, R. S., Harrison, L. C., Muggeo, M., Gorden, P., Lahn, R. C., and Roth, J., 1979, Regulation of insulin receptors in normal and abnormal physiology in humans, *Adv. Intern. Med.* **24:**23–52.

Beachey, E. H. (ed.), 1980, *Bacterial Adherence,* Chapman & Hall, London.

Beachey, E. H., 1981, Bacterial adherence: Adhesin–receptor interactions mediating the attachment of bacteria to mucosal surfaces, *J. Infect. Dis.* **143:**325–345.

Beachey, E. H., Eisenstein, I., and Ofek, I., 1982, *Bacterial Adherence in Infectious Diseases: Current Concepts,* The Upjohn Co., Kalamazoo, Mich.

Beachey, E. H., Simpson, W. A., Ofek, I., Hasty, D. K., Dale, J. B., and Whitneck, E., 1983, Attachment of *Streptococcus pyogenes* to mammalian cells, *Rev. Infect. Dis.* **5**(Suppl. 4)**:**5670–5677.

Berkeley, R. C. W., Lynch, J. H., Melling, J., Rutter, P. R., and Vincent, B. (eds.), 1980, *Microbial Adhesion to Surfaces,* Horwood, Chichester.

Bernard, E. A., 1979, Visualisation and counting of receptors at the light and electron microscopic levels, in: *The Receptors,* Volume 1 (R. D. O'Brien, ed.), Plenum Press, New York, pp. 247–310.

Bitton, G., and Marshall, K. C. (eds.), 1980, *Adsorption of Microorganisms to Surfaces,* Wiley, New York.

Blumenthal, R., and Shamoo, A. E., 1979, Incorporation of transport molecules into black lipid membrane, in: *The Receptors,* Volume 1 (R. D. O'Brien, ed.), Plenum Press, New York, pp. 215–245.

Briles, E. B., 1982, Lectin-resistant cell surface variants of eukaryotic cells, *Int. Rev. Cytol.* **75:**101–165.

Burridge, K., 1978, Direct identification of specific glycoproteins and antigens in sodium dodecyl sulfate gels, *Methods Enzymol.* **50:**54–64.

Callies, R., Schwarzmann, G., Radsak, K., Siegert, R., and Wiegandt, H., 1977, Characterization of the cellular binding of exogenous gangliosides, *Eur. J. Biochem.* **80:**425–432.

Cherry, R. J., 1979, Rotational and lateral diffusion of membrane proteins, *Biochim. Biophys. Acta* **559:**289–327.

Clamp, J., 1974, Analysis of glycoproteins, *Biochem. Soc. Symp.* **40:**3–16.

Critchley, D. R., 1979, Glycolipids as membrane receptors important in growth regulation, in: *Surfaces of Normal and Malignant Cells* (R. O. Hynes, ed.), Wiley, New York, pp. 63–101.

Critchley, D. R., Streuli, C. H., Kellie, S., Ansell, S., and Patel, B., 1982, Characterization of the cholera toxin receptor on Balb/c 3T3 cells as a ganglioside similar to, or identical with, ganglioside GM$_1$: No evidence for galactoproteins with receptor activity, *Biochem. J.* **204:**209–219.

Cuatrecasas, P., 1974, Membrane receptors, *Annu. Rev. Biochem.* **43:**169–214.

Das, M., and Fox, C. F., 1979, Chemical cross-linking in biology, *Annu. Rev. Biophys. Bioeng.* **8:**165–193.

Duk, M., Lisowska, E., Kordowicz, M., and Wasniowska, K., 1982, Studies on the specificity of the binding site of *Vicia graminea* anti-N lectin, *Eur. J. Biochem.* **123:**105–112.

Edidin, M., 1974, Rotational and translational diffusion in membranes, *Annu. Rev. Biophys. Bioeng.* **3:**179–201.

Elbein, A. D., Rolf, R., Dorling, P. R., and Vosbeck, K., 1981, Swainsonine, an inhibitor of glycoprotein processing, *Proc. Natl. Acad. Sci. USA* **78:**7393–7397.

Elliott, K., O'Connor, M., and Whelan, J. (eds.), 1981, *Adhesion and Microorganism Pathogenicity,* Pitman Press, London.

Eytan, G. D., 1982, Use of liposomes for reconstruction of biological function, *Biochim. Biophys. Acta* **694:**185–202.

Fader, R. C., and Davis, C. P., 1980, Effect of piliation on *Klebsiella pneumoniae* infection in rat bladders, *Infect. Immun.* **30:**554–561.

Feldner, J., Bredt, W., and Kahane, I., 1979, Adherence of erythrocytes to *Mycoplasma pneumoniae, Infect. Immun.* **25:**60–67.

Findlay, J. B. C., 1974, The receptor proteins for concanavalin A and *Lens culinaris* phytohemagglutinin in the membrane of the human erythrocyte, *J. Biol. Chem.* **249:**4398–4403.

Finean, J. B., Coleman, R., and Michell, R. H., 1978, *Membranes and Their Cellular Function,* 2nd ed., Blackwell, Oxford.

Finne, J., 1980, Identification of the blood-group ABH-active glycoprotein components of human erythrocyte membrane, *Eur. J. Biochem.* **104:**181–189.

Fishman, P. H., 1982, Role of membrane gangliosides in the binding and action of bacterial toxins, *J. Membr. Biol.* **69:**85–97.

Flowers, H. M., and Sharon, N., 1979, Glycosidases—Properties and application to the study of complex carbohydrates and cell surfaces, *Adv. Enzymol.* **48:**29–95.

Fujita-Yamaguchi, Y., Choi, S., Sakamoto, Y., and Itakura, K., 1983, Purification of insulin receptor with full binding activity, *J. Biol. Chem.* **258:**5045–5049.

Fukuda, M., and Fukuda, M. N., 1978, Changes in cell surface glycoproteins and carbohydrate structures during the development and differentiation of human erythroid cells, *J. Supramol. Struct.* **8:**313–324.

Furlan, M., Perret, B. A., and Beck, E. A., 1979, Staining of glycoproteins in polyacrylamide and agarose gels with fluorescent lectins, *Anal. Biochem.* **96:**208–214.

Gahmberg, C. G., 1977, Cell surface proteins: Changes during cell growth and malignant transformation, in: *Cell Surface Reviews* (G. Poste and G. L. Nicolson, eds.), North-Holland, Amsterdam, pp. 371–421.

Gahmberg, C. G., 1981, Membrane glycoproteins and glycolipids: Structure, localization and function of carbohydrates, in: *Membrane Structure* (J. B. Finean and R. H. Michell, eds.), Elsevier/North-Holland, Amsterdam, pp. 127–160.

Gahmberg, C. G., and Andersson, L. C., 1982, Surface glycoproteins of malignant cells, *Biochim. Biophys. Acta* **651:**65–83.

Garoff, H., 1979, Structure and assembly of the Semliki Forest virus membrane, *Biochem. Soc. Trans.* **7:**301–306.

Geiger, B., 1983, Membrane cytoskeleton interactions, *Biochim. Biophys. Acta* **737:**305–341.

Gibbons, R. J., 1977, Adherence of bacteria to host tissue, in: *Microbiology 1977* (D. Schlessinger, ed.), American Society for Microbiology, Washington, D.C., pp. 395–406.

Grunberger, G., Taylor, S. I., Dons, R. F., and Gorden, P., 1983, Insulin receptors in normal and diabetic states, *Clin. Endocrinol. Metab.* **12:**191–219.

Hakomori, S., 1981, Glycosphingolipids in cellular interaction, differentiation, and oncogenesis, *Annu. Rev. Biochem.* **50:**733–764.

Hay, D.E., 1981, *Cell Biology of Extracellular Matrix,* Plenum Press, New York.

Hynes, R. O., and Yamada, K. M., 1982, Fibronectins: Multifunctional molecular glycoproteins, *J. Cell Biol.* **95:**369–377.

Izhar, M., Nuchamowitz, Y., and Mirelman, D., 1982, Adherence of *Shigella flexneri* to guinea pig intestinal cells is mediated by a mucosal adhesin, *Infect. Immun.* **35:**1110–1118.

Jaffe, C. L., Lis, H., and Sharon, N., 1979, Identification of peanut agglutinin receptors on human erythrocyte ghosts by affinity crosslinking using a new cleavable heterobifunctional reagent, *Biochem. Biophys. Res. Commun.* **91:**402–409.

Jaffe, C. L., Lis, H., and Sharon, N., 1980, New cleavable photoreactive heterobifunctional cross-linking reagents for studying membrane organization, *Biochemistry* **19:**4423–4429.

Jain, M. K., and Wagner, R. C., 1980, *Introduction to Biological Membranes,* Wiley, New York.

Jones, G. W., 1977, The attachment of bacteria to the surface of animal cells, in: *Microbial Interactions* (J. L. Reissig, ed.), Chapman & Hall, London, pp. 139–176.

Källenius, G., Svensson, S. B., Mollby, R., Cedergren, B., Hultberg, H., and Winberg, J., 1981, Structure of carbohydrate part of receptor on human uroepithelial cells for pyelonephritogenic *Escherichia coli*, *Lancet* **2:**604–606.

Källenius, G., Svensson, S. B., Hultberg, H., Mollby, R., Winberg, J., and Roberts, J. A., 1983, P-Fimbria of pyelonephritogenic *Escherichia coli*: Significance for reflux and renal scarring—A hypothesis, *Infection* **11:**73–76.

Karnovsky, M. J., and Unanue, E. P., 1973, Mapping and migration of lymphocyte surface macromolecules, *Fed. Proc.* **32:**55–59.

Kelleher, D. J., Rashidbaigi, A., Ruoho, A. E., and Johnson, G. L., 1983, Rapid vesicle reconstitution of alprenolol-Sepharose purified β_1-adrenergic receptors: Interaction of the purified receptor with N_3, *J. Biol. Chem.* **258:**12881–12885.

Kościelak, J., Miller-Podraza, H., Krauze, R., and Piasek, A., 1976, Isolation and characterization of poly(glycosyl) ceramides (megaloglycolipids) with A, H and I blood-group activities, *Eur. J. Biochem.* **71:**9–18.

Leffler, H., and Svanborg-Edén, C., 1981, Glycolipid receptors for uropathogenic *Escherichia coli* on human erythrocytes and uroepithelial cells, *Infect. Immun.* **34:**920–929.

Levitzki, A., 1980, Quantitative aspects of ligand binding to receptors, in: *Cellular Receptors for Hormones and Neurotransmitters* (D. Schulster and A. Levitzki, eds.), Wiley, New York, pp. 9–28.

Lichtenberg, D., Robson, R. J., and Dennis, E. A., 1983, Solubilization of phospholipids by detergents: Structural and kinetic aspects, *Biochim. Biophys. Acta* **737:**285–304.

Lodish, H. F., Braell, W. A., Schwartz, A. L., Strous, G. J. A. M., and Zilberstein, A., 1981, Synthesis and assembly of membrane and organelle proteins, *Int. Rev. Cytol. Suppl.* **12:**247–307.

Lotan, R., and Nicolson, G. L., 1981, Plasma membranes of eukaryotes, in: *Advanced Cell Biology* (L. M. Schwartz and M. M. Azar, eds.), Van Nostrand–Reinhold, Princeton, N.J., pp. 129–154.

Macara, I. G., and Cantley, L. C., 1983, The structure and function of band 3, in: *Cell Membranes, Models and Reviews,* Volume I (E. Elson, W. Frazier, and L. Glaser, eds.), Plenum Press, New York, pp. 41–87.

Magnani, J. L., Smith, D. F., and Ginsburg, V., 1980, Detection of gangliosides that bind cholera toxin: Direct binding of ^{125}I-labeled toxin to thin-layer chromatograms, *Anal. Biochem.* **109:**399–402.

Marchesi, V. T., Furthmayr, H., and Tomita, M., 1976, The red cell membrane, *Annu. Rev. Biochem.* **45:**667–698.

Marinetti, G. V., and Crain, R. C., 1978, Topology of amino-phospholipids in the red cell membrane, *J. Supramol. Struct.* **8:**191–213.

Miller, C., and Racker, E., 1979, Reconstitution of membrane transport functions, in: *The Receptors,* Volume 1 (R. D. O'Brien, ed.), Plenum Press, New York, pp. 1–31.

Momoi, T., Tojunaga, T., and Nagai, Y., 1982, Specific interaction of peanut agglutinin with the glycolipid asialo GM1, *FEBS Lett.* **141:**6–10.

Nicolson, G. L., 1976, Trans-membrane control of the receptors on normal and tumor cells. I. Cytoplasmic influence on cell surface components, *Biochim. Biophys. Acta* **457:**57–108.

Nicolson, G. L., 1979, Topographic display of cell surface components and their role in transmembrane signaling, *Curr. Top. Dev. Biol.* **3:**305–338.

O'Brien, R. D. (ed.), 1979, *The Receptors,* Volume 1, Plenum Press, New York.

Ofek, I., and Beachey, E. H., 1980, Bacterial adherence, *Adv. Intern. Med.* **25:**503–531.

Pan, Y. T., Schmitt, J. W., Sanford, B. A., and Elbein, A. D., 1979, Adherence of bacteria to mammalian cell: Inhibition by tunicamycin and streptovirudin, *J. Bacteriol.* **139:**507–514.

Pricer, W. E., Jr., Hudgin, R. L., Ashwell, G., Stockert, R. J., and Morell, A. G., 1974, A membrane receptor protein for asialoglycoproteins, *Methods Enzymol.* **34:**688–691.

Prives, J. M., 1980, Nicotinic acetylcholine receptors, in: *Cellular Receptors for Hormones and Neurotransmitters* (D. Schulster and A. Levitzki, eds.), Wiley, New York, pp. 331–351.

Raff, M. C., and de Petris, S., 1973, Movement of lymphocyte surface antigens and receptors: The fluid nature of the lymphocyte plasma membrane and its immunological significance, *Fed. Proc.* **32:**48–54.

Roth, J., 1980, The use of lectins as probes for carbohydrates—Cytochemical techniques and their application in studies on cell surface dynamics, *Acta Histochem. Suppl.* **22:**113–121.

Ruoslahti, E., Pierschbacher, M., Hayman, E. G., and Engvall, E., 1982, Fibronectin: A molecule with remarkable structural and functional diversity, *Trends Biochem. Sci.* **7:**188–190.

Schrevel, J., Gros, D., and Monsigny, M., (eds.), 1981, *Cytochemistry of Cell Glycoconjugates, Prog. Histochem. Cytochem.* **14,** No. 2.

Schulster, D., and Levitzki, A. (eds.), 1980, *Cellular Receptors for Hormones and Neurotransmitters,* Wiley, New York.

Schwarz, R. T., and Datema, R., 1982, The lipid pathway of protein glycosylation and its inhibitors: Tools in the study of the biological significance of protein-bound carbohydrates, *Adv. Carbohydr. Chem. Biochem.* **40:**287–379.

Sedlacek, H. H., Stark, J., Seiler, F. R., Ziegler, W., and Wiegandt, H., 1976, Cholera toxin induces redistribution of sialoglycolipid receptor at the lymphocyte membrane, *FEBS Lett.* **61:**272–276.

Sharon, N., 1975, *Complex Carbohydrates: Their Chemistry, Biosynthesis, and Functions,* Addison–Wesley, Reading, Mass.

Sharon, N., 1979, Some biological functions of cell surface sugars, in: *Structure and Function of Biomembranes* (K. Yagi, ed.), Japan Scientific Societies Press, Tokyo, pp. 63–82.

Sharon, N., 1981, Glycoproteins in membranes, in: *Membranes and Intercellular Communication* (R. Balian, M. Chabre, and P. F. Devaux, eds.), North-Holland, Amsterdam, pp. 117–182.

Sharon, N., and Lis, H., 1975, Use of lectins for the study of membranes, *Methods Membr. Biol.* **3:**147–200.

Sharon, N., and Lis, H., 1981, Glycoproteins: Research booming on long-ignored, ubiquitous compounds, *Chem. Eng. News* **59**(13):21–44.

Sharon, N., and Lis, H., 1982, Glycoproteins, in: *The Proteins,* Volume V (H. Neurath and R. L. Hill, eds.), 3rd ed., Academic Press, New York, pp. 1–144.

Shorr, R. G. L., Lefkowitz, R. J., and Caron, M. G., 1981, Purification of the β-adrenergic receptor: Identification of the hormone binding subunit, *J. Biol. Chem.* **256:**5820–5826.

Sim, E., 1982, *Membrane Biochemistry,* Chapman & Hall, London.

Singer, S. J., 1974, The molecular organization of membranes, *Annu. Rev. Biochem.* **43:**805–833.

Singer, S. J., 1981, The cell membrane, in: *Membranes and Intercellular Communication* (R. Balian, M. Chabre, and P. F. Devaux, eds.), North-Holland, Amsterdam, pp. 1–16.

Singer, S. J., and Nicolson, G. L., 1972, The fluid mosaic model of cell membranes, *Science* **175:**720–731.

Steck, T. L., 1978, Band 3 protein of the human red cell membrane: A review, *J. Supramol. Struct.* **8:**311–324.

Svanborg-Edén, C., Freter, R., Hagberg, L., Hull, R., Hull, S., Leffler, H., and Schoolnik, G., 1982, Inhibition of experimental ascending urinary tract infection by an epithelial cell surface receptor analogue, *Nature (London)* **298:**560–562.

Tanford, C., 1980, *The Hydrophobic Effect: Formation of Micelles and Biological Membranes,* 2nd ed., Wiley, New York.

Tanner, M. J. A., 1978, Erythrocyte glycoproteins, *Curr. Top. Membr. Transp.* **11:**279–325.

Tollefsen, S. E., and Kornfeld, R., 1983, The B_4 lectin from *Vicia villosa* interacts with *N*-acetylgalactosamine residues α-linked to serine or threonine residues in cell surface glycoproteins, *J. Biol. Chem.* **258:**5172–5176.

Tulsiani, D. R. P., Harris, T. M., and Touster, O., 1982, Swainsonine inhibits the biosynthesis of complex glycoproteins by inhibition of Golgi mannosidase II, *J. Biol. Chem.* **257:**7936–7939.

Wicken, A. J., and Knox, K. W., 1981, Composition and properties of amphiphiles, in: *Chemistry and Biological Activities of Bacterial Surface Amphiphiles* (G. D. Shockman and A. J. Wicken, eds.), Academic Press, New York, pp. 1–7.

Wiegandt, H., Kanda, S., Inoue, K., Utsumi, K., and Nojima, S., 1981, Studies on the cell association of exogenous glycolipids, *Adv. Exp. Med. Biol.* **152:**3433–3352.

Yamakawa, T., and Nagai, Y., 1978, Glycolipids at the cell surface and their biological functions, *Trends Biochem. Sci.* **3:**128–131.

4

Characteristics of Plant Surfaces

J. GOULD and D. H. NORTHCOTE

1. INTRODUCTION

This chapter will examine the chemistry of epidermal cell walls and associated materials of higher plants. It is primarily a review of work on the specialized external surfaces of intact vegetative plants with some reference to epidermal polymers involved in plant defense.

The plant surface has a fundamental protective role as a barrier between plant and environment (Martin and Juniper, 1970; Holloway, 1982a). The strength of this barrier is the result of its physicochemical properties which provide resistance to desiccation, leaching, abrasion, and an ability to withstand microbial attack. Protection cannot be absolute as a certain degree of translucency and permeability is required by leaves for photosynthesis and by roots for water and inorganic nutrient uptake. In addition, cell–cell recognition required for tissue organization, reproduction, and interactions with symbionts and some pathogens needs receptive, compatible surfaces for successful development (Smith, 1977; Heslop-Harrison, 1978; Dazzo, 1980; Hughes and Pena, 1981; Kosuge, 1981; Edelman, 1983). Thus, the various competing functions of epidermal cell walls impose constraints upon their construction.

Plant surfaces differ because of genetically determined distinctions between species (see, for example, Chapter 14, Section 2.1.3), but even on any one plant, differences occur at various sites and between young, mature, and senescing tissues. Differences also arise because of the polydisperse and microheterogeneous nature of some cell wall polymers and because in natural environments, plant surfaces are host to a specific nonpathogenic flora. These distinctions may not be apparent from an examination of cell wall composition as such data generally reflect an average value for the fraction or type of tissue analysed. Therefore, differences between opposite faces of a cell or stratification of a particular constituent will be concealed.

The cell wall is thus a dynamic structure whose composition and properties constantly respond to the growth, stage of differentiation, and environment of the cell. Its formation and maintenance is a continuous part of a plant's growth processes (Northcote, 1972).

J. Gould and D. H. Northcote • Department of Biochemistry, University of Cambridge, Cambridge CB2 1QW, England.

2. MATERIAL COMMON TO ALL EPIDERMAL CELL WALLS: THE PRIMARY CELL WALL

Primary cell walls are laid down by actively dividing cells. These walls may contribute to control of the rate and extent of cell expansion (Cook and Stoddart, 1973). All other cell wall types (secondary cell walls), including epidermal cell surfaces, are derived by continual modification and additions either to the inside or outer surface of the primary cell wall after expansion has slowed or ceased. This differentiation represents a development by cells toward their mature, specialized functions (Northcote, 1972).

The primary cell wall is predominantly polysaccharide with water and proteins as other important constituents. The wall components have been compared with man-made composites such as glass-fiber reinforced plastic, i.e., consisting of a dispersed, but highly oriented, phase of microfibrils entangled and keyed within a complex, continuous, hydrophilic matrix. The primary wall has considerable tensile strength but little rigidity (Northcote, 1972).

The polysaccharides of the wall are conveniently classified on the basis of selective extraction procedures, monosaccharide composition, and linkage into α-cellulose, hemicelluloses, and pectins (Cook and Stoddart, 1973; Aspinall, 1980). Absolute definition of many polysaccharide groups is not possible as these polymers, which are secondary gene products, are generally polydisperse and microheterogeneous (Northcote, 1972; Cook and Stoddart, 1973). Because cell wall polysaccharides often contain many different sugars, there is much opportunity for structural variation by changes in monosaccharide, linkage, or branching (Cook and Stoddart, 1973). Recent reviews of this field include those by Aspinall (1980), Darvill *et al.* (1980), and Kato (1981).

2.1. Microfibrillar Phase

α-*Cellulose* is the insoluble fraction of cell walls which remains after delignification, depectination, and strong alkali extraction. It constitutes the microfibrillar phase of the walls and is almost totally composed of cellulose, a long-chain β(1→4)glucan (Aspinall, 1970; Bikales and Segal, 1971). Estimates for the degree of polymerization (d.p.) vary with plant source and isolation method from about 6000 to 15,000 (Preston, 1974). The glucan molecule has a twofold screw axis (Preston, 1974). The conformation of the linear chains is maintained by inter- and intramolecular hydrogen bonds. These chains are grouped to form semicrystalline microfibrils. The microfibril is the fundamental unit in which cellulose is found and is about 10 nm in diameter and very long (Preston, 1974).

It has been suggested by Gardner and Blackwell (1974) that the glucan chains within the microfibrils of the alga *Valonia* may be parallel, instead of antiparallel (Meyer and Misch, 1937). Although there is no conclusive evidence for either conformation in higher plants, the parallel arrangement is probably present (Blackwell, 1982). Nonglucan polymers are consistenly found associated with microfibrils. They are possibly normal constituents of the glucan or perhaps hemicelluloses tightly adsorbed to the surface of the microfibrils and may provide a keying material for the entanglement of the microfibrils within the matrix (Northcote, 1972; Preston, 1974; Darvill *et al.*, 1980).

During initial stages of cell growth when the matrix is semifluid the microfibrils are found in loosely arranged bands which are at a large angle to the long axis of the cell. In secondarily thickened cells microfibrils are much more closely packed in distinct laminae, and they are oriented at a smaller angle to the long axis of the cell (Shafizadeh and McGinnis, 1971). Lamination ensures efficient microfibril packing and the maintenance of strength in more than one direction (Northcote, 1972).

2.2. Matrix Phase

Matrix polysaccharides consist of linear oriented polymers synthesized throughout development (hemicelluloses) and others formed only at specific stages of growth (pectins). The polysaccharide composition varies between plant species and between different organs as the composition of cell wall polysaccharides is related to the functions of various tissues (Northcote, 1972). Structural features of polysaccharides are insufficiently specific to provide a basis for chemotaxonomy (Stephen, 1980).

2.2.1. Hemicelluloses

Hemicelluloses are a diverse polysaccharide group, defined empirically by extraction method. They are alkaline soluble.

2.2.1a. Xylans.   Xylans are the most abundant noncellulosic polysaccharides of angiosperm secondary cell walls. The polymers comprise a linear $\beta(1 \rightarrow 4)$-D-xylopyranose [$\beta(1 \rightarrow 4)$-O-Xyl$p$] backbone of about 150 to 200 residues long. Short side chains on every 7 to 10 xylose residues are of three types: (1) α-4-O-Me-D-GlcAc$p(1 \rightarrow 2)$ or α-D-GlcAc$p(1 \rightarrow 2)$; (2) $\alpha(1 \rightarrow 3)$-L-arabinofuranose [α-L-Ara$f$], sometimes additionally $(1 \rightarrow 2)$; and (3) extended side chains where α-L-Ara$f$ is substituted (Siddiqui and Wood, 1977; Wilkie and Woo, 1977; Shimizu *et al.*, 1978; Wilkie, 1979). Xylose groups may be acetylated at positions C-3 and/or C-2 (rarely the latter). This renders the polymer more water soluble as acetyl substituents are sufficiently numerous to prevent alignment and aggregation with other xylan chains. Xylan molecules can be associated as hydrated, linear, semicrystals (Nieduszynski and Marchessault, 1971). Their degree of hydration is dependent on the number of uronic acid and arabinose side chains. The linear xylan chains have also been shown to be adsorbed strongly by hydrogen bonds to cellulose fibers. In addition to these associations the chains of any loosely arranged xylans can be noncovalently associated with other matrix polymers. In a cell wall these associations made by hydrogen bonds between polysaccharides would be dependent on the water content of the wall (Northcote, 1972).

2.2.1b. Xyloglucans.   Xyloglucans were originally identified as a common component of seeds but they have now been found to be widespread as structural wall polymers in primary walls (Aspinall, 1980; Darvill *et al.*, 1980). Xyloglucans contain a $\beta(1 \rightarrow 4)$Glc$p$ backbone with short side chains on C-6 of approximately half of the glucose residues. The three most common side chains are: (1) β-D-Xyl$p(1 \rightarrow 6)$, (2) β-D-Gal$p(1 \rightarrow 2)$-O-α-D-Xyl$p(1 \rightarrow 6)$, and (3) α-L-Fuc$p(1 \rightarrow 2)$-O-β-D-Gal$p(1 \rightarrow 2)$-O-α-D-Xyl$p(1 \rightarrow 6)$ (Aspinall, 1980; O'Neill and Selvendran, 1983). Purified xyloglucans strongly bind to cellulose by hydrogen bonds (Kato, 1981; O'Neill and Selvendran, 1983). In a cell wall the glucan chains of cellulose and xyloglucans may become aligned and side chains of the latter key into the matrix.

2.2.1c. Glucomannans and Galactoglucomannans.   These polysaccharides are the main hemicelluloses of gymnosperms. They consist of $\beta(1 \rightarrow 4)$-linked chains of randomly arranged D-Glc$p$ and D-Man$p$ units in the approximate molar ratio of 1 : 3, respectively. They may be acetylated on C-2 or C-3 of D-mannose residues (Aspinall, 1980). In galactoglucomannans terminal, single, α-D-Gal$p(1 \rightarrow 6)$ units are present as irregularly spaced side chains (Aspinall, 1980). The molar ratio of sugar residues in galactoglucomannans is approximately 1 : 1 : 3, respectively. These polymers have a d.p. of about 100. Glucomannans are larger with a d.p. of 200 (Northcote, 1972). In angiosperms glucomannans (Glc : Man molar ratio 1 : 2) occur in smaller amounts. The chains are shorter (d.p. 70) than their gymnosperm counterparts (Northcote, 1972; Aspinall, 1980).

2.2.1d. Noncellulosic Mixed β-Glucans.

Linear β-D-glucans with repeating sequences of β(1→3) and β(1→4) linkages are characteristic of the hemicellulosic fraction of cell walls of monocotyledons (Nevins *et al.*, 1977; Darvill *et al.*, 1980). Linkage molar ratios β(1→3) : β(1→4) vary with tissue, age, and source from 1 : 2 to 1 : 3 (Aspinall, 1980).

2.2.2. Pectins

Pectic polysaccharides comprise a large part (up to 35%) of the primary cell wall of dicotyledons (Darvill *et al.*, 1980). Pectins are a mixture of neutral (arabinogalactans) and acidic (rhamnogalacturonans) polysaccharides which are soluble in water and aqueous solutions of chelating agents and are able to form gels. Gelation properties are important in their role in the cell wall as a comparatively fluid hydrophilic filler (Rees and Welsh, 1977). Their presence allows relative movement of the microfibrillar and matrix components during wall expansion (Northcote, 1972).

2.2.2a. Rhamnogalacturonans.

The acidic moieties of pectin contain long ribbonlike chains (d.p. 2000) of α(1→4)D-GalAc*p* with a threefold screw axis. There may be extended regions of homogalacturonan or more commonly the domains of uronic acid are interspersed, on average every 10 uronide residues, by β-L-Rha*p*. The inclusion of (1→2)-linked rhamnose makes a kink in an otherwise linear molecule. The rhamnose linkage has been assigned β but this has not been unambiguously confirmed (Rees and Wight, 1971). A variable proportion of the carboxyl groups are methyl esterified (Siddiqui and Wood, 1976). Acetylation of uronide hydroxyl groups is less frequent (Aspinall, 1980). Acidic groups on rhamnogalacturonans form ionic links with cations. Polymer conformation will differ in the presence of divalent cations from that in monovalent ones (Northcote, 1972; Preston, 1979). Nondegradative purification procedures show that the majority of the acidic polymers are heterosaccharides and that the acidic backbones carry side chains. Large, branched side chains of arabinose and galactose are most common but L-fucose, D-xylose, D-glucuronic acid, and, more rarely, the sugars 2-*O*-Me-D-xylose, 2-*O*-Me-L-fucose, and D-apiose are also present (Darvill *et al.*, 1978; Aspinall, 1980; McNeil *et al.*, 1982). Each additional structural feature of the basic polygalacturonic acid backbone prevents a tendency of chains to form ordered conformations (Rees and Wight, 1971) and enables the polysaccharide to maintain fluidity of the wall during cell expansion.

2.2.2b. Arabinogalactans.

Associated with the polyrhamnogalacturonans are high-molecular-weight polymers largely composed of the neutral sugars L-Ara*f* and D-Gal*p* (Barrett and Northcote, 1965). These "blocks" may be attached, within the wall, by transglycosylation to independently assembled rhamnogalacturonan chains during the later stages of pectin biosynthesis (Stoddart and Northcote, 1967).

The basic structure of arabinogalactans is a linear core of β(1→4)galactopyranose [although β(1→6) and other linkages are occasionally found] with a d.p. of 33 to 50 (Toman *et al.*, 1972) to which is attached side chains composed of arabinose (Aspinall, 1980; Kato, 1981). The arabinan moieties vary in size from 34 to 100 units. The arabinose residues are linked α-L-Ara*f*(1→2), (1→3), and (1→5) and may be branched through positions 2, 3, and 5 (Karacsonyi *et al.*, 1975; Joseleau *et al.*, 1977). The varied glycosyl linkage compositions of arabinans have been reviewed by Aspinall (1980) and Darvill *et al.* (1980). The molar ratio of arabinose to galactose varies from 9 : 1 respectively in rape to 1 : 9 in larch (Darvill *et al.*, 1980). Galacturonic and glucuronic acids, fucose, xylose, and rhamnose are minor constituents of some arabinogalactans (Darvill *et al.*, 1980).

2.3. Hydroxyproline-Rich Glycoproteins

The primary cell wall of higher plants contains from 2 to 10% (w/w) protein (Lamport, 1970; Burke *et al.*, 1974). The total protein is made up of at least three types of hydroxyproline-rich glycoproteins (insoluble glycoproteins associated with cellulose, soluble lectins and agglutinins, and soluble high-molecular-weight arabinogalactan glycoproteins) in addition to various cell wall enzymes such as peroxidase, phosphatases, and carbohydrases (Yung and Northcote, 1975; Mäder, 1976; Mäder *et al.*, 1977; Crasnier *et al.*, 1980; Gibson and Liu, 1981).

In sycamore, more than 90% of the total hydroxyproline-rich glycoprotein is insoluble and remains with α-cellulose after extraction of all other wall polymers (Heath and Northcote, 1971; Pope, 1977). The glycoprotein may play a role in disease resistance (Esquerre-Tugaye *et al.*, 1979; Esquerre-Tugaye and Lamport, 1979; Stuart and Varner, 1980). It is suggested that soluble glycoprotein monomers (lectins or bacterial agglutinins; see Chapter 9, Section 2.3.2) are secreted from the cytoplasm into the wall where they are probably cross-linked by oxidative coupling of tyrosine to form isodityrosine and a protein matrix (Fry, 1982; Cooper and Varner, 1983). The glycoprotein matrix may be covalently linked to polysaccharides of the wall (Monro *et al.*, 1976; O'Neill and Selvendran, 1980) perhaps by cross-links such as diferulic acid esters.

Soluble hydroxyproline-rich glycoproteins which resemble (by composition) the insoluble wall matrix are of two types: lectins and bacterial agglutinins. They may both occur in the same tissue [e.g., potato (Allen *et al.*, 1978; Leach *et al.*, 1982)] and may be distinguished by their binding properties for different sugars and action on erythrocytes (Mellon and Helgeson, 1982).

The chemistry of the lectin and bacterial agglutinin from potato has been investigated and the general pattern of composition and structure for these glycoproteins is probably similar (Allen *et al.*, 1978; Murray and Northcote, 1978; Ashford *et al.*, 1982; Leach *et al.*, 1982; Selvendran and O'Neill, 1982; Matsumoto *et al.*, 1983). They are basic [isoelectric point (pI) 9.5 to 11] and contain about 50% (w/w) carbohydrate. The carbohydrate is made up predominantly of tri- and tetra-arabinofuranosides β-linked onto hydroxyproline. The terminal arabinose unit may be α-linked [α-L-Ara*f*(1→3)-*O*-β-L-Ara*f*(1→2)-*O*-β-L-Ara*f*(1→2)-*O*-β-L-Ara*f*-1→Hyp]. Galactose (5 to 8% of neutral sugars) is present as a single unit α-linked onto serine.

The role of these lectins and bacterial agglutinins is unresolved. It has been suggested, however, that they may function in disease resistance, as determinants of host–symbiont specificity, or as toxins (Dazzo, 1980; Lamport, 1980; Sequeira, 1980; Barondes, 1981; Graham, 1981; see Chapter 9, Sections 2.3 and 4.2; Chapter 14, Section 2.1).

Arabinogalactan proteins have been identified in most higher plants. For example, they have been found in seeds, in extracellular polysaccharides and exudations of tissue cultured cells and also as major components of stigma surfaces and mucilages of style canals (A. E. Clarke *et al.*, 1979; Fincher *et al.*, 1983). The location of arabinogalactan proteins on receptive parts of flowers, and an ability to bind β-glycosyl residues with a single sugar-binding site (Yariv *et al.*, 1962; Jermyn and Yeow, 1975), has led to the suggestion that they may have a role in cell–cell recognition (Clarke, 1981; Lamport and Catt, 1981). Arabinogalactan proteins are heterogeneous, high-molecular-weight (M_r about 300K) polymers composed largely of carbohydrate (90 to 98%) (A. E. Clarke *et al.*, 1979; Fincher *et al.*, 1983). The carbohydrate portion is typically a β-D-Gal*p*(1→3) backbone with oligo-D-galactoside side chains at C-6. Arabinofuranosyl residues are linked by (1→6) or (1→3) bonds onto these side chains (A. E. Clarke *et al.*, 1979; Lamport and Catt, 1981). Uronic acids (at nonreducing termini), mannose, and rhamnose are also present and amino sugars, glucose, fucose, and xylose, occur in trace amounts (Akiyama and Kato, 1981; van Holst *et al.*, 1981). The large polysaccharide groups may be linked to the hydroxyproline-rich protein core either as Gal-Hyp or less frequently as Ara-Hyp (Pope, 1977; Akiyama and Kato, 1981). Galactose has been found linked to serine (Pope, 1977; A. E. Clarke *et al.*, 1979; Fincher *et al.*, 1983).

2.4. Interpolymeric Linkages

No primary cell wall model accurately describes the nature and extent of covalent and noncovalent linkages between the major cell wall polymers (cellulose, hemicelluloses, pectins, and glycoproteins) and their absolute and relative contributions to the structure of the wall at any developmental stage. The glycosidic bonds of the polysaccharides are essentially monovalent and undirectional and therefore cannot act as true, multifunctional, cross-linking agents, as can peptidoglycans of bacterial walls. Comparison of plant cell walls with other materials such as connective tissue proteoglycans suggests that noncovalent bonding is sufficient for the stable association of their polymeric constituents (Lindahl and Höök, 1978).

2.4.1. Noncovalent Bonding

One of the more variable features of plant cell walls is their degree of hydration. The water content of walls is to some extent determined by its polymeric constituents. In the matrices of expanding primary cell walls the major components are hydrophilic pectins which form gels. In the rigid secondary wall, water is replaced by the deposition of hydrophobic lignin.

When present water can (1) play a structural role as part of a gel, (2) control hydrogen bonding between microfibrillar and hemicellulosic components, (3) determine polysaccharide conformation and aggregation with various cations, (4) aid noncovalent aggregation of dissimilar polysaccharides, and (5) affect wall permeability.

2.4.2. Covalent Bonding

The primary cell walls of many, if not all, higher plants contain two types of polymer-bound phenols: (1) tyrosine residues of cell wall matrix protein and (2) cinnamate derivatives esterified to polysaccharides (Fry, 1983).

The hydroxyproline-rich glycoprotein of the cell wall may be cross-linked by isodityrosine bonds. The degree of coupling may be controlled by an extracellular peroxidase/ascorbate oxidase system in the wall which can provide a means for control of wall extensibility and growth (Fry, 1982; Cooper and Varner, 1983). Cell wall protein when extracted is sometimes found to cochromatograph with polysaccharides. It has been suggested that a small proportion of the cell wall protein could be covalently linked to polysaccharides (Monro *et al.*, 1976; O'Neill and Selvendran, 1980).

Pectin-bound phenylpropanoids may occur as frequently as one every 50 to 100 sugar residues probably at nonreducing ends of the polysaccharide chains (Fry, 1983; Tanner and Morrison, 1983). The extent of peroxidase-catalyzed cross-linkage of pectic polymers, possibly by diferulic acid ester linkages, may provide control of primary cell wall extension (Fry, 1983). However, the relative importance of these cross-linkages for control of cell wall expansion is difficult to assess.

In secondary walls there are covalent linkages between lignins and hemicelluloses and between protein and lignins (Morrison, 1974; Whitmore, 1978a). These cross-links are produced by the formation of phenylpropanoid free radicals generated by the peroxidase reaction (Whitmore, 1978b). Polymer-bound phenylpropanoids of the primary wall may act as initiation sites for lignification during secondary wall formation (Whitmore, 1978a; Tanner and Morrison, 1983).

Bauer *et al.* (1973) showed that in sycamore, hemicelluloses may be covalently linked to pectins. However, attempts by these workers to isolate these linked polymers have been unsuccessful (Darvill *et al.,* 1980). Ring and Selvendran (1980) were unable to find glycosidic bonds which linked pectins and xyloglucans (hemicellulose) in potato tissue. The nature and existence of possible glycosidic linkages between hemicelluloses and pectins remain undetermined.

3. SUPERFICIAL COVERINGS OF PLANTS

The structural components of plant skins differ according to the specialization of the underlying tissue especially with regard to water permeability.

Aerial parts of plants and older decorticated root surfaces (which are less active in water and inorganic nutrient uptake) are covered by hydrophobic skins. These coverings vary in thickness, complexity, and composition. For example, there are delicate, translucent membranous cuticles on photosynthetic tissues and receptive surfaces of reproductive organs, thicker skins on fruit surfaces, and elaborate surface incrustations of tree barks (periderms) (Martin and Juniper, 1970; Heslop-Harrison and Heslop-Harrison, 1982; Holloway, 1982a). The structural components which characterize these outer coverings are the chemically related components cutin and suberin which are ester-linked polymers built of hydroxy-fatty acids. Cuticles contain cutin and periderms contain suberin (Martin and Juniper, 1970). These materials provide a structural network of long-chain hydroxy-fatty acid esters which are made waterproof by intussusception of a complex mixture of comparatively nonpolar lipids collectively called *wax* (Espelie *et al.,* 1980a; Kolattukudy, 1980b; Schönherr and Ziegler, 1980; Schönherr, 1982). This diffusion barrier offers some degree of protection from pathogens, allows stomatal control of transpiration and gaseous exchange, and restricts passive losses of nutrients by leaching (Martin and Juniper, 1970).

In contrast to more impermeable tissues, the surfaces of intact root apical regions, which are comparatively active in water and inorganic nutrient adsorption, are covered by gels or mucilages. Mucilages are the major product of the root that forms the interface between a plant and the soil. In addition to mucilages the root exudes a great variety of other materials. These materials together with a number of accompanying physiochemical modifications to the soil form the complex ecological system intimately associated with roots termed the *rhizosphere* (Harley and Scott-Russell, 1979).

3.1. Mucilages

Young root epidermal cell walls may be shown by selective microscopic techniques to have up to four wall layers (Fig. 1). These consist of an outer covering which stains with osmium, a mucilagenous sheath of variable thickness derived from the root cap, a thick

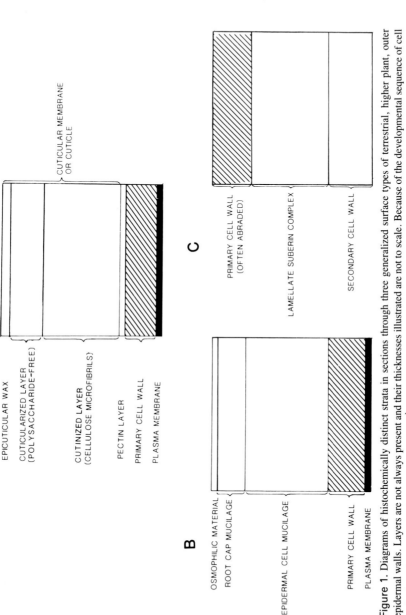

Figure 1. Diagrams of histochemically distinct strata in sections through three generalized surface types of terrestrial, higher plant, outer epidermal walls. Layers are not always present and their thicknesses illustrated are not to scale. Because of the developmental sequence of cell walls the primary walls of epidermal surfaces are stratified into an outer pectin-enriched layer [called *primary wall* by Foster and Bowen (1982)] and an inner layer that contains the majority of cellulose microfibrils [called *secondary wall* by Foster and Bowen (1982)]. (A) Surface of an aerial plant tissue that is ephemeral or annual (after Martin and Juniper, 1970; Hallam, 1982; Holloway, 1982a). (B) Young root epidermal wall (after Oades, 1978; K. J. Clarke *et al.*, 1979; Miki *et al.*, 1980; Vermeer and McCully, 1982). (C) Periderm wall found on healed wounds (irrespective of former surface type) and perennating tissues. The plasma membrane is omitted from the bottom of this illustration as suberized cells are often dead (after Martin and Juniper, 1970; Wattendorff, 1974; Sitte, 1975; Dean and Kolattukudy, 1976; Dean *et al.*, 1977; Barckhausen, 1978; Robards *et al.*, 1979; Soliday *et al.*, 1979; Schönherr and Ziegler, 1980).

mucilagenous layer of epidermal cell origin interspersed with cellulose microfibrils, and a multilamellate cellulosic layer (Werker and Kislev, 1978; K. J. Clarke *et al.*, 1979; Miki *et al.*, 1980; Vermeer and McCully, 1981, 1982; Foster, 1982; Foster and Bowen, 1982). There are little substantive data correlating microscopically observable strata with biochemical composition. In older parts of a root much of the mucilage may be of microbial (see Chapter 14, Section 5.1) rather than plant origin (Rovira *et al.*, 1979). The extent and perhaps the type of mucilages produced by roots depend on plant species, soil type, and region of the root (Foster, 1982; see also Chapter 14, Section 5.1).

Mucilages protect the delicate meristematic tissues of the root apex and epidermal cells by their physical presence, by chelation of toxins, and by lubricating the elongation of a root through the soil (Rougier, 1981; Horst *et al.*, 1982; Allen, 1983). As water and ions diffuse through root mucilages in a manner not significantly different from that in free water (Greenland, 1979) and as root mucilages can maintain their water content over a wide range of water potentials (i.e., they are very hydrophilic) their presence may aid nutrient diffusion (Stephen, 1980; Kennedy and White, 1983). In periods of drought, mucilages, and soil particles bound by them, are drawn close to the roots surface. This forms a "soil sheath" which preserves a rhizosphere moisture film and prevents desiccation (Nambiar, 1976; Greenland, 1979; Vermeer and McCully, 1982).

Maize root-cap slime is the most thoroughly investigated of root mucilages. It contains approximately 30% (w/w) protein; the remainder is carbohydrate mainly as polysaccharide (Wright and Northcote, 1975). Whether this protein has a function in slime or merely represents contents of autolysing root cap cells is not known.

The polysaccharide moiety can be fractionated by glass-fiber paper electrophoresis into a neutral polymer and weakly and strongly acid polymers (Wright and Northcote, 1974). The neutral polymer is largely glucose as a $\beta(1\rightarrow4)$glucan (Wright and Northcote, 1976). The major monosaccharides of the acidic polymers are fucose, arabinose, galactose, and a variable proportion of galacturonic acid. The difference in uronic acid content distinguishes the weakly from the strongly acidic polymers. Xylose, glucose, and mannose are present in minor amounts (Paull *et al.*, 1975; Wright, 1975). Absolute sugar proportions are variable but the general compositional features are constant.

Unlike pectins, the acidic polymers do not seem to have long chains of galacturonic acid but consist of either mixed chains of glucose and uronic acids carrying neutral sugars as side chains or as glucan chains with acidic and neutral sugars as side chains (Wright and Northcote, 1975; Wright *et al.*, 1976). The polymers of slime are probably arranged as a central $\beta(1\rightarrow4)$glucan core rendered water soluble by a coating of the hydrophilic acidic polysaccharides linked both covalently and noncovalently (Wright and Northcote, 1976; Northcote, 1982). Sandhu *et al.* (1981) suggest that L-Ara*f* and D-GalAc*p* bind water at the surface and within the gels which are stabilized by the more rigid core.

3.2. Cutins

Aerial epidermal surfaces of ephemeral or annual tissues are covered by a cuticle or cuticular membrane. Cuticles, like suberin-complexes, are generally stratified (Fig. 1) (Kolattukudy *et al.*, 1981; Hallam, 1982; Holloway, 1982a). These strata are an epicuticular wax layer, a polysaccharide-free cuticularized layer, a cutinized layer with cellulose fibrils, and a pectin-rich layer (Holloway, 1982a). Homologous strata (histochemically defined) may be present to varying degrees on different plants or tissues, or may be entirely absent. The sequence of primary cell wall formation is that in which matrix material, notably pectin, is initially deposited into the cell plate. Into this are woven cellulose microfibils. This means that on outer surfaces of epidermal walls or between adjacent cells there is a concentration of pectin material. The cuticle is attached to epidermal cell walls by pectinaceous materials (Fig. 1) (Kolattukudy, 1980b; Kolattukudy *et al.*, 1981; Holloway, 1982a). Incorporation of new cutin occurs predominantly in cuticular strata adjacent to the primary wall (Lendzian and Schönherr, 1983).

Cutins are insoluble high-molecular-weight polyesters composed of various long-chain, hydroxy-fatty acids. Additional substituents in the long chains are commonly carboxyl and hydroxyl but aldehyde, ketone, epoxide, and unsaturated groups may also be present (Table I) (Kolattukudy, 1980a,b; Holloway, 1982b).

Depolymerized cutins have been classified into three categories on the basis of the predominating chain length of the monomer: C_{16}, C_{18}, and a mixed C_{16} and C_{18} type (Kolattukudy, 1980b; Holloway, 1982b). There is no obvious chemotaxonomic relationship between these categories but gymnosperms generally lack C_{18} monomers (Holloway, 1982b). Claims of associations of cutin classes with tissues of particular growth rates (e.g., Kolattukudy, 1980b; Kolattukudy *et al.*, 1981) are considered by Holloway (1982b) to be inadequately defined in physiological and morphological terms.

Major components of the C_{16} class of cuticle are palmitic, 16-hydroxypalmitic, 9,16-dihydroxy- and 10,16-dihydroxypalmitic acids. In the C_{18} family, stearic, oleic, linoleic, 18-hydroxyoleic, 18-hydroxy-9,10-epoxystearic, 9,10,18-trihydroxystearic and their corresponding Δ^{12} unsaturated analogs are the common acids (Kolattukudy, 1980b; Kolattukudy *et al.*, 1981; Holloway, 1982b). The minor components differ from the major ones in their degree and type of substitution, oxidation state, and they have shorter or longer chain lengths (Holloway, 1982b).

Various physical analyses of the intact polymer suggest that cutin is amorphous (Wilson and Sterling, 1976; Kolattukudy *et al.*, 1981; Holloway, 1982b). Different structures for cutins are possible as the numbers and positions of hydroxyl groups which may be involved in ester linkages vary in different plants (Holloway, 1982b). Little is known, however, of the detail of the intermolecular structure of cutins. Partial degradation is difficult because all the ester linkages of the polymer have similar lability. Labeling of free hydroxyl groups within cutin followed by depolymerization indicates that most primary hydroxyls are esterified (Deas and Holloway, 1977; Kolattukudy, 1977). These results concur with observations that treatment of cutin with an enzyme specific for primary alcohol esters (mammalian pancreatic lipase) causes extensive depolymerization (Brown and Kolattukudy, 1978). Kolattukudy (1980a) has proposed models for polymeric cutin and suberin. Holloway (1982b) suggests, however, that such models have little value in the absence of sufficient substantive data.

Cuticles are strongly fluorescent. This is due to the presence in them of condensed tannins and simple phenolic acids (Hunt and Baker, 1980). These compounds may account for 20% of the total depolymerization products in the cuticle (Martin and Juniper, 1970). Small amounts of these phenylpropanoids are esterified to cutin (Riley and Kolattukudy, 1975).

3.3. Suberin-Complex

There is disagreement over usage of the term *suberin*. Martin and Juniper (1970) and Holloway (1972a,b) define *suberin* as the aliphatic polyester component of periderm walls. However, Kolattukudy (1978) defines it as the total polymeric material of the suberin-containing wall layer; this includes both the polyester and also an uncharacterized polyphenylpropanoid polymer. The term *suberin-complex* is perhaps better fitted for the latter usage. In this chapter, *suberin* refers solely to the aliphatic polyester moiety.

Table I. Aliphatic Monomers of Cutin[a] and Suberin[b]

Class of component	Cutin		Suberin	
	Homologs	Most common examples	Homologs	Most common examples
Monobasic acids	C_{12} to C_{36}[c]	16:0, 18:0, 18:1, 18:2	C_{16} to C_{28}	22:0, 24:0
α,ω-Dibasic acids	C_{16} and C_{18}	16:0, 18:0, 18:1, 18:2	C_{10} to C_{28}[d]	16:0, 18:0, 18:1, 20:0, 22:0
Monohydroxymonobasic acids	C_{14}, C_{15}, and C_{18}	14:1, 15:0		
ω-Hydroxymonobasic acids	C_{16} and C_{18}[d]	16:0, 18:0, 18:1, 18:2	C_{16} to C_{28}[d]	16:0, 18:0, 18:1, 20:0, 22:0
Monohydroxy-α,ω-dibasic acids	C_{15} and C_{16}	16:0	—	
Monohydroxyoxomonobasic acids	C_{16}	9- and 10-hydroxy-16-oxo-16:0 and ketone analogs	—	
Dihydroxymonobasic acids	C_{15}, C_{16}, and C_{18}[d]	7,16-, 8,16-, 9,16-, and 10,16-dihydroxy-16:0	C_{16} and C_{18}[d]	7,16-, 8,16-, 9,16-, and 10,16-dihydroxy-16:0 and 18:0
Epoxymonohydroxymonobasic and trihydroxymonobasic acids	C_{18}[d]	cis-9,10-epoxy-18-hydroxy-18:0, 9,10,18-trihydroxy-18:0 and Δ12 analogs	C_{18}	9,10,18-trihydroxy-18:0
Epoxy-α,ω-dibasic and dihydroxy-α,ω-dibasic acids	C_{18}	9,10-epoxy-18:0 and 9,10-dihydroxy-18:0	C_{18}[d]	9,10-epoxy-18:0 and 9,10-dihydroxy-18:0
Fatty alcohols	—[e]		C_{18} to C_{30}	18:0, 20:0, 22:0
α,ω-Diols	—		C_{18} to C_{24}	18:0, 20:0, 22:0

[a]Data from Kolattukudy (1980b), Kolattukudy et al. (1981), Holloway (1982b).
[b]Data from Holloway (1972a,b), Holloway and Deas (1973), Kolattukudy et al. (1975), Dean and Kolattukudy (1976), Dean et al. (1977), Espelie et al. (1979), Espelie et al. (1980b).
[c]Homologs have chains of carbon atoms that are preponderantly even-numbered.
[d]Denotes classes of components which are often quantitatively dominant in depolymerized cutin or suberin.
[e]Signifies rare, minor, or absent components.

The skins and periderms of perennating organs, the scars of all healed wounds irrespective of the original covering of the tissue, and decorticated plant root systems contain suberin (Martin and Juniper, 1970; Waid, 1974; Dean and Kolattukudy, 1976; Dean et al., 1977; Henry and Deacon, 1981). The suberin-complex is itself stratified microscopically into light and dark zones or lamellae (Sitte, 1975). The suberin-complex, in contrast to the cuticle, is laid down on proximal faces of the primary cell wall and distal to the plasma membrane (Fig. 1). It may partially or completely enclose the cell. This protective barrier can be thickened by the activity of a meristem or cork cambium that produces many more suberized cells and this constitutes a periderm. The outer cell wall is often abraded (Martin and Juniper, 1970; Kolattukudy et al., 1981). Cork cells lose their protoplasm after differentiation and may then be filled with air, colored resins, or tannin-rich compounds. The periderm is compressible, light, buoyant, and impervious to decay (Martin and Juniper, 1970; Gibson et al., 1981).

Analyses of periderm components have not been as extensive as those for cuticles (Kolattukudy et al., 1981). Suberin monomers comprise 10 to 50% (by weight) of the periderm and cannot therefore form such an extensive interesterified matrix as does cutin in cuticles (Kolattukudy, 1978). The periderm may also contain large quantities of phenylpropanoids (up to 70% by weight), some of which may be in a lignin-like matrix, and a smaller proportion of waxes (up to 4% by weight) (Martin and Juniper, 1970; Kolattukudy, 1978). Suberin-associated waxes are the major diffusion barrier of the periderm. However, there is no relation between periderm thickness and porosity (Schönherr, 1982). Soliday et al. (1979) selectively inhibited wax synthesis and thereby prevented lamellation of the suberin-complex. This observation coupled with data on the different permeabilities of light and dark lamellae is evidence for the lighter lamellae being wax-enriched (Sitte, 1975).

The nature and composition of the polyphenylpropanoid matrix of the suberin-complex are unknown. However, the aliphatic monomers of suberin have been investigated. Their general composition is shown in Table I. Major components are often ω-hydroxymonobasic acids and α,ω-dibasic acids. These are commonly present with chain lengths between C_{10} and C_{28}. The predominant homologs are C_{16}, C_{18}, C_{20}, C_{22} and for monoenoic acids C_{18} (Holloway, 1972a,b; Kolattukudy et al., 1975; Kolattukudy, 1978). Other fatty acids which may be major components in some suberins include epoxyoctadecanoic (Holloway and Deas, 1973) and variously substituted monobasic acids (Kolattukudy, 1978). There is a small proportion of neutral lipid in the suberin-complex and alkan-1-ols (C_{16}–C_{30}) are often the largest component of this fraction. However, α,ω-diols and cyclic alcohols are also found (Holloway, 1972b; Kolattukudy, 1978; Kolattukudy et al., 1981).

3.4. Waxes

The cuticles of all higher plants carry a partial or continuous covering of amorphous epicuticular wax (Fig. 1). Crystalline waxes or blooms are frequently superimposed upon these amorphous layers. Wax skins are generally thinner on rapidly expanding tissues and thicker on fruits and xerophytes (Martin and Juniper, 1970). The genetically determined epicuticular wax morphology differs between species, among organs, and within the life cycle of a plant (von Wettstein-Knowles, 1979; Baker, 1982). In addition to epicuticular wax deposits there are other waxes of different composition, more intimately associated with cutin and suberin (Baker, 1982). More is known of cuticular than suberin-associated waxes. Waxes are composed of complex mixtures of several classes of long-chain aliphatic compounds. The striking differences in wax composition between species and organs arise from the relative abundance of particular wax classes and their homolog range (Kolattukudy, 1980b; Baker, 1982). Waxes decrease transpiration and wettability and increase resistance to disease and frost (Tulloch, 1976).

Espelie *et al.* (1980a) suggest that although the same classes of waxes present in cuticles are also found in root periderms, in the latter there is a preponderance of even-numbered, short-chain hydrocarbons of relatively broader homolog distribution. The great variety of cyclic, branched, and long-chain aliphatic constituents of plant surface waxes has been reviewed by Tulloch (1976), Kolattukudy (1980b), Kolattukudy *et al.* (1981), and Baker (1982). Only a tiny proportion of the waxes of plant species have been analyzed. When a higher proportion has been examined new types of components will undoubtedly be found and others now rare or unusual may be shown to be common (Tulloch, 1976).

Although one wax may contain over 50 distinct compounds, commonly one or a few alkanes of the C_{21} to C_{37} or primary alcohols of the C_{22} to C_{32} range predominate (Table II) (Martin and Juniper, 1970; Tulloch, 1976; Kolattukudy, 1980b; Baker, 1982). Fatty acids are major components of intracuticular wax. In epicuticular deposits, secondary alcohols, ketones, and β-diketones are generally minor constituents but these compounds may comprise a major proportion of waxes of some plants (Baker, 1982). The cyclic components of waxes such as the steroids and terpenes are commonly associated with the periderm and rarely found in cuticular waxes (Tulloch, 1976; Bianchi *et al.*, 1976). Monoesters between long-chain alcohols and acids commonly occur but they are minor constituents whereas linear or cyclic polyesters of ω-hydroxy acids (estolides) are often major components of the waxes in gymnosperms (Martin and Juniper, 1970; Tulloch, 1976; Corrigan *et al.*, 1978; Franich *et al.*, 1978).

Various phenylpropanoids have been identified both free and esterified to waxes (Adamovics *et al.*, 1977). Their concentration is generally greater in comparable extracts from abaxial leaf surfaces and suberin-associated wax than from adaxial leaf surfaces and cutin-associated waxes (Kolattukudy *et al.*, 1981; Baker, 1982). Nothing is known of the function of these phenylpropanoids. One of their possible roles in the various coverings of aerial organs such as leaves is protection from ultraviolet radiation from the sun, as these delicate organs have to be exposed for photosynthesis.

4. PLANT DEFENSE: POLYMERS SYNTHESIZED IN RESPONSE TO DAMAGE

The various epidermal cells all have defensive roles irrespective of individual specialized functions. Both biochemical and mechanical defense against herbivores and microbial pathogens occur (see Chapter 9, Section 4.1; Chapter 14, Section 4). There are no simple, single determinants in the establishment of compatible plant–pathogen or plant–herbivore relationships, as interactions between parasite and host are multifactorial (Smith, 1977). Plant defense mechanisms are customarily divided into components which are preformed (Campbell *et al.*, 1980; Misaghi, 1982) and those which are actively synthesized in response to wounding (Kahl, 1978; Wood, 1981; Bailey and Mansfield, 1982).

4.1. Preformed Defensive Components

The epidermal wall is a preformed defensive barrier. Associations of these walls with preformed antimicrobial compounds such as various phenylpropanoids (Friend, 1979; Frossard, 1981), terpenoids (Blakeman and Atkinson, 1981), enzymes, and H_2O_2 (Misaghi, 1982) generally ensure that plants remain uninfected. Plants may also contain unpalatable material such as silicon (Kaufman *et al.*, 1981; Sangster and Parry, 1981), lignin (Gross, 1979), and cyanogenic glycosides (Conn, 1981) or glucosinolates (Larsen, 1981) to discourage herbivore activity.

Table II. Common Major Components of Waxes from Cuticles and Periderms[a]

Compound	Homolog range	Most common examples
Hydrocarbons (n-alkanes)	Odd No. of C chains from C_{17} to C_{35}	Nonacosane (C_{29}), hentriacontane (C_{31})
Primary alcohols	Even No. of C chains from C_{22} to C_{34}	Hexacosanol (C_{26}), octacosanol (C_{28})
Secondary alcohols	Odd No. of C chains from C_{21} to C_{33}	Nonacos-10-ol (C_{29}), hentriacontan-9-ol (C_{31})
Ketones	Odd No. of C chains from C_{23} to C_{33}	Nonacosanone (C_{29}), hentriacontanone (C_{31})
β-Diketones	Odd No. of C chains from C_{29} to C_{33}	Nonacos-12,13-dione (C_{29}), hentriacontan-14,16-dione (C_{31}), tritriacontan-16,18-dione (C_{33})
Triterpenoids		Ursolic acid, oleanolic acid
Monoesters of long-chain fatty acids and alcohols	Even No. of C chains from C_{36} to C_{72}	Esters with total C chains of C_{44}, C_{46}, C_{48}, C_{50}
Polyesters of ω-hydroxy acids[b]	Esters are made largely of 4 to 6 molecules of ω-hydroxy acids. The acids have an even No. of C chains from C_{12} to C_{18}	Linear and cyclic esters

[a]Data from Tulloch (1976, 1978), Corrigan et al. (1978), Tulloch and Hoffman (1979), Kolattukudy et al. (1981), Baker (1982).
[b]Commonly occur in the waxes of gymnosperms.

4.2. Postformed Defensive Components

Active defense mechanisms include physical sealing of wounded sites by callose (Eschrich, 1975; Aist, 1976), by phloem proteins (Read and Northcote, 1983), and by lignin and suberin-complexes which may occur with periderm development (Barckhausen, 1978; Ride, 1978; Vance et al., 1980; Loebenstein et al., 1982).

Phytoalexins are biocidal or biostatic compounds synthesized by plants in response to infection or stress (see Chapter 14, Section 4.3.2). The chemistry of these compounds is generally specific to a plant family. Many major groups of secondary metabolites are represented—phenylpropanoids, flavonoids, stilbenes, terpenes, and polyacetylenes. These compounds have been reviewed by Grisebach and Ebel (1978), Bailey (1981), and Bailey and Mansfield (1982). Mechanisms of phytoalexin elicitation and biosynthesis have been reviewed by Albersheim and Valent (1978), Rhodes and Wooltorton (1978), West (1981), and Lamb et al. (1983).

4.2.1. Lignification and Suberization after Wounding

Lignins, the insoluble constituents of the cell wall, are high-molecular-weight polymers composed of phenylpropanoids. They differ in composition in different plants, e.g., between angiosperms and gymnosperms and between sclerenchyma and tracheid lignins of the same plant (Stafford, 1974; Higuchi, 1981). Ligninlike material is present in epidermal cell walls either as a component of a localized response to wounding (Aist, 1976; Barckhausen, 1978; Ride, 1978; Vance et al., 1980; Loebenstein et al., 1982; Touze and Esquerre-Tugaye, 1982) or as a normal constituent of suberized cells (Kolattukudy, 1978). The identification of lignin in epidermal layers has been shown simply by selective color reactions (reviewed by Barckhausen, 1978; Vance et al., 1980; Touze and Esquerre-Tugaye, 1982) and not by chemical analysis of the polymeric structure. From such incomplete studies it is unclear whether phenylpropanoid polymers present in a wounding response represent lignification similar to that in secondarily thickened internal tissue, or whether it is suberization with a high phenylpropanoid content (Dean and Kolattukudy, 1976; Dean et al., 1977; Espelie et al., 1979, 1980b). The structure and definition of lignin are based exclusively on models derived from chemical analysis of standardized preparations made from xylem and tracheids of woody plants rather than epidermal tissue. However, there are probably similarities between phenylpropanoid material from these different sites.

In secondary thickening, lignin penetrates the wall from outside (primary wall) inward. It is a hydrophobic filler that replaces water and encrusts microfibrillar and matrix polysaccharides. Covalent bonds are formed between lignin and polysaccharide (hemicelluloses) (Morrison, 1974; Whitmore, 1978b; Tanner and Morrison, 1983). In secondarily thickened walls the linear carbohydrate polymers are enclosed in a cross-linked lignin matrix which greatly aids mechanical strength, impermeability, and rigidity (Northcote, 1972). Additionally, this biological composite is extremely resistant to microbial decay (Higuchi, 1981) and is therefore admirably suited to a role as a component of a wound sealant or of the periderm.

Lignins are formed by enzymatic oxidation of the phenolic group and subsequent, essentially random polymerization of mesomeric phenoxy radicals of coumaryl, coniferyl, and sinapyl alcohols. These phenoxy radicals are joined by ether and carbon–carbon linkages (reviewed by Gross, 1979; Grisebach, 1981; Higuchi, 1981). In addition, up to about 10% by weight of hydroxycinnamic acids (p-hydroxybenzoate, ferulate, and p-coumarate) may be esterified to the main polymer core. Lignin polymers possess similar structures but are distinguished by differences in precursor composition, variation of linkage types, methoxyl content, degree of aromatic acid substitution, and association with cell wall polysaccharides and/or aliphatic suberin polymers.

4.2.2. Callose Deposition after Wounding

Callose is a linear $\beta(1\rightarrow3)$-linked glucan. It is not very soluble in dilute acids, dilute alkali, or cuprammonium solutions (Clarke and Stone, 1963). About 2% uronic acid is present in many preparations which may account for the affinity of callose for the basic blue dyes of the triphenylmethane series such as aniline blue (Aspinall and Kessler, 1957). It is characteristically found in wounded or dormant phloem sieve tubes or in various cells as a component of a response to fungal infection (Aist, 1976; Brett, 1978). It acts as a seal for damaged tissue and may very quickly be deposited or removed (Eschrich, 1975).

5. SUMMARY

All terrestrial higher plant surfaces are partly composed of the external primary walls of epidermal tissues. The sequential development of the primary wall makes outer layers pectin-enriched while inner ones contain the majority of microfibrils. Because of the specialized functions of epidermal cells for protection, photosynthesis, reproduction, and water and inorganic nutrient uptake, the primary walls are adapted in particular ways. Modifications take several forms either by overproduction and modification of an existing fraction of the cell wall such as the pectinlike mucilages of young roots or by specific synthesis of additional materials as in the hydrophobic coverings which may be apposed to either the outer or the inner surfaces (cuticles and suberin-complexes, respectively) of the primary wall.

ACKNOWLEDGMENTS. We thank Mrs. Ann Lombardi for typing the manuscript and the Natural Environment Research Council for a research studentship (J.G.).

REFERENCES

Adamovics, J. A., Johnson, G., and Stepmitz, F. R., 1977, Ferrulates from cork layers of *Solanum tuberosum* and *Pseudotsuga menziesii, Phytochemistry* **16**:1089–1090.

Aist, J. R., 1976, Papillae and related wound plugs of plant cells, *Annu. Rev. Phytopathol.* **14**:145–163.

Akiyama, Y., and Kato, K., 1981, An extracellular arabinogalactan-protein from *Nicotiana tabacum, Phytochemistry* **20**:2507–2510.

Albersheim, P., and Valent, B. S., 1978, Host pathogen interactions in plants: Plants, when exposed to oligosaccharides of fungal origin, defend themselves by accumulating antibiotics, *J. Cell Biol.* **78**:627–643.

Allen, A., 1983, Mucus—A protective secretion of complexity, *Trends Biochem. Sci.* **8**:169–173.

Allen, A. K., Desai, N. N., Neuberger, A., and Creeth, J. M., 1978, Properties of potato lectin and the nature of its glycoprotein linkages, *Biochem. J.* **171**:665–674.

Asford, D., Desai, N. N., Allen, A. K., Neuberger, A., O'Neill, M. A., and Selvendran, R. R., 1982, Structural studies of the carbohydrate moieties of lectins from potato (*Solanum tuberosum*) tubers and thorn apple (*Datura stramonium*) seeds, *Biochem. J.* **201**:199–208.

Aspinall, G. O., 1970, *Polysaccharides,* Pergamon Press, Elmsford, N.Y.

Aspinall, G. O., 1980, Chemistry of cell wall polysaccharides, in: *The Biochemistry of Plants: A Comprehensive Treatise,* Volume 3 (J. Preiss, ed.), Academic Press, New York, pp. 473–500.

Aspinall, G. O., and Kessler, G., 1957, The structure of callose from the grape vine, *Chem. Ind. (London)* **39**:1296.

Bailey, J. A., 1981, Physiological and biochemical events associated with the expression of resistance to disease, in: *Active Defense Mechanisms in Plants* (R. K. S. Wood, ed.), Plenum Press, New York, pp. 39–65.

Bailey, J. A., and Mansfield, J. W. (eds.), 1982, *Phytoalexins,* Blackie, Glasgow.

Baker, E. A., 1982, Chemistry and morphology of plant epicuticular waxes, in: *The Plant Cuticle* (D. F. Cutler, K. L. Alvin, and C. E. Price, eds.), Academic Press, New York, pp. 139–165.

Barckhausen, R., 1978, Ultrastructural changes in wounded plant storage tissue cells, in: *Biochemistry of Wounded Plant Tissues* (G. Kahl, ed.), de Gruyter, Berlin, pp. 1–42.

Barondes, S. H., 1981, Lectins: Their multiple endogenous cellular functions, *Annu. Rev. Biochem.* **50:**207–231.

Barrett, A. J., and Northcote, D. H., 1965, Apple fruit pectic substances, *Biochem. J.* **94:**617–627.

Bauer, M. D., Talmadge, K. W., Keegstra, K., and Albersheim, P., 1973, The structure of plant cell walls. II. The hemicellulose of the walls of suspension-cultured sycamore cells, *Plant Physiol.* **51:**174–187.

Bianchi, G., Avato, P., Bertorelli, P., Mariani, G., 1978, Epicuticular waxes of two sorghum varieties, *Phytochemistry* **17:**999–1001.

Bikales, N. M., and Segal, L., 1971, *Cellulose and Cellulose Derivatives,* Wiley, New York.

Blackwell, J., 1982, The macromolecular organisation of cellulose and chitin, in: *Cellulose and Other Natural Polymer Systems* (R. M. Brown, ed.), Academic Press, New York, pp. 403–428.

Blakeman, J. P., and Atkinson, P., 1981, Antimicrobial substances associated with the aerial surfaces of plants, in: *Microbial Ecology of the Phylloplane* (J. P. Blakeman, ed.), Academic Press, New York, pp. 245–263.

Brett, C. T., 1978, Synthesis of β(1→3)-glucan from extracellular UDP-Glc as a wound response in suspension cultured soybean cells, *Plant Physiol.* **62:**377–382.

Brown, A. J., and Kolattukudy, P. E., 1978, Evidence that pancreatic lipase is responsible for the hydrolysis of cutin, *Arch. Biochem. Biophys.* **190:**17–26.

Burke, D., Kaufman, P., McNeil, M., and Albersheim, P., 1974, The structure of plant cell walls. VI. A survey of the walls of suspension-cultured monocots, *Plant Physiol.* **54:**109–115.

Campbell, C. L., Huang, J. S., and Payne, G. A., 1980, Defense of the perimeter: The outer walls and the gates, in: *Plant Disease: An Advanced Treatise,* Volume V (J. G. Horsfall and E. B. Cowling, eds.), Academic Press, New York, pp. 103–120.

Clarke, A. E., 1981, Defined components involved in pollination, in: *Plant Carbohydrates, II. Extracellular Carbohydrates* (W. Tanner and F. A. Loewus, eds.), Springer-Verlag, Berlin, pp. 577–583.

Clarke, A. E., and Stone, B. A., 1963, Chemistry and biochemistry of β-1,3-glucans, *Rev. Pure Appl. Chem.* **13:**134–156.

Clarke, A. E., Anderson, R. L., and Stone, B. A., 1979a, Form and function of arabinogalactans and arabinogalactan proteins, *Phytochemistry* **18:**521–540.

Clarke, K. J., McCully, M. E., and Miki, N. K., 1979b, A developmental study of the epidermis of young roots of *Zea mays* L., *Protoplasma* **98:**283–309.

Conn, E. E., 1981, Cyanogenic glycosides, in: *The Biochemistry of Plants: A Comprehensive Treatise,* Volume 7 (E. E. Conn, ed.), Academic Press, New York, pp. 479–601.

Cook, G. M. W., and Stoddart, R. W., 1973, *Surface Carbohydrates of the Eukaryotic Cell,* Academic Press, New York.

Cooper, J. B., and Varner, J. E., 1983, Insolubilization of hydroxyproline-rich cell wall glycoprotein in aerated carrot root slices, *Biochem. Biophys. Res. Commun.* **112:**161–167.

Corrigan, D., Timoney, R. F., and Donnelly, D. M. X., 1978, n-Alkanes + ω-hydroxyalkanoic acids from the needles of 28 *Picea* species, *Phytochemistry* **17:**907–910.

Crasnier, M., Noat, G., and Ricard, J., 1980, Purification and molecular properties of acid phosphatases from sycamore cell walls, *Plant Cell Environ.* **3:**217–224.

Darvill, A. G., McNeil, M., and Albersheim, P., 1978, Structure of plant cell walls. VIII. A new pectic polysaccharide, *Plant Physiol.* **62:**418–422.

Darvill, A., McNeil, M., Albersheim, P., and Delmer, D. P., 1980, The primary cell walls of flowering plants, in: *The Biochemistry of Plants: A Comprehensive Treatise,* Volume 1 (N. E. Tolbert, ed.), Academic Press, New York, pp. 92–162.

Dazzo, F. B., 1980, Adsorption of microorganisms to roots and other plant surfaces, in: *Absorption of Microorganisms to Surfaces* (G. Bitton and K. C. Marshall, eds.), Wiley, New York, pp. 253–316.

Dean, B. B., and Kolattukudy, P. E., 1976, Synthesis of suberin during wound-healing in jade leaves, tomato fruits and bean pods, *Plant Physiol.* **58:**411–416.

Dean, B. B., Kolattukudy, P. E., and Davis, R. W., 1977, Chemical composition and ultrastructure of suberin from hollow heart tissue of potato tubers (*Solanum tuberosum*), *Plant Physiol.* **59:**1008–1010.

Deas, A. H. B., and Holloway, P. J., 1977, The intermolecular structure of plant cutins, in: *Lipids and Lipid Polymers in Higher Plants* (M. Trevinc and H. K. Lichtenthaler, eds.), Springer-Verlag, Berlin, pp. 293–299.

Edelman, G. M., 1983, Cell adhesion molecules, *Science* **219:**450–457.

Eschrich, W., 1975, Sealing systems in phloem, in: *Encyclopedia of Plant Physiology,* New Series, Volume I (M. H. Zimmerman and J. A. Milburn, eds.), Springer-Verlag, Berlin, pp. 40–56.

Espelie, K. E., Dean, B. B., and Kolattukudy, P. E., 1979, Composition of lipid derived polymers from different anatomical regions of several plant species, *Plant Physiol.* **65:**1089–1093.

Espelie, K. E., Sadek, N. Z., and Kolattukudy, P. E., 1980a, Composition of suberin-associated waxes from the subterranean storage organs of seven plants [parsnip, carrot, rutabaga, turnip, red beet, sweet potato and potato], *Planta* **148:**468–476.

Espelie, K. E., Davis, R. W., and Kolattukudy, P. E., 1980b, Composition, ultrastructure and function of the cutin and suberin containing layers in the leaf, fruit peel, juice-sac and inner seed coat of grapefruit (*Citrus paradisi* Macfed.), *Planta* **149:**498–511.

Esquerre-Tugaye, M.-T., and Lamport, D. T. A., 1979, Cell surfaces in plant microorganism interactions. I. A structural investigation of cell wall hydroxyproline-rich glycoproteins which accumulate in fungus-infected plants, *Plant Physiol.* **64:**314–319.

Esquerre-Tugaye, M.-T., Lafitte, C., Mazau, D., Toppan, A., and Toze, A., 1979, Cell surfaces in plant–microorganism interactions, *Plant Physiol.* **64:**320–326.

Fincher, G. B., Stone, B. A., and Clarke, A. E., 1983, Arabinogalactan-proteins: Structure, biosynthesis and function, *Annu. Rev. Plant Physiol.* **34:**47–70.

Foster, R. C., 1982, The fine structure of epidermal cell mucilages of roots, *New Phytol.* **91:**727–740.

Foster, R. C., and Bowen, G. D., 1982, Plant surfaces and bacterial growth: The rhizosphere and rhizoplane, in: *Phytopathogenic Prokaryotes* (M. S. Mount and G. H. Lacy, eds.), Academic Press, New York, pp. 159–185.

Franich, R. A., Wells, L. G., and Holland, P. T., 1978, Epicuticular wax of *Pinus radiata* needles, *Phytochemistry* **17:**1617–1623.

Friend, J., 1979, Phenolic substances and plant disease, *Recent Adv. Phytochem.* **12:**557–588.

Frossard, R., 1981, Effect of guttation fluids on growth of microorganisms on leaves, in: *Microbial Ecology of the Phylloplane* (J. P. Blakeman, ed.), Academic Press, New York, pp. 213–226.

Fry, S. C., 1982, Isodityrosine, a new cross linking amino acid from plant cell wall glycoprotein, *Biochem. J.* **204:**449–455.

Fry, S. C., 1983, Feruloylated pectins from the primary cell wall: Their structure and possible functions, *Planta* **157:**111–123.

Gardner, K. H., and Blackwell, N., 1974, The structure of native cellulose, *Biopolymers* **13:**1975–2001.

Gibson, D. M., and Liu, E. H., 1981, Inhibition of cell wall-associated peroxidase isoenzymes with British anti-Lewiste, *J. Exp. Bot.* **32:**419–426.

Gibson, L. J., Easterling, K. E., and Ashby, M. F., 1981, The structure and mechanics of cork, *Proc. R. Soc. London Ser. A* **377:**99–117.

Graham, T. L., 1981, Recognition in *Rhizobium*–legume symbioses, *Int. Rev. Cytol. Suppl.* **13:**127–148.

Greenland, D. J., 1979, The physics and chemistry of the soil root interface, in: *The Soil–Root Interface* (J. L. Harley and R. Scott-Russell, eds.), Academic Press, New York, pp. 83–98.

Grisebach, H., 1981, Lignins, in: *The Biochemistry of Plants: A Comprehensive Treatise,* Volume 7 (E. E. Conn, ed.), Academic Press, New York, pp. 457–478.

Grisebach, H., and Ebel, J., 1978, Phytoalexins, chemical defense substances of higher plants, *Angew. Chem. Int. Ed. Engl.* **17:**635–647.

Gross, G. G., 1979, Recent advances in the chemistry and biochemistry of lignin, *Recent Adv. Phytochem.* **12:**177–220.

Hallam, N. D., 1982, Fine structure of the leaf cuticle and the origin of leaf waxes, in: *The Plant Cuticle,* (D. F. Cutler, K. L. Alwin, and C. E. Price, eds.), Academic Press, New York, pp. 197–214.

Harley, J. L., and Scott-Russell, R. (eds.), 1979, *The Soil–Root Interface,* Academic Press, New York.

Heath, M. F., and Northcote, D. H., 1971, Glycoprotein of the wall of sycamore tissue-culture cells, *Biochem. J.* **125:**953–961.

Henry, C. M., and Deacon, J. W., 1981, Natural (nonpathogenic) death of the cortex of wheat, and barley seminal roots, as evidenced by nuclear staining with acridine orange, *Plant Soil* **60:**255–274.

Heslop-Harrison, J., 1978, *Cellular Recognition Systems in Plants,* Arnold, London.

Heslop-Harrison, J., and Heslop-Harrison, Y., 1982, The specialised cuticles of the receptive surfaces of angiosperm stigmas, in: *The Plant Cuticle* (D. F. Cutler, K. L. Alvin, and C. E. Price, eds.), Academic Press, New York, pp. 99–119.

Higuchi, T., 1981, Biosynthesis of lignin, in: *Plant Carbohydrates. II. Extracellular Carbohydrates* (W. Tanner and F. A. Loewus, eds.), Springer-Verlag, Berlin, pp. 194–224.

Holloway, P. J., 1972a, The composition of suberin from the corks of *Quercus suber* L. and *Betula pendula* Roth, *Chem. Phys. Lipids* **9:**158–170.

Holloway, P. J., 1972b, The suberin composition of the cork layers of some *Ribes* species, *Chem. Phys. Lipids* **9:**171–179.

Holloway, P. J., 1982a, Structure and histochemistry of plant cuticular membranes: An overview, in: *The Plant Cuticle* (D. F. Cutler, K. L. Alvin, and C. E. Price, eds.), Academic Press, New York, pp. 1–32.

Holloway, P. J., 1982b, The chemical constitution of plant cutins, in: *The Plant Cuticle* (D. F. Cutler, K. L. Alvin, and C. E. Price, eds.), Academic Press, New York, pp. 45–85.

Holloway, P. J., and Deas, A. H. B., 1973, Epoxyoctadecanoic acids in plant cutins and suberins, *Phytochemistry* **12:**1721–1735.

Horst, W. J., Wagner, A., and Marschner, H., 1982, Mucilage protects root meristems from aluminium injury, *Z. Pflanzenphysiol. B.D.* **105S:**435–444.

Hughes, R. G., and Pena, S. D. J., 1981, The role of carbohydrates in cellular recognition and adhesion, in: *Carbohydrate Metabolism and Its Disorders* (P. J. Randle, D. F. Stainer, and W. J. Whelan, eds.), Academic Press, New York, pp. 362–423.

Hunt, G. M., and Baker, E. A., 1980, Phenolic constituents of tomato fruit cuticles, *Phytochemistry* **19:**1415–1419.

Jermyn, M. A., and Yeow, Y. M., 1975, A class of lectins present in the tissues of seed plants, *Aust. J. Plant Physiol.* **2:**501–531.

Joseleau, J.-P., Chambat, G., Vignon, M., and Barnoud, F., 1977, Chemical and ^{13}C N.M.R. studies of two arabinans from the inner bark of young stems of *Rosa glauca, Carbohydr. Res.* **58:**165–175.

Kahl, G. (ed.), 1978, *Biochemistry of Wounded Plant Tissues,* de Gruyter, Berlin.

Karacsonyi, S., Toman, R., Janecek, F., and Kubachkova, M., 1975, Polysaccharides from the bark of the white willow (*Salix alba* L.): Structure of an arabinan, *Carbohydr. Res.* **44:**285–290.

Kato, K., 1981, Ultrastructure of the plant cell wall; biochemical viewpoint, in: *Plant Carbohydrates. II. Extracellular Carbohydrates* (W. Tanner and F. A. Loewus, eds.), Springer-Verlag, Berlin, pp. 29–46.

Kaufman, P. B., Dayanandan, P., Takeoka, Y., Bigelow, W. C., Jones, J. D., and Iler, P., 1981, Silica in shoots of higher plants, in: *Silicon and Siliceous Structures in Biological Systems* (T. L. Simpson and B. E. Volcani, eds.), Springer-Verlag, Berlin, pp. 409–449.

Kennedy, J. F., and White, C. A., 1983, *Bioactive Carbohydrates in Chemistry, Biochemistry and Biology,* Horwood, Chichester.

Kolattukudy, P. E., 1977, Lipid polymers and associated phenols, their chemistry, biosynthesis, and role in pathogenesis, in: *The Structure, Biosynthesis and Degradation of Wood* (F. A. Loewus and V. C. Runeckles, eds.), Plenum Press, New York, pp. 165–246.

Kolattukudy, P. E., 1978, Chemistry and biochemistry of the aliphatic components of suberin, in: *Biochemistry of Wounded Plant Tissues* (G. Kahl, ed.), de Gruyter, Berlin, pp. 43–84.

Kolattukudy, P. E., 1980a, Biopolyester membranes of plants: Cutin and suberin, *Science* **208:**990–1000.

Kolattukudy, P. E., 1980b, Cutin, suberin and waxes, in: *The Biochemistry of Plants: A Comprehensive Treatise,* Volume 4 (P. K. Strumpf, ed.), Academic Press, New York, pp. 571–645.

Kolattukudy, P. E., Kronman, K., and Poulose, A. J., 1975, Determination of structure and composition of suberin from the roots of carrot, parsnip, rutabaga, turnip, red beet and sweet potato by combined gas–liquid chromatography and mass spectrometry, *Plant Physiol.* **55:**567–573.

Kolattukudy, P. E., Espelie, K. E., and Soliday, C. L., 1981, Hydrophobic layers attached to cell walls: Cutin, suberin and associated waxes, in: *Plant Carbohydrates. II. Extracellular Carbohydrates* (W. Tanner and F. A. Loewus, eds.), Springer-Verlag, Berlin, pp. 225–254.

Kosuge, T., 1981, Carbohydrates in plant–pathogen interactions, in: *Encyclopedia of Plant Physiology,* New Series, Volume 13B (W. Tanner and F. A. Loewus, eds.), Springer-Verlag, Berlin, pp. 584–623.

Lamb, C., Bell, J., Norman, P., Lawton, M., Dixon, R., Rowell, P., and Bailey, J., 1983, Early molecular events in the phytoalexin defense response, in: *Structure and Function of Plant Genomes,* Vol. 63 (O. Ciferri and L. Dure, eds.), NATO–ASI Series A, Plenum Press, New York, pp. 313–327.

Lamport, D. T. A., 1970, Cell wall metabolism, *Annu. Rev. Plant Physiol.* **21:**235–270.

Lamport, D. T. A., 1980, Structure and function of plant glycoproteins, in: *The Biochemistry of Plants: A Comprehensive Treatise,* Volume 3 (J. Preiss, ed.), Academic Press, New York, pp. 501–541.

Lamport, D. T. A., and Catt, J. W., 1981, Glycoproteins and enzymes of the cell wall, in: *Plant Carbohydrates. II. Extracellular Carbohydrates* (W. Tanner and F. A. Loewus, eds.), Springer-Verlag, Berlin, pp. 133–165.

Larsen, P. O., 1981, Glucosinolates, in: *The Biochemistry of Plants: A Comprehensive Treatise,* Volume 7 (E. E. Conn. ed.), Academic Press, New York, pp. 502–525.

Leach, J. E., Cantrell, M. A., and Sequeira, L., 1982, Hydroxyproline-rich bacterial agglutinin from potato: Extraction, purification and characterization, *Plant Physiol.* **70:**1353–1359.

Lendzian, K. J., and Schönherr, J., 1983, In vivo study of cutin synthesis in leaves of *Clivia miniata* Reg., *Planta* **158:**70–75.

Lindahl, V., and Höök, M., 1978, Glycosaminoglycans and their binding to biological macromolecules, *Annu. Rev. Biochem.* **47**:385–417.

Loebenstein, G., Spiegel, S., and Gera, A., 1982, Localised resistance and barrier substances, in: *Active Defense Mechanisms in Plants* (R. K. S. Wood, ed.), Plenum Press, New York, pp. 211–230.

McNeil, M., Darvill, A. G., and Albersheim, P., 1982, Structure of plant cell walls. XII. Identification of seven differently linked glycosyl residues attached to O4 of the 2-4 linked L-rhamnosyl residue of rhamnogalacturonan I, *Plant Physiol.* **70**:1586–1591.

Mäder, M., 1976, Localisation of peroxidase-isoenzyme group G1 in the cell wall of tobacco tissues, *Planta* **131**:11–15.

Mäder, M., Nessel, A., and Bopp, M., 1977, On the physiological significance of the isoenzyme groups of peroxidase from tobacco demonstrated by biochemical properties. II. pH optima, Michaelis constants, maximal oxidation rates, *Z. Pflanzenphysiol.* **82**:247–260.

Martin, J. T., and Juniper, B. E., 1970, *The Cuticles of Plants,* Arnold, London.

Matsumoto, I., Jimbo, A., Mizuno, Y., Seno, N., and Jeanloz, R. W., 1983, Purification and characterization of potato lectin, *J. Biol. Chem.* **258**:2886–2891.

Mellon, J. E., and Helgeson, J. P., 1982, Interaction of a hydroxyproline-rich glycoprotein from tobacco cells with potential pathogens, *Plant Physiol.* **70**:401–405.

Meyer, K. H., and Misch, L., 1937, Positions des atomes dans le nouveau modele spatial de la cellulose, *Helv. Chim. Acta* **20**:232–244.

Miki, N. K., Clarke, K. J., and McCully, M. E., 1980, A histological and histochemical comparison of the root tips of several grasses, *Can. J. Bot.* **58**:2581–2593.

Misaghi, I. J., 1982, *Physiology and Biochemistry of Plant Pathogen Interactions,* Plenum Press, New York.

Monro, J. A., Penny, D., and Bailey, R. W., 1976, The organisation and growth of primary cell walls of lupin hypocotyl, *Phytochemistry* **15**:1193–1198.

Morrison, I. M., 1974, Structural investigation on the lignin–carbohydrate complexes of *Lolium perenne, Biochem. J.* **139**:197–204.

Murray, R. H. A., and Northcote, D. H., 1978, Oligoarabinosides of hydroxyproline isolated from potato lectin, *Phytochemistry* **17**:623–629.

Nambiar, G. K. S., 1976, Uptake of ^{65}Zn from dry soil by plants, *Plant Soil* **44**:267–271.

Nevins, D. J., Huber, D. J., Vamamoto, R., and Loescher, W. H., 1977, β-D-glucan of *Avena* coleoptile cell walls, *Plant Physiol.* **60**:617–621.

Nieduszynski, I., and Marchessault, R. H., 1971, Structure of β-D-(1-4) xylan hydrate, *Nature (London)* **232**:46–47.

Northcote, D. H., 1972, Chemistry of the plant cell wall, *Annu. Rev. Plant Physiol.* **23**:113–132.

Northcote, D. H., 1982, The synthesis and transport of some plant glycoproteins, *Philos. Trans. R. Soc. London Ser. B* **300**:195–206.

Oades, J. M., 1978, Mucilages at the root surface, *J. Soil Sci.* **29**:1–16.

O'Neill, M. A., and Selvendran, R. R., 1980, Glycoproteins from the cell wall of *Phaseolus coccineus, Biochem. J.* **187**:53–63.

O'Neill, M. A., and Selvendran, R. R., 1983, Isolation and partial characterisation of a xyloglucan from the cell walls of *Phaseolus coccineus, Carbohydr. Res.* **111**:239–255.

Paull, R. G., Johnson, C. M., and Jones, R. L., 1975, Studies on the secretion of maize root cap slime. I. Some properties of the secreted polymer, *Plant Physiol.* **56**:300–306.

Pope, D. G., 1977, Relationships between hydroxyproline-containing proteins secreted into the cell wall and medium of suspension cultured *Acer pseudoplatanus* cells, *Plant Physiol.* **59**:894–900.

Preston, R. D., 1974, *The Physical Biology of Plant Cell Walls,* Chapman & Hall, London.

Preston, R. D., 1979, Polysaccharide conformation and cell wall function, *Annu. Rev. Plant Physiol.* **30**:55–78.

Read, S. M., and Northcote, D. H., 1983, Chemical and immunological similarities between the phloem proteins of three genera of the Cucurbitaceae, *Planta* **158**:119–127.

Rees, D. A., and Welsh, E. J., 1977, Secondary and tertiary structure of polysaccharides in solutions and gels, *Angew. Chem. Int. Ed. Engl.* **16**:214–234.

Rees, D. A., and Wight, A. W., 1971, Polysaccharide conformation. Part VII. Model building computations for α-1,4galacturonan and the kinking function of L-rhamnose residues in pectic substances, *J. Chem. Soc. B* **1971**:1366–1372.

Rhodes, M. J., and Wooltorton, L. S. C., 1978, The biosynthesis of phenolic compounds in wounded plant storage tissues, in: *Biochemistry of Wounded Plant Tissues* (G. Kahl, ed.), de Gruyter, Berlin, pp. 243–286.

Ride, J. P., 1978, The role of cell wall alterations in response to fungi, *Ann. Appl. Biol.* **89**:302–306.

Riley, G., and Kolattukudy, P. E., 1975, Evidence for covalently attached β-coumaric and ferulic acid in cutins and suberins, *Plant Physiol.* **56**:650–654.

Ring, S. G., and Selvendran, R. R., 1980, An arabinogalactan xyloglucan from the cell wall of *Solanum tuberosum*, *Phytochemistry* **20:**2511–2519.

Robards, A. W., Clarkson, D. T., and Sanderson, J., 1979, Structure and permeability of the epidermal/hypodermal layers of the sand sedge (*Carex arenaria* L), *Protoplasma* **101:**331–347.

Rougier, M., 1981, Secretory activity of the root cap, in: *Encyclopedia of Plant Physiology*, New Series, Volume 13B (W. Tanner and F. A. Loewus, eds.), Springer-Verlag, Berlin, pp. 542–576.

Rovira, A. D., Foster, R. C., and Martin, J. K., 1979, Note on terminology, origin, nature and nomenclature of the organic materials in the rhizosphere, in: *The Soil–Root Interface* (J. L. Harley, ed.), Academic Press, New York, pp. 1–4.

Sandhu, J. S., Hudson, G. J., and Kennedy, J. F., 1981, The gel nature and structure of the carbohydrate of ispaghula husk *ex Plantago ovata* Forsk, *Carbohydr. Res.* **93:**247–259.

Sangster, A. G., and Parry, D. W., 1981, Ultrastructure of silica in higher plants, in: *Silicon and Siliceous Structures in Biological Systems* (T. L. Simpson and B. E. Volcani, eds.), Springer-Verlag, Berlin, pp. 383–407.

Schönherr, J., 1982, Resistance of plant surfaces to water loss: Transport properties of cutin, suberin and associated lipids, in: *Encyclopedia of Plant Physiology*, New Series, Volume 12B (O. L. Lange, P. S. Nobel, C. B. Osmond, and H. Ziegler, eds.), Springer-Verlag, Berlin, pp. 153–179.

Schönherr, J., and Ziegler, H., 1980, Water permeability of *Betula* periderm, *Planta* **147:**345–354.

Selvendran, R. R., and O'Neill, M. A., 1982, Plant glycoproteins, in: *Encyclopedia of Plant Physiology*, New Series, Volume 13A (F. A. Loewus and W. Tanner, eds.), Springer-Verlag, Berlin, pp. 515–583.

Sequeira, L., 1980, Defenses triggered by the invader: Recognition and compatibility phenomena, in: *Plant Disease*, Volume 5 (J. G. Horsfall and E. B. Cowling, eds.), Academic Press, New York, pp. 179–200.

Shafizadeh, F., and McGinnis, G. D., 1971, Morphology and biogenesis of cellulose and plant cell walls, *Adv. Carbohydr. Chem. Biochem.* **26:**297–349.

Shimizu, K., Hashi, M., and Sakurai, K., 1978, Isolation from a soft wood xylan of oligosaccharides containing two 4-*O*-methyl-D-glucuronic acid residues, *Carbohydr. Res.* **62:**117–126.

Siddiqui, I. R., and Wood, P. J., 1976, Structural investigation of oxalate-soluble rapeseed (*Brassica campestris*) polysaccharides. Part IV. Pectic polysaccharides, *Carbohydr. Res.* **50:**97–107.

Siddiqui, I. R., and Wood, P. J., 1977, Structural investigation of an acidic xylan from rapeseed, *Carbohydr. Res.* **54:**231–236.

Sitte, P., 1975, Significance of the laminated fine structure of cork cell walls, *Biochem. Physiol. Pflanz.* **168S:**287–297.

Smith, H., 1977, Microbial surfaces in relation to pathogenicity, *Bacteriol. Rev.* **41:**475–500.

Soliday, C. L., Kolattukudy, P. E., and Davis, R. W., 1979, Chemical and ultrastructural evidence that waxes associated with the suberin polymer constitute the major diffusion barrier to water vapour in potato tuber (*Solanum tuberosum* L), *Planta* **146:**607–614.

Stafford, H. A., 1974, Possible multienzyme complexes regulating the formation of C6–C3 phenolic compounds and lignins in higher plants, *Recent Adv. Phytochem.* **8:**53–79.

Stephen, A. M., 1980, Plant carbohydrates, in: *Encyclopedia of Plant Physiology*, New Series, Volume 8 (E. A. Bell and B. V. Charlwood, eds.), Springer-Verlag, Berlin, pp. 555–584.

Stoddart, R. W., and Northcote, D. H., 1967, Metabolic relationships of the isolated fractions of the pectin substances of actively growing sycamore cells, *Biochem. J.* **105:**45–59.

Stuart, D. A., and Varner, J. E., 1980, Purification and characterization of a salt extractable hydroxyproline rich glycoprotein from aerated carrot discs, *Plant Physiol.* **66:**787–792.

Tanner, G. R., and Morrison, I. M., 1983, Phenolic-carbohydrate complexes in the cell walls of *Lolium perenne*, *Phytochemistry* **22:**1433–1439.

Toman, R., Karácsonyi, S., Kováčik, V., 1972, Polysaccharide from the bark of the white willow (*Salix alba* L.): Structure of a galactan, *Carbohydr. Res.* **25:**371–378.

Touze, A., and Esquerre-Tugaye, M.-T., 1982, Defense mechanisms of plants against varietal non-specific pathogens, in: *Active Defense Mechanisms in Plants* (R. K. S. Wood, ed.), Plenum Press, New York, pp. 103–117.

Tulloch, A. P., 1976, Chemistry of waxes of higher plants, in: *Chemistry and Biochemistry of Natural Waxes* (P. E. Kolattukudy, ed.), Elsevier, Amsterdam, pp. 235–279.

Tulloch, A. P., 1978, Epicuticular wax of *Poa ampla* leaves, *Phytochemistry* **17:**1613–1615.

Tulloch, A. P., and Hoffman, L. L., 1979, Epicuticular waxes of *Andropogon hallii* and *A. scoparius*, *Phytochemistry* **18:**267–271.

Vance, C. P., Kirk, T. K., and Sherwood, R. T., 1980, Lignification as a mechanism of disease resistance, *Annu. Rev. Phytopathol.* **18:**259–288.

van Holst, G. J., Klis, F. M., de Wilot, P. J. M., Hazenberg, C. A. M., Buijs, J., and Stegwee, D., 1981,

Arabinogalactan protein from a crude cell organelle fraction of *Phaseolus vulgaris* L., *Plant Physiol.* **68:**910–913.

Vermeer, J., and McCully, M. E., 1981, Fucose in the surface deposits of axenic and field grown roots of *Zea mays* L., *Protoplasma* **109:**233–248.

Vermeer, J., and McCully, M. E., 1982, The rhizosphere in *Zea*: New insight into its structure and development, *Planta* **156:**45–61.

von Wettstein-Knowles, P., 1979, Genetics and biosynthesis of plant epicuticular waxes, in: *Advances in the Biochemistry and Physiology of Plant Lipids* (L.-A. Appelquist and C. Liljenberg, eds.), Elsevier/North-Holland, Amsterdam, pp. 1–33.

Waid, J. S., 1974, Decomposition of roots, in: *Biology of Litter Decomposition*, Volume 1 (C. H. Dickinson and G. J. F. Pugh, eds.), Academic Press, New York, pp. 175–211.

Wattendorff, J., 1974, Ultrahistochemical reactions of the suberized cell walls in *Acorus, Acacia*, and *Larix*, *Z. Pflanzenphysiol.* **73:**214–225.

Werker, E., and Kislev, M., 1978, Mucilage on the root surface and root hairs of sorghum: Heterogeneity in structure, manner of production and site of accumulation, *Ann. Bot. (London)* **42:**809–816.

West, C. A., 1981, Fungal elicitors of the phytoalexin response in higher plants, *Naturwissenschaften* **68:**447–457.

Whitmore, F. W., 1978a, Lignin–protein complex catalysed by peroxidase, *Plant Sci. Lett.* **13:**241–245.

Whitmore, F. W., 1978b, Lignin–carbohydrate complex formed in isolated cell walls of callus, *Phytochemistry* **17:**421–425.

Wilkie, K. C. B., 1979, The hemicelluloses of grasses and cereals, *Adv. Carbohydr. Chem. Biochem.* **36:**215–264.

Wilkie, K. C. B., and Woo, S.-L., 1977, A heteroxylan and hemicellulosic materials from bamboo leaves, and a reconsideration of the general nature of commonly occurring xylans and other hemicelluloses, *Carbohydr. Res.* **57:**145–162.

Wilson, L. A., and Sterling, C., 1976, Studies on the cuticle of tomato fruit: Fine structure of the cuticle, *Z. Pflanzenphysiol.* **77:**359–371.

Wood, R. K. S. (ed.), 1981, *Active Defense Mechanisms in Plants*, Plenum Press, New York.

Wright, K., 1975, Polysaccharides of root cap slime from five maize varieties, *Phytochemistry* **14:**759–763.

Wright, K., and Northcote, D. H., 1974, The relationship of root-cap slimes to pectins, *Biochem. J.* **139:**525–534.

Wright, K., and Northcote, D. H., 1975, An acidic oligosaccharide from maize slime, *Phytochemistry* **14:**1793–1798.

Wright, K., and Northcote, D. H., 1976, Identification of $\beta(1\rightarrow4)$ glucan chains as part of a fraction of slime synthesized within the dictyosomes of maize root cap cells, *Protoplasma* **88:**225–239.

Wright, K., Northcote, D. H., and Davey, R. M., 1976, Preparation of rat epididymal α-L-fucosidase free from other glycosidases: Its action on root cap slime from *Zea mays* L, *Carbohydr. Res.* **47:**141–150.

Yariv, J., Rapport, M. M., and Graf, L., 1962, The interaction of glycosides and saccharides with antibody to the corresponding phenylazo glycosides, *Biochem. J.* **85:**383–388.

Yung, K. H., and Northcote, D. H., 1975, Some enzymes present in the walls of mesophyll cells of tobacco leaves, *Biochem. J.* **151:**141–144.

5

The Properties of Nonbiological Surfaces and Their Characterization

GEORGE I. LOEB

1. INTRODUCTION

The forces involved in adhesion of cells to other surfaces may include any of the interactions possible in aqueous physical systems (see Chapter 6). Thus, a wide variety of surface properties and measurements may have some relevance to adhesion problems. The classical work on colloid stability, known as the DLVO theory, considered the balance between the London–van der Waals attractive forces and the electrostatic repulsive interaction between the two interacting surfaces [see Section 3.1, and Rutter and Vincent (1980) for a short introduction and references]. Thus, appreciation of the electrostatic charge on surfaces and of the London-type, or dispersion, interactions characteristic of the surfaces involved are required. In addition, the polarity of biological material means that polar interactions must also be taken into account. The measurement of surface charge in aqueous systems, and of estimation of the London and polar components of the surface energies of solid surfaces, have consequently been closely associated with research into bioadhesion. By measuring and assigning experimental values to these various parameters, it becomes possible to examine experimentally the theoretical basis of adhesion processes and to test the validity of such theories.

Attempts to systematize the results of bioattachment data have led to postulated correlations with such general surface properties, e.g., with the degree of surface hydrophilicity or hydrophobicity (Chapter 6, Section 4.5). For example, the tendency for hydrophilic containers to accelerate blood clotting and to enhance tissue attachment and spreading has been recognized for many years (Baier, 1970), while other cell types attach preferentially to hydrophobic surfaces (e.g., Fletcher and Loeb, 1979). Attempts to enhance or inhibit bioattachment to materials without degrading their bulk properties have resulted in a variety of physical and chemical surface treatments, ranging from simple abrasion to ultraviolet and electrical discharge treatments, surface modification with wet or gaseous chemical reagents,

George I. Loeb • Naval Research Laboratory, Washington, D. C. 20375. *Present address:* David Taylor Naval Ship R & D Center, Annapolis, Maryland 21402.

and grafting of polymeric chains containing desired functionalities to surfaces. Methods for controlling and characterizing such modifications are still required.

The effects of adsorbed material greatly complicate the picture. The degree to which adsorbed layers on cells and on immersed surfaces interact and the tendency of various adsorbed materials and solvent components to compete for adsorption sites on cells and surfaces and to form permanent bonds are subjects of current active research (Chapter 6, Section 4.8). The production of extracellular material by cells leads to similar phenomena and questions (Chapter 6, Section 6). These questions emphasize the importance of analytical techniques for detection and characterization of small amounts of materials on surfaces, while the need for surface characterization in terms which allow understanding of the more general physicochemical properties of surfaces remains.

Understanding the interaction of cells with surfaces requires an understanding of the cellular side of the interface as well as the "inert" side. The complex considerations required to understand the dynamic surfaces of cells have been discussed (Weiss, 1972; Fletcher *et al.*, 1980; Rutter and Vincent, 1980; Pethica, 1980; Tadros, 1980; Ward and Berkeley, 1980), and efforts continue to increase knowledge of their structures. However, the cell or bacterial surface, as distinct from the "inert" surface, is discussed in Chapter 2 and not dealt with in this chapter.

2. SURFACE ENERGY AND CONTACT ANGLE

Thermodynamic treatments of adhesion or attachment deal with the change in free energy of materials as they form interfaces, and the underlying premise is that adhesion is energetically favored if the formation of the adhesion interface is accompanied by a decrease in the free energy of the system. The general expression for the free energy change of such processes may be expressed in terms of the surface free energies of the two materials of interest, e.g., 1 and 2, initially in contact with a surrounding fluid phase, 3, and then coming together to form a 1–2 interface. Thus,

$$\Delta F^a = \Delta F_{1,2} - \Delta F_{1,3} - \Delta F_{2,3} \tag{1}$$

where ΔF^a is the free energy of adhesion and $\Delta F_{1,2}$, $\Delta F_{1,3}$, and $\Delta F_{2,3}$ are the free energies per unit area of interface between phases 1 and 2, 1 and 3, and 2 and 3, respectively, expressed in ergs per square centimeter or millijoules per square meter. The surface free energy of a fluid phase boundary is numerically equal to its surface tension, which is usually measured mechanically. Thus, understanding of the thermodynamics of attachment and adhesion requires a knowledge of the surface energies or surface tensions of the materials at the interfaces.

Although the surface tension of a fluid–fluid (e.g., liquid–gas, liquid–liquid) interface is, at least in principle, easily measured by controlled deformation of the interface, solids cannot be handled in this way because their viscoelasticity prevents reversible deformation in a reasonable time. The methods used are therefore indirect and involve interaction of the solid with fluids, and a number of approaches have been employed.

2.1. Theoretical Treatments of Thermodynamic Parameters

2.1.1. Contact Angles and Critical Surface Tension

The contact angle was analyzed in terms of Young's equation in the last century, as shown in Fig. 1. The property observed is the angle θ, which the boundary between two fluid

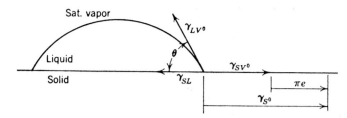

Figure 1. Vector representation of vapor–liquid–solid equilibrium of a sessile liquid drop on a solid surface in a vapor atmosphere. Symbols: γ_{SL}, γ_{LV}, and γ_{SV} are the solid–liquid, liquid–vapor, and solid–vapor interfacial energies, respectively; γ_S is the solid surface free energy; π_e is the spreading pressure, i.e., the difference between γ_{SV} (surface energy of the solid in equilibrium with the vapor of the liquid to form the drop) and γ_S; it is close to 0 for liquids which form nonzero contact angles. From Kaelble (1971).

phases forms with the surface of a smooth flat solid. In the case where one of the fluid phases is air and the other a liquid, the angle is usually measured in the liquid phase. Inasmuch as the liquid–air interfacial free energy is numerically equivalent to the surface tension of the fluid phase boundary, it has become customary to describe the situation in the notation of surface tension, although the solid surface tension is not as easily thermodynamically defined. Thus, Young's equation states:

$$\gamma_{SV} = \gamma_{SL} + \gamma_{LV}\cos \theta \qquad (2)$$

where γ denotes the surface tension or free energy of the phase boundary, and S, V, and L denote solid and two fluid phases (the former is frequently vapor), respectively, e.g., γ_{SV} is the surface free energy at the solid–vapor phase boundary. Understanding adhesion from the thermodynamic point of view comes about by relating the composition and properties of the phases in contact to the free energy of adhesion.

The concept of the critical surface tension, or γ_c, was developed primarily by Zisman and co-workers (Zisman, 1964) as an empirical tool for organizing a body of data developed through wetting measurements. It was defined as the surface tension of the liquid of highest surface tension which completely wets the solid being studied, and liquids of higher surface tension form drops with nonzero contact angle. The graph of cosine of contact angle of a liquid drop on a surface in air, for a series of liquids, versus the surface tension of the liquid has become known as a "Zisman plot." The intercept of the plot at $\cos \theta = 1$ (where $\theta = 0$) is taken as γ_c for the solid concerned.

The γ_c concept allowed correlation of the available data, and surfaces were ranked and classified in terms of γ_c. Changes in chemical composition were reflected in γ_c values. For example, it was found that substitution of hydrogen on paraffin or polyethylene by halogens led to pronounced changes in γ_c, with fluorine causing reductions, and chlorine, bromine, and iodine leading to increases in γ_c (Zisman, 1964). Extensive studies led to ranking of many polymers with respect to their γ_c values, and a compilation by Shafrin (1966) contains many references. Substances could be conveniently classified as high- or low-γ_c materials, and the relationship between surface tension and surface free energy has encouraged the terms *high-* and *low-energy materials* to be used more or less interchangeably with the γ_c concept. Values of γ_c above 10^2 mJ/m^2 are generally considered high, and water, whose surface tension is 73 mJ/m^2, spreads on such surfaces.

The γ_c concept was presented as an empirical means of systematizing a large body of data. However, the need for more fundamental understanding has led to efforts to bridge the gap between the γ_c concept and the surface free energy, despite the ambiguity presented by the solid surface.

2.1.2. Equation of State Approach

If the surface free energy of a solid is considered to be a thermodynamically significant property of a well-prepared surface, then thermodynamic treatments similar to those applied to polymers and other liquids may be applied to solids as well. The validity of such treatments will be judged by the degree to which a consistent picture emerges. Young's equation contains two terms involving the surface energy of the solid phase: γ_{SV} and γ_{SL}. Therefore, another relation between them is required if the two unknown values for the interfacial energies are to be calculated using the contact angle measurement from a single liquid. Two relations for this purpose have been suggested, based on observed empirical relationships among γ_{SL}, γ_{SV}, and γ_{LV}, which have been derived from a large body of data in the literature, resulting from measurements on many solids with many liquids. The first, which depends upon a linear correlation, is that of Neumann et al. (1974b):

$$\cos \theta = \frac{(0.015\gamma_{SV} - 2.00)(\gamma_{LV}\gamma_{SV})^{1/2} + \gamma_{LV}}{\gamma_{LV}[0.015(\gamma_{LV}\gamma_{SV})^{1/2} - 1.00]} \tag{3}$$

A computer-generated table of solutions to this equation, which interpolates through the singularity region, is available (Neumann et al., 1980).

The second formulation (Gerson 1982), based on an exponential relation, is

$$\frac{\gamma_{LV}(1 + \cos \theta)}{2(\gamma_{LV}\gamma_{SV})^{1/2}} = \exp[(\gamma_{SV}\text{-}\gamma_{LV}\cos \theta)(6.5 \times 10^{-5}\gamma_{SV} - 0.010)] \tag{4}$$

Each of these treatments yields a value of γ_{SV} from the contact angle obtained with a single liquid of known surface tension, and the value of γ_{SL} may then be obtained from Young's equation. If data can be obtained with more than one liquid, however, a better judgment of the validity of these assumptions for the particular solid considered will be possible.

2.1.3. The Interaction Parameter ϕ and the Separability of Interfacial Interaction Components

The need for greater understanding of the nature of the interactions between the phases making up the interface has led to another approach, in which the different types of interaction forces acting at the interface are explicitly taken into account. Girifalco and Good (1957) and Good and Girifalco (1960) formulated the concept of the interaction parameter ϕ, which evaluates departure from the simple geometric mean formulation of the interfacial free energy between phases 1 and 2. Thus,

$$\gamma_{12} = \gamma_1 + \gamma_2 - 2\phi(\gamma_1\gamma_2)^{1/2} \tag{5}$$

In their treatment it is shown that $\phi = 1$ when the two phases in contact have similar ratios of intermolecular interaction types, while differences in the ratio of the types of interactions in the contacting phases result in $\phi < 1$. Specific attractive interfacial interactions may lead to $\phi > 1$. ϕ has thus been termed an interfacial bonding efficiency factor (Kaelble, 1971, p. 174).

The value of ϕ for any liquid–solid system is related to the fractional polarities of the liquid and solid and the volumes of their interacting species. Thus, if the fractional London force, polar force, and induction force, contributions to the surface free energy are defined as L, P, and I, respectively, and the molar volumes are not very unusual, then

$$L + P + I = 1 \tag{6}$$

for each phase, and

$$\phi = (L_S L_L)^{1/2} + (P_S P_L)^{1/2} + (I_S I_L)^{1/2} \tag{7}$$

(Good, 1979), while

$$\gamma_i = \gamma_i L_i + \gamma_i P_i + \gamma_i I_i \tag{8}$$

The terms are defined by the ionization potentials, the polarizabilities, and the dipole moments of the constituents of the phases in contact.

Fowkes (1962, 1964) had also pointed out the separability of the components of the surface energy. For his evaluation of the components of the surface energy of materials, a large number of liquid–solid contact angles were determined. Hydrocarbon and fluorocarbon low-energy surfaces were used to characterize the London components of liquids, assuming that only London forces were significant in the interfacial interactions. (Results of this analysis indicated discrepancies which were accounted for by assuming small non-London force components to the fluorocarbon surface free energies.) Then, the residual component of γ_L was considered to be of non-London force origin. The treatment (summarized by Good, 1979) considers the non-London force components to be the sum of a "polar" term and an "induction" term. Because the induction term is smallest, it has been neglected in most analyses based on these concepts, with the result that the London component, and a residual component usually designated "polar," are evaluated by treatments of contact angle data.

The work of adhesion, W_a, defined by the free energy change on bringing two separated phases into contact, is

$$W_a = \gamma_{LV} + \gamma_{SV} - \gamma_{LS} \tag{9}$$

Using Young's equation (eq. 2)

$$W_a = \gamma_{LV} (1 + \cos \theta) \tag{10}$$

Using (9) and (5)

$$W_a = 2\phi(\gamma_L \gamma_S)^{1/2} \tag{11}$$

and (7) yields (neglecting the induction component):

$$W_a = 2(\gamma_L \gamma_S)^{1/2}[(L_S L_L)^{1/2} + (P_S P_L)^{1/2}] \tag{12}$$

Using equations (8), (10), and (12),

$$\frac{\gamma_L}{2}(1 + \cos\theta) = (\gamma_L{}^L\gamma_S{}^S)^{1/2} + (\gamma_L{}^P\gamma_S{}^P)^{1/2} \tag{13}$$

Measurement of $\cos\theta$ for a single liquid is not sufficient to determine $\gamma_S{}^L$ and $\gamma_S{}^P$, even when values of $\gamma_L{}^L$ and $\gamma_L{}^P$ are known, unless one of the empirical relationships discussed above (Section 2.1.2) is adopted. Measurement using two liquids, however, is sufficient, in principle. If several liquids are used, the system is overdetermined, and a set of values of $\gamma_S{}^L$ and $\gamma_S{}^P$ may be chosen to give the best fit to the data. One may take the average over a number of pairs of liquids, or, alternatively, make a plot of

$$\frac{\gamma_L(1 + \cos\theta)}{2(\gamma_L{}^P)^{1/2}} \quad \text{vs.} \quad (\gamma_L{}^L/\gamma_L{}^P)^{1/2}$$

The slope will yield $(\gamma_S{}^L)^{1/2}$ and the intercept $(\gamma_S{}^P)^{1/2}$. The best values may be obtained via a linear least-squares fit.

The geometric mean formulation of the interfacial tension (eq. 5) has been criticized by Wu (1971), who prefers a reciprocal mean form for low-energy systems, i.e.,

$$\gamma_{1,2} = \gamma_1 + \gamma_2 - \frac{4(\gamma_1{}^L\gamma_2{}^L)}{\gamma_1{}^L + \gamma_2{}^L} - \frac{4(\gamma_1{}^P\gamma_2{}^P)}{\gamma_1{}^P + \gamma_2{}^P} \tag{14}$$

This formulation, when combined with Young's equation (eq. 2), yields:

$$\frac{\gamma_L}{4}(1 + \cos\theta) = \frac{\gamma_L{}^L\gamma_S{}^L}{\gamma_L{}^L + \gamma_S{}^L} + \frac{\gamma_L{}^P\gamma_S{}^P}{\gamma_L{}^P + \gamma_S{}^P} \tag{15}$$

Wu has presented some data which indicate better consistency between surface energies derived from contact angle measurements and interfacial tensions of molten polymers when the reciprocal mean rather than the geometric mean is used (Wu, 1971).

Determination of polar and nonpolar components of the surface tension of liquids by means of contact angle against hydrocarbons, or by interfacial tension with other liquids, leads to different values of γ_L components when using the reciprocal mean than when using the geometric mean. Thus, the value obtained by Fowkes (1962) using the geometric mean approach for the nonpolar component for water is 21.8 mJ/m^2, and, therefore, the remainder of 51.0 mJ/m^2 would be assigned to the polar component. Wu, using the reciprocal mean, obtained very similar values, 22.1 and 50.7, respectively. The values for methylene iodide, however, are 48.1 and 2.3 according to Fowkes's geometric mean, and 44.1 and 6.7 according to Wu's treatment (Wu, 1971). A list of liquids, with the nonpolar components of the surface energy pertinent to the geometric mean treatment, was given by Kaelble and Moacanin (1977).

Both approaches have been criticized in that only two components of surface free energy are considered pertinent because all types of nondispersion interactions are lumped into the "polar" term. Good (1977, 1979) and Fowkes and Mostafa (1978) have pointed out the omission of other factors, especially hydrogen bonding and Lewis acid–base interactions, and that the induction term may not be separable. However, evaluation of additional parameters for solid surfaces will require evaluation of corresponding parameters in series of test liquids. Current reported values of $\gamma_S{}^L$ and $\gamma_S{}^P$, evaluated on the assumption that such separation is valid, may have to be reevaluated as better understanding of other factors is achieved. Thus, Kitazaki and Hata (1972) reported a study in which three separate series of liquids were used to determine three components of the γ_S. They were (1) hydrocarbons

(London force), (2) esters and halogenated hydrocarbons (polar), and (3) hydrogen-bonding liquids. Penn and Bowler (1981) suggested that a simple diagram which shows both advancing and receding contact angles (see Section 2.2.2) for a number of liquids on a solid could be inspected visually; they used a set of seven liquids. Like Kitazaki and Hata (1972), they optimized adhesion by matching the adhesive and adherend which seemed most similar in wetting properties over the broad range of liquids used, but felt that calculations of γ_S^P and γ_S^L made on the basis of advancing contact angles only are misleading because of hysteresis. The significance of contact angle hysteresis is still an active subject of investigation, and a significantly greater sample of data will be needed before this approach can be evaluated.

The Zisman plot approach (Section 2.1.1), making use of advancing contact angles formed with a particular set of liquids (including aliphatic hydrocarbons, aromatic and halogenated organic liquids, hydrogen-bonding organic liquids, and water), has been used for a large number of solid materials. γ_c is closely related to γ_S^L because the low γ_L members of the series are hydrocarbon and interact primarily via London forces, while the slope of the plot has been correlated with γ_S^P (Zisman, 1964; Baier, 1980). The position of water on this plot is often anomalous (probably because it is present in air, penetrates the air-equilibrated surfaces, and is strongly hydrogen bonding) and is not given much weight in the analysis. (This is in contrast to the bubble and two-liquid approach, in which water is frequently used as one of the liquids in contact with a water-equilibrated surface, as discussed in Section 2.2.2.)

The Zisman plot was developed for empirical correlation of contact angle data. If, however, the definition of the interaction parameter (eq. 5) is combined with equation (2), then the relation between them can be seen. Thus,

$$\cos \theta = 2\phi \, (\gamma_S / \gamma_L)^{1/2} - 1 \qquad (16)$$

and, for a series of homologous liquids, a plot of $\cos \theta$ vs. $(\gamma_L)^{-\frac{1}{2}}$ can be extrapolated to $\theta = 0$ for evaluation of γ_c. This plot has been described as more consistent than the Zisman plot (Good, 1979). Equation 16 also predicts a horizontal line for a plot of $\gamma_{LV} (1 + \cos \theta)^2 / 4$ vs. γ_{LV} (Good, 1977). An alternative plot for determination of γ_S^L and γ_S^P for solids and adsorbed films has also been presented by Birdi (1981).

2.1.4. Hydrated Materials

Despite the problems encountered with measurements on highly hydrated materials in air, van Oss and Gillman (1972a,b) and Neumann et al. (1974a) reported contact angles upon bacterial films obtained with aqueous sessile drops in air. They report that careful control of the exposure of the films to the atmosphere leads to a stable condition lasting long enough for reproducible measurements (see also Chapter 6, Section 4.5.2).

Baier (1980) and Weisberg and Dworkin (1983) have also reported contact angles obtained in air by the sessile drop technique on hydratable materials but use nonaqueous liquid drops, immiscible with water. The convenience of varying the contacting liquid easily allows limited samples to be characterized rapidly in this manner (Baier, 1980).

2.2. Techniques

2.2.1. Vertical Plate

A very comprehensive review of techniques of contact angle measurement and surface preparation has been published (Neumann and Good, 1979). These authors have pointed out that the height of the meniscus of a liquid at a vertical surface is a very sensitive method of determining contact angle and more precise than the more widely used method of observing the profile of a sessile drop. Both advancing and receding contact angles may be obtained by raising or lowering the vertical surface at known rates. The method was sensitive enough to allow heats of wetting to be determined from the usually unobserved temperature dependence of the contact angle, and transitions in the polymer were also evident in the contact angle temperature dependence (Neumann, 1978).

2.2.2. Captive Bubble and Two-Liquid Contact Angles

If the properties of a material which is normally in contact with a liquid medium are of interest, the possibility that the surface properties of the dry material might be misleading arises; the molecular arrangement at the wet surface might reflect the nature of fluid contacting the solid. Thus, to preserve the immersed configuration of the solid surface and to avoid evaporation of volatile liquids such as water or light hydrocarbons, contact angles at bubbles formed against immersed surfaces may be observed. Allowing bubbles, of a gas or an immiscible liquid, to form under a horizontal immersed surface of interest or holding a bubble on the end of a capillary against the material are possible. However, the presence of a layer of liquid which is slow to drain from the interface is a possible complication (Davies and Rideal, 1963, p. 422) which must be considered.

Results of bubble measurements differ from measurements made on dry materials using an advancing liquid front (Section 2.1.1). The advancing angle technique, on which a great amount of data are based, measures the interaction of the liquid with a solid which has been exposed to air, a low-energy environment. On exposure to water, or indeed any test fluid, the material surface may undergo changes, whose progress will depend on the mobility of the kinetic units within the solid, the degree of swelling, and the relative affinity of the various components of the material for the liquids in contact. Such changes induced in the solid are one cause of the hysteresis common to contact angle measurements, in that the values obtained for advancing and receding drops may not agree. Measurements made with underwater bubbles may be analogous to receding sessile drop measurements, in that a previously wet surface is under investigation, but in both cases, the kinetics of reorientation and diffusion and the difference in humidity near the edge of the drop prevent general statements concerning the equivalence of the methods.

Young's equation (eq. 2) for contact angle is as valid in solid–liquid–liquid systems as in solid–liquid–gas systems. The substitution of an immiscible liquid for the gas phase requires equilibration of the two liquid phases with each other and measurement of their interfacial tension in the equilibrated state. Hysteresis of contact angle is common in the two-liquid system and is usually at least as serious as in gas–liquid systems (Neumann and Good, 1979).

These techniques are especially valuable for samples such as hydrogels or layers of adsorbed highly hydrated substances which may be located on the surface of other materials, because the use of air bubbles or immiscible liquid droplets in a larger volume of aqueous medium will maintain the hydration state. Two-liquid systems allow greater flexibility in resolving the components of solid surface energy, and the realization that the dispersion force components of the surface tensions of water and octane are equal has led to the use of the water–octane–solid contact angle for evaluating polar force contributions to surface free energies of several polymers (Hamilton, 1974). Also, macromolecular adsorbed layers and bacterial films have been investigated by air bubble-in-water contact angle measurement in relation to bioattachment of bacteria (Fletcher and Marshall, 1982) and bryozoan larvae (Mihm *et al.*, 1981).

3. ELECTROKINETIC PROPERTIES AND MEASUREMENTS

3.1. Zeta Potentials

Electrical properties of surfaces may be expected to have significant effects upon interfacial interactions. The well-known DLVO theory of colloid stability (see Rutter and Vincent, 1980) places very strong emphasis upon the electrostatic term of the interaction (see also Chapter 11, Section 1.2). Its application to surfaces of plant and animal cells, microorganisms, and so-called "inert surfaces" is mentioned (at least in passing) in every work on the subject of bioadhesion or bioattachment (see, e.g., Pethica, 1980; Rutter and Vincent, 1980; see also Chapter 6, Section 3, and Chapter 7, Section 6.1).

Briefly, the DLVO theory proposes that the interaction between two particles is made up of two additive components. These are (1) an attractive component due to van der Waals forces and (2) a repulsive component (if the two particles bear the same charge sign), which is due to the overlap between the electrical double layers associated with the charge groups on the two particles. There are two particle separation distances at which attraction may occur. These are called the *primary minimum,* at small interparticle distances and where attraction forces are strong, and the *secondary minimum,* at relatively large interparticle distances and where attraction forces are weaker. These are separated by a repulsion barrier at intermediate distances. The total interaction at any given distance depends on the particle radii, surface potentials, Hamaker constants (a constant whose value depends on the compositions of the particles and the separating medium), and electrolyte concentrations. As electrolyte concentration is increased, the repulsion barrier decreases and disappears at high electrolyte concentrations.

The Stern concept of the electrical "double layer" evolved from previous simpler concepts. The concept recognizes that preferential adsorption properties and ionization of functional groups lead to a net charge distribution at the solid surface, and that electrical neutrality in a finite region requires that ions of opposite charge (counterions) in the liquid be more concentrated in its neighborhood than ions of the same sign of charge. Some of the counterions are held very closely to the surface, forming the Stern layer. The rest of the counterions form a diffuse ionic atmosphere, in which the positions of the ions are much less restricted with relatively free motion as a result of thermal agitation. The distribution of ions in this diffuse region is consistent with the requirements of the Gouy–Chapman approach to the diffuse double layer, and ions of both signs are present, although with more of the counterions, as required by the Poisson–Boltzmann equation.

The statistical thickness of the double layer is commonly signified by $1/\kappa$. When a charged particle in liquid is subjected to an electric field, as in electrophoresis, it moves toward the electrode of opposite sign. The region of solvent nearest the particle surface moves with it, carrying along the associated ions. Conversely, if a pressure gradient moves fluid along a charged interface, the fluid nearest the wall is retarded. The charge associated with the surface in flow is thus determined by the extent to which the ionic atmosphere extends beyond the "shearing plane," or effective boundary between the solution associated with the solid surface and the bulk liquid phase. The potential at the shearing plane is signified by ζ and called the zeta potential. The potential decay curves corresponding to a positively charged surface with, and without, adsorbed cations are shown in Fig. 2. Here, ψ_0, ψ_δ, and ζ signify the potentials at the particle surface, at the boundary between the fixed Stern layer and the diffuse part of the ionic atmosphere, and the potential at the shearing plane, respectively. If counterions are tightly bound to the extent that the particle sign is changed, as may be true if the bound counterions have higher valence than the particle

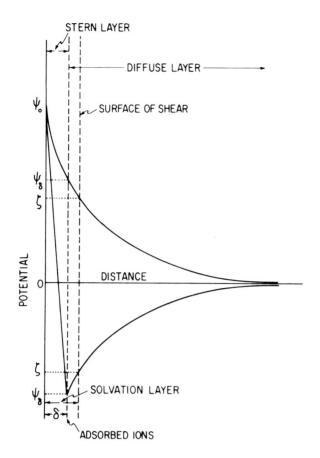

Figure 2. Potential decay curves for an electrical double layer associated with a particle of surface potential ψ_0. The lower curve results from strong adsorption within the Stern layer. For explanation, see text. From James (1979).

surface groups, then charge reversal within the Stern layer occurs, and the potentials are as shown in the lower curve of Fig. 2.

The Stern layer is normally one molecule thick, and the shearing plane outside it. Therefore, ζ is normally less than ψ_δ. Because the ionic atmosphere contracts as the electrolyte concentration increases, the value of ζ also decreases with electrolyte concentration, because less of the ionic imbalance remains outside the shear plane. The exact position of shear will also be dependent on surface roughness.

The measurement of the electrical charge on immersed surfaces, or of the electrical potential, and properties of the electrical double layer are generally approached differently for conductors than for insulating materials. Conducting metal electrodes may be treated by the powerful methods of traditional electrochemistry, as the potential within the electrode out to the surface may be well defined. Insulating materials do not have this property and are treated by methods dependent upon electrokinetics.

The two general methods for analysis of the surface charge and potential of nonconductors are microelectrophoresis (for microparticulates) and generation of an electric current or potential due to streaming of a liquid through a porous specimen or past a solid surface.

Particle electrophoresis requires particles small enough that sedimentation is not a significant factor. The theoretical treatments of electrophoretic mobility involve the ratio of particle radius, α, to ionic atmosphere radius, $1/\kappa$, as described in more detail in several reviews (Brinton and Lauffer, 1959; James,1979). The treatments of Smoluchowski, Henry,

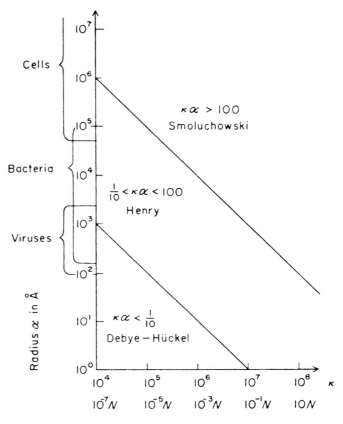

Figure 3. The product $\kappa\alpha$ plotted as a function of κ and α to show the regions where the equations of Smoluchowski, Henry, and Debye-Hückel are applicable. From Brinton and Lauffer (1959).

and Debye–Hückel are valid at different values of this ratio (Brinton and Lauffer, 1959). These regimes are shown in Fig. 3, where the lines show the limits of applicability of the treatments.

The relevant relations are:

1. The Smoluchowski Equation:

$$m = \frac{v}{x} = \frac{\epsilon \zeta}{4\pi \eta} \qquad (17)$$

where v = velocity, x = field strength, ϵ = the permittivity, η = the dynamic viscosity, and m = the mobility.

2. The Debye–Hückel Equation:

$$m = \frac{c\epsilon \zeta}{\eta} \qquad (18)$$

where c is a shape factor, with value $1/6\ \pi$ for spheres, and $1/8\ \pi$ or $1/4\ \pi$ for cylinders, depending on orientation.

3. The Henry Equation:

$$m = \frac{\epsilon\zeta}{6\pi\eta}f\,(\kappa\alpha) \tag{19}$$

where $f\,(\kappa\alpha)$ is a power series whose terms depend on shape and orientation with respect to the field. The Henry equation reduces to the Smoluchowski equation when $\kappa\alpha > 100$. These equations must be corrected for surface conductance, K_s, which leads to a significant correction if $K\alpha > 100$; neglecting K_s can lead to spuriously low values for ζ (Fricke and Curtis, 1936). The effect is to cause a lower potential gradient across the particle than expected. A similar effect occurs in streaming potential measurements where small capillary radii, or fine passages in pores, lead to low values of ζ. The specific conductance of the solution within the pores is different from that in the bulk solution, and fine capillary spaces allow overlap of the ionic atmospheres from opposite walls of the channel. The conductivity should be measured in the capillary system and determined with respect to the same capillary system containing solution of KC1 which is so concentrated that $1/\kappa$ is smaller than the capillary radius and the surface conductance K_s is negligible compared to that of the bulk solution, K_0 (Bull, 1964).

Henry (1931) showed that for an insulating particle

$$m = \frac{\epsilon\zeta}{4\pi\eta}\left[\alpha K_0(\alpha K_0 + K_s)^{-1}\right]$$

The equations predict that mobility should be a function of particle radius of curvature. As rather large cells and protein-coated particles are observed to have mobilities similar to much smaller structures, it has been suggested that the topography of cells incorporates a rough contour on a microscale, with projections that determine the effective radius of curvature. Similarly, one can postulate that adsorbed polymer coatings have sufficient textural structure to have a similar effect on mobility (Brinton and Lauffer, 1959; Bull, 1964).

3.2. Surface Charge Densities

The mobility measurement in electrophoresis, and the streaming potential, are transformed into ζ potential values using the relations (17) to (20). The ζ values are useful in themselves, in that comparisons may be made of charge properties of different materials contacting the same medium. However, the need to evaluate the charge on an absolute basis, so as to characterize materials (Chapter 7, Section 2.1), has led to use of the ζ potential to estimate charge, despite some questionable assumptions. The correct values of the viscosity and permittivity near the interface, and the position of the plane of shear between the fixed and the diffuse parts of the double layer, are somewhat uncertain. Nevertheless, assuming bulk values of viscosity and permittivity, and that the potential ζ is closely related to ψ_δ (the value at the inner edge of the diffuse part of the double layer), a calculation may be made. Thus, the surface charge density is given by the Chapman–Gouy treatment as

$$\sigma_D{}^2 = (N_a\epsilon kT/2000\pi)\,\{\Sigma\ c_i\,[\exp(-z_ie\zeta/kT) - 1]\} \tag{21}$$

where c_i and z_i are the concentration and charge of the ions of species i, respectively, N_a is Avogadro's number, and k is Boltzmann's constant. Ions tightly adsorbed are considered part of the charged particle itself in this treatment. Haydon (1961a,b) pointed out that the charge may be underestimated by a factor which can be as much as 2 in this treatment, because overlap of the fixed charge and counterion charge layers is not accounted for. As James (1979) points out: ''On account of these uncertainties, most experimenters now prefer to record and report mobility values and to discuss their results in terms of these experimentally measured quantities." Many experimental considerations are presented by James for the microelectrophoretic method, with references to previous workers. An appendix in which SI units are discussed may be of interest also.

3.3. Streaming Measurements

Streaming measurements may be made in channels whose walls consist of the material of interest. When streaming potentials are measured, turbulence becomes a complicating factor to be avoided. Groves and Sears (1975) have presented a streaming current method which allows turbulence in the flow system to be tolerated, but this method must be standardized via the streaming potential. The channels or pores must therefore be small enough and the driving pressure low enough to ensure laminar flow.

Most streaming potential measurements have been made in porous beds or in capillary tubes. Recently, however, equipment to allow measurements to be made on macroscopic flat surfaces has been designed. The convenience of the flat plate for other additional measurements, such as wettability or reflectance spectroscopy, allows a convenient series of measurements to be carried out on the same sample.

The ζ potential may be calculated for capillary flow in an insulator cylinder or porous plug from the equation (Bull, 1964, p. 297):

$$\zeta = (4\pi\eta/D) \ [E/(\Delta P)]K_c \tag{22}$$

where E = field strength, K_c = specific conductance of the liquid, η = liquid viscosity, D = double layer dielectric constant, and ΔP = pressure difference generating this flow in the capillary.

3.4. Conducting Solids and Liquids

Electrokinetic analyses have been primarily directed toward nonconducting solids, or gas bubbles, because charge transport through both the liquid medium and a conducting solid complicates the interpretation of results. Charge transport through a conducting solid is the result of three processes: (1) charge transfer across the liquid–solid interface, (2) electric current within the conductive solid, and (3) subsequent charge transfer across the solid–liquid interface. The interface processes, then, determine whether charge transfer occurs with conductors. Several experimental studies have shown that conductive metals behave similarly to insulators in electrophoresis, probably because they are quickly polarized so that no appreciable current flows through them. Successful measurements have been made using gold, silver, platinum, and steel in natural waters and in artificial analogs to natural waters (Hurd and Hackerman, 1955; Harrison and Elton, 1959; Benton and Sparks, 1966; Loeb and Niehof, 1977). Systems are also known where reversible charge transfer occurs, e.g., silver/silver nitrate, and, therefore, only very low electrophoretic mobility is observed (Henry, 1931).

The effect of solution conductivity upon electrokinetic measurements is manifested in two ways. First, the ionic atmosphere becomes thinner with increasing salt concentration. The result of this is a lessening of excess counterion charge outside the shear plane, and so lower ζ. A second effect is increased conductivity of the electrolyte, which makes it more difficult to maintain the potential gradients and charge separation required to perform the measurements. However, in the case of seawater, the palladium electrodes developed by Neihof and Schuldiner (1960) allowed the required current density to be achieved without gas generation during electrophoresis.

For streaming potential results to be meaningful, a flow profile over $> 90\%$ of the test area must be characteristic of well-developed laminar flow. This hydrodynamic requirement was discussed by DePalma (1980) and Van Wagenen and Andrade (1980) in the context of cells of both cylindrical and rectangular cross-sections. As Groves and Sears (1975) point out, however, this requirement may be relaxed when streaming current is measured, if proper calibrations are made (Sears and Groves, 1978). The streaming current method allows use of a low-resistance shunt to alleviate the problem of short-circuiting the streaming potential when high-conductivity solutions and capillaries are used, as pointed out by Carius and Dobias (1980), following on work of Eversole and Deardorff (1941), Hurd and Hackerman (1955), and Pravdic (1963). Hurd and Hackerman (1955) were able to use this stratagem to measure streaming currents with metal capillaries.

3.5. Surface Films

Modification of surface charge due to the presence of adsorbed films has been a subject of much electrokinetic study. Bull (1964) summarizes much work on the electrokinetic consequences of protein adsorption to particulates; the electrophoretic behavior of the coated particles becomes similar to the behavior of the protein. Thus, Neihof and Loeb (1972) found that the electrophoretic mobilities of different marine and estuarine natural particulates were similar. Later work, utilizing artificial seawater and particulates of a number of materials of widely different charge, demonstrated that dissolved organic matter adsorbed to immersed particulates and caused their mobilities to converge to a relatively small range of mobilities indicating that all had a similar moderate negative charge (Neihof and Loeb, 1974; Loeb and Neihof, 1975).Studies by Hunter and Liss (1979) on marine and estuarine particulates indicated similar negative mobilities, which were also attributed to surface coatings. Similar findings have been reported using a marine aquarium and DePalma's streaming potential technique after adsorption to flat germanium plates occurred (DePalma and Baier, 1977), although in a study near an industrial area (Pravdic, 1970, and private communication) and in certain sediments measured by plug streaming potential (Hunter and Liss, 1979), positive charges were found.

Investigations of flat-plate specimens of a number of polymer films (Van Wagenen *et al.*, 1981) indicated that the ζ potential followed the change in charge expected from the chemical composition but with a bias toward $\zeta < 0$. An explanation could perhaps be preferential adsorption of anions, although the bias is not well understood, and it was suggested that an alternative charge transfer process proposed by Fowkes and Heilscher (1980) might be operating.

4. SURFACE CHEMICAL ANALYSIS

Surface analysis, by methods developed relatively recently, is being introduced into general use at a spectacular rate. A review (Turner and Colton, 1982) which did not completely cover a 2-year period, but emphasized only what the authors considered significant developments, contained more than 900 references. Although it is not possible to discuss these methods in this limited space, they should be briefly mentioned, as they are potentially extremely valuable in the surface analysis of the various components involved in bioadhesion phenomena.

The principal methods for surface analysis are (1) electron spectroscopy for chemical analysis (ESCA), also called X-ray photoelectron spectroscopy (XPS); (2) Auger electron spectroscopy (AES); (3) secondary ion mass spectroscopy (SIMS); and (4) ion-scattering spectroscopy (ISS). Figure 4 indicates the basic features of these methods. They are sensitive

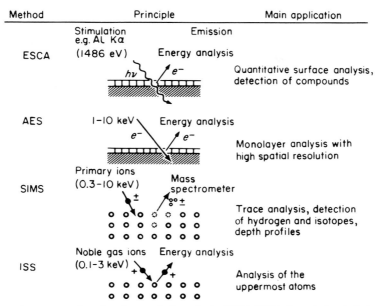

Figure 4. Schematic representation of surface analysis methods. ESCA (XPS) X-ray-induced photo- and Auger-electron spectroscopy), AES (electron-induced Auger electron spectroscopy), SIMS (secondary ion mass spectroscopy), and ISS (ion-scattering spectrometry). From Holm and Storp (1980).

techniques which allow analysis of elemental composition of surfaces and provide some information on the types of bonds present. Although powerful techniques, which can provide much useful information on solid surface properties, they are all normally carried out in vacuum chambers. This limits the samples which can be handled and unfortunately thus far, excludes study of the normal attachment surface, i.e., one suspended in aqueous solution.

5. ADSORPTION AND CONTAMINATION

Adsorption of soluble materials on substrata may affect adhesion and attachment, and studies of the effect of adsorption on wettability and charge have frequently been made (e.g., Loeb and Niehof, 1975; Mihm et al., 1981; Baier, 1982; Fletcher and Marshall, 1982). The adsorbed layers can serve as surfaces for attachment of subsequently arriving organisms, or may themselves be displaced or diluted by later arrivals (Chapter 6, Section 4.8).

The effect of the substratum surface properties upon the adsorbed molecules can be significant and, therefore, be expected to influence subsequent interactions. Thus, Mattson and Jones (1976), using infrared transparent internal reflection elements of the semiconductor germanium, showed the quantity of the protein fibrinogen adsorbed increased sharply when the semiconductor's electrical potential was raised beyond the "onset potential," and the infrared spectrum of the adsorbed protein indicated a conformational change as the potential was increased. Morrissey et al. (1976) confirmed both the existence of the onset potential and a conformational change in the adsorbed protein by ellipsometry on a platinum substratum. The sudden onset of heavy adsorption raises questions concerning long-range electrostatic attraction, and the spectral and ellipsometric changes are clear evidence that the nature of the adsorbed protein changed with charging. Infrared evidence of the dependence of adsorbed protein conformation on substratum surface energy was also obtained by surface pretreatment of internal reflection elements before adsorption of β-lactoglobulin from solution (Baier *et al.*, 1971).

A recent overview introduced several studies of adsorbed proteins by Fourier transform infrared spectroscopy, intrinsic fluorescence, and circular dichroism, among other techniques, at a symposium on interfacial phenomena on biomaterials (Horbett, 1982).

Because the difference between adsorption and modification of surfaces on the one hand and contamination on the other is predominantly in the experimenter's perception, studies of the phenomena are similar. As modern methods of surface analysis become more widely available, control of surface contamination and procedures for cleaning may be evaluated to a much greater extent than previously. As examples of the insight which can now be gained into contamination, studies of surfaces to be used for tissue culture and their preparation are instructive. Cleaning of high-energy surfaces by plasma discharge techniques is very convenient, as no extensive rinsing is required to remove conventional chemical agents at the last step and observation in the vacuum system by ESCA, for example, is possible (Mittal, 1979). Samples cleaned by this method have been exposed for controlled times to laboratory environments and reanalyzed (Ratner *et al.*, 1979). The conclusion of these studies was that even very short exposure to laboratory atmospheres results in more than a monolayer of carbon-containing contaminants. There was, however, no correlation between the amount of this material and mouse embryo fibroblast adhesion, despite the fact that monolayers of protein are known to be inhibitory. A similar effect was observed for adsorption of ambient dissolved matter from an artificial seawater aquarium (Mihm *et al.*, 1981). Attachment of invertebrate larvae (bryozoans) was not correlated with changes in contact angle after exposure of substrata to seawater from which particulate matter had been removed, although correlation with wettability had been established for fresh substrata. Among other possible reasons for these results, one may suggest that some adventitious contaminants may be easily displaced by exudates of the organisms.

6. OVERVIEW

The wide variety of colonized surfaces and the degrees of preference shown by attaching organisms indicate a corresponding variety of bioattachment and surface attachment processes. General concepts, such as the work of adhesion and interfacial free energy, have validity, but their evaluation using simple probes may not adequately reflect the many possibilities present at a complex organism's surface. The separability of nonpolar forces, and the availability of nonpolar materials, allowed attempts to explain differences in bioadhesion to low-energy surfaces (Baier, 1970, 1982), predominantly in terms of nonpolar interactions (Schrader, 1982). Dexter (1979) has presented a formulation which takes explicit account of the chemistry of the attaching organisms, adsorbed materials, and substrata, as reflected in their interaction parameters, ϕ, in evaluating the system work of adhesion. The values assigned to ϕ will have to be determined in each case and reflect the variety of interactions involved.

The newer methods of surface characterization coming into general use, as well as the improvements in techniques familiar to most workers in the field of bioattachment, will certainly raise doubt about the validity of old correlations as more parameters characterizing surfaces are added to descriptions of materials. However, the fact that a more complete characterization for any given material sample may now be obtained should allow increasing control in isolating the variables in future studies and give greater guidance to theoretical studies of bioadhesion.

ACKNOWLEDGMENTS. I am grateful to Drs. R. Jones, N. Neihof, and N. Turner for critical reading of portions of the manuscript and their suggestions.

REFERENCES

Baier, R. E., 1970, Surface properties influencing biological adhesion, in: *Adhesion in Biological Systems* (R. S. Manley, ed.), Academic Press, New York, pp. 15–48.

Baier, R. E., 1980, Substrata influences on adhesion of microorganisms and their resultant new surface properties, in: *Adsorption of Microorganisms to Surfaces* (G. Bitton and K. C. Marshall, eds.), Wiley, New York, pp. 59–104.

Baier, R. E., 1982, Conditioning surfaces to suit the biomedical environment, *J. Biomech. Eng.* **104:**257–271.

Baier, R. E., Loeb, G. I., and Wallace, G. T., 1971, Role of an artificial boundary in modifying blood proteins, *Fed. Proc. Fed. Am. Soc. Exp. Biol.* **30:**1523–1538.

Benton, D. P., and Sparks, B. D., 1966, Adsorption from aqueous solutions of ionic surfactants by gold, *Trans. Faraday Soc.* **62:**3244–3252.

Birdi, K. S., 1981, Cell adhesion on solids and the role of surface forces, *J. Theor. Biol.* **93:**1–5.

Brinton, C. C., and Lauffer, M. A., 1959, The electrophoresis of viruses, bacteria, and cells and the microscope method of electrophoresis, in: *Electrophoresis* (M. Bier, ed.), Academic Press, New York, pp. 427–492.

Bull, H. B., 1964, *An Introduction to Physical Biochemistry,* Davis, Philadelphia.

Carius, W., and Dobias, B., 1980, Semi-automatic apparatus for the measurement of streaming potential and streaming current, *Colloid Polym. Sci.* **259:**470–471.

Davies, J. T., and Rideal, E. K., 1963, *Interfacial Phenomena,* 2nd ed., Academic Press, New York.

DePalma, V. A., 1980, Apparatus for zeta-potential measurement of rectangular flow cells, *Rev. Sci. Instrum.* **51:**1390–1395.

DePalma, V., and Baier, R. E., 1977, Microfouling of metallic and coated metallic flow surfaces in model heat exchange cells, in: *Proceedings of Ocean Thermal Energy Biofouling and Corrosion Symposium, Seattle,* U.S. Department of Energy, pp. 89–106.

Dexter, S. C., 1979, Influence of substratum critical surface tension on bacterial adhesion, *J. Colloid Interface Sci.* **70:**346–354.

Eversole, W. G., and Deardorff, D. L., 1941, Flow potentials through metals, *J. Phys. Chem.* **45:**236–241.

Fletcher, M., and Loeb, G. I., 1979, The influence of substratum characteristics on the attachment of a marine pseudonomad to solid surfaces, *Appl. Environ. Microbiol.* **37:**67–72.

Fletcher, M., and Marshall, K. C., 1982, Bubble contact angle method for evaluating substratum-interfacial characteristics and its relevance to bacterial attachment, *Appl. Environ. Microbiol.* **44:**184–192.

Fletcher, M., Latham, M. J., Lynch, J. M., and Rutter, P. R., 1980, The characteristics of interfaces and their role in microbial attachment, in: *Microbial Adhesion to Surfaces* (R. C. W. Berkeley, J. M. Lynch, J. Melling, P. R. Rutter, and B. Vincent, eds.), Horwood, Chichester, pp. 67–78.

Fowkes, F. M., 1962, Determination of interfacial tensions, contact angles and dispersion forces in surfaces by assuming additivity of intermolecular interactions in surfaces, *J. Phys. Chem.* **66:**382–388.

Fowkes, F. M., 1964, Dispersion-force contributions to surface and interfacial tensions, contact angles, and heats of immersion, *Adv. Chem. Ser.* **43:**99–111.

Fowkes, F. M., and Heilscher, F. H., 1980, Electron injection from water into hydrocarbons and polymers, *Polym. Prepr. Am. Chem. Soc. Div. Polym. Chem.* **42:**164–174.

Fowkes, F. M., and Mostafa, M. A., 1978, Acid–base interactions in polymer adsorption, *Ind. Eng. Chem. Prod. Dev.* **17:**3–7.

Fricke, H., and Curtis, H. J., 1936, Determination of surface conductance from measurements on suspensions of spherical particles, *J. Phys. Chem.* **40:**715–722.

Gerson, D. F., 1982, An empirical equation-of-state for solid–fluid interfacial free energies, *Colloid Polym. Sci.* **260:**539–544.

Girifalco, L. A., and Good, R. J., 1957, A theory for estimation of surface and interfacial energies, *J. Phys. Chem.* **61:**904–909.

Good, R. J., 1977, Surface-free energy of solids and liquids: Thermodynamics, molecular forces and structure, *J. Colloid Interface Sci.* **59:**398–419.

Good, R. J., 1979, Contact angles and the surface free energy of solids, in: *Surface and Colloid Science,* Volume II (R. J. Good and R. R. Stromberg, eds.), Plenum Press, New York, pp. 1–29.

Good, R. J., and Girifalco, L. A., 1960, A theory for estimation of surface and interfacial energies. III. Estimation of surface energies of solids from contact angle data, *J. Phys. Chem.* **64:**561–570.

Groves, J. N., and Sears, A. R., 1975, Alternating streaming current measurements, *J. Colloid Interface Sci.* **53:**83–89.

Hamilton, W. C., 1974, Measurement of the polar force contribution to adhesive bonding, *J. Colloid Interface Sci.* **47:**672–675.

Harrison, J. T., and Elton, G. A. H., 1959, Electrophoresis of gold and silver particles, *J. Chem. Soc.* **162:**3838–3843.

Haydon, D. A., 1961a, The surface charge of cells and some other small particles as indicated by electrophoresis. I. Zeta-potential–surface charge relationships, *Biochim. Biophys. Acta* **50:**450–456.

Haydon, D. A., 1961b, The surface charge of cells and some other small particles as indicated by electrophoresis. II. Interpretation of electrophoretic charge, *Biochim. Biophys. Acta* **50:**457–462.

Henry, D. C., 1931, The cataphoresis of suspended particles, *Proc. R. Soc. London Ser. A* **133:**106–129.

Holm, R., and Storp, S., 1980, Surface and interface analysis in polymer technology: A review, *Surf. Interf. Anal.* **2:**96–106.

Horbett, T. A., 1982, Protein adsorption on biomaterials, in: *Biomaterials: Interfacial Phenomena and Applications* (S. L. Cooper, N. A. Pappas, A. S. Hoffman, and B. D. Ratner, eds.), American Chemical Society, Washington, D.C., pp, 233–244.

Hunter, K. A., and Liss, P. S., 1979, The surface charge of suspended particles in estuarine and coastal waters, *Nature (London)* **282:**823–825.

Hurd, R. M., and Hackerman, N., 1955, Electrokinetic potentials of bulk metals by streaming current measurements, *J. Electrochem. Soc.* **102:**594–597.

James, A. M., 1979, Electrophoresis of particles in suspension, in: *Colloid and Surface Science,* Volume 11 (R. J. Good and R. R. Stromberg, eds.), Plenum Press, New York, pp. 121–185.

Kaelble, D. H., 1971, *Physical Chemistry of Adhesion,* Wiley, New York.

Kaeble, D. H., and Moacanin, J., 1977, A surface energy analysis of bioadhesion, *Polymer* **18:**475–482.

Kitazaki, Y., and Hata, T, 1972, Surface chemical criteria for adhesion, *J. Adhes.* **4:**123–132.

Loeb, G. I., and Niehof, R. A., 1975, Marine conditioning films, in: *Applied Chemistry at Protein Interfaces* (R. E. Baier, ed.), American Chemical Society, Washington, D.C., pp. 319–335.

Loeb, G. I., and Neihof, R. A., 1977, Adsorption of an organic film at the platinum–seawater interface, *J. Mar. Res.* **5:**283–291.

Mattson, J. S., and Jones, T. T., 1976, Infrared spectrophotometric observations of the absorption of fibrinogen from solution at optically transparent carbon film electrode surfaces, *Anal. Chem.* **48:**2164–2167.

Mihm, J. W., Banta, W. C., and Loeb, G. I., 1981, Effects of adsorbed organic and primary fouling films on bryozoan settlement, *J. Exp. Mar. Biol. Ecol.* **54:**167–179.

Mittal, K. L., 1979, Surface contamination: An overview, in: *Surface Contamination: Genesis, Detection, and Control* (K. L. Mittal, ed.), Plenum Press, New York, pp. 3–46.

Morrissey, B. W., Smith, L. E., Stromberg, R. R., and Fenstermaker, C. A., 1976, Ellipsometric investigation of the effect of potential on blood protein conformation and absorbance, *J. Colloid Interface Sci.* **56:**557–563.

Neihof, R. A., and Loeb, G. I., 1972, The surface charge of particulate matter in seawater, *Limnol. Oceanogr.* **17:**7–16.

Neihof, R. A., and Loeb, G. I., 1974, Dissolved organic matter in seawater and the electric charge at immersed surfaces, *J. Mar. Res.* **32:**5–12.

Neihof, R. A., and Schuldiner, S., 1960, Simple non-gassing electrodes for use in electrophoresis, *Nature (London)* **185:**526.

Neumann, A. W., 1978, Contact angles: Introductory lecture, in: *Wetting, Spreading and Adhesion* (J. F. Padday, ed.), Academic Press, New York, pp. 3–33.

Neumann, A. W., and Good, R. J., 1979, Techniques of measuring contact angles, in: *Surface and Colloid Science,* Volume 11 (R. J. Good and R. R. Stromberg, eds.), Plenum Press, New York, pp. 31–91.

Neumann, A. W., Gillman, C. F., and van Oss, C. J., 1974a, Phagocytosis and surface free energies, *J. Electroanal. Chem.* **49:**393–400.

Neumann, A. W., Good, R. J., Hope, C. J., and Sejpal, M., 1974b, An equation of state approach to determine surface tensions of low-energy solids from contact angles, *J. Colloid Interface Sci.* **49:**291–304.

Neumann, A. W., Absolom, D. R., Francis, D. W., and van Oss, C. J, 1980, Conversion tables of contact angles to surface tensions for use in determining the contribution of the van der Waals attraction or repulsion to various separation properties, *Sep. Purif. Methods* **9:**69–163.

Penn, L.S., and Bowler, E. R., 1981, A new approach to surface energy characterization, *Surf. Interface Anal.* **3:**161–164.

Pethica, B. A., 1980, Microbial and cell adhesion, in: *Microbial Adhesion to Surfaces* (R. C. W. Berkeley, J. M. Lynch, J. Melling, P. R. Rutter, and B. Vincent, eds.), Horwood, Chichester, pp. 19–46.

Pravdic, V., 1963, Electrokinetic Studies in disperse systems. VI. Streaming current measurements, *Croat. Chem. Acta* **35**:233–237.

Pravdic, 1970, Surface charge characterization of sea sediments, *Limnol. Oceanogr.* **15**:230–233.

Ratner, B. D., Rosen, J. J., Hoffman, A. S., and Scharpen, L. H., 1979, An ESCA study of surface contaminants on glass substrates for cell adhesion, in: *Surface Contamination: Genesis, Detection, and Control* (K. L. Mittal, ed.), Plenum Press, New York, pp. 669–686.

Rutter, P. R., and Vincent, B., 1980, The adhesion of microorganisms to surfaces: Physico-chemical aspects, in: *Microbial Adhesion to Surfaces* (R. C. W. Berkeley, J. M. Lynch, J. Melling, P. R. Rutter, and B. Vincent, eds.), Horwood, Chichester, pp. 79–92.

Schrader, M. E., 1982, On adhesion of biological substances to low energy solid surfaces, *J. Colloid Interface Sci.* **88**:296–297.

Sears, A. R., and Groves, J. N., 1978, The use of oscillating laminar-flow streaming potential measurements to determine the zeta potential of a capillary surface, *J. Colloid Interface Sci.* **65**:479–482.

Shafrin, E., 1966, Critical surface tension of polymers, in: *Polymer Handbook* (J. Brandrup and E. H. Immergut, eds.), Wiley, New York, pp. III-113–III-114.

Tadros, H. F., 1980, Particle–surface adhesion, in: *Microbial Adhesion to Surfaces* (R. C. W. Berkeley, J. M. Lynch, J. Melling, P. R. Rutter, and B. Vincent, eds.), Horwood, Chichester, pp. 93–116.

Turner, N., and Colton, R. J., 1982, Surface analysis, *Anal. Chem. Rev.* **54**:293R–322R.

van Oss, C. J., and Gillman, C. F., 1972a, Phagocytosis as a surface phenomenon. I. Contact angles and phagocytosis of nonopsonised bacteria, *J. Reticuloendothel. Soc.* **12**:283–292.

van Oss, C. J., and Gillman, C. F., 1972b, Phagocytosis as a surface phenomenon. II. Interfacial energies and phagocytosis, *J. Reticuloendothel. Soc.* **12**:424–425.

Van Wagenen, R. A., and Andrade, J. D., 1980, Flat plate streaming potential investigations: Hydrodynamics and electrokinetic equivalency, *J. Colloid Interface Sci.* **76**:305–314.

Van Wagenin, R. A., Coleman, D. L., King, R. N., Triolo, P., Brostrom, L., Smith, L. M., Gregonis, D. E., and Andrade, J. D., 1981, Streaming potential investigations: Polymer thin films, *J. Colloid Interface Sci.* **84**:155–162.

Ward, J. B., and Berkeley, R. C. W., 1980, The microbial cell surface and adhesion, in: *Microbial Adhesion to Surfaces* (R. C. W. Berkeley, J. M. Lynch, J. Melling, P. R. Rutter, and B. Vincent, eds.), Horwood, Chichester, pp. 47–66.

Weisberg, D. S., and Dworkin, M., 1983, Method for measuring changes in surface tension on agar, *Appl. Environ. Microbiol.* **45**:1378–1381.

Weiss, L., 1972, Interactions between normal and malignant cells, in: *The Chemistry of Biosurfaces*, Vol. 2 (M. L. Hair, ed.), Marcel Dekker, New York, pp. 377–448.

Wu, S., 1971, Calculation of the interfacial tensions in polymer systems, *J. Polym. Sci.* **C34**:19–30.

Zisman, W. A., 1964, Relation of the equilibrium contact angle to liquid and solid constitution, in: *Contact Angle, Wettability and Adhesion* (R. F. Gould, ed.), American Chemical Society, Washington, D.C., pp. 1–51.

II

Mechanisms of Adhesion

6

Mechanisms of Bacterial Adhesion at Solid–Water Interfaces

KEVIN C. MARSHALL

1. INTRODUCTION

Any consideration of the mechanisms whereby bacteria adhere to solid surfaces must take account of the surface properties of the bacteria as well as those of the substrata concerned. Properties of bacterial surfaces have been dealt with by Wicken (Chapter 2) and those of apparently inert substratum surfaces by Loeb (Chapter 5). Bacteria can be considered as living colloidal particles (Marshall, 1973, 1976). They range in size from about 0.2 μm to several micrometers in length, and the majority of bacteria are about 1.0 μm in length or diameter. The density of a bacterial cell is only slightly greater than that of water. Stable suspensions of bacteria in dilute electrolytes (e.g., nutrient media) result, in part, from a mutual electrostatic repulsion between like charges on the bacterial surfaces. As revealed by electrophoretic measurements (Fig. 1), bacteria possess a net negative surface charge at pH values found in most natural habitats. A charge reversal at low pH values is indicative of the presence of some charged basic (amino) groups, which are revealed when dissociation of acidic (carboxyl, phosphate) groups is suppressed at the low pH values (Plummer and James, 1961). The positively charged groups appear to be evenly distributed over the surface of *Flexibacter aurantiacus* CW7 (Marshall and Cruickshank, 1973). Bacteria also exhibit variations in overall surface free energy of the cells, with some bacteria possessing relatively hydrophilic surfaces and others relatively hydrophobic surfaces (Mudd and Mudd, 1924; Magnusson *et al.*, 1977; Dahlbäck *et al.*, 1981). The distribution of hydrophobic sites on bacterial cells is not necessarily uniform and may result in a preferred orientation of certain bacteria at interfaces (Marshall and Cruickshank, 1973).

Bacteria are not inert colloidal particles, but are living organisms capable of metabolism, growth, and, in some instances, independent motion. An expression of the metabolism of bacteria is the production of variable amounts of polymers that are bound to the exterior of the cell envelope (Costerton *et al.*, 1981). These polymers, along with flagella, pili, and

Kevin C. Marshall • School of Microbiology, University of New South Wales, Kensington, New South Wales 2033, Australia.

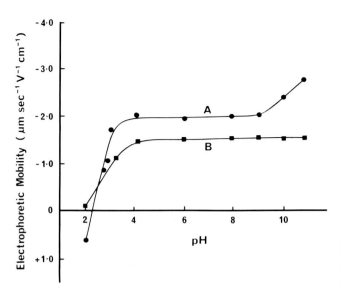

Figure 1. Electrophoretic mobility versus pH for (A) *Rhizobium trifolii* SU297A and (B) *R. lupini* UT12. Adapted from Marshall (1967).

fimbriae (Chapter 2), are important in the final process of adhesion of bacteria to solid surfaces. The formation of these extracellular polymers by bacteria varies with species, age, and growth conditions and, hence, can have a bearing on the efficiency and selectivity of adhesion occurring between different species in natural habitats.

2. TRANSPORT OF BACTERIA TO SURFACES

Before addressing the question of how bacteria adhere to solid surfaces, it is relevant to consider the means by which the bacteria are transported from the bulk liquid to the vicinity of the solid–water interface.

2.1. Conditions in the Aqueous Phase

Adhesion of bacteria to solid surfaces occurs both in quiescent waters and in conditions of turbulent flow. The mechanisms of transport of bacteria to surfaces are certainly different under these different conditions. Bryers and Characklis (1981) have shown that the rate of overall bacterial film development on surfaces in turbulent flow conditions increases with the biomass concentration dispersed in the bulk phase. This indicates that the particle flux to a surface is directly proportional to the bulk particle concentration as predicted by mass transport and particle deposition theories. Consequently, in very dilute suspensions of bacteria (as in open ocean waters) transport is probably the rate-limiting step in the process of bacterial deposition at the surface (Characklis, 1981a). That the growth conditions of the bacteria in the dispersed phase have an important effect on adhesion characteristics was reported by Leech and Hefford (1980). The initial rates of deposition of inert polystyrene latex particles and of cells of *Streptococcus sanguis* grown slowly ($D = 0.04$/hr) increased with increasing flow rate, yet with cells grown at faster rates ($D = 0.2$ and 0.5/hr) the deposition rate decreased with increasing flow rate. It is obvious from such results that the physiological state of the bacteria plays an important role in the final adhesion process.

Under turbulent flow conditions a zone of relatively still water exists near a solid surface. This is termed the *viscous* (or boundary) *sublayer* and its thickness is dependent upon the magnitude of the fluid shear rate. For a shear stress of 6.5 to 7.9 N/m², the thickness of the viscous sublayer is approximately 40 μm (Characklis, 1981b). Bacteria in the bulk liquid must penetrate this viscous sublayer in order to be deposited at the solid surface.

2.2. Transport Mechanisms

Transport of molecules and small particles (0.01 to 0.1 μm) may be described satisfactorily in terms of diffusion. Particles of bacterial size are transported by a number of mechanisms, but in a turbulent flow regime, transport is largely by fluid dynamic forces (Characklis, 1981a).

2.2.1. Sedimentation

This process is only of significance in low-shear systems and with relatively large particles, such as very large bacteria or aggregates (flocs) of normal bacteria. As stated in Section 1, most bacteria form stable suspensions and only sediment when the system becomes destabilized as, for example, in floc formation. In a flowing system, for instance, Leech and Hefford (1980) observed the deposition of single cells on both the top and the bottom of a rectangular glass capillary tube, but aggregates only appeared on the bottom surface. According to Characklis (1981b), sedimentation of bacteria is unlikely to occur under turbulent flow conditions.

2.2.2. Chemotaxis

Many bacteria are motile in the bulk aqueous phase as a result of the propulsive action of flagella. A maximum velocity of 56 μm/sec has been recorded for *Pseudomonas aeruginosa* by Vaituzis and Doetsch (1969). Motile bacteria are capable of displaying a positive chemotactic response to certain nutrient sources (Adler, 1969) and, as such, can respond to a nutrient concentration gradient that exists at a solid–water interface. Thus, bacteria may actively swim toward a surface and ultimately be held by the attractive forces operative near the surface (Marshall *et al.*, 1971a; see also Chapter 15, Section 2.2.1). The role of chemotactic responses in the formation of primary surface films by bacteria has been considered by Young and Mitchell (1973a). Bacteria may also exhibit a negative chemotactic response where inhibitory substances, such as hydrogen ions or antibiotics, accumulate at the solid–water interface (Doetsch and Seymour, 1970; Seymour and Doetsch, 1973; Young and Mitchell, 1973b). Chemotactic responses are probably not of significance in turbulent flow conditions (Characklis, 1981b), but are probably important in transport through the viscous sublayer (Characklis, 1981a).

It should be obvious that nonmotile bacteria are incapable of chemotaxis and these bacteria must be transported to the surface by other mechanisms.

2.2.3. Brownian Motion

The magnitude of displacement of a particle by Brownian motion is dependent upon particle size. The Brownian displacement (Δ) of a particle is given by the equation

$$\Delta = 2Dt \tag{1}$$

where t is the time in seconds and D is the diffusion coefficient given by

$$D = RT/6\pi\eta aN \tag{2}$$

Table I. Diffusion Coefficients (*D*) and Brownian
Displacement (Δ) Values Calculated for Uncharged
Spheres in Water at 20°C[a]

Particle radius	Diffusion coefficient (cm^2/sec)	Brownian displacement (cm/hr)
1 nm	2.1×10^{-6}	1.23×10^{-1}
10 nm	2.1×10^{-7}	3.90×10^{-2}
0.1 μm	2.1×10^{-8}	1.23×10^{-2}
1.0 μm	2.1×10^{-9}	3.90×10^{-3}

[a]Adapted from Shaw (1966).

where R is the gas constant, T the absolute temperature, η the viscosity of the medium, a the effective radius of the particle, and N is Avogadro's number. Values for D and Δ are presented in Table I. Most bacteria have an effective radius of less than 1.0 μm and exhibit a significant degree of Brownian displacement when viewed under quiescent conditions under a microscope. Larger bacteria do not show this form of motion. Brownian motion probably contributes little to the transport of bacteria in turbulent flow (Characklis, 1981b), but should be a significant form of transport within the viscous sublayer.

2.2.4. Cell Surface Hydrophobicity

Mudd and Mudd (1924) demonstrated that bacteria can vary appreciably in their degree of cell surface hydrophobicity, and these observations have been confirmed and extended by more recent studies on the relative hydrophobicities of bacterial surfaces (Marshall and Cruickshank, 1973; Magnusson *et al.*, 1977; Rosenberg *et al.*, 1980; Dahlbäck *et al.*, 1981). According to Marshall and Cruickshank (1973), hydrophobic bacteria could reasonably be considered as being rejected from the aqueous phase and attracted toward any nonaqueous phase, including solid surfaces. Consequently, hydrophobicity could provide a means whereby bacteria are attracted toward solid surfaces and, in certain instances, would provide a means for a particular orientation of some bacteria at interfaces (Marshall and Cruickshank, 1973).

2.2.5. Fluid Dynamic Forces

Bacteria in turbulent flow systems are dispersed by eddy diffusion in the turbulent core region, thus maintaining a uniform concentration in the bulk of fluid. Such eddy diffusion may only transport the bacteria to the region of the viscous sublayer. If the bacteria are traveling faster than the fluid in the region of the wall, a lift force directs the bacteria toward the wall (Characklis, 1981a). In the viscous sublayer, the bacteria encounter significant frictional drag forces, which gradually slow a bacterium down as it approaches the surface. There is an additional fluid drainage force resulting from the resistance a bacterium encounters near the wall due to pressure in the draining fluid film between the wall and the approaching bacterial surface (Characklis, 1981a). Obviously, lift, drainage, and drag forces need to be included in any general model for bacterial deposition at a solid surface from a turbulent flow system.

Apart from lift forces, another mechanism for directing particles of bacterial size through the viscous sublayer to the wall is that of turbulent "downsweeps" (Lister, 1981). These spontaneous bursts of turbulence penetrate the viscous sublayer and provide a significant fluid mechanical force to direct the bacteria to the solid surface.

The above mechanisms provide the means for transporting bacteria from the bulk aqueous phase to the vicinity of the wall. The final adhesion of bacteria to solid surfaces involves a consideration of various aspects of colloid chemistry along with an understanding of some of the physiological responses of different bacteria to the conditions existing both in the bulk liquid and at the solid–water interface.

3. REVERSIBLE ADHESION

Reversible adhesion is an instantaneous attraction by long-range forces holding bacteria near a surface, so that the bacteria continue to exhibit Brownian motion and can be readily removed from the surface by the shearing effects of a water jet or by the violent rotational movements of motile bacteria (Marshall *et al.*, 1971a). This initial phase of adhesion was first described by ZoBell (1943) and has been studied in some detail by Marshall *et al.* (1971a), van Houte and Upeslacis (1976), Hamada (1977), and Jones *et al.* (1980, 1981).

3.1. Long-Range Forces

Most solid surfaces found in natural habitats possess a net negative surface charge. Superficially, there appears to be a problem in explaining how negatively charged bacteria can be attracted to a surface of similar charge. Because of the charge on these surfaces, a potential exists between the surfaces and the bulk aqueous phase (Chapter 5, Section 3). To counterbalance the surface charge, ions of the opposite charge (counter- or gegen ions) are loosely attracted to the surface to form a diffuse double layer of ions. When two negatively charged bodies are in close association they may be repelled from or attracted to each other. This effect depends on the thickness of the double layer which, in turn, is dependent on the valency and concentration of the electrolyte. The extent of interaction between particles under such conditions is predicted by the colloid stability, or DLVO, theory, developed independently by Derjaquin and Landau (1941) and Verwey and Overbeek (1948) (see also Chapter 11, Section 1.2). Potential problems in the application of the DLVO theory to biological systems have been considered in some detail by Pethica (1980) and Tadros (1980). The basic principles arising from the DLVO theory, however, do help explain the observed behavior of bacteria when they arrive at the vicinity of the solid–water interface.

The potential energy of interaction with increasing distance between a bacterium and a solid surface has been estimated from the magnitude of both the London–van der Waals attractive energies and the electrical repulsion energies in the overlapping double layers surrounding the bacterial and substratum surfaces (Marshall *et al.*, 1971a; Jones *et al.*, 1981). The derivation of the DLVO theory employed (Weiss and Harlos, 1972) is applicable only to the analysis of weak long-range interactions of the reversible type. Marshall *et al.* (1971a) obtained the energies of interaction between *Achromobacter* R8 and a glass surface for different double layer thicknesses ($1/\kappa$) from the following values: radius of curvature of glass $= 10^4$ nm and of the bacterium $= 0.4$ μm; surface potential of glass $= -15$ mV and of the bacterium $= -25$ mV; $1/\kappa$ varied from 0.645 nm at 2×10^{-1} M NaCl to 20 nm at 2×10^{-4} M NaCl; the dielectric constant (of water) $= 81.07$; Hamaker constant $= 5 \times 10^{-16}$ μJ (as suggested by Weiss, 1971). Although a repulsion barrier is evident at all values of $1/\kappa$ (Fig. 2), a secondary attraction minimum is apparent at the lower $1/\kappa$ values (higher electrolyte concentrations). The shapes of the curves in Fig. 2 suggest that bacteria should be attracted reversibly to the secondary attraction minimum at high electrolyte concentrations, but should be repulsed from the glass surface at lower electrolyte concentrations.

Direct observations (Marshall *et al.*, 1971a) of the attraction of *Achromobacter* R8 to

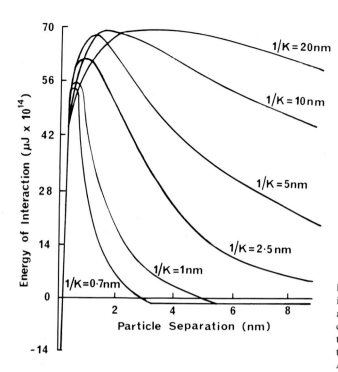

Figure 2. Curves depicting the energy of interaction between *Pseudomonas* R3 and glass surfaces at different interparticle distances for electrolyte concentrations from 2×10^{-1} M ($1/\kappa = 0.7$ nm) to 2×10^{-4} M ($1/\kappa = 20$ nm) NaCl. Adapted from Marshall *et al.* (1971a).

glass surfaces showed that repulsion of the bacteria from the surface increases as the electrolyte concentration decreases (Fig. 3). The concentrations at which complete repulsion occurred were $\sim 5 \times 10^{-4}$ M for NaCl (a uni-univalent electrolyte) and $\sim 5 \times 10^{-5}$ M for $MgSO_4$ (a di-divalent electrolyte). This difference is related to the greater double layer compression in divalent systems at comparable electrolyte concentrations. For both types of electrolyte, the concentrations at which bacteria are repelled from the surface correspond to theoretical $1/\kappa$ values of ~ 20 nm.

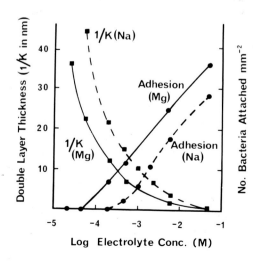

Figure 3. The theoretical double-layer thickness ($1/\kappa$) and reversible adhesion of *Achromobacter* R8 in relation to electrolyte valency and concentration. Adapted from Marshall *et al.* (1971a).

Using a similar analysis for the interaction of *Salmonella typhimurium* to HeLa cells, Jones *et al.* (1981) reported a distance of separation between the surfaces of 26, 14, and 3 nm when the values for $1/\kappa$ were 4.12, 2.5, and 0.72 nm, respectively. Accounting for variations in the radius of curvature of HeLa surfaces due to filamentous protrusions and for variations in the dielectric constant did not alter these values appreciably. However, assignment of different values for the Hamaker constant did alter the calculated distances of separation at a particular value for $1/\kappa$.

3.2. Some Consequences of Reversible Adhesion

3.2.1. Positioning of the Bacterium at the Solid–Water Interface

The most important aspect arising from the interaction between a negatively charged bacterium and a negatively charged surface is that the bacterium itself is unable to overcome the repulsion barrier evident at very small interparticle distances (Fig. 2). Consequently, the main body of the bacterial cell is held at some small, but finite, distance from the surface as reported in Section 3.1. In the absence of strong shear forces, the bacterium being held by forces at the secondary minimum is then ideally positioned to make use of other means to bind irreversibly to the surface (see Section 4). Observations of bacteria immediately after attraction to a surface reveal that they continue to exhibit Brownian motion (Marshall *et al.,* 1971a) and that some show a preferential polar orientation as a result of nonhomogeneous distribution of hydrophobic sites on the bacteria (Marshall and Cruickshank, 1973).

3.2.2. Behavior of Motile Bacteria

Depending on their orientation when they approach a surface, motile bacteria may exhibit either a continuous gyratory motion around the pole of the cell or a propellerlike motion (Marshall *et al.,* 1971a; Jones *et al.,* 1981). In some instances bacteria appear to be held at a surface by the flagella (Meadows, 1971; Sjoblad and Doetsch, 1982; Belas and Colwell, 1982). Some of the motile bacteria were able to overcome the attractive forces at the secondary minimum and swim away, often becoming reversibly attached at other sites (Marshall *et al.,* 1971a). This behavior may be of significance in the search for nutrients at the solid–water interface. The fact that a motile bacterium is able to rotate at and break away from the surface suggests that it has not made contact with the actual surface in its initial approach. For the motile *Pseudomonas* R3, Marshall *et al.* (1971a) observed a maximum velocity of 33 μm/sec. Assuming an average mass of 10^{-12} g, the calculated kinetic energy of 5.45×10^{-19} μJ is not sufficient to overcome the repulsion barrier at any value for $1/\kappa$ shown in Fig. 2. Consequently, some other mechanism is necessary to account for firm adhesion to a surface.

From a study of the dual occurrence of sheathed, polar flagella and lateral flagella in species of *Vibrio,* Belas and Colwell (1982) proposed a model for the possible events taking place in the adhesion of such vibrios to solid surfaces. Their model assumes that initial reversible adhesion results from contact of the sheathed polar flagellum with the surface as suggested by Meadows (1971). After initial contact with the surface, the production of lateral flagella begins within about 3 hr and may increase the forces holding the bacterium to the surface, yet still allow for movement of the bacterium on the substratum. According to this model, lateral flagella would serve as an intermediate structure between reversible adhesion, mediated by the polar flagellum, and irreversible adhesion involving polymer bridging to the surface (see Section 4.1).

3.2.3. Nutrient Scavenging

Are bacteria located at the secondary minimum (reversibly adhering) capable of utilizing nutrient molecules localized at a solid surface? Kefford *et al.* (1982) have shown that the saprophytic *Leptospira interrogans (biflexa patoc* 1) is able to efficiently scavenge labeled fatty acids bound to glass surfaces. This organism possesses the ability to "crawl" on a surface (Cox and Twigg, 1974) and return to the aqueous phase, and Kefford *et al.* found that the leptospires in the aqueous phase had accumulated labeled substrate. Whether more conventional bacteria reversibly adhering to surfaces are able to scavenge nutrients bound at the surface remains to be established. Bacteria irreversibly bound to surfaces certainly scavenge labeled fatty acids (Kefford *et al.,* 1982).

An interesting adaptation to reversible adhesion is that of dinoflagellates grazing on microbial films adhering to glass surfaces (Marshall, 1976, p. 60). One flagellum of the dinoflagellate was observed being held as a trailing anchor at the level of the secondary attraction minimum, with the second flagellum providing the motive force to enable the dinoflagellate to move across the surface. This arrangement ensures that the dinoflagellate remains as close as possible to its nutrient source as it grazes across the surface. The dinoflagellates readily detached from the surface and often attached at other nearby sites.

3.2.4. Reversible Adhesion in Natural Habitats

When bacteria are added to saline sediments, very few are recovered following shaking of the sediment with water (Rubentschick *et al.,* 1936; Roper and Marshall, 1974). Desorption of such added bacteria from sediments was shown by Roper and Marshall (1974) to be dependent on the electrolyte concentration. Similarly, some of the natural population of the sediment can be desorbed by lowering the electrolyte concentration (Fig. 4). At high salinities (or low $1/L_s$ values), few bacteria are recovered, but, following successive washings in distilled water, large numbers of bacteria appear in the supernatant at an electrolyte concentration coinciding with the dispersion of colloids and, hence, an increase in turbidity of the supernatant. Such results suggest that many bacteria in saline sediments are merely reversibly adhering to sediment particulates and are repelled from these surfaces at low electrolyte concentrations.

It may be that many bacteria in natural habitats lack the ability to adhere irreversibly to surfaces but are still capable of reversible adhesion at such surfaces. Most isolation or

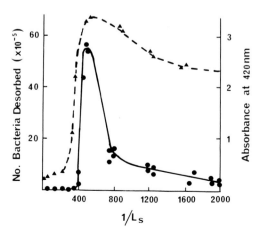

Figure 4. Numbers of bacteria desorbed from a saline sediment (solid line) and turbidity of the supernatant (dotted line) following successive washings of the sediment in distilled water. The values of $1/L_S$ represent the reciprocal of the specific conductance of the supernatant liquid after each washing. Adapted from Roper and Marshall (1974).

observation techniques involve rigorous washing of surfaces prior to further study. The shear forces involved in such procedures would certainly remove any reversibly adhering bacteria.

4. IRREVERSIBLE ADHESION

4.1. Polymer Bridging

ZoBell (1943) suggested that, following attraction to a surface, bacteria become firmly attached to the surface as a result of the synthesis of extracellular adhesive materials. Hirsch and Pankratz (1970) described a variety of amorphous, granular, and fibrous "holdfast" structures produced by bacteria firmly adhering to electron microscope grids that had been immersed in aquatic habitats. Following a study of the adhesion of marine bacteria to surfaces, Marshall *et al.* (1971a) defined *irreversible adhesion* as a time-dependent firm adhesion whereby the bacteria no longer exhibited Brownian motion and could not be removed by washing. On the basis of extracellular polymer production by bacteria and, more particularly, of polymer "footprints" remaining on surfaces following removal of bacteria, Marshall *et al.* (1971a) suggested that polymer bridging was responsible for the firm anchoring of bacteria to a surface. It was suggested that this polymer was produced by the bacteria in response to nutrients accumulated at the surface. On the other hand, Pourdjabber and Russell (1979) concluded that instantaneous irreversible adhesion of some oral bacteria occurs on tooth surfaces without the need for a period of active bacterial metabolism. Fletcher (1980a) has also indicated that spontaneous attachment of bacteria to certain substrata is a distinct possibility.

Direct evidence for the involvement of polymer bridging between bacteria and solid surfaces was obtained by using solid substrata amenable to sectioning at the adhesion surface (Marshall and Cruickshank, 1973; Fletcher and Floodgate, 1973). Bacteria were allowed to attach to araldite or membrane filter surfaces, and the surfaces were washed thoroughly, then embedded and sectioned. Such techniques showed the presence of ruthenium red-reactive polymer between the bacterium and the solid attachment surface and suggested the presence of acidic polysaccharide material (Fletcher and Floodgate, 1973). Any fine polymeric component external to the cell envelope, such as extracellular polymers, pili, fimbriae, and flagella (Chapters 2 and 11) may be responsible for the bridging of bacteria to solid surfaces. More recent studies of a wide variety of surfaces have confirmed the universality of polymer bridging as the means whereby bacteria bind irreversibly to surfaces (Costerton *et al.*, 1981; Chapter 1).

4.2. Short-Range Forces

How do extracellular polymers bind to the solid surface? Bangham and Pethica (1960) noted that contact between animal cells is facilitated energetically by a reduction of the radii of curvature of appropriate regions of the cells. On this basis, Tadros (1980) presented calculations on the interaction energy at various radii of curvature for a cell approaching another with a fixed radius of curvature of 10^3 nm. It was shown that, for a surface potential of -25 mV at 1.5-nm separation, only probes (such as extracellular polymers) with radii of less than 2 nm can make contact under Brownian motion. Once these filaments are deposited in the primary minimum they become firmly bound as a result of the high attractive energies at this point.

At short separations between the polymers and the surface, short-range forces are particularly important. These forces may be divided into (Tadros, 1980): (1) chemical bonds, such as electrostatic, covalent, and hydrogen bonds, (2) dipole interactions, such as dipole–dipole (Keesom), dipole-induced dipole (Debye), and ion–dipole interactions, and (3) hydrophobic bonding. Because of the obvious variability in the extracellular polymers produced by different bacteria (see Section 6.3), different interactions will occur during firm attachment of these bacteria and it is possible that various types of interactions take place simultaneously with a single bacterial type.

It is relevant here to note the model proposed by Doyle *et al.* (1982) to describe the observed positive cooperativity in the adhesion of *Streptococcus sanguis* to saliva-coated (pellicle) hydroxyapatite beads. Positive cooperativity is interpreted to mean that when one bond in the cell–pellicle interaction occurs, subsequent bond formation is more energetically favored. The model envisages an initial interaction via chemical or dipole means and that this interaction is stabilized by hydrophobic groups. Once the initial cell is bound, adjacent pellicle proteins may change configuration to create new receptor sites for other cells (i.e., positive cooperativity). This potential situation is relevant to so-called inert surfaces which, in natural habitats, rapidly become conditioned by adsorbed macromolecular species (see Chapter 5, Sections 3.5 and 5). Under these circumstances, extracellular polymers of adhering bacteria may interact with surface conditioning molecules rather than the actual substratum surface (see Section 4.8 and Chapter 10, Section 4.1).

The possibility of hydrophobic forces being involved in the adhesion of bacteria to solid surfaces was first alluded to by Marshall and Cruickshank (1973). Subsequently, a variety of methods for determining bacterial cell surface hydrophobicity have been devised (see Chapter 2, Section 5.2). Hydrophobic bacteria appear to adhere more firmly to solid surfaces than hydrophilic bacteria (Rosenberg, 1981; Kefford *et al.*, 1982; Kjelleberg *et al.*, 1983), although other factors may modify this relationship (see Section 4.5.3).

4.3. Adhesion Strength

To measure the strength of adhesion of bacteria to surfaces it is necessary to measure the force required to remove the bacteria from the surface. According to Tadros (1980), the most obvious method is to apply a centrifugal force (F_c) which would balance the adhesion force (F_a) of the bacteria. Such an approach was employed by Zvyagintsev *et al.* (1971), who reported adhesion values ranging from 4×10^{-12} to 4×10^{-9} N/cell for several gram-positive and gram-negative bacteria and a yeast. Fisher *et al.* (1980) reported a method that may be suitable for the measurement of bacterial adhesion.

The problem with any measurement of bacterial adhesion is to define just what is being removed during the process. As demonstrated by Marshall *et al.* (1971a), mechanical shearing of bacteria from a surface left a polymer "footprint" on the solid surface. This suggests that the cohesive strength of the bridging polymer is substantially less than the true adhesive strength between the polymer and the surface. Consequently, any measurement of bacterial adhesive strengths is likely to provide a gross underestimate of the true adhesive forces. It is not known whether the adhesive strength of a single bacterial type to different surfaces varies according to the properties of those surfaces.

4.4. Passive versus Active Adhesion

Active adhesion implies the need for some physiological activity on the part of the bacterium when it is in the vicinity of the surface. Bacterial motility, for instance, can aid attachment by increasing the statistical chances of a bacterium approaching the surface (Fletcher, 1980a), but does not necessarily overcome the repulsion barrier holding the bacterium at some distance from the surface (see Sections 3.1 and 3.2.2). ZoBell (1943) and

Marshall *et al.* (1971a) suggested that a time-dependent phase of irreversible adhesion resulted from the need for polymer synthesis by the bacteria following initial attraction to the surface. The addition of chloramphenicol to a seawater sample inhibited the irreversible adhesion of most, but not all, of the bacteria to glass surfaces (Marshall, 1973). It was suggested the inhibitor of protein synthesis prevented the production of enzymes necessary for polymer synthesis by some of the bacteria. Such bacteria may be in a state of starvation in seawater (Novitsky and Morita, 1978) and respond to nutrients accumulated at a surface by the production of extracellular polymer. Bacteria adhering to the surface despite the presence of chloramphenicol may possess constitutive enzymes capable of synthesizing polymer in the low-nutrient conditions of the bulk seawater. Takakuwa *et al.* (1979) demonstrated that *Thiobacillus thiooxidans* attachment to sulfur granules was prevented by a range of inhibitors. A more extensive study by Fletcher (1980a) revealed complex interactions between the effects of the bacterial species, the inhibitor, and the substratum on firm adhesion. In some instances, the inhibitors had their maximum effect when added before the bacteria encountered the substratum, indicating that bridging polymers were synthesized by the bacteria in the bulk liquid phase. Other combinations revealed that time-dependent physiological activity at the surface is required for firm attachment. As pointed out by Fletcher, the indication that the attachment process of a particular bacterium might vary with different substrata makes the study of attachment mechanisms much more complicated.

The attachment of killed (by UV, formaldehyde, or heat) bacteria to surfaces (Meadows, 1971; Ørstavik, 1977; Fletcher, 1980a) does suggest some form of passive attachment. There is no guarantee, however, that damage to cell membranes by these treatments may release internal polymers (such as DNA) that may enhance the "stickiness" of the killed bacteria. If bacterial adhesion to certain substrata is a passive process, then it should be possible to remove the organisms from the surface with a suitable eluant (Fletcher, 1980a). For instance, Minato and Suto (1976) found that certain rumen bacteria could be eluted from starch grains with formalin and others with a salt solution. The same authors later found that species of *Bifidobacterium* could be eluted from starch grains by the use of dextrin or partially hydrolyzed hydroxylpropyl starch (Minato and Suto, 1979). Similarly, Ørstavik (1977) has eluted *Streptococcus faecium* from glass with Tween 80, and Fletcher (personal communication) has eluted *Pseudomonas* NCMB 2021 from polystyrene with hexane.

Obviously, the question of active versus passive adhesion of bacteria to surfaces requires much more detailed investigation. Such studies should provide a greater insight into the complexities involved in the processes of bacterial adhesion.

4.5. Surface Energy Approach to Adhesion

4.5.1. Effects of Substratum Wettability

Substratum wettability has been described in terms of (1) surface free energy (γ_S), approximated by determining the critical surface tension of γ_C (Zisman, 1964), by calculating the sum of polar and dispersion ($\gamma_S^p + \gamma_S^d$) components of surface free energy (Wu, 1980), or by determining γ_S from a single contact angle measurement using the equation-of-state approach (Neumann *et al.*, 1974), or (2) hydrophobicity, expressed as the water contact angle, Θ_W (Fletcher and Loeb, 1979), or as the work of adhesion between water and the surface, W_A (Pringle and Fletcher, 1983; see also Chapter 5, Section 2).

On the basis of experience with other biological systems, Baier (1973) predicted that minimal bacterial adhesion in aquatic habitats should occur on substrata within a critical surface tension range of 20 to 30 mN/m. Such substrata are relatively hydrophobic, and

Baier (1980) has termed this γ_C range the "minimally bioadhesive" range. Several studies on a range of substrata immersed in seawater have shown minimal adhesion by bacteria on hydrophobic substrata, that is, in the "minimally bioadhesive" range predicted by Baier, and maximal adhesion on high-energy (or hydrophilic) substrata (Dexter et al., 1975; Dexter, 1979). On the other hand, other studies (Loeb, 1977; Carson and Allsopp, 1980) have indicated more rapid colonization of hydrophobic surfaces than of hydrophilic surfaces immersed in seawater.

Fletcher and Loeb (1976, 1979) reported that a marine pseudomonad exhibited maximal attachment to hydrophobic surfaces and lower attachment to hydrophilic surfaces, with a direct relationship between increasing adhesion and increasing Θ_W of the test substrata (Fletcher and Loeb, 1979). A more recent laboratory study with a range of freshwater bacteria (Pringle and Fletcher, 1983) did not reveal any generic pattern of attachment to test surfaces, although the majority of isolates showed a preferential attachment to hydrophobic surfaces. These workers expressed the hydrophobicity of substrata in terms of W_A, a value determined from the Young–Dupre equation:

$$W_A = \gamma_{LV} (1 + \cos \Theta_E) + \pi \tag{3}$$

where γ_{LV} is the surface tension of the liquid, Θ_E is the equilibrium contact angle for water (assumed to be equivalent to the advancing contact angle, Θ_A), and π is the spreading pressure of the adsorbed vapor on the solid (assumed to be negligible; Chapter 5, Section 2). Pringle and Fletcher reported that *Pseudomonas fluorescens* and *Aeromonas hydrophila* showed maximum attachment on poly(vinyl chloride) (W_A of 85.5 mJ/m²) and *Acinetobacter* sp. on Nylon 6.6 (W_A of 97.7 mJ/m²). By comparison the *Pseudomonas* NCMB 2021 (Fletcher and Loeb, 1979) showed maximal attachment on polystyrene (W_A of 72.8 mJ/m²).

Fletcher and Pringle (1984) indicate that W_A, derived from Θ_W, relates better to attachment data than say γ_C, derived from contact angles of a variety of liquids on the test surface. For adhesion to occur between hydrated bacterial and solid surfaces, water must be displaced as the two surfaces move together, and the greater W_A between water and the surface, the more difficult will this displacement of water become as shown in the results of Pringle and Fletcher (1983).

4.5.2. Reasons for Apparent Discrepancies

There are several possible explanations for the conflicting results reported in Section 4.5.1, namely: (1) the effects of bacterial surface free energy and liquid surface tension were not considered in these studies, (2) the variable effects of competition between different colonizing bacteria at different test sites and the lack of such competition in pure culture studies, and (3) the variable effects of different conditioning films (Fletcher and Marshall, 1982a) that might be encountered in different experimental conditions.

4.5.3. The Thermodynamic Model

A thermodynamic model has been described in tissue cell adhesion studies (Neumann et al., 1979; Absolom et al., 1980) that accounts for the interfacial tensions between all three interacting components of the system, namely, the cell surface, the solid substratum, and the liquid phase (see also Chapter 5, Section 2). According to this thermodynamic model, bacterial adhesion will be favored if the process itself causes the free energy to decrease. For systems in which electric charges and specific receptor–ligand interactions may be neglected, the change in free energy is given by

$$\Delta F^{\text{adh}} = \gamma_{BS} - \gamma_{BL} - \gamma_{SL} \qquad (4)$$

where ΔF^{adh} is the change in free energy of adhesion, and γ_{BS}, γ_{BL}, and γ_{SL} are the bacterium–substratum, bacterium–liquid, and substratum–liquid interfacial tensions, respectively. Values for these interfacial tensions have recently been derived via the application of Young's equation (Neumann *et al.*, 1980):

$$\gamma_{SV} - \gamma_{SL} = \gamma_{LV} \cos \Theta \qquad (5)$$

where γ_{SV}, γ_{SL}, and γ_{LV} are the substratum–vapor, substratum–liquid, and liquid–vapor interfacial tensions, respectively, and Θ is the contact angle of the liquid on the substratum. Only γ_{LV} and Θ are readily determined experimentally. Using an equation-of-state approach, γ_{SV} and γ_{SL} can be calculated from a computer program presented by Neumann *et al.* (1980). Similarly, it is possible to calculate all of the interfacial tensions in equation (4) required to determine ΔF^{adh}.

Gerson and Scheer (1980) used essentially this approach to study the adhesion of three bacteria to each of five substrata. Solid–liquid interfacial free energies were determined from contact angles of water and growth media on samples of substrata and lawns of bacteria on membrane filters (van Oss *et al.*, 1975). Surface tensions of the liquids were determined, the interfacial free energies were calculated by the equation-of-state approach (Neumann *et al.*, 1974), and the ΔF^{adh} was derived from these values. Gerson and Scheer (1980) described a straight-line relationship between ΔF^{adh} and the logarithm of the number of cells attached per unit area for *Staphylococcus aureus* (bacterium–vapor interfacial energy, or γ_{BV}, of 69.7 mN/m). A generally linear relationship was established for *Staphylococcus epidermidis* (γ_{BV} of 67.9 mN/m) and *Serratia marcescens* (γ_{BV} of 66.3 mN/m) except for the high-molecular-weight polyethylene substratum, with a high γ_{SV} (64.6 mN/m). These discrepancies for a high-γ_{SV} substratum may be explained by the recent results of Absolom *et al.* (1983) reported below. According to Gerson and Scheer (1980), their results show that thermodynamics of partitioning of suspended particles (bacteria) between two immiscible liquid phases (Gerson, 1980; Gerson and Akit, 1980) also apply to the partitioning of the particles between a liquid and a solid phase. In a study on the effects of alcohols on the attachment of a marine pseudomonad to hydrophilic and hydrophobic surfaces, Fletcher (1983) reported a consistent minimum attachment at medium γ_{LV} values of 64 to 69 mN/m for both substrata employed.

Absolom *et al.* (1983) have determined the effect of γ_{SV} and γ_{LV} on the degree of adhesion of five bacteria to a range of substrata. Theoretical calculations of ΔF^{adh} for the attachment of a bacterium to various substrata as a function of γ_{SV} (Fig. 5) lead to three possible situations according to these authors: (1) where $\gamma_{LV} < \gamma_{BV}$, ΔF^{adh} decreases with increasing γ_{SV} and predicts increasing bacterial adhesion with increasing γ_{SV}; (2) where $\gamma_{LV} > \gamma_{BV}$, ΔF^{adh} increases with increasing γ_{SV} and predicts decreasing bacterial adhesion with increasing γ_{SV}; and (3) where $\gamma_{LV} = \gamma_{BV}$, $\Delta F^{\text{adh}} = 0$ independently of the value of γ_{SV} and implies that bacterial adhesion does not depend on γ_{SV} and should be zero if other effects, such as electrostatic interactions, are not operative.

Solutions of different γ_{LV} values were obtained by Absolom *et al.* (1983) by the addition of various levels of dimethylsulfoxide to Hanks' balanced salt solution (HBSS). This gave a range of γ_{LV} from 72.8 to 64.0 mN/m. Values of γ_{SV} and γ_{BV} were obtained by measuring contact angles of water or HBSS on the substrata or lawns of bacteria deposited on agar (van Oss *et al.*, 1975), respectively, and calculating the interfacial tensions by the equation-of-state approach (Neumann *et al.*, 1974). Although objections have been raised to the method of

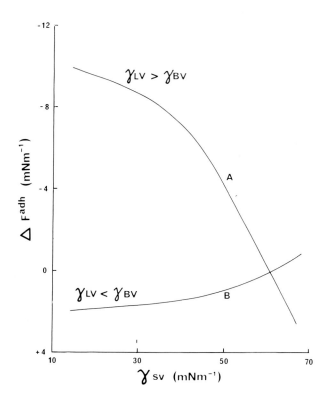

Figure 5. Values of ΔF^{adh} calculated for *Escherichia coli* 2627 (γ_{BV} = 67.8 mN/m) at different values of the substratum surface tension (γ_{SV}) for the following conditions: (A) γ_{LV} = 72.8 mN/m and (B) γ_{LV} = 64.0 mN/m. Reproduced with permission from Absolom *et al.* (1983).

obtaining values for γ_{BV} from contact angles on bacterial lawns, particularly in relation to leakage of cellular materials into the water droplets being measured (Pethica, 1980), Absolom *et al.* (1983) obtained very good agreement between results using this method and results obtained by several other methods (Table II). These authors found that the number of bacteria adhering per unit surface area varied with γ_{SV} in accordance with the predictions of the thermodynamic model, i.e., adhesion increased or decreased with increasing γ_{SV} where γ_{LV} < γ_{BV} or γ_{LV} > γ_{BV}, respectively. It was also demonstrated that values of γ_{BV} could be determined from a plot of the slope derived from (number of bacteria adhering over γ_{SV}) against γ_{LV}. γ_{BV} was taken to be the same as the γ_{LV} value at the point at which the slope was equal to zero. Values of γ_{BV} derived from these adhesion results are included in Table II.

Fletcher and Pringle (1984) have also examined bacterial adhesion at different γ_{SV} values using a freshwater isolate of *Pseudomonas fluorescens* grown in both carbon- and nitrogen-limiting conditions in a chemostat. They also observed a change in the adhesion rate with increasing γ_{SV} when values of γ_{LV} vary, as predicted by the above thermodynamic model. Using the same adhesion method to derive values of γ_{BV} as employed by Absolom *et al.* (1983), Fletcher and Pringle obtained values of 66.4 and 63.6 mN/m for carbon- and nitrogen-limited bacteria, respectively. It is obvious from such values that the surface free energy of a single bacterial type may be extensively modified by the particular growth conditions.

These results of Absolom *et al.* (1983) and Fletcher and Pringle (1984) partially explain the apparent discrepancies found in earlier studies of bacterial adhesion to substrata of varying surface free energies (see Section 4.5.1). By accounting for all components of the system, different degrees of bacterial attachment are observed for different relationships

Table II. Comparison of Surface Free Energy Values (γ_{BV}) of Various Bacteria
Obtained Using a Range of Methods[a]

| | | γ_{BV} obtained by | | |
| | Contact angle by equation-of-state (mN/m) | Phagocytic ingestion | | |
Bacteria		Granulocytes (mN/m)	Platelets (mN/m)	Adhesion (mN/m)
E. coli 055	69.7	69.6	69.3	69.9
S. aureus 049	69.1	68.7	68.9	69.3
E. coli 2627	67.8	ND[b]	ND	67.8
S. epidermidis	67.1	66.9	67.3	66.9
L. monocytogenes	66.3	66.1	65.8	65.6

[a]From Absolom *et al.* (1983).
[b]ND, not determined.

between γ_{SV}, γ_{LV}, and γ_{BV}. Absolom *et al.* (1983) have also pointed out that complications arise when ΔF^{adh} or adhesion values are plotted against γ_{LV} rather than against γ_{SV}. For a high-energy surface, such as sulfonated polystyrene (SPS; $\gamma_{SV} = 66.7$ mN/m), the adhesion behavior is predicted by the thermodynamic model. For values of γ_{LV} greater than 66.7 mN/m, adhesion increases with increasing bacterial hydrophobicity (decreasing γ_{BV}), but when γ_{LV} is less than 66.7 mN/m, adhesion decreases with increasing bacterial hydrophobicity (Fig. 6). With a lower energy surface, such as polystyrene ($\gamma_{SV} = 25.6$ mN/m), adhesion follows the thermodynamic prediction when γ_{LV} is greater than γ_{BV}. At lower values of γ_{LV}, the curve flattens out and adhesion becomes independent of γ_{LV} (Fig. 6). According to Absolom *et al.* (1983), this discrepancy may result from electrostatic interactions mediated by cationic bridging. It may be that errors are introduced into the measure-

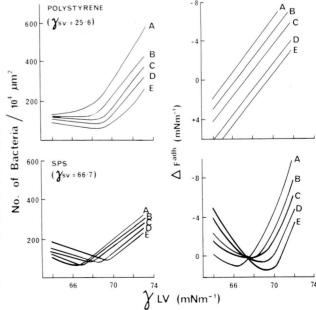

Figure 6. Calculated values of ΔF^{adh} as a function of the liquid surface tension (γ_{LV}) for five species of bacteria (right-hand side) and the experimentally determined extent of bacterial adhesion under the same conditions (left-hand side). (A) *Listeria monocytogenes* ($\gamma_{BV} = 66.3$ mN/m), (B) *Staphylococcus epidermidis* ($\gamma_{BV} = 67.1$ mN/m), (C) *Escherichia coli* 2627 ($\gamma_{BV} = 67.8$ mN/m), (D) *S. aureus* ($\gamma_{BV} = 69.1$ mN/m), and (E) *E. coli* 055 ($\gamma_{BV} = 69.7$ mN/m). Reproduced with permission from Absolom *et al.* (1983).

ment of γ_{LV} by adsorption of dimethylsulfoxide or bacterial excretion products at surfaces leading to erroneous estimates of γ_{SV} and γ_{BV} (Fletcher and Marshall, 1982a).

4.6. Colloid Stability versus Surface Energy Approaches

In Sections 3.1 and 4.5 the feasibility of the colloid stability (DLVO) theory or the surface energy approach in explaining bacterial adhesion has been discussed as though they were separate entities. As pointed out by Pethica (1980), the connections between these two approaches have not been fully appreciated. Pethica emphasizes the problems existing with both approaches and considers that a general descriptive framework is required for biological adhesion which takes account of experimentally accessible variables and which provides a basic critique for the DLVO and surface energy approaches. In considering a thermodynamic description of cell adhesion, Pethica presents a detailed analysis of the publications of Hall (1972) and Everett and Radke (1975), which render more apparent the relationships between the DLVO and surface energy approaches to cell adhesion. Pethica (1980) derived several equations to show that the total change in surface tension of opposing surfaces can be calculated from the interaction forces and notes that this feature is automatically built into the DLVO theory.

4.7. Adsorption Isotherms

Adsorption isotherms define the relationship between the adsorption capacity of an absorbent and the concentration of bacterial cells at equilibrium (Marshall, 1976). Type I (Langmuir type) isotherms exhibit a rapid rise in the numbers of attached bacteria (q) with increasing concentration (C) up to a limiting value, and such isotherms indicate that adsorption is restricted to a monolayer (Fig. 7A). The sigmoid type II isotherm represents multilayer adsorption to nonporous solids, with the inflection point representing the formation of a complete monolayer (Fig. 7B).

4.7.1. Monolayer Attachment

Most workers studying adsorption isotherms of bacteria have employed the Langmurian model presented by Gibbons *et al.* (1976). This model is based on the following assumptions: (1) a finite number of identical sites are available for attachment per unit area of adsorbent, (2) bacteria approach the adsorbent surface without any steric hindrance (see Maroudas, 1975), and (3) adsorption is reversible up to and at the equilibrium state. The Langmuir adsorption isotherm can be expressed mathematically as:

$$q = CN/C + K_d \tag{6}$$

where C is the bacterial concentration at equilibrium (cells/ml), q is the number of bacteria adsorbed per unit area of adsorbent material, N is the adsorption capacity of the adsorbent material (binding sites per unit area), K_d is the dissociation constant (cells/ml), and the reciprocal of K_d is the affinity constant, K_a. One linearization (Lupton and Marshall, 1981) of the Langmuir adsorption isotherm is

$$C/q = C/N + K_d/N \tag{7}$$

and a plot of C/q versus C results in a straight line with x and y intercepts of $-K_d$ and K_d/N, respectively (Fig. 7C).

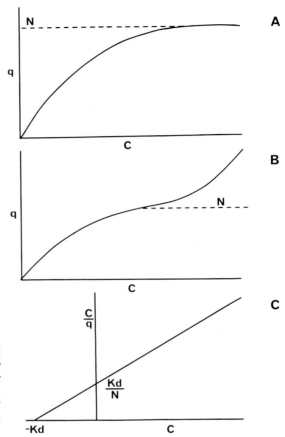

Figure 7. Examples of adsorption isotherms. (A) Type 1 (Langmuir) isotherm, (B) type II (multilayer) isotherm, and (C) reciprocal plot of the Langmuir isotherm, where q = number of bacteria adhering per unit area, C = bacterial concentration in the bulk phase at equilibrium, N = number of binding sites per unit area, and K_d = equilibrium constant.

Using this approach, Gibbons *et al.* (1976) demonstrated a greater number of adsorption sites for *Streptococcus mitis (S. miteor)* on saliva-coated hydroxyapatite (HA) than on untreated HA. A more detailed study by Clark *et al.* (1978) revealed the following results: (1) a relatively similar level of adhesion of *S. mutans, S. salivarius, S. sanguis, S. mitis, Actinomyces viscosus,* and *A. naeslundii* to untreated HA, suggesting a common mechanism of adhesion, possibly Ca^{2+} bridging, to this adsorbent, and (2) a more complex pattern on saliva-treated HA with lower adhesion of *S. mutans, S. salivarius,* and *A. naeslundii* and higher adhesion of *S. sanguis, S. mitis,* and *A. viscosus,* suggesting that salivary components impart a high order of specificity. Wheeler *et al.* (1979) found that the affinity of the virulent strain *A. viscosus* T14V for saliva-treated HA was 10-fold greater than that of the avirulent strain T14AV. Pretreatment of strain T14V with proteolytic enzymes or with heat inhibited adhesion to saliva-treated HA, indicating that adhesion receptors on the bacterial surface are protein in nature. Similarly, Celesk and London (1980) found that adhesion of oral *Cytophaga* species to HA was partially inhibited by saliva, by pretreatment of the bacteria with proteinase K or phospholipases C and D, or by exposure to high temperatures.

Fletcher (1977) did not observe desorption of a marine pseudomonad from a surface following firm adhesion and attributed this to the efficiency of polymer bridging where there are a number of adsorption sites and where it is statistically improbable that all of these anchor sites could be desorbed simultaneously. Consequently, Fletcher modified the Langmuir equation to eliminate the desorption term. She demonstrated that adhesion was more rapid for log phase than for stationary and death phase cells, and that adhesion increased with increasing temperature from 3 to 20°C.

From derived values of K_d and N, Applebaum *et al.* (1979) chose the product K_aN as an index of the binding of a given strain to an adsorbent, for this product combines the affinity constant (K_a) and the number of sites (N) into a single factor. As $K_aN = q(1/C + K_a)$, then K_aN is a linear function of q (the number of cells bound), is modulated by the relationship between the free cell concentration and the affinity constant K_a, and is independent of N, irrespective of the range of bacterial concentrations used in the adhesion experiments (Applebaum *et al.*, 1979). As K_aN represents the reciprocal of the y intercept (Fig. 7C), its units represent the number of bound cells per unit area adsorbent divided by the number of free cells per milliliter. It was found that *S. sanguis* serotype 1 strains adhered to saliva-coated HA better [$K_aN = ((187 \pm 72) \times 10^{-2}$] than serotype 2 strains [$K_aN = (97 \pm 84) \times 10^{-2}$]. Other oral streptococci and *Caprocytophaga* gave lower K_aN values.

Lupton and Marshall (1981) used a similar approach to examine the specific adhesion of bacteria to heterocysts of *Anabaena* species. A study of monolayer adhesion showed that, although the number of binding sites (N) for *Zoogloea* SL20 was greater than for *Pseudomonas* SL10, the adhesion index (K_aN) for *Pseudomonas* SL10 was greater than that for *Zoogloea* SL20 (0.42 and 0.14, respectively). The same approach was employed by Belas and Colwell (1982) to examine the adsorption kinetics of laterally and polarly flagellated *Vibrio* species. These authors observed that, when conditions were favorable for the production of lateral flagella, adsorption of *Vibrio parahaemolyticus* followed the Langmuir adsorption isotherm, but proportional adsorption kinetics were observed when conditions did not favor lateral flagellation. K_aN values obtained from Langmuir isotherms indicated maximal adsorption of *V. parahaemolyticus* to chitin surfaces at 25°C, and relatively constant K_aN values were obtained over salinity values of 1.0 to 2.0% NaCl.

4.7.2. Multilayer Attachment

Instances where adsorption kinetics do not conform to the Langmuir model have been reported by Applebaum *et al.* (1979), Lupton and Marshall (1981), and Belas and Colwell (1982). In these cases, a plot of q versus C indicates that bacterial adhesion remains proportional to cell concentration and that surface saturation kinetics do not seem to apply. This is usually the result of multilayer adhesion, and a detailed examination of adhesion at lower cell concentrations should reveal a sigmoid type II isotherm (Lupton and Marshall, 1981) as shown in Fig. 7B. The first phase represents bacterial adhesion to the adsorbent (cell–adsorbent interaction), which should show saturation kinetics, followed by an aggregation of cells away from the surface (cell–cell interaction). Lupton and Marshall (1981) clearly showed the aggregation of *Zoogloea* SL20 on cyanobacterial heterocysts (multilayer adsorption), compared with the perpendicular orientation of *Pseudomonas* SL10 adhering to heterocysts (monolayer adsorption).

4.7.3. Competition Studies

Applebaum *et al.* (1979) employed a fixed concentration of a radiolabeled reference strain and increasing concentrations of a second strain to test the hypothesis that bacteria having the same surface receptor on an adsorbent should compete for the same binding site. Using *S. sanguis* G9B, serotype 1, as the reference strain, these authors observed that the serotype 1 strains competed for the same, or similar, binding sites on HA, whereas serotype 2 strains increased binding of the reference strain to HA. Scanning electron microscopy revealed that this enhanced binding may have resulted from aggregation between the two bacterial types. Other oral bacteria showed little or no competition with the reference strain.

Using a similar method, Lupton (1982) found that *Zoogloea* SL20 competed with *Pseudomonas* SL10 for available binding sites on cyanobacterial heterocysts, but that *Pseudomonas* SL10 did not compete with the adhesion of *Zoogloea* SL20. Lupton concluded that the inhibition of *Pseudomonas* SL10 by *Zoogloea* SL20 was due to steric hindrance of heterocyst adhesins for the pseudomonad by the *Zoogloea* SL20 polymer and probably did not reflect direct competition for a common binding site. Belas and Colwell (1982) used dual-label adsorption competition experiments to demonstrate that laterally flagellated *V. parahaemolyticus* inhibited the adsorption of polarly flagellated *V. cholerae*. In contrast, polarly flagellated bacteria enhanced the adsorption of *V. cholerae* and of *V. parahaemolyticus*. These authors suggest that lateral flagella represent a component of certain vibrios that is important in their adsorption to surfaces.

4.8. Conditioning Films

The alteration in substratum surface properties resulting from the adsorption of macromolecules at the surface has been considered by Loeb (Chapter 5, Section 5) and Fletcher and Marshall (1982b). The most intensively studied effect of a conditioning film on bacterial adhesion is that of the salivary pellicle formed on tooth surfaces. The modifying effect of saliva on the specificity of bacterial attachment to HA surfaces was dealt with in Section 4.7.1.

Previous studies have revealed that bacterial adhesion may be enhanced, inhibited, or unaffected by the presence of macromolecules in the aqueous phase (Meadows, 1971; Fletcher, 1976; Ørstavik, 1977). It has generally been assumed that these macromolecules adsorb to the substratum surface, thereby altering the surface free energy and possibly creating steric exclusion effects (Maroudas, 1975). Fletcher and Marshall (1982a) described a bubble contact angle method to determine changes in surface free energy characteristics in the presence of the aqueous phase when macromolecules were introduced into the system. Bubble contact angles were found to vary with the composition and concentration of the macromolecular solutions, as well as with the substratum characteristics. Modification by Pronase of the conditioning effect of proteins varied with the protein, the substratum, and the time of addition of the enzyme relative to the conditioning of the substratum. Fletcher and Marshall demonstrated that the effects of protein-conditioning films and modification with Pronase were consistent with the attachment of *Pseudomonas* NCMB 2021 to the treated substrata. For instance, Pronase reduced the inhibition of attachment of bacteria to hydrophobic polystyrene by proteins. The results on hydrophilic tissue culture plastic, however, were not entirely consistent.

Little is known about the effects of natural conditioning films on the adhesion of and competition between bacteria colonizing such surfaces. At some stage, this aspect needs to be considered in any development of the thermodynamic model of adhesion discussed in Section 4.5.3.

5. TEMPORARY ADHESION IN GLIDING BACTERIA

Because gliding bacteria are capable of translational motion across a surface and are capable of breaking away from the attachment surface, the mode of adhesion can be described as temporary (Crisp, 1973). A temporary adhesive holds surfaces together by the

Table III. Calculated Values of the Force Preventing
Separation (F_x) and Horizontal Drag (F_y) for *Flexibacter*
BH3[a]

Fluid	F_x (μN)	Fy (μN)
Water	4.71×10^{-3}	2.16×10^{-8}
Slime	1.41×10^{-1}	6.48×10^{-7}

[a]Adapted from Humphrey *et al.* (1979).

work done against viscosity when the surfaces are separated, yet allows translational motion across the surface. This has been termed *Stefan adhesion* and describes the situation where the force required to separate two surfaces is very much greater than the horizontal drag (Crisp, 1973). The force preventing separation, F_x, and the horizontal drag, F_y, are given by

$$F_x = 1.5 \ \pi\eta R^4 \ (dx/dt)x^{-3} \tag{8}$$

$$F_y = \pi\eta R^2 \ (dy/dt)x^{-1} \tag{9}$$

where η is the Newtonian viscosity, R the effective radius, x the distance of separation between the two surfaces, dx/dt the rate of separation, and dy/dt the rate of horizontal motion.

Humphrey *et al.* (1979) obtained two extracellular fractions from *Flexibacter* BH3; one was a hydrophobic material identified as lipopolysaccharide and the second was a water-soluble glycoprotein slime. The slime exhibited viscous properties characteristic of a linear colloid and was considered suitable as a Stefan adhesive. This slime was only produced by *Flexibacter* BH3 when grown on solid medium. Although the precise concentration of slime separating the bacterium from the surface is unknown, Humphrey *et al.* (1979) employed a highly conservative viscosity value of 0.03 Nsm^{-2} at 26°C in calculating forces relevant to temporary adhesion of *Flexibacter* BH3. The values reported for F_x and F_y where the fluid separating the bacterium from the surface was water or the slime are given in Table III. It is obvious that the viscous slime greatly increased the force preventing separation compared to that for water and, therefore, should provide an effective means of adhesion. On the other hand, the horizontal drag in the presence of slime remains some five orders of magnitude lower than the force preventing separation. Provided a suitable form of locomotion is present, such as the generation of rhythmic contractions of the outer cell membrane (Humphrey *et al.*, 1979), the bacterium should be held near the surface by the slime but be able to glide across the surface (O'Brien, 1981).

6. ROLE OF POLYMERS

6.1. Polymers as Adhesives

As discussed in Section 4.1, extracellular polymers play a crucial role in the adhesion of bacteria to surfaces by bridging across the repulsion barrier (Section 3.1) and thereby anchoring the cell to the surface. ZoBell (1943) first suggested that extracellular polymers were responsible for the firm adhesion of bacteria to surfaces following microscope observations of crystal violet-stained slides that had been immersed in seawater. Subsequently, Corpe (1970a,b) isolated a number of strains of *Pseudomonas atlantica* from immersed surfaces, and demonstrated the production of large quantities of an acidic polysaccharide that he considered might be involved in adhesion of these bacteria to surfaces. Hirsch and Pankratz (1970) and Marshall *et al.* (1971a, b) immersed electron microscope grids in aquatic habitats and in laboratory cultures and observed that extracellular polymer appeared to anchor bacteria to the support surface. Marshall *et al.* (1971a) reported polymer "footprints" remaining on the grid surface after an adherent marine pseudomonad had been sheared from the surface. Direct demonstration of bridging of the space between the bacterium and the solid surface by polymer was presented by Marshall and Cruickshank (1973), Marshall (1973), and Fletcher and Floodgate (1973). To achieve this, bacteria were allowed to adhere to suitable substrata (araldite, membrane filter), which were then washed, embedded, and ultrathin sections cut perpendicularly to the original solid–water interface. Subsequently, polymer bridging between bacteria and surfaces in numerous natural habitats has been observed, and excellent representative electron micrographs can be seen in Corpe (1980) and Costerton *et al.* (1981; Chapter 1). Marshall (1980) has presented a series of photographs illustrating the progressive buildup of bridging polymer with time between *Flexibacter* CW7 and the attachment surface.

Fletcher and Floodgate (1973) recognized two alcian blue- and ruthenium red-reactive polymers produced by a marine pseudomonad attached to surfaces. One was a compact layer on the outer surface of the cell wall, which the authors suggested was involved in the adhesion process, and a second loose fibrous polysaccharide, which may have been produced by the bacterium after attachment. Similarly, Costerton *et al.* (1981) describe the production of additional polymeric material after adhesion of the bacteria, followed by the replication of the bacteria within this polymer matrix to form microcolonies. Costerton *et al.* (1981) caution that normal subculturing techniques used in the laboratory often select for fast-growing mutants lacking the mucoid extracellular polymer (glycocalyx) found in natural habitats (Chapter 1, Section 1.1). Consequently, other polymers are exposed, and the adhesive properties may be very different from those of the wild-type cells. The possession of mucoid properties, however, is not necessarily synonymous with adhesiveness. For example, Brown *et al.* (1977) and Wardell *et al.* (1980) demonstrated widespread adhesion from mixed populations in a carbon-limited culture, despite no obvious evidence of extracellular polymer production. In nitrogen-limited culture, however, little adhesion was observed, yet abundant extracellular polymer production was observed. Similarly, Pringle *et al.* (1983) used an adsubble process (Lemlich, 1972) to foam fractionate the wild type of a freshwater strain of *Pseudomonas fluorescens* from a fermenter during flow conditions. The wild type, which concentrated in the foam, produced very little extracellular polysaccharide. Under these conditions, the aqueous phase became dominated by a mucoid mutant that produced an alginate polymer and was far less adhesive than the wild type. A second mutant, with crenated colony morphology, was found on the vessel walls and produced very little extracellular polymer but exhibited increased adhesiveness compared to the wild type.

6.2. Polymers as Dispersants

Excessive polymer production by bacteria may serve to effectively disperse bacteria in suspension and prevent them from adhering to solid surfaces. The phenomenon of steric hindrance is well known in colloid chemistry (LaMer and Healy, 1963; Vincent, 1974) and occurs when there is a high degree of coverage of two interacting surfaces by a polymer. This results in a mutual repulsion between the two surfaces because of the exclusion volume of the polymer essentially saturating both surfaces. Such steric hindrance has been observed in studies of bacterial aggregation (Harris and Mitchell, 1973) and of animal cell fusion (Maroudas, 1975).

It is more than likely that the poor adhesion observed by Brown *et al.* (1977) in nitrogen-limited bacterial cultures and by Pringle *et al.* (1983) in a mucoid mutant of *P. fluorescens* was the result of steric hindrance between the polymer on the bacterial surface and the same polymer that had adsorbed to the substratum surface. This possibility suggests that some bacteria may produce a polymer that is solely for adhesive purposes and, following adhesion, produce a secondary polymer (Fletcher and Floodgate, 1973), the function of which is to either reinforce the attachment process, render the surface more amenable to colonization by other microorganisms (Corpe, 1970b; Tosteson and Corpe, 1975), or protect the bacteria from predators and toxins (Costerton *et al.*, 1981). To speak of the bacterial "glycocalyx" as being adhesive (Costerton *et al.*, 1981) is potentially simplistic and misleading, and requires extensive qualification.

6.3. Polymer Composition

Probably the major problem in accurately defining the mechanisms whereby bacteria attach irreversibly to inert surfaces is the paucity of knowledge concerning the detailed structure of the extracellular polymers responsible for the anchoring process. Final definition of the forces involved in polymer binding at the surface will have to await the availability of detailed analyses of the different polymers produced by a wide range of bacteria. Such analyses should ultimately reveal the actual group or groups on the polymer molecules that interact directly with different substratum surfaces. It is certain that for a single bacterial type, different modes of interaction exist with different substrata (Fletcher, 1980a). Examples of the physically different polymers produced by bacteria are those illustrated in Marshall (1981) showing amorphous (*Hyphomicrobium*), random fibrillar (*Pseudomonas*), and radiating fibrillar (*Flexibacter*) polymer "footprints."

Few chemical analyses of extracellular polymers produced by bacteria initially isolated from inert surfaces have been reported. Corpe (1970b) found the polymer of *P. atlantica* to contain mannose, glucose, and galactose, together with glucuronic acid and pyruvate. This polymer is notable for its high content of uronic acid, with a ratio of hexose to uronic acid of 1 : 1. According to Sutherland (1980), most microbial acidic polysaccharides contain about 20 to 25% uronic acid. It must be emphasized that Corpe's (1970b) polymer was obtained from bacteria grown in an unrealistically rich medium. A polysaccharide isolated from *Sphaerotilus natans* by Gaudy and Wolfe (1962) contained approximately equal amounts of fucose, glucose, galactose, and glucuronic acid. A similar monosaccharide composition was reported by Humphrey *et al.* (1979) for the extracellular slime produced by the gliding bacterium, *Flexibacter* BH3, but the uronic acid levels were very low (3%). The physical properties of this slime suggested it was a linear polymer and suitable as a temporary (Stefan) adhesive (see Section 5).

Sutherland (1980) reported the chemical composition of extracellular polymers extracted from 17 seawater and marine bacteria capable of attaching to surfaces. Most of the polysaccharides contained D-mannose, D-glucose, and D-galactose. Alternatively, D-mannose was present with either D-glucose or D-galactose. In the freshwater strains, galacturonic or glucuronic acids were sometimes present. Fucose was found in some of the polymers from marine strains. Sutherland noted that an interesting feature was the apparent absence of *O*-acetyl groups from these polymers, whereas this substituent is often found in polysaccharides from nonaquatic bacteria. Sutherland predicted that these polymers may have a high incidence of 1,3- or 1,4-linkages which would confer some degree of rigidity to the molecules. It was found that the apparent viscosity of some of the polymers decreased with increasing shear rate and that the apparent viscosity of the polymer from the marine strain 5.2 decreased with increasing levels of calcium ions.

As reported in Section 6.1, Fletcher and Floodgate (1973) recognized the presence of two different acidic polysaccharides produced at a surface by a marine pseudomonad (*Pseudomonas* NCMB 2021). The extracellular polysaccharides of this bacterium are currently undergoing detailed analysis by Bjørn E. Christensen and Olav Smidsrød of the University of Trondheim. According to Christensen (personal communication), the nondialyzable fraction of exudates of NCMB 2021 consists of approximately 65% carbohydrate and 35% protein. The carbohydrate fraction consists of equal amounts of the high-molecular-weight "A-polysaccharide" and "B-polysaccharide." The former consists mainly of glucose, galactose, and other unidentified sugars (some with acidic groups), whereas the latter is somewhat hydrophobic and contains *N*-acetylglucosamine and several other components (probably a 6-deoxy-sugar and *O*-acetyl groups). Both polysaccharides bind to anion exchangers and both are precipitated by ferric-iron. A-polysaccharide gives viscous solutions and forms a gel at high concentrations, whereas B-polysaccharide is less viscous, more easily soluble in water, and is partly soluble in 90% phenol. The roles of these two polysaccharides in the adhesion process remain to be determined.

That proteins may be part of the adhesive polymers produced by some bacteria is suggested by the results of Danielsson *et al.* (1977), who reported the successful removal of a marine pseudomonad from surfaces by treatment with proteolytic enzymes. Marshall (1973) and Fletcher (1980b) failed to remove marine bacteria from glass surfaces with proteolytic enzymes, but Fletcher and Marshall (1982a) noted that Pronase removed *Pseudomonas* NCMB 2021 from polystyrene surfaces but not from glass surfaces, suggesting that different mechanisms of polymer bridging are involved in the attachment of this bacterium to such surfaces.

The evidence for two distinct polymers produced by NCMB 2021 introduces an element of caution in considering analytical data of crude polymer extracts. It is obvious that under certain growth conditions, and particularly in broth culture, the polymers produced by the bacteria may bear little resemblance to those that are actually responsible for adhesion to solid surfaces (e.g., the polymer described by Corpe, 1970b). In many natural habitats, bacteria in the aqueous phase are starving (Novitsky and Morita, 1978; Dawson *et al.*, 1981), and bridging polymers produced under such conditions may be unique. It is obvious that many more detailed studies of the composition and physical properties of adhesive polymers of bacterial origin are required.

6.4. Genetic Aspects of Polymer Production

The fact that the adhesive properties of many bacteria are lost or modified under *in vitro* conditions (Costerton *et al.*, 1981) suggests either a chromosomal mutation or the loss of a plasmid. Rosenberg *et al.* (1982) have shown that nonadhesive mutants of *Acinetobacter calcoaceticus* RAG-1 fail to produce the thin fimbriae that appear to be responsible for adhesion to polystyrene and to hexadecane. The thin fimbriae were observed on partial revertants derived from the nonadhesive mutants. The lower adhesion observed by Pringle *et al.* (1983) for mucoid mutants of *P. fluorescens* emphasizes the fact that the mere possession of extracellular polymer in mutants does not ensure enhanced adhesiveness.

Both chromosomal and plasmid control have been implicated in the adhesion of *Agrobacterium tumefaciens* to host plants (Whatley *et al.*, 1978; Matthysse *et al.*, 1978;

Chapter 9, Section 3.2). More recently, Jones *et al.* (1982) have presented evidence associating an autonomous 60-megadalton plasmid with adhesiveness, invasiveness, and virulence in *Salmonella typhimurium* (Chapter 15, Section 2.2.3). Morris *et al.* (1982) have demonstrated that transfer of the Vir plasmid from *Escherichia coli* S5 into other strains of this organism resulted in increased adhesiveness in the recipient strains. The possibility of plasmids with a wide host range and capable of coding for the production of adhesive extracellular polymers could be of major significance in the ecology of bacteria in natural habitats.

Bacteria obtain many advantages by residing at surfaces (Costerton *et al.*, 1981), and, consequently, it is likely that there would be a natural selection pressure toward the production of adhesive polymers in bacteria in natural habitats. Obviously, detailed studies on genetic aspects of the adhesion of bacteria to inert surfaces are urgently required.

7. CONCLUSIONS

Despite a great deal of research currently under way on adhesion of bacteria to inert surfaces, there are many areas where much more information is required before we have a satisfactory understanding of the overall processes involved. The mechanisms whereby bacteria are transported to surfaces are not entirely clear. Much of the emphasis in current work is on irreversible adhesion, and it is obvious that we know little of the bacteria that can be washed off surfaces (reversibly adhering) prior to our examination of bacteria remaining attached to the surfaces. It is quite likely that reversible adhesion is a more significant factor in microbial ecology than is generally recognized.

In terms of irreversible adhesion, our knowledge of the actual events occurring at the solid–liquid interface is limited by the paucity of information on the chemical composition and physical characteristics of the bridging polymers produced by bacteria. More detailed studies are required on the question of passive versus active adhesion for a range of different substrata. The surface energy approach to adhesion is beginning to provide some sensible approximations as to the factors involved in the adhesion process, but this approach needs to be reconciled with aspects of the DLVO theory, the role of bridging polymers, and the modifying effects of conditioning films.

Too many studies on adhesion are carried out on single cultures and do not reflect the various competitive effects that exist between different bacteria in natural habitats. An extension of the competition studies dealt with in Section 4.7.3 is certainly warranted. Finally, studies on genetic aspects of adhesion may provide useful insights into the control of polymer production by bacteria and into the functional groups of particular polymers that may be involved in the adhesion process.

REFERENCES

Absolom, D. R., van Oss, C. J., Genco, R. J., Francis, D. W., Zingg, W., and Neumann, A. W., 1980, Surface thermodynamics of normal and pathological granulocytes, *Cell Biophys.* **2:**113–126.

Absolom, D. R., Lamberti, F. V., Policova, Z., Zingg, W., van Oss, C. J., and Neumann, A. W., 1983, Surface thermodynamics of bacterial adhesion, *Appl. Environ. Microbiol.* **46:**90–97.

Adler, J., 1969, Chemoreceptors in bacteria, *Science* **166:**1588–1597.

Applebaum, B., Golub, E., Holt, S. C., and Rosan, B., 1979, In vitro studies of dental plaque formation: Adsorption of oral streptococci to hydroxyapatite, *Infect. Immun.* **25:**717–728.

Baier, R. E., 1973, Influence of the initial surface condition of materials on bioadhesion, in: *Proceedings, 3rd International Congress on Marine Corrosion Fouling* (R. F. Acker, B. F. Brown, J. R. DePalma, and W. P. Iverson, eds.), Northwestern University Press, Evanston, Ill., pp. 633–639.

Baier, R. E., 1980, Substrate influence on adhesion of microorganisms and their resultant new surface properties, in: *Adsorption of Microorganisms to Surfaces* (G. Bitton and K. C. Marshall, eds.), Wiley–Interscience, New York, pp. 59–104.

Bangham, A. D., and Pethica, B. A., 1960, The adhesiveness of cells and the nature of chemical groups at their surfaces, *Proc. R. Phys. Soc. Edinburgh* **28:**43–52.

Belas, M. R., and Colwell, R. R., 1982, Adsorption kinetics of laterally and polarly flagellated *Vibrio, J. Bacteriol.* **151:**1568–1580.

Brown, C. M., Ellwood, D. C., and Hunter, J. R., 1977, Growth of bacteria at surfaces: Influence of nutrient limitation, *FEMS Microbiol. Lett.* **1:**163–166.

Bryers, J. D., and Characklis, W. G., 1981, Kinetics of initial biofilm formation within a turbulent flow system, in: *Fouling of Heat Transfer Equipment* (E. F. C. Somerscales and J. G. Knudsen, eds.), Hemisphere, Washington, D.C., pp. 313–333.

Carson, J., and Allsopp, D., 1980, The enumeration of marine periphytic bacteria from a temporal sampling series, in: *Biodeterioration, Proceedings of the 4th International Biodeterioration Symposium, Berlin* (T. A. Oxley, G. Becker, and D. Allsopp, eds.), Pitman, London, pp. 193–198.

Celesk, R. A., and London, J., 1980, Attachment of oral *Cytophaga* species to hydroxyapatite-containing surfaces, *Infect. Immun.* **29:**768–777.

Characklis, W. G., 1981a, Bioengineering report: Fouling biofilm development: A process analysis, *Biotech. Bioeng.* **23:**1923–1960.

Characklis, W. G., 1981b, Microbial fouling: A process analysis, in: *Fouling of Heat Transfer Equipment* (E. F. C. Somerscales and J. G. Knudsen, eds.), Hemisphere, Washington, D.C., pp. 251–291.

Clark, W. B., Bammann, L. L., and Gibbons, R. J., 1978, Comparative estimates of bacterial affinities and adsorption sites on hydroxyapatite surfaces, *Infect. Immun.* **19:**846–853.

Corpe, W. A., 1970a, Attachment of marine bacteria to solid surfaces, in: *Adhesion in Biological Systems* (R. S. Manly, ed.), Academic Press, New York, pp. 73–87.

Corpe, W. A., 1970b, An acid polysaccharide produced by a primary film-forming marine bacterium, *Dev. Ind. Microbiol.* **11:**402–412.

Corpe, W. A., 1980, Microbial surface components involved in the adsorption of microorganisms onto surfaces, in: *Adsorption of Microorganisms to Surfaces* (G. Bitton and K. C. Marshall, eds.), Wiley–Interscience, New York, pp. 105–144.

Costerton, J. W., Irvin, R. T., and Cheng, K.-J., 1981, The bacterial glycocalyx in nature and disease, *Annu. Rev. Microbiol.* **35:**299–324.

Cox, P. J., and Twigg, G. I., 1974, Leptospiral motility, *Nature (London)* **250:**260–261.

Crisp, D. J., 1973, Mechanisms of adhesion of fouling organisms, in: *Proceedings, 3rd International Congress on Marine Corrosion Fouling* (R. F. Acker, B. F. Brown, J. R. DePalma, and W. P. Iverson, eds.), Northwestern University Press, Evanston, Ill., pp. 691–709.

Dahlbäck, B., Hermansson, M., Kjelleberg, S., and Norkrans, B., 1981, The hydrophobicity of bacteria—An important factor in their initial adhesion at the air–water interface, *Arch. Microbiol.* **128:**267–270.

Danielsson, A., Norkrans, B., and Björnsson, A., 1977, On bacterial adhesion—The effect of certain enzymes on adhered cells of a marine *Pseudomonas* sp., *Bot. Mar.* **20:**13–17.

Dawson, M. P., Humphrey, B. A., and Marshall, K. C., 1981, Adhesion, a tactic in the survival strategy of a marine vibrio during starvation, *Curr. Microbiol.* **6:**195–198.

Derjaguin, B. V., and Landau, L., 1941, Theory of the stability of strongly charged lyophobic sols and of the adhesion of strongly charged particles in solutions of electrolytes, *Acta Physicochim. URSS* **14:**633–662.

Dexter, S. C., 1979, Influence of substratum critical surface tension on bacterial adhesion—*In situ* studies, *J. Colloid Interface Sci.* **70:**346–354.

Dexter, S. C., Sullivan, J. D., Jr., Williams, J., III, and Watson, S. W., 1975, Influence of substrate wettability on the attachment of marine bacteria to various surfaces, *Appl. Microbiol.* **30:**298–308.

Doetsch, R. N., and Seymour, F. W. K., 1970, Negative chemotaxis in bacteria, *Life Sci.* **9:**1029–1037.

Doyle, R. J., Nesbitt, W. E., and Taylor, K. G., 1982, On the mechanism of adherence of *Streptococcus sanguis* to hydroxylapatite, *FEMS Microbiol. Lett.* **15:**1–5.

Everett, D. H., and Radke, C. J., 1975, Thermodynamics of adsorption and interparticle forces, in: *Adsorption at Interfaces* (K. L. Mittal, ed.), American Chemical Society, Washington, D.C., pp. 1–15.

Fisher, L. R., Israelachvili, J. N., Parker, N. S., and Sharples, F., 1980, Adhesion measurement, in: *Microbial Adhesion to Surfaces* (R. C. W. Berkeley, J. M., Lynch, J. Melling, P. R. Rutter, and B. Vincent, eds.), Horwood, Chichester, pp. 515–517.

Fletcher, M., 1976, The effects of proteins on bacterial attachment to polystyrene, *J. Gen. Microbiol.* **94:**400–404.

Fletcher, M., 1977, The effects of culture concentration and age, time, and temperature on bacterial attachment to polystyrene, *Can. J. Microbiol.* **23:**1–6.

Fletcher, M., 1980a, The question of passive versus active attachment mechanisms in non-specific bacterial adhesion, in: *Microbial Adhesion to Surfaces* (R. C. W. Berkeley, J. M. Lynch, J. Melling, P. R. Rutter, and B. Vincent, eds.), Horwood, Chichester, pp. 197–210.

Fletcher, M., 1980b, Adherence of marine microorganisms to smooth surfaces, in: *Bacterial Adherence* (E. H. Beachey, ed.), Chapman & Hall, London, pp. 345–374.

Fletcher, M., 1983, The effects of methanol, ethanol, propanol and butanol on bacterial attachment to surfaces, *J. Gen. Microbiol.* **129:**633–641.

Fletcher, M., and Floodgate, G. D., 1973, An electron-microscopic demonstration of an acidic polysaccharide involved in the adhesion of a marine bacterium to solid surfaces, *J. Gen. Microbiol.* **74:**325–334.

Fletcher, M., and Loeb, G. I., 1976, The influence of substratum surface properties on the attachment of a marine bacterium, in: *Colloid and Interface Science,* Volume III (M. Kerker, ed.), Academic Press, New York, pp. 459–469.

Fletcher, M., and Loeb, G. I., 1979, Influence of substratum characteristics on the attachment of a marine pseudomonad to solid surfaces, *Appl. Environ. Microbiol.* **37:**67–72.

Fletcher, M., and Marshall, K. C., 1982a, Bubble contact angle method for evaluating substratum interfacial characteristics and its relevance to bacterial attachment, *Appl. Environ. Microbiol.* **44:**184–192.

Fletcher, M., and Marshall, K. C., 1982b, Are solid surfaces of ecological significance to aquatic bacteria?, in: *Advances in Microbial Ecology,* Volume 6 (K. C. Marshall, ed.), Plenum Press, New York, pp. 199–236.

Fletcher, M., and Pringle, J. H., 1984, The effect of surface free energy and medium surface tension on bacterial attachment to solid surfaces, *J. Colloid Interface Sci.,* in press.

Gaudy, E., and Wolfe, R. S., 1962, Composition of an extracellular polysaccharide produced by *Sphaerotilus natans, Appl. Microbiol.* **10:**200–205.

Gerson, D. F., 1980, Cell surface energy, contact angles and phase partition. I. Lymphocyte cell lines in biphasic aqueous mixtures, *Biochim. Biophys. Acta* **602:**269–280.

Gerson, D. F., and Akit, J., 1980, Cell surface energy, contact angles and phase partition. II. Bacterial cells in biphasic aqueous mixtures, *Biochim. Biophys. Acta* **602:**281–284.

Gerson, D. F., and Scheer, D., 1980, Cell surface energy, contact angles and phase partition. III. Adhesion of bacterial cells to hydrophobic surfaces, *Biochim. Biophys. Acta* **602:**506–510.

Gibbons, R. J., Moreno, E. C., and Spinell, D. M., 1976, Model delineating the effects of a salivary pellicle on the adsorption of *Streptococcus miteor* onto hydroxyapatite, *Infect. Immun.* **14:**1109–1112.

Hall, D. G., 1972, Thermodynamic treatment of some factors affecting the interaction between colloidal particles, *J. Chem. Soc. Faraday Trans. 2* **68:**1269–1282.

Hamada, S., 1977, New glucan synthesis as a prerequisite for adherence of *Streptococcus mutans* to smooth glass surfaces, *Microbios Lett.* **5:**141–146.

Harris, R. H., and Mitchell, R., 1973, The role of polymers in microbial aggregation, *Annu. Rev. Microbiol.* **27:**27–50.

Hirsch, P., and Pankratz, S. H., 1970, Studies of bacterial populations in natural environments by use of submerged electron microscope grids, *Z. Allg. Mikrobiol.* **10:**589–605.

Humphrey, B. A., Dickson, M. R., and Marshall, K. C., 1979, Physicochemical and in situ observations on the adhesion of gliding bacteria to surfaces, *Arch. Microbiol.* **120:**231–238.

Jones, G. W., Richardson, L. A., and Vanden Bosch, J. L., 1980, Phases in the interaction between bacteria and animal cells, in: *Microbial Adhesion to Surfaces* (R. C. W. Berkeley, J. M. Lynch, J. Melling, P. R. Rutter, and B. Vincent, eds.), Horwood, Chichester, pp. 211–219.

Jones, G. W., Richardson, L. A., and Uhlman, D., 1981, The invasion of HeLa cells by *Salmonella typhimurium:* Reversible and irreversible bacterial attachment and the role of bacterial motility, *J. Gen. Microbiol.* **127:**351–360.

Jones, G. W., Rabert, D. K., Svinarich, D. M., and Whitfield, H. J., 1982, Association of adhesive, invasive and virulent phenotypes of *Salmonella typhimurium* with autonomous 60-megadalton plasmids, *Infect. Immun.* **38:**476–486.

Kefford, B., Kjelleberg, S., and Marshall, K. C., 1982, Bacterial scavenging: Utilization of fatty acids localized at a solid–liquid interface, *Arch. Microbiol.* **133:**257–260.

Kjelleberg, S., Humphrey, B. A., and Marshall, K. C., 1983, Initial phases of starvation and activity of bacteria at surfaces, *Appl. Environ. Microbiol.* **46:**978–984.

LaMer, V. K., and Healy, T. W., 1963, Adsorption flocculation of macromolecules at the solid–liquid surface, *Rev. Pure Appl. Chem.* **13:**112–133.

Leech, R., and Hefford, R. J. W., 1980, The observation of bacterial deposition from a flowing suspension, in: *Microbial Adhesion to Surfaces* (R. C. W. Berkeley, J. M. Lynch, J. Melling, P. R. Rutter, and B. Vincent, eds.), Horwood, Chichester, pp. 544–545.

Lemlich, R., 1972, Adsubble processes: Foam fractionation and bubble fractionation, *J. Geophys. Res.* **77**:5204–5210.

Lister, D. H., 1981, Corrosion products in power generating systems, in: *Fouling of Heat Transfer Equipment* (E. F. C. Somerscales and J. G. Knudsen, eds.), Hemisphere, Washington, D.C., pp. 135–200.

Loeb, G. I., 1977, The Settlement of Fouling Organisms on Hydrophobic Surfaces, Naval Research Laboratory Memorandum Report 3665, Washington, D.C.

Lupton, F. S., 1982, The adhesion of bacteria to heterocysts of *Anabaena*: An example of specific microbial attachment of ecological significance, Ph.D. thesis, University of New South Wales, Kensington, Australia.

Lupton, F. S., and Marshall, K. C., 1981, Specific adhesion of bacteria to heterocysts of *Anabaena* spp. and its ecological significance, *Appl. Environ. Microbiol.* **42**:1085–1092.

Magnusson, K. E., Stendahl, O., Tagesson, C., Edebo, L., and Johansson, G., 1977, The tendency of smooth and rough *Salmonella typhimurium* bacteria and lipopolysaccharide to hydrophobic and ionic interactions as studied in aqueous polymer two-phase system, *Acta Pathol. Microbiol. Scand. Sect. B* **85**:212–218.

Maroudas, N. G., 1975, Polymer exclusion, cell adhesion and membrane fusion, *Nature (London)* **254**:695–696.

Marshall, K. C., 1967, Electrophoretic properties of fast- and slow-growing species of *Rhizobium, Aust. J. Biol. Sci.* **20**:429–438.

Marshall, K. C., 1973, Mechanism of adhesion of marine bacteria to surfaces, in: *Proceedings, 3rd International Congress on Marine Corrosion Fouling* (R. F. Acker, B. F. Brown, J. R. DePalma, and W. P. Iverson, eds.), Northwestern University Press, Evanston, Ill., pp. 625–632.

Marshall, K. C., 1976, *Interfaces in Microbial Ecology*, Harvard University Press, Cambridge, Mass.

Marshall, K. C., 1980, Microorganisms and interfaces, *BioScience* **30**:246–249.

Marshall, K. C., 1981, Bacterial behavior at solid surfaces—A prelude to microbial fouling, in: *Fouling of Heat Transfer Equipment* (E. F. C. Somerscales and J. G. Knudsen, eds.), Hemisphere, Washington, D.C., pp. 305–312.

Marshall, K. C., and Cruickshank, R. H., 1973, Cell surface hydrophobicity and the orientation of certain bacteria at interfaces, *Arch. Mikrobiol.* **91**:29–40.

Marshall, K. C., Stout, R., and Mitchell, R., 1971a, Mechanism of the initial events in the sorption of marine bacteria to surfaces, *J. Gen. Microbiol.* **68**:337–348.

Marshall, K. C., Stout, R., and Mitchell, R., 1971b, Selective sorption of bacteria from seawater, *Can. J. Microbiol.* **17**:1413–1416.

Matthysse, A. G., Wyman, P. M., and Holmes, K. V., 1978, Plasmid-dependent attachment of *Agrobacterium tumefaciens* to plant tissue culture cells, *Infect. Immun.* **22**:516–522.

Meadows, P. S., 1971, The attachment of bacteria to solid surfaces, *Arch. Mikrobiol.* **75**:374–381.

Minato, H., and Suto, T., 1976, Technique of fractionation of bacteria in rumen microbial ecosystem. I. Attachment of rumen bacteria to starch granules and elution of bacteria attached to them, *J. Gen. Appl. Microbiol.* **22**:259–276.

Minato, H., and Suto, T., 1979, Technique for fractionation of bacteria in rumen microbial ecosystem. III. Attachment of bacteria isolated from bovine rumen to starch granules *in vitro* and elution of bacteria attached therefrom, *J. Gen. Appl. Microbiol.* **25**:71–93.

Morris, J. A., Thorns, C. J., Scott, A. C., and Sojka, W. J., 1982, Adhesive properties associated with the Vir plasmid: A transmissible pathogenic characteristic associated with strains of invasive *Escherichia coli, J. Gen. Microbiol.* **128**:2097–2103.

Mudd, S., and Mudd, E. B. H., 1924, Certain interfacial tension relations and the behavior of bacteria in films, *J. Exp. Med.* **40**:647–660.

Neumann, A. W., Good, R. J., Hope, C. J., and Sejpal, M., 1974, An equation-of-state approach to determine surface tensions of low-energy solids from contact angles, *J. Colloid Interface Sci.* **49**:291–304.

Neumann, A. W., Absolom, D. R., van Oss, C. J., and Zingg, W., 1979, Surface thermodynamics of leukocyte and platelet adhesion to polymer surfaces, *Cell Biophys.* **1**:79–92.

Neumann, A. W., Hum, O. S., Francis, D. W., Zingg, W., and van Oss, C. J., 1980, Kinetic and thermodynamic aspects of plaetelet adhesion from suspension to various substrates, *J. Biomed. Matter. Res.* **14**:499–509.

Novitsky, J. A., and Morita, R. Y., 1978, Possible strategy for the survival of marine bacteria under starvation conditions, *Mar. Biol.* **48**:289–295.

O'Brien, R. W., 1981, The gliding motion of a bacterium: *Flexibacter* strain BH3, *J. Aust. Math. Soc. Ser. B* **23**:2–16.

Ørstavik, D., 1977, Sorption of streptococci to glass: Effects of macromolecular solutes, *Acta Pathol. Microbiol. Scand.* **85**:47–53.

Pethica, B. A., 1980, Microbial and cell adhesion, in: *Microbial Adhesion to Surfaces* (R. C. W. Berkeley, J. M. Lynch, J. Melling, P. R. Rutter, and B. Vincent, eds.), Horwood, Chichester, pp. 19–45.

Plummer, D. T., and James, A. M., 1961, Some physical investigations of the behaviour of bacterial surfaces. III. The variation of the electrophoretic mobility and capsule size of *Aerobacter aerogenes* with age, *Biochim. Biophys. Acta* **53**:453–460.

Pourdjabber, F., and Russell, C., 1979, Factors affecting adhesion of bacteria to a tooth *in vitro*, *Microbios* **26**:73–84.

Pringle, J. H., and Fletcher, M., 1983, Influence of substratum wettability on attachment of freshwater bacteria to solid surfaces, *Appl. Environ. Microbiol.* **45**:811–817.

Pringle, J. H., Fletcher, M., and Ellwood, D. C., 1983, Selection of attachment mutants during the continuous culture of *Pseudomonas fluorescens* and relationship between attachment ability and surface composition, *J. Gen. Microbiol.* **129**:2557–2569.

Roper, M. M., and Marshall, K. C., 1974, Modification of the interaction between *Escherichia coli* and bacteriophage in saline sediment, *Microb. Ecol.* **1**:1–13.

Rosenberg, M., 1981, Bacterial adherence to polystyrene: A replica method of screening for bacterial hydrophobicity, *Appl. Environ. Microbiol.* **42**:375–377.

Rosenberg, M., Gutnik, D., and Rosenberg, E., 1980, Adherence of bacteria to hydrocarbons: A simple method for measuring cell-surface hydrophobicity, *FEMS Microbiol. Lett.* **9**:29–33.

Rosenberg, M., Bayer, E. A., Delarea, J., and Rosenberg, E., 1982, Role of thin fimbriae in adherence and growth of *Acinetobacter calcoaceticus* RAG-1 on hexadecane, *Appl. Environ. Microbiol.* **44**:929–937.

Rubentschick, L., Roisin, M. B., and Bieljansky, F. M., 1936, Adsorption of bacteria in salt lakes, *J. Bacteriol.* **32**:11–31.

Seymour, F. W. K., and Doetsch, R. N., 1973, Chemotactic responses by motile bacteria, *J. Gen. Microbiol.* **78**:287–296.

Shaw, D. J., 1966, *Introduction to Colloid and Surface Chemistry*, Butterworths, London.

Sjoblad, R. D., and Doetsch, R. N., 1982, Adsorption of polarly flagellated bacteria to surfaces, *Curr. Microbiol.* **7**:191–194.

Sutherland, I. W., 1980, Polysaccharides in the adhesion of marine and freshwater bacteria, in: *Microbial Adhesion to Surfaces* (R. C. W. Berkeley, J. M. Lynch, J. Melling, P. R. Rutter, and B. Vincent, eds.), Horwood, Chichester, pp. 329–338.

Tadros, T. F., 1980, Particle–surface adhesion, in: *Microbial Adhesion to Surfaces* (R. C. W. Berkeley, J. M. Lynch, J. Melling, P. R. Rutter, and B. Vincent, eds.), Horwood, Chichester, pp. 93–116.

Takakuwa, S., Fujimori, T., and Iwasaki, H., 1979, Some properties of cell–sulphur adhesion in *Thiobacillus thiooxidans*, *J. Gen. Appl. Microbiol.* **25**:21–29.

Tosteson, T. R., and Corpe, W. A., 1975, Enhancement of adhesion of the marine *Chlorella vulgaris* to glass, *Can. J. Microbiol.* **21**:1025–1031.

Vaituzis, A., and Doetsch, R. N., 1969, Motility tracks: Technique for quantitative study of bacterial movement, *Appl. Microbiol.* **17**:584–588.

van Houte, J., and Upeslacis, V. N., 1976, Studies on the mechanism of sucrose-associated colonization of *Streptococcus mutans* on teeth of conventional rats, *J. Dent. Res.* **55**:216–222.

van Oss, C. J., Gillman, C. F., and Neumann, A. W., 1975, *Phagocytic Engulfment and Cell Adhesiveness as Cellular Surface Phenomena*, Dekker, New York.

Verwey, E. J. W., and Overbeek, J. T. G., 1948, *Theory of the Stability of Lyophobic Colloids*, Elsevier, Amsterdam.

Vincent, B., 1974, The effect of adsorbed polymers on dispersion stability, *Adv. Colloid Interface Sci.* **4**:193–277.

Wardell, J. N., Brown, C. M., and Ellwood, D. C., 1980, A continuous culture study of the attachment of bacteria to surfaces, in: *Microbial Adhesion to Surfaces* (R. C. W. Berkeley, J. M. Lynch, J. Melling, P. R. Rutter, and B. Vincent, eds.), Horwood, Chichester, pp. 221–230.

Weiss, L., 1971, Biophysical aspects of initial cell interactions with solid surfaces, *Fed. Proc. Fed. Am. Soc. Exp. Biol.* **30**:1649–1657.

Weiss, L., and Harlos, J. P., 1972, Short term interaction between cell surfaces, *Prog. Surf. Sci.* **1**:355–405.

Whatley, M. H., Margot, J. B., Schell, J., Lippincott, B. B., and Lippincott, J. A., 1978, Plasmid and chromosomal determination of *Agrobacterium* adherence specificity, *J. Gen. Microbiol.* **107**:395–398.

Wheeler, T. T., Clark, W. B., and Birdsell, D. C., 1979, Adherence of *Actinomyces viscosus* T14V and T14AV to hydroxyapatite surfaces in vitro and human teeth in vivo, *Infect. Immun.* **25**:1066–1074.

Wu, S., 1980, Surface tension of solids: Generalization and reinterpretation of critical surface tension, in: *Adhesion and Adsorption of Polymers* (L.-H. Lee, ed.), Plenum Press, New York, pp. 53–65.

Young, L. Y., and Mitchell, R., 1973a, The role of chemotactic responses in primary microbial film formation, in: *Proceedings, 3rd International Congress on Marine Corrosion Fouling* (R. F. Acker, B. F. Brown, J. R. DePalma, and W. P. Iverson, eds.), Northwestern University Press, Evanston, Ill., pp. 617–624.

Young, L. Y., and Mitchell, R., 1973b, Negative chemotaxis of marine bacteria to toxic chemicals, *Appl. Microbiol.* **25:**972–975.

Zisman, W. A., 1964, Relation of equilibrium contact angle to liquid and solid constitution, in: *Contact Angle, Wettability and Adhesion* (R. F. Gould, ed.), American Chemical Society, Washington, D.C., pp. 1–51.

ZoBell, C. E., 1943, The effect of solid surfaces upon bacterial activity, *J. Bacteriol.* **46:**39–56.

Zvyagintsev, D. G., Pertsovskaya, A. F., Yakhnin, E. D., and Averbach, E. I., 1971, Adhesion value of microorganism cells to solid surfaces, *Microbiology (Engl. Transl.)* **40:**889–893.

7

Mechanisms of Bacterial Adhesion at Gas–Liquid Interfaces

S. KJELLEBERG

1. INTRODUCTION

The air–water interface is a simple and useful model system for work on bacterial adhesion, and the similarities between gas–liquid and solid–liquid interfaces have been recognized. With respect to its chemical constituents as a result of adsorption at the air–water interface, the surface microlayer, as "seen" by bacteria, may often be indistinguishable from that of a solid substratum (Baier, 1975).

It is easy to conclude, from the paucity of data and contradictory arguments in the literature, that there is a considerable amount of information to be collected regarding the interaction of microorganisms at gas–liquid interfaces in natural systems. The surface micro-layers in aquatic environments, both marine and limnic systems, are of ecological importance (Norkrans, 1980), and the interfaces of gas bubbles rising through water columns are covered with bacteria being transported to the air–water surface and ejected into the atmosphere via jet and film drops from bursting bubbles (Blanchard, 1983). At any given instant, 1% of the ocean is covered with bubbles (Cipriano and Blanchard, 1982), and approximately 10^{18} of these bubbles break every second (MacIntyre, 1974a). The information presented from systems other than those we traditionally define as aquatic is, to say the least, scarce. However, terrestrial systems include several microenvironments containing boundary water films, and soil particles, litter, and plants are known to have a special microflora in surface films. Also, useful data on the mechanisms for bacterial adhesion at gas–liquid interfaces have been obtained from the formation of aerosols from wastewater treatment facilities.

A model system approach using an air–water interface becomes especially valuable in at least three research areas, each demanding very precise studies: (1) the nature of interaction between bacteria and substratum components (e.g., Kjelleberg and Stenström, 1980; Chapters 6 and 12), (2) behavior of copiotrophic and oligotrophic bacteria utilizing very small amounts of nutrients in a substrate-deficient environment (e.g., Kjelleberg et al., 1982), and (3) activity measurements and exchange processes revealing exact differences in

S. Kjelleberg • Department of Marine Microbiology, University of Göteborg, S-413 19 Göteborg, Sweden.

scavenging ability of surface-bound versus free-living bacteria (e.g., Hermansson and Dahlbäck, 1983).

The utilization of the air–water interface for these studies is based on the high flexibility of experimental design. The interface can be modified in numerous ways, and the amounts of applied surface-active substances are known. As will be considered in Section 4.4, physicochemical parameters, such as surface pressure and surface potential, can be controlled, which leads, for example, to an understanding of the importance of the architecture of the film (surface topography) and allows bacteria and bacterial surface groups to interact in various, controlled ways with molecules residing at the boundary. In addition to the change in surface free energy, analogous with substratum wettability of solid surfaces (Chapter 5), the charge characteristics of the interface can be manipulated by spreading of appropriate substances.

The scavenging of surface-localized nutrients is of enormous importance in oligotrophic environments (Chapter 12). Changes in metabolism and morphology of small starved cells of copiotrophic bacteria can be recorded following uptake of minute amounts of nutrients (see Section 3). Meaningful comparsions with bulk-phase systems can only be carried out knowing the composition at the surface, and this is easily achieved using the air–water interface. The range of substrates that can be used varies from highly surface-active long-chain fatty acids to more hydrophilic amino acids, held at the interface by virtue of relatively hydrophobic groups. The system may be used for several easy-to-perform, easy-to-evaluate studies of bacterial activity (Section 5.1).

In this chapter, Section 2 describes the surface microlayers of aquatic systems, summarizes the state of art in bubble, jet, and film drop research, and comments upon other related gas–liquid systems. Section 3 deals with prerequisites for bacterial adhesion, with emphasis on changes in bacterial surface structure following altered environmental parameters. The bulk of information found in the literature bearing some kind of relationship to the topic "mechanisms" is collected in Section 4, which starts with a short summary of transport and accumulation processes. Although the purpose of this chapter is to present information and ideas concerning mechanisms for bacterial adhesion, it is important to have at least some knowledge of the consequences of bacterial and chemical enrichments at an air–water interface. Thus, Section 5 presents an up-to-date summary of bacterial activity and exchange at the surface microlayers and the possible food chains in and underneath this environment.

2. CHARACTERIZATION OF NATURAL GAS–LIQUID INTERFACES

In order to appreciate the mechanisms for bacterial adhesion and the subsequent enrichment and activity at gas–liquid interfaces, the formation, composition, and behavior of the various strata that make up the so-called surface microlayers will be outlined. This presentation deals almost exclusively with the air–water interface of aquatic systems. References to informative reviews or overview-type articles will be given in most instances, and this outline will be restricted to the shortest possible logical chain of events. The gas–liquid interfaces of rising bubbles, terrestrial systems, and specialized applied areas, such as sewage treatment plants (with respect to transfer of pathogens), are introduced and summarized separately. No attempt is made to discuss gas–liquid boundaries in higher organisms.

2.1. Surface Microlayers of Aquatic Systems

2.1.1. Stratification

Much information concerning the air–water interface of aquatic systems has been detailed, in its basic form, with genetic precision in succeeding articles. Few attempts have

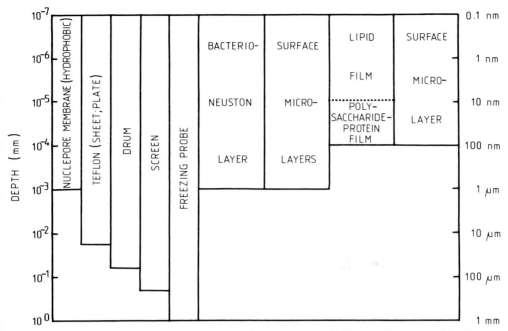

Figure 1. Methods for sampling (left) and possible definitions of the strata (right) in the 1-mm layer of the air–water interface. Samplers operating at various depths have resulted in the concept of a 1-μm bacterioneuston layer, also defined as surface microlayers (SMs). The organic chemical strata, i.e., the lipid and polysaccharide–protein complex films, are equivalent to the surface microlayer (SM). After Norkrans (1980).

been made to systematize the pool of information, and this has very often led to widely different ways of interpreting available data. The first attempt to compile a substantial part of published results, subdivide this into some order, and deduce from the information what could possibly be a complete picture of the air-water interface was made by Norkrans (1979, 1980). The approach she used to deal with much of the information that was available is presented in Fig. 1. The lipid film of so-called dry surfactants (e.g., fatty acids, alcohols, and phospholipids), being 1–2 nm thick in its monomolecular form and approximately 16 nm thick when multilayered, is combined with a lower stratum of wet surfactants, i.e., those with associated water, to make up the surface microlayer (SM) (Sieburth *et al.*, 1976; Norkrans, 1980). The wet surfactants, which form a more irregular stratum than the dry ones, are essentially hydrophilic but stick to the surface by virtue of their few hydrophobic chains. These protein–polysaccharide complexes range in thickness from 10 to 30 nm (Baier *et al.*, 1974; Norkrans, 1980).

An interaction between these SM strata and microorganisms has been demonstrated both in natural habitats and in model systems and shown to result in an accumulation of cells (see, e.g., Lion and Leckie, 1981). This event is described in terms of an enrichment factor (EF), i.e., the ratio between the number of microorganisms per milliliter of the SMs and the number per milliliter of subsurface water. Microorganisms associated with the interface are referred to as neuston (Naumann, 1917). With the use of neuston samplers operating at various depths at the surface, i.e., collecting different amounts of subsurface water, it has been concluded that bacteria interact strongly with, and form a distinct layer in or just below, the organic constituents of the surface molecular film (Daumas *et al.*, 1976; Kjelleberg *et al.*, 1979; Syzdek, 1982). It seems, thus, very plausible to recognize a bacterioneuston layer, extending over 1 μm and forming the SMs (MacIntyre, 1974a; Kjelleberg *et al.*, 1979; Norkrans, 1980). This differs from the SM described above, which is purely molecular (Fig. 1). Figure 1 includes some sampling devices for collecting the SMs. The purpose of using samplers that collect a relatively thin stratum is to remove as nearly as possible only those organic molecules and organisms situated at the air–water interface, leaving behind subsurface water whose chemical constituents and biological inhabitants do not contribute to the modification of the interface. By analogy, samplers that select for foams or collapsed surface films, i.e., those of Szekielda *et al.* (1972) and Pellenbarg (1976), obtained metal concentrations in the SM which were 10^4 times higher than those obtained from other surface sampling devices, in which some dilution by subsurface water had occurred. Surface sampling of chemical constituents cannot always be performed using the most efficient bacterioneuston samplers, and vice versa (Kjelleberg *et al.*, 1979; van Vleet and Williams, 1980b; Carlson, 1982). The sensitivity of methods employed to detect the chemical components and amounts of SMs, as well as the volumes of bulk water withdrawn by a given sampler, has to be taken into account.

Informative introductions to the formation of the SMs and to interrelated processes with bulk water and the atmosphere can be found in Wallace *et al.* (1972), Hatcher and Parker (1974), and Duce and Hoffman (1976).

2.1.2. Nature of Accumulated Chemicals

The nature of chemical constituents in the upper two strata can be predicted on the basis of thermodynamics of surface adsorption. A solute species which decreases surface tension will adsorb at the air–water interface. Most organic solutes decrease the surface tension of water and would therefore be expected to positively and spontaneously adsorb at the air–water interface. All naturally occurring air–water interfaces will consequently differ from the bulk phase in their chemical constituents. Furthermore, competitive adsorption phenomena also occur at the air–water interface (Garrett, 1967). Field observations of surface-active organics are thoroughly summarized in Lion and Leckie (1981). Norkrans (1980) discusses the major constituents of the lipid film, i.e., free fatty acids, glycerides, phospholipids, hydrocarbons, sterols, etc., and considers the stability and coherence of this upper stratum layer. Lipids might not be, however, the quantitatively most important fraction of the SM (Sieburth *et al.*, 1976; Carlson and Mayer, 1980; Gucinski *et al.*, 1981). Surface potential and surface film pressure measurements were made on films of known composition, and the results of these measurements were compared to those generated by films from natural seawater (van Vleet and Williams, 1980a). The surface potential results from interactions between a surface film and the bulk water, which reflect the dipolar characteristics of the film, and is measured as a difference in potential between the liquid and a probe placed above the liquid surface. The surface pressure is defined as the interfacial (surface) tension of a pure liquid minus the interfacial tension in the presence of a surface film. van Vleet and Williams concluded that (1) natural seawater films do not appear to be primarily composed of free fatty acids, free fatty alcohols, or hydrocarbons but resemble films composed of proteins, polysaccharides, humic-type materials, and waxes, and (2) relatively small concentrations of certain lipids in films composed primarily of protein and carbohydrates can determine the resultant surface potential and film pressure of the films. These conclusions are compatible with similar determinations of physical parameters, where monomolecular films of SM samples were spread on synthetic seawater in a surface balance apparatus (Jarvis *et al.*, 1967; Dragcevic *et al.*, 1979). Comparison of the shape of the curves for surface potential, surface tension, and surface viscosity resembled those of high-molecular-weight saturated, and unsaturated, fatty acids and their esters.

2.1.3. High Concentrations of Materials at Interfaces in Natural Waters

Natural surface films often form slicks, i.e., multilayers of surface-active material. This occurs not only in lakes and in coastal regions of the sea but, under certain conditions, also on the open oceans. Much of the accumulation processes at the air–water surface are connected with a convergent flow, and the surface-active compounds so enriched are compressed by horizontal flow toward a convergence. At the frontal zone there will be a production of slicks and foam lines in the area of convergence. This is, in turn, the result of a downward transport at the frontal zones between water masses of different densities (Szekielda et al., 1972). If the wind speed remains low, natural slicks can persist for hours or even days (Hühnerfuss et al., 1977). Slicks modify air–water processes in different ways, e.g., by capillary wave dumping (Larsson et al., 1974) and retardation of evaporation (LaMer, 1962), gas exchange (Hawke and Alexander, 1962), heat transfer (Jarvis, 1962), and light transmission and reflection (Garrett, 1969). With respect, however, to most open ocean locations, Hunter and Liss (1977) point out, on the basis of force–area plots and low film pressures recorded at field stations, that the organic matter at the surface is unlikely to form coherent films.

There are numerous reports on the accumulation of dissolved and particulate organic material (e.g., Williams, 1967; Nishizawa and Nakajima, 1971; Daumas et al., 1976; Dietz et al., 1976; Sieburth et al., 1976; Hunter and Liss, 1977; Carlson, 1982), as well as trace metals (Piotrowicz et al., 1972; Elzerman and Armstrong, 1979; Hunter, 1980; Lion et al., 1982) and organic toxic substances (as reviewed by Norkrans, 1980) at the surfaces of natural waters.

2.1.4. Interrelation between Organisms and Substrates

The concentrations of utilizable substrates at an interface may be greater than those of the bulk aqueous phase if it is oligotrophic. Thus, selective enrichment at the surface of so-called copiotrophic bacteria (Poindexter, 1981a), which require higher amounts of nutrients for growth than oligotrophic bacteria, can be expected (see Section 3.1 and Chapter 12). The concentration of dissolved organic carbon in subsurface water of oligotrophic aquatic habitats is often less than 1 mg carbon/liter, to be compared with a mean concentration of 1427 mg/liter calculated for the 0.1-μm SMs in the North Atlantic (Sieburth et al., 1976). The latter corresponds to the amount of nutrients in media generally used to grow bacteria in the laboratory. Evidence for close correlations between organisms and possible substrates in the SM can be found in the literature. For example, surface enrichment of nitrate, nitrite, phosphate, and silicate has clearly been shown to be associated with phytoplankton growth (Lyons et al., 1980). There are also, based on in situ measurements of sea-surface film potential and pressure: (1) a correlation of total organic carbon (TOC) and total bacteria with high surface pressure values, indicating a condensed surface film (Section 4.4) and (2) an increase in TOC and total bacteria in the surface film, after bubbling air through the water beneath the film (Williams et al., 1980) (see Section 2.2).

The occurrence of high numbers of microorganisms, predominantly bacteria, at air–water interfaces is well established (e.g., Roy et al., 1970; Harvey and Burzell, 1972; Hatcher and Parker, 1974; Crow et al., 1975; Kjelleberg et al., 1979), although patchiness in cell distribution seems to exist (Tsiban, 1975; Dietz et al., 1976; Young, 1978). Among other things, bacteria seem to be responsible, through aggregation, for the formation of particles (Barber, 1966; Batoosingh et al., 1969), even though other hypotheses for this process, based on, for example, convection currents, have been suggested (Wangersky, 1972, 1976; Wheeler, 1975; Johnson, 1976). The observed levels of partitioning of dissolved, particulate, and total metals, and the correlation between the partitioning of particulate metals and particle number, strongly suggest that microlayer enrichment of trace metals is related to the behavior of particles, e.g., of bacteria.

2.2. Gas Bubbles in Aquatic Systems—Adsubble Processes

The compositions of the SMs, the surfaces of bubbles rising through the water, and the jet and film drops formed when the bubbles burst at the air–water interface are closely interrelated and interdependent. Thus, the SM is a dynamic milieu.

The term *adsubble processes* is a contraction of "adsorptive bubble separation processes." These processes are defined as those phenomena and means by which dissolved or suspended material are segregated within or removed from a liquid by adsorption or attachment at the surfaces of rising bubbles (Lemlich, 1972). It is known that these bubbles collect, for example, surface-active compounds that are released by phytoplankton (Wilson and Collier, 1972) and concentrate these materials in the surface film. Droplets that rise from the sea in the surf zone similarly carry a highly compressed surface-active film (Blanchard, 1964).

The original work on effects of bubbling (Baylor *et al.*, 1962) was, however, concerned with the transport of particulate, rather than dissolved, material to the surface film, and it has since become evident that any mechanism moving particles into the surface layer will move bacteria as well. The collection of bacteria on the surfaces of bubbles, with subsequent transfer to the surface film, has been demonstrated numerous times (e.g., Blanchard and Syzdek, 1970, 1972; Bezdek and Carlucci, 1972). Bubbles breaking at the air–water interface act as effective microtomes, skimming material from the bubble surface and ejecting it as jet drops into the atmosphere (MacIntyre, 1968, 1970, 1974b). As the bubble reaches the surface, the surface free energy of the bubble is converted into kinetic energy causing a burst of jet drops and film drops, which are smaller and more numerous, into the atmosphere. Figure 2 illustrates this process schematically. A number of studies report, not surprisingly, the presence of marine microorganisms in the atmosphere (ZoBell and Mathews, 1936; Woodcock, 1948; Stevenson and Collier, 1962; Maynard, 1968), a subject that is excellently reviewed by Duce and Hoffman (1976).

2.2.1. Bubble Scavenging of Bacteria

It is known that bacterial concentrations in jet drops from bursting bubbles may be 1000-fold greater than concentrations in the bulk liquid (e.g., Bezdek and Carlucci, 1972;

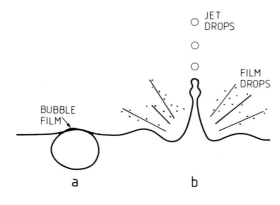

Figure 2. Bubbles rise to (a) and burst at the surface of water to produce both film and jet drops (b). After Blanchard and Syzdek (1982).

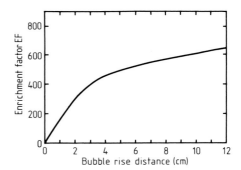

Figure 3. Effect of distance a bubble rises through a bacterial suspension of *Serratia marcescens* in pond water on enrichment factor (EF) for bacteria in top jet drop. After Blanchard *et al.* (1981).

Blanchard and Syzdek, 1972). Where do these bacteria come from? Blanchard (1983) puts all the pieces of information together and his review forms the basis of the following summary. Bacteria reaching the surface of the bubble do not do so by either diffusion or inertial effects. They do it by interception, a collection mechanism that comes into play because the bacteria have a finite size. As inertial effects are negligible, the bacteria follow the fluid streamlines past the bubble. But due to their finite size, bacteria following streamlines very close to the bubble surface make contact with the bubble and are scavenged. The collection efficiency (E_c) is written as the product of a collision efficiency (E) and an attachment efficiency (Ea): $E_c = E \cdot E_a$. E may be related to hydrodynamic effects, as described above, and E_a to surface chemical effects, e.g., the hydrophobic nature of the bacterial surface (Section 4.3). The number of bacteria collected by a bubble surface is also a function of the distance the bubble rises through water. Figure 3 gives the enrichment factor (EF) (i.e., the number of cells per drop divided by the number of cells per milliliter in the bulk suspension) for the top jet drop from a bubble of 380-μm diameter, rising through a suspension of *Serratia marcescens* in pond water (Blanchard *et al.*, 1981). During the first few centimeters of rise, a bubble moves as a fluid sphere with a mobile surface. The collection efficiency has been reported to decrease after about 3 cm of rise, and due to adsorption of surfactants, including bacteria, the mobility of the surface decreases, letting it rise as a solid sphere (Blanchard, 1983). A bubble rise distance of 2 cm in a bacterial suspension (2×10^6 per ml) was enough to produce an EF of about 300 for the bacteria in the top jet drop.

Finally, the EF of bacteria in jet drops is also controlled by the drop size. The thickness of the microlayer that is skimmed off the collapsing bubble to produce the jet drops is in proportion to the size of the rising bubble. The thickness of the layer giving the maximum EF is reported to be the width of a bacterial cell, about 0.5 μm. There is a maximum in EF at a bubble drop size of about 60 μm (e.g., Hejkal *et al.*, 1980). The reasons for this are quite simple: for smaller bubbles, the microlayer that is skimmed is less than the size of the cells. This leaves many bacteria in the suspension when the microlayer of the bubble is transformed into jet drops. Bubbles larger than 60 μm, on the other hand, have a boundary layer that extends deeper into the bulk suspension, approaching a bacterial concentration equal to that of the bulk suspension, thus resulting in a relatively low EF.

There appears to be no mechanism to transfer bacteria from the SMs to the jet drops. Cells will not move fast enough into the bursting bubble cavity, which is necessary in order to reach the descending capillary wave that produces the jet drops. Bacteria in jet drops come from the bubble scavenging in the bulk, not the enriched populations at the SMs. Surface-active materials in the SM can, however, be transported into the atmosphere on ejected jet drops. The transfer from the SM into the bubble cavity is fast enough, as has been shown using radiolabeled material (Bezdek and Carlucci, 1974). It is possible, although this has not been experimentally verified, that film drops can carry bacteria from the SMs into the atmosphere (Blanchard, 1983).

2.2.2. Transfer of Pathogens

The transfer of bacteria from water to air could also constitute a health problem. Woodcock (1955) suggested, with respect to sewage treatment plants, that bacteria and SM material might become airborne in jet drops from bursting bubbles. Previous to this it was known that aerosols caused distribution of infectious, as well as noninfectious, bacteria (Wells and Stone, 1934; ZoBell, 1942). The transport of bacteria by aerosols could be induced by bubbling or splashing procedures in the laboratory (Anderson et al., 1952). It is now well documented that the air around sewage treatment plants is rich in microorganisms (Napolitano and Rowe, 1966; Adams and Spendlove, 1970; Goff et al., 1973), and that bacterial aerosols from various agricultural, municipal, and industrial sources could be public health hazards (Spendlove, 1974; Hickey and Reist, 1975; Adams et al., 1982). Certain bacteria have been shown to be emitted in relatively high concentrations from wastewater treatment facilities (Randall and Ledbetter, 1966), and some investigations have suggested that coliform bacteria could serve well as indicators of airborne contamination by wastewater treatment plants (Adams and Spendlove, 1970; Goff et al., 1973). A definite relation between a disease and ejection of the infectious agent into the atmosphere by bubbling and splashing processes has been established for the pathogen Legionella pneumophila. Bacteria were associated with the aerosols produced by air-conditioning cooling systems (Dondero et al., 1980).

The small film drops which are produced as bubbles burst at the air–water interface will naturally remain in the atmosphere for longer periods than the relatively large jet drops. Blanchard and Syzdek (1982) point out that the size range of 1 to 10 μm for film drops allows for penetration into the lungs and that it might be important, from a health standpoint, to carefully examine the enrichment factors and bacterial species composition in these drops.

2.3. Terrestrial Systems

Very little is known about the microbiology of gas–liquid interfaces in terrestrial systems. Marshall (1975) states that soil microorganisms normally function in the aqueous phase and, at the solid– or gas–water interfaces, within the voids or pore spaces. High numbers of many bacterial types have been seen in surface films from soil crumbs and particles (Harris, 1972; Waid, 1973). Aqueous films on terrestrial litter are inhabited by numerous microorganisms. The surfaces of such films are covered by monolayer-forming substances, of which the spreading pressure can result in a considerable lateral transport of floating and submerged organisms to adjacent water films which have no surface monolayers (Bandoni and Koske, 1974). In 1938, Schulman and Teorell showed that a molecular monolayer moving over water carries with it a considerable volume of associated water, and they determined that the thickness of a boundary layer (consisting of the monolayer and associated water) extends about 0.03 mm into the bulk water, thus including the SM strata defined in Section 2.1. Even though there are a few articles published on this topic, it is clear that more emphasis should be paid to organisms in surface films of water in terrestrial systems. As a further example, Bandoni (1972) examined the surface film of water covering leaves and other plant debris and found this interface to contain large numbers of floating conidia (see also Chapter 9, Section 4.3).

3. BACTERIAL NUTRITIONAL AND CELL SURFACE PROPERTIES WHICH AFFECT ADHESION

3.1. Survival Tactics of Bacteria

3.1.1. The Relationship between Survival and Adhesion

In natural environments, low-nutrient conditions can occur frequently (Morita, 1982), and nutrient availability in these systems is often in flux, rather than occurring at constant substrate levels (Poindexter, 1981a). Jannasch (1974) argued convincingly that the concept of steady state has no bearing on natural ecosystems. This approach, which has been experimentally verified (e.g., Azam and Hodson, 1981), has led to definition of nutritionally distinct groups of bacteria in relation to their response to substrate availability. Obligate and facultative oligotrophic bacteria (Yanagita *et al.*, 1978; Ishida *et al.*, 1982) grow and reproduce under low-nutrient flux, while copiotrophic bacteria demand a flux of nutrient input many times higher for similar energy-demanding processes (Poindexter, 1981a; Chapter 12, Section 4.2.1). Of these, much knowledge is currently being collected about copiotrophs, which are often able to survive in low-nutrient, starvation conditions. Some do so by means of initiation of an active and rapid process that leads to small, starved, metabolically and morphologically altered cells (e.g., Novitsky and Morita, 1976, 1977, 1978; Dawson *et al.*, 1981; Torella and Morita, 1981; Grossman *et al.*, 1982; Kjelleberg *et al.*, 1982, 1983; Humphrey *et al.*, 1983). The existence of starvation-induced responses has naturally been known for a longer period (e.g., Postgate and Hunter, 1962; Harrison and Lawrence, 1963; Druilhet and Sobek, 1976), but has not until recently been related to the context of population dynamics and ecology (Hirsch, 1979; Morita, 1982). Nor has its significance with respect to microbial processes at surfaces previously been highlighted (Fletcher and Marshall, 1982).

Marshall *et al.* (1971) observed that small, dwarflike bacteria were primary colonizers of surfaces immersed in seawater, and increased adhesion to surfaces and higher activity of surface-bound bacteria in nutrient-limited situations have been reported (Jannasch and Pritchard, 1972; Brown *et al.*, 1977). Hydrophobic bacteria were found to be more efficient in nutrient uptake of firmly bound material, as compared with more hydrophilic cells (Kjelleberg *et al.*, 1983), and some irreversibly bound bacteria seemed to act as the "surface specialists." In contrast, reversibly bound cells could not be considered to take advantage of surface-localized substrates (Kefford *et al.*, 1982). It is possible that firmly bound cocci and rod-shaped bacteria are metabolically better adapted to the bulk liquid phase. It was suggested by Dawson *et al.* (1981) that adhesion is a survival tactic of copiotrophic bacteria under low-nutrient stress. The increase in irreversible binding of cells of a marine *Vibrio*, exposed to a starvation regime, has since been confirmed to be a process that can be applicable to some other strains (Kjelleberg and Hermansson, 1984). This degree of irreversible binding has been related to the degree of hydrophobicity of the organisms (Fig. 4), possibly implying that the degree of polarity (i.e., water structure-breaking ability) is a crucial physicochemical factor affecting adhesion at interfaces (Chapters 5 and 6).

In summary, it is important to understand the survival strategies of copiotrophic and oligotrophic bacteria in order to successfully deal with "bacteria–interfaces phenomena." Some small, starved, naturally occurring copiotrophic bacteria can adapt to the environmental stress of low-nutrient conditions by changes in, among other characteristics, morphology and cell surface physicochemical properties (Dawson *et al.*, 1981). The most important such changes, from an adhesion point of view, are increased roughness, decreased polarity of surface groups, and a high degree of irreversible binding.

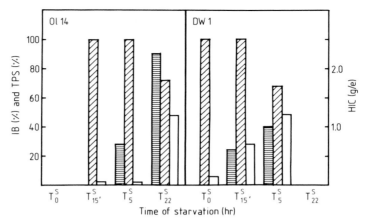

Figure 4. Degree of irreversible binding of bacteria to glass (IB; ▤), two-phase separation in a hexadecane–water system (TPS: ▨), both expressed as the percentage of cells that remain in the aqueous phase, and hydrophobic interaction chromatography results (HIC; ☐), expressed as the ratio of cells retained in the gel to those in the eluate (g/e). With TPS, 100% denotes hydrophilic bacteria. However, with HIC, 100% denotes hydrophobic bacteria, as the higher g/e values, the greater is the tendency for hydrophobic interaction. The bacteria, *Spirillum* 01 14 and *Vibrio* DW1, were examined at various times during exposure to a starvation regime. T_0^S and T_n^S denote the start of the starvation process and the starvation time (n) in minutes, respectively.

3.1.2. Oligotrophs and Copiotrophs at the Air–Water Interface

In many natural environments, as in most experiments where cells are washed, diluted, and suspended in a diluent, the aqueous phase resembles a starvation regime. Wangersky (1977) hypothesized that free-living bacteria exist in a perennial state of malnutrition, and Stevenson (1978) presented a case for dormancy as the natural situation. This forms the basis for interpreting many reports on distribution of bacterioneuston in SMs, as compared with bacterial numbers and types in the water column. Sieburth (1971a) found that a 10-fold dilution of the sample resulted in a 90-fold increase in colonial development from SMs at some oceanic stations. This he considered to result from dilution of inhibitory substances, but the observed effect could also depend on a reduced carryover of SM nutrients if this resulted in increased clumping of the cells. Such bacterial aggregation in extreme oligotrophic conditions was described by Hirsch (1980), where oligotrophic conditions in an aquatic system resulted in extreme polymer production by many diverse strains. The supplement of only 0.005% peptone or yeast extract to lake water media yielded 16 different strains of gram-negative bacteria. These grew dispersed in the bulk phase at nutrient levels typical of oligotrophic conditions (0.005% yeast extract) but formed thin surface pellicles on lake water with even lower nutrient concentrations, i.e., 0.0005% glucose or peptone. It was concluded that extreme oligotrophic conditions resulted in extensive polymer production which allowed for the development of neustonlike pellicles of cell chains. Crawford *et al.* (1982) conducted a survey of lakes and found a relationship between the trophic state of the lake and bacterial enrichments in surface films. The more oligotrophic lakes had a greater proportion of the bacteria occurring in their SMs.

It is, however, also possible to observe the predominance of copiotrophs at the air–water interface. The abundance of nutrients assembled at the SMs in some waters (Sieburth *et al.*, 1976) would not give oligotrophs a suitable environment. In such conditions, they are likely to be outcompeted by copiotrophs in the SMs, and the drawbacks associated with the physiotype of an ideal oligotroph (Poindexter, 1981b) would be accentuated at this interface as well. An ideal oligotroph risks, for example, the occurrence of an uncontrolled uptake of unbalanced nutrients or toxic substances (Poindexter, 1981b), known to be accumulated at the air–water interface (Norkrans, 1980). A larger fraction of copiotrophic, as compared with oligotrophic, bacteria occurs in the SMs, according to results presented by Dahlbäck *et al.* (1982) and Harvey and Young (1980a). Ratios between viable plate counts and total counts were higher in the SMs than in the bulk liquid. Salt marsh samples contained many chains of rod-shaped bacteria (oligotrophs would tend to form spherical cells), which were generally associated with particulate matter (Harvey and Young, 1980a,b). The proportion of large cells found in the latter study (more than 50% > 0.5 μm; 84% ≥ 0.4 μm) also indicated that the dominant nutritional group was the copiotrophs. Using a froth flotation method, the concentration of green algae, e.g., *Chlorella* sp. and *Chlamydomonas* sp., was shown to depend on the food concentration, among other factors, in a similar way (Levin *et al.*, 1962).

However, the sequence of competitive processes at the SMs is likely to be more complex than the simple dominance of one, or the other, of the bacterial types, as outlined above. A. Södergren, G. Odham, B. Stehn, and B. Norén (personal communication) followed the composition of the lipid surface film and associated microorganisms in a static freshwater laboratory system. A succession of bacterial types in the film was noted between the 10th and the 18th day of incubation at 20°C, and appendaged bacteria, resembling *Caulobacter,* were dominant. At 5°C, slime-producing rods in encapsulated aggregates of two to four cells were more common. The occurrence of *Caulobacter* at later stages of the experiment at 20°C could coincide with the production of a more oligotrophic environment as nutrients were depleted, and fatty acids in the lipid surface film were noted to be consumed. Kjelleberg *et al.* (1982) have also shown the efficient scavenging of nutrients by copiotrophic bacteria at an air–water interface. The genus *Caulobacter* has been suggested to represent an ideal oligotroph by Poindexter (1981b), and as such would be outcompeted by copiotrophs should the nutrient availability in the SM be high. The succession from copiotrophs to oligotrophs seems to be delayed at the lower temperature of 5°C.

3.2. The Influence of Surface Properties of Bacteria on Their Distribution at the Air–Water Interface

Of the seven problems related to jet drop research posed by Blanchard and Syzdek (1978), three deal with bacteria. There may presently be a satisfactory answer to at least two of them: (1) "Why is there sometimes an increase in jet drop bacteria with time?" and (2) "Why do we find a relative inability of the nonpigmented cells of *Serratia marcescens* to attach to air bubbles?"

With respect to the first problem, the authors then asked what may be a more appropriate question: "What was happening in the bulk suspension to make the surface of the bacteria more hydrophobic with time?" The experiment lasted for several hours, using a mixture of 4 ml of a 22-hr nutrient broth culture of *S. marcescens* in 1200 ml of sterile distilled water. That mixture resulted in a nutrient concentration of approximately 5 mg organic carbon per liter and contained a starting bacterial density of $\sim 5 \times 10^6$ per ml. This is a perfect condition for the immediate start of events leading to the formation of small starved cells (Section 3.1.1). Furthermore, the initial dwarfing period of mesophilic, copiotrophic bacteria, including rapid division without growth (i.e., fragmentation; Novitsky and Morita, 1977) and continuous size reduction of fragmented cells (Kjelleberg *et al.,* 1983), is known to last approximately 5 hr (Humphrey *et al.,* 1983). Figure 5 shows that a

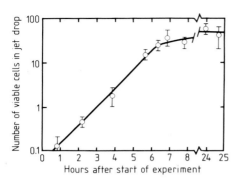

Figure 5. Number of viable *Serratia marcescens* cells in top jet drop as a function of the incubation time when the jet drop was produced. After Blanchard and Syzdek (1978).

plateau in the number of cells in the top jet drop is reached after 6 hr, which would be concurrent with the end of the dwarfing period. Further changes during the subsequent long-term starvation phase are known to take place at a slower rate and are relatively constant between 5 and 24 hr of exposure to a starvation regime (Kjelleberg *et al.*, 1982, 1983; Amy *et al.*, 1983). A similar stabilization is shown in Fig. 5, as there is only a small increase in concentration of bacteria in the top jet drop in this time interval. The number of cells in the bulk phase was checked in an attempt to account for the slope (Fig. 5). Constant values were obtained, however, for counts taken at 2.5, 8, and 25 hr, which is in agreement with the fact that the fragmentation process (Novitsky and Morita, 1977) is completed in 2.5 hr, followed by a continuous size reduction of the fragmented cells (Kjelleberg *et al.*, 1983). Blanchard and Syzdek (1978) should have recorded an increase in numbers due to fragmentation, had samples been taken at intervals within 2.5 hr. The above explanation seems further supported by the fact that the rapid increase of bacteria in top jet drops with time disappeared when the bacterial suspension was prepared with pond water rather than with distilled water. Sufficient amounts of organic material in the pond water presumably supported growth or prevented the initiation of an active starvation-induced response.

The second problem (Blanchard and Syzdek, 1978) could possibly be answered exclusively in terms of cell surface hydrophobicity. The results that puzzled the authors showed enormous differences in concentration values of different types of bacteria in top jet drops. The concentration value for red-pigmented *Serratia* cells was over 17 times that of pink *Serratia* cells, which have a smaller amount of the surface-localized tetrapyrrole-containing pigment, prodigiosin (Purkayastha and Williams, 1960; Williams, 1973), and nearly 750 times that of white nonpigmented cells. It is prodigiosin, or its interaction with surface-localized molecules, which is responsible for various degrees of surface hydrophobicity and degree of attachment of *Serratia* sp. For example, Hermansson *et al.* (1979) found a much higher enrichment of the red wild type of *S. marcescens* than of the pigmentless mutant in the SMs. Also in a quantitative technique for determining bacterial cell surface hydrophobicity using the partitioning of ^{14}C-labeled dodecanoic acid between cells and the diluent, Kjelleberg *et al.* (1980) reported a higher number of hydrophobic binding sites on late-log-phase red cells than on early log-phase pink cells of *S. marcescens*, while the binding constants of these sites were of similar strength. Kefford *et al.* (1982) found a higher degree of irreversible binding and better scavenging ability of surface-localized nutrients for the wild type compared with the mutant, and unpublished results (S. Kjelleberg) on the retention of *Serratia* cells in hydrophobic-interaction chromatography columns revealed a much higher degree of hydrophobicity of pigmented cells than of nonpigmented ones. In a recent survey by Humphrey *et al.* (1983) investigating starvation effects on hydrophobic and hydrophilic bacteria at surfaces, the red-pigmented wild type responded similarly to other hydrophobic organisms, while the mutant displayed a typical "hydrophilic behavior." Wallace *et al.* (1972) were one of the first to emphasize the importance of surface properties of bacteria, especially their degree of hydrophobicity, in enabling the cells to float.

There are other factors, in addition to levels of nutrients in the surrounding medium, that influence cell surface characteristics and thereby alter the enrichment factors and partitioning of cells of gas–liquid interfaces. For example, the surface properties change during the growth cycle. Hejkal *et al.* (1980) concluded that age effects in several bacterial species could possibly affect their ability to partition between the SMs and the bulk phase. Wallace *et al.* (1972) found that rising bubbles most efficiently fractioned cells of *Cyclotella nana* (a diatom) in the post-stationary growth phase. It could be debated, however, as done by Wallace *et al.* (1972) and Lion and Leckie (1981), as to whether an increased recovery by flotation should be attributed to adsorption of surfactants onto the cell surface or to an intrinsic change in the nature of the cell surface itself.

Lion and Leckie (1981) summarize the use of bubbles to concentrate bacteria at surfaces and to separate algae and bacteria from dilute suspensions (Boyles and Lincoln, 1958; Levin *et al.*, 1962; Rubin *et al.*, 1966; Grieves and Wang, 1967; Wallace *et al.*, 1972). Flotation, as seen in these experiments, can be influenced by factors directly related to the physicochemical properties of the interfaces involved, e.g., the presence of surfactants, salt concentrations, pH, age of bacterial components, cell type, and cell surface functional groups.

Thus, it appears that many environmental factors influence the bacterial cell surface and hence the degree of interaction and concentration at a gas–liquid interface. Of these, the naturally and frequently occurring starvation-induced responses, which lead to large changes in cell volumes and morphology, are by far the best understood. Even though this knowledge has been obtained from different experimental studies, its significance in natural systems seems very plausible indeed.

4. MECHANISMS OF BACTERIAL ADHESION AT GAS–LIQUID INTERFACES

"Mechanisms for bacterial adhesion at gas–liquid interfaces" refers to any means of holding, for any length of time, a bacterial cell at a gas–liquid boundary. Special bacterial features such as morphological adaptations (including, for example, holdfasts and fibrillar structures), high degree of surface-localized pigments, and proteins promoting hydrophobicity are all means of increasing bacterial enrichment factors at gas–liquid interfaces. The interfacial adsorption of all such structures and surface components, as well as any adsorbed surface molecular film that is advantageous for bacterial adhesion, is an indication of the basic binding forces involved. Such components may reduce or contribute to the short- and long-range binding forces and, thus, influence bacterial adhesion at the air–water interface. Due to the relatively meager amount of data resulting from actual physicochemical-oriented experiments, this section will deal primarily with differences observed in bacterial populations at the gas–liquid interfaces compared to those in the bulk phase, as seen in both field and laboratory studies. Some of these observations are subsequently interpreted in light of the few model system experiments designed to reveal more precisely physicochemical factors involved in bacterial adhesion mechanisms.

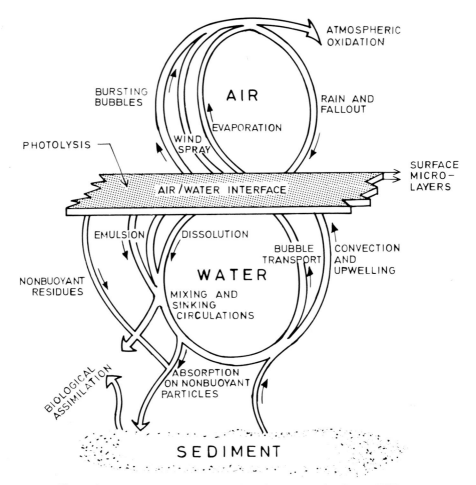

Figure 6. Interrelated processes in aquatic environments. After Garrett (1972).

4.1. Transport and Accumulation Processes

The role of transport and accumulation processes in the formation of the SMs will not be dealt with in depth here, but a short summary and relevant references are given. Several interrelated processes were described by Garrett (1972). His diagram (Fig. 6) summarizes some of the diverse processes that may bring matter into contact with the interface. Additional ways for bacteria to be transported to air–water interfaces have been suggested by Hatcher and Parker (1974) and Marshall (1972). Of the many factors listed in their papers, chemotaxis, aerotaxis, thermotaxis, geotaxis, and phototaxis could all play a part. Turbulent eddy flow, buoyancy, bacteria adsorbed to organic detritus rising to the SMs by virtue of its low density, and surface geometry must also be considered. No doubt, transport over long distances depends on various forms of fluid flow, Langmuir circulation (Sutcliffe *et al.*, 1963), and so on, while short-distance effects for bacterial accumulation at the interface would, to a greater extent, depend on bacterial characteristics, e.g., structures that render the bacteria more buoyant. It seems likely that the upward transport of particulate organic matter (also in the form of bacteria) by bubbles may be more efficient than its removal by downward turbulent diffusion under high wind conditions. Even though insoluble monolayers are mixed downward at wind speeds above 6 m/sec, (Scott, 1972), Tsiban (1971) observed that bacteria are enriched in the SMs even during stormy weather. An excellent review on the primary processes for concentration of organisms, surfactants, and their associated matter, further stressing the dynamics of the SM milieu, can be found in Duce and Hoffman (1976).

4.2. Field Observations on Selective Enrichments

There is substantial evidence for differences between bacteria at gas–liquid interfaces and those in the adjacent bulk liquid phase. All of these observations are not primarily concerned with the mechanisms for bacterial accumulation and adhesion, and general statements such as "the predominant bacteria at a number of stations in the SMs were often found to vary from both bulk and air samples" are easy to find in the literature. As for information of more precise character, bacterioneuston in marine systems have been characterized as follows: (1) primarily *Pseudomonas* spp. (Sieburth, 1965); (2) gram-negative motile rods, presumptively identified as pseudomonads (Crow *et al.*, 1975, 1976) with predominance of either pigmented (Crow *et al.*, 1976) or nonchromogenic (Crow *et al.*, 1975) isolates; (3) isolates consisting of 89.5% *Pseudomonas*, 6.5% *Alcaligenes*, 4% *Enterobacter* species (Sieburth, 1971b); (4) a selective enrichment of pigmented, as compared with nonpigmented, bacteria in aquatic foam (Carlucci and Williams, 1965); (5) 64.5% *Pseudomonas* sp. and 56% pigmented cells in the SMs of an estuary (Fehon and Oliver, 1979); and (6) a prevalence of colored forms and, at numerous stations, a predominance of the genera *Pseudomonas* and *Chromobacterium* (Tsiban, 1975). Carlucci and Williams (1965) employed four bacterial strains in order to compare concentration factors in collected foam from bubbling experiments using nearshore-sampled seawater. Air-bubbling through filtered seawater gave, over a 4-hr period, average concentrations of cells in the foam of 17, 9, 4, and 2 times those in the liquid for *Achromobacter aquamarinus, Serratia marinorubra, Micrococcus infirmis,* and *Vibrio haloplanktis,* respectively. A markedly higher concentration in the 0.5-hr fraction in unfiltered seawater containing particulate organic matter supports observations that high concentrations of microorganisms are associated with particulate matter (cf., e.g., ZoBell, 1946; Young, 1978; Maki and Remsen, 1983). It is not unusual to find bacteria in marine SMs that require no salt for growth, indicating their transport from terrestrial or freshwater origins (Fehon and Oliver, 1979). There is also in marine waters an enrichment of nitrifying bacteria, particularly at night (Horrigan *et al.*, 1981). The observed enhancement of nitrate in SMs was not, however, due to bacterial nitrification, as shown by tracer (^{15}N) experiments.

Investigations of several freshwater environments revealed that many health-indicator bacteria were frequently at higher concentrations in the SMs than in the subsurface waters (Hatcher and Parker, 1975). Michaels *et al.* (1976) have studied the enrichment of pollution-indicator and pathogenic bacteria. The greatest numbers of *Aeromonas hydrophila*, fecal coliforms, fecal streptococci, *Klebsiella pneumoniae, Pseudomonas aeruginosa, Salmonella* sp., and *Staphylococcus aureus* were isolated from the SMs of polluted sites. Some of these species were selectively enriched in the SMs in a pattern different from that in clean sampling sites. The selective enrichment of *Salmonella* and *A. hydrophila* appears to be in agreement with results obtained by Cook *et al.* (1977).

The general concept of a different microflora or a bacterial population with surface-related differences at an air–water interface, as compared with the bulk phase, was supported by the work of DiSalvo (1973), in which the extent of irreversible sorption on interface-exposed marbles indicated that the SM bacteria were capable of rapid attachment. The number of cells that attached to the solid surface was 1000 times higher in the SMs than in subsurface samples, despite the fact that the total number of cells in the SMs was only about 10 times higher.

"Rare" bacteria tend to be found more easily in the SMs than in the water column (Romanenko *et al.*, 1978), and bacteria possessing holdfasts are often found. *Leptothrix* sp. dominated the neuston population of a small pond in springtime (Frølund, 1976), and on some occasions the water film was completely covered with holdfasts. Various forms of the genus *Caulobacter* have been observed by several authors to inhabit the SMs (Romanenko *et al.*, 1978; Romanenko, 1979; Hirsch, 1980). Large numbers of prosthecate bacteria partition readily at the air–water interface (Babenzien and Schwartz, 1970; Young, 1978; Hirsch, 1980; Wyndham and Costerton, 1982), although highly eutrophic waters might not have such bacteria in high enough concentrations to be easily traced (Maki and Remsen, 1983). Marshall and Cruickshank (1973) demonstrated that the mother cell was more hydrophobic than the attached hypha and budding daughter cell, and the perpendicular mode of orientation of *Hyphomicrobium* was related to an increased degree of hydrophobicity at the attaching pole (S. Kjelleberg and K. C. Marshall, unpublished observations). The importance of cell surface and extracellular polymer materials has been stressed for bacterioneuston (Young, 1978). However, a membrane adsorption/SEM technique used for observing neuston organisms revealed extracellular material only when bacteria formed microcolonylike groups (Maki and Remsen, 1983).

4.3. Experiments Investigating the Relationship between Selective Enrichments and Bacterial Surface Characteristics

In addition to the results given in bubbling experiments of *S. marcescens* at various stages of prodigiosin production (Section 3.2.2), work (quoted in Blanchard, 1983) with mixed suspensions of *S. marcescens* and *Escherichia coli* revealed that the more hydrophobic the bacterial surface, the more easily the cell became attached to air bubbles (see also Bezdek and Carlucci, 1972). Differences in jet drop concentrations of 1.5- and 5-day-old cultures of *E. coli* were suggested to be due to involvement of components of the cell itself (Hejkal *et al.*, 1980), and it was further assumed that an increased lipid content in the cell envelope could have promoted the higher concentrations of cells in SMs. Results and conclusions similar to those of Blanchard were also reached using a foam separation process (Boyles and Lincoln, 1958). Cells of *S. marcescens* were more easily collected than various strains of *Brucella suis*, while *Francisella (Pasteurella) tularensis* could not be collected.

Some additional examples can be given to emphasize the relation between hydrophobicity of the cell surface and adhesion at gas–liquid interfaces. Employing *Pseudomonas putida (ovalis),* the resulting layer of enriched bacteria in SMs resulted in retardation of mass transfer rates of glucose and oxygen into the film (Bungay and Masak, 1981). The authors concluded that organisms with greater surface-active properties may produce a greater bacterioneuston layer thickness. *Mycobacterium tuberculosis* is an excellent example of a bacterium that grows only at the air–water interface in normal liquid media. That it grows as a homogeneous suspension in the presence of a detergent such as Tween 80 (Dubos and Middlebrook, 1948) indicates that its outer surface is hydrophobic (Marshall, 1975). This characteristic may even present a health hazard, as field experiments gave evidence for the existence of a water-to-air transfer of nontuberculosis mycobacteria biochemically resembling those isolated from humans (Wendt *et al.*, 1980).

Rubin *et al.* (1966) concluded that *E. coli* behaves like a typical hydrophilic colloid and that surface hydration is the primary mechanism that determines whether an organism will adsorb a collector (any salting-out solute) and subsequently separate in a foam. This was based on the high concentrations of salt needed to promote flotation of the bacteria. The more effective foam-fractionating agent gives a greater decrease in entropy of the system as water is displaced from the bacterial surface. [Salting-out was recently suggested as a fast method for determining degree of bacterial cell surface hydrophobicity (Lindahl et al., 1981).] Hejkal et al. (1980) suggested that concentration factors (ratios of the number of cells at the surface to the number in the bulk phase) less than 1 for cells of *Pseudomonas bathycetes* were due to migration of bacteria away from the surface of rising bubbles. However, an alternative explanation for such concentration factors may be found in the outcome of the interception mechanism for collection of bacteria at bubbles (Section 2.2.1), as bacteria would not move fast enough, compared to the bubble rise speed, to migrate away from the bubble surface (D. C. Blanchard, personal communication). Surface balance experiments indicated that the cells of *P. fluorescens* could not ''scatter'' the monolayer of lipids at an air–water interface but formed a layer underneath the surface (Kjelleberg et al., 1976). These cells could be considered relatively hydrophilic and not able to react with the surface-active molecules in the SM. Similarly, smooth, gliding movements of hydrophilic (non-acid-fast) bacteria along air–water interfaces were observed by Mudd and Mudd (1924a). However, a high interfacial tension resulted in jerky and spasmodic movements of bacteria at the interface, indicating binding attempts. Mudd and Mudd (1924a,b) explained these observations in terms of a kinetic mechanism, dependent primarily upon interfacial surface tension forces in the boundary layer between two wholly or partially immiscible fluids. The membrane surface of ordinary, or non-acid-fast, bacteria was said to contain many active or polar groups. Hence, the authors concluded that ordinary bacteria, i.e., hydrophilic ones, should be wet by water, the polar liquid, and thus remain in the aqueous phase. In their oil–water systems, acid-fast (hydrophobic) microorganisms were ''wet'' by the nonpolar organic liquids (Mudd and Mudd, 1924b).

There are, as seen from these examples, strong indications that bacterial populations in natural SMs should consist of higher percentages of hydrophobic cells than can be found in subsurface waters. In fact, this seems to be the case. Bacterial isolates, from four stations along the Swedish west coast, were assessed for their degree of hydrophobicity using hydrophobic interaction chromatography (Dahlback et al., 1981). Isolates were labeled with [3H]leucine and run on columns packed with octyl-Sepharose Cl-4B gels, whereafter the partitioning between gel (g) and eluate (e) was examined. A high g/e value corresponded to a high degree of hydrophobicity. The results revealed a positive relationship between the degree of accumulation of bacteria at the surface and their hydrophobicity. Figure 7 shows the frequency of observation of the bacterial hydrophobicity parameter, log g/e, for isolates sampled at one station from (a) the air–water interface and (b) the bulk water (0.5-m depth). The enrichment value for bacteria in the SMs at this station was 360 and there was concomitantly a significantly higher degree of hydrophobicity of the bacteria at the interface, compared to subsurface water.

4.4. Model System Experiments to Determine Physicochemical Properties

Model systems seem well justified for interface studies, for the initial events of interaction are governed by physicochemical forces where microorganisms behave as colloid particles, their biological activity being manifested only in subsequent stages. The following refers to model system studies based on the simple idea of applying an appropriate film-forming substance to the air–water interface of an aqueous phase containing suspended

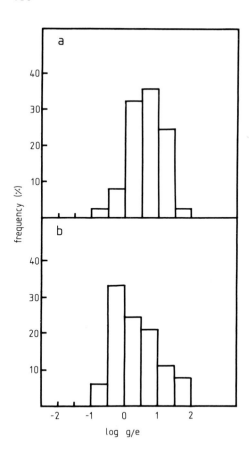

Figure 7. Distribution of degree of hydrophobicity (log g/e) among marine bacteria sampled at one station at the Swedish west coast; from (a) air–water interface, (b) subsurface water. The isolates were labeled with [^{3}H]-L-leucine and separated on octyl-Sepharose C1-4B gel. After Dahlbäck et al. (1981).

bacteria. Both a static system and the more flexible approach of surface balance studies will be considered. In the latter system, the surface tension was measured with a glass plate suspended from a torsion wire, i.e., the Wilhelmy technique, while the surface film was subjected to a continuously increasing compression with a Teflon barrier (Kjelleberg and Stenström, 1980). The film was thereby changed from a gaseous state, if less than a monolayer of surface-active substances was spread, via an expanded liquid state, to a condensed monolayer (Gaines, 1966) (Fig. 8). Penetration of enzyme molecules and a lipopolysaccharide preparation from the envelope of *Salmonella typhimurium* has been demonstrated and shown to alter the pressure–area isotherm of, for example, phospholipid monolayers (Shah and Schulman, 1967; Rothfield and Romeo, 1971; Verger *et al.*, 1973). The adhesion of bacteria at the surface is similarly indicated by an increase in the surface pressure at a given area per molecule. Various aggregation states of a lipid microlayer, that alter penetration of bacterial surface groups, naturally exist, depending basically on the magnitude of lateral cohesive forces between, and the concentration of, film molecules. Some films will exist in clusters, separated by large empty spaces (Odham *et al.*, 1978), whereas stratified multilayered films can be formed at higher concentrations (Larsson, 1973). Charge effects can furthermore be evaluated, depending on the type of film which is spread.

    The number of adhered bacteria in the static model system was determined by withdrawing the surface with Teflon surface samplers (Norkrans and Sörensson, 1977; Kjelleberg *et al.*, 1979). The latter authors found *Serratia marinorubra* [known to be hydrophobic

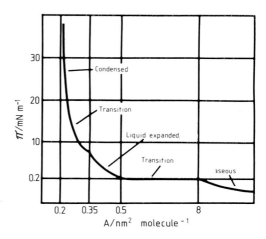

Figure 8. Schematic representation of the π–A (surface pressure–area) curve for myristic acid spread on
0.1 mole/dm³ HCl at 14°C, revealing the different
states of a monolayer film. These represent various
degrees of molecular freedom or order, resulting from
the intermolecular forces in the film and between film
and subphase. The gaseous state has a low density
with film molecules moving almost independently of
each other. Condensed film molecules are closely
packed, approaching a crystalline state. After Shaw
(1970).

(Kjelleberg *et al.*, 1980)] to give enrichment values of 100, 10 times higher than those
obtained for *Aeromonas* and *Pseudomonas* species. Hermansson *et al.* (1979) showed a
much higher enrichment for pigmented cells of *S. marcescens* than for nonpigmented cells of
this species at an air–water interface covered with a monomolecular layer of stearic acid.
Norkrans and Sörensson (1977) further observed that a multilayer (10 molecules thick) of
oleic acid gave twice the bacterial enrichment of a monolayer, and the number of bacteria
found in SMs varied with the cell density in the subsurface waters. Syzdek (1982) could not,
however, find a limiting number of binding sites for cells in the lipid surface film, as was
reported by Norkrans and Sörensson (1977).

A delicate balance between different adsorption and dispersion interactions, dependent
upon bacterial surface structures, was discovered by Hermansson *et al.* (1982), who performed static model experiments with strains of *Serratia* and *Salmonella*. The air–water
model system revealed hydrophobic interactions. As phospholipids spread as a lipid surface
film, hydrocarbon tails will be oriented partly horizontally at the interface and in contact with
each other, the remainder being oriented away from the aqueous phase. The bacteria that
interact at the air–water interface thus "see" a hydrophobic surface. *Salmonella typhimurium* was cultivated under static conditions to produce fimbriated cells. The fimbria
filaments are hydrophobic proteins exhibiting acidic isoelectric points (Korhonen, 1980; see
also Chapter 11, Section 3.2). Such fimbriated *Salmonella* cells showed high degrees of
hydrophobicity (as assessed by hydrophobic interaction chromatography) and enrichment at
the interface. These values were high for fimbriated cells, regardless of whether the cells had
polysaccharide strands on the lipopolysaccharide (LPS) or not, as compared with nonfimbriated cells, Smooth strains were, however, less hydrophobic than rough strains (lacking the
O-side chain of the LPS) when fimbriae were present, and the interaction at the air–water
interface was reduced for both nonfimbriated and fimbriated smooth cells, as compared with
the corresponding rough cells. The importance of LPS in determining cell surface properties
was also shown by the use of various rough mutants of *S. typhimurium*. There was a
tendency toward increased hydrophobic characteristics accompanying the sequential loss of
the O-side chain and core oligosaccharide components (Fig. 9). This relationship between
fimbriation and hydrophobicity, illustrated by *S. typhimurium,* probably resembles that of
many aquatic bacteria, but was not shown to be valid for *Serratia marcescens* (Hermansson
et al., 1982). Fimbriated cells were less hydrophobic and adhered to a lesser degree at the

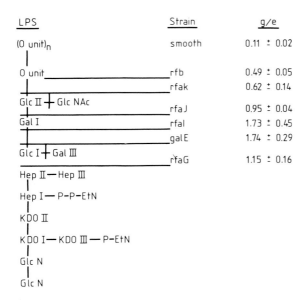

LPS	Strain	g/e
(0 unit)n	smooth	0.11 ± 0.02
O unit ——————————— rfb		0.49 ± 0.05
——————————— rfak		0.62 ± 0.14
Glc II ┼ Glc NAc		
——————————— rfaJ		0.95 ± 0.04
Gal I	rfal	1.73 ± 0.45
——————————— galE		1.74 ± 0.29
Glc I ┼ Gal III		
——————————— rfaG		1.15 ± 0.16
Hep II—Hep III		
Hep I — P—P—EtN		
KDO II		
KDO I— KDO III — P—EtN		
Glc N		
Glc N		

Figure 9. Degree of hydrophobicity (*g/e*) assessed by hydrophobic interaction chromatography of *Salmonella typhimurium*, including a smooth strain and an isogenic series of rough mutants displaying a sequential loss of the O-side chain and core oligosaccharide components of the lipopolysaccharide (LPS). Abbreviations of rough mutants according to Mäkelä and Stocker (1981). After Hermansson *et al.* (1982).

air–water interface than their nonfimbriated counterparts. Many investigations (Hermansson *et al.*, 1979; Kjelleberg *et al.*, 1980; Rosenberg *et al.*, 1980; Hermansson and Dahlbäck, 1983) of *Serratia* cells have shown them to possess strong adhesion properties and a high degree of hydrophobicity, probably due to the production of the surface-localized pigment, prodigiosin (see Section 3.2).

Bacteria have been shown by surface balance experiments to interact at sites in, or just beneath, the surface lipid film (Kjelleberg *et al.*, 1976; Kjelleberg and Stenström, 1980) (Fig. 10). A bacterial interaction is indicated by an increase in the surface pressure at a given

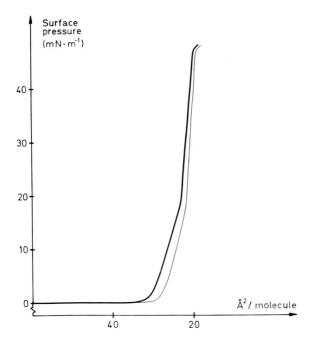

Figure 10. Surface pressure as a function of molecular area for a monolayer of palmitic acid on phosphate-buffered solution with (——) or without (—) cells of *Serratia marinorubra*. After Kjelleberg and Stenström (1980).

area per molecule. Cells of *Serratia marinorubra* scattered the lipid film to a greater extent than did cells of the more hydrophilic *Pseudomonas halocrenaea,* which formed a layer beneath the film. The mycoplasma *Acholeplasma laidlawii,* which has its lipid bilayer plasma membrane exposed, produced the largest increase in surface pressure. The surface films used were long-chain fatty acids and phospholipids.

In these studies, factors beside the hydrophobic interaction shown to be of significance were the architecture of the film and the surface charge, of both the monolayered film and the bacteria. With respect to film structure, the lateral cohesive forces between the hydrocarbon chains have a strong influence on the microbial adhesion. If the surface molecules are sufficiently separated to prevent interchain reactions, the possibility for bacterial interaction is greatly enhanced. There was, similarly, a stronger bacterial influence on condensed films than on expanded ones (Kjelleberg and Stenström, 1980). Saturated fatty acids forming condensed film layers result more easily in island formation, which effectively lowers the surface density at less than complete monomolecular concentrations. As the surface is compressed, i.e., the film substances are forced into contact with each other, bacteria actually interacting with the surface-active molecules are not forced out of the film and produce an increased surface pressure at a given area per surface molecule.

Most bacteria carry a negative charge on their outer layer (Longton *et al.,* 1975). By replacing negatively charged fatty acid films with octadecylamine films, which are positively charged at the experimental pH 7.4, it was found that interacting bacteria provoked a much larger increase in surface pressure (Kjelleberg and Stenström, 1980). Furthermore, *S. marinorubra,* which is characterized by a greater negative charge than *P. halocrenaea,* caused a larger surface pressure increase with the positively charged film.

Syzdek (1982) recently concluded that lower bacterial enrichment values obtained with greater thicknesses of collected SMs were a result of dilution of highly enriched SMs in the upper few micrometers by the less enriched subsurface water. This supports data given by Kjelleberg *et al.* (1979) suggesting that bacteria occupy sites in or just below the SM, and is consistent with the surface balance experiments presented here. The idea of a strong interaction between bacteria and the surface-active molecules likewise finds support in the literature. For example, bacteria did not bind to the same extent at the air–water interface using a lipid-free surface (Syzdek, 1982), and the initial rate of increase in enrichment for undisturbed, uncleaned surfaces was about a factor of 14 per hr, while for cleaned surfaces it was a factor of ≤ 7.5 per hr. Williams *et al.* (1980) performed some interesting *in situ* measurements of sea-surface films. There was a correlation between the total number of bacteria and high surface pressure values, implying a direct interaction between bacteria and surface-active molecules, similar to the surface balance studies.

5. EFFECTS OF BACTERIAL ENRICHMENT AT THE AIR–WATER INTERFACE

What are the results of the sometimes massive bacterial enrichment in SMs, and is it possible to speculate on the ecosystem at these interfaces? For example, is it possible to recognize a distinct food chain within the SM? Our understanding of bacterial interactions with air–water interfaces allows us to perform reasonably sound experimental investigations of questions that deal with bacterial activity, as information on mechanisms for bacterial adhesion is a prerequisite for studies regarding bacterial activity at an interface.

First, it is advisable to introduce some words of caution with respect to the presumed similarities between various interfaces and their influence on bacterial behavior. As mentioned in Section 3.1.1, there appears to be a relationship between degree of binding at solid surfaces by the hydrophobic bacteria and their metabolic activity. Let us compare this with a liquid–liquid interface situation, e.g., that of oil degradation by bacteria. This interfacial activity is carried out by relatively hydrophilic bacteria (Lupton and Marshall, 1979), which remain essentially in the aqueous phase. Hydrophobic bacteria will not remain in a stable position at the oil–water interface where oil degradation can be maintained, but will tend to pass into the hydrocarbon phase (Mudd and Mudd, 1924a). Furthermore, Hermansson and Dahlbäck (1983) described a nonpreferential uptake of surface-localized fatty acids at an air–water interface by hydrophobic and hydrophilic bacteria (see Section 5.1). Thus, it seems that conclusions drawn about bacterial activity at interfaces should apply to the specific interface studied.

5.1. Bacterial Activity

Hydrophobic bacteria and small starved cells displaying increased degrees of hydrophobicity and irreversible binding can take advantage of the presence of interfaces. Air– and solid–water interfaces, lacking sufficient amounts of nutrients for growth of adhered bacteria, induced the associated bacteria (except for those which were very hydrophobic) to produce even smaller cells upon starvation than those produced by bacteria in the bulk phase (Kjelleberg et al., 1982; Humphrey et al., 1983; Chapter 12, Section 3.4.3). Very hydrophobic bacteria showed the opposite effect, i.e., cells in the bulk phase formed apparently smaller cells as compared with those at the interface. This has been interpreted as efficient scavenging of the contaminating nutrients that are adsorbed at the interface (resulting in less size reduction), although the surface free energy of the substratum might well be the decisive factor explaining the trigger mechanism. Adding nutrients in very low concentrations to the interface after 24 hr of starvation led to an immediate increase in size of bacteria at the interface. However, the small amount of nutrients at the interface was obviously consumed in 2 hr, whereafter starvation effects reappeared.

In a similar study, low protein concentrations added to the SMs *in situ* resulted in a more rapid degradation by bacteria at the surface than in the bulk (Hermansson and Dahlbäck, 1983). In contrast, high amounts of supplied proteins gave similar rates of breakdown of the substrate in surface and in bulk. Not only did more efficient uptake and/or metabolism occur at the interface at low protein concentrations, but the authors also concluded that interfaces might be necessary for degradation of large surface-active molecules in aquatic environments.

Several authors have dealt with bacterial activity at the air–water interface. The methodological approaches have considered bacterial uptake and respiration of various types of substrates (Dietz et al., 1976; Romanenko et al., 1978; Passman et al., 1979; Mitamura and Matsumoto, 1981; Bell and Albright, 1982; Crawford et al., 1982; Hermansson and Dahlbäck, 1983), biochemical activities of bacterial isolates (Sieburth, 1971a; Tsiban, 1975; Crow et al., 1976; Ahearn et al., 1977; Fehon and Oliver, 1977, 1979; Hoppe, 1977; Kjelleberg and Haåkransson, 1977; Dahlbäck et al., 1982), and fraction of respiring cells (Harvey and Young, 1980b). It is not easy to form a coherent picture from these results. Results presented on biochemical activities are difficult to interpret, for comparable bulk phase figures are not always supplied. Measurements of bacterial uptake of labeled substrates, however, have often shown higher activities for cells in the aqueous phase than in the SMs (Dietz et al., 1976; Romanenko et al., 1978; Passman et al., 1979; Bell and Albright, 1982). Explanations for such results often refer to detrimental factors as being important for determining bacterial activity in the SMs. Toxic substances, intense solar radiation, and stress resulting from quick changes in environmental factors, such as temperature and salinity, would reduce the activity (Norkrans, 1980). Enrichment of nutrients in the SMs would, on the other hand, act in the opposite way (Drachev et al., 1965). Other factors of possible importance, such as fluctuations (of unknown orders of magnitude) in water activity, pH, E_h, and so on, are difficult to take into account.

In most investigations the SMs are sampled and transferred for subsequent incubation in the laboratory. This results in a disruption of the stratified SMs and the mixing of surface bacteria into a different environment. Moreover, short-chain and low concentrations of long-chain fatty acids, i.e., free, soluble forms of these molecules, appear to be toxic to bacteria (Helprin and Hiatt, 1957; Maxcy and Dill, 1967; Henderson, 1973; Maczulak et al., 1981), but are nontoxic and constitute a suitable substrate when immobilized at surfaces. Much care should therefore be taken in the interpretation of results that are generated from such experiments. The approach of studying relatively undisturbed interfacial strata was first applied by Gallagher (1975). He transferred a surface sampler, after collection of the SMs in a salt marsh [using the so-called "screen" sampler (Garrett, 1967)] to a seawater solution, and allowed the sampler to line the wall of the container. The subsequent incubation revealed that 37% of the photosynthetic activity was localized in the surface. Gallagher (1975) concluded that the SM is one of the centers of metabolic activity in salt marshes.

Hermansson and Dahlbäck (1983) measured bacterial activity using amino acids, a fatty acid, and a protein mixture as substrates in the SMs and in the bulk liquid. The aim of these experiments was to keep the stratification of the interface in its original form. In situ incubations revealed only small differences in the fraction of active cells between surface and bulk samples, and an exchange of matter between the surface and the bulk was said to affect the results. By spreading [³H]palmitic acid on the surface of seawater samples, such an exchange of bacteria was established (Hermansson and Dahlbäck, 1983). Addition of different bacteria with thoroughly defined surface characteristics led the authors to conclude that a bacterial strain showing a high accumulation is most likely firmly bound, with little or no exchange of cells between interface and bulk (Table I). The accumulation of cells at the interface is, to a certain extent, inversely related to the ratio between the total number of labeled bacteria in the bulk and the total number of labeled bacteria in the surface (ACB/ACS), as seen in Table I. These data agree well with the surface balance studies carried out by Kjelleberg and co-workers (Kjelleberg et al., 1976; Kjelleberg and Stenström, 1980;

Table I. Scavenging of [³H]-Palmitic Acid at the Air–Water Interface[a]

| | Active bacteria %[b] | | | |
	Surface	Bulk	EF[c]	ACB/ACS[d]
Seawater samples	20	28	12	189
Test bacteria				
Serratia marcescens (m)[e]	54	0.5	98	1.8
Serratia marcescens (w)[e]	42	2.0	7.0	145
Pseudomonas halocrenaea	65	0.9	1.0	412

[a]Data from Hermansson and Dahlbäck (1983).
[b]A combined microautoradiography–epifluorescence method makes it possible to determine the fraction of the bacterial population that scavenges tritiated material.
[c]Enrichment factors of bacteria at the air–water interface.
[d]Ratios between the total number of labeled bacteria in the bulk and the total number of labeled bacteria in the surface.
[e]The wildtype (w) has a surface-localized pigment that increases its tendency for accumulation at the interface compared to the mutant (m) that lacks this pigment.

Section 4.4). Cells of *Serratia* penetrate the surface film while cells of *Pseudomonas* are squeezed out and form a layer of loosely bound bacteria underneath the interface.

The study carried out by Hermansson and Dahlbäck (1983) furthermore supplied valuable information on the ways of assessing and interpreting uptake and respiration of substrates at interfaces. It was clear from the use of a fatty acid that the surface-active substances alter to some extent the composition of the adsorbed molecular layer at the interface by competing with less surface-active substances. Normal Michaelis–Menten kinetics cannot be used above certain concentrations of added surface-active substrates.

Through the action of extracellular enzymes, short peptide fragments might be distributed into the bulk from protein originally spread at the interface (Rogers, 1961). Enzymes could either be more efficient at interfaces as a result of a higher collision rate between the enzyme and the substrate (ZoBell, 1943), or inactivated following unfavorable changes in conformation at the adsorption event (Marshall, 1976; Burns, 1980; Chapter 12). Substrate uptake may also be affected by conditions at an interface. No differences in uptake of glutamic acid and leucine were found in the bulk phase (Hermansson and Dahlbäck, 1983), but higher percentages of active cells using leucine were recorded at the interface, as compared with the bulk. The more hydrophobic leucine was assumed to accumulate more efficiently at an interface.

5.2. Possible Food Chains

The abundance of nutrient material and microorganisms in the SMs suggests the possibility of the development of a complex biotic system. Several reports on "higher" organism types, thriving in upper surface waters, and in some occasions as members of the SM microcommunity, make it legitimate to consider the existence of a different type of food chain at the air–water interfaces of oceans, lakes, and rivers, as compared with the adjacent water column.

Brockmann *et al.* (1976) found an active and dense accumulation of phytoplankton in large slicks west of the island of Sylt (West Germany). At the maximum buildup of slick material, a dinoflagellate, *Prorocentrum micans,* make up more than 90% of the calculated biomass in the upper 250-μm surface layer. It should also be mentioned that these authors found a direct correlation between the increased numbers of *P. micans* cells and the chemical composition of the lipid film. A concomitant increase in palmitic acid, being the main component of *Prorocentrum* lipids, was especially noted. It seems clear that the phytoneuston community, of at least the upper 50 μm, is greater in density and different in species composition, as compared with the plankton community at a depth of some centimeters (Harvey, 1966; Harvey and Burzell, 1972; Hardy, 1973; Sieburth *et al.,* 1976; Wandschneider, 1979). Moreover, Wotton (1982) concluded, based on model and field limnic system experiments, that surface films constitute suitable substrates for lotic animals. Harvey (1966) deduced from microscopical studies and analysis of biomass indicators that a different type of food chain with a higher ratio of herbivores to primary producers occurred in the SMs, as compared with the water below. Such consumer organisms seem to have special morphological and feeding adaptations in order to use the surface film as a habitat (David, 1965; Wotton, 1982). Many periphytic "benthic" organisms and typical surface film-hangers have been described (Norris, 1965; Maynard, 1968). Furthermore, it has been found that the SMs are important to fish, shellfish, and crustaceans, including those of commercial importance, whose eggs and larval stages often concentrate in the neustonic layers (Hardy and Valett, 1981, and work cited therein).

The existence of food chains in or near the SMs is likewise implied by studies on the fate of heavy metals and pollutants. A description of the entry of these materials into food chains via the SMs can be found in Duce *et al.* (1972). However, there is still no thorough study to elucidate the patterns in the possibly unique food chains that are part of the stratified surface layers.

6. CONCLUDING REMARKS

The mechanisms for bacterial adhesion seem similar in character at gas–, liquid–, and solid–liquid interfaces. Strata in the SM of the air–water interface and conditioning films of solid surfaces appear similar, and some methods for determining physicochemical properties of the bacterial cell surface employ the various types of interfaces and reveal information of general significance. It is scientifically correct and convenient from an experimental standpoint to utilize the gas–liquid interface as a model system for research into broad aspects of bacterial adhesion. Much useful data have been obtained from defined experiments using both the air–water interface of an aqueous subphase and the gas bubbles that rise through the water column and subsequently produce jet and film drops, as the bubble bursts at the surface. Valuable information has similarly been obtained in relation to bacterial survival tactics and activity at interfaces. With respect to bacterial activity, however, differences between various interfacial types might well exist. Bacteria that are merely loosely adsorbed at the air–water interface, displaying a relatively hydrophilic cell surface, scavenge surface-localized nutrients as efficiently as do hydrophobic firmly bound cells. Thus, the situation resembles that at liquid–liquid interfaces, for example as occurs in oil degradation. This condition appears not to be the case at solid surfaces, at which hydrophobic irreversibly bound bacteria may play a dominant role in scavenging of surface-localized molecules. However, there may be exceptions, as illustrated by reversibly attached, gliding *Leptospira* cells, which can scavenge nutrients on surfaces (Kefford *et al.*, 1982).

Bearing these similarities and differences between interfaces of dissimilar phases in mind, it is possible to successfully apply, transfer, and compare a broad range of data and observations concerning bacterial adhesion and behavior.

ACKNOWLEDGMENTS. I am grateful to Professor B. Norkrans for stimulating discussions concerning this manuscript. Work by the author was supported by the Swedish Natural Science Research Council and the Australian Research Grants Committee.

REFERENCES

Adams, A. P., and Spendlove, J. C., 1970, Coliform aerosols emitted by sewage treatment plants, *Science* **169**:1218–1220.

Adams, D. J., Spendlove, J. C., Spendlove, R. S., and Barnett, B. B., 1982, Aerosol stability of infectious and potentially infectious reovirus particles, *Appl. Environ. Microbiol.* **44**:903–908.

Ahearn, D. G., Crow, S. A., and Cook, W. L., 1977, Microbial interactions with pesticides in estuarine surface slicks, *U.S. Environ. Prot. Agency Rep. Ecol. Res. Ser.* 600-3.77.050.

Amy, P. S., Pauling, C., and Morita, R. Y., 1983, Starvation—survival processes of a marine *Vibro*, *Appl. Environ. Microbiol.* **45**:1041–1048.

Anderson, R. E., Sein, L., Moss, M. L., and Gross, N. H., 1952, Potential health hazards of common bacteriological techniques, *J. Bacteriol.* **64**:473–481.

Azam, F., and Hodson, R. E., 1981, Multiphastic kinetics for D-glucose uptake by assemblages of natural marine bacteria, *Mar. Ecol. Prog. Ser.* **6**:213–222.

Babenzien, H.-D., and Schwartz, W., 1970, Studien zur Mikrobiologie des Neustons, *Limnologica* **7**:247–272.

Baier, R. E., 1975, Applied chemistry at protein interfaces, in: *Applied Chemistry at Protein Interfaces* (R. F. Gould, ed.), American Chemical Society, Washington, D.C., pp. 1–25.

Baier, R. E., Goupil, D. W., Perlmutter, S., and King, R., 1974, Dominant chemical composition of sea-surface films, natural slicks, and foams, *J. Rech. Atmos.* **8:**571–600.

Bandoni, R. J., 1972, Terrestial occurrence of some aquatic hyphomyctes, *Can. J. Bot.* **50:**2283–2288.

Bandoni, R. J., and Koske, R. E., 1974, Monolayers and microbial dispersal, *Science* **183:**1079–1081.

Barber, R. T., 1966, Interaction of bubbles and bacteria in the formation of organic aggregates in sea-water, *Nature (London)* **211:**257–258.

Batossingh, E., Riley, G. A., and Keshwar, B., 1969, An analysis of experimental methods for producing particulate organic matter in sea water by bubbling, *Deep-Sea Res.* **16:**213–219.

Baylor, E. R., Sutcliffe, W. H., and Hirschfeld, D. S., 1962, Adsorption of phosphate onto bubbles, *Deep-Sea Res.* **9:**120–124.

Bell, C. R., and Albright, L. J., 1982, Bacteriological investigation of the neuston and plankton in the Fraser River estuary, British Columbia, *Estuarine Coastal Shelf Sci.* **15:**385–394.

Bezdek, H. F., and Carlucci, A. F., 1972, Surface concentration of marine bacteria, *Limnol. Oceanogr.* **17:**566–569.

Bezdek, H. F., and Carlucci, A. F., 1974, Concentration and removal of liquid microlayers from a seawater surface by bursting bubbles, *Limnol. Oceanogr.* **19:**126–132.

Blanchard, D. C., 1964, Sea-to-air transport of surface active material, *Science* **146:**396–397.

Blanchard, D. C., 1983, The production, distribution and bacterial enrichment of the sea-salt aerosol, in: *Air–Sea Exchange of Gases and Particles* (P. S. Liss and W. G. N. Slinn, eds.), Reidel, Dordrecht, pp. 407–454.

Blanchard, D. C., and Syzdek, L. D., 1970, Mechanism for the water-to-air transfer and concentration of bacteria, *Science* **170:**626–628.

Blanchard, D. C., and Syzdek, L. D., 1972, Concentration of bacteria in jet drops from bursting bubbles, *J. Geophys. Res.* **77:**5087–5099.

Blanchard, D. C., and Syzdek, L. D., 1978, Seven problems in bubble and jet drop researches, *Limnol. Oceanogr.* **23:**389–400.

Blanchard, D. C., and Syzdek, L. D., 1982, Water-to-air transfer and enrichment of bacteria in drops from bursting bubbles, *Appl. Environ. Microbiol.* **43:**1001–1005.

Blanchard, D. C., Syzdek, L. D., and Weber, M. E., 1981, Bubble scavenging of bacteria in freshwater quickly produces bacterial enrichment in airborne drops, *Limnol. Oceanogr.* **26:**961–964.

Boyles, W. A., and Lincoln, R. E., 1958, Separation and concentration of bacterial spores and vegetative cells by foam flotation, *Appl. Microbiol.* **6:**327–334.

Brockmann, U. H., Kattner, G., Hentzschel, G., Wandschneider, K., Junge, H. D., and Hühnerfuss, H., 1976, Natürliche Oberflächenfilme im Seegebiet vor Sylt, *Mar. Biol.* **36:**135–146.

Brown, C. M., Ellwood, D. C., and Hunter, J. R., 1977, Growth of bacteria at surfaces: Influence of nutrient limitation, *FEMS Microbiol. Lett.* **1:**163–166.

Bungay, H. R., and Masak, R. D., 1981, Estimation of thickness of bacterial films at an air–water interface, *Biotechnol. Bioeng.* **23:**1155–1157.

Burns, R. G., 1980, Microbial adhesion to soil surfaces: Consequences for growth and enzyme activities, in: *Microbial Adhesion to Surfaces* (R. C. W. Berkeley, J. M. Lynch, J. Melling, P. R. Rutter, and B. Vincent, eds.), Horwood, Chichester, pp. 249–262.

Carlson, D. J., 1982, A field evaluation of plate and screen microlayer sampling techniques, *Mar. Chem.* **11:**189–208.

Carlson, D. J., and Mayer, L. M., 1980, Enrichment of dissolved phenolic material in the surface microlayer of coastal waters, *Nature (London)* **286:**482–483.

Carlucci, A. F., and Williams, P. M., 1965, Concentration of bacteria from sea water by bubble scavenging. *J. Cons. Cons. Int. Explor. Mer.* **30:**28–33.

Cipriano, R. I., and Blanchard, D. C., 1982, Reply, *J. Geophys. Res.* **87:**5869–5870.

Cook, W. L., Michaels, G. B., Crow, S. A., and Rector, D. D., 1977, Indicator bacteria in surface microlayers of Lake Sidney Lanier, *Dev. Ind. Microbiol.* **19:**473–476.

Crawford, R. L., Johnson, L., and Martinson, M., 1982, Numbers and metabolic activities of bacteria in surface films of freshwater lakes, in: *Abstr. Annu. Meet. Am. Soc. Microbiol.* No. 2, p. 178.

Crow, S. A., Ahearn, D. G., Cook, W. L., and Bourquin, A. W., 1975, Densities of bacteria and fungi in coastal surface films as determined by a membrane-adsorption procedure, *Limnol. Oceanogr.* **20:**644–646.

Crow, S. A., Cook, W. L., Ahearn, D. G., and Bourquin, A. W., 1976, Microbial populations in coastal surface slicks, in: *Proceedings, 3rd International Biodegradation Symposium* (J. M. Sharpley and A. M. Kaplan, eds.), Applied Science Publishers, London, pp. 93–98.

Dahlbäck, B., Hermansson, M., Kjelleberg, S., and Norkrans, B., 1981, The hydrophobicity of bacteria—An important factor in their initial adhesion at the air–water interface, *Arch. Microbiol.* **128:**267–270.

Dahlbäck, B., Gunnarsson, L. A. H., Hermansson, M., and Kjelleberg, S., 1982, Microbial investigations of surface microlayers, water column, ice and sediment in the Arctic Ocean, *Mar. Ecol. Prog. Ser.* **9:**101–109.

Daumas, R. A., LaBorde, P. L., Marty, J. C., and Saliot, A., 1976, Influence of sampling method on the chemical composition of water surface film, *Limnol. Oceanogr.* **21:**319–326.

David, P. M., 1965, The surface fauna of the ocean, *Endeavour* **24:**95–100.

Dawson, M. P., Humphrey, B. A., and Marshall, K. C., 1981, Adhesion: A tactic in the survival strategy of a marine *Vibrio* during starvation, *Curr. Microbiol.* **6:**195–198.

Dietz, A. S., Albright, L. J., and Tuominen, T., 1976, Heterotrophic activities of bacterioneuston and bacterioplankton, *Can. J. Microbiol.* **22:**1699–1709.

DiSalvo, L. H., 1973, Contamination of surfaces by bacterial neuston, *Limnol. Oceanogr.* **18:**165–168.

Dondero, T. J., Jr., Rendtorff, R. C., Mallison, G. F., Weeks, R. M., Levy, J. S., Wong, E. W., and Schaffner, W., 1980, An outbreak of Legionnaires' disease associated with a contaminated air-conditioning cooling tower, *N. Engl. J. Med.* **302:**365–370.

Drachev, S. M., Bylinkina, V. A., and Sosunova, I. N., 1965, The importance of adsorption and surface phenomena in the self-cleaning of lakes and ponds [abstract, Engl. Transl.], *Biol. Abstr.* **46:**101500.

Dragcevic, D., Vukovic, M., Cukman, D., and Pravdic, V., 1979, Properties of the seawater–air interface: Dynamic surface tension studies, *Limnol. Oceanogr.* **24:**1022–1030.

Druilhet, R. E., and Sobek, J. M., 1976, Starvation survival of *Salmonella enteritidis*, *J. Bacteriol.* **125:**119–124.

Dubos, R. J., and Middlebrook, G., 1948, The effect of wetting agents on the growth of tubercle bacilli, *J. Exp. Med.* **88:**81–88.

Duce, R. A., and Hoffman, E. J., 1976, Chemical fractionation at the air/sea interface. *Annu. Rev. Earth Planet. Sci.* **4:**187–228.

Duce, R. A., Quinn, J. G., Olney, C. E., Piotrowicz, S. R., Ray, B. I., and Wade, T. L., 1972, Enrichmen of heavy metals and organic compounds in the surface layer of Narragansett Bay, Rhode Island, *Science* **176:**161–163.

Elzerman, A. W., and Armstrong, D. E., 1979, Enrichment of Zn, Cd, Pb and Cu in the surface microlayer of Lakes Michigan, Ontario and Mendota, *Limnol. Oceanogr.* **24:**133–144.

Fehon, W. C., and Oliver, J. D., 1977, Degradation of crude oil by mixed populations of bacteria from the surface microlayer in an estuarine system, *J. Elisha Mitchell Sci. Soc.* **93:**72–73.

Fehon, W. C., and Oliver, J. D., 1979, Taxonomy and distribution of surface microlayer bacteria from two estuarine sites, *Estuaries* **2:**194–197.

Fletcher, M., and Marshall, K. C., 1982, Are solid surfaces of ecological significance to aquatic bacteria?, in: *Advances in Microbial Ecology,* Volume 6 (K. C. Marshall, ed.), Plenum Press, New York, pp. 199–236.

Frølund, A., 1976, The seasonal variation of the neuston of a small pond, *Bot. Tidsskr.* **72:**45–56.

Gaines, G. L., 1966, *Insoluble Monolayers at Liquid–Gas Interfaces,* Wiley–Interscience, New York.

Gallagher, J. L., 1975, The significance of the surface film in salt marsh plankton metabolism, *Limnol. Oceanogr.* **20:**120–123.

Garrett, W. D., 1967, The organic chemical composition of the ocean surface, *Deep-Sea Res.* **14:**221–227.

Garrett, W. D., 1969, Surface-chemical modification of the air–sea interface, *Ann. Meteorol. N.F.* **4:**25–29.

Garrett, W. D., 1972, Impact of natural and man-made surface films on the properties of the air–sea interface, in: *The Changing Chemistry of the Oceans* (D. Dyrssen and D. Jagner, eds.), Almqvist & Wiksells, Stockholm, pp. 75–91.

Goff, G. D., Spendlove, J. S., Adams, A. P., and Nicholes, P. S., 1973, Emission of microbial aerosols from sewage treatment plants that use trickling filters, *Health Serv. Rep.* **88:**640–652.

Grieves, R. B., and Wang, S. L., 1967, Foam separation of bacteria with a cationic surfactant, *Biotechnol. Bioeng.* **9:**187–194.

Grossman, N., Ron, E. C., and Wodlringh, C. L., 1982, Changes in cell dimensions during amino acid starvation of *Escherichia coli, J. Bacteriol.* **152:**35–41.

Gucinski, H., Goupil, D. W., and Baier, R. E., 1981, Sampling and composition of the surface microlayer, in: *Atmospheric Pollutants in Natural Waters* (S. J. Eisenreich, ed.), Ann Arbor Science, Ann Arbor, Mich., pp. 165–180.

Hardy, J. T., 1973, Phytoneuston ecology of a temperate marine lagoon, *Limnol. Oceanogr.* **18:**525–533.

Hardy, J. T., and Valett, M., 1981, Natural and microcosm phytoneuston communities of Sequim Bay, Washington, *Estuarine Coastal Shelf Sci.* **12:**13–22.

Harris, P. J., 1972, Micro-organisms in surface films from soil crumbs, *Soil Biol. Biochem.* **4:**105–106.

Harrison, A. P., and Lawrence, F. R., 1963, Phenotypic, genotypic, and chemical changes in starving populations of *Aerobacter aerogenes, J. Bacteriol.* **85:**742–750.

Harvey, G. W., 1966, Microlayer collection from the sea surface: A new method and initial results, *Limnol. Oceanogr.* **11:**608–613.

Harvey, G. W., and Burzell, L. A., 1972, A simple microlayer method for small samples, *Limnol. Oceanogr.* **17:**156–157.

Harvey, R. W., and Young, L. Y., 1980a, Enrichment and association of bacteria and particulates in salt marsh surface water, *Appl. Environ. Microbiol.* **39:**894–899.

Harvey, R. W., and Young, L. Y., 1980b, Enumeration of particle-bound and unattached respiring bacteria in the salt marsh environment, *Appl. Environ. Microbiol.* **40:**156–160.

Hatcher, R. F., and Parker, B. C., 1974, Laboratory comparisons of four surface microlayer samplers, *Limnol. Oceanogr.* **19:**162–165.

Hatcher, R. F., and Parker, B. C., 1975, Concentration of coliform organisms at freshwater surfaces and their transfer into the atmosphere, *Va. J. Sci.* **26:**141–143.

Hawke, J. G., and Alexander, A. E., 1962, The influence of surface-active compounds upon the diffusion of gases across the air–water interface, in: *Retardation of Evaporation by Monolayers: Transport Processes* (V. K. LaMer, ed.), Academic Press, New York, pp. 67–73.

Hejkal, T. W., LaRock, P. A., and Winchester, J. W., 1980, Water-to-air fractionation of bacteria, *Appl. Environ. Microbiol.* **39:**335–338.

Helprin, J. J., and Hiatt, C. W., 1957, The effect of fatty acids on the respiration of *Leptospira icterohemmor- rhagiae, J. Infect. Dis.* **100:**136–140.

Henderson, R. E., 1973, The effects of fatty acids on pure cultures of rumen bacteria, *J. Agric. Sci.* **81:**107–112.

Hermansson, M., and Dahlbäck, B., 1983, Bacterial activity at the air/water interface, *Microb. Ecol.* **9:**317–328.

Hermansson, M., Kjelleberg, S., and Norkrans, B., 1979, Interaction of pigmented wildtype and pigment-less mutant of *Serratia marcescens* with lipid surface film, *FEMS Microbiol. Lett.* **6:**129–132.

Hermansson, M., Kjelleberg, S., Korhonen, T. K., and Stenström, T.-A., 1982, Hydrophobic and electrostatic characterization of surface structures of bacteria and its relationship to adhesion at an air–water surface, *Arch. Microbiol.* **131:**308–312.

Hickey, J. L. S., and Reist, P. C., 1975, Health significance of airborne microorganisms from wastewater treatment processes. Part 1. Summary of investigations, *J. Water Pollut. Control Fed.* **4:**2741–2757.

Hirsch, P., 1979, Life under conditions of low nutrient concentrations, in: *Strategies of Microbial Life in Extreme Environments* (M. Shilo, ed.), Verlag Chemie, Weinheim, pp. 357–372.

Hirsch, P., 1980, Neuston microorganisms: Observation, isolation and studies of surface growth, in: *Abstr. Annu. Meet. Am. Soc. Microbiol.* No. 53, p. 172.

Hoppe, H. G., 1977, Analysis of actively metabolizing bacterial populations with the autoradiographic method, in: *Microbial Ecology of a Brackish Water Environment* (G. Reinheimer, ed.), Springer-Verlag, Berlin, pp. 179– 197.

Horrigan, S. G., Carlucci, A. F., and Williams, P. M., 1981, Light inhibition of nitrification in sea-surface films. *J. Mar. Res.* **39:**557–565.

Hühnerfuss, H., Walter, W., and Kruspe, G., 1977, On the variability of surface tension with mean wind speed, *J. Phys. Oceanogr.* **7:**567–571.

Humphrey, B. A., Kjelleberg, S., and Marshall, K. C., 1983, Responses of marine bacteria under starvation conditions at a solid–water interface, *Appl. Environ. Microbiol.* **45:**43–47.

Hunter, K. A., 1980, Processes affecting particulate trace metals in the sea surface microlayer, *Mar. Chem.* **9:**49– 70.

Hunter, K. A., and Liss, P. S., 1977, The input of organic material to the oceans: Air–sea interactions and the organic chemical composition of the sea surfaces, *Mar. Chem.* **5:**361–379.

Ishida, Y., Imai, I., Miyagaki, T., and Kadota, H., 1982, Growth and uptake kinetics of a facultatively oligotrophic bacterium at low nutrient concentrations, *Microb. Ecol.* **8:**23–32.

Jannasch, H. W., 1974, Steady state and the chemostat in ecology, *Limnol. Oceanogr.* **19:**716–720.

Jannasch, H. W., and Pritchard, P. H., 1972, The role of inert particulate matter in the activity of aquatic microorganisms. *Mem. Ist. Ital. Idrobiol. Suppl.* **29:**289–308.

Jarvis, N. L., 1962, The effect of monomolecular films on surface temperature and convective motion at the water– air interface, *J. Colloid Sci.* **17:**512–522.

Jarvis, N. L., Garrett, W. D., Scheiman, M. A., and Timmons, C. O., 1967, Surface chemical characterization of surface active material in seawater, *Limnol. Oceanogr.* **12:**88–96.

Johnson, B. D., 1976, Nonliving organic particle formation from bubble dissolution, *Limnol. Oceanogr.* **21:**444– 446.

Kefford, B., Kjelleberg, S., and Marshall, K. C., 1982, Bacterial scavenging: Utilization of fatty acids localized at a solid/liquid interface, *Arch. Microbiol.* **133:**257–260.

Kjelleberg, S., and Hermansson, M., 1984, Starvation induced effects on bacterial surface characteristics, *Appl. Environ. Microbiol.* **48:**497–503.

Kjelleberg, S., and Håkansson, N., 1977, Distribution of lipolytic, proteolytic, and amylolytic marine bacteria between the lipid film and the subsurface water, *Mar. Biol.* **39:**103–109.

Kjelleberg, S., and Stenström, T.-A., 1980, Lipid surface films: Interaction of bacteria with free fatty acids and phospholipids at the air–water interface, *J. Gen. Microbiol.* **116:**417–423.

Kjelleberg, S., Norkrans, B., Löfgren, H., and Larsson, K., 1976, Surface balance study of the interaction between microorganisms and lipid monolayer at the air/water interface, *Appl. Environ. Microbiol.* **31:**609–611.

Kjelleberg, S., Stenström, T.-A., and Odham, G., 1979, Comparative study of different hydrophobic devices for sampling lipid surface films and adherent microorganisms, *Mar. Biol.* **53:**21–25.

Kjelleberg, S., Lagercrantz, C., and Larsson, T., 1980, Quantitative analysis of bacterial hydrophobicity studied by the binding of dodecanoic acid, *FEMS Microbiol. Lett.* **7:**41–44.

Kjelleberg, S., Humphrey, B. A., and Marshall, K. C., 1982, The effects of interfaces on small starved marine bacteria, *Appl. Environ. Microbiol.* **43:**1166–1172.

Kjelleberg, S., Humphrey, B. A., and Marshall, K. C., 1983, Initial phases of starvation and activity of bacteria at surfaces, *Appl. Environ. Microbiol.* **46:**978–984.

Korhonen, T. K., 1980, Purification of pili from *Escherichia coli* and *Salmonella typhimurium*, *Scand. J. Infect. Dis. Suppl.* **24:**154–157.

LaMer, V. K. (ed), 1962, *Retardation of Evaportation by Monolayers*, Academic Press, New York.

Larsson, K., 1973, Lipid multilayers, *Surf. Colloid Sci.* **6:**261–285.

Larsson, K., Odham, G., and Södergren, A., 1974, On lipid films on the sea. I. A simple method for sampling and studies of composition, *Mar. Chem.* **2:**49–57.

Lemlich, R., 1972, Adsubble processes: Foam fractionation and bubble fractionation, *J. Geophys. Res.* **77:**5204–5209.

Levin, G. V., Clendenning, J. R., Gibor, A., and Bogar, F. D., 1962, Harvesting of algae by froth flotation, *Appl. Microbiol.* **10:**169–175.

Lindahl, M., Faris, A., Wadström, T., and Hjertén, S., 1981, A new test based on "salting out" to measure relative surface hydrophobicity of bacterial cells, *Biochim. Biophys. Acta* **677:**471–476.

Lion, L. W., and Leckie, J. O., 1981, The biogeochemistry of the air–sea interface, *Annu. Rev. Earth Planet. Sci.* **9:**449–486.

Lion, L. W., Harvey, R. W., and Leckie, J. O., 1982, Mechanisms for trace metal enrichment at the surface microlayer in an estuarine salt marsh, *Mar. Chem.* **11:**235–244.

Longton, R. W., Cole, J. S., III, and Quinn, P. F., 1975, Isoelectric focusing of bacteria: Species location within an isoelectric focusing column by surface charge, *Arch. Oral Biol.* **20:**103–106.

Lupton, F. S., and Marshall, K. C., 1979, Effectiveness of surfactants in the microbial degradation of oil, *Geomicrobiol. J.* **1:**235–247.

Lyons, W. M., Pybus, M. J. S., and Coyne, J., 1980, The seasonal variation in the nutrient chemistry of the surface microlayer of Galway Bay, Ireland, *Oceanol. Acta* **3:**151–155.

MacIntyre, F., 1968, Bubbles: A boundary-layer "microtome" for micron-thick samples of liquid surface, *J. Phys. Chem.* **72:**589–592.

MacIntyre, F., 1970, Geochemical fractionation during mass transfer from sea to air by breaking bubbles, *Tellus* **22:**451–461.

MacIntyre, F., 1974a, The top millimeter of the ocean, *Sci. Am.* **230:**62–77.

MacIntyre, F., 1974b, Chemical fractionation and sea-surface microlayer processes, in: *The Sea*, Volume 5 (E. D. Goldberg, ed.), Wiley–Interscience, New York, pp. 245–299.

Maczulak, A. E., Dehority, B. A., and Palmquist, D. L., 1981, Effects of long-chain fatty acids on growth of rumen bacteria, *Appl. Environ. Microbiol.* **42:**856–862.

Mäkelä, P. H., and Stocker, F. A. D., 1981, Genetics of the bacterial cell surface, in: *Genetics as a Tool in Microbiology* (S. W. Glover and D. A. Hopwood, eds.), Cambridge University Press, London, pp. 219–264.

Maki, J. S., and Remsen, C. C., 1983, A membrane adsorption-SEM technique for observing neuston organisms, *Microb. Ecol.* **9:**177–183.

Marshall, K. C., 1972, Mechanism of adhesion of marine bacteria to surfaces, in: *Proceedings, 3rd International Congress on Marine Corrosion and Fouling* (R. F. Acker, B. F. Brown, J. R. DePlama, and W. P. Iverson, eds.), Northwestern University Press, Evanston, Ill., pp. 625–632.

Marshall, K. C., 1975, Clay minerology in relation to survival of soil bacteria, *Annu. Rev. Phytopathol.* **13:**357–373.

Marshall, K. C., 1976, *Interfaces in Microbial Ecology*, Harvard University Press, Cambridge, Mass.

Marshall, K. C., and Cruickshank, R. H., 1973, Cell surface hydrophobicity and the orientation of certain bacteria at interfaces, *Arch. Microbiol.* **91:**29–40.

Marshall, K. C., Stout, R., and Mitchell, R., 1971, Selective sorption of bacteria from seawater, *Can. J. Microbiol.* **17:**1413–1416.

Maxcy, R. B., and Dill, C. W., 1967, Adsorption of free fatty acids on cells of certain microorganisms, *J. Dairy Sci.* **50:**472–476.

Maynard, N. G., 1968, Significance of airborne algae, *Z. Allg. Mikrobiol.* **8:**119–126.

Michaels, G. B., Rector, D. D., and Cook, W. L., 1976, Densities of water quality indicator microorganisms in the surface microlayer, in: *Abstr. Annu. Meet. Am. Soc. Microbiol.* No. 15, p. 172.

Mitamura, O., and Matsumoto, K., 1981, Uptake rate of urea nitrogen and decomposition rate of urea carbon at the surface microlayer in Lake Biwa, *Verh. Int. Ver. Theor. Angew. Limnol.* **21:**556–564.

Morita, R. Y., 1982, Starvation survival of heterotrophs in the marine envirnoment, in: *Advances in Microbial Ecology*, Volume 6 (K. C. Marshall, ed.), Plenum Press, New York, pp. 171–198.

Mudd, S., and Mudd, E. B. H., 1924a, The penetration of bacteria through capillary spaces. IV. A kinetic mechanism at interfaces, *J. Exp. Med.* **40:**633–646.

Mudd, S., and Mudd, E. B. H., 1924b, Certain interfacial tension relations and the behaviour of bacteria in films, *J. Exp. Med.* **40:**647–660.

Napolitano, P. J., and Rowe, D. R., 1966, Microbial content of air near sewage treatment plants, *Water Sewage Works*, 113:480–483.

Naumann, E., 1917, Beiträge zur Kenntnis des Teichnannoplanktons. II. Uber das Neuston des Süsswassers, *Biol. Zentralbl.* **37:**98–106.

Nishizawa, S., and Nakajima, K., 1971, Concentration of particulate organic material in the sea surface layer, *Bull. Plankton Soc. Jpn.* **18:**12–19.

Norkrans, B., 1979, Role of surface microlayers, in: *Microbial Degradation of Pollution in Marine Environments* (A. W. Bourquin and P. H. Pritchard, eds.), U.S. Environmental Protection Agency, EPA-60/9-79-012, pp. 201–213.

Norkrans, B., 1980, Surface microlayers in aquatic environments, in: *Advances in Microbial Ecology*, Volume 4 (M. Alexander, ed.), Plenum Press, New York, pp. 51–85.

Norkrans, B., and Sörensson, F., 1977, On the marine lipid surface microlayer: Bacterial accumulation in model systems. *Bot. Mar.* **20:**473–478.

Norris, R. E., 1965, Neustonic marine *Creaspedomonodales* (choano flagellates) from Washington and California, *J. Protozool.* **12:**589–602.

Novitsky, J. A., and Morita, R. Y., 1976, Morphological characterization of small cells resulting from nutrient starvation of a psychrophilic marine vibrio, *Appl. Environ. Microbiol.* **32:**617–622.

Novitsky, J.A., and Morita, R. Y., 1977, Survival of a psychrophilic marine vibrio under long-term nutrient starvation, *Appl. Environ. Microbiol.* **33:**635–641.

Novitsky, J. A., and Morita, R. Y., 1978, Possible strategy for the survival of marine bacteria under starvation conditions, *Mar. Biol.* **48:**289–295.

Odham, G., Norén, B., Södergren, A., and Löfgren, H., 1978, Biological and chemical aspects of the aquatic lipid surface microlayer, *Prog. Chem. Fats Other Lipids* **16:**31–44.

Passman, F. J., Novitsky, T. J., and Watson, S. W., 1979, Surface microlayers of the North Atlantic: Microbial populations, heterotrophic and hydrocarbonoclastic activities, in: *Microbial Degradation of Pollution in Marine Environments* (A. W. Bourquin and P. H. Pritchard, eds.), U.S. Environmental Protection Agency, EPA-60/0-79-012, pp. 214–226.

Pellenbarg, R. E., 1976, The aqueous surface microlayer and trace metals in the salt marsh, Ph.D. thesis, University of Delaware, Newark.

Piotrowicz, S. R., Ray, B. J., Hoffman, G. L., and Duce, R. A., 1972, Trace metal enrichment in the sea-surface microlayer, *J. Geophys. Res.* **77:**5243–5254.

Poindexter, J. S., 1981a, The caulobacters: Ubiquitous unusual bacteria, *Microbiol. Rev.* **45:**123–179.

Poindexter, J. S., 1981b, Oligotrophy: Feast and famine existence, in: *Advances in Microbial Ecology*, Volume 5 (M. Alexander, ed.), Plenum Press, New York, pp. 63–89.

Postgate, J. R., and Hunter, J. R., 1962, The survival of starved bacteria, *J. Gen. Microbiol.* **21:**233–306.

Purkayastha, M., and Williams, R. P., 1960, Association of pigment with the cell envelope of *Serratia marcescens (Chrymobacterium prodigiosum)*, *Nature (London)* **187:**349–350.

Randall, C. W., and Ledbetter, J. O., 1966, Bacterial air pollution from activated sludge units *Am. Ind. Hyg. Assoc. J.* **27:**506–509.

Rogers, H. J., 1961, The dissimilation of higher molecular weight substances, in: *The Bacteria*, Volume II (I. C. Gunsalus and R. Y. Stanier, eds.), Academic Press, New York, pp. 257–318.

Romanenko, V. I., 1979, Bacterial growth on slides and electron microscope grids in surface water films and ooze deposits, *Microbiology (Engl. Transl.)* **48:**105–109.

Romanenko, V. I., Pubienes, M. A., and Daukshta, A. S., 1978, Growth and activity of bacteria on the surface film of water under experimental conditions, *Microbiology (Engl. Transl.)* **47:**119–124.

Rosenberg, M., Gutnick, D., and Rosenberg, E., 1980, Adherence of bacteria to hydracarbons: A simple method for measuring cell-surface hydrophobicity, *FEMS Microbiol. Lett.* **9:**29–33.

Rothfield, L., and Romeo, D., 1971, Role of lipids in the biosynthesis of the bacterial cell envelope, *Bacteriol. Rev.* **35:**14–38.

Roy, W. M., Dupuy, J. L., MacIntyre, W. G., and Harrison, W., 1970, Abundance of marine phytoplankton in surface films: A method of sampling, in: *Hydrobiology Bioresources of Shallow Water Environments* (W. G. Weist, Jr., and P. E. Greeson, eds.), University of Illinois Press, Urbana, pp. 371–380.

Rubin, A. J., Cassel, E. A., Henderson, O., Johnson, J. D., and Lamb, J. C., III, 1966, Microflotation: New low gas-flow rate foam separation technique for bacteria and algae, *Biotechnol. Bioeng.* **8:**135–151.

Schulman, J. H., and Teorell, T., 1938, On the boundary layer at membrane and monolayer interfaces, *Trans. Faraday Soc.* **34:**1337–1342.

Scott, J. C., 1972, The influence of surface-active contamination in the initiation of wind waves, *J. Fluid Mech.* **56:**591–606.

Shah, D. O., and Schulman, J. H., 1967, Enzymic hydrolysis of various lecithin monolayers employing surface pressure and potential technique, *J. Colloid Interface Sci.* **25:**107–119.

Shaw, D. J., 1970, *Introduction to Colloid and Surface Chemistry,* 2nd ed., Butterworths, London.

Sieburth, J. M., 1965, Bacteriological samplers for air–water and water–sediment interfaces, in: *Ocean Science and Ocean Engineering,* Volume 2, Transactions of the Joint Conference on Marine Technology Society and American Society of Limnology and Oceanography, Washington, D.C., pp. 1064–1068.

Sieburth, J. M., 1971a, An instance of bacterial inhibition in oceanic surface water, *Mar. Biol.* **11:**98–100.

Sieburth, J. M., 1971b, Distribution and activity of oceanic bacteria, *Deep-Sea Res.* **18:**1111–1121.

Sieburth, J. M., Willis, P.-J., Johnson, K. M., Burney, C. M., Lavoie, D. M., Hinga, K. R., Caron, D. A., French, F. W., III, Johnson, P. W., and Davis, P. G., 1976, Dissolved organic matter and heterotrophic microneuston in the surface microlayers of the North Atlantic, *Science* **194:**1415–1418.

Spendlove, J. C., 1974, Industrial, agricultural, and municipal microbial aerosol problems, *Dev. Ind. Microbiol.* **15:**20–27.

Stevenson, L. H., 1978, A case for dormancy in aquatic systems, *Microb. Ecol.* **4:**127–133.

Stevenson, R. E., and Collier, A., 1962, Preliminary observations on the occurrence of air-borne marine phytoplankton, *Lloydia* **25:**89–93.

Sutcliffe, W. H., Baylor, E. R., Jr., and Menzel, D. W., 1963, Sea surface chemistry and Langmuir circulation, *Deep-Sea Res.* **10:**233–243.

Syzdek, L. D., 1982, Concentration of *Serratia* in the surface microlayer, *Limnol. Oceanogr.* **27:**172–177.

Szekielda, K.-H., Kupperman, S. L., Klemas, V., and Polis, D. F., 1972, Element enrichment in organic films and foam associated with aquatic frontal systems, *J. Geophys. Res.* **77:**5287–5282.

Torella, F., and Morita, R. Y., 1981, Microcultural study of bacterial size changes and microcolony and ultramicrocolony formation by heterotrophic bacteria in seawater, *Appl. Environ. Microbiol.* **41:**518–527.

Tsiban, A. V., 1971, Marine bacterioneuston, *J. Oceanogr. Soc. Jpn.* **27:**56–66.

Tsiban, A. V., 1975, Bacterioneuston and problem of degradation in surface films of organic substances released into the sea, *Prog. Water Technol.* **7:**793–799.

van Vleet, E. S., and Williams, P. M., 1980a, Surface potential and film pressure measurement in seawater systems, FCRG Annual Report, University of California, San Diego, pp. 359–378, unpublished manuscript.

van Vleet, E. S., and Williams, P. M., 1980b, Sampling sea surface films: A laboratory evaluation of techniques and collecting materials, *Limnol. Oceanogr.* **25:**764–770.

Verger, R., Mieras, M. C. E., and de Haas, G. H., 1973, Action of phospholipase A at interfaces, *J. Biol. Chem.* **248:**4023–4034.

Waid, J. S., 1973, A method to study microorganisms on surface films from soil particles with the aid of the transmission electron microscope, *Bull. Ecol. Res. Comm. NFR (Swedish Natural Science Research Council)* **17:**103–108.

Wallace, G. T., Jr., Loeb, G., and Wilson, D. F., 1972, On the flotation of particulates in sea water by rising bubbles, *J. Geophys. Res.* **77:**5293–5301.

Wandschneider, K., 1979, Vertical distribution of phytoplankton during investigations of a natural surface film, *Mar. Biol.* **52:**105–111.

Wangersky, P. J., 1972, The cycle of organic carbon in seawater, *Chimia* **26:**559–564.

Wangersky, P. J., 1976, The surface film as a physical environment, *Annu. Rev. Ecol. Syst.* **7:**161–176.

Wangersky, P. J., 1977, The role of particulate matter in the productivity of surface waters, *Helgol. Wiss. Meeresunters.* **30:**546–564.

Wells, W. F., and Stone, W. R., 1934, On air-borne infection. Study III. Viability of droplet nuclei under infection, *Am. J. Hyg.* **20:**619–627.

Wendt, S. L., George, K. L., Parker, B. C., Gruft, H., and Falkinham, J. O., III, 1980, Epidemiology of infection by nontubercolous mycobacteria. III. Isolation of potentially pathogenic mycobacteria from aerosols, *Am. Rev. Respir. Dis.* **122:**259–263.

Wheeler, J. R., 1975, Formation and collapse of surface films, *Limnol. Oceanogr.* **20:**338–342.

Williams, P. M., 1967, Sea surface chemistry: Organic carbon and organic and inorganic nitrogen and phosphorus in surface films and subsurface waters, *Deep-Sea Res.* **14:**791–800.

Williams, P. M., van Vleet, E. S., and Booth, C. R., 1980, *In situ* measurements of sea-surface film potentials, *J. Mar. Res.* **38:**193–204.

Williams, R. P., 1973, Biosynthesis of prodigiosin, a secondary metabolite of *Serratia marcescens, Appl. Microbiol.* **25:**396–402.

Wilson, W. B., and Collier, A., 1972, The production of surface-active material by marine phytoplankton cultures, *J. Mar. Res.* **30:**15–25.

Woodcock, A. H., 1948, Note concerning human respiratory irritation associated with high concentrations of plankton and mass mortality of marine organisms, *J. Mar. Res.* **7:**56–62.

Woodcock, A. H., 1955, Bursting bubbles and air polution, *Sewage Ind. Wastes* **27:**1189–1192.

Wotton, R. S., 1982, Does the surface film of lakes provide a source of food for animals living in lake outlets?, *Limnol. Oceanogr.* **27:**959–960.

Wyndham, R. C., and Costerton, J. W., 1982, Bacterioneuston involved in the oxidation of hydrocarbons at the air–water interface, *J. Great Lakes Res.* **8:**316–322.

Yanagita, T., Ichikawa, T., Tsuji, T., Kamata, Y., Ito, K., and Sasaki, M., 1978, Two trophic groups of bacteria, oligotrophs and eutrophs: Their distributions in fresh and sea water areas in the central northern Japan, *J. Gen. Appl. Microbiol.* **24:**59–88.

Young, L. Y., 1978, Bacterioneuston examined with critical point drying and transmission electron microscopy, *Microb. Ecol.* **4:**267–277.

ZoBell, C. E., 1942, Microorganisms in marine air, *Publ. Am. Assoc. Adv. Sci.* **17:**55–68.

ZoBell, C. E., 1943, The effect of solid surfaces upon bacterial activity, *J. Bacteriol.* **46:**39–54.

ZoBell, C. E., 1946, *Marine Microbiology,* Chronica Botanica, Waltham, Mass.

ZoBell, C. E., and Mathews, H. M., 1936, A qualitative study of the bacterial flora of sea and land breezes, *Proc. Natl. Acad. Sci. USA* **22:**55–60.

8

Mechanisms of Adhesion to Clays, with Reference to Soil Systems

G. STOTZKY

1. INTRODUCTION

Soil is undoubtedly the most complex of all microbial habitats. Primarily because of this complexity, there is insufficient information on how and where most microbial events— including adhesion—occur in soil *in situ* and which microbes are numerically and physiologically the most important participants in these events. The abiotic components of soil have been relatively well defined, both qualitatively and quantitatively. However, the micro-geographic distribution and the geometric relations of abiotic components to each other, and to the microbiotic component, and the interactions among and between the abiotic and microbiotic components are not clearly defined. Most of what is known about the composition of the abiotic components has been obtained by dispersing soil (either chemically or physically, including sonically) and, after careful fractionation, conducting detailed chemical (both inorganic and organic) and physical analyses on the individual fractions (see Marshall, 1964; Rich and Kunze, 1964; Black, 1965; Schnitzer and Kahn, 1972; Gieseking, 1975; Dixon and Weed, 1977; Greenland and Hayes, 1978; Stevenson, 1982).

Most of what is known about the composition of the microbiotic component has been obtained by a variety of techniques, both direct (e.g., visible light, infrared, transmission and scanning electron, epifluorescence, and fluorescent-antibody microscopy of intact soil, of hardened soil sections, and of various types of surfaces that have been immersed in soil) and indirect (e.g., dilution or replica plating of soil on a spectrum of selective media incubated under a variety of cultural conditions, metabolic studies utilizing different types of organic and inorganic substrates and which reflect the enzymatic and, therefore, the genetic diversity of the microbiota, resistance to inhibitors) (see Stotzky, 1973, 1974; Atlas and Bartha, 1981; Burns and Slater, 1982). The information obtained about the microbiotic component has not been as definitive as that for the abiotic components, primarily as a result of the greater difficulties involved in completely extracting the microbiotic component in a viable state, concocting the appropriate media, and establishing cultural conditions that mimic the natural

G. Stotzky • Laboratory of Microbial Ecology, Department of Biology, New York University, New York, New York 10003.

soil habitat in the laboratory, and of the differences in the degree of sensitivity between physicochemical and microbiological techniques.

These differences in the relative sensitivity and precision between physicochemical and microbiological methodologies are probably not as important as the fact that these techniques usually disperse, fractionate, and then study each component of soil individually. Although this "first approximation" is undoubtedly critical to defining the system (i.e., the soil) that is being studied, it is an unwarranted and unsupportable "inductive leap" from this approximation to assumptions of what actually occurs in soil *in situ,* i.e., the actual microgeographic and geometric location of the various abiotic and biotic components and any and all interactions among and between them. Furthermore, *in vitro* studies on interactions between presumed "major and active" abiotic soil components and portions of the microbiotic component are based on assumptions that may not reflect the *in situ* situation. This, however, is the present state-of-the-art of studying soil–microbe interactions.

Another unwarranted assumption is that the microbiological responses obtained when individual particulate soil components—most frequently, clay minerals—are added to microbial systems are the result of adhesion between the soil components and the microbial cells. Although there is considerable evidence that some particulate soil components, especially some types of clay minerals, significantly affect microbial events, these effects are primarily indirect and do not appear to involve surface interactions between soil particulates and microbes. For example, the stimulatory effects of montmorillonite on fungi (Filip, 1975) and actinomycetes (Martin *et al.,* 1976) and the ability of clays to protect microbes against the inhibitory effects of hypertonic osmotic pressures (G. Stotzky, unpublished data) were the same when the clays and microbes were separated or not separated by dialysis tubing. Clay minerals apparently exert their primary influence on microbial events by modifying the physicochemical characteristics of microbial habitats, and this either enhances or attenuates the growth and metabolic activities of individual microbial populations, which, in turn, influences the growth and activities of other populations (see Stotzky, 1971, 1974, 1980; Filip, 1973, 1975; Hattori, 1973; Hattori and Hattori, 1976; Stotzky and Burns, 1982).

In contrast to these *indirect* effects of clay minerals, relatively little is known about *direct* surface interactions (e.g., adhesion) between clays, and other soil components, and microbes and viruses in soil *in situ.* Although there is empirical evidence to suggest that such surface interactions occur *in situ* (e.g., lack of movement of large numbers of microbes and viruses from the surface to underlying soil layers and then to groundwaters during heavy rains, snow melts, floods, or irrigation; failure to wash substantial numbers of microbes and viruses from soil columns in perfusion or leaching experiments; partial removal of microbes and viruses from waste waters in percolation beds; increased release of microbes from soil by sonication, surfactants, and other methods used to enhance the numbers of microbes enumerated), there are few hard data to substantiate this, and even fewer on the mechanisms involved. There are, however, considerable "mystiques" and unsupported assumptions.

One of these assumptions is that in soils and other environments with low levels of organic substrates, particulates (especially surface-active ones, such as clay minerals) concentrate these substrates at the solid–liquid interface, where they presumably reach levels high enough to support the growth of microbes which adhere to these surfaces in response to the nutrient enrichment. The available evidence, however, does not support this. For example, samples of mined montmorillonite, attapulgite (palygorskite), and kaolinite that contained 0.24, 0.24, and 0.15% organic carbon, respectively (the montmorillonite and attapulgite also contained 0.024 and 0.008% organic nitrogen, respectively; no organic nitrogen was detected in the kaolinite) (F. Andreux and G. Stotzky, unpublished data), did not support microbial growth in the absence of exogenous substrates (see Stotzky and Rem,

1966, 1967; Stotzky, 1974; Dashman and Stotzky, 1985b). Furthermore, when organic substrates, such as proteins (Pinck and Allison, 1951; Harter and Stotzky, 1971, 1973; see Stotzky, 1974), peptides, amino acids (Sørensen, 1972, 1975; Dashman and Stotzky, 1982, 1985a,b), polysaccharides (Olness and Clapp, 1972; see Hayes and Swift, 1978), nucleic acids (Goring and Bartholomew, 1952; Greaves and Wilson, 1969, 1970), and nucleotides (Greaves and Wilson, 1973; Ivarson et al., 1982), have been adsorbed to clay minerals, their availability to microbes was, in general, reduced. Consequently, even if clay surfaces in soil are "nutrient enriched," these nutrients may not be available to support microbial growth. Furthermore, once microbes have adhered to these surfaces in response to a presumed nutrient enrichment, the transfer of new nutrients to the surfaces must be continuous and rapid to be of benefit to the microbes (Ellwood et al., 1982). The enrichment of other inorganic particles in soil (e.g., sand, silt) with organic substrates has apparently not been studied, but as discussed below, these particles may not be too important to microbial life in soil. Although clay minerals may not be a good source of organic nutrients, exchangeable inorganic nutrients appear to be readily available to microbes (Stotzky and Rem, 1966).

A corollary to the assumption that organic substrates are concentrated at solid–liquid interfaces is that organic inhibitors are also concentrated there. Although this assumption may be valid, as there is some evidence that especially clay minerals bind antibiotics and other inhibitors and reduce their antimicrobial activity (e.g., Skinner, 1956; Soulides, 1969; Martin et al., 1976; Campbell and Ephgrave, 1983; see Stotzky, 1974), a paradox is apparent, which is an example of the mystiques that are associated with the adhesion of microbes to surfaces, especially in soil. If adhesion of microbes to soil particulates, especially to clay minerals, occurs because of the concentration of organic substrates by these particulates, then this adhesion could be detrimental to microbes if the same or closely adjacent particulates also bind antibiotics and other inhibitors. Inasmuch as inert surfaces probably do not distinguish between "good" and "bad" adsorbates, the same mechanism cannot be invoked for the direct influence of both organic substrates and organic inhibitors on the adhesion of microbes to particulates.

As with inorganic nutrients, inorganic inhibitors bound to clay appear to be readily exchangeable. For example, the protection of microbes against the toxic effects of soluble Cd, Pb, and Ni by clay minerals and particulate organic matter was correlated with the cation exchange capacity (CEC) of the particles (Babich and Stotzky, 1977a, 1979, 1980b, 1983a,b). However, when Cd was added only as either montmorillonite or kaolinite homoionic to Cd (i.e., no free ionic Cd was added), the toxicity of Cd was also correlated with the CEC of the clays, with Cd-montmorillonite (with the higher CEC) being more toxic than Cd-kaolinite at comparable or lower concentrations (Babich and Stotzky, 1977a). Both the nutritional and the toxic effects of exchangeable cations appeared to be the result of indirect effects of the clays, rather than of adhesion of the cells to the clays, and probably involved exchange between cations on the clays and protons produced during metabolism of the cells (Stotzky and Rem, 1966; Babich and Stotzky, 1980b, 1983a,b).

A further consideration—regardless of whether substrates or inhibitors are involved—is that there may actually be a dilution, rather than a concentration, of these materials at the solid–liquid interface as a result of negative adsorption. Negative adsorption can occur if the adsorbate is negatively charged and is repelled from a highly negatively charged adsorbent (e.g., some inorganic and organic anions from montmorillonite) (see Theng, 1974, 1979). Similarly, a highly hydrophobic molecule may be repelled from the surface of a clay mineral with a high CEC, as a result of the hydration of the exchangeable cations. The importance of negative adsorption of molecules or of cells in adhesion of microbes to soil particulates has apparently not been studied.

A major factor contributing to these mystiques and assumptions is the variability and heterogeneity of soil. Even over small distances, e.g., less than 1 mm, the composition and size of the particulates, the amounts and types of water, nutrients, and gases, and the pH, E_h, ionic strength, and other physicochemical characteristics can vary. This variability in abiotic factors is reflected by the heterogeneity of the microbiota, which is demonstrated by the simultaneous occurrence in the same soil sample of autotrophs and heterotrophs [both oligotrophs and copiotrophs (Poindexter, 1981)], aerobes and anaerobes, vegetative cells and spores, prokaryotes and eukaryotes, and cells with different requirements for and tolerances to osmotic pressure, pH, E_h, temperature, etc. (e.g., Stotzky, 1974; Hattori and Hattori, 1976).

Not only is the microbiota in soil heterogeneous, but the inanimate surfaces in soil are also heterogeneous as a result of being coated (partially or completely) with clay minerals, hydrous metal oxides, and organic matter, of pH-dependent changes, and of fluctuations in the types and concentrations of ambient electrolytes, etc. This heterogeneity precludes the realistic application to soil of many concepts of the mechanisms of adhesion (e.g., DLVO and critical surface tension theories; see Chapters 5, 6, and 11), as these concepts assume homogeneous and clean surfaces (Tadros, 1980).

The complexity and heterogeneity of soil is probably responsible, in part, for the lack of the elegant types of studies on microbial adhesion in soil that have been conducted in waters (see Chapters 6 and 7), in animals (see Chapters 1, 10, 11, and 15), on plants (see Chapters 3, 9, and 14), and in other natural microbial habitats that are more homogeneous in their microbial populations, surfaces, and physicochemical characteristics. Because these studies have not been conducted in soil, it is difficult—and unwarranted—to assume that adhesion of microbes occurs in soil and that, if adhesion occurs, the mechanisms are similar to those in other habitats, even though "intuition" tends to skew the assumptions in that direction.

The purposes of this chapter are to review briefly what is and is not known about surface interactions between abiotic soil components and microbes, the mechanisms involved, and the importance of these interactions to microbial events and to the survival of viruses in soil, as well as to illuminate the "mystiques," expose the unsupported assumptions, and indicate directions for future realistic and productive studies. This chapter is not intended to be an exhaustive review of the various subject areas that impact on this topic—a herculean task that would not only exceed the limits of a chapter but also the energies and capabilities of the author—but rather an exploration of the topic. Consequently, relevant reviews are cited more than the original literature that is the foundation for these and the present review. The author apologizes for those research reports that have not been cited: this was the result primarily of constraints in space and time, and references to these can be found in the research reports of the author and his co-workers and of other investigators in this area.

2. PROPERTIES OF SOIL PARTICLES

Soil differs from most other microbial habitats in that it is dominated by a solid phase consisting of particulates of different sizes which is surrounded by liquid and gaseous phases that fluctuate markedly in time and space. The solid phase is a tripartite system composed of (1) finely divided minerals (both primary and secondary), (2) plant, animal, and microbial residues in various stages of decay, and (3) a living and metabolizing microbiota. These particulates exist as both independent entities and mixed conglomerates.

2.1. Inorganic Components

In most surface soils, inorganic particulates occupy approximately 50% of the soil volume. These inorganic particulates vary in size from stones (> 20 mm in diameter), to gravel (2–20 mm), to coarse sand (0.2–2.0 mm), to fine sand (0.02–0.2 mm), to silt (0.002–0.02 mm), and to clay (< 0.002 mm in diameter) (White, 1979). This classification is based solely on size (there are several classifications of particle-size classes) and does not reflect completely the mineralogical composition of each size fraction. For example, primary minerals, such as quartz, may be present in all size fractions, including the clay fraction. Conversely, some clay minerals, such as vermiculite, may be larger than 0.002 mm (2.0 μm) (Brown et al., 1978; Farmer, 1978). Nevertheless, the clay-sized fraction is composed primarily of clay minerals and the larger fractions of weathered primary minerals.

One result of the differences in particle diameter is that the surface area per unit mass (specific surface) increases logarithmically as the particle diameter decreases (Hattori and Hattori, 1976). For example, clay-sized particles with an average diameter of 2 μm have 50 to 100 times more surface area than an equivalent amount of silt- or fine sand-sized particles. These values are based on highly dispersed systems, and comparable values may not occur in soil in situ because of aggregation (see Section 2.2). Nevertheless, the large differences in specific surface among soil particulates not only influence where microbial activities, including adhesion, occur in soil, but also distinguish surfaces in soil from those in most other microbial habitats.

Particulates larger than the clay-sized fraction do not long retain water films because of their low specific surface and relatively inert surfaces (see Section 3), and therefore, they probably do not maintain large or permanent microbial populations (Stotzky, 1974). Consequently, the major inorganic particulates that affect microbial events in soil are within the clay-sized fraction and consist primarily of clay minerals and polymeric hydrous oxides of mainly Fe^{3+}, Al^{3+}, and Mn^{4+}.

2.1.1. Clays

Clay minerals are primarily crystalline hydrous aluminosilicates composed of two basic structural units: a tetrahedron of oxygen atoms (O) surrounding a central cation, usually tetravalent silicon (Si^{4+}); and an octahedron of O and hydroxyl ions (OH^-) surrounding a central cation, ususally trivalent aluminum (Al^{3+}) or divalent magnesium (Mg^{2+}). The geometry of these units is dominated mainly by O (radius = 0.140 nm) or OH^- (radius = 0.146 nm), as Si^{4+} (radius = 0.041 nm), Al^{3+} (radius = 0.050 nm), and Mg^{2+} (radius = 0.065 nm) are considerably smaller. The sharing of O or OH^- between neighbor tetrahedra or octahedra results in the formation of sheets called tetrahedral (or silica) and octahedral (or alumina or magnesia) sheets, respectively. The almost identical symmetry and dimensions of the tetrahedral and octahedral sheets permit the sharing of O between the sheets. In two-layer clays, such as kaolinite, the sheets are associated in a 1:1 (silica–alumina) ratio, and in three-layer clays, such as montmorillonite, the sheets are associated in a 2:1 (silica–alumina–silica) ratio. Some clay minerals, such as attapulgite, do not form such platelike structures but form fibrious needlelike crystals.

The individual 1:1 or 2:1 associations are called unit layers, and these are stacked to give packets of different sizes. The successive unit layers of 1:1 clays are held together tightly by hydrogen (H) bonds that are formed between the OH groups of the octahedral sheet of one layer and the O of the tetrahedral sheet of another layer. Consequently, packets of 1:1 clays do not normally expand upon wetting and do not expose any internal surface. In contrast, direct H-bonding in 2 : 1 clays is not possible, as the outer sheets of the unit layer are silica tetrahedra and only O surfaces face each other, and the unit layers are held together by van der Waals forces (it is actually the hydration energy of the interlayer cations that forces the layers apart). As these forces are relatively weak, water and other polar molecules can penetrate between the unit layers and increase the basal (c) spacing and the electrostatic interaction of interlayer cations. Water molecules become adsorbed on the O surfaces or on the hydration layers of the charge-compensating cations by forming H-bonds, and a stable configuration is obtained with two to four monomolecular layers of water between the unit layers, which, depending on the associated cations, can increase the c-spacing on the order of 1.25 to 3.00 nm and expose a considerable amount of internal surface area. It has been suggested that, depending on the type of charge-neutralizing cation present (e.g., Li^+, H^+, Na^+), the increase in c-spacing could exceed 13 nm, which would result in essentially complete dispersion of the packets into individual unit layers (Norrish, 1954). If such dispersion occurs in soil *in situ*, then surface interactions between clays and microbial cells would be between small, highly negatively charged unit layers of clay and net negatively charged cells, which would result in repulsion of the unit layers from the cells or, under certain conditions (see Section 7.2), in the coating of the cells with individual unit layers.

In montmorillonite, the total specific surface can be as high as 800 m^2/g oven-dry clay, with the external area contributing only about 40 to 80 m^2/g. In kaolinite, the specific surface ranges from 15 to 50 m^2/g and consists almost entirely of external area. In attapulgite, the specific surface is approximately 125 m^2/g and consists primarily of open narrow channels in the needlelike crystals that comprise the fibrous mass.

In some clay minerals, especially in 2 : 1 clays, some structural cations are replaced by cations of similar size but of lower valence with an almost imperceptible modification in the crystalline structure. This process is called *isomorphous substitution* and usually involves the replacement of some Si^{4+} in the tetrahedron by Al^{3+} or Fe^{3+} and some Al^{3+} in the octahedron by Mg^{2+}, Fe^{2+}, or other divalent cations. This isomorphous substitution imparts a net negative charge on the unit layers, which is either permanently neutralized by K^+ in minerals with a high amount of substitution (especially in the tetrahedral layers) and a high surface charge deficiency, resulting in essentially noncharged and nonexpandable clays (e.g., mica), or transiently neutralized in clays with less isomorphous substitution by cations (e.g., Ca^{2+}, Mg^{2+}, K^+, Na^+, H^+, NH_4^+) present in the soil solution. In the latter case, the charge-neutralizing cations are readily exchanged by other cations, depending on their valence, hydration, and concentration in the soil solution. These cations are hydrated, and their associated water forms H-bonds with the clay-associated water and with organic molecules and cell surfaces that may be present.

The amount of neutralizing cations that can be retained by and exchanged from a clay mineral is termed the CEC and is expressed in units of milliequivalents (meq) per gram or per 100 g of oven-dry clay. In some 2 : 1 clays, the surface charge deficiency, which is located primarily between the unit layers, is neutralized by a positively charged sheet of $Al_2(OH)_6$ (gibbsite) or $Mg_3(OH)_6$ (brucite) (as in chlorite) or by smaller interlayers of Al or Fe hydroxides (as in various interstratified clays), which reduces not only the CEC but also the ability of the clays to swell. In general, the greater the amount of isomorphous substitution, the more tenaciously will charge-neutralizing cations be retained by the clays. The differences in the characteristics (e.g., CEC, specific surface) of the various platelike 2 : 1 layer clays (phyllosilicates) are the result primarily of differences in the amount of substitution.

In addition to the negative charges that result from isomorphous substitution, some negative charges also result from dissociation of OH groups on the surfaces and edges of clays. This dissociation, which probably constitutes the primary source of the negative charge on 1 : 1 clay minerals and on various hydrous metal oxides, is dependent on the pH of and the type and concentration of ions in the soil solution. In 2 : 1 minerals, the negative charge is essentially pH-independent, as more than 80% of the CEC results from iso-morphous substitution, whereas in 1 : 1 clays and hydrous metal oxides, most of the negative charge is pH-dependent and increases with an increase in pH.

The CEC of 2 : 1 phyllosilicates ranges from 60 to 200 meq/100 g clay, of 2 : 1 fibrous clays from 3 to 38 meq/100 g clay, and of 1 : 1 phyllosilicates from 2 to 10 meq/100 g clay. These values, as those for specific surface, apply only to dispersed "clean" clays and probably overestimate actual values in soil, where aggregation, interactions between clays, organic matter, and hydrous metal oxides (e.g., interlayer formation), and other phenomena (e.g., cutan formation on larger particulates) occur (see Section 2.2). Furthermore, these negative charges are not uniformly distributed over the clay surfaces but occur as discrete point charges.

The differences in the specific surface area and CEC between clays are reflected in their mean surface charge density, which is a rough indication of the closeness of the negative charges and is probably important in adhesion. For a montmorillonite with a total specific surface of 800 m^2/g and a CEC of 0.98 meq/g, the derived mean surface charge density is 1.23×10^{-3} meq/m^2, and for a kaolinite with a specific surface of 10 m^2/g and a CEC of 0.06 meq/g, it is 6×10^{-3} meq/m^2. Consequently, even though both the specific surface and the CEC are vastly greater for montmorillonite than for kaolinite, the negative charges are closer together on kaolinite (i.e., the mean area per charge is greater on montmorillonite).

The importance of the distance between negative charges in surface interactions be-tween microbes and clays has not been established. However, the distance between point charges if they, in fact, exist, appears to affect the adsorption of organic cations to clays. The adsorption of paraquat and diquat (organic, divalent, cationic herbicides) to various clay minerals was related to the apparent distance between the positive charges on the herbicides and the surface charge density of the minerals. Paraquat, with a distance of 0.7 to 0.8 nm between the positive charges, was preferentially adsorbed to clays with a relatively low surface charge density (e.g., montmorillonites) whereon the distance between surface charge sites was similar to the charge separation on the herbicide, whereas diquat, with a charge separation of 0.3 to 0.4 nm, was preferentially adsorbed to minerals with a higher surface charge density (e.g., micas) (Philen *et al.*, 1971).

Clays also have an anion exchange capacity (AEC), which is located primarily at the edges of the clay packets and results from breakage at the edges to expose Al^{3+} or Si^{4+} or from the exchange of OH groups. Some anions, such as phosphate, arsenate, or borate, which have a size and geometry similar to the silica tetrahedron, can be adsorbed by fitting on the edges of the tetrahedral sheet, whereas other anions commonly present in the soil solution but which differ in size and geometry from the silica tetrahedron, such as sulfate, nitrate, and chloride, do not fit and are not adsorbed. The AEC is difficult to determine accurately, but it is essentially similar for 2 : 1 and 1 : 1 phyllosilicates and ranges from 7 to 30 meq/100 g. Consequently, the CEC : AEC ratio is probably as important a parameter in adhesion and other microbial events as are the CEC and the AEC, as this ratio will determine the relative net negativity of different clays and the ability of net negatively charged mi-crobes, viruses, and organic molecules to approach the clays. The average CEC : AEC ratio is about 6.7 for montmorillonite and 0.5 for kaolinite. In soil *in situ,* however, the actual CEC : AEC ratios may be considerably different from those in dispersed clay suspensions (especially of clean clays), as many of the positively charged edge sites may be unavailable as a result of face-to-edge aggregation of the clay particles or blockage by negatively charged inorganic or organic molecules.

The reader is referred to review articles and books for more details on clay mineralogy (e.g., Marshall, 1964; Rich and Kunze, 1964; Mysels, 1967; Grim, 1968; Swartzen-Allen and Matijevic, 1974; Gieseking, 1975; Dixon and Weed, 1977; van Olphen, 1977; Brown *et al.*, 1978; van Olphen and Fripiat, 1979).

2.1.2. Other Inorganic Components

In addition to crystalline hydrous aluminosilicates, many soils also contain allophane, imogolite, and polymeric hydrous oxides of Fe^{3+}, Al^{3+}, Mn^{4+}, Ti^{4+}, and Si^{4+} (*hydrous oxide* refers here to any oxide, oxyhydroxide, or hydroxide). Allophane and imogolite are present primarily in young soils derived from recent volcanic activity, whereas hydrous oxides are usually present in highly weathered and old soils, especially in tropical soils. Although these materials are not as well characterized as are the crystalline clay minerals, they may be as or more important in microbial events—especially in adhesion—in soil, as they can be either positively, negatively, or neutrally charged. The pH at which the surface of these materials is uncharged (i.e., has a net neutral charge) is called the zero point of charge (ZPC). The ZPC is distinguished from the isoelectric point (pI) of amphoteric organic molecules (and of cells and viruses containing such molecules on their surfaces) in that at the pI, the total charge is the highest but the negative and positive charges are equal and there is no net movement in an electrical field. [This distinction between the ZPC and the pI of hydrous oxides and organic molecules is not accepted by all mineralogists, and considerable controversy exists in this area (e.g., Gast, 1977; Arnold, 1978; Greenland and Mott, 1978; Sposito, 1981; Escudey and Galindo, 1983).] The approximate ZPC of hydrous oxides that are probably most important in soils are: SiO_2, pH 2; MnO_2, pH 4; TiO_2, pH 4.5; Fe_2O_3, between pH 6.5 and 8; and Al_2O_3, between pH 7.5 and 9.5 (Parks, 1965). At pH values above its ZPC, a hydrous oxide is negatively charged as a result of dissociation of protons, and at pH values below the ZPC, it is positively charged as a result of association of protons. Because of the presence of impurities, the state of hydration, the different coordination numbers of some cations, and other mediating factors in the soil environment, the actual ZPC can vary *in situ*. Hydrous metal oxides also form polymers, the stability and charge of which are also pH-dependent.

Although it has often been stated that hydrous metal oxides form coatings on the surfaces of crystalline clays, this concept has been questioned, as the CEC of soils can usually be related directly to the clay mineralogy of soils [e.g., the CEC after treatment to remove these oxides was not significantly different from that of untreated soils (see Greenland and Mott, 1978)]. However, with respect to microbial events, these measurements involved bulk soil samples and may not have been sensitive enough to determine the effects of metal oxides in the microhabitats. Just as the bulk pH of soil samples was not correlated with the rate of spread of *Fusarium* wilt of banana (Stotzky *et al.*, 1961; Stotzky and Martin, 1963) or with the worldwide geographic distribution of *Histoplasma capsulatum* (Stotzky and Post, 1967; Stotzky, 1970) in soils *in situ*, subsequent studies (see Stotzky, 1967, 1971, 1974) under more defined conditions showed that montmorillonite, the clay type that was correlated with the slow spread of the wilt and with the apparent failure of *H. capsulatum* to colonize, exerted its effect primarily by influencing the pH of the microhabitats (Section 7.4.1). Consequently, the presence of even small amounts of these hydrous metal oxides in a soil may have a significant effect on adhesion and other microbial events, as these oxides appear to be concentrated in the microenvironments where they probably interact with clay minerals and alter the surface characteristics of the clays. Furthermore, the CEC of various minerals and soil fractions was decreased at pH values below the ZPC of Fe- and Al-hydroxides, whereas the CEC was increased at pH values above the ZPC of these hydroxides (Hendershot and Lavkulich, 1983).

2.2. Aggregates

Clay minerals in soil probably do not exist as individual packets, as the individual packets would, because of their small size (< 2 μm), be washed to underlying soil layers and then to ground waters. Rather, the clays exist as cutans (skins) on larger soil particulates (Brewer, 1964) and as domains (oriented clusters of individual packets), which, together with silt and fine sand particles, form microaggregates (Emerson, 1959; see Arnold, 1978; Greenland and Hayes, 1978). The clustering of clay packets into cutans and domains occurs, in part, from electrostatic interactions between oppositely charged faces and edges of the packets, resulting in a "card-house" (van Olphen, 1977) or a "honeycomb" structure (Lockhart, 1980a,b). These microaggregates are stabilized by organic matter and precipitated inorganic materials. One effect of this clustering of clay packets into stable microaggregates is a reduction in both the specific surface area and the cation and anion exchange capacities of the clays. Consequently, data obtained with dispersed and clean clays that relate specific surface and exchange capacities to adhesion of microbes and viruses, to binding of organics, and even to the indirect effects of clays on microbial events may have limited relevance to what occurs in soil *in situ*.

The microaggregates, in turn, cluster together and, in combination with larger sand particles and cemented by organic matter and inorganic precipitates, form aggregates that can range from 0.5 to 5 mm. These aggregates retain water, the thickness and permanence of which depend on the amount and location of the clay fraction and organic matter within the aggregates (see Section 3), and this water may form bridges with adjacent aggregates close enough to retain water by surface tension. These aggregates, or clusters of aggregates, with their adjacent water comprise the microhabitats in soil wherein microbes presumably function (see Stotzky, 1974; Hattori and Hattori, 1976). The space between the microhabitats constitutes the pore space, which is filled with N_2, O_2, CO_2, and other gases and volatiles, the amounts of which depend on the water content of the soil and the metabolic activity of the microbes at any given time (see Babich and Stotzky, 1974; Stotzky, 1974; Stotzky and Schenck, 1976; Burns, 1983).

2.3. Organic Components

In addition to clay minerals, particulate organic matter is undoubtedly important in microbial events, including adhesion, in soil. In contrast to clay minerals, soil organic matter is relatively poorly defined, primarily as a result of its complexity. Particulate soil organic matter can be separated into two major groups: (1) nonhumic substances consisting of plant, animal, and microbial debris in the early stages of physical and chemical disintegration and decay, processes that are mediated, in part, by soil macro- and microorganisms, and (2) transformed products that bear no morphological resemblance to the materials from which they were derived and which consist of amorphous, polymeric (i.e., macromolecular), dark-colored humic materials and of chemically definable polymers, such as polysaccharides, polypeptides, and altered lignins. The transformed products result both from modifications of similar structures in the original debris and from *de novo* synthesis by the soil microbiota or by abiotic means. The amorphous humic substances are differentiated, on the basis of their solubility, into humic acids (soluble in alkali but precipitated by acid), fulvic acids (soluble in both alkali and acid), and humins (insoluble in both alkali and acid).

The chemical composition of the humic substances is not completely known and varies from soil to soil. Elemental analysis has shown that the carbon, nitrogen, and hydrogen contents of fulvic acids are usually lower than those of humic acids, whereas the reverse is true for the oxygen content. The principal functional groups are carboxyl, phenolic, and alcoholic hydroxyl, carbonyl, methoxyl, and amino. The carboxyl groups are perhaps the most important, as they dissociate at the pH of most soils and, hence, contribute to the CEC of soils. Depending on the pH, the type and concentration of cations in the soil solution, and the composition of the humic polymers, the CEC of humic substances ranges from 200 to 600 meq/100 g. Consequently, even in soils with low organic matter contents, these materials may be as significant to microbial events, especially through indirect effects, as clay minerals. Particulate organic matter is net negatively charged at the pH of most soils, as are most clay minerals.

In addition to particulate organic matter, soluble organic matter, composed of smaller and usually chemically definable compounds (e.g, carbohydrates, proteinaceous materials, organic acids), is also present in soils. Although these materials are mostly labile and, therefore, transient, they constitute the major carbon and energy sources for microbial activities in soil. The binding of these components to the clay fraction, and probably also to particulate organic matter, appears to render them less available for microbial utilization.

Although the composition, age, turnover rates, and other properties of soil organic matter, both particulate and soluble, have been and are being extensively studied (e.g., Bremner, 1967; Schnitzer and Khan, 1972; Flaig *et al.*, 1975; Parsons and Tinsley, 1975; Hayes and Swift, 1978; Stevenson, 1982), no studies have apparently been conducted on the role of particulate organic matter in the adhesion of bacteria, other microbes, or viruses in soil. This deficiency is probably the result of the difficulty in extracting organic matter from soil without extensive alteration in its structure and surface properties, from the vagueness and contradictory data about its chemical composition and structure, and from the reluctance both of investigators interested in adhesion to work with this complex and ill-defined particulate fraction and of specialists in soil organic matter to devote time from their studies of the properties of this fraction to studies of adhesion. This is an example of the still-existing lack of significant interaction between the various specialities—e.g., chemistry, physics, mineralogy, microbiology—in soil science.

Even the indirect effects of particulate soil organic matter on microbial events have been sparsely studied. Commercially available humic acids have been shown to protect microbes against the toxic effects of some heavy metals (Babich and Stotzky, 1979, 1980b, 1983a; Stotzky and Babich, 1980), and other studies have focused on the chemistry of organic matter–metal interactions (see Stevenson, 1982). In contrast, numerous studies have investigated the interactions between clay minerals and organic matter (e.g., Greenland, 1965, 1971; Mortland, 1970; Theng, 1974, 1979; Schnitzer and Kodama, 1977; Hayes and Swift, 1978), but with few exceptions (see Sections 5 and 7.2.3), these studies have not been concerned with the effects—either indirect or direct—of these interactions on microbial events in soil.

In view of the paucity of data on the role of particulate organic matter in microbial events in soil—especially in adhesion—this discussion will focus on the role of the clay mineral fraction. Not only is there more known about clay minerals and their effects on microbes, but this fraction is more stable than the organic matter fraction, which is constantly being modified by the activities of the soil microbiota, including the microfauna, as well as by some macrofauna and abiological reactions.

3. IMPORTANCE OF WATER

Although numerous physicochemical factors affect microbial life in soil (see Stotzky, 1974), perhaps the most important is an adequate supply of utilizable water. Microbes are aquatic organisms and, even in soil, require a sufficiently high water activity (a_w) for growth. Except for short periods following rain, snow melts, or irrigation, a significant portion of the volume of most arable soils contains insufficient available water to support microbial activity. Most sand- and silt-sized particles do not retain water against gravitational pull, and it is primarily the clay minerals that, because of their charged surfaces, retain sufficient water to sustain growth. Consequently, the apparent correlation of microbial activity with the clay fraction in soil may be the result primarily of the availability of water associated with this fraction rather than of the presumed concentration of nutrients, removal of inhibitors, and surface interactions with cells by this fraction. In aquatic environments, colonization of nonclay inorganic particles appears to be common (e.g., Meadows and Anderson, 1966; DeFlaun and Mayer, 1983), inasmuch as the a_w surrounding these particles in such environments is not limiting. Furthermore, the absence, or near-absence, of negative charges on these particles facilitates the adhesion of net negatively charged bacteria, viruses, and organic molecules in aquatic habitats. Similar situations probably prevail in trickling filters, where the particles of the filter bed are large and have a low surface charge density, and water and nutrients are not limiting. In such biofilms, the diffusion rate of oxygen usually becomes the limiting factor to sustained growth.

The relatively constant occurrence of water around clay-containing microaggregates in soil probably also bestows a degree of permanence to the associated bacteria and may explain, in part, the limited vertical movement of bacteria through soil after rain, irrigation, perfusion, etc. Furthermore, the slow lateral spread of bacteria through soils (e.g., in soil replica plating studies with soils maintained at their 1/3 bar water tension) indicates that the spread is the result of growth (i.e., cell division) rather than of motility and suggests that bacteria, even if flagellated in soil, do not have sufficient energy to overcome the surface tension of the water associated with these microaggregates. In contrast, fungi, which grow by apical and lateral extension of hyphae from a food- and water-base and are able both to translocate water internally and to retain water externally on the hyphae, can grow across pore spaces devoid of water and in areas with a low a_w and, thereby, ramify rapidly through soil (Stotzky, 1965, 1973, 1974; Rosenzweig and Stotzky, 1979, 1980).

The retention of water by clays is presumably a result primarily of their high surface charge and of the coordination of water molecules to the charge-compensating cations, which apparently induces an ordering of adjacent water molecules to form a quasicrystalline structure within the water (Low, 1960, 1961, 1962, 1979; Farmer, 1978). The extent of this structure varies with the type and arrangement of the clay particles and the exchangeable cations (Low, 1960, 1979; Forslind and Jacobsson, 1970) This adsorbed water is less dense, more viscous, and freezes at a lower temperature than water not associated with clays (Low, 1960). Furthermore, the interlayer water of montmorillonite differs from the bulk water in specific volume, specific heat capacity, heat of compression, specific expansibility, and viscosity (Low, 1979). Because of this ordering, water close to the surfaces of clays is probably not available to microbes, as its a_w is too low to sustain growth. Consequently, it is difficult to accept the assumption that microbes are metabolizing within this highly ordered water, which further mitigates against the concept that, in soil, microbes—especially bacteria—are attached to domains or cutans of clay. It is more likely that microbes are growing in clay-associated water some distance from the clay surfaces. This water is probably still under the attraction of the clays and, thereby, sufficiently ordered to resist gravitational pull, but it has a sufficiently high a_w to enable its use by microbes. Furthermore, it is probable that available organic nutrients are concentrated in this outer water and, perhaps, retained by surface tension rather than being bound to the clay surfaces where the availability of the nutrients to microbes is reduced (e.g., Gerard and Stotzky, 1973; see Stotzky, 1974; Dashman and Stotzky, 1985b).

Particulate organic matter also retains ordered water, primarily as a result of the polar groups of the organic matter, and this water appears also to be tenaciously held. For example, the water-holding capacity of a muck soil (85% organic matter) was 262% (w/w), but only when the water content was above 157% (w/w) (the 1/3 bar water tension) was the a_w sufficiently high to support microbial activity (Stotzky and Mortensen, 1957, 1958).

Although the structure (primarily the tertiary structure) of particulate organic matter is such that it will expand upon wetting, excessive drying can render the organic matter difficult to rewet as a result of its structural reorientation and the exposure of hydrophobic regions during drying. These hydrophobic regions are the result of the presence in the organic matter of lipids, waxes, and hydrophobic moieties in proteinaceous materials. These hydrophobic components can also render inorganic particles—especially clay-containing microaggregates—hydrophobic when complexes between these inorganic and organic components are formed (Farmer, 1978). The importance of the relative hydrophobicity of inanimate surfaces to microbial events—including adhesion—in soil has not been clarified. However, as regions of some microbial surfaces are also hydrophobic (e.g., some surface protein antigens and nonpolar amino acid residues), the importance of "hydrophobic interactions" in adhesion of microbes to surfaces in soil requires study and clarification (see also Chapter 5, Section 2; Chapter 6, Section 4.5). Furthermore, the occurrence in soil of chaotropic ions (which decrease the structure of water and tend to disrupt hydrophobic interactions by increasing the accommodation of nonpolar compounds in aqueous solutions) and of anti-chaotropic ions (which increase the structure of water and, thereby, hydrophobic interactions by reducing the ability of aqueous solutions to accomodate nonpolar groups) (Dandiker *et al.*, 1967; Hatefi and Hanstein, 1969, 1974; Farrah *et al.*, 1981; Shields and Farrah, 1983) and the influence that these compounds have on microbial events—especially in the microhabitats near reactive surfaces—should be investigated.

The microbial cell also retains water, and the extent and physicochemical characteristics of this water are probably influenced by the chemical nature of and the charges on the outer coverings of the cell. The source of this water is not only the environment but also cellular metabolism. Little appears to be known about the interactions and exchanges between nonadsorbed water and that on the surfaces of cells and soil particulates and associated with adsorbed ions. However, despite the apparently highly ordered nature of clay-associated water, exchange of individual water molecules appears to occur constantly and at rates measured in fractions of a second (M. H. B. Hayes, personal communication).

The water associated with highly charged clay particles appears to be more tightly bound than that to microbial cells, as cellular water can be removed by drying at approximately 100°C, whereas higher temperatures are necessary to remove all clay-associated water. Furthermore, increasing the concentration of swelling clays, such as montmorillonite, resulted in gels that interfered with the diffusion of oxygen and reduced the growth of aerobic bacteria (Stozky and Rem, 1966) and filamentous fungi (Stotzky and Rem, 1967) and caused a shift in metabolism in faculative bacteria from aerobic to anaerobic pathways (Lee *et al.*, 1970). Consequently, it is difficult to ascribe the apparent protective effect of such clays against the desiccation of bacteria in soils under extreme water stress to the availability of water retained by the clays (Marshall and Roberts, 1963; Bushby and Marshall, 1977). In fact, it has been suggested that the apparent tolerance of some rhizobia to desiccation in the presence of montmorillonite, but not of illite or kaolinite, was the result of the greater and more rapid dehydration of the cells because of the higher affinity for water of the montmorillonite than of the cells. The survival of these rhizobia during desiccation appeared to be enhanced by rapid drying to low water contents, as the loss in viability was less when the cells were dried at a relative humidity of 70% or lower than at 90% (Bushby and Marshall, 1977; Osa-Afiana and Alexander, 1982).

Although too little is known about how the physicochemical characteristics of soil water affect microbial events, the type of water, the type of soil surfaces, and the interactions between water and these surfaces probably affect adhesion and metabolism in soil (Fletcher and Marshall, 1983). The effects of these interactions on microbial events may again reflect only an indirect effect of clays rather than a direct surface interaction between clays and microbes. Furthermore, it is obvious that it is not possible to extrapolate results obtained in aquatic systems (see Chapters 12 and 13) (wherein water is not limiting and, although it may be ordered to some extent near surfaces, is continuous) to soil (wherein the more permanent water is contiguous to highly charged clay and, probably, organic surfaces and, therefore, is highly ordered). Similarly, in other microbial habitats (e.g., the oral cavity, the gastrointestinal and genitourinary tracts, plant surfaces), the water associated with surfaces is usually more continuous and less ordered than in soil because of the relative low charge of these surfaces.

4. "CONDITIONING FILMS" AND EXTRACELLULAR SLIME LAYERS

Another area in which extrapolation from aquatic environments to the soil environment may not be valid is the role of conditioning films of organic molecules on soil surfaces in the subsequent adhesion of bacteria and other organisms to these surfaces (see Chapter 6, Section 4.8). There appears to be convincing evidence that the binding of such molecules— probably of microbial origin—to surfaces in aquatic systems renders these surfaces more susceptible to adhesion, apparently as a result of the reduction in their surface free energy (see Fletcher et al., 1980; Rutter and Vincent, 1980; Chapters 5 and 6). Comparable data for surfaces in soil are lacking. However, studies on the adsorption of viruses to clays in vitro indicated that organic molecules either competed with hydrophilic viruses, i.e., reovirus (Lipson and Stotzky, 1984b) and bacteriophages (Schiffenbauer and Stotzky, 1985b), for adsorption sites (both negatively and positively charged) on clays or did not affect the adsorption of hydrophobic viruses, i.e., herpes hominis type I (Yu and Stotzky, 1979; Stotzky et al., 1981). The amount of competition was related to the size and pI of the molecules and to the type of clay, the type of cations on the clay, and the sites of adsorption on the clays (Lipson and Stotzky, 1984b; Schiffenbauer and Stotzky, 1985).

4.1. Organic and Inorganic Conditioning Films

When proteins were bound on montmorillonite homoionic to various cations (Harter and Stotzky, 1971), both the bulk pH and the electrophoretic mobility of the clays were altered to either higher or lower values depending on the type of protein bound (Harter and Stotzky, 1973). Changes to higher values were observed when a synthetic humic acid polycondensate (composed of catechol and glycylglycylglycine, average $M_r = 30,000$) was bound on various homoionic clays (F. Andreux and G. Stotzky, unpublished data). These observations suggest that if binding of organic molecules to clays occurs in soil in situ, the surfaces of clays will be modified (conditioned), which may affect their subsequent surface interactions with microbes and viruses. Changes in surface charge properties of solids in aqueous environments have also been observed as a result of the adsorption of organic molecules (Neihof and Loeb, 1972, 1974; Marshall, 1980b), but, as emphasized above, conditions (e.g., continuous water, size, and physicochemical properties of the surfaces) are different in aquatic habitats and in soil.

In contrast, when amino acids and small peptides were bound on various homoionic montmorillonites and kaolinites (Dashman and Stotzky, 1982, 1984), there were no significant differences in the pH or electrophoretic mobility of the clay–amino acid or clay–peptide complexes as compared with the homoionic clays alone (Dashman and Stotzky, 1984), even though the amounts of amino acids and peptides bound were greater than the amounts of proteins bound (Harter and Stotzky, 1971, 1973). In contrast to the macromolecular proteins, the smaller amino acids and peptides apparently did not cover sufficient portions of the surface of the clays to alter significantly their bulk pH or electrokinetic potentials.

Although changes in the electrokinetic properties of clays reacted with organic molecules other than proteins and amino acids have apparently not been studied directly, flocculation (which reflects changes in surface charge) of clays by polysaccharides (i.e., rhizobial gums) and microbial metabolites (i.e., culture filtrates) occurred only when the clays were homoionic to polyvalent cations (Santoro and Stotzky, 1967b). Consequently, the type of organic molecules and the cation status of the clays appear to be important in the conditioning of clay surfaces.

In soils, inorganic materials may be more important as conditioning agents than organic materials, as not only are precursors and metabolic energy required for synthesis of the latter, but they are also susceptible to biodegradation. The formation of polymeric hydrous oxides of Fe^{3+}, Al^{3+}, and Mn^{4+} is an abiotic process, and their interaction with clays can result in marked changes in the electrokinetic properties of clays (see Section 2.1.2). These inorganic polymers have their own electrokinetic properties, and if coverage of the clays is adequate, these properties will dominate. For example, the net charge of montmorillonite homoionic to Al^{3+} in a dilute $AlCl_3$ solution (ionic strength $= 3 \times 10^{-3}$) was positive from about pH 3.5 to 8.2, whereas in a solution of NaCl of the same ionic strength, the net charge of montmorillonite homoionic to Al^{3+} was negative at all pH values measured (i.e., pH 2.0 to 9.5) (Santoro and Stotzky, 1967c). The interaction of clays with such inorganic polymers also greatly increased the binding on the clays of a synthetic humic acid composed of catechol and triglycine (F. Andreux and G. Stotzky, unpublished data).

In addition to changes in the net surface charge of clays caused by the adsorption of polymeric hydrous metal oxides, some heavy metals (e.g., Cu, Ni) also caused a reversal in the charge of kaolinite and montmorillonite from their normal net negative charge to a net positive charge at alkaline pH values. Although the mechanisms responsible for these charge reversals and the relevance of these reversals to clays in soil *in situ* are not clear, they occurred at concentrations of the metals (1×10^{-4} M) that may be present in soils exposed to heavy metal pollution and, therefore, may be important in adhesion processes (Collins and Stotzky, 1982, 1983). Consequently, inorganic materials may be important in the conditioning of surfaces in soil.

Despite, or perhaps because of, the paucity of data about the occurrence of organic conditioning films in soil, several questions must be asked. If such organic molecules are produced, do they bind to the highly net negatively charged surfaces of clays, especially if the molecules are also net negatively charged? If they do bind, are they rendered resistant to microbial degradation, especially by bacteria and other microbes that presumably would subsequently be involved in surface interactions with such conditioned clays? If the organic molecules adsorb to less highly charged soil particulates (e.g., sand and silt particles), are they bound sufficiently tightly to prevent their biodegradation during the brief periods when these particulates are surrounded by adequate water to support microbial growth? Why would microbes under the starvation conditions that are presumed to be the norm in soils waste precious energy and structural components to synthesize and excrete molecules that may never reach appropriate surfaces to become conditioning films because of their utilization by other microbes?

Here, again, the differences between aquatic and soil environments must be considered when attempting to relate data on conditioning films in aquatic habitats to what may occur in soils, where data on the occurrence of such films are lacking. Aquatic habitats have continuous water, generally lower population densities, and surfaces that are large and mostly uncharged (e.g., ship bottoms, pilings, bulk headings, pipes, sediments that generally—but not always—are composed of large particulates, experimentally submersed surfaces, such as glass, metal, plastic, and Formvar grids). Soil habitats have restricted and highly ordered water and higher population densities, and the microbiologically active particulates are small and have highly charged surfaces. Similar caution must be exercised in extrapolating to soil data obtained from other microbial habitats that share many of the characteristics of aquatic habitats but which may have more energy and carbon sources (e.g., oral cavity, gastrointestinal and genitourinary tracts).

4.2. Bacterial Extracellular Slime Layers

Similar questions must be raised about the possible role of extracellular slime layers in the adhesion of bacteria to surfaces in soil. Such slime layers and capsules are presumably synthesized widely in aquatic and other habitats, and they have been suggested to be a primary mechanism by which bacteria attach to surfaces in these environments (e.g., Fletcher and Floodgate, 1973; Corpe, 1980; Marshall, 1980b; Costerton, 1980; Costerton et al., 1981). The synthesis of these extracellular materials in soil and their involvement in the adhesion of cells to soil particulates have not been unequivocally demonstrated.

In environments where slime layers around bacteria have been observed, the slimes appear to be predominantly acidic (e.g., they stain with ruthenium red, crystal violet, and alcian blue), and hence, their net charge is negative (see Corpe, 1980; Costerton, 1980; Sutherland, 1980, 1983; Costerton et al., 1981; Chapter 1, Section 1; Chapter 6, Section 6). Consequently, even if such materials are synthesized in soil, the mechanisms whereby these net negatively charged polymers bind on net negatively charged clays require clarification. In fact, highly acidic polysaccharides are apparently not adsorbed to clays (see Hayes and Swift, 1978). Furthermore, it would appear that because of the charge relations, encapsulated cells would preferentially bind to nonclay particulates (e.g., sand and silt) in soil—as they do in aquatic habitats—only to die from lack of water. It has been suggested that in aquatic environments, encapsulated cells adhere to sand and silt grains, and the slimes then entrap clay particles from the surrounding water (DeFlaun and Mayer, 1983).

Although most extracellular polymeric materials appear to be predominantly polysaccharide in nature, some also contain proteins, including glycoproteins (Corpe, 1980; Sutherland, 1980, 1983). Consequently, even though many polysaccharides, especially very acidic ones, may not bind on clay minerals, the protein components of slimes may be involved in the adhesion between bacteria and clay minerals. Some proteins have been shown to bind tenaciously to clay minerals (Harter and Stotzky, 1971, 1973; Stotzky, 1974), and some proteases, but not saccharases, were able to remove attached bacteria (Fletcher and Marshall, 1983).

Nevertheless, many species of bacteria are genetically incapable of synthesizing extracellular materials, and many that have the genetic capability do so only under specific nutritional conditions (e.g., even under laboratory conditions, a rich medium, such as one containing skim milk, is necessary to establish whether a bacterial isolate is capable of producing capsules and/or a more extensive slime layer). Not all autochthonous soil bacteria are able to synthesize these materials. Consequently, if extracellular materials are critical to adhesion and adhesion is critical to the survival of bacteria in natural habitats, then all bacteria in soil (and in other natural habitats as well) should be able to synthesize them, as nonsynthesizers would have been selected against. This does not appear to be the situation.

Furthermore, bacterial endospores do not produce extracellular slimes, yet these spores are retained in soil.

Even for those bacteria that are genetically able to synthesize slime materials, the nutritional conditions in soil may be too poor to provide the structural components for the synthesis of the monomers and the energy necessary for the formation of the glycosidic linkages of the polymers. As with the synthesis of conditioning films, it would appear to make little sense, teleologically at least, for an organism to squander its limited resources to synthesize extracellular polymers, especially if these do not provide specificity in terms of sticking the cell to surfaces in microenvironments that will provide the water and nutrients necessary for growth and survival. The observation that little or no extracellular materials were synthesized by bacteria in continuous culture under carbon-limiting conditions (Wardell *et al.*, 1980; Ellwood *et al.*, 1982) adds support to the suggestion that these materials are not formed under the usual oligotrophic conditions in soil. Furthermore, the types of carbon and energy sources available in soil at any given time may determine whether appropriate slimes for adhesion are synthesized. This would be similar to the observations that *Streptococcus mutans* did not adhere to teeth when growing on glucose but did adhere when growing on sucrose (see Gibbons, 1980), that the specificity of adhesion of *Streptococcus sanguis* to hydroxyapatite was different for cells grown on glucose than on fructose (Rosan *et al.*, 1982), and that species of *Rhizobium*, which produced copious amounts of polysaccharides in pure culture, showed the least amount of adsorption to glass of seven genera of bacteria tested (Zvyagintsev, 1973).

However, in the continuous culture studies of Wardell *et al.* (1980) and Ellwood *et al.* (1982), the numbers and species diversity of the bacterial population attached to glass slides placed into the chemostats were greater under carbon-limited conditions than under conditions of excess carbon (but limited nitrogen), where large quantities of extracellular materials were produced. Consequently, the possibility that adhesion to soil particulates may be enhanced by "starvation conditions" in soil must be considered. Marshall and others (see Marshall, 1980b) have indicated that in aqueous systems, adhesion to surfaces (e.g., glass beads) occurred only at extremely low nutrient concentrations and was inhibited by higher concentrations. Perhaps bacteria have no choice under low nutrient conditions but to utilize their endogenous carbon and energy sources to attach to surfaces, otherwise they die. Furthermore, the nutritional conditions in soil are seldom as poor as they are in many natural waters, and yet bacteria appear to form copious quantities of extracellular polymers in such waters. However, in many waters, there is a more continuous supply of carbon and energy sources that are provided by photosynthetic algae. Obviously, the questions of whether bacteria form exopolymers in soil and, if so, whether they have a role in adhesion and survival of the cells require study.

Because of the opaque nature of soil, it is difficult to view undisturbed soil microhabitats with electron microscopy to determine directly whether extracellular materials, especially thin capsules that presumably surround most bacteria (Costerton, 1980), are formed *in situ*. Most electron micrographs of disturbed soil samples do not suggest the presence of such materials (e.g., Nikitin *et al.*, 1966), but these micrographs have been of samples that have not been stained specifically for extracellular polysaccharides or other compounds. Light micrographs of bacteria in soil, including *Agromyces ramosus*, which may constitute the majority of the bacterial population in soil and may represent true oligotrophs, suggest the presence of a capsule around the cells (see Casida, 1965, 1971, 1977). However, these capsules may only be apparent rather than real, as they may be similar to the clear "ring" seen by electron microscopy around bacterial species placed into soil and which do not appear to contain any extracellular slime (Labeda *et al.*, 1976). Even if the presence of extracellular materials around bacteria can be demonstrated in soil, it will still have to be resolved whether these materials are involved in adhesion or whether they are formed after adhesion has occurred by some other mechanism, as has been suggested to occur in aquatic habitats (Ward and Berkeley, 1980). The slime layers could then serve the adhered cell by trapping nutrients, reducing desiccation, decreasing susceptibility to ingestion by predators, providing a reserve source of nutrients, etc. (Stotzky, 1974).

Regardless of whether the extracellular materials are formed before or after adhesion of the producer cells to surfaces in soil, the stability of the exopolymers, especially against degradation by other microbes, must be considered. The number of bacteria in soil usually range from 10^6 to 10^9/g oven-dry soil, and the number in the microhabitats are undoubtedly higher, as soil is not uniformly colonized (Gray *et al.,* 1968). Consequently, these exopolymers are probably under constant enzymatic attack. Although the binding of polysaccharides on clay minerals and humic substances appears to protect many polysaccharides against rapid degradation, this protection is not absolute (see Hayes and Swift, 1978). Although a similar cycle of synthesis and degradation probably occurs in aquatic and other habitats (even though the population densities may be lower than in soil) where the production of exopolymers has been extensively studied, there appear to be no data on the rates of degradation of these extracellular materials. If the synthesis of extracellular materials is expensive for a cell in terms of its carbon and energy reserves, and the materials are then rapidly degraded by other organisms, thereby negating any positive effects of the exopolymers, including adhesion, then the evolutionary value of the production of these extracellular materials must be questioned.

Nevertheless, studies of the adhesion of bacteria in various natural habitats indicate that both conditioning films and extracellular slime layers are formed and that these may be important in the adhesion of cells to surfaces and in the subsequent pathogenicity of some species (e.g., Costerton *et al.,* 1981). Furthermore, many fungi excrete polysaccharides with "adhesive qualities" *in vitro* (Martin and Adams, 1965), and such materials have been implicated in the aggregation of clays around fungal structures (Dorioz and Robert, 1982). However, the same reservations, stated above, concerning the production, persistence, and importance *in vivo* of extracellular bacterial polysaccharides can be applied to fungi. Consequently, comparable studies should be conducted in soil to determine whether these films and layers have a similar function in the adhesion of cells to soil particulates and in the pathogenicity of soil-borne plant pathogens. The structural and physicochemical complexity of soil will make these studies difficult.

5. METABOLIC ACTIVITY OF PARTICLE-BOUND CELLS IN SOIL

There is considerable interest in whether the adhesion of bacteria to surfaces affects their metabolic activity (see Chapter 12). However, studies to determine this have been conducted essentially only in the laboratory with model systems or in aquatic systems, usually with synthetic surfaces (e.g., Hattori and Hattori, 1976; Bright and Fletcher, 1983). There appear to be no comparable studies with soil, probably because of the lack of techniques with which to separate quantitatively attached and unattached cells from soil. The use of a series of filter membranes with different pore sizes, which is extensively used with aquatic systems to separate presumably attached and unattached cells, is of little value in studies of soil, wherein the wide range of particle sizes of inorganic and organic particulates would obscure any differentiation between free and bound cells. Furthermore, any physical and chemical disruption of a soil sample in preparation for separating attached and unattached cells might detach cells that were associated with particles *in situ*.

The results obtained with model and aquatic systems have been extremely variable, with some studies indicating a higher metabolic activity of attached cells, some indicating a lower activity of attached cells, and others indicating no difference in activity between attached and unattached cells (Hattori and Hattori, 1976; Kirchman, 1983; see Chapter 12). Bacteria that attached to glass slides submersed in a chemostat grew about twice as fast as unattached cells (Ellwood *et al.*, 1982), and the nitrification and respiratory rates of *Nitrobacter* attached to baked expanded macroporous clays were greater than those of unattached cells (Audic *et al.*, 1984).

Variable results have also been reported with respect to the size of attached cells: some investigators have reported that small (''dwarf'') cells attached more than larger (''normal'') cells (see Marshall 1980b; Kjelleberg *et al.*, 1982, 1983), whereas others have reported that the volume of free-living cells was vastly smaller than that of attached cells (e.g., Hodson *et al.*, 1981; Pedros-Alio and Brock, 1983). The relative size of attached and unattached bacteria in freshwater appeared to be temperature-dependent, with larger cells being predominant in the summer, when the percentage of attached bacteria and the number of attached cells per particle were also highest (Kirchman, 1983; Chapter 12, Section 2).

No comparable data are available for soils, although the studies of Peele (1936) indicated that the degradation of glucose by *Azotobacter chroococcum* was markedly reduced in a soil saturated with Fe^{3+} or Al^{3+}, presumably because the bacteria were adsorbed to soil components, whereas no comparable adsorption or reduction in glucose utilization occurred in the same soil saturated with Na^+ or NH_4^+. Unfortunately, these interesting studies have apparently not been repeated. It has also been suggested that cells of *Agromyces ramosus* in soil are metabolically inactive dwarfs (Casida, 1977; Marshall, 1980a). However, there appears to be no evidence for the adhesion of these cells to surfaces in soil *in situ*.

Until unequivocal evidence for the adhesion of cells to soil particles has been obtained, speculation as to the relative metabolic activity of attached and unattached cells in soil is premature. Although there is ample evidence that some soil particulates, especially clay minerals, affect many metabolic activities in soil (see Stotzky, 1967, 1974, 1980; Filip, 1973, 1975; Hattori and Hattori, 1976), caution must again be exercised in attributing these effects to the adhesion of cells to these particulates. In most instances, these effects can be explained by *indirect* rather than by *direct* effects of the soil particulates (see Stotzky, 1974, 1980). Furthermore, nothing appears to be known about the desorption of cells that may attach to soil particles, especially about the effects that different substrates and their subsequent metabolic products may have on the desorption process. This aspect also appears to have received insufficient attention in aquatic habitats (Roper and Marshall, 1974), although such desorption, or the shedding of cells from attached developing colonies, must occur to ensure colonization of new sites and perpetuation of the species.

6. PHYSICOCHEMICAL PROPERTIES AND MECHANISMS THAT CAN AFFECT ADHESION

Although the properties of clay minerals that affect surface interactions with microbes have not been studied extensively, attempts have been made to identify the properties that affect the adsorption and binding of soluble organic materials (see Mortland, 1970; Theng, 1974, 1979; Harter, 1977). Inasmuch as some soluble organic compounds are similar to components of the surface structures of microbial cells and viruses, these studies have relevance to the adhesion of particulate biological entities. However, the differences in the spatial and conformational constraints of a soluble organic compound and of the same compound as part of a more rigid surface structure must be considered when making extrapolations from adsorption and binding studies with soluble compounds to the adhesion of microbes and viruses (see Section 7.1).

6.1. Charge Interactions

Although the CEC and AEC of clays affect their coulombic interaction with net positively and net negatively charged organic compounds, respectively, the CEC : AEC ratio may be more important in their surface interactions with microbes and viruses, as this ratio affects the net negativity of the clays and, therefore, the ability of net negatively charged biological particulates to come close enough to the clays for mechanisms of adhesion to operate. Similarly, the surface charge density and the distribution, location, and source of the negative charges on the clays probably affect the relative repulsion : attraction ratio at clay–liquid interfaces. If the surface charge density is low or the charges are spatially located primarily in a cluster or on only one side of a clay unit [e.g., as has been suggested for kaolinite (McLaren, 1954; McLaren *et al.*, 1958)], then the probability of surface interactions between net negatively charged clays and cells is increased. If the source of the charge of 2 : 1 clays is primarily in the tetrahedral sheets (i.e., if the majority of the isomorphous substitution is in these sheets), the electrostatic force and the surface acidity will be greater than in clays with predominantly octahedral charges, and the repulsion of cells may be greater. These aspects have not been evaluated sufficiently in surface interactions between clays and microbes.

One property of clay minerals that has received considerable attention with respect to their surface interactions with microbes, viruses, and organic compounds is the composition of the cationic counterions on the clays. In general, if a net positively charged biological entity is involved, surface interactions are usually greater when the charge-compensating cations have a low charge (e.g., monovalent) and a small ionic radius, presumably as the exchange of monovalent inorganic cations by organic cations is easier than the exchange of multivalent cations. For example, the amount of lysozyme (which was net positively charged at the pH of the experiments) adsorbed on montmorillonite homoionic to different cations was inversely related to the valence and the nonhydrated ionic radius of the cations saturating the exchange complex of the clay (Harter, 1977). In contrast, when net negatively charged biological entities are involved (a much more usual situation, as most microbes, viruses, and organic compounds are net negatively charged at the pH of most soils), surface interactions between these entities and clays are usually greater when the valence of the exchangeable cations is higher. This enhancement is usually the result primarily of the reduction of the extent of the diffuse double layer (i.e., of the net negative charge) of the clays (and also of cells if there is an exchange between cations on the clays and on the cells), which enables the clays and the biological entities to approach each other more closely. This phenomenon, which is an example of the Schultze–Hardy rule, is discussed in greater detail below (see Section 6.3). In addition, cations with a higher valence appear to form stronger bonds, at least with soluble organic compounds (Lailach *et al.*, 1968; Mortland, 1970). Multivalent cations, however, also reduce the expansibility of swelling 2 : 1 clays and the dispersibility of all clays. Although these reductions in surface area can reduce the binding of some soluble macromolecules (e.g., proteins) (Harter and Stotzky, 1971, 1973), the surface interactions of microbes (see Stotzky, 1974) and viruses (Stotzky *et al.*, 1981; Lipson and Stotzky, 1983) with clay increased as the valency of the cation on the exchange complex of the clays increased. Whether the enhancement of these surface interactions was a direct result of the reduction of the extent of the diffuse double layer (Chapter 5, Section 3; Chapter 6, Section 3.1) or an indirect result of the flocculation of the clays and a concomitant decrease in the expression of the interlayer charge on swelling clays (most of the negative charge on these clays is located between the unit layers) has not been determined. This distinction, however, may not be necessary, as both these direct and indirect results cause a reduction in the net negative charge on the clays and, therefore, a reduction in the electrokinetic repulsion between the clays and the microbes and viruses.

The characteristics of organic molecules that are important in their adsorption and binding on clays have been studied extensively (e.g., Greenland, 1965; Mortland, 1970; Theng, 1974, 1979; Harter, 1977; Hayes and Swift, 1978; Stevenson, 1982), but again, the differences between a soluble molecule and the same molecule covalently bound in a relatively rigid surface structure on a microbe or virus must be considered. Even with soluble organic molecules, the characteristics of different molecules—and, therefore, the mechanisms—that are involved in surface interactions between the molecules and clays have not been unequivocally defined. If the soluble or particulate biological entity is net positively charged (a rare occurrence in most soils), the initial reaction (e.g., reversible adsorption) with net negatively charged clays is probably coulombic, and subsequent interactions (e.g., irreversible binding) probably involve additional reactions that render the entities resistant to cation exchange. The "two-stage" attachment concept of Marshall (1976, 1980a) may be applicable to interactions between oppositely charged biological entities and clays.

More usually, however, the soluble or particulate biological entity is also net negatively charged at the pH of most soils. The net negativity of the biological entity is dependent on the pK of its dissociable functional groups and the ambient pH. For amphoteric entities (i.e., those having both acidic and basic functional groups, which is characteristic of most bacteria and viruses), the pI (which is an empirical summation of the pK values of all components capable of accepting or releasing protons) and the pH of the solution surrounding the entity will determine its net charge. However, as discussed below (see Section 7.5.1), the pI appears to be important only in the adsorption of relatively small molecules, as large molecules (e.g., proteins) and particulate entities (e.g., microbes and viruses) may have positively charged sites even at pH values above their pI (when their net charge is negative). Furthermore other mechanisms, such as H-bonding and van der Waals forces, may be involved in the surface interactions of these molecules and particulates with the net negatively charged surfaces of clays (e.g., Harter and Stotzky, 1971, 1973; Dashman and Stotzky, 1982; Lipson and Stotzky, 1983; see Stotzky, 1974, 1980).

If the molecule is uncharged or if the molecule is large and the distribution of charges is relatively localized, adsorption can result from differences in the polarity of the molecule, either from permanent dipole moments or from induced dipole moments caused by the presence of the charge field around clays. The size (i.e., molecular weight) and conformation of the molecule, whether charged or uncharged, can influence its ability to intercalate expanding clays and, through steric hindrance, may influence its ability to interact with surface or edge charges, as has been shown for paraquat and diquat (Philen et al., 1971) and some proteins (e.g., Harter and Stotzky, 1971, 1973).

These characteristics of clays and biological entities, in conjunction with the physicochemical properties of their environment, will influence the mechanisms of surface interactions between them. As already indicated, if the biological entity has a net positive charge, the major interaction will be exchange with inorganic charge-compensating cations on the negative sites on the clay, with the exchange being mediated by such factors as the valence and ionic radius of the inorganic cation and the conformation (i.e., steric hindrance) of the biological entity. If the biological entity has a net negative charge, its interaction with net negatively charged clays can be mediated by a variety of mechanisms. These mechanisms are briefly described below, and the interested reader is referred to more extensive treatments of these mechanisms (e.g., Greenland, 1965; Mortland, 1970; Theng, 1974, 1979; Harter, 1977; Stevenson, 1982).

6.2. Hydrogen Bonding

Perhaps the most important mechanism involves some type of H-bonding. Although H-bonds are relatively weak bonds, in contrast to coulombic or covalent bonds, the interaction resulting from numerous H-bonds functioning in concert and over small distances can be extremely strong. For example, the resistance of 1 : 1 clay minerals (e.g., kaolinite) to intercalation by polar molecules and, thereby, to swelling is presumably the result primarily of H-bonding between the surface OH groups of the octahedral sheet and the surface O of the tetrahedral sheet. Similarly, the binding of some proteins and viruses to clays appears to result primarily from H-bonding (Harter and Stotzky, 1971; Lipson and Stotzky, 1983).

The formation of H-bonds between some functional groups (e.g., COH, COOH, NH) on organic entities and the surface O of the tetrahedral sheet (Si—O—Si or siloxane surfaces) of clays is presumably unfavorable energetically, as the basal O are poor electron donors (Farmer, 1978; Greenland and Mott, 1978; Theng, 1979). However, there is some evidence that such H-bonds do form between clays and some organic molecules (see Harter, 1977), and the structure of kaolinite packets also indicates that such H-bonds are formed. The formation of H-bonds between structural OH groups of clays and functional groups (e.g., CO, N) on biological entities is probably of minor importance, as exposed OH groups are located primarily at the broken edges of clays and are usually dissociated at the pH of most soils. In 2 : 1 clays, the OH groups are in the middle octahedral sheet, and therefore, they are never exposed at the surface of the unit layers or the clay packets. In 1 : 1 clays, the octahedral sheet within the packet are not exposed because of the nonexpansion of these clays, and only OH groups on one outer side of the packet are exposed. These OH groups appear to be involved in surface interactions [e.g., with proteins (McLaren, 1954; McLaren *et al.,* 1958) and some viruses (Lipson and Stotzky, 1984b)], but as the relative amount of OH surface per clay packet is small, these interactions may be of limited importance in soil.

The major form of H-bonding between clays (at least, dispersed ''clean'' clays) and biological entities appears to be the result of protonation. Protons from the clay surface—or, more probably, from clay-associated water—are transferred to the biological entity (usually to atoms that have lone pair electrons, such as O and N and, to a lesser extent, S and P), rendering the entity either net neutrally or net positively charged, and surface interaction with the net negatively charged clay then occurs either by H-bonding or van der Waals forces or by coulombic interaction, respectively. Protonation is a function of the surface pH (pH_s) of the clay, which is presumably several pH units lower than the bulk pH (pH_b) of the adjacent solution as a result of hydrolysis and the formation of a Stern layer around the clay (see also Chapter 5, Section 3). The magnitude of the difference between the pH_s and the pH_b (i.e., the ΔpH) depends on the type of charge-compensating cations on the clay. For a detailed discussion of protonation of biological entities, see Greenland (1965), McLaren and Skujins (1968), Mortland (1970), Theng (1974, 1979), and Harter (1977).

In water bridging, another form of H-bonding, appropriate anionic groups on the biological entity are H-bonded to water molecules in the primary hydration shell of a charge-compensating cation on the clay. This type of H-bond is most effective when the water content is low, which suggests that if cells are attached to clays in soil by this mechanism, they are probably not metabolizing because of a low a_w.

In soils *in situ,* where clays are not dispersed but are aggregated into domains and cutans and are probably complexed, in part, with organic matter and hydrous metal oxides, H-bonding between biological entities and these organic and inorganic coatings (''conditioning films'') may be as, or more important than, either protonation or water bridging. Consequently, if organic conditioning films are formed in soil, they may be important in adhesion by being intermediaries in H-bonding between clays and cells.

6.3. Other Interactions

Cation bridging, wherein multivalent exchangeable cations on the surface of the clay act as ''bridges'' between the clay and an anionic biological entity, has often been suggested as a mechanism of adhesion. The evidence for this type of direct cation bridging is not convincing and only apparent, as under natural conditions, the cation would be hydrated, and adhesion would more likely result indirectly from protonation or water bridging. However, multivalent cations reduce the extent of the diffuse double layer, thereby allowing a net negatively charged entity to approach closer to the clays and for other mechanisms of adhesion, such as H-bonding and van der Waals interactions, to become effective.

van der Waals forces (or, more correctly, London–van der Waals forces; see also Chapter 6, Section 3.1) result from the polarization of one molecule or atom by the fluctuations in the charge distribution of a nearby molecule or atom. The attractive energy between individual molecules is small, and the attractive force decays with the seventh power of the distance between interacting molecules. However, as the attractive forces are approximately additive, the attractive energy between two particles or macromolecules is the sum of the attractive forces between all component molecules, and the decay of the force decreases to about the second power of the distance between the particles. Consequently, the importance of van der Waals attractive forces in surface interactions between clays and biological entities increases as the molecular size of the biological entity increases. van der Waals interactions occur between permanent dipole–permanent dipole, permanent dipole–induced dipole, and induced dipole–induced dipole. The exchangeable cations on clays can induce dipoles, and consequently, ion–dipole interactions are probably very important in surface interactions between clays and biological entities. As with H-bonds, numerous adjacent van der Waals forces acting in concert can result in strong interactions, and the range of attractive energies between clays and biological entities can be much larger than that predicted from the component atoms or molecules of the entities.

Another type of interaction is coordination, although it is sometimes difficult to distinguish this mechanism from ion–dipole interactions and water bridge formation (Harter, 1977; Theng, 1979). In coordination, appropriate functional groups of the biological entity form a complex (e.g., via lone pair electrons on carboxylate groups) with charge-compensating metal cations on the clay (usually by competing with the cation-associated water) or with aluminol (Al–OH) groups on the edges of the clay. This type of interaction is sometimes referred to as ligand exchange, especially when the entity exchanges for OH groups on hydrous oxides of Fe^{3+} and Al^{3+}. A special type of coordination is the chelation by amino acids, through adjacent amine and carboxylate groups, of exchangeable cations, especially of transition metal ions that have a high affinity for electrons (e.g., Jang and Condrate, 1972a,b; Dashman and Stotzky, 1984). Transition metal cations on clays can also form π bonds with organic molecules through donation of π electrons to the molecule from the d orbitals of the metal ions (Mortland, 1970; Harter, 1977). The importance of these types of mechanisms in surface interactions between clays and biological entities is not well defined, but as they occur primarily in dehydrated systems and depend on competition with cation-associated water, they may not be too important in adhesion in soil *in situ*.

Although both ''primary'' (e.g., anion exchange, ligand exchange, protonation, cation or water bridging) and "secondary" (e.g., H-bonding, van der Waals forces) interactions appear to be involved in the surface interactions between net negatively charged biological entities and clays, the indirect effect of the increase in entropy that results from the displacement of many water molecules from the clay surface by a single macromolecule also favors adsorption (e.g., Greenland, 1965; Parfitt and Greenland, 1970; Theng, 1979).

6.4. Factors Determining Dominant Interactions

As is apparent from this brief overview, numerous mechanisms have been suggested to explain surface interactions between clay minerals and organic compounds. Notably absent is covalent bonding, which apparently does not normally occur between clay minerals and organic molecules (Theng, 1979), and the occurrence and importance of hydrophobic interactions between clays and biological entities has received insufficient attention. Despite this plethora of possible mechanisms, it is seldom clear which mechanism(s) is operative in a specific interaction, and different mechanisms have often been invoked by different investigators to explain the same interaction. Furthermore, it is seldom clear why one mechanism has been suggested to be responsible for the adsorption of one molecule and another mechanism for another molecule, even though both molecules have the same or similar types of functional groups. Moreover, it is probable that several mechanisms function simultaneously. Perhaps most important for the purposes of this chapter is the difficulty in extrapolating from results obtained from studies on surface interactions between clays and soluble organic molecules to those between clays and larger and more rigid microbial cells and viruses.

Nevertheless, interactions with both soluble and particulate entities are affected by the physicochemical characteristics of the ambient soil environment, as these factors influence the reactivity of both the clays and the biological entities. Among these factors are pH, ionic strength, concentration, and size relations. The pH will determine the net charge on the biological entity (depending on the pK values of the constituent dissociable groups and the pI of amphoteric functional groups) and on some clays (i.e., those with a pH-dependent charge) and associated hydrous metal oxides, as well as the pH_s. The ionic strength will affect the extent of the diffuse double layer, with an increase in ionic strength and in the valency of the cations decreasing the thickness and reducing the repulsion between net negatively charged clays and biological entities. If the biological entity is net positively charged, however, an increase in ionic strength and in the valency of the cations can reduce the interaction, as the inorganic cations will compete with the organic cation for negative sites on the clays. A similar reduction can occur if the interaction between the clay and the biological entity involves anion exchange at positively charged sites on the clay or associated hydrous metal oxides. An increase in the concentration of either the clay or the biological entity will usually result in an increase in surface interactions, both because of an increase in the frequency of the components coming close enough together for an interaction to occur and because there is a partitioning of the biological entity between the soil solution and the surfaces of the clay. Both the frequency of interaction and the partitioning will depend on the relative size of the biological entity and the clays, e.g., whether the clays are present as individual packets or as components (e.g., domains or cutans) of aggregates.

In addition, the distribution of the charges on the biological entity, especially on microbes and viruses, will affect the orientation (e.g., end-on, side-on, capsid vs. tail) of the entity with respect to the clays. The orientation may also be affected by the types of organic solutes (e.g., hydrophilic vs. hydrophobic) present in the soil solution, which will also affect the surface tension of particulate-associated water (Daniels, 1980).

INTERACTIONS BETWEEN CLAY MINERALS
JLOGICAL ENTITIES

eral Considerations

Relatively few studies have investigated the adhesion of bacteria and other biological entities in soil, and essentially all of these studies have been with model systems using disperse, monospecific, defined, and "clean" clay minerals, laboratory-cultured microbes and viruses, and usually single, defined organic compounds. Consequently, the results of these studies must be carefully evaluated with respect to their relevance to surface interactions between biological and abiological components in soil *in situ*. As has already been emphasized, clays in soil are seldom disperse and exist primarily as aggregates that form domains and cutans in conjunction with other soil particulates. They are seldom monospecific and defined and probably consist of mixtures of clay types (e.g., 1:1 and interstratified 2:1 clays) in the same domains and cutans. They are seldom "clean," as they are coated, at least partially, with various hydrous metal oxides and organic matter. The surfaces of microbes grown in laboratory media containing adequate or, as more usual, surplus carbon and energy sources may be different from those of microbes growing under the oligotrophic conditions that presumably prevail in soil *in situ* (e.g., Ellwood *et al.*, 1982; Rosan *et al.*, 1982) and wherein the generation times greatly exceed those in normal laboratory media (see Stotzky, 1974). In soil *in situ*, numerous organic molecules, both monomers and polymers, may be simultaneously present in the soil solution, and competition for sites on the clays will occur not only among the molecules but also between the molecules and microbes and viruses. The relative affinity of the molecules, microbes, and viruses for sites on the clays will ultimately determine which will be bound, but there have been few studies, even in model systems, on binding affinities and competition between these biological entities (Krumins and Stotzky, 1980, 1982, 1983; Schiffenbauer and Stotzky, 1983; Lipson and Stotzky, 1984b; 1985b; see Stotzky, 1980).

Even when using model systems, the purposes of the studies and the orientation of the investigators have been different: the physical scientist is usually interested primarily in the mechanism of adhesion, whereas the microbiologist is more concerned about the stability of the complexes formed between clay minerals and the biological entities. For example, the microbiologist wants to know whether and how the metabolism of microbes complexed with clay differs from that of noncomplexed microbes, whether the adhesion of viruses to clays alters their infectivity and their susceptibility to inactivation by biological and abiological factors, and whether the availability of organic compounds as nutrients for or toxicants to microbes is different in a complexed than in the noncomplexed form. However, to determine the stability of these complexes, the microbiologist must use the techniques of the physical scientist to quantitate the amount of the biological entity stably complexed with the clays (e.g., by the construction of adsorption and binding isotherms), to determine the location of the entity (e.g., by X-ray diffraction analysis and electron microscopy), and to ensure that the complexes will not be disrupted during the experiments and that the biological activity of the complexes—rather than of the noncomplexed biological entity in the presence of the clay—is really being studied (Harter and Stotzky, 1971; Stotzky, 1973, 1980). Some of the techniques used in these types of studies have been described (see Stotzky and Burns, 1982).

Because of these differences in purposes and orientation between disciplines, some commonly used terminology is unilaterally defined for the purpose of this portion of this chapter. Clays will be considered as *adsorbents* and the biological entities (whether microbial cells, viruses, or organic molecules) as *adsorbates*. Even though the particle size of dispersed clays is usually smaller than that of microbial cells—but not of viruses and soluble organic molecules—clays are usually aggregated in soil and the size of the aggregates probably approaches or exceeds that of most cells. Because the direction of adhesion (e.g., does clay attach to a microbial cell or vice versa?) and the mechanisms involved are seldom known, and because some soluble organics may intercalate clays, *sorption* and *surface interaction* will be used primarily when discussing complex formation between clays and biological entities. Because of the necessity, as discussed above, of ensuring stability of the complex between the clay and the biological entity, *adsorption,* which is usually loosely defined as the collection or concentration of an adsorbate at an interface of an adsorbent, is too vague a term. Consequently, the terms *equilibrium adsorption* and *binding* will be used.

Equilibrium adsorption is reversible, exhibits low specificity between the adsorbent and the adsorbate, may result in multilayer sorption, and involves primarily physical forces, such as London–van der Waals interactions, simple H-bonds, protonation, water bridging, and coordination bonds (Theng, 1979; Section 6). Equilibrium is attained when the rate of adsorption equals the rate of desorption. As coverage of the adsorbent surface increases, it becomes increasingly difficult for adsorption to proceed, and mutual repulsion of the adsorbate molecules may result in a reduction in the bond strength and in lower values for the energy of adsorption.

Binding, sometimes referred to as chemical adsorption or chemisorption, can, in general, be either reversible or irreversible (irreversibility is a relative, rather than an absolute, phenomenon), may exhibit high specificity of the adsorbate for binding sites on the adsorbent, usually results in only single-layer adsorption, as only the reactive groups of the adsorbate and the adsorbent enter into the reaction, and usually involves the formation of a chemical bond. However, there appear to be significant differences in chemisorption between systems containing a free surface of a solid and those not involving a solid. Reactive groups of molecules in a liquid or gas phase are usually unrestrained and can form conventional chemical bonds, whereas the reactive groups of a surface are components of surface molecules which, in turn, are part of the solid, and therefore, these molecules are restrained (Clark, 1974).

These restraints, in the form of bonds in three directions to neighboring molecules, may influence the properties of the reactive groups of the adsorbent and, consequently, their ability to form bonds with the adsorbate. If the adsorbate is also a solid, such as a microbe or a virus, whose surface molecules are also restrained in three directions, the effect of these restraints on surface interactions between clays and such biological entities may be exacerbated. Consequently, experimental models using relatively unrestrained (or flexible) adsorbates (e.g., proteins, peptides, amino acids, polysaccharides, lipids) and restrained (or rigid) adsorbents (e.g., clays) may not be predictive of sorptive reactions between clays and solid biological adsorbates, even if the reactive surface groups of the restrained adsorbates (e.g., the proteins, polysaccharides, glycoproteins, lipids, lipopolysaccharides, lipoproteins, teichoic acids, etc. of microbes and viruses) are similar to the models of unrestrained adsorbates.

Binding is often the result of chemical bonds, but it may also result from the attachment of an adsorbate by numerous physical forces, the sum of which exceeds the tendency of the adsorbate to detach from the adsorbent. For example, one end of a biological entity attached to a clay surface by H-bonds may detach, but as additional H-bonds are disrupted along the length of the entity, the initial detached end again forms H-bonds with the clay, and the entity remains attached (e.g., the "zipper effect"). This alternate detachment–reattachment may explain why small ions and molecules are usually ineffective in removing large organic molecules from clays, whereas such removal can sometimes be effected by molecules equivalent to or larger in size than the attached molecule (Soulides, 1969). When the bonds attaching the original molecule to the clay are broken, the large exogenous molecule inserts like a wedge along the length of the clay surface between the detaching molecule and the surface and forms bonds with the surface, thereby preventing the re-formation of H-bonds with the initial molecule, which is released (Stotzky, 1980). If these physical forces are sufficiently numerous and close, they can, especially in conjunction with hydrophobic interactions (e.g., expulsion of the biological entity from the aqueous phase), result in binding that is even less reversible than that resulting solely from chemical bonds. Because both chemical and physical forces are probably involving in binding, it would be incorrect to refer to the binding between biological entities and clays as chemisorption. Consequently, *adsorption* will be used herein to describe equilibrium adsorption (which is reversible and probably the result primarily of physical forces), and *binding* will be used to describe the condition in which the biological entity cannot be separated from the clay by repeated washings, with appropriate solutions, of the complex of the clay and biological entity (Chassin, 1969; Harter and Stotzky, 1971).

Although microbial cells, viruses, and organics differ greatly in size, shape, chemical composition, solubility, etc., some factors are common in their surface interactions with clays. Most clay minerals, as well as microbes, viruses, and most naturally occurring organics, have a net negative charge at the pH of most soils (i.e., the pH_b is above the pI or the pK of the biological entities). Consequently, the prelude to any surface interaction between clays and biological entities must involve a sufficient reduction in the electrokinetic potentials (EKP) of the components to enable them to come close enough together for attractive forces, either or both physical and chemical, to overcome electrostatic repulsion. These surface interactions are probably restricted to physical and ionic mechanisms, as clay minerals are not normally capable of participating in covalent bonding (Theng, 1979). Clay minerals are rigid, and their charge distribution is relatively stable, when compared to the fluidity of the lipid bilayer membranes of living cells, most of which can rapidly and greatly alter their conformation and that of their associated integral and peripheral proteins in response to changes in their ambient environment. This fluidity, the apparent specificity of some receptor sites (e.g., for hormones, lectins, certain drugs, and viruses; antigen–antibody reactions), the ability to form covalent bonds, and the pH-dependence of the surface charge of cells as contrasted to the predominantly pH-independent charge of most clays distinguish living cells from clay minerals as adsorbents (e.g., Krumins and Stotzky, 1980, 1982, 1983).

Because of the paucity of published information on surface interactions between clay minerals and biological entities—especially in soil—this portion of the chapter emphasizes work conducted in the author's laboratory. Unfortunately, most of these studies have also been conducted in model systems with ''clean'' clays, laboratory-cultured microbes and viruses, and defined organic molecules.

7.2. Bacteria

7.2.1. Charge Interactions and Sorption to Clays

To determine whether a reduction in the electrokinetic potential (EKP) of clays and bacteria was necessary before surface interactions between them could occur, sorption between bacteria and either montmorillonite or kaolinite homoionic to different cations was studied by comparing the particle size distributions of mixtures of bacteria and clays to those of each population of particles alone (Santoro and Stotzky, 1967a, 1968). Sorption between these populations [and flocculation of clays by microbial metabolites (Santoro and Stotzky, 1967b)] occurred only when the EKP of the clays and cells were sufficiently decreased by multivalent cations, and the amount of sorption increased as the valency of the cations increased. Sorption also occurred at pH values below the pI of the bacteria, when the net charge on the cells was positive but that of the clays remained negative and ionic interactions apparently occurred.

When surface interactions between the same particulates were studied by transmission electron microscopy, the adhesion to bacteria of different size fractions of various clays made homoionic to Na$^+$ followed the sequence: kaolinite (2 to 5 μm > 0.2 to 2 μm) > attapulgite (> 5 μm) > vermiculite (> 5 μm) > montmorillonite (0.2 to 2 μm > 0.2 μm). This sequence reflected the net negativity (i.e., the extent of the diffuse double layer) of the clays, which increased with this sequence and with a decrease in particle size. Furthermore, attapulgite was attached primarily by the tips of the clay fibers, which presumably have a net positive charge. When the 0.2- to 2-μm fraction of montmorillonite was made homoionic to different cations, the sequence of adhesion of the clay to the bacteria was tetravalent > trivalent > divalent > monovalent cations (Stotzky, 1974). These sequences of the adhesion of the different types of clay with different particle sizes and different cation saturation were the same with gram-positive and gram-negative bacteria, both in the particle size distribution and electron microscopy studies, suggesting that reducing the EKP of even one of the reacting populations of particulates was sufficient to bring the populations close enough together for surface interactions to occur.

These observations were in agreement with those of Lahav (1962) and Marshall (1968, 1969), who suggested that clays sorb to net negatively charged bacteria by the positively charged sites at the broken edges of the clays. This suggestion was substantiated by blocking the positive sites on the clays with hexametaphosphate (Michaels, 1958), which reduced or eliminated, depending on the chemical composition of the bacterial surface, sorption. When the surface charge of the species of *Rhizobium* used was derived primarily from the ionization of carboxyl groups, the edge (clay)-to-face (bacterium) association was eliminated by hexametaphosphate. When the surface charge of the bacteria was derived from both carboxyl and amino groups, treatment of the clays (montmorillonite and illite homoionic to Na$^+$) with hexametaphosphate only reduced the amount of sorption, presumably as some face-to-face association, in addition to edge-to-face association, occurred between the negatively charged faces of the clay and the amino groups on the bacterial surface, which were apparently protonated at the pH of the experiments. Furthermore, the amount of clay sorbed per unit area of bacterial surface was almost twice as much to rhizobia with carboxyl-type surfaces than with carboxyl-amino-type surfaces (Marshall, 1968, 1969, 1976). The reasons for this difference are not clear, as it would be expected that more clay would be sorbed where the possibility for both edge-to-face and face-to-face associations existed than where only edge-to-face association was possible, as with the carboxyl-type surfaces. Furthermore, despite the presence of some positively charged edge sites on these 2:1 clays, their CEC:AEC ratio was large, and it is not clear how the net negatively charged clays were able to avoid the repulsion from the net negatively charged bacteria and how the diffuse double layers of each type of particle were able to overlap sufficiently for the relatively few positively charged sites, both on the edges of the clays and on the amino groups on some rhizobial species, to interact. Moreover, the net surface charge density of the rhizobial species was not related either to the types of surface components or to the amount of clay sorbed.

The presumed changes in EKP that resulted in surface interactions between clays and bacteria were verified by direct measurement of the electrophoretic mobility of the clays and bacteria in the presence of different cations (Santoro and Stotzky, 1967c; Stotzky, 1974).

The negativity of the bacteria decreased, at pH levels above their pI, as the valency of the cations in the ambient medium increased and also as the concentration of the cations, even of monovalent ones, increased. These observations, in agreement with those obtained by Peele (1936) with mixed soil components, were consistent with a presumed reduction in the extent of the diffuse double layer surrounding the cells. Although the net charge of the bacteria was positive at pH values below their pI, the measured pI of some bacterial species was dependent on the type of cations present: e.g., the pI in the presence of low concentrations (ionic strength = 3×10^{-4}) of the chloride salts of mono- or divalent cations was between pH 2.5 and 3.5, whereas in solutions of the same ionic strength but containing $LaCl_3$ or $CrCl_3$, the pI was approximately at pH 5.0, and in the presence of $FeCl_3$ or $AlCl_3$, the pI was shifted to approximately pH 7.0, probably reflecting the effect of pH on the polymeric hydrous oxides of Fe^{3+} and Al^{3+}, which may have coated the cells. Similar shifts in the pI with an increase in the valence of ambient cations have been observed with normal human lymphoid and transformed Burkitt lymphoma cell lines and with frog kidney cells (Kiremidjian and Stotzky, 1975, 1976). Reductions in negativity and shifts in measured pI to higher pH values also occurred when bacteria were suspended in dilute soil extracts (1 : 5 soil : water), dilute seawater (1 : 100 seawater : distilled water), solutions of various microbial metabolites, or culture media containing organic constituents (Santoro and Stotzky, 1967c; Kiremidjian and Stotzky, 1973).

When the electrophoretic mobility of kaolinite and montmorillonite was measured at the same ionic strength used with bacteria, the clays remained net negatively charged at all pH values studied (pH 2 to 9), regardless of the valence of the ambient cations (Santoro and Stotzky, 1967c; Stotzky, 1974). However, the negativity decreased as the valency of the cations, both on the exchange complex of the clays and in the ambient solution, was increased. In the presence of higher concentrations of tri- and tetravalent cations (e.g., when clays homoionic to Al^{3+} or Th^{4+} were suspended in solutions containing multivalent cations), some charge reversal occurred between approximately pH 4 and 8, which probably reflected the pH-dependent ionization of the polymeric hydrous oxides of these cations that were associated with the surface of the clays. When suspended in dilute soil extracts or seawater, the negativity of the clays decreased but no charge reversal occurred. In solutions of microbial metabolites or culture media, the negativity either increased or decreased, depending on the source of the metabolites or the type of medium, but there was no charge reversal. Similar modifications in the electrophoretic mobility of clays complexed with proteins (Harter and Stotzky, 1973), peptides, amino acids (Dashman and Stotzky, 1985a), a synthetic humic acid (F. Andreux and G. Stotzky, unpublished data), or algal products (Kiremidjian and Stotzky, 1973) have been observed.

These studies confirmed that there must be a reduction in the EKP (of like charge) of reacting particulates to enable them to approach each other closely enough for the mechanisms involved in adsorption and binding to become effective. However, these *in vitro* studies also indicated that surface interactions between clays and bacteria occur essentially only at pH values below the pI of the cells or in the presence of multivalent inorganic cations or certain organic compounds. In most natural soils, however, such surface interactions appear to occur (if, indeed, they do occur) at pH values above the apparent pI of cells and in the usual absence of large quantities of soluble organics and tri- and tetravalent cations (e.g., the dominant cations in most soils are K^+, Na^+, Ca^{2+}, and Mg^{2+}).

7.2.2. Factors Affecting Interactions

As indicated in Section 1, the empirical evidence for surface interactions between bacteria and soil particulates *in situ* is strong (e.g., lack of movement of large numbers of microbes from the surface to underlying soil layers during heavy rains, snow melts, or flooding; failure to wash substantial numbers of microbes from soil columns in perfusion or leaching experiments), but this retention of bacteria may be the result of their entrapment in capillaries and narrow channels between soil particulates, inside soil crumbs, and by surface tension within the water associated with the reactive surfaces of clay minerals and particulate organic matter. However, it is difficult to assign too large a role to these phenomena, as, for example, channels are constantly altered, crumbs are slaked, and water is rearranged during heavy rains, perfusion, and mechanical disturbance of soil. It has been suggested (see Parfitt and Greenland, 1970; Greenland and Mott, 1978; Hayes and Swift, 1978; Theng, 1979; Daniels, 1980; Burns, 1983) that precipitation or coflocculation of microbial cells with soluble components might occur in soil. However, the effects of such reactions on cell viability and whether—and how—the reactions are reversible do not appear to be known. Consequently, physicochemical surface interactions with clay minerals and other soil particulates must still be considered as a major mechanism for the retention of bacteria in soil, and the apparent paradox between the *in vitro* and the *in situ* situations requires clarification.

7.2.2a. Charged Solutes.

Some soluble multivalent cations and organic macromolecules are undoubtedly present in the microhabitats in soil, probably at concentrations higher than those measured on a bulk soil basis. Bacteria in these habitats may be positively charged or, at least, have a lower net negative charge at the prevailing pH than at comparable pH values *in vitro,* which would enable their surface interactions with net negatively charged clays. Another factor that requires consideration is the possible role that heavy metals may have in altering the charge of clays and cells. Some heavy metals (i.e., Cd, Cu, Cr, Ni, and Zn, but not Pb and Hg) caused a reversal in the charge of bacterial cells in pure culture to positive values at pH levels above neutrality (Collins and Stotzky, 1982, 1983). This charge reversal occurred at an ionic strength of 3×10^{-4} (other ionic strengths have not been evaluated), which is equivalent to about 10 to 20 ppm (mg/liter) cation, depending on the heavy metal. Some heavy metals (e.g., Cu, Cr, Ni, and Zn) are present at these and even higher concentrations in nonpolluted soils (e.g., Babich and Stotzky, 1982) and at considerably higher concentrations in polluted soils (the concentration of each metal will depend on the type of pollution, and elevated levels of even Cd occur in some soils) (Babich and Stotzky, 1978, 1980b). Some heavy metals (i.e., Cu, Cr, and Ni, but not Cd, Hg, Pb, and Zn) also caused a reversal in the charge of montmorillonite and kaolinite. However, heavy metals are freely exchangeable from the clays, whereas they appear to be retained more strongly by microbial cells. Inasmuch as some heavy metal cations (i.e., "soft" ions), in contrast to essential divalent cations (i.e., "hard" ions), tend to form bonds of a more covalent nature with organic entities because of their easily deformable outer electron shells (Christie and Costa, 1983), they may be preferentially retained by microbial cells, even though the levels of essential cations, for example, Ca^{2+} and Mg^{2+}, are vastly higher in most soils. The importance of heavy metals in surface interactions between clays and bacteria requires further study.

7.2.2b. pH$_s$ and Surface Charge.

Another consideration is that the actual pH at the surface of charged particulates (i.e., the pH$_s$) may be lower than that measured in the bulk suspension (i.e., the pH$_b$), and consequently, the pH$_s$ of clays may actually be near or below the pI of bacterial cells, even though the pH$_b$ is above the pI (see Stotzky, 1974, and Section 6). Such pH relations could result directly in ionic interactions between clays and bacteria or indirectly after protonation of the cells. Furthermore, bacteria may not interact directly with the surfaces of clays but, rather, with the polymeric hydrous metal oxides which coat or extend from the clay surfaces and have amphoteric properties.

The observation (Peele, 1936) that *Azotobacter chroococcum* sorbed extensively (90% of the input population) to the electrophoretically separated, positively charged fractions of a soil, whereas there was no sorption to the negatively charged fractions, requires confirmation. Even if confirmed, the composition of the net positively charged fractions must be determined to establish whether these fractions are "biologically active" *in situ* (e.g., if they retain sufficient water for metabolic activity). Nevertheless, these observations support the concept that both positive and negative charges can coexist in some soils, presumably as a result of steric factors that prevent the complete cancellation of all opposite charges (see Arnold, 1978). The importance of these positive charges to the adhesion of microbes in soil requires study.

7.2.2c. Bacterial Surface Properties. As mentioned in Section 4.2, the importance of extracellular slimes in the adhesion of bacteria in soil also needs clarification. Similarly, the involvement of pili or fimbriae in the adhesion of bacteria in soil *in situ* has not been demonstrated. Because of their small diameter (3 to 40 nm), long length (several μm), apparent hydrophobicity, and probable positively charged tips, pili could penetrate the negatively charged double layer of clays and thereby cause an initial adhesion of the cell (Marshall, 1976; Corpe, 1980; Pethica, 1980). That pili are present on bacteria in soil has been shown not only by electron microscopy (Nikitin *et al.*, 1966), but also by the demonstration that conjugation, during which DNA is transferred from a donor to a recipient bacterium through the sex pili, occurs in soil (Weinberg and Stotzky, 1972; Stotzky and Krasovsky, 1981; Stotzky and Babich, 1984).

7.2.3. Physiological and Ecological Factors

In addition, although the model studies discussed above have indicated some of the factors that may be involved in surface interactions between clays and bacteria in soil *in situ,* these studies have not sufficiently addressed such questions as the stability of clay–bacteria complexes, the metabolic activity and competitive ability of complexed bacteria, and the susceptibility of complexed bacteria to predation, amensalism, and infection by bacteriophages and predators (e.g., Roper and Marshall, 1974, 1978a,b; Keya and Alexander, 1975; Habte and Alexander, 1978a,b) relative to those of noncomplexed bacteria. For example, montmorillonite specifically enhanced nitrification (Kunc and Stotzky, 1980; Macura and Stotzky, 1980) and aldehyde degradation (Kunc and Stotzky, 1974, 1977) by bacteria in soil and the nitrate reductase activity of two species of *Stachybotrys* in culture (Bondietti *et al.*, 1971). Although these enhancements may have resulted from some indirect effects of the clay, the specificity of the stimulation (e.g., neither ammonification nor utilization of the acids or alcohols homologous to the aldehydes was enhanced by the presence of the clay; $NaNO_3$, but not reduced organic nitrogen sources, inhibited glucose utilization by the fungi in the absence of the clay) suggests that surface interactions may have occurred between the clay and the microbial cells. Such surface interactions could enhance the activity (and, perhaps, the synthesis) of the enzymes involved in these diverse metabolic reactions, as a result of changes in the distribution and density of the charges on and the permeability of the microbial membranes (Pethica, 1980) and by an increase in ATP production by cells in contact with a surface because of a change in the chemiosmotic proton gradient (Ellwood *et al.*, 1982). However, before such direct effects can be invoked, unequivocal evidence for sorption between the clay and the specific microbes involved in these transformations must be obtained.

It must again be emphasized that most of the studies described above were conducted *in vitro*, primarily with disperse, fractionated, monospecific, and "clean" clays and with monospecific laboratory-grown bacteria. Consequently, the relevance of the results of these studies to what occurs in soil *in situ* must be questioned. Although some studies with intact soil have shown both the retention of bacteria and that this retention ususally increases with increasing concentrations of clay (e.g., Krasil'nikov, 1958; Hattori, 1973; Zvyagintsev, 1973; Marshall, 1976), the mechanisms responsible for this retention, i.e., whether adhesion or mechanisms not involving surface interactions, have not been clarified. More studies on the occurrence and mechanisms of surface interactions between soil particulates, especially complex clay-containing aggregates, and bacteria *in situ* are obviously needed. The major limitation for the conduct of such studies is the availability of adequate experimental techniques.

7.3. Viruses

Inasmuch as some organics were protected to various degrees against biodegradation when bound on clays (see Section 7.5.2), the adhesion of viruses to clays may also protect viruses against both biological and abiological inactivation, thereby enabling their persistence in soils, sediments, and waters in the absence of their hosts. This protection may be a contributing factor to the apparent increase in viral diseases of humans, primarily by waterborne enteric viruses (see Berg, 1967; Berg *et al.*, 1974; Stotzky *et al.*, 1981; Lipson and Stotzky, 1984a), and may affect policies on the use of soils as repositories for sewage sludge and wastewater (see Bitton, 1980a). Viruses may also be good models with which to study surface interactions between biological entities and clays: e.g., viruses share with bacteria a relatively rigid structure, although viruses are approximately 10 to 100 times smaller; the molecular weight of viruses is approximately 100 to 10,000 times greater than that of most proteins, but the outer covering (capsid) of many viruses is composed primarily of protein subunits (capsomeres), although some animal viruses have an envelope composed primarily of lipid.

Studies have been conducted with (1) two bacteriophages of *Escherichia coli*, coliphages T7 and T1, which contain double-stranded DNA, are approximately 50 nm in diameter, and have a molecular weight of 20×10^6 to 25×10^6, (2) reovirus type 3, which contains double-stranded RNA, is 75 to 80 nm in diameter, and has a molecular weight of 70×10^6, and (3) herpesvirus hominis type 1 (HSV 1), which contains double-stranded DNA, is approximately 100 nm in diameter, has a molecular weight of 100×10^6, and has a lipid envelope.

7.3.1. Coliphages

Coliphages T1 and T7 had a greater affinity for montmorillonite than for kaolinite, and T7 had a greater affinity for both clays than T1 (Schiffenbauer and Stotzky, 1982). The primary adsorption sites on the clays for the two types of coliphages differed: T1 adsorbed primarily to positively charged sites on kaolinite and to both positively and negatively charged sites on montmorillonite; T7 adsorbed primarily to negatively charged sites on both clays, as the increase in its adsorption was related to an increase in the CEC of the systems. Pretreatment of the clays with sodium metaphosphate, which blocked positively charged sites on the clays (Michaels, 1958; Lahav, 1962; Marshall, 1968, 1969), reduced the adsorption of T1 more than that of T7, and there was no significant competition for adsorption sites when the two phages were added simultaneously or sequentially after preadsorption of the heterologous phage. Pretreatment of the clays with nutrient broth (with an average pI of 4.2) or egg albumin (with a pI of 4.6) reduced the adsorption of both phages to the clays, with the reduction in adsorption to kaolinite being greater than that to montmorillonite, probably because the CEC : AEC ratio of kaolinite was lower, and the reduction being greater with T1 than with T7. Conversely, lysozyme (with a pI of 11) enhanced the adsorption of both phages (T1 more than T7) to both clays. Inasmuch as the pH_b of the clay–phage systems was 6.9, these results supported the conclusion that the phages, especially T1, adsorbed primarily to the positively charged sites on the clays, as these were blocked by the net negatively charged nutrient broth and egg albumin. In contrast, the net positively charged lysozyme adsorbed to the negatively charged sites on the clays where it did not compete with the adsorption of the coliphages to positively charged sites, and reduced the overall negativity of the clays, thereby facilitating the approach of the net negatively charged coliphages (the pI of which was about 4.0, although the tips of the tail fibers were probably positively charged), which enhanced the probability of their interacting with the positively charged sites on the clays. Furthermore, a nondialyzable but filterable (<0.22 μm) component released from montmorillonite, which was apparently a negatively charged hydrolysis product of this clay, reduced the adsorption of the phages to kaolinite but not to montmorillonite. No comparable component was released from kaolinite (M. Schiffenbauer and G. Stotzky, unpublished data).

The adsorbed and bound phages [i.e., those not desorbed from the clays after 30 (for T1) and 66 (for T7) washings with a 1 : 1 mixture of Tris buffer: distilled water at pH 7] retained their lytic capabilities, and even after six cycles of exposure to the appropriate bacterial hosts, the clay–phage mixtures were as lytic as phages in the absence of the clays (Schiffenbauer and Stotzky, 1984). Inasmuch as only phage DNA is injected into host cells, it is not clear whether the bound phages detached from the clays and then attached to the host cells or whether—and how—DNA was injected into the host cells by phages bound to clays, especially as the phages appeared, in transmission electron micrographs, to be bound to the clays primarily by their tails (Bystricky *et al.,* 1975). If infection was by bound phages, then the empty capsids were apparently rapidly replaced on the clays by intact virions, as empty capsids were seldom observed on the clays. If infection was by virions that detached from the clays, then the affinity of the phages for their host cells was greater than for the clays, even though the extensive washings with Tris buffer: water did not desorb bound phages. This apparent greater affinity for host cells may have resulted from the desorption of the bound phages from the clays by bacterial metabolites, as washing the clay–T1 and clay–T7 complexes with nutrient broth or solutions of casein or fetal bovine serum after 30 and 66 washings with Tris buffer: water, respectively, caused desorption of these apparently bound phages.

The clays protected the coliphages against abiological inactivation. For example, the infectivity of the phages at 4, 24, and 37°C persisted significantly longer in the presence than in the absence of the clays, and the clays protected the phages against inactivation by increasing concentrations of sodium metaphosphate and lysozyme (M. Schiffenbauer and G. Stotzky, unpublished data). There are insufficient data on the effects of clays on biological inactivation of phages, although some clays (attapulgite > vermiculite > montmorillonite > kaolinite) extended the survival time of phage φ11M15 of *Staphylococcus aureus* in nonsterile lake water (Babich and Stotzky, 1980a).

Studies on the adsorption of coliphages T1 and T7 to microbes suggested that microbial cells are not significant reservoirs for phages in natural habitats. Adsorption of the coliphages to stationary cultures (20 to 24 hr old) of 13 species of yeasts and bacteria, including actinomycetes and the host strains of *E. coli,* was low (0 to 20%) when compared to the adsorption to montmorillonite and kaolinite (65 to 95%). Even the adsorption of the phages to log-phase cultures (e.g., 3 hr old) of the host bacteria was generally less (e.g., 68% to unwashed cells and 50% to washed cells) than to the clays (Schiffenbauer and Stotzky, 1983).

7.3.2. Reovirus

In contrast to the coliphages, the adsorption of reovirus type 3 to kaolinite and montmorillonite was highly correlated with the CEC of the clays, indicating that although this virus had a pI of 3.8 and, therefore, was net negatively charged at the pH_b of these studies (7.0 to 7.2), it adsorbed to the negatively charged sites on the clays (Lipson and Stotzky, 1983). Not only was a single adsorption isotherm obtained when the amount of reovirus adsorbed was plotted against the CEC of the systems, but neither sodium metaphosphate nor the supernatant from montmorillonite reduced the adsorption. There was no correlation between adsorption of the virus and the specific surface area or the surface charge density of the clay systems. Furthermore, when chymotrypsin (pI of 8.1 to 8.6), lysozyme (pI of 11), or ovalbumin (pI of 4.6) was bound on the clays before exposure to the virus, adsorption of the virus to montmorillonite and, with the exception of lysozyme, to kaolinite was reduced, probably as a result of the blockage of negatively charged sites on the clays by the proteins (Lipson and Stotzky, 1984a,b). Inasmuch as lysozyme does not appear to bind on kaolinite by cation exchange (Albert and Harter, 1973), the failure of lysozyme to block the adsorption of the virus to kaolinite further confirmed that the virus adsorbed to negatively charged sites on the clays. Although the pI of ovalbumin was below the pH_b of the systems, this protein probably adsorbed on the clays by H-bonding and van der Waals interactions (Albert and Harter, 1973) and blocked the access of the virus to surface sites on the clays.

The adsorption of the net negatively charged virus to negatively charged sites on the clays was probably the result of protonation of the capsomere proteins of the virus, followed by cation exchange between the net negatively charged clays and the now net positively charged virus. In addition, H-bonding and van der Waals forces may have been involved in holding the virus on the clay surfaces.

The importance of the EKP of the clays and the reovirus in their surface interactions was demonstrated by the greater adsorption to lower concentrations of the clays in synthetic estuarine than in distilled water (Lipson and Stotzky, 1983) and by the increased adsorption, especially to kaolinite, as the pH_b of the systems approached or was reduced below the pI of the virus (Lipson and Stotzky, 1985a). That the higher ionic strength of the estuarine water reduced the EKP was substantiated by the enhanced adsorption of the virus when individual chloride salts of the major cations present in estuarine water were added: Ca^{2+} and Mg^{2+} were more effective than Na^+ and K^+, and a salt concentration of 10^{-2} M was more effective than 10^{-3} M in causing adsorption. However, KCl, at both 10^{-2} and 10^{-3} M, almost completely inhibited the adsorption of the reovirus to montmorillonite, probably because K^+ collapsed the clay lattices and prevented expression of the interlayer-derived CEC. Furthermore, in studies with homoionic clays, the adsorption to montmorillonite increased as the valency of the cation on the exchange complex of the clay increased (Al^{3+} > Ca^{2+} > Mg^{2+} > Na^+ > K^+; there was again essentially no adsorption to montmorillonite homoionic to K^+). With kaolinite, the valency of the exchangeable cation was not as important as with montmorillonite (the sequence of adsorption with kaolinite was Na^+ > Al^{3+} > Ca^{2+} > Mg^{2+} > K^+) (Lipson and Stotzky, 1983).

Competitive adsorption studies indicated that the reovirus and coliphage T1 did not share common adsorption sites on the clays. Coliphage T1 did not interfere with the adsorption of the reovirus to either montmorillonite or kaolinite in synthetic estuarine water or to montmorillonite in distilled water (kaolinite was not studied in distilled water). Reovirus did not affect the adsorption of T1 to kaolinite in estuarine water or to montmorillonite in distilled water, but it appeared to suppress the adsorption of T1 to montmorillonite in estuarine water. This apparent suppression of the adsorption of T1 by the reovirus may have been caused by the enhanced adsorption of reovirus to montmorillonite in estuarine water, which resulted in a steric interference by the relatively larger reovirus of the adsorption of T1 (the diameter and molecular weight of reovirus are approximately 1.5 and 3 times greater, respectively). This steric interference did not occur in distilled water, as fewer reovirus particles were adsorbed. Furthermore, as T1 adsorbed primarily to positively charged sites on the clays, and as the CEC : AEC ratio of kaolinite is considerably less than that of montmorillonite, the net negatively charged coliphage particles were able to approach kaolinite more easily than montmorillonite.

The organic materials (fetal bovine serum, amino acids, penicillin, streptomycin, and amphotericin B) present in the maintenance medium (Hanks' minimal essential medium) of the L-929 mouse fibroblast cells that were used to titer the reovirus totally blocked the adsorption of T1 to kaolinite and reduced the adsorption to montmorillonite by more than 90%, but they did not affect the adsorption of the reovirus to the clays (Lipson and Stotzky, 1984a, 1985b). These results again indicated that coliphage T1 adsorbed primarily to positively charged sites and reovirus to negatively charged sites on the clays. Furthermore, these results also indicated that "conditioning films"—even if they are produced in soil and cause, as they presumably do in aquatic environments, a "convergence" in the surface charge of different particulates because of the adsorption of similar films (Neihof and Loeb, 1972; Baier, 1980; see Marshall, 1980b)—will differentially affect the adhesion of biological entities, depending on the surface characteristics and relative affinity of the entities for positively or negatively charged sites on clays.

The reovirus was not bound tightly to the clays, as most of the virus particles were desorbed by three or four washes with distilled water, but the binding appeared to be stronger to montmorillonite than to kaolinite (Lipson and Stotzky, 1985a). Furthermore, the amount of virus bound was influenced by the type of charge-compensating cations on the clays. Approximately 100% of the input virus population was recovered from both clays homoionic to Na^+ and from kaolinite either homoionic to Mg^{2+} or containing a mixed complement of cations, which consisted primarily of Na^+ and Ca^{2+}. However, only 78% of the input population was recovered from montmorillonite with a mixed complement of cations, but 150% was apparently recovered from montmorillonite homoionic to Mg^{2+}. This apparent increase in the recovery of the virus was caused by Mg^{2+} and was probably the result of several phenomena (e.g., enhanced infectivity of the testor cells), none of which appeared to be related to surface interactions between the clay and the virus (see Lipson and Stotzky, 1985a). However, as montmorillonite, because of its greater CEC, had a higher exchangeable concentration of Mg^{2+} than kaolinite, more Mg^{2+} was released from montmorillonite during the desorption washings.

As with the coliphages, the bound reoviruses retained their infectivity. However, it is also not known whether the reoviruses detached from the clays before infecting the testor cells or whether the clay–reovirus complexes underwent viropexis (i.e., engulfment of the complexes by the cells). The infectivity of the virus in the absence of clay decreased as the pH_b approached its pI, probably as the result of the aggregation of the virus (Lipson and Stotzky, 1985a). Chymotrypsin and lysozyme also markedly decreased the infectivity of the virus in the absence of clay, but when these basic proteins were bound to clay, their inhibition of viral infectivity was almost completely eliminated. Conversely, ovalbumin increased the infectivity of the virus in the absence of clay, probably because this net negatively charged protein dispersed aggregates of the virus, but this enhancement in infectivity did not occur when the protein was bound to clay (Lipson and Stozky, 1984b).

The clays reduced the rate of inactivation of the reovirus in both distilled and synthetic estuarine waters: e.g., at 23°C, the virus could not be detected in either water in the absence of clay after 8 weeks, whereas it was still detected after 22 and 18 weeks in estuarine and distilled water, respectively, containing 1% montmorillonite or kaolinite (Lipson and Stotzky, 1985a).

7.3.3. Herpesvirus

In contrast to the coliphages and the reovirus, the adsorption of HSV 1 was essentially the same to montmorillonite and to kaolinite (Yu and Stotzky, 1979; Stotzky et al., 1981). There was no correlation between the adsorption of HSV 1 and the CEC of the clay systems, as two distinct isotherms, one for each clay, were obtained when the amount of sorption was plotted against CEC, indicating that this virus did not adsorb primarily to negatively charged sites on the clays. There was also no correlation between adsorption and the specific surface area of the systems. Drying the clays at 180°C did not affect the amount of virus subsequently adsorbed.

Both clays also bound essentially the same amounts of HSV 1, and 28 to 30 washings with water were necessary to desorb completely loosely bound virus from the clays. The virus particles that remained bound after these washings were capable of infecting the testor cells (human embryonic lung fibroblasts), but the rate of infection was significantly slower with the montmorillonite–HSV 1 complex than with the free virus or with the kaolinite–virus complex. A similar relation was also noted in the rate of infection of the different clay–reovirus complexes (Lipson and Stotzky, 1985a). As with the coliphages and reovirus, it is also not known whether bound HSV 1 desorbed from the clays before infecting the testor cells or whether the clay–virus complexes underwent viropexis.

The preadsorption of bovine serum albumin (with a pI of 4.9) to the clays did not affect subsequent adsorption of HSV 1 but did increase the amounts of virus bound. This lack of effect of the protein on adsorption of the virus, the apparent lack of correlation between the characteristics of the clays (e.g., CEC, specific surface area) and adsorption, and the fact that HSV 1 has a lipid envelope suggest that hydrophobic interactions may have been involved in the adsorption of this virus to the clays.

The stability of HSV 1 at 4, 24, and 37°C was greater in the presence of montmorillonite than of kaolinite, and the addition of bovine serum albumin increased the stability in both the absence and the presence of the clays. The clays, however, did not protect the virus against inactivation by human IgG, even when the protein was bound on the clays (B. H. Yu and G. Stotzky, unpublished data).

7.3.4. Virus Adsorption Characteristics

The studies on surface interactions between viruses and clay minerals showed that different viruses, even closely related ones such as coliphages T1 and T7, adhered to different sites on the clays. These differences may have been related to differences in the amino acid composition of the capsid proteins of the reovirus (Pett et al., 1973) and the coliphages (Stent, 1963), although differences in the composition of the proteins of the tail fibers of the coliphages may also have been involved. The outer lipid envelope and the associated glycoprotein peplomers of HSV 1 were different from the capsids of the other viruses.

Despite these differences, there were some similarities in the adhesion of these viruses. When the concentration of the viruses was maintained constant and the concentration of the clays was varied, the adsorption isotherms were of the L (Langmuir) type, according to the classification system of Giles et al. (1960, 1974). However, when the concentration of the clays was maintained constant and that of the viruses was varied, the isotherms were of the C (constant partition) type, in that they did not show a plateau, even at clay : virus ratios that resulted in plateaus (indicating saturation-type kinetics) in the L-type isotherms. Although the C-type isotherm has been interpreted as indicating the penetration of an adsorbate into the structure of an adsorbent (Giles *et al.*, 1960, 1974), none of the viruses, probably because of their size, intercalated the clays, as shown by X-ray diffraction analyses (Stotzky *et al.*, 1981). It is not clear why C-type isotherms—which indicated that as the virus concentration was increased, there was a concomitant increase in available adsorption sites—were obtained when the concentration of the viruses was varied and L-type isotherms were obtained when the concentration was maintained constant. However, these differences emphasize the importance of how adsorption studies are conducted and the care that must be taken in comparing the results obtained, especially results from different laboratories.

The C-type isotherms also indicated that a fixed proportion of the viruses was adsorbed, regardless of the concentration of virus added. This constant proportion may have been more apparent than real and may have resulted from an inhomogeneity in the viral populations. Inasmuch as viral assays only detect those viruses capable of infecting testor cells, noninfectious but noninactivated virions may have competed with infectious virions for sites on the clays. For example, more than 90% of reovirus particles that were not adsorbed initially by either montmorillonite or kaolinite were adsorbed when the supernatants were treated with fresh clay (Lipson and Stotzky, 1983, 1984a). The viral particles adsorbed in the initial reaction with the clays were apparently noninfectious (or defective interfering) particles, which have been identified as components of populations of reovirus by both genetic analysis and their reaction to proteolytic enzymes (Ahmed and Fields, 1981; Adams *et al.*, 1982). These defective interfering particles are genetically heterogenous, contain multiple mutations, and, most importantly, have a mutation on the S4 RNA segment (Ahmed and Fields, 1981). This segment encodes for the $\sigma 3$ outer capsid polypeptide, which is necessary for the infectivity of reovirus type 3 (Drayna and Fields, 1982).

The heterogeneity of viral populations was also observed with HSV 1 (Stotzky *et al.*, 1981). Approximately 10% of the viral population was not adsorbed, even when fresh clay was added. Although the nonadsorbable HSV 1 particles were capable of infecting testor cells, in contrast to the defective interfering reovirus particles, their failure to adsorb to the clays indicated that their surface was different from that of the other 90% of the HSV 1 population. Inhomogeneities in bacterial populations have also been observed (e.g., Lahav, 1962).

The results of these studies indicated that the sorption of viruses to clay minerals is a function of the morphological, biochemical, and biophysical characteristics of the viruses (e.g., pI, composition of the capsomeres, absence or presence of a tail or a lipid envelope), the physiochemical characteristics of the clays (e.g., structure, CEC, CEC : AEC ratio, type of cation on the exchange complex), and the composition of the ambient environment (e.g., ionic strength, the presence of other viruses or soluble organic compounds). Surface interactions between viruses and clays appeared to protect viruses against abiological inactivation (essentially nothing is known about biological inactivation), but they did not impair the ability of the viruses to infect appropriate host cells. Furthermore, these studies indicated that viruses are good models with which to study basic aspects of surface interactions between clay minerals and biological entities.

This section on the adhesion of viruses to clay minerals has concentrated primarily on studies conducted in the author's laboratory. References to studies conducted by others can be found in Berg (1967), Berg *et al.* (1974), Roper and Marshall (1978a), Sykes and Williams (1978), Duboise *et al.* (1979), Bitton (1980a,b), Stotzky *et al.* (1981), Taylor

(1981), and Lipson and Stotzky (1984a), which also contain references to the relatively few studies that have been concerned with the movement and survival of viruses in soil *in situ*.

7.4. Fungi

7.4.1. Bacteria–Fungi–Clay Interactions

There have been few studies on surface interactions between clay minerals—or any other soil particulates—and fungi, whether filamentous or yeasts. The *in situ* observations that the rate of spread of *Fusarium* wilt of banana was faster in soils not containing a clay mineral having the characteristics of montmorillonite (Stotzky *et al.*, 1961; Stotzky and Martin, 1963; Stotzky, 1974) and that the discrete geographic distribution of *Histoplasma capsulatum* was highly correlated with the absence of this clay mineral in soils in which this fungus pathogenic to humans was able to establish (Stotzky and Post, 1967; Stotzky, 1971, 1974) were initially assumed to be the result primarily of indirect effects of this clay. One apparent indirect effect was that in soils containing montmorillonite, the bacterial populations were able to develop to a greater extent as a result of (1) the buffering capacity of this clay (i.e., because of its high CEC, it was able to maintain a suitable pH for continued bacterial development by the exchange of metabolically produced protons for basic cations on its exchange complex) (Stotzky, 1966a,b, 1974; Stotzky and Rem, 1966), (2) protection by the clay against hypertonic osmotic pressures (Stotzky, 1966a), which also appeared to be related to the higher CEC of this clay (Nanfara and Stotzky, 1979), (3) ability of the clay to adsorb heavy metals that were more toxic at the same concentration to bacteria than to fungi (Babich and Stotzky, 1977a, 1978, 1979, 1980b, 1982, 1983a,b), and (4) other mechanisms, which ultimately resulted not only in more rapid depletion of substrates in soils containing this clay, but apparently also in other antagonistic effects of bacteria on the fungi (Stotzky, 1974, 1980). A second apparent indirect affect of this 2 : 1 expanding clay was that as a result of its ability to swell by trapping water between its lattices, it interfered with the movement of O_2 to sites of metabolism (Stotzky and Rem, 1967) and spore germination (Santoro *et al.*, 1967) of these obligately aerobic fungi. These assumptions were based primarily on pure culture studies with bacteria, fungi, and mined clay minerals (see Stotzky, 1974, 1980).

When these variables were studied in sterile soil systems *in vitro*, similar results were obtained: i.e., bacteria grew about twice as fast in soils containing montmorillonite (either naturally or amended with mined clay), whereas fungi grew about twice as fast in soils not containing this clay (the addition of mined kaolinite had little effect on growth) (Stotzky, 1974). The enhanced growth of bacteria was assumed to be the result of the same factors by which montmorillonite stimulated bacterial development in pure culture. However, as the soils were maintained at their 1/3 bar water tension, where O_2 diffusion should not have been impaired, the slower growth of fungi in soils containing montmorillonite could not be attributed to limitations in O_2. No mechanisms for the reduced growth of fungi in soils containing montmorillonite were apparent at that time.

Subsequent studies, wherein sterile soils were inoculated with mixtures of fungi and bacteria, either into the same or separate sites in soil, indicated that the antagonistic effects of bacteria against fungi were enhanced by the addition of montmorillonite but not of kaolinite. The antagonism resulted primarily from the better competition by the bacteria for available carbon in soils containing montmorillonite, apparently as a result of the ability of the clay to maintain the pH of the microenvironments at levels conducive to bacterial metabolism (Rosenzweig and Stotzky, 1979, 1980). In addition, some bacteria that did not produce diffusible inhibitors of fungi in culture appeared to produce them in soil containing montmorillonite, whereas other bacteria, e.g., species of *Bacillus,* that produced antifungal inhibitors in culture either did not produce them in soil or they were inactivated, perhaps by adsorption on the clay (Skinner, 1956; Soulides, 1969; Stotzky, 1974; Rosenzweig and Stotzky, 1979; Campbell and Ephgrave, 1983). Although the rate of growth of most fungi studied was slower in soils containing montmorillonite (Stotzky, 1974), some exceptions have been noted, e.g,, *Penicillium asperum* (Babich and Stotzky, 1977b) and *Gaeumannomyces graminis* var. *tritici* (Campbell and Ephgrave, 1983). Furthermore, some studies in pure culture have indicated that montmorillonite enhances growth of (Bondietti *et al.,* 1979; Filip *et al.,* 1972a; McCormick and Wolf, 1979a,b) and synthesis of humic-type polymers by fungi (Filip *et al.,* 1972b; Filip, 1975), but these enhancements probably resulted from the adsorption by the clay of inhibitory materials, such as heavy metals.

7.4.2. Fungi–Clay Interactions

7.4.2a. Adsorption.
Although these observations were initially attributed to indirect effects of montmorillonite, subsequent studies have indicated that montmorillonite, as well as other clay minerals, may interact directly, i.e., undergo surface interactions, with fungi. The respiration of *Histoplasma capsulatum* was reduced in pure culture by montmorillonite and, to a lesser extent, by attapulgite and kaolinite (Lavie and Stotzky, 1981, and unpublished data), similar to what was observed previously with 27 other fungal species (Stotzky and Rem, 1967). However, in contrast to these previous studies, the reduction in respiration of *H. capsulatum* already occurred at low concentrations of clay (0.01 to 0.5% w/v), and higher concentrations (1 to 8%) resulted in only slightly more inhibition (i.e., the pattern of reduction in respiration suggested saturation-type kinetics). Furthermore, the reduction was not the result of increases in viscosity (i.e., there was no apparent impairment in the diffusion of O_2). Scanning electron microscopy indicated that the clays were bound to the surface of the fungal mycelium even after washing with water, with fixative (5% glutaraldehyde in sodium cacodylate buffer), and then 10 times in a graded series of ethanol for dehydration before critical point drying.

The mechanisms by which the clays adhered to the mycelium are not known. The cell wall of fungi is composed of four intergrading regions, which in *Neurospora crassa* (the fungus most studied with respect to the structure of the cell wall) are: an outermost layer (80 to 90 nm thick) of amorphous glucans containing β-1,3 and β-1,6 linkages; a glycoprotein reticulum (40 to 50 nm thick), in which the glucans merge into the protein layer; a discrete protein layer (8 to 10 nm thick); and an innermost layer (about 20 nm thick) of chitin microfibrils embedded in protein (Bartnicki-Garcia, 1968; Hunsley and Burnett, 1970; Smith and Berry, 1974; Burnett, 1976). The composition of the cell wall of *H. capsulatum,* as well as that of most septate fungi (i.e., Ascomycetes, Basidiomycetes, and Deuteromycetes), is presumably similar to that of *N. crassa.* The composition of the cell walls of Phycomycetes differs and shows a wide diversity (e.g., cellulose–glycogen, cellulose–β-glucan, cellulose–chitin, chitin–chitosan), as does that of yeasts (e.g., mannan–β-glucan, chitin–mannan) (Bartnicki-Garcia, 1968).

Because the outermost layer of the cell wall of *H. capsulatum* is composed of glucan (a polymer of glucose), there is ample opportunity for the formation of H-bonds between the hyphae and clay minerals. In addition, many fungi excrete polysaccharides with ''adhesive'' qualities (Martin and Adams, 1965) and which can cause aggregation of clays around fungal structures (Dorioz and Robert, 1982), and these may also have been responsible for the apparent tight adhesion of the clays to the surface of *H. capsulatum.* These mechanisms may have also been involved in the adhesion of montmorillonite and kaolinite to the hyphae of

Gaeumannomyces graminis (Campbell, 1983; Campbell and Ephgrave, 1983), although there may also have been some hydrophobic effects (Campbell, 1983).

7.4.2b. Siderophores.

In the case of *H. capsulatum*, the reduction in the respiration of the fungus in the presence of the clays was attributed, in part, to the reduction in the amount of mycelial surface available for the movement of nutrients, waste products, and gases across the cell wall as the result of the adhesion of the clays. However, the adhesion of the clays did not explain the differences in the amount of reduction in respiration caused by the same maximum concentration (8%) of each clay (18, 32, and 40% reduction by kaolinite, attapulgite, and montmorillonite, respectively). Even though the reduction in respiration showed a high correlation with an increase in both the CEC and the specific surface area of the liquid culture medium as the concentration of the clays was increased, the amount of kaolinite and montmorillonite bound to the hyphae appeared to be similar (the adhesion of attapulgite was not studied by electron microscopy), and none of the clays interfered with the availability of essential nutrients (S. Lavie and G. Stotzky, unpublished data).

H. capsulatum, like many fungi (Emery, 1971, 1978; Emery and Emery, 1973; Neilands, 1973, 1981a,b), requires siderophores for the transport of iron from the environment into the cell (Burt *et al.,* 1981; Burt, 1982). The siderophore of *H. capsulatum* is a trihydroxamic acid (deferricoprogen B), which, upon reductive hydrolysis, yielded ornithine, dimerumic acid, and fusarinine (Burt, 1982). Although various trihydroxamic acid siderophores differ in their structure, they are basically cyclic hexapeptides containing three residues of glycine and/or serine and three residues of ornithine, the terminal nitrogen atoms of which have been oxidized to hydroxylamine and combined with three acyl groups (which can differ in length) to yield the hydroxamate group that binds Fe^{3+} more tightly than most other substances (see Emery, 1982). Siderophores appear to be produced by fungi in soil (Powell *et al.,* 1980, 1983), wherein they may be adsorbed by clays and humic acids (Powell *et al.,* 1981).

To determine whether the reduction in the respiration of *H. capsulatum* in the presence of the clays was the result of the adsorption of its siderophore on the clays, various concentrations of a commercially available siderophore, Desferal Mesylate [deferrioxamine mesylate (DFOM), Ciba-Geigy Pharmaceutical Co.], were added to systems containing the fungus and various concentrations of the clays. The addition of DFOM reversed, in part, the inhibition in respiration caused by montmorillonite but not that caused by kaolinite and attapulgite. The reversal increased with increasing concentrations of DFOM, but the amount of respiration restored by DFOM never exceeded 40% of the reduction caused by montmorillonite. These observations indicated that the reduction in respiration caused by montmorillonite was a result of both the adhesion of the clay to the hyphae and the adsorption of the fungal siderophore, which hampered the iron-nutrition of the fungus, whereas the reduction caused by kaolinite and attapulgite was the result of only the adhesion of these clays to the hyphae. Because of the low CEC and specific surface area of the kaolinite used (approximately 6 meq/100 g and 16 m²/g, respectively), not much siderophore was apparently adsorbed. Although the attapulgite used had a higher CEC and specific surface area (approximately 34 meq/100 g and 125 m²/g, respectively) than the kaolinite, although considerably less than that of the montmorillonite used (approximately 98 meq/100 g and 800 m²/g total area, respectively), the structure of the fibrous attapulgite may have precluded access of the siderophore to the narrow, nonexpandable, internal channels where the exchange sites are located.

The mechanisms by which siderophores adsorb on montmorillonite have not been studied. The apparent pI of DFOM is around 10 (the pK of the protonated amino group is at approximately pH 11 and the pK of the hydroxamic acid group is at approximately pH 9) (H. Stober, personal communication), and based on the similarity in structure of most hydroxamate siderophores, the fungal siderophore had a net positive charge at the pH of these studies (pH 6.8 to 7.0). Consequently, coulombic interactions between the net negatively charged montmorillonite and the net positively charged siderophore probably occurred, in addition to H-bonding and van der Waals forces (based on the structure of hydroxamate siderophores). Although the correlation between the reduction in respiration and the increase in both the CEC and the specific surface area of the clay-containing systems supported this assumption, critical studies (e.g., construction of equilibrium adsorption and binding isotherms of the siderophore on clays) are necessary. Furthermore, it is not known whether the added DFOM served as the iron-transport siderophore or replaced the natural siderophore of *H. capsulatum* from the clay.

In addition to being an iron-transport system, some hydroxamic acids exhibit antibacterial activity (Emery and Emery, 1973), although other siderophores appear to stimulate both bacterial and plant growth (Emery, 1982). Consequently, montmorillonite may exert both direct and indirect effects on the growth of *H. capsulatum*—and, perhaps, other fungi—in soil: the direct effect results from the adhesion of the clay to the mycelial cell wall, which reduces the surface area for transmembrane transport; the indirect effects result from the binding of siderophores, which not only reduces the iron-nutrition of the fungus but also reduces their antibacterial activity and, therefore, further enables bacteria to antagonize the growth of the fungus. This interesting example of a clay exerting both direct and indirect effects on the activity, ecology, and population dynamics of a portion of the soil microbiota requires more study for confirmation, detail, and broadness of applicability. A reduction in the iron-nutrition of fungi by siderophores produced by bacteria has also been suggested as being involved in the "suppressiveness" of some soils to fusarial wilt diseases (Kloepper *et al.*, 1980; Schroth and Hancock, 1982; Scher and Baker, 1982; Vandenbergh *et al.*, 1983).

7.4.2c. Fungal Surface Charge.

The net surface charge of filamentous fungi has apparently not been measured. However, studies on the electrophoretic mobility of yeasts (Santoro and Stotzky, 1967c; Collins and Stotzky, 1982, 1983) and on changes in particle size distributions of mixtures of yeasts and clays homoionic to cations of different valence and at different pH values (Santoro and Stotzky, 1967a, 1968)—in addition to the commonality in the composition of the outermost layer of the cell wall of most filamentous fungi and yeasts (i.e., polymers of either glucose, mannose, or galactose) (Bartnicki-Garcia, 1968)—indicate that fungi have a net negative charge at pH values normally encountered in soils. Because the polymeric sugars that comprise the outermost layer of the wall are not amphoteric, a pI and a net positive charge at pH values below the pI are possible only if the proteins in the underlying layers are exposed. Such an exposure does not appear to occur with *Saccharomyces cerevisiae,* as cells of this yeast remained net negatively charged at pH values from 1.5 to 10, whereas cells of *Candida albicans,* which apparently has the same type of outermost layer in its wall, had a pI at pH 2.3 and was net positively charged at lower pH values. However, as with bacteria, some heavy metals caused a reversal to a net positive charge in both yeasts at alkaline pH values (Collins and Stotzky, 1983).

Although there have been some studies on the attachment of fungi to plant, animal, and some inanimate surfaces (see Corpe, 1980), there is a dearth of studies on surface interactions between fungal propagules and soil particulates, both *in vitro* and *in situ*. Coatings of montmorillonite protected blastospores of *Beauveria bassiana* against destruction by the soil microbiota (Reisinger *et al.,* 1977), similar to the apparent protection of *E. coli* from bacteriophages as a result of the coating of the bacteria by this clay (Roper and Marshall,

1978a). Although these studies were conducted *in vitro* with "clean" clays, they and the studies described above on the indirect and direct effects of clays on fungi suggest that studies on surface interactions between fungi and soil particulates might provide valuable information for the development of methods for the control in soil of fungi pathogenic to both plants and animals, including humans. Furthermore, the effects of clays that were initially assumed to be indirect (e.g., in the reduction of respiration of 27 fungal species and the increase in time required for spore germination when montmorillonite was added) may, in fact, have been the result of direct surface interactions between the clays and fungal propagules, as was observed with *H. capsulatum* and *G. graminis*. These and other studies need to be repeated and extended with both "clean" and "dirty" clays.

7.5. Proteinaceous Compounds

7.5.1. Adsorption and Binding of Proteinaceous Compounds on Clays

7.5.1a. Proteins. In contrast to results obtained with microbes and viruses, the amounts of various proteins—as examples of amphoteric macromolecules—adsorbed and bound on montmorillonite homoionic to different cations were generally inversely proportional to the valence of the cation saturating the clay (Harter and Stotzky, 1971), indicating that a reduction in the EKP was not a prerequisite for surface interactions between clays and these water-soluble polymeric adsorbates. Adsorption and binding occurred with every combination of clay (i.e., homoionic to H^+, Na^+, Ca^{2+}, La^{3+}, Al^{3+}, or Th^{4+}) and protein (i.e., casein, catalase, chymotrypsin, edestin, lactoglobulin, lysozyme, ovalbumin, ovomucoid, and pepsin) studied, even when the pH_b of the clay suspensions (which ranged from approximately pH 3 to 7) was above the pI of the proteins (which ranged from approximately pH 1 to 11) and the net charge on both clays and proteins was negative. The binding isotherms of these proteins were either of the L or H (high affinity) type, with the exception of catalase which was of the C type (Giles *et al.*, 1960, 1974; see Section 7.3.4).

These results suggested (Harter and Stotzky, 1971, 1973) that:

1. The pH_s of the clays was lower than the pH_b, and therefore, the pH that a protein with a low pI encountered at the clay surface may not have been as unfavorable for surface interactions as indicated by the measured pH_b. This possibility, however, is questionable, as montmorillonite homoionic to Na^+ (pH_b 6.2) adsorbed and bound almost as much of proteins with pI values between pH 3.8 and 5.7 as did clay homoionic to H^+ (pH_b 2.7)
2. As these studies were conducted in unbuffered systems to prevent changes in the cation saturation of the clays, alterations in either or both the clay surface and the proteins by the prevailing ambient pH could have occurred
3. As the pI indicates only the pH at which the net charge of an ampholyte is zero, the proteins may have had positive charges at some loci, even at pH values above their pI. If these positively charged loci were sufficiently numerous and close, a relatively flexible adsorbate, such as a protein, might have been able to approach negatively charged sites on the clay for ionic interactions to occur
4. A predominantly net negatively charged protein may be sorbed to positively charged sites on the clays, as appeared to be the situation with pepsin, the pI of which has been reported to range between 0 and 2 and which, on the basis of X-ray diffraction analysis and transmission electron microscopy, was apparently bound to positively charged sites
5. If the sorption of proteins on clays is a cation exchange reaction (possibly after

protonation of acidic proteins), then the proteins competed better for exchange sites with monovalent cations, which have a lower energy of binding and can be more easily replaced, than with cations of higher valence. The possible involvement of cation exchange in sorption of the proteins was reinforced by the observations that the pH_b of the clays, especially of the acidic homoionic clays, increased as the amounts of various proteins bound increased and that the electrophoretic mobility of the clays toward the anode decreased, presumably as a result of the covering of negatively charged sites on the clay surfaces by the proteins

6. The amount of clay surface available to proteins was lower with clays homoionic to multivalent cations, as these clays were better flocculated and more resistant to dispersion (Santoro and Stotzky, 1967a,b).

The molecular weight, shape, and number of binding sites on the proteins also appeared to be important in their surface interactions with the clays (Harter and Stotzky, 1971). Although the amount (i.e., weight) of protein bound generally increased as the molecular weight of the protein increased, the number of moles bound was generally inversely related to molecular weight. The increase in weight or the decrease in moles of protein bound was not always in direct proportion to molecular weight, indicating that some steric hindrance occurred. For example, the molecular weight of casein was about five times greater than that of chymotrypsin and ovomucoid, but the amount of casein bound on montmorillonite homo-ionic to H^+ was only about 15% more than that of the other two proteins. The importance of the number of binding sites on the proteins, resulting from differences in the composition and sequence of amino acids and in tertiary structures, was shown, for example, by the twofold greater binding of chymotrypsin than of lysozyme, even though their molecular weights were essentially the same. Lysozyme apparently had more binding sites, and fewer molecules were necessary to saturate binding sites on the clay.

Although the mechanisms involved in the adsorption and binding of these complex molecules on clays are not clearly defined, the pI of the proteins and the pH_b of the systems were apparently not as important as the other factors discussed above. Nevertheless, some proteins showed maximum adsorption and binding at a pH_b near their pI, possibly because (1) repulsive forces among protein molecules, both in solution and on the clay, were at a minimum, (2) the rate of collision of proteins having a net neutral charge with charged clay surfaces was at a maximum (MacRitchie and Alexander, 1963), (3) proteins with a net neutral charge contributed less to the surface potential than did charged proteins, and there-fore, neutral proteins were at a lower energy level and in a more stable state when attached to clay surfaces than when in the bulk solution (Rao, 1972), and (4) other factors, such as solubility (see Theng, 1979), were also involved.

On the basis of X-ray diffraction and electron microscopic analyses, the proteins, with the exception of catalase and pepsin, appeared to intercalate the homoionic montmorillonites (Harter and Stotzky, 1973). In general, the extent of expansion of the clays caused by the proteins was proportional to the amounts of protein added rather than to the amounts bound, suggesting that with higher added concentrations, the protein molecules were initially dis-tributed more uniformly across the interlayer surfaces rather than being restricted to the edges of the interlayers as with more dilute protein solutions. The X-ray diffraction data also indicated that, in general, several layers of protein were intercalated and that denaturation (i.e., uncoiling) of the proteins did not occur.

Even though the amount of catalase bound was greater than that of the other proteins, it caused essentially no expansion of the clays, probably because catalase was too large ($M_r \sim$ 250,000) to be intercalated. Distinct globules of catalase were observed, by electron microscopy, to be located relatively uniformly and densely over the surface of the clay packets, indicating that binding sites for this protein were distributed over the clay surface. Pepsin, in contrast, was concentrated around the edges of the clay particles and adjacent to "folds" in the surface rather than being distributed randomly, as were some proteins, or uniformly, as was catalase, on the surface, suggesting that pepsin was bound to positively charged sites (Harter and Stotzky, 1973).

7.5.1b. Amino Acids.

Proteins are composed of numerous amino acids with different pK values, and the relative proportion of negative and positive charges varies within a single protein molecule at different pH values. Furthermore, the large size of proteins facilitates the formation of numerous physical bonds with clays. Consequently, the interpretation of adsorption and binding isotherms of proteins on clays and definition of the mechanisms involved in sorption are difficult. Therefore, the adsorption and binding of amino acids and small peptides on montmorillonite and kaolinite homoionic to various cations were studied (Dashman and Stotzky, 1982, 1984).

Aspartic acid, cysteine, proline, and arginine adsorbed on clays homoionic to H^+, Ca^{2+}, Zn^{2+}, or Al^{3+} but not on clays homoionic to Na^+ or La^{3+}. Glycine adsorbed only on clays homoionic to H^+ and was completely removed by a single wash with water. Aspartic acid was bound only on montmorillonite homoionic to Ca^{2+} or Zn^{2+}, cysteine only on montmorillonite or kaolinite homoionic or Zn^+, proline only on montmorillonite homoionic to H^+, Ca^{2+}, or Zn^{2+} and on kaolinite homoionic to H^+, and arginine only on montmorillonite homoionic to H^+ or Al^{3+}. Both the adsorption and binding isotherms were mostly of the C type, with some being of the S or H type (Giles *et al.*, 1960, 1974). The mechanisms of binding were related to the structure of the amino acid and the cation saturating the clays and appeared to involve either cation or anion exchange or the formation of π- or chelate-complexes.

The peptides studied were adsorbed and bound more on montmorillonite than on kaolinite, and the amounts increased, in general, as the molecular weight and chain length of the peptides increased. Peptides of glycine, with chain lengths ranging from two to four monomers, were not bound, with the exception of glycylglycine which bound on kaolinite homoionic to Na^+ or Ca^{2+}. The acidic peptide, L-aspartylglycine, and the basic peptide, alanyl-L-lysine, were bound only on montmorillonite homoionic to H^+. Of the heterocyclic peptides, L-histidylglycine was bound only on montmorillonite homoionic to Al^{3+}, Zn^{2+}, or Na^+, and L-prolyl-L-phenylalanylglycyl-L-lysine was bound only on montmorillonite homoionic to Al^{3+}, Zn^{2+}, H^+, La^{3+}, or Na^+ and on kaolinite homoionic to Al^{3+}, Zn^{2+}, Na^+, H^+, or Ca^{2+} (the sequence of cations indicates the sequence of the amounts bound).

Several amino acids and peptides intercalated montmorillonite after equilibrium adsorption, but after removal of weakly adsorbed molecules by washing with water, only arginine resulted in detectable expansion of montmorillonite homoionic to Al^{3+}. In contrast to the results obtained with proteins (Harter and Stotzky, 1973), the pH_b and the electrophoretic mobility of the clay–amino acid or clay–peptide complexes were not significantly different from those of the clays alone. The amounts of amino acids or peptides bound on the clays were apparently insufficient to cover enough surface sites to affect significantly the negativity or pH_b of the clays (Dashman and Stotzky, 1985a).

7.5.1c. Other Polymers.

In an attempt to simulate more closely the conditions that presumably exist in soil *in situ*, studies were initiated on the adsorption and binding of a synthetic humic acid polycondensate (avg. M_r 30,000, prepared by the chemical oxidation of catechol in the presence of triglycine) on clays not only homoionic to cations of different valence but also on clays complexed with polymeric hydrous oxides of Al^{3+} (F. Andreux and G. Stotzky, unpublished data). The sequence of adsorption of the humic acid polycondensate on the clays was montmorillonite > illite > attapulgite > kaolinite, indicating that adsorption increased as the CEC and specific surface area increased. In contrast to the results obtained with proteins but in agreement with those obtained with bacteria and viruses (see Sections 7.2.1 and 7.3.2), the adsorption of the polycondensate increased as the valency of the charge-neutralizing cations increased. Similarly, the amounts of the polycondensate bound after extensive washing of the clay–polycondensate complexes were greater as the valency of the cations increased. Most significantly, the greatest amounts of the synthetic humic acid were adsorbed and bound on clays complexed with the polymeric hydrous oxides of Al^{3+}.

When lactoglobulin was added to complexes of clay and the humic acid polymer, the polymer was not desorbed, but the protein was bound, albeit less than on clays without the polymer. The same relations were observed when the humic acid polymer was added to complexes of clay and lactoglobulin. When the two adsorbates were added simultaneously, greater amounts of both were apparently bound than when added individually, which probably reflected an interaction between the polymer and lactoglobulin as well as their binding on the clays.

The binding of the humic acid polymer increased both the acidity (i.e., decreased the pH_b) and the net negativity (as measured electrophoretically) of the clays. When lactoglobulin was subsequently bound to the clay–polymer complexes, the acidity and the net negativity were decreased and approached those of clay–lactoglobulin complexes, which, in general, were similar to those of the clays alone.

These observations, as well as those on the adsorption of viruses to protein-coated clays (see Section 7.3), introduce another confounding aspect: i.e., the reactivity of clay surfaces changes when organic—and inorganic—molecules are bound on the clays. These changes in reactivity undoubtedly alter the surface interactions between the clays and microbial cells, viruses, and other organics. This, again, emphasizes the probability that studies on surface interactions between "clean" clays and biological entities do not accurately reflect the interactions that may occur *in situ*, wherein various types of "conditioning" molecules are present (see Sections 4 and 7.3.2).

7.5.2. Microbial Utilization of Complexes between Clays and Proteinaceous Compounds

7.5.2a. Protein–Clay Complexes.   When the montmorillonite–protein complexes were presented as a sole source of carbon to microbes that had been cultured on the individual proteins for several weeks, the relative availability of the complexed proteins appeared to be dependent on the cation saturating the clay, the location of the protein on the clay, the amount of protein bound, and the characteristics of the individual protein (Gerard and Stotzky, 1973; Stotzky, 1974). Of the five proteins studied in detail, catalase and pepsin were not utilized when complexed, although they were rapidly utilized when not complexed. Casein, chymotrypsin, and lactoglobulin were utilized when complexed, but considerably less than when not complexed. The latter three proteins intercalated the clay and, in general, were present in multilayers (Harter and Stotzky, 1973). Only protein molecules attached to the protein molecules bound on the clay were apparently utilized, and once the multilayered molecules were degraded, there was no apparent further utilization of molecules bound on the clays.

Even though the location of catalase and pepsin on the outer surfaces of the clay packets should have rendered them more available to proteolytic enzymes than the intercalated proteins, their lack of susceptibility to proteolysis suggested that (1) the terminal amino acid residues necessary for initiation of cleavage of the polypeptide chain by exopeptidases were involved in binding or were "hidden" because of binding, and/or (2) binding so altered the configuration or conformation of the proteins that it rendered them insusceptible to the action of endopeptidases. These factors may also have been involved in the lack of utilization of protein molecules bound on the clay interlayers, even though the multilayer molecules were susceptible to degradation.

The enzymatic activity of catalase bound in some homoionic montmorillonites was at least four times greater than that of free catalase, further indicating that binding resulted in a structural change in some proteins (Stotzky, 1974). Although this enhanced enzymatic activity could have resulted from a concentration of the enzyme on clay surfaces (e.g., Burns, 1983), both the inability of proteolytic enzymes to degrade catalase and its greater enzymatic activity when bound suggested that binding altered the shape of the molecule so that more "active centers" were exposed but sites for attachment and/or activation by proteolytic enzymes were obscured. The retention of enzymatic activity by bound catalase further indicated that the protein was not denatured as a result of binding.

The reduction in utilization of complexed proteins was not a result of the binding on the clays of either the microbial exoenzymes or the products of proteolysis. Control experiments, which included the addition of clay to noncomplexed protein and of noncomplexed protein (the same as in the complex) to clay–protein complexes just before the start of the experiments, showed that any binding of proteolytic enzymes or of their products did not account for the amount of reduction in the utilization of the complexed proteins (Gerard and Stotzky, 1973; Stotzky, 1973, 1974).

7.5.2b. Amino Acid–, Peptide–, and Synthetic Humic Acid Polycondensate Clay Complexes.

When the complexes of homoionic montmorillonite or kaolinite and amino acids or peptides were presented to bacteria and fungi as a sole source of carbon, nitrogen, or both carbon and nitrogen, only some complexes were utilized, and then only by some microbes, even though the noncomplexed amino acids and peptides were rapidly utilized (Dashman and Stotzky, 1985b). For example, neither aspartic acid nor cysteine complexed with clays was utilized, whereas complexed proline and arginine were. In more detailed studies with the bacterium, *Agrobacterium radiobacter,* proline bound to montmorillonite homoionic to H^+ or Ca^{2+} was utilized only as a source of nitrogen (when glucose was added), but it was not utilized as a source of carbon or of both carbon and nitrogen. Arginine bound to montmorillonite homoionic to H^+ was available as a source of carbon, nitrogen, and both carbon and nitrogen, but when bound to montmorillonite homoionic to Al^{3+}, arginine was available only as a source of either nitrogen (when glucose was added) or carbon (when NH_4NO_3 was added) but not of both carbon and nitrogen. Glycylglycine bound on kaolinite homoionic to Na^+ was utilized as a source of nitrogen in the presence of glucose. These differences in utilization suggested different types and/or energies of binding of the amino acids to the various homoionic clays.

The apparent selective utilization of portions of amino acids that were external to the cell was in contrast to the action of extracellular proteases that cleave proteins into small peptides or amino acids, which are then transferred, presumably intact, into cells and catabolized intracellularly. The mechanisms whereby a microorganism appears to use selectively either the carbon or the nitrogen moieties of an extracellular amino acid are not clear, especially as no extracellular enzymes capable of degrading amino acids have been reported. Consequently, the amino acids were probably dissociated intact from the clays and then transferred into the cells.

The release of proline and arginine (as well as glycylglycine) from the clays probably involved a specific transport or permease system in the membrane of the bacteria (Ardeshi and Ames, 1980; Rogers *et al.*, 1980). The clay–amino acid complex may have been adsorbed on or came sufficiently close to the bacterial cell in such a configuration that the amino acid was "recognized" by a region of the membrane containing a specific transport protein, which then complexed with the amino acid. The permease–amino acid complex was then transported to the inner side of the membrane, and the amino acid was released into the cytoplasm where it was metabolized.

For a permease system to complex successfully with an amino acid bound on clay, the affinity of the amino acid for the permease system must be greater than its affinity for the clay. If it is assumed that the selectivity coefficient (K_s) of an amino acid for a clay approximates the affinity constant of the amino acid for the clay and that the K_m (the concentration of amino acid to produce one-half saturation of the maximum rate of transport) of the amino acid–permease system approximates the affinity constant of the amino acid for its permease, then the relative affinity of amino acids for clays and permeases can be compared. The K_s of proline bound on montmorillonite homoionic to Ca^{2+} or H^+ was 7 and 37 mM, respectively, and for arginine complexed on montmorillonite homoionic to Al^{3+} or H^+, it was 113 and 142 mM, respectively (Dashman and Stotzky, 1982, 1984c). The average K_m of the permeases of various bacteria for several amino acids is approximately 1 μM (Oxender, 1972), indicating that the affinity of these amino acids for transport proteins was vastly greater than their affinity for the clays. Consequently, it was possible for permeases to complex with the amino acids on the clays and facilitate their transfer into the bacterial cell.

The different affinities of the various amino acids for the homoionic clays might also explain the differences in the amount of utilization of the bound amino acids. Proline complexed on montmorillonite homoionic to Ca^{2+} or H^+ was used only as a source of nitrogen, arginine complexed on montmorillonite homoionic to Al^{3+} was used only as a source of either carbon or nitrogen, but arginine complexed on montmorillonite homoionic to H^+ was simultaneously utilized as a source of carbon and nitrogen. Inasmuch as the normal metabolism of either 1 mole of arginine or proline yields 1 mole of glutamate (Mahler and Cordes, 1969) and as bacteria were capable of utilizing free proline and arginine as both a carbon and a nitrogen source, the differences in the amount of utilization of the bound amino acids may have been related to differences in the amounts of energy necessary for their transport from the clay into the cell.

On the basis of the K_s values, the affinity of arginine for montmorillonite homoionic to H^+ was 4- to 20-fold higher than the affinity of proline for montmorillonite homoionic to H^+ or Ca^{2+}, respectively (Dashman and Stotzky, 1982, 1985b). If the net energy for the dissociation of the same amount of proline and arginine from montmorillonite homoionic to H^+ was greater for arginine and assuming that the energy required for the transport of proline and arginine into the cell was the same, then, perhaps, the energy produced from the incorporation of arginine into the urea cycle, which was then linked to the citric acid cycle, may have provided additional energy not available from proline to sustain dissociation, transport, and growth. However, when additional energy in the form of glucose was present, the nitrogen of proline was used for growth. Although the amount of arginine bound on montmorillonite homoionic to Al^{3+} was greater than that on montmorillonite homoionic to H^+ (2.33 and 1.33 mg/30 mg clay, respectively), sufficient arginine to yield a detectable increase in *d* spacing remained intercalated only in the clay homoionic to Al^{3+} after extensive washing (Dashman and Stotzky, 1985a), and therefore, the amount of arginine accessible to its permease from the Al–clay–arginine complex was apparently inadequate to provide sufficient energy for dissociation, transport, and growth. However, when the medium was supplemented with glucose or NH_4NO_3, sufficient energy and nutrients were apparently available for growth.

Cysteine was not utilized when bound on any homoionic montmorillonite or kaolinite, possibly because the amount of energy required for the dissociation of cysteine from the clays and its subsequent transport into the cell was greater and its affinity for its permease was less than those for proline or arginine. The K_s for the binding of cysteine on montmorillonite homoionic to Zn^{2+} and on kaolinite homoionic to H^+ or Zn^{2+} was 93, 1.4, and 0.6 mM, respectively (Dashman and Stotzky, 1982). The microbes studied may not have been able to utilize bound cysteine because the K_m for transport may have required too much energy [e.g., the K_m for the transport of cysteine is about 10 times greater than that for arginine in *E. coli* strain K_{12} W (Rogers *et al.*, 1980)].

A comparison of the affinities of aspartic acid, which was not utilized when bound on the clays, with cysteine, proline, and arginine was not possible, because K_s values for aspartic acid could not be calculated as the binding isotherms of aspartic acid were not of the C type (Dashman and Stotzky, 1982).

Although the catechol–triglycine polycondensate was recalcitrant to degradation by microbes that rapidly utilized proteins, peptides, and amino acids, a *Streptomyces* sp. that could degrade, albeit slowly, this synthetic humic acid was isolated from soil (F. Andreux and G. Stotzky, unpublished data). The utilization of the polycondensate bound on montmorillonite homoionic to various cations was, in general, less than that of the polymer alone. However, the synthetic humic acid bound on montmorillonite complexed with polymeric hydrous oxides of Al^{3+} was utilized more than that bound on the homoionic clays. Similarly, the utilization of lactoglobulin bound on various homoionic montmorillonites was considerably less than that of the free protein, but the utilization of the protein bound on montmorillonite complexed with the polymeric hydrous oxides of Al^{3+} was, in general, greater than that of the other clay–lactoglobulin complexes. The humic acid polycondensate appeared to exert some toxicity to the *Streptomyces* sp., but its toxicity was reduced or eliminated when bound on the various homoionic clays, as the addition of various homoionic clay–polycondensate complexes did not inhibit the utilization of free lactoglobulin. However, when the polycondensate bound on clay complexed with polymeric hydrous oxides of Al^{3+} was added to free lactoglobulin, the utilization of the protein was inhibited, suggesting that the humic acid polycondensate was easily released from this clay–hydrous oxide complex.

A detailed discussion of the microbial utilization of other organic compounds complexed with clays is beyond the scope of this chapter, but references to these types of studies and to studies on the utilization of complexes of clays and humic matter and of humic matter and soluble organics, including pesticides, can be found in the reviews of Schnitzer and Khan (1972), Theng (1974, 1979), Filip (1975), Marshall (1976), Greenland and Hayes (1978), Stotzky and Burns (1982), Burns (1983), and others. In general, the complexation of organic compounds with clays or particulate humic matter decreases their degradation by microbes. However, the experimental procedures used in these studies must be carefully evaluated to be certain that stable complexes were formed and that the data obtained were not just the result of some indirect effects of the clays or particulate humic matter on the microbial systems (e.g., maintenance of a favorable pH, provision of inorganic nutrients, removal of toxic materials). Unfortunately, few studies on the microbial utilization of complexed organic compounds have constructed adsorption and binding isotherms, have attempted to define the mechanisms involved, or have related these mechanisms to the reduction (or increase) in the utilization of the complexed organic compounds.

7.5.3. Implications of the Binding of Organics on Clays to Adhesion and Ecology of Microbes in Soil

Inasmuch as the binding of organics on clays appears to decrease their availability as nutrient and energy sources for microbes *in vitro,* the differential effects of different clay minerals saturated with different cations on the availability of organic nutrients could have a marked influence on the activity, ecology, and population dynamics of microbes in soil. Furthermore, surface interactions between organics and clays may prevent the migration of organics through soil and may explain, in part, the lower concentrations of organic matter in deeper than in surface horizons. The lower content of organic matter in deeper horizons may be the result of the restricted mobility of organics rather than of their degradation by microbes, especially as microbial activity is less in these deeper horizons, primarily because of limitations in available oxygen.

Although several *in vitro* studies have demonstrated that clays bind organics and that the degradation of these organics is thereby reduced, caution must be used in extrapolating the results of such studies to soil *in situ* until another apparent paradox is resolved: if clays bind organic molecules and these are then relatively resistant to microbial degradation, then essentially all clays in soil *in situ* should be saturated with such molecules and, with the exception of newly formed clay minerals (a relatively infrequent occurrence), no clay surfaces should be available for further binding. Consequently, either mechanisms exist for the removal of bound organics, which would provide suitable clay surfaces for the binding of newly formed organics, or such binding no longer occurs in soil *in situ.*

Inasmuch as enzymatic degradation of some bound organic molecules (e.g., proteins) is apparently not a major removal mechanism, mechanical methods of removal, which probably have energies greater than those involved in binding, might be involved: e.g., (1) abrasion, such as is exerted in the guts of earthworms (or filter-feeders in aquatic sediments), during cultivation, or during disturbance of soil by growing roots and micro- and macrofauna, or (2) release during cycles of soil wetting and drying or freezing and thawing. However, there are many areas of the world where the soils neither contain significant numbers of earthworms, are cultivated, nor have cycles of freezing and thawing, and yet turnover of clay–organic complexes appears to occur.

Perhaps a better explanation is that organic molecules in soil *in situ* are not bound directly on the surfaces of "clean" clays or to their charge-neutralizing cations, as in these *in vitro* studies, but rather to polymeric metal hydrous oxides that may be associated with clays in soil. The preliminary studies of F. Andreux and G. Stotzky, discussed in Section 7.5.2b, tend to support this possibility. Both the catechol–triglycine polycondensate and lactoglobulin were not only bound more on clays complexed with polymeric hydrous oxides of Al^{3+}, but the utilization of both compounds was greater when bound on these clays than on clays homoionic to nonpolymeric cations of different valence. Inasmuch as the integrity of these inorganic polymers is pH-dependent and their binding of organic molecules probably involves chemical bonds, the release of such bound organics may occur frequently as the pH of natural microhabitats fluctuates, primarily as a result of microbial metabolism. Once the organics have been released from the clay–hydrous oxide complexes, they could be either degraded by the soil microbiota or again bound on the complexes. Furthermore, newly formed organics may also bind to organics directly bound on clays, and these multilayer organics may be more susceptible to enzymatic degradation, as suggested by the partial utilization *in vitro* of the proteins of certain clay–protein complexes (Gerard and Stotzky, 1973; Stotzky, 1974; see Section 7.5.2a).

There has been little study of the mechanisms involved in the binding of organics on natural (i.e., aggregated, mixed, and "dirty") clays and other soil particulates and on the subsequent utilization of these organics by microbes. Until such mechanisms have been defined, the importance of the binding of organics on clays and other particulates to microbial events or to the migration of organics in soil *in situ* remains speculative.

Nevertheless, *in vitro* studies on the binding of organic molecules on "clean" clays and on defined clays complexed with polymeric hydrous metal oxides may provide information about the importance of organic "conditioning films" to the adhesion of microbes and viruses to clays in soil. Inasmuch as many organics, depending on their size, appear to intercalate expanding 2 : 1 clays, the direct value of these organics as conditioning films on such clays may be limited, as microbial cells and viruses are too large to enter between the unit layers. However, the primary location of the negative charges on 2 : 1 clays is between the unit layers, and therefore, the covering of these negative charges may reduce the net negativity of these clays—even when present as domains in and cutans on aggregates—and, thereby, enable net negatively charged cells and viruses to approach close enough to the clays for adsorption and then binding to occur. The net negative charge on nonexpanding 1 : 1 clays is probably too low to be of much importance in the retention of water, the concentration of nutrients, the removal of inhibitors, the repulsion of cells, etc., and therefore, the binding of organic conditioning films to these clays may not be too important to microbial events in soil. Nevertheless, if these bound organics (i.e., conditioning films) are as recalcitrant to microbial degradation in soil *in situ* as they were in these *in vitro* studies with "clean" clays, then this may explain, to a great extent, the apparent surface interactions *in situ* between net negatively charged clays and net negatively charged microbial cells, viruses, and soluble organics. This is obviously a speculation that requires extensive investigation—once appropriate techniques for these types of studies have been developed.

8. CONCLUSIONS

Soil is an extremely complex habitat for microbes because of the heterogeneity of soil particulates, the constantly changing water and gas regimes, the intermittent supply of nutrients, etc. This complexity of soil is reflected in the heterogeneity of its microbial inhabitants and in their fluctuations—both in time and in space—in numbers and activity. This complexity also makes it extremely difficult both to determine whether the adhesion of microbes, viruses, and organic compounds to soil particulates occurs in soil *in situ* and, if so, to clarify the mechanisms involved. Although the addition to soil of surface-active particulates (e.g., clay minerals) may result in changes in microbial numbers and activities, this is not proof that surface interactions between the added particulates and the microbiota (i.e., *direct* effects) have occurred, as the effects of the added particulates could be, and usually are, the result of modifications in the physicochemical characteristics of the microhabitats of the microbes (i.e., *indirect* effects). Because of the present inability to distinguish between direct and indirect effects in soil directly, model systems are often used. However, these model systems are too often unrealistic both in their simplicity and in their assumptions, and their relevance to what really occurs in soil *in situ* must be constantly and critically questioned. Furthermore, more attempts must be made to test and verify the results obtained from model systems in actual soil systems. Unfortunately, the state-of-the-art of studying surface interactions in soil has not progressed to this stage.

In addition to using more realistic models (e.g., using "dirty" rather than "clean" clays), care must be exercised in extrapolating to soil results obtained from studies of other, often quite dissimilar and less complex, habitats. Furthermore, it should be realized that many of the conclusions about adhesion in these other habitats are also based, to a great extent, on model systems. Because of the extensive use of model systems, which often vary from laboratory to laboratory and with the habitat being modeled, a considerable amount of the literature—including this chapter—on adhesion and other interactions between inanimate surfaces and biological entities is confusing, contradictory, insufficiently specific, and unrealistic.

For example, thermodynamic concepts (e.g., changes in entropy) may be applicable to model systems, but they may have less applicability to natural, and therefore, open (i.e., nonadiabatic) systems. Similarly, the DLVO and other theories that are based essentially on the presence of homogeneous and clean surfaces (and microbial populations) may not be relevant to the complex, heterogeneous, and "dirty" conditions in soil and most other natural microbial habitats.

Although there are more holes than cloth in the fabric of what is known about adhesion and other interactions between soil surfaces and biological entities—both in model and in *in situ* systems—some areas appear to have received less study than others. For example, the occurrence and relative importance of hydrophobic "interactions" in soil need study. Not only are many microbial cells amphipathic (i.e., they contain both polar and nonpolar regions), but so also may be some soil particulates (e.g., clays partially coated with hydrophobic organic matter, long stretches of siloxane surfaces in clay tetrahedra with little or no isomorphous substitution, possibly even the highly ordered water associated with clay surfaces and their charge-compensating cations). Furthermore, both chaotropic (disrupting water structure) and antichaotropic substances are variably present in the soil solution, which can affect the structure of water, especially in the microhabitats near clay surfaces, and thereby, the accommodation of hydrophobic entities.

If microbes stick to soil particulates—and the evidence for this adhesion is not conclusive—does such adhesion affect the metabolism (e.g., the activity and the externalization/internalization of enzymes), gene expression, and other physiological functions of the microbes? Do microbes and other biological entities stick in mono- or multilayers? What is the importance of the relative size of the adsorbents (e.g., clay domains) and adsorbates (e.g., bacteria vs. viruses vs. soluble macromolecules) to surface interactions in soil? Are the concepts of "conditioning films" applicable to soil (e.g., are such films formed in soil)? If so, do they affect the subsequent adhesion of particulate biological entities, and if so, do they confer any specificity with respect to which entities stick? How resistant are such films to biodegradation (i.e., what is their retention time)? Similar questions must be asked about the importance and production of extracellular structures (e.g., slime layers, fimbriae, holdfasts) in the adhesion of microbes in soil, especially as many microbes appear to stick in model systems without the production of such structures. Although this volume focuses primarily on the adhesion of bacteria to surfaces, other types of microbes, as well as viruses, are as important as bacteria in soil, and their interactions with surfaces in soil should be studied more extensively.

Perhaps most important is the necessity to relate adhesion and other surface interactions of bacteria and other biological entities to their activity, ecology, and population dynamics in soil. A potential interaction with a surface is only one of the multitude of interactions that microbes and viruses encounter in the complex soil habitat. Consequently, even though the study of surface interactions is intellectually stimulating and rewarding—albeit frustrating at times—its relative importance to the survival and other functions of microbes and viruses in soil—and in other microbial habitats—must be kept in perspective.

REFERENCES

Adams, D. J., Ridinger, D. N., Spendlove, R. S., and Barnett, B. B., 1982, Protamine precipitation of two reovirus particle types from polluted waters, *Appl. Environ. Microbiol.* **44:**589–596.

Ahmed, R., and Fields, B. N., 1981, Reassortment of genome segments between reovirus defective interfering particles and infectious virus: Construction of temperature-sensitive and attenuated viruses by rescue of mutants from DI particles, *Virology* **111:**351–363.

Albert, J. T., and Harter, R. D., 1973, Adsorption of lysozyme and ovalbumin by clay: Effect of clay suspension pH and clay mineral type, *Soil Sci.* **115:**130–136.

Ardeshi, F., and Ames, G. F. L., 1980, Cloning of the histidine transport genes from *Salmonella typhimurium* and characterization of an analogous transport system in *Escherichia coli,* in: *Membrane Transport and Neuroreceptors* (D. L. Oxender, A. Blume, I. Diamond, and C. F. Fox, eds.), Liss, New York, pp. 51–64.

Arnold, P. W., 1978, Surface–electrolyte interactions, in: *The Chemistry of Soil Constituents* (D. J. Greenland and M. H. B. Hayes, eds.), Dekker, New York, pp. 355–404.

Atlas, R. M., and Bartha, R., 1981, *Microbial Ecology,* Addison–Wesley, Reading, Mass.

Audic, J. M., Faub, G. M., and Navarro, J. M., 1984, Specific activity of *Nitrobacter* through attachment on granular media, *Water Res.* **18:**745–750.

Babich, H., and Stotzky, G., 1974, Air pollution and microbial ecology, *Crit. Rev. Environ. Control* **4:**353–421.

Babich, H., and Stotzky, G., 1977a, Reductions in the toxicity of cadmium to microorganisms by clay minerals, *Appl. Environ. Microbiol.* **33:**696–705.

Babich, H., and Stotzky, G., 1977b, Effect of cadmium on fungi and on interactions between fungi and bacteria in soil: Influence of clay minerals and pH, *Appl. Environ. Microbiol.* **33:**1059–1066.

Babich, H., and Stotzky, G., 1978, Effects of cadmium on the biota: Influence of environmental factors, *Adv. Appl. Microbiol.* **23:**55–117.

Babich, H., and Stotzky, G., 1979, Abiotic factors affecting the toxicity of lead to fungi, *Appl. Environ. Microbiol.* **38:**506–513.

Babich, H., and Stotzky, G., 1980a, Reductions in inactivation rates of bacteriophages by clay minerals in lake water, *Water Res.* **14:**185–187.

Babich, H., and Stotzky, G., 1980b, Environmental factors that influence the toxicity of heavy metals and gaseous pollutants to microorganisms, *Crit. Rev. Microbiol.* **9:**99–145.

Babich, H., and Stotzky, G., 1982, Gaseous and heavy metal air pollutants, in: *Experimental Microbial Ecology* (R. G. Burns and J. H. Slater, eds.), Blackwell, Oxford, pp. 631–670.

Babich, H., and Stotzky, G., 1983a, Influence of chemical speciation on the toxicity of heavy metals to the microbiota, in: *Aquatic Toxicology* (J. O. Nriagu, ed.), Wiley, New York, pp. 1–46.

Babich, H., and Stotzky, G., 1983b, Toxicity of nickel to microbes: Environmental aspects, *Adv. Appl. Microbiol.* **29:**195–265.

Baier, R. E., 1980, Substrata influences on adhesion of microorganisms and their resultant new surface properties, in: *Adsorption of Microorganisms to Surfaces* (G. Bitton and K. C. Marshall, eds.), Wiley, New York, pp. 59–104.

Bartnicki-Garcia, S., 1968, Cell wall chemistry, morphogenesis and taxonomy in fungi, *Annu. Rev. Microbiol.* **22:**67–108.

Berg, G. (ed.), 1967, *Transmission of Viruses by the Water Route,* Wiley, New York.

Berg, G., Bodily, H. L., Lennette, E. H., Melnick, J. L., and Metecalf, T. G. (eds.), 1974, *Viruses in Water,* American Public Health Association, Washington, D.C.

Bitton, G., 1980a, *Introduction to Environmental Virology,* Wiley, New York.

Bitton, G., 1980b, Adsorption of viruses to surfaces: Technological and ecological implications, in: *Adsorption of Microorganisms to Surfaces* (G. Bitton and K. C. Marshall, eds.), Wiley, New York, pp. 331–374.

Black, C. A. (ed.), 1965, *Methods of Soil Analyses,* Parts 1 and 2, American Society of Agronomy, Madison, Wisc.

Bondietti, E., Martin, J. P., and Haider, K., 1971, Influence of nitrogen source and clay on growth and phenolic polymer production by *Stachybotrys* species, *Hendersonula toruloidea,* and *Aspergillus sydowi, Soil Sci. Soc. Am. Proc.* **35:**917–922.

Bremmer, J. M., 1967, Nitrogenous compounds, in: *Soil Biochemistry,* Volume 1 (A. D. McLaren and G. H. Peterson, eds.), Dekker, New York, pp. 19–66.

Brewer, R., 1964, *Fabric and Mineral Analysis of Soils,* Wiley, New York.

Bright, J. J., and Fletcher, M., 1983, Amino acid assimilation and electron transport system activity in attached and free-living marine bacteria, *Appl. Environ. Microbiol.* **45:**818–825.

Brown, G., Newman, A. C. D., Rayner, J. H., and Weir, A. H., 1978, The structure and chemistry of soil clay minerals, in: *The Chemistry of Soil Constituents* (D. J. Greenland and M. H. B. Hayes, eds.), Dekker, New York, pp. 29–178.

Burnett, J. H., 1976, *Fundamentals of Mycology,* Arnold & Arnold, New York.

Burns, R. G., 1983, Extracellular enzyme–substrate interactions in soil, in: *Microbes in Their Natural Environments* (J. H. Slater, R. Whittenbury, and J. W. T. Wimpenny, eds.), Cambridge University Press, London, pp. 249–298.

Burns, R. G., and Slater, J. H. (eds), 1982, *Experimental Microbial Ecology,* Blackwell, Oxford.

Burt, W. R., 1982, Identification of coprogen B and its breakdown products from *Histoplasma capsulatum, Infect. Immun.* **35:**990–996.

Burt, W. R., Underwood, A. L., and Appleton, G. L., 1981, Hydroxamic acid from *Histoplasma capsulatum* that displays growth factor activity, *Appl. Environ. Microbiol.* **42:**560–563.

Bushby, H. V. A., and Marshall, K. C., 1977, Water status of rhizobia in relation to their susceptibility to desiccation and to their protection by montmorillonite, *J. Gen. Microbiol.* **99:**19–27.

Bystricky, V., Stotzky, G., and Schiffenbauer, M., 1975, Electron microscopy of T1-bacteriophage adsorbed to clay minerals: Application of the critical point drying method, *Can. J. Microbiol.* **21:**1278–1282.

Campbell, R., 1983, Ultrastructural studies of *Gaeumannomyces graminis* in the waterfilms on wheat roots and the effect of clay on the interaction between this fungus and antagonistic bacteria, *Can. J. Microbiol.* **29:**39–45.

Campbell, R., and Ephgrave, J. M., 1983, Effect of bentonite clay on the growth of *Gaeumannomyces graminis* var. *tritici* and on its interactions with antagonistic bacteria *J. Gen. Microbiol.* **129:**771–777.

Casida, L. E., Jr., 1965, Abundant microorganisms in soil, *Appl. Microbiol.* **13:**327–334.

Casida, L. E., Jr., 1971, Microorganisms in unamended soil as observed by various forms of microscopy and staining, *Appl. Microbiol.* **21:**1040–1045.

Casida, L. E., Jr., 1977, Small cells in pure cultures of *Agromyces ramosus* and in natural soils, *Can. J. Microbiol.* **23:**214–216.

Chassin, P., 1969, Adsorption du glycolle par la montmorillonites, *Bull. Groupe Fr. Argiles* **21:**71–88.

Christie, N. T., and Costa, M. 1983, *In vitro* assessment of the toxicity of metal compounds. III. Effects on DNA structure and function in intact cells, *Biol. Trace Elem. Res.* **5:**55–71.

Clark, A., 1974, *The Chemisorptive Bond,* Academic Press, New York.

Collins, Y. E., and Stotzky, G., 1982, Influence of heavy metals on the electrokinetic properties of bacteria, *Abstr. Annu. Meet. Am. Soc. Microbiol.* 229.

Collins, Y. E., and Stotzky, G., 1983, Heavy metals alter the electrokinetic properties of bacteria, yeasts, and clay minerals, *Abstr. 3rd Int. Symp. Microb. Ecol.* p. 76.

Corpe, W. A., 1980, Microbial surface components involved in adsorption of microorganisms onto surfaces, in: *Adsorption of Microorganisms to Surfaces* (G. Bitton and K. C. Marshall, eds.), Wiley, New York, pp. 105–144.

Costerton, J. W., 1980, Some techniques involved in study of adsorption of microorganisms to surfaces, in: *Adsorption of Microorganisms to Surfaces* (G. Bitton and K. C. Marshall, eds.), Wiley, New York, pp. 403–424.

Costerton, J. W., Irwin, R. T., and Cheng, K.-J., 1981, The role of bacterial surface structures in pathogenesis, *Crit. Rev. Microbiol.* **8:**303–338.

Dandiker, W. B., Alonso, R., de Saussure, V. A., Kierszenbaum, F., Levison, S. A., and Schapiro, H. C., 1967, The effect of chaotropic ions on the dissociation of antigen-antibody complexes, *Biochemistry* **6:**1460–1467.

Daniels, S. L., 1980, Mechanisms involved in sorption of microorganisms to solid surfaces, in: *Adsorption of Microorganisms to Surfaces* (G. Bitton and K. C. Marshall, eds.), Wiley, New York, pp. 7–58.

Dashman, T., and Stotzky, G., 1982, Adsorption and binding of amino acids on homoionic montmorillonite and kaolinite, *Soil Biol. Biochem.* **14:**447–456.

Dashman, T., and Stotzky, G., 1984, Adsorption and binding of peptides on homoionic montmorillonite and kaolinite, *Soil Biol. Biochem.* **16:**51–55.

Dashman, T., and Stotzky, G., 1985a, Physical properties of homoionic montmorillonite and kaolinite complexed with amino acids and peptides, *Soil Biol. Biochem.* **17:**1–7.

Dashman, T., and Stotzky, G., 1985b, Microbial utilization of amino acids and a peptide bound to homoionic montmorillonite and kaolinite, *Soil Biol. Biochem.,* in press.

DeFlaun, M. F., and Mayer, L. M., 1983, Relationship between bacteria and grain surfaces in intertidal sediments, *Limnol. Oceanogr.* **28:**873–881.

Dixon, J. B., and Weed, S. B. (eds.), 1977, *Minerals in Soil Environments,* Soil Science Society of America, Madison, Wisc.

Dorioz, J.-M., and Robert, M., 1982, Etude experimentale de l'interaction entre Champignons et argile: Consequences sur la microstructure des sols, *C.R. Acad. Sci. Ser. II* **295:**511–516.

Drayna, D., and Fields, B. N., 1982, Biochemical studies on the mechanism of chemical and physical inactivation of reovirus, *J. Gen, Virol.* **63:**161–170.

Duboise, S. M., Moore, B. E., Sorber, C. A., and Sagik, B. P., 1979, Viruses in soil systems, *Crit. Rev. Microbiol.* **7:**245–285.

Ellwood, D. C., Keevil, C. W., Marsh, P. D., Brown, C. M., and Wardell, J. N., 1982, Surface-associated growth, *Philos. Trans. R. Soc. London* **297:**517–532.

Emerson, W. W., 1959, The structure of soil crumbs, *J. Soil Sci.* **10:**235–244.

Emery, T., 1971, Hydroxamic acids of natural origin, *Adv. Enzymol.* **35:**135–185.

Emery, T., 1978, The storage and transport of iron, in: *Metal Ions in Biological Systems,* Volume 7 (H. Sigel, ed.), Dekker, New York, pp. 77–126.

Emery, T., 1982, Iron metabolism in humans and plants, *Am. Sci.* **70:**626–632.

Emery, T., and Emery, L., 1973, The biological activity of some siderochrome derivatives, *Biochem. Biophys. Res. Commun.* **50:**670–675.

Escudey, M., and Galindo, G., 1983, Effect of iron oxide coatings on electrophoretic mobility and dispersion of allophane, *J. Colloid Interface Sci.* **93:**78–83.

Farmer, V. C., 1978, Water on particle surfaces, in: *The Chemistry of Soil Constituents* (D. J. Greenland and M. H. B. Hayes, eds.), Dekker, New York, pp. 405–408.

Farrah, S. R., Shah, D. O., and Ingram, L. O., 1981, Effect of chaotropic and antichaotropic agents on elution of poliovirus adsorbed on membrane filters, *Proc. Natl. Acad. Sci. U.S.A.* **78:**1229–1232.

Filip, Z., 1973, Clay minerals as a factor influencing the biochemical activity of soil microorganisms, *Folia Microbiol. (Prague)* **18:**56–74.

Filip, Z., 1975, Wechselbeziehungen zwischen Mikroorganismen und Tonmineralen und ihre Auswirkung auf die Bodendynamik, Habilitationsschrift, Justus Liebig Universitat, Giessen.

Filip, Z., Haider, K., and Martin, J. P., 1972a, Influence of clay minerals on growth and metabolic activity of *Epicoccum nigrum* and *Stachybotrys chartarum, Soil Biol. Biochem.* **4:**135–145.

Filip, Z., Haider, K., and Martin, J. P., 1972b, Influence of clay minerals on the formation of humic substances by *Epicoccum nigrum* and *Stachybotrys chartarum, Soil Biol. Biochem.* **4:**147–154.

Flaig, W., Beutelspacher, H., and Rietz, E., 1975, Chemical composition and physical properties of humic substances, in: *Soil Components,* Volume 1 (J. E. Gieseking, ed.), Springer-Verlag, Berlin, pp. 1–211.

Fletcher, M., and Floodgate, G. D., 1973, An electron-microscopic demonstration of an acidic polysaccharide involved in adhesion of a marine bacterium to solid surfaces, *J. Gen. Microbiol.* **74:**325–334.

Fletcher, M., and Marshall, K. C., 1983, Are solid surfaces of ecological significance to aquatic bacteria? in: *Advances in Microbial Ecology,* Volume 6 (K. C. Marshall, ed.), Plenum Press, New York, pp. 199–236.

Fletcher, M., Latham, M. J., Lynch, J. M., and Rutter, P. R., 1980, The characteristics of interfaces and their role in microbial attachment, in: *Microbial Adhesion to Surfaces* (R. C. W. Berkeley, J. M. Lynch, J. Melling, P. R. Rutter, and B. Vincent, eds.), Horwood, Chichester, pp. 67–78.

Forslind, E., and Jacobsson, A., 1970, Clay–Water Interactions: An Experimental Study of Interface Phenomena, European Research Office, U.S. Army, Frankfurt, Germany, DAJA 37-69-C-0657.

Gast, R. G., 1977, Surface and colloid chemistry, in: *Minerals in Soil Environments* (J. B. Dixon and S. B. Weed, eds.), Soil Science Society of America, Madison, Wisc., pp. 27–74.

Gerard, J. F., and Stotzky, G., 1973, Smectite–protein complexes vs. non-complexed proteins as energy and carbon sources for bacteria, *Agron. Abstr.* p. 91.

Gibbons, R. J., 1980, Adhesion of bacteria to the surfaces of the mouth, in: *Microbial Adhesion to Surfaces* (R. C. W. Berkeley, J. M. Lynch, J. Melling, P. R. Rutter, and B. Vincent, eds.), Horwood, Chichester, pp. 351–388.

Gieseking, J. E. (ed.), 1975, *Soil Components,* Volume 2, Springer-Verlag, Berlin.

Giles, C. H., MacEwan, T. H., Nakhwa, S. N., and Smith, D., 1960, Studies on adsorption. XI. A system of classification of solution adsorption isotherms, and its use in diagnosis of adsorption mechanisms and in measurement of specific surface areas of solids, *J. Chem. Soc.* **786:**3973–3993.

Giles, C. H., Smith, D., and Huitson, A., 1974, A general treatment and classification of the solute adsorption isotherm. I. Theoretical, *J. Colloid Interface Sci.* **47:**755–765.

Goring, C. A. I., and Bartholomew, W. V., 1952, Adsorption of mononucleotides, nucleic acids, and nucleoproteins by clays, *Soil Sci.* **74:**149–164.

Gray, T. R. G., Baxby, P., Hill, J. R., and Goodfellow, M., 1968, Direct observation of bacteria in soil, in: *The Ecology of Soil Bacteria* (T. R. G. Gray and D. Parkinson, eds.), Liverpool University Press, Liverpool, pp. 171–192.

Greaves, M. P., and Wilson, M. J., 1969, The adsorption of nucleic acids by montmorillonite, *Soil Biol. Biochem.* **1:**317–323.

Greaves, M. P., and Wilson, M. J., 1970, The degradation of nucleic acids and montmorillonite–nucleic acid complexes by soil microorganisms, *Soil Biol. Biochem.* **2:**257–268.

Greaves, M. P., and Wilson, M. J., 1973, Effects of soil microorganisms on montmorillonite–adenine complexes, *Soil Biol. Biochem.* **5:**275–276.

Greenland, D. J., 1965, Interactions between clays and organic compounds in soils, II, *Soils Fert.* **28:**521–532.

Greenland, D. J., 1971, Interactions between humic and fulvic acids and clays, *Soil Sci.* **111:**34–41.

Greenland, D. J., and Hayes, M. H. B. (eds), 1978, *The Chemistry of Soil Constituents,* Wiley, New York.

Greenland, D. J., and Mott, C. J. B., 1978, Surfaces of soil particles, in: *The Chemistry of Soil Constituents* (D. J. Greenland and M. H. B. Hayes, eds.), Wiley, New York, pp. 321–354.

Grim, R. E., 1968, *Clay Mineralogy,* McGraw-Hill, New York.

Habte, M., and Alexander, M., 1978a, Mechanisms of persistence of low numbers of bacteria preyed upon by protozoa, *Soil Biol. Biochem.* **10:**1–6.

Habte, M., and Alexander, M., 1978b, Protozoan density and the coexistence of protozoan predators and bacterial prey, *Ecology* **59:**140–146.

Harter, R. D., 1977, Reactions of minerals with organic compounds in the soil, in: *Minerals in Soil Environments* (J. B. Dixon and S. B. Reed, eds.), Soil Science Society of America, Madison, Wisc., pp. 709–739.

Harter, R. D., and Stotzky, G., 1971, Formation of clay–protein complexes, *Soil Sci. Soc. Am. Proc.* **35:**383–389.

Harter, R. D., and Stotzky, G., 1973, X-ray diffraction, electron microscopy, electrophoretic mobility, and pH of some stable smectite–protein complexes, *Soil Sci. Soc. Am. Proc.* **37:**116–123.

Hatefi, Y., and Hanstein, W. G., 1969, Solubilization of particulate proteins and nonelectrolytes by chaotropic agents, *Proc. Natl. Acad. Sci. U.S.A.* **62:**1129–1136.

Hatefi, Y., and Hanstein, W. G., 1974, Destabilization of membranes with chaotropic ions, *Methods Enzymol.* **31:**770–790.

Hattori, T., 1973, *Microbial Life in the Soil,* Dekker, New York.

Hattori, T., and Hattori, R., 1976, The physical environment in soil microbiology: An attempt to extend principles of microbiology to soil microorganisms, *Crit. Rev. Microbiol.* **4:**423–461.

Hayes, M. H. B., and Swift, R. S., 1978, The chemistry of soil organic colloids, in: *The Chemistry of Soil Constituents* (D. J. Greenland and M. H. B. Hayes, eds.), Dekker, New York, pp. 179–320.

Hendershot, W. H., and Lavkulich, L. M., 1983, Effect of sesquioxide coatings on surface charge of standard mineral and soil samples, *Soil Sci. Soc. Am. J.* **47:**1252–1260.

Hodson, R. E., Maccubin, A. E., and Pomeroy, L. R., 1981, Dissolved adenosine triphosphate utilization by free-living and attached bacterioplankton, *Mar. Biol.* **64:**43–51.

Hunsley, D., and Burnett, J. H., 1970, The ultrastructural architecture of the walls of some hyphal fungi, *J. Gen Microbiol.* **66:**203–218.

Ivarson, K. C., Schnitzer, M., and Cortez, J., 1982, The biodegradability of nucleic acid bases adsorbed on inorganic and organic soil components, *Plant Soil* **64:**343–353.

Jang, S. D., and Condrate, R. A., Sr., 1972a, The IR spectra of lysine adsorbed on several cation-substituted montmorillonites, *Clays Clay Miner.* **20:**79–82.

Jang, S. D., and Condrate, R. A., Sr., 1972b, Infared spectra of α-alanine adsorbed on Cu-montmorillonite, *Appl. Spectrosc.* **26:**102–104.

Keya, S. D., and Alexander, M., 1975, Factors affecting growth of *Bdellovibrio* on *Rhizobium, Arch. Microbiol.* **103:**37–43.

Kirchman, D., 1983, The production of bacteria attached to particles suspended in a freshwater pond, *Limnol. Oceanogr.* **28:**858–872.

Kiremidjian, L., and Stotzky, G., 1973, Effects of natural microbial preparations on the electrokinetic potential of bacterial cells and clay minerals, *Appl. Microbiol.* **25:**964–971.

Kiremidjian, L., and Stotzky, G., 1975, Influence of mono- and multivalent cations on the electrokinetic properties of adult *Rana pipiens* kidney cells, *J. Cell. Physiol.* **85:**125–134.

Kiremidjian, L., and Stotzky, G., 1976, Influence of mono- and multivalent cations on the electrokinetic properties of normal human lymphoid and Burkitt lymphoma cells, *Experientia* **32:**312–314.

Kjelleberg, S., Humphrey, B. A., and Marshall, K. C., 1982, Effect of interfaces on small, starved marine bacteria, *Appl. Environ. Microbiol.* **43:**1166–1172.

Kjelleberg, S., Humphrey, B. A., and Marshall, K. C., 1983, Initial phases of starvation and activity of bacteria at surfaces, *Appl. Environ. Microbiol.* **46:**978–984.

Kloepper, J. W., Leong, J., Teintze, M., and Schroth, M. N., 1980, *Pseudomonas* siderophores: A mechanism explaining disease-suppressive soils, *Curr. Microbiol.* **4:**317–320.

Krasil'nikov, N. A., 1958, *Soil Microorganisms and Higher Plants,* Akad. Nauk SSSR, Moscow (translated by Israel Program Sci. Transl., Washington, D.C., 1961).

Krumins, S., and Stotzky, G., 1980, Protein–membrane interactions: Equilibrium adsorption and binding of proteins and polyamino acids to erythroblasts transformed by Friend virus, *Cell Biol. Int. Rep.* **4**:1131–1141.

Krumins, S., and Stotzky, G., 1982, Scanning electron microscopy studies of interactions of proteins and polyamino acids with erythroblasts transformed by Friend virus, *Cell Biol. Int. Rep.* **6**:443–453.

Krumins, S., and Stotzky, G., 1983, Protein–membrane interactions: Specific vs. non-specific adsorption and binding of proteins and a polyamino acid on erythroblasts transformed by Friend virus, *Cell Biol. Int. Rep.* **7**:625–635.

Kunc, F., and Stotzky, G., 1974, Effect of clay minerals on heterotrophic microbial activity in soil, *Soil Sci.* **118**:186–195.

Kunc, F., and Stotzky, G., 1977, Acceleration of aldehyde decomposition in soil by montmorillonite, *Soil Sci.* **124**:167–172.

Kunc, F., and Stotzky, G., 1980, Acceleration by montmorillonite of nitrification in soil, *Folia Microbiol. (Prague)* **25**:106–125.

Labeda, D. P., Liu, K.-C., and Casida, L. E., Jr., 1976, Colonization of soil by *Arthrobacter* and *Pseudomonas* under varying conditions of water and nutrient availability as studied by plate counts and transmission electron microscopy, *Appl. Environ. Microbiol.* **31**:551–561.

Lahav, N., 1962, Adsorption of sodium bentonite particles on *Bacillus subtilis*, *Plant Soil* **17**:191–208.

Lailach, G. E., Thompson, T. D., and Brindley, G. W., 1968, Adsorption of pyrimidines, purines, and nucleosides by Li-, Na-, Mg-, and Ca-montmorillonite (clay–organic studies XII), *Clays Clay Miner.* **16**:285–293.

Lavie, S., and Stotzky, G., 1981, Effects of clay minerals on respiration and growth of *Histoplasma capsulatum*, *Abstr. Annu. Meet. Am. Soc. Microbiol.* p. 316.

Lee, K. W., Davey, B. G., and Low, P. F., 1970, Effect of the sol–gel transformation in clay–water systems on biological activity. I. Seed germination and bacterial thermogenesis, *Soil Sci. Soc. Am. Proc.* **34**:45–49.

Lipson, S. M., and Stotzky, G., 1983, Adsorption of reovirus to clay minerals: Effects of cation exchange capacity, cation saturation, and surface area, *Appl. Environ. Microbiol.* **46**:673–682.

Lipson, S. M., and Stotzky, G., 1984a, Adsorption of viruses to particulates: Possible effects on virus survival, in: *Viral Ecology* (A. H. Misra and H. Polasa, eds.), South Asian Publishers, New Delhi, pp. 165–178.

Lipson, S. M., and Stotzky, G., 1984b, Effect of proteins on reovirus adsorption to clay minerals, *Appl. Environ. Microbiol.* **48**:525–530.

Lipson, S. M., and Stotzky, G., 1985a, Infectivity of reovirus adsorbed to homoionic and mixed cation clays, *Water Res.*, in press.

Lipson, S. M., and Stotzky, G., 1985b, Specificity of virus adsorption to clay minerals, *Can. J. Microbiol.*, in press.

Lockhart, N. C., 1980a, Electrical properties and the surface characteristics and structure of clays. I. Swelling clays, *J. Colloid Interface Sci.* **74**:509–519.

Lockhart, N. C., 1980b, Electrical properties and the surface characteristics and structure of clays. II. Kaolinite—a nonswelling clay, *J. Colloid Interface Sci.* **74**:520–529.

Low, P. F., 1960, Viscosity of water in clay systems, *Clays Clay Miner.* **8**:170–182.

Low, P. F., 1961, Physical chemistry of clay–water interactions, *Adv. Agron.* **13**:269–327.

Low, P. F., 1962, Effect of quasi-crystalline water on rate processes involved in plant nutrition, *Soil Sci.* **93**:6–15.

Low, P. F., 1979, Nature and properties of water in montmorillonite–water systems, *Soil Sci. Soc. Am. J.* **43**:651–658.

McCormick, R. W., and Wolfe, D. C., 1979a, Effect of montmorillonite and trace elements on the growth of *Penicillium frequentans*. I. Ammonium nitrogen source, *Soil Sci. Soc. Am. J.* **43**:1114–1120.

McCormick, R. W., and Wolf, D. C., 1979b, Effect of montmorillonite and trace elements on the growth of *Penicillium frequentans*. II. Nitrate nitrogen source, *Soil Sci. Soc. Am. J.* **43**:1120–1124.

McLaren, A. D., 1954, The adsorption and reactions of enzymes and proteins on kaolinite, I, *J. Phys. Chem.* **58**:129–137.

McLaren, A. D., and Skujins, J., 1968, The physical environment of microorganisms in soil, in: *The Ecology of Soil Bacteria* (T. R. G. Gray and D. Parkinson, eds.), Liverpool University Press, Liverpool, pp. 3–24.

McLaren, A. D., Peterson, G. H. and Barshad, I., 1958, The adsorption and reactions of enzymes and proteins on clay minerals. IV. Kaolinite and montmorillonite, *Soil Sci. Soc. Am. Proc.* **22**:239–244.

MacRitchie, F., and Alexander, A. E., 1963, Kinetics of adsorption of proteins at interfaces. III. The role of electrical barriers in adsorption, *J. Colloid Sci.* **18**:464–469.

Macura, J., and Stotzky, G., 1980, Effect of montmorillonite and kaolinite on nitrification in soil, *Folia Microbiol. (Prague)* **25**:90–105.

Mahler, H. R., and Cordes, E. H., 1969, *Basic Biological Chemistry,* Harper & Roe, New York.

Marshall, C. E., 1964, *The Physical Chemistry and Mineralogy of Soils,* Wiley, New York.

Marshall, K. C., 1968, Interaction between colloidal montmorillonite and cells of *Rhizobium* species with different ionogenic surfaces, *Biochim. Biophys. Acta* **156**:179–186.

Marshall, K. C., 1969, Studies by microelectrophoretic and microscopic techniques of the sorption of illite and montmorillonite to rhizobia, *J. Gen. Microbiol.* **56**:301–306.

Marshall, K. C., 1976, *Interfaces in Microbial Ecology*, Harvard University Press, Cambridge, Mass.

Marshall, K. C., 1980a, Adsorption of microorganisms to soils and sediments, in: *Adsorption of Microorganisms to Surfaces* (G. Bitton and K. C. Marshall, eds.), Wiley, New York, pp. 317–330.

Marshall, K. C., 1980b, Bacterial adhesion in natural environments, in: *Microbial Adhesion to Surfaces* (R. C. W. Berkeley, J. M. Lynch, J. Melling, P. R. Rutter, and B. Vincent, eds.), Horwood, Chichester, pp. 187–196.

Marshall, K. C., and Roberts, J. F. J., 1963, Influence of fine particle materials on survival of *Rhizobium trifolii* in sandy soil, Nature (London) **198**:410–411.

Martin, J. P., Filip, Z., and Haider, K., 1976, Effect of montmorillonite and humate on growth and metabolic activity of some actinomycetes, *Soil Biol. Biochem.* **8**:409–413.

Martin, S. M., and Adams, G. A., 1965, A survey of fungal polysaccharides, *Can. J. Microbiol.* **2**:715–719.

Meadows, P. S., and Anderson, J. G., 1966, Microorganisms attached to marine and freshwater sand grains, *Nature (London)* **212**:1059–1060.

Michaels, A. S., 1958, Deflocculation of kaolinite by the alkali polyphosphates, *Ind. Eng. Chem.* **50**:951–958.

Mortland, M. M., 1970, Clay–organic complexes and interactions, *Adv. Agron.* **22**:75–117.

Mysels, K. J., 1967, *Introduction to Colloid Chemistry*, Wiley, New York.

Nanfara, M., and Stotzky, G., 1979, Protection of microorganisms by clay minerals against hypertonic osmotic pressures, *Abstr. Annu. Meet. Am. Soc. Microbiol.* p. 189.

Neihof, R. A., and Loeb, G. I., 1972, The surface charge of particulate matter in seawater, *Limnol. Oceanogr.* **17**:7–16.

Neihof, R. A., and Loeb, G. I., 1974, Dissolved organic matter in seawater and the electric charge of immersed surfaces, *J. Mar. Res.* **32**:5–12.

Neilands, J. B., 1973, Microbial iron transport compounds (siderochromes), in: *Inorganic Biochemistry* (G. L. Eichorn, ed.), Elsevier, Amsterdam, pp. 167–202.

Neilands, J. B., 1981a, Iron absorption and transport in microorganisms, *Annu. Rev. Nutr.* **1**:27–46.

Neilands, J. B., 1981b, Microbial iron compounds, *Annu. Rev. Biochem.* **50**:715–731.

Nikitin, D. I., Vasil'eva, L. V., and Lokhmacheva, R. A., 1966, *New and Rare Forms of Soil Microorganisms*, Nauka, Moscow.

Norrish, K., 1954, The swelling of montmorillonite, *Discuss. Faraday Soc.* **18**:120–134.

Olness, A., and Clapp, C. E., 1972, Microbial degradation of a montmorillonite–dextran complex, *Soil Sci. Soc. Am. Proc.* **36**:179–181.

Osa-Afiana, L. O., and Alexander, M., 1982, Clays and the survival of *Rhizobium* in soil during desiccation, *Soil Sci. Soc. Amer. J.* **46**:285–288.

Oxender, D. L., 1972, Amino acid transport in microorganisms, in: *Metabolic Pathways*, Volume VI (L. E. Hokin, ed.), Academic Press, New York, pp. 133–172.

Parfitt, R. L., and Greenland, D. J., 1970, Adsorption of polysaccharides by montmorillonite, *Soil Sci. Soc. Am. Proc.* **34**:862–866.

Parks, G. A., 1965, The isoelectric points of solid oxides, solid hydroxides, and aqueous hydroxo complex systems, *Chem. Rev.* **65**:177–198.

Parsons, J. W., and Tinsley, J., 1975, Nitrogenous substances, in: *Soil Components*, Volume 1 (J. E. Gieseking, ed.), Springer-Verlag, Berlin, pp. 263–304.

Pedros-Alio, C., and Brock, T. D., 1983, The importance of attachment to particles for planktonic bacteria, *Arch. Hydrobiol.* **98**:354–379.

Peele, T. C., 1936, Adsorption of bacteria by soils, *Cornell Univ. Exp. Sta. Mem.* 197.

Pethica, B. A., 1980, Microbial and cell adhesion, in: *Microbial Adhesion to Surfaces* (R. C. W. Berkeley, J. M. Lynch, J. Melling, P. R. Rutter, and B. Vincent, eds.), Horwood, Chichester, pp. 19–46.

Pett, D. M., Vanaman, T. C., and Joklik, W. K., 1973, Studies on the amino and carboxyl terminal amino acid sequences of reovirus capsid polypeptides, *Virology* **52**:174–186.

Philen, O. D., Jr., Weed, S. B., and Weber, J. B., 1971, Surface charge characterization of layer silicates by competitive adsorption of two organic divalent cations, *Clays Clay Miner.* **19**:295–302.

Pinck, L. A., and Allison, F. E., 1951, Resistance of a protein–montmorillonite complex to decomposition by soil microorganisms, *Science* **114**:131.

Poindexter, J. S., 1981, Oligotrophy: Fast and famine existence, in: *Advances in Microbial Ecology*, Volume 5 (M. Alexander, ed.), Plenum Press, New York, pp. 63–89.

Powell, P. E., Cline, G. R., Reid, C. P. P., and Szaniszlo, P. J., 1980, Occurrence of hydroxamate siderophore iron chelators in soils, *Nature (London)* **287**:833–834.

Powell, P. E., Cline, G. R., Reid, C. P. P., and Szaniszlo, P. J., 1981, Factors affecting the concentration of hydroxamate siderophores in soil solution, *Abstr. Annu. Meet. Am. Soc. Microbiol.* p. 184.

Powell, P. E., Szaniszlo, P. J., and Reid, C. P. P., 1983, Confirmation of occurrence of hydroxamate siderophores in soil by a novel *Escherichia coli* bioassay, *Appl. Environ. Microbiol.* **46:**1080–1083.

Rao, S. R., 1972, *Surface Phenomena,* Hutchinson, London.

Reisinger, O., Fargues, J., Robert, P., and Arnold, M.-F., 1977, Effet de l'argile sur la conservation des micro-organismes. I. Etude ultrastructurale de la biodegradation dans le sol de l'hyphomycete entomopathogene *Beauveria bassiana* (Bals.) Vuill., *Ann. Microbiol.* **128B:**271–287.

Rich, C. I., and Kunze, G. W. (eds.), 1964, *Soil Clay Mineralogy,* University of North Carolina Press, Chapel Hill.

Rogers, H. J., Perkins, H. R., and Ward, J. B., 1980, *Microbial Cell Walls and Membranes,* Chapman & Hall, London.

Roper, M. M., and Marshall, K. C., 1974, Modification of the interaction between *Escherichia coli* and bacterio-phage in saline sediment, *Microb. Ecol.* **1:**1–13.

Roper, M. M., and Marshall, K. C., 1978a, Effect of clay particle size on clay–*Escherichia coli*–bacteriophage interactions, *J. Gen. Microbiol.* **106:**187–189.

Roper, M. M., and Marshall, K. C., 1978b, Effects of a clay mineral on microbial predation and parasitism of *Escherichia coli, Microbiol. Ecol.* **4:**279–289.

Rosan, B., Appelbaum, B., Campbell, L. K., Knox, K. W., and Wicken, A. J., 1982, Chemostat studies of the effect of environmental control on *Streptococcus sanguis* adherence to hydroxyapatite, *Infect. Immun.* **35:**64–70.

Rosenzweig, W. D., and Stotzky, G., 1979, Influence of environmental factors on antagonism of fungi by bacteria in soil: Clay minerals and pH, *Appl. Environ. Microbiol.* **38:**1120–1126.

Rosenzweig, W. D., and Stotzky, G., 1980, Influence of environmental factors on antagonism of fungi by bacteria in soil: Nutrient levels, *Appl. Environ. Microbiol.* **39:**354–360.

Rutter, P. R., and Vincent, B., 1980, The adhesion of micro-organisms to surfaces: Physicochemical aspects, in: *Microbial Adhesion to Surfaces* (R. C. W. Berkeley, J. M. Lynch, J. Melling, P. R. Rutter, and B. Vincent, eds.), Horwood, Chichester, pp. 79–92.

Santoro, T., and Stotzky, G., 1967a, Effect of electrolyte composition and pH on the particle size distribution of microorganisms and clay minerals as determined by the electrical sensing zone method, *Arch. Biochem. Biophys.* **122:**664–669.

Santoro, T., and Stotzky, G., 1967b, Influence of cations on flocculation of clay minerals by microbial metabolites as determined by the electrical sensing zone particle analyzer, *Soil Sci. Soc. Am. Proc.* **31:**761–765.

Santoro, T., and Stotzky, G., 1967c, Effect of cations and pH on the electrophoretic mobility of microbial cells and clay minerals, *Bacteriol. Proc.* A15.

Santoro, T., and Stotzky, G., 1968, Sorption between microorganisms and clay minerals as determined by the electrical sensing zone particle analyzer, *Can. J. Microbiol.* **14:**299–307.

Santoro, T., Stotzky, G., and Rem, L. T., 1967, The electrical sensing zone particle analyzer for measuring germination of fungal spores in presence of other particles, *Appl. Microbiol.* **15:**935–939.

Scher, F. M., and Baker, R. R., 1982, Effect of *Pseudomonas putida* and a synthetic iron chelator on induction of soil suppressiveness to *Fusarium* wilt pathogens, *Phytopathology* **72:**1567–1573.

Schiffenbauer, M., and Stotzky, G., 1982, Adsorption of coliphages T1 and T7 to clay minerals, *Appl. Environ. Microbiol.* **43:**590–596.

Schiffenbauer, M., and Stotzky, G., 1983, Adsorption of coliphages T1 and T7 to host and nonhost microbes and to clay minerals, *Curr. Microbiol.* **8:**245–259.

Schiffenbauer, M., and Stotzky, G., 1984, Adsorption and desorption of coliphages T1 and T7 to and from kaolinite and montmorillonite and the lytic capabilities of clay–coliphage complexes, *Can J. Microbiol.,* submitted for publication.

Schnitzer, M., and Khan, S. U., 1972, *Humic Substances in the Environment,* Dekker, New York.

Schnitzer, M., and Kodama, H., 1977, Reactions of minerals with soil humic substances, in: *Minerals in Soil Environments* (J. B. Dixon and S. B. Weed, eds.), Soil Science Society of America, Madison, Wisc., pp. 741–770.

Schroth, M. N., and Hancock, J. G., 1982, Disease-suppressive soil and root-colonizing bacteria, *Science* **216:**1376–1381.

Shields, P. A., and Farrah, S. R., 1983, Influence of salts on electrostatic interactions between poliovirus and membrane filters, *Appl. Environ. Microbiol.* **45:**526–531.

Skinner, F., 1956, The effect of adding clays to mixed cultures of *Streptomyces albidoflavus* and *Fusarium culmorum, J. Gen. Microbiol.* **14:**393–405.

Smith, J. B., and Berry, D. R., 1974, *An Introduction to Biochemistry of Fungal Development,* Academic Press, New York.

Sørensen, L. H., 1972, Stabilization of newly-formed amino acid metabolites in soil by clay minerals, *Soil Sci.* **114:**5–11.

Sørensen, L. H., 1975, The influence of clay on the rate of decay of amino acid metabolites synthesized in soil during decomposition of cellulose, *Soil Biol. Biochem.* **7:**171–177.

Soulides, D. A., 1969, Antibiotic tolerance of the soil microflora in relation to type of minerals, *Soil Sci.* **107:**105–107.

Sposito, G., 1981, The operational definition of zero point of charge in soils, *Soil Sci. Soc. Am. J.* **45:**292–297.

Stent, G. S., 1963, *Molecular Biology of Bacterial Viruses,* Freeman, San Francisco.

Stevenson, F. J., 1982, *Humus Chemistry,* Wiley, New York.

Stotzky, G., 1965, Replica plating technique for studying microbial interactions in soil, *Can. J. Microbiol.* **11:**629–636.

Stotzky, G., 1966a, Influence of clay minerals on microorganisms. II. Effect of various clay species, homoionic clays, and other particles on bacteria, *Can. J. Microbiol.* **12:**831–848.

Stotzky, G., 1966b, Influence of clay minerals on microorganisms. III. Effect of particle size, cation exchange capacity, and surface area on bacteria, *Can. J. Microbiol.* **12:**1235–1246.

Stotzky, G., 1967, Clay minerals and microbial ecology, *Trans. N.Y. Acad. Sci. II* **20:**11–21.

Stotzky, G., 1970, Further observations on the apparent relation between clay mineralogy and geographic distribution of human pathogens in soil, *10th Int. Congr. Microbiol.* (abstract).

Stotzky, G., 1971, Ecologic eradication of fungi—Dream or reality?, in: *Histoplasmosis, Proceedings of the Second National Conference* (M. L. Furcolow and E. W. Chick, eds.), Thomas, Springfield, Ill., pp. 477–486.

Stotzky, G., 1973, Techniques to study interactions between microorganisms and clay minerals *in vivo* and *in vitro,* *Bull. Ecol. Res. Comm. (Stockholm)* **17:**17–28.

Stotzky, G., 1974, Activity, ecology, and population dynamics of microorganisms in soil, in: *Microbial Ecology* (A. I. Laskin and H. Lechevalier, eds.), Chemical Rubber Co., Cleveland, pp. 57–135.

Stotzky, G., 1980, Surface interactions between clay minerals and microbes, viruses, and soluble organics, and the probable importance of these interactions to the ecology of microbes in soil, in: *Microbial Adhesion to Surfaces* (R. C. W. Berkeley, J. M. Lynch, J. Melling, P. R. Rutter, and B. Vincent, eds.), Horwood, Chichester, pp. 231–249.

Stotzky, G., and Babich, H., 1980, Physicochemical factors that affect the toxicity of heavy metals to microbes in aquatic habitats, in: *Proceedings of the ASM Conference on Aquatic Microbial Ecology* (R. R. Colwell and J. Foster, eds.), University of Maryland, College Park, pp. 181–203.

Stotzky, G., and Babich, H., 1984, Fate of genetically-engineered microbes in natural environment, *Recomb. DNA Tech. Bull.* **7:**167–191.

Stotzky, G., and Burns, R. G., 1982, The soil environment: Clay–humus–microbe interactions, in: *Experimental Microbial Ecology* (R. G. Burns and J. H. Slater, eds.), Blackwell, Oxford, pp. 105–133.

Stotzky, G., and Krasovsky, V. N., 1981, Ecological factors that affect the survival, establishment, growth, and genetic recombination of microbes in natural habitats, in: *Molecular Biology, Pathogenicity, and Ecology of Bacterial Plasmids* (S. B. Levy, R. C. Clowes, and E. L. Koenig, eds.), Plenum Press, New York, pp. 31–42.

Stotzky, G., and Martin, R. T., 1963, Soil mineralogy in relation to the spread of *Fusarium* wilt of banana in Central America, *Plant Soil* **18:**317–338.

Stotzky, G., and Mortensen, J. L., 1957, The effect of crop residue and nitrogen additions on the decomposition of an Ohio muck soil, *Soil Sci.* **83:**165–174.

Stotzky, G., and Mortensen, J. L., 1958, Effect of addition level and maturity of rye tissue on the decomposition of a muck soil, *Soil Sci. Soc. Am. Proc.* **22:**521–524.

Stotzky, G., and Post, A. H., 1967, Soil mineralogy as possible factor in geographic distribution of *Histoplasma capsulatum, Can. J. Microbiol.* **13:**1–7.

Stotzky, G., and Rem, L. T., 1966, Influence of clay minerals on microorganisms. I. Montmorillonite and kaolinite on bacteria, *Can. J. Microbiol.* **12:**547–563.

Stotzky, G., and Rem, L. T., 1967, Influence of clay minerals on microorganisms. IV. Montmorillonite and kaolinite on fungi, *Can. J. Microbiol.* **13:**1535–1550.

Stotzky, G., and Schenck, S., 1976, Volatile organic compounds and microorganisms, *Crit. Rev. Microbiol.* **4:**333–382.

Stotzky, G., Dawson, J. E., Martin, R. T., and ter Kuile, G. H. H., 1961, Soil mineralogy as a factor in the spread of *Fusarium* wilt of banana, *Science* **133:**1483–1485.

Stotzky, G., Schiffenbauer, M., Lipson, S. M., and Yu, B. H., 1981, Surface interactions between viruses and clay minerals and microbes: Mechanisms and implications, in: *Viruses and Wastewater Treatment* (M. Goddard and M. Butler, eds.), Pergamon Press, Elmsford, N.Y., pp. 199–204.

Sutherland, I. W., 1980, Polysaccharides in the adhesion of marine and freshwater bacteria, in: *Microbiol Adhesion to Surfaces* (R. C. W. Berkeley, J. M. Lynch, J. Melling, P. R. Rutter, and B. Vincent, eds.), Horwood, Chichester, pp. 329–338.

Sutherland, I. W., 1983, Microbial exopolysaccharides—Their role in microbial adhesion in aqueous systems, *Crit. Rev. Microbiol.* **10:**173–200.

Swartzen-Allen, L. S., and Matijevic, E., 1974, Surface and colloid chemistry of clays, *Chem. Rev.* **74:**385–400.

Sykes, I. K., and Williams, S. T., 1978, Interactions of actinophage and clays, *J. Gen. Microbiol.* **108:**97–102.

Tadros, T. F., 1980, Particle–surface adhesion, in: *Microbial Adhesion to Surfaces* (R. C. W. Bekeley, J. M. Lynch, J. Melling, P. R. Rutter, and B. Vincent, eds.), Horwood, Chichester, pp. 93–116.

Taylor, D. H., 1981, Interpretation of the adsorption of viruses by clays from their electrokinetic properties, in: *Chemistry of Water Reuse* (W. J. Cooper, ed.), Ann Arbor Science, Ann Arbor, Mich., pp. 595–612.

Theng, B. K. G., 1974, *The Chemistry of Clay–Organic Reactions,* Wiley, New York.

Theng, B. K. G., 1979, *Formation and Properties of Clay–Polymer Complexes,* Elsevier, Amsterdam.

Vandenbergh, P. A., Gonzales, C. F., Wright, A. M., and Kunka, B. S., 1983, Iron-chelating compounds produced by soil pseudomonads: Correlation with fungal growth inhibition, *Appl. Environ. Microbiol.* **46:**128–132.

van Olphen, H., 1977, *An Introduction to Clay Colloid Chemistry,* 2nd ed., Wiley, New York.

van Olphen, H., and Fripiat, J. J. (eds.), 1979, *Data Handbook for Clay Materials and Other Non-Metallic Minerals,* Pergamon Press, Elmsford, N.Y.

Ward, J. B., and Berkeley, R. C. W., 1980, The microbial cell surface and adhesion, in: *Microbial Adhesion to Surfaces* (R. C. W. Berkeley, J. M. Lynch, J. Melling, P. R. Rutter, and B. Vincent, eds.), Horwood, Chichester, pp. 47–66.

Wardell, J. N., Brown, C. M., and Ellwood, D. C., 1980, A continuous culture study of the attachment of bacteria to surfaces, in: *Microbial Adhesion to Surfaces* (R. C. W. Berkeley, J. M. Lynch, J. Melling, P. R. Rutter, and B. Vincent, eds.), Horwood, Chichester, pp. 221–230.

Weinberg, S. R., and Stotzky, G., 1972, Conjugation and genetic recombination of *Escherichia coli* in soil, *Soil Biol. Biochem.* **4:**171–180.

White, R. E., 1979, *Introduction to the Principles and Practice of Soil Science,* Wiley, New York.

Yu, B. H., and Stotzky, G., 1979, Adsorption and binding of herpes-virus hominis type 1 (HSV 1) by clay minerals, *Abstr. Annu. Meet. Am. Soc. Microbiol.* p. 188.

Zvyaginstsev, D., 1973, *Interaction between Microorganisms and Solid Surfaces,* Moscow University Press, Moscow.

9

Mechanisms of Bacterial Adhesion to Plant Surfaces

ANN G. MATTHYSSE

1. INTRODUCTION

1.1. General Scope

In comparison to the large amount of information available on mechanisms of bacterial adhesion to animal cells, relatively little is known about bacterial adhesion to the plant cell surface. Indeed, the role of adhesion in symbiotic and pathogenic interactions with plants is still not completely understood. The most detailed studies of bacterial adhesion to plant cells have been made with three systems—the interaction of rhizobia with legume roots resulting in the formation of nitrogen-fixing root nodules, the interaction of agrobacteria with dicot cells resulting in the formation of crown gall tumors, and the interaction of several species of phytopathogenic gram-negative bacteria with plants (which are not susceptible hosts for the bacteria) resulting in a hypersensitive response by the plant. Although occasional studies have been made on other bacterial interactions with the plant cell surface and a considerable literature exists on the microbial ecology of the surface of leaves and roots (Foster, 1981; Blakeman, 1982), these three examples are the only ones about which sufficient information is available to permit discussion of the mechanisms of adhesion; this chapter will consider only these three cases. The consequences for plant cells of bacterial adhesion are considered in Chapter 14.

1.2. Comparison of Some Aspects of Bacterial Adhesion to Plant and Animal Cells

Attachment of bacteria to plant cells differs in several respects from attachment of bacteria to animal cells. Plant cells have a substantial cell wall; thus, attachment of bacteria, unless accompanied by degradation of the host cell wall, does not bring the bacterial cells into contact with the host cell plasmalemma. (The surface of the plant cell is described in Chapter 3.) In general, bacterial attachment to plant cells appears to be slower than bacterial

Ann G. Matthysse • Department of Biology, University of North Carolina, Chapel Hill, North Carolina 27514.

attachm   *desiccation 干燥*   m attachment of *Agrobacterium tumefaciens* to
carrot t   *hypersensitive*   it 20 min of incubation (Matthysse *et al.*, 1981).
Bacteri   *过敏的*   dom includes the entire bacterial population and
may be   *plasmalemma*   b of the bacterial inoculum is attached (Bal *et al.*,
1978;   *原膜*   do not possess circulatory or digestive systems
similar to those   ot need to adhere to the plant cell surface to avoid
being carried with the bulk of the fluid. Bacteria on the surface of a plant must adhere to
avoid removal by wind or rain. By contrast, bacteria in the intercellular spaces of the
mesophyll of a leaf need not adhere to the surface of the cells to remain in the area. However,
bacteria may adhere to cell surfaces in such places to avoid desiccation. Adhesion to the
surface of conducting tissue in the phloem or xylem would prevent bacteria from being
carried with the stream of fluid. Fluid movement in these tissues is much slower, nevertheless, than in the bloodstream. Thus, the role of adhesion in most bacterial diseases of
plants is not obvious and is a subject of much current research.

1.3. Methods of Studying Bacterial Adhesion to Plant Cells

The attachment of bacteria to the plant cell surface has been studied using several
different techniques. The most direct method is microscopy. Interaction of bacteria with
living plant surfaces has been observed with the light microscope using living material,
particularly Fåhraeus's (1957) slide technique for observing the interaction of aseptic roots of
small legumes with *Rhizobium*, and the interaction of tissue culture cells with *Agrobacterium*
(Matthysse *et al.*, 1978; Douglas *et al.*, 1982). Fixed materials have been studied in the light
microscope and in the scanning and transmission electron microscopes. The location, appearance, and number of attached bacteria are generally recorded as is any apparent host cell
response. For attached bacteria to be visible in the microscope, relatively high bacterial
concentrations (10^5 to 10^{10} bacteria/ml) usually must be used.

Numbers of attached bacteria have been assessed with a procedure in which radiolabeled bacteria are bound to the plant surface. Unbound bacteria are removed by washing
of roots or leaves or by filtration of suspension culture cells (Chen and Phillips, 1976;
Ohyama *et al.*, 1979; Suhayda and Goodman, 1981). This technique generally also requires
rather large numbers of bacteria. In the case of *Agrobacterium* where attached bacteria
remain viable, the numbers of free and attached bacteria in suspensions of bacterial and
cultured plant cells can be estimated with viable cell counts, after free bacteria are separated
by filtration from plant cells and attached bacteria. The plant cells are homogenized before
plating for the viable cell count. This technique allows the use of small concentrations of
bacteria (10^3 bacteria/ml) (Matthysse *et al.*, 1978). Attempts have been made to estimate by
viable count the numbers of free and attached bacteria in the intercellular spaces of leaves
after the free bacteria are washed out. The procedures give minimum estimates of free
bacteria in the leaf tissue; many difficulties are inherent in removing free bacteria from tissue
with numerous convolutions in which the bacterial cells can be trapped (Atkinson *et al.*,
1981).

Bacterial attachment to the plant cell surface has also been measured with indirect
assays. If *Agrobacterium* must attach to a specific site on the host plant surface in order to
induce tumors, then anything that occupies those sites will prevent the bacteria from inducing
the tumors. The number of tumors formed is measured with either a pinto bean leaf assay or a
potato disc assay. In these assays, the number of tumors formed increases with increasing
numbers of bacteria inoculated (Lippincott and Heberlein, 1965; Anand and Heberlein,
1977; Glogowski and Galsky, 1978). When heat-killed virulent *A. tumefaciens* or certain

avirulent strains of *A. tumefaciens* are inoculated onto wounded bean leaves prior, but not subsequent, to the inoculation of virulent *A. tumefaciens,* the dead or avirulent bacteria inhibit tumor formation by the virulent bacteria (Lippincott and Lippincott, 1969). This inhibition is dependent on the relative numbers of virulent and avirulent bacteria inoculated in a manner consistent with the hypothesis that the inhibition is due to the occupation of receptor sites on the plant cell surface by the avirulent bacteria (Lippincott and Lippincott, 1969). This assay suffers from the obvious difficulty that it is indirect. There may be steps between bacterial attachment and the appearance of visible tumors which could be affected by the dead or avirulent bacteria. However, *Agrobacterium* normally infects wound sites in whole plants; wound sites are difficult material for direct microscopic assays of attachment. The binding of *Agrobacterium* to plant cells in liquid suspension cultures can be observed directly in the microscope or estimated with viable counts or with radiolabeled bacteria as described above. Although the interaction leads to the formation of crown gall tumor cells, there is no guarantee that the bacteria interacting with tissue culture cells are identical to the bacteria interacting with wound sites. Nevertheless, results of studies of attachment of *A. tumefaciens* using the indirect and direct assays with tissue cultures are generally similar, suggesting that the bacteria interact with tissue culture cells and with wound sites on leaves in similar ways. Most studies of attachment of agrobacteria use one of these systems.

Rhizobia appear to interact with roots at the surface of root hairs formed by epidermal cells slightly behind the growing tip of the root. As root hairs are formed only by epidermal cells of roots and not by tissue culture cells, studies of the attachment of rhizobia to the root hair surface generally utilize whole roots or plants grown under aseptic conditions.

The interactions of incompatible bacteria with plants which lead to the hypersensitive response and of compatible bacteria which lead to disease have been studied most frequently by infiltrating or injecting bacterial suspensions into the intercellular space in leaves. Working with this type of material is difficult; therefore, most studies rely heavily on microscopic observations of fixed material. Recently, some researchers have utilized tissue culture cells which provide obvious advantages for biochemical studies. It is by no means certain, however, that the hypersensitive or susceptible response as observed in the leaf can be mimicked in tissue culture. These responses may depend upon the presence of organized tissue. In particular, the restriction of bacterial growth to the site of inoculation seen in the hypersensitive response is difficult to obtain in callus cultures growing on agar and generally is not obtained with suspension cultures. The tendency in these situations is for the bacteria to take over the culture probably because they grow so much faster than the plant cells.

The use of various species and strains of bacteria and species and cultivars of plants has also aided in the study of bacterial attachment to plant cells. The study of bacterial mutants with altered attachment properties is just beginning and holds great promise for the elucidation of the mechanisms of attachment. Bacterial mutants with altered attachment properties have been difficult to obtain, because only a fraction of the wild-type bacterial population typically attaches to the plant cell surface. As a consequence, no direct method is available for selecting for attachment mutants. The attachment mutants which exist have been obtained largely by screening populations mutated in some character other than adhesion for mutants with defects in attachment. This procedure is rather laborious.

In many studies, closely related species of bacteria which give different responses in the plant host have been compared. The interactions of rhizobia belonging to different cross-inoculation groups with the same plant are often compared, as are the interactions of compatible and incompatible species of *Pseudomonas, Xanthomonas,* or *Erwinia* on one host plant.

Although it is difficult to obtain bacterial mutants with altered attachment properties, it is far more difficult to obtain mutant plants with altered responses to the bacteria. In some cases different cultivars of the same species of plant differ in their response to *Rhizobium* or to *Pseudomonas, Xanthomonas,* or *Erwinia.* The study of these plants can be helpful in understanding the surface interactions between the bacteria and the plant. In the case of *Agrobacterium,* for example, some dicots are not susceptible to tumor formation by some strains or biotypes of the bacteria; in those examples which have been examined, however, the block to tumor formation is at a later stage than bacterial attachment (Matthysse and Gurlitz, 1982; Binns *et al.,* 1982; Kao *et al.,* 1982). The molecular genetic analysis of plants, or at least of dicots, should become possible when *A. tumefaciens* can be used to insert selected genes into the plant chromosomes. Such analysis may aid considerably in elucidating the genetic control of the response of the plant host to potentially pathogenic or symbiotic bacteria.

2. INTERACTIONS OF RHIZOBIA WITH THE ROOT SURFACE

2.1. The Infection Process

The initiation by rhizobia of nitrogen-fixing nodules on the roots of leguminous plants involves a complex series of interactions (see also Chapter 14, Section 2). These events appear to be similar for interactions of many rhizobia with temperate-zone legumes, and have been characterized microscopically. Perhaps the best studied is the interaction of *Rhizobium trifolii* with *Trifolium repens* (white clover) (Callaham and Torrey, 1981). The bacteria are inoculated into clover plants growing in Fåhraeus (1957) slide cultures, and bind to the surface of root hairs. By 5 hr postinoculation the originally straight root hairs begin to curl at the tip or branch along their length. At 16–24 hr the root hairs are fully deformed, but no large accumulations of attached rhizobia are seen. Infection threads generally arise from areas where there are tight folds in the root hair created by curling or branching or accidental apposition of one root hair with another. In these regions the rhizobia apparently digest the host cell wall and form an entry site into the host cell (Callaham and Torrey, 1981). *Rhizobium* has been shown to possess pectolytic and cellulose- and hemicellulose-degrading enzymes (Hubbell *et al.,* 1978; Martinez-Molina *et al.,* 1979; Chapter 3, Sections 2.1.1 and 2.2.1). These enzymes may hydrolyze the root hair cell wall in a localized region where bacteria are attached. The host cell responds by synthesizing a new thin layer of cell wall over the advancing zone of hydrolysis (Callaham and Torrey, 1981). Thus, the infection thread is formed. The infection thread with proliferating bacteria within it develops along the root hair cell and then invades the cortex of the root where the events leading to the formation of a nitrogen-fixing root nodule take place (Callaham and Torrey, 1981; Dazzo and Hubbell, 1981).

The interaction of rhizobia with legumes exhibits a considerable degree of species specificity. Typically one species of *Rhizobium* will form nodules on only one or a few species of legumes. Thus, *R. trifolii* infects and nodulates roots of *Trifolium* (clover) but not alfalfa or beans. *R. phaseoli* infects and nodulates beans (*Phaseolus*) but not clover or alfalfa. It appears that this host specificity is generally expressed at a relatively early stage in the infection process prior to the formation of the infection thread (Napoli and Hubbell, 1975). Considerable effort has been expended in attempts to elucidate the mechanism of this host specificity.

Although the rhizobia which nodulate clover, alfalfa, beans, peas, etc. are distinct from each other, they are rather closely related and can be grouped together as the "fast-growing" rhizobia. The "slow-growing" rhizobia such as *Rhizobium japonicum* which nodulates soybeans appear to be a genetically distinct group, although as far as is known the initial stages of nodule formation in soybeans proceed as described above for clover (Turgeon and Bauer, 1982). In soybeans only cells of a particular age in the growing root are susceptible to nodulation. These are epidermal cells which initiate root hairs shortly after the inoculation of the bacteria (Bhuvaneswari *et al.*, 1980). Bacteria attach to these newly formed root hairs which then curl. Infection threads are formed in the tightly curled root hairs. The infection threads grow down the root hairs and penetrate the host root cortex where the nitrogen-fixing root nodule is formed (Turgeon and Bauer, 1982).

2.2. Role of Bacterial Adhesion in the Formation of Nodules

The host specificity observed in the interaction of rhizobia with legumes could result from either the failure of the nonnodulating rhizobia to attach to the root hair surface or the failure of attached rhizobia to initiate an infection thread (Li and Hubbell, 1969). Non-nodulating mutants of rhizobia may be blocked at either stage. Thus, the results of different studies have given different answers to the question of the stage involved in host specificity. Chen and Phillips (1976) found that *Rhizobium leguminosarum* did not bind in greater numbers to pea roots than did other species of *Rhizobium*. They also observed that *R. leguminosarum* bound to root hairs of species such as clover and *Medicago* which these bacteria do not nodulate. On the other hand, Jansen van Rensburg and Strijdom (1982) found that a greater proportion of the rhizobia present are bound more tightly to hosts in which they can form nodules than to hosts in which they do not form nodules. Thus, at the present time it is unclear which stage in the infection process is primarily responsible for host specificity.

It appears likely, however, that rhizobia must bind to root hairs in order to initiate infection threads and nodules. Mutants of *R. meliloti* and *R. japonicum* which do not bind to alfalfa or soybean root hairs, respectively, do not form nodules (Hirsch *et al.*, 1982; Stacey *et al.*, 1982). These mutants were isolated as nodulation-minus mutants, however, and then shown not to bind to root hairs, rather than vice versa. Therefore, the evidence is not strong that the bacteria must adhere to the root hairs to be able to induce nodules.

2.3. Role of Plant Lectins in Adhesion of Rhizobia to Root Hairs

2.3.1. R. trifolii and Clover

The hypothesis that plant lectins play an important role in the binding of rhizobia to the root hair surface has received the most support from studies of the interaction of *R. trifolii* with white clover. Lectins are proteins or glycoproteins with multiple sugar-binding sites. The sugar(s) bound by a particular lectin is usually quite specific; binding of a particular lectin can generally be inhibited by including the specific mono- or oligosaccharide in the reaction mixture (Goldstein and Hayes, 1978). Clover contains a lectin, trifoliin, which binds specifically to *R. trifolii* and is found on clover root hair tips as well as elsewhere in the plant (Dazzo *et al.*, 1978). The binding of *R. trifolii* to clover root hairs proceeds in two stages (Dazzo, 1981). In the first stage, the bacteria attach at the tips and then along the length of root hairs. After about 12 hr of incubation, attached bacteria and fibrillar material, probably bacterial cellulose, are seen to cover more extensive areas on the root surface (Dart, 1971; Napoli *et al.*, 1975; Dazzo, 1981). The sugar, 2-deoxyglucose, specifically inhibits the initial stage of binding of *R. trifolii* to clover root hairs; it has no effect on the binding of *R. meliloti* to alfalfa (Dazzo *et al.*, 1976). Binding of purified trifoliin to *R. trifolii* and *R.*

trifolii capsular polysaccharide to clover root hairs are also inhibited by 2-deoxyglucose, suggesting that trifoliin and *R. trifolii* capsular polysaccharide are involved in the binding of the bacteria to clover root hairs. In addition, transformation of *Azotobacter vinelandii* with DNA from *R. trifolii* gives rise to transformants which have acquired the ability simultaneously to bind both trifoliin and clover root hairs. This binding is inhibited by 2-deoxyglucose (Bishop *et al.,* 1977; Dazzo and Brill, 1979).

Antibody prepared against clover roots cross-reacts with *R. trifolii.* The reaction of this antibody with the bacteria is inhibited by 2-deoxyglucose (Dazzo and Brill, 1979). Both trifoliin and this cross-reactive antibody appear to react with similar sugars on the surface of clover roots and of *R. trifolii.* Dazzo and Hubbell (1975) have proposed a model to explain these results in which the clover lectin serves as a bridging molecule to link the cross-reactive antigens on the surface of the *R. trifolii* to those on the surface of the clover root hair.

The location and biochemistry of the *R. trifolii* saccharides which bind trifoliin have been examined. This has not been an easy task; rhizobia typically secrete large quantities of extracellular and capsular polysaccharides. Trifoliin reacts with both a purified acidic capsular polysaccharide and the lipopolysaccharide of *R. trifolii* (Dazzo and Brill, 1979). The reaction of the lectin with both of these polysaccharides is inhibited by 2-deoxyglucose.

The ability of *R. trifolii* to bind to clover roots and to bind trifoliin varies with the age of the bacterial culture; cells in the early stationary phase of growth show the most binding. The composition and binding to trifoliin of the bacterial lipopolysaccharide vary in parallel with culture age (Hrabek *et al.,* 1981). The clover host possesses enzymes which degrade the bacterial capsular polysaccharide and destroy its ability to bind to trifoliin (Dazzo *et al.,* 1982). For reasons which are as yet unclear, these enzymes apparently tend first to remove the capsular polysaccharide from the middle of the bacteria, leaving the trifoliin-binding activity at one or both poles of the bacteria. Dazzo *et al.* (1982) have proposed that this reaction is responsible for the observation that, while rhizobia appear initially to be attached in a random orientation to the root hair surface, after a long incubation most of the bacteria are attached in a polar orientation.

Rhizobia do not form root nodules or infection threads on plants growing in combined nitrogen. The bacteria also do not bind to clover root hairs of plants grown in 15 mM nitrate. The decrease in bacterial binding to root hairs with increasing levels of nitrate from 1 to 15 mM is paralleled by a decrease in the amount of trifoliin detected (Dazzo and Brill, 1978). The ability of clover roots to bind exogeneous trifoliin also decreases when the roots are grown in nitrate (Dazzo, 1981). Thus, it appears that neither the bacterial nor the plant component of the binding interaction is constitutive and that the expression of these components is regulated by the growth conditions of the organisms.

2.3.2. R. japonicum and Soybeans

Soybean lectin labeled with flourescein isothiocyanate binds to *R. japonicum,* but not to other species of rhizobia which do not nodulate soybeans (Bohlool and Schmidt, 1974). This lectin binds galactose and *N*-acetylgalactosamine; both of these compounds inhibit the binding of *R. japonicum* to soybean root hairs (Stacey *et al.,* 1980). The detection of soybean lectin on the root hair surface, however, has proved to be a matter of some controversy (Stacey *et al.,* 1980; Gade *et al.,* 1981; Pueppke *et al.,* 1981; Bhuvaneswari, 1981). Some inbred lines of soybeans lack detectable seed lectin (Orf *et al.,* 1978). These soybeans have an insertion sequence in the major seed lectin that effectively prevents synthesis of this protein (Goldberg *et al.,* 1983). However, *R. japonicum* forms normal root nodules on these plants (Pull *et al.,* 1978). Ananlysis of the DNA of these plants reveals a second lectinlike gene the function of which is unknown (Goldberg *et al.*, 1983). Thus, the major seed lectin of soybeans must not be required for nodulation by *R. japonicum*. The possible role of other minor lectins is uncertain.

Although most infective strains of *R. japonicum* bind soybean lectin, some strains do not (Bohlool and Schmidt, 1974; Pueppke *et al.*, 1980). There are also strains of *R. japonicum* which synthesize lectin-binding capsules only when soybean roots are present (Bhuvaneswari and Bauer, 1978; Law *et al.*, 1982). Bal *et al.* (1978) have suggested that for one strain of *R. japonicum* only a small fraction (less than 1%) of the bacterial population may synthesize capsule material, bind soybean lectin, and bind to soybean root hairs; nevertheless, this bacterial strain does form nodules on soybeans.

The binding site for soybean lectin on the surface of *R. japonicum* appears to be both capsular and extracellular polysaccharide (Bal *et al.*, 1978; Mort and Bauer, 1980). Although some bacterial mutants which synthesize less extracellular polysaccharide may still retain the ability to nodulate soybean, it is unclear whether these mutants continue to synthesize particular extracellular or capsular polysaccharides required for bacterial binding to root hairs while failing to synthesize other polysaccharides (Sanders *et al.*, 1981). Two mutants of *R. japonicum* have been isolated which fail to bind to soybean roots and which do not nodulate soybeans (Stacey *et al.*, 1982). The exact nature of the defect in these mutants is not yet known; they should prove to be helpful, however, in elucidating the bacterial components required for binding to soybean roots.

As is the case for *R. trifolii*, the ability of *R. japonicum* to bind to soybean lectin depends on the stage of growth of the bacterial culture. Cells in the early and mid-log phases of growth bind the most lectin (Bhuvaneswari *et al.*, 1977). The ability of the cells to bind soybean lectin disappears as 4-O-methylgalactose replaces galactose in the capsular polysaccharide (Mort and Bauer, 1980). Soybean lectin and galactose residues in the bacterial capsular or exopolysaccharides are the best candidates for the molecules involved in the binding of *R. japonicum* to soybean roots. However, the conclusion that these are the plant receptor and bacterial binding site is far from certain.

2.3.3. R. leguminosarum and Pea, and R. meliloti and Alfalfa

Much less information is available on the mechanism of bacterial adhesion to the root hair surface and the possible role of lectins for rhizobia other than *R. trifolii* and *R. japonicum*. Pea lectin binds 3-O-methyl-D-glucose and methyl-D-mannopyranoside and both of these sugars inhibit the binding of *R. leguminosarum* to pea roots (Kato *et al.*, 1981). Pea lectin can be found on the surface of seedling pea roots but is more difficult to detect on older roots (Rougé and Labroue, 1977; Kijne *et al.*, 1980). The bacterial binding site for pea lectin appears to be the capsular polysaccharide (Kato *et al.*, 1981).

A protein which agglutinates *R. meliloti* but not other rhizobia has been purified from alfalfa seeds (Paau *et al.*, 1981). Nonnodulating mutants of *R. meliloti* which fail to bind to alfalfa root hairs also fail to be agglutinated by this protein. Nevertheless, nonnodulating mutants which do bind to root hairs are agglutinated by this protein. Thus, the protein is strongly implicated as being important in the binding of *R. meliloti* to alfalfa roots. The nature of the mutations which cause the bacteria to fail to bind to alfalfa roots is not yet known (Hirsch *et al.*, 1982).

Thus, the role of plant lectins in the specific binding of rhizobia to root hairs remains unclear. The best documented case is that of *R. trifolii* which appears to be bound to clover root hairs via the clover lectin, trifoliin. The role of soybean lectin, as described above, is uncertain, and we have only fragmentary, although suggestive, information regarding the lectins of other species of legumes and their interaction with rhizobia.

2.4. Genetics of Attachment of *Rhizobium*

Although the rhizobia were originally divided into species on the basis of their ability to form nodules on particular species of legumes, recent research makes it appear that this distinction does not reflect a fundamental genetic difference between some of these *Rhizobium* species. The genes involved in the nodulation of legumes by *R. leguminosarum, R. trifolii, R. phaseoli,* and *R. meliloti* are located on large plasmids (100–500 kb in size) (Benyon *et al.*, 1980; Hirsch *et al.*, 1980; Banfalvi *et al.*, 1981; Hooykaas *et al.*, 1981; Rosenberg *et al.*, 1981). The ability of *R. leguminosarum* to nodulate peas can be transferred to *R. trifolii* or *R. phaseoli*, which ordinarily form nodules only on clover or beans, respectively, by transfer of one of the plasmids found in *R. leguminosarum* (Johnston *et al.*, 1978). The ability to form nodules on particular cultivars of peas which are not nodulated by most strains of *R. leguminosarum* can also be transferred with a plasmid from a strain able to nodulate the cultivar to a nonnodulating mutant of *R. leguminosarum* (Brewin *et al.*, 1980). The genes responsible for host range of *R. leguminosarum* have been cloned as a 10-kb fragment of the plasmid DNA. This DNA fragment when transferred to *R. phaseoli* confers on it the ability to nodulate peas (Downie *et al.*, 1983). Thus, the genes in these rhizobia which determine host range appear to be located on a plasmid and to be rather few in number. The chromosomes of these fast-growing rhizobia seem to be somewhat homologous (Ausubel, 1982). It is difficult, therefore, to determine from these experiments the extent of the chromosomal component involved in the nodulation process.

Nevertheless, the genes required for the initial binding of *R. trifolii* to clover roots can be expressed in other chromosomal backgrounds as well. The ability to bind trifoliin and to bind to clover roots can be transferred by DNA transformation from *R. trifolii* to *Azotobacter vinelandii* (Bishop *et al.*, 1977; Dazzo and Brill, 1979) and by plasmid conjugation to *Agrobacterium tumefaciens* (Dazzo *et al.*, 1983). This suggests that at least the portion of the host range genes required for bacterial binding can be expressed in diverse chromosomal backgrounds. When the *R. leguminosarum* plasmid containing host range genes was transferred to *R. japonicum*, a member of the slow-growing group of rhizobia, however, the plasmid apparently was not expressed (Sadowsky and Bohlool, 1983). This result is consistent with the apparent genetic distinctiveness of the fast-growing and slow-growing rhizobia.

2.5. Model for Attachment of *Rhizobium* to the Root Hair Surface

Given our present information we can construct a model for the attachment of rhizobia to the root hair surface. This model is, of necessity, incomplete and probably incorrect in some respects. Moreover, as we have the most information about *R. trifolii* and binding to clover root hairs, any model will naturally be slanted toward that particular interaction.

The first step in nodule formation by rhizobia is the binding of the bacteria to the root hair surface. This binding involves polysaccharides on the surface of the bacterium; for various species these may include extracellular polysaccharide, capsular polysaccharide, and lipopolysaccharide. The particular carbohydrate required for bacterial binding may be present only transiently during the growth of the bacteria and may be synthesized only in the presence of the appropriate host plant. The receptor on the plant root hair surface is a protein, most probably a lectin, which binds the bacterial polysaccharide. The plant lectin may also bind polysaccharides on the surface of the root hair forming a bridge between the root hair and the bacterium. The plant protein receptor is also not constitutive, and its synthesis may be repressed by combined nitrogen. The bacteria are originally bound in a random orientation but in some cases the plant may possess enzymes which degrade the bacterial polysaccharide-binding site leaving only a residue at the ends of the bacteria, thus resulting in polar bacterial binding. The bound bacteria are viable; they can divide to form microcolonies. They synthesize fibrils probably composed of cellulose which may aid them in adhering firmly to the root hair surface. The bacteria also synthesize enzymes which degrade the root hair cell wall and thus begin the formation of the infection thread by a process of bacterial degradation of the host cell wall balanced by the host's attempt to synthesize new cell wall in the area of degradation.

3. INTERACTIONS OF AGROBACTERIA WITH THE PLANT CELL SURFACE

3.1. Bacterial Induction of Crown Gall Tumors

Infection of wound sites on dictoyledonous plants by *Agrobacterium tumefaciens* results in the formation of crown gall tumors (see also Chapter 14, Section 3). A wound is required for entrance of the bacteria on most dicots, although some plants can be infected via natural openings such as lenticles. The host range of the bacterium is wide including most dicots and some gymnosperms (Lippincott and Lippincott, 1975). Naturally occurring and laboratory mutant strains of *A. tumefaciens* exist, however, which have a more limited host range (Anderson and Moore, 1979; Loper and Kado, 1979; Thomashow *et al.*, 1980b). Except for a few specific cases, the reasons for the limited host range are not known (Kao *et al.*, 1982).

Once they have entered the wound site, the bacteria attach to the host cell surface. This attachment is probably required for tumor formation. The prior but not subsequent inoculation of avirulent strains of the bacteria inhibits tumor formation in a fashion consistent with a hypothesis that receptor sites are occupied by the avirulent bacteria, thus blocking attachment of the virulent bacteria (Lippincott and Lippincott, 1969). When avirulent transposon mutants of *A. tumefaciens* are examined, some of these mutants are found to be unable to attach to dicot cells (Douglas *et al.*, 1982). That these mutants are in fact altered in a single gene (or operon) was shown in one case by marker exchange of the DNA segment carrying the transposon into a virulent strain of *A. tumefaciens*. The transposon and the avirulent and nonattaching phenotypes were all transferred together (Douglas *et al.*, 1982). In addition, revertants of the avirulent nonattaching mutants selected for recovery of the ability to bind to dicot cells also recover virulence at the same time (Matthysse, unpublished observation). It appears likely, therefore, that bacterial attachment is required for tumor formation by *A. tumefaciens*.

Once the bacteria have attached to the plant cell surface, bacterial Ti (tumor inducing) plasmid DNA is transferred to the host cell. A portion of this plasmid DNA, the T region, becomes integrated into host chromosomal DNA where it is transcribed and is responsible for maintaining the tumor phenotype (Matthysse and Stump, 1976; Chilton *et al.*, 1977; Gurley *et al.*, 1979; Thomashow *et al.*, 1980a). Very little information is available on the mechanism of DNA transfer from the bacterium to the plant cell or on the subsequent steps of DNA integration.

3.2. General Characteristics of Binding of *A. tumefaciens* to Plant Cells

Two different methods are in current use for measuring the attachment of *A. tumefaciens* to plant host cells. The first and older method is indirect, and involves assaying how various treatments interfere with tumor formation on wounded bean leaves (or other tissues). The number of tumors formed on bean leaves is a linear function of the number of bacteria inoculated in the range of 10^7 and 10^9 bacteria/ml (Lippincott and Heberlein, 1965). The inoculation of avirulent or heat-killed *Agrobacterium* prior to or simultaneously with inoculation of virulent *A. tumefaciens* interferes with tumor formation. Adding the same avirulent or heat-killed bacteria subsequent to adding the virulent bacteria has no effect on tumor formation (Lippincott and Lippincott, 1969).

As an assay for factors involved in attachment, the number of tumors eventually formed has been estimated after various treatments of the bacteria, or the effects of adding various substances prior or subsequent to adding the virulent bacteria to the wounded bean leaves have been compared (Lippincott and Lippincott, 1976). In this system, binding to the receptor on the plant cell surface is specific for agrobacteria. All virulent strains of *A. tumefaciens* bind as do some avirulent strains. Some avirulent strains of *A. radiobacter* also bind. *Escherichia coli*, *Pseudomonas aeruginosa*, *P. savastoni*, *Corynebacterium fasciens*, and *Bacillus megaterium* do not bind to the plant receptor site (Lippincott and Lippincott, 1969). Thus, this assay of bacterial binding is specific for those bacteria which might be expected to bind to the receptor of *A. tumefaciens*.

The second method used to measure the attachment of *A. tumefaciens* to plant cells is a direct one. Bacteria are inoculated into a suspension culture of plant cells; the resulting attachment is either observed directly in the microscope (Matthysse *et al.*, 1978, 1981; Ohyama *et al.*, 1979; Douglas *et al.*, 1982) or measured using viable cell counts of the bacteria (Matthysse *et al.*, 1978). The specificity of each bacterial strain for attaching as assessed with the direct assay is generally the same as seen with the indirect assay (Matthysse *et al.*, 1978). Bacterial attachment to tissue culture cells appears to be rather slow; half-maximum attachment is reached only after about 20 to 60 min of incubation, depending upon the bacterial strain and species of plant used (Matthysse *et al.*, 1978, 1981). Bacterial attachment to wound sites as measured using the indirect assay appears to be more rapid; avirulent bacteria added 15 min after inoculating the virulent bacteria do not interfere with tumor formation (Lippincott and Lippincott, 1969).

A number of substances appear to have no effect on attachment of agrobacteria as examined with either assay method. No sugar is known which inhibits the attachment, nor do lectins inhibit attachment in general, suggesting that unlike *Rhizobium*, *Agrobacterium* attachment does not involve plant lectins (Lippincott and Lippincott, 1976; Ohyama *et al.*, 1979; Gurlitz *et al.*, 1984). Divalent cations also are not required for attachment of agrobacteria to tissue culture cells (Ohyama *et al.*, 1979; Gurlitz *et al.*, 1984). High ionic strength (0.25 M NaCl) also does not inhibit their attachment to such cells, suggesting that hydroxyproline-rich glycoproteins are not involved in the attachment (Mellon and Helgeson, 1982; Gurlitz *et al.*, 1984; Chapter 3, Sections 2.3 and 2.4.2).

The direct and indirect assays yield differing results in tests of the ability of pectin or polygalacturonic acid to inhibit bacterial attachment and of the ability of certain avirulent strains of *A. tumefaciens* to bind to plant cells. The explanation for the first difference is unclear at present and will be discussed further below. The second difference appears to be explained by differing bacterial concentrations used in the two assays (10^8 or more bacteria/ml in the indirect assay and 10^3 to 10^4 bacteria/ml in the direct assay). At least one avirulent strain of *A. tumefaciens* (NTI) binds to plant cells only at high bacterial concentrations when binding is measured using the direct assay (Matthysse and Lamb, 1981).

3.3. Plant Receptors for Binding of *A. tumefaciens*

The plant cell appears to play a passive role in the binding of *A. tumefaciens*. Bacteria bind to heat-killed or glutaraldehyde-fixed carrot suspension culture cells with unaltered strain specificity and only slightly altered kinetics (Matthysse *et al.*, 1981). This finding suggests that the receptor sites on the plant cell surface to which the bacteria attach are constitutive rather than induced by some interaction with the bacteria. Thus, it is possible to examine the effects on attachment of treatments of the plant cell surface prior to adding the bacteria, or the effects on bacterial induction of tumors of treatments of the bacteria with various extracts of the plant cell surface.

The inhibition of tumor formation on wounded bean leaves by the pectin fraction of plant cell walls has been studied extensively. Addition of pectin to the wound site prior, but not subsequent, to inoculating virulent bacteria inhibits tumor formation (Lippincott and Lippincott, 1980). However, plant cell membrane preparations have no effect on tumor formation (Lippincott and Lippincott, 1976). The inhibitory effect of the cell wall preparations can be reversed by pretreating them with a heat-killed strain of *A. tumefaciens* capable of blocking tumor formation by virulent strains, and presumably able to bind to the plant receptor site. Treatment of the inhibitory cell wall preparations by strains of *Agrobacterium* which do not bind to the plant receptor site has no effect on the inhibitory activity of the preparations (Lippincott *et al.*, 1977). Treatment of plant cell wall preparations with pectinase, macerase, cellulase, EDTA and Triton X-100, or sodium desoxycholate does not affect their inhibitory activity. In fact, hot acid (0.05 M H_2SO_4 at 100°C) is required to destroy the inhibitory activity of these preparations (Lippincott and Lippincott, 1980). Treatment of cell wall preparations with pectin methyltransferase also destroys their inhibitory activity; this reaction can be reversed by treatment with pectin esterase. Cell wall preparations from crown gall tumor cells, monocots, or embryonic dicot tissues do not inhibit tumor formation on wounded bean leaves by virulent *A. tumefaciens*. These cell wall preparations become inhibitory after treatment with pectin esterase (Lippincott and Lippincott, 1976; Rao *et al.*, 1982).

These observations are interpreted by the authors to suggest that pectin may be the receptor (or a portion of the receptor) on the plant cell surface. However, other interpretations are possible. The assay used to measure bacterial attachment to the plant cell surface is indirect; thus, treatments that interfere with steps in tumor formation subsequent to bacterial attachment may be confused with treatments which interfere with bacterial attachment per se. Plant cell wall fractions can have effects of their own when inoculated into plant tissues. Hahn *et al.* (1981) have shown that certain plant cell wall fractions can act as elicitors of phytoalexins. The effects of phytoalexins on *Agrobacterium* and on crown gall tumor formation are not known.

When a more direct assay for bacterial attachment involving suspension culture cells is used, pectin or polygalacturonic acid has no effect on attachment of virulent *A. tumefaciens* to carrot cells (Matthysse *et al.*, 1982). In fact, *A. tumefaciens* binds to carrot cell protoplasts from which all detectable cell wall components have been removed. The bacteria bind both to living protoplasts and to protoplasts fixed with glutaraldehyde immediately after preparation to prevent cell wall resynthesis. Attachment to protoplasts shows the same *Agrobacterium* strain specificity as attachment to intact cells (Matthysse *et al.*, 1982). It is unclear whether the receptor on the intact plant cell and the receptor on plant protoplasts are different or identical. Microscopic observations of bacteria attaching to plasmalysed plant cells show bacteria attached to the cell wall in regions in which the plasmalemma is no longer in contact with the plant cell wall; thus, there may be receptors for *A. tumefaciens* located in the plant cell wall itself as well as on the surface of protoplasts (Gurlitz *et al.*, 1984).

The receptor for the binding of *A. tumefaciens* to intact carrot suspension culture cells can be removed by brief treatment of the cells with detergent, $CaCl_2$, or proteolytic enzymes (Gurlitz *et al.*, 1984). These treatments do not damage the carrot cells irreversibly; if the cells are returned to their culture medium, they recover the ability to bind *A. tumefaciens* after about 6 hr. This recovery is inhibited by cycloheximide (Gurlitz *et al.*, 1984). These results suggest that a protein is one component of the receptor on the carrot cell surface. When the material extracted from the carrot cell surface with detergent or $CaCl_2$ is examined by gel electrophoresis, more than 30 distinct polypeptides are found to be present. At the present time it is not clear which, if any, of these peptides may be the receptor for binding *A. tumefaciens* (Gurlitz *et al.*, 1984).

Agrobacterium does not induce tumors on monocots or embryonic dicots (Lippincott and Lippincott, 1975; Matthysse *et al.*, 1982). Moreover, as assessed with the indirect assay, cell wall preparations from monocots and embryonic dicots do not inhibit tumor formation (Lippincott and Lippincott, 1978). As determined with the direct assay, *A. tumefaciens* does not bind to monocot tissue culture cells nor do the bacteria bind to carrot tissue culture cells 24 hr after they have been induced to form embryos. [At that time the carrot cells are less than two cell divisions into the process of embryo formation (Matthysse and Gurlitz, 1982).] These results suggest that one factor limiting the host range of *A. tumefaciens* may be a receptor which binds the bacteria to the surface of the plant host cell.

3.4. The Bacterial Binding Site

Although the plant cell is apparently a passive partner in the binding of *A. tumefaciens* to tissue culture cells, the bacterium appears to play a more active role. Live bacteria bind to dead carrot cells; dead bacteria do not bind to living or dead carrot cells (Matthysse *et al.*, 1981). However, bacterial protein synthesis is not required for attachment to carrot cells (Matthysse *et al.*, 1983).

In contrast to the observation that heat-killed bacteria fail to bind to tissue culture cells, such bacteria do interfere with tumor formation by live virulent bacteria (Lippincott and Lippincott, 1969). The reasons are unknown why live bacteria are required for attaching to tissue culture cells and are not necessary for the inhibition of tumor formation. As bacterial protein synthesis is not required in either case, the bacterial binding site for attaching to the plant cell surface must be made constitutively and is not induced by the plant cells.

The binding site appears to be located on the surface of the bacteria. Live bacterial spheroplasts show a time delay in attaching to carrot cells, similar to the time required for the bacteria to lose their osmotic fragility, and thus presumably to resynthesize their cell wall and outer membrane (Matthysse *et al.*, 1983). Treatment of the bacteria with trypsin also delays bacterial attachment, although for a shorter time than does removing the cell wall and outer membrane in the preparation of bacterial spheroplasts (Matthysse *et al.*, 1983).

Bacterial lipopolysaccharide (LPS; Chapter 2, Section 3.3.1) purified from virulent strains of *A. tumefaciens* inhibits the binding of *A. tumefaciens* to carrot and tobacco tissue culture cells (Matthysse *et al.*, 1978) and interferes with tumor formation on bean leaves (Whatley *et al.*, 1976). LPS purified from an avirulent *A. radiobacter* strain, which does not itself bind to tissue culture cells or inhibit tumor formation on bean leaves, has no effect on binding of virulent *A. tumefaciens* to tissue culture cells or on tumor formation (Whatley *et al.*, 1976; Matthysse *et al.*, 1978). Avirulent mutants of *A. tumefaciens* which fail to bind to tissue culture cells show altered phage absorption. The chemical basis of this alteration is not known (Douglas *et al.*, 1982). The results suggest that the site for binding bacteria to plant cells is constitutive and contains LPS and possibly also proteins.

The genetics of the ability of *A. tumefaciens* to bind to the plant cell surface are not well understood at the present time. Some avirulent strains of *A. radiobacter* which lack the Ti plasmid do not bind to plant cells (Lippincott and Lippincott, 1969; Matthysse *et al.*, 1978).

When a Ti plasmid is introduced into these strains, they become virulent and acquire the ability to bind to plant cells (Whatley *et al.*, 1978). These results suggest that some genes required for bacterial attachment are located on the Ti plasmid. When various strains of *A. tumefaciens* are cured of their Ti plasmids, however, they may still bind to plant cells (Whatley *et al.*, 1978; Matthysse, 1983b). One cured strain (NT1) shows a dependence on bacterial concentration; the bacteria bind to tissue culture cells when present at high numbers and fail to bind when present at low numbers (Matthysse and Lamb, 1981). Avirulent transposon mutants of virulent strain C58 have been isolated which fail to bind to tissue culture cells. These mutants all appear to be chromosomal mutants (Douglas *et al.*, 1982; Matthysse, unpublished observations).

Thus, the genetics of bacterial attachment appear to be confusing. Some attachment genes are apparently chromosomal while others are found on the Ti plasmid. The distribution of attachment genes between the chromosome and the Ti plasmid may vary with different strains of *A. tumefaciens*.

3.5. Role of Cellulose Fibrils in Attachment of *A. tumefaciens*

Observations with the scanning electron microscope of *A. tumefaciens* bound to tissue culture cells reveal fibrils surrounding the attached bacteria and covering the surface of the tissue culture cells (Fig. 1). These fibrils are visible as early as 90 min after the bacteria are added to carrot suspension culture cells. By 19 hr of incubation, the fibrils form networks that surround the plant cell surface and entrap additional bacteria, attaching them indirectly

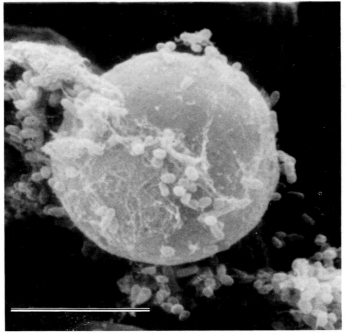

Figure 1. Scanning electron micrograph of *A. tumefaciens* strain A6 incubated with killed carrot protoplasts for 8 hr. The bacteria are surrounded by fibrillar material. Large clusters of bacteria held together by fibrils are visible. Most of the bacteria in these clusters are only indirectly attached to the plant cell surface. Some individually attached bacteria are visible. Scale bar: 10 μm.

to the plant cell surface. Eventually the bacteria and fibrils form a large mass linking all the plant cells in the culture flask together into a few large aggregates (Matthysse *et al.*, 1981).

Similar fibrils have been observed in transmission electron microscope studies of *A. tumefaciens* attaching to *Datura* tissue culture cells and in scanning electron microscope studies of *A. tumefaciens* attaching to zinnia cells (Ohyama *et al.*, 1979; Douglas *et al.*, 1982). The fibrils are also seen when *A. tumefaciens* is attached to tobacco callus cultures and leaves. However, the number of fibrils present appears to be inversely related to the wetness of the environment; many fibrils are present when the organisms are attached to tobacco suspension culture cells; the fewest fibrils are seen with the bacteria attached to tobacco leaves. Bacteria attached to tobacco callus show an intermediate number of fibrils (Deasey and Matthysse, 1983).

These fibrils are synthesized by the bacteria rather than the plant cells; they are made even when the microorganisms adhere to heat-killed or glutaraldehyde-fixed carrot cells. The bacteria can be induced to synthesize fibrils, in the absence of the plant cells, by including 0.02% soytone in the bacterial culture medium. Fibrils have been purified from these cultures. They are resistant to digestion by proteases and 2 N trifluoroacetic acid at 121°C for 3 hr. The fibrils show a fluorescent stain with calcafluor which stains β-linked polysaccharides such as chitin and cellulose. Treatment with 6 N HCl at 100°C or with purified cellulase digests the fibrils. The sole product of prolonged digestion is glucose. Intermediate digestion times with cellulase produce both cellobiose and glucose. The fibrillar material has been purified and its infrared spectrum compared with that of cellulose. The two spectra are almost identical (Matthysse *et al.*, 1981). Thus, the bacterial fibrils appear to be composed of cellulose. Synthesis of such fibrils can be regarded as a clever strategy on the part of the

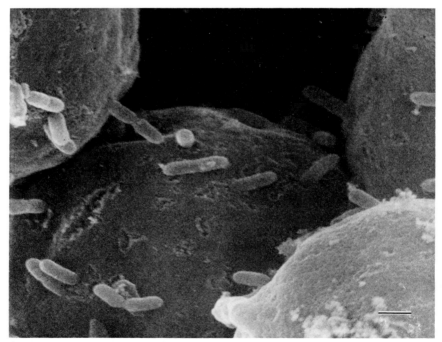

Figure 2. Scanning electron micrograph of a cellulose-minus mutant of *A. tumefaciens* strain ACH_5C_3 incubated with killed carrot protoplasts for 19 hr. Note the absence of fibrillar material surrounding the bacteria. Scale bar: 1 μm.

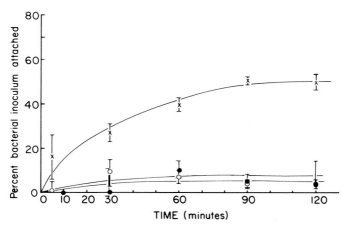

Figure 3. Time course of attachment of *A. tumefaciens* strain A6 to carrot suspension culture cells. X, untreated carrot cells; ○, carrot cells treated with 0.1% trypsin for 60 min prior to addition of the bacteria; ●, carrot cells treated with 0.1% chymotrypsin for 60 min prior to addition of the bacteria.

bacterium: even should the plant be able to recognize these cellulose fibrils as foreign material, it would be unable to digest them without damaging its own cell wall.

The role of the cellulose fibrils in infection has been studied with *A. tumefaciens* transposon mutants which were constructed to be unable to synthesize cellulose fibrils. These mutants synthesize no cellulose detectable by calcafluor staining. They do not form aggregates as do the parent wild-type bacteria when grown in the presence of 0.02% soytone. The mutant bacteria attach to the surface of carrot suspension culture cells, but do not form large aggregates on the plant cell surface as does the parent strain. In the scanning electron microscope the mutant bacteria are seen to be attached individually. No fibrillar material is visible (Fig. 2). Wild-type strains form large aggregates of carrot cells after they are incubated with the plant cells for 1 to 2 days. The mutant bacteria never form such aggregates. When cellulose fibril synthesis is absent, the kinetics of attachment also are altered. The parent strain shows a biphasic attachment curve which never saturates (Fig. 3). The mutant bacteria show only the first phase of the attachment curve, suggesting that the second phase is due to indirect attachment of bacteria to fibrils synthesized by directly attached bacteria (Matthysse, 1983a).

Cellulose synthesis is not required for bacterial virulence under laboratory conditions. The cellulose-minus mutants induce tumors on tobacco, *Bryophyllum,* and beans. For most of the mutants, only cellulose-minus bacteria can be recovered from the tumors, indicating that the mutants themselves and not cellulose-positive revertants had caused the tumors. The ability of the parent strain to produce tumors on *Bryophyllum* leaves is not affected by washing the inoculation site with water 2 hr after inoculating the bacteria. However, the ability of the cellulose-minus bacteria to induce tumors is much reduced by washing the inoculation site with water (Matthysse, 1983a). Revertants of the cellulose-minus mutants which regain the ability to synthesize cellulose regain at the same time all of the other properties of the parent strain. Thus, a major role of the cellulose fibrils synthesized by *A. tumefaciens* appears to be to anchor large numbers of bacteria to the surface of the host cells, thereby preventing bacterial removal by factors such as rain, and aiding in the production of tumors.

3.6. Attachment of *A. rhizogenes* to Plant Cells

The studies described above all refer to the attachment of *A. tumefaciens* to plant host cells. Another species of *Agrobacterium, A. rhizogenes,* is also pathogenic in plants, causing hairy root disease. The pathogenic mechanism of *A. rhizogenes* is similar to that of *A. tumefaciens* and involves transfer of plasmid DNA of the Ri (root inducing) plasmid from the bacterium to the plant host cell (Chilton *et al.,* 1982). Physical contact between the bacterium and the plant host is required for development of the disease (Moore *et al.,* 1979).

These observations might suggest that the attachment of *A. rhizogenes* to carrot suspension culture cells would have similar properties to the attachment of *A. tumefaciens.* However, this does not appear to be the case. *A. rhizogenes* makes very few cellulose fibrils. The attachment to plant cells of the organism is much weaker than that of *A. tumefaciens* and can be disrupted by sheer forces such as in vortexing a mixture of the bacteria and carrot cells. Only a small fraction of the *A. rhizogenes* population (less than 1%) is apparently able to attach to carrot cells. These bacteria appear to bind to the carrot cell surface, however, as assessed by both light and electron microscopy (Council, 1983). Thus, even rather closely related bacteria with similar pathogenic mechanisms may differ with respect to the mechanism by which they adhere to the host cell surface.

3.7. Model for Attachment of *A. tumefaciens* to the Plant Cell Surface

Using the information just detailed, we can construct a model of the attachment of *A. tumefaciens* to the plant cell surface (Fig. 4). Both the bacterial binding site and the plant receptor site are constitutive and exist prior to any interaction between the bacteria and the plant cells. The bacteria bind to a receptor on the plant cell surface. The plant receptor is composed of protein and possibly pectin. The binding site is located on the bacterial cell surface, probably on the outer membrane, and is composed of LPS and protein. Once the

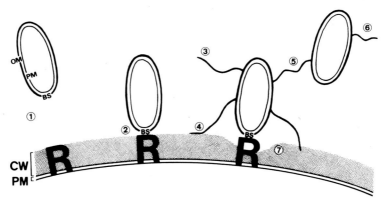

Figure 4. A model for the attachment of *A. tumefaciens* to carrot cells. *A. tumefaciens* with a constitutive binding site on the bacterial outer membrane (OM) encounters a plant cell with a receptor site (R) on its surface (step 1). The bacterial binding site (BS) attaches to the plant receptor site (R) (step 2). The attached bacterium begins to elaborate cellulose fibrils (step 3). These fibrils anchor the bacteria to the surface of the plant cell wall (CW) (step 4) and entrap additional bacteria (step 5). These entrapped bacteria may themselves begin to synthesize cellulose fibrils (step 6) which may entrap additional bacteria. This facilitated attachment and multiplication of attached bacteria result in the development of a bacterial colony attached to the plant cell wall via binding sites and cellulose fibrils. The attached bacteria secrete enzymes such as pectinase which digest the plant cell wall (step 7) and may aid in establishing contact of the bacteria with the plant cell plasmalemma (PM) to facilitate transfer of the Ti plasmid from the bacteria to the plant host cell.

bacteria have bound to the plant cell surface, they begin to elaborate cellulose fibrils. These fibrils serve to anchor the bacteria firmly to the host cell surface and are responsible for the extreme difficulty encountered in attempts to remove bound *A. tumefaciens* from the plant cell surface. Some of the fibrils are free in the medium and entrap and entangle free bacteria, thus binding them indirectly to the plant cell surface. The entrapment of free bacteria and the division of attached and entrapped bacteria result in the formation of large bacterial aggregates on the plant cell surface. The bacteria may also possess enzymes such as pectinase (Matthysse *et al.*, 1981) which alter the host cell wall and allow a closer association of the bacteria with the host cell. Thus, although not all of the steps in the adhesion of *A. tumefaciens* to its host cells are known, the adhesion can be seen to be a complicated multistep rather than single simple process.

4. INTERACTIONS OF *PSEUDOMONAS, ERWINIA,* AND *XANTHOMONAS* WITH THE PLANT CELL SURFACE

4.1. Bacterial Attachment and the Hypersensitive Response

The majority of the bacterial pathogens of plants need not attach to host cells in order to cause disease. Indeed, evidence suggests that attachment of these bacteria to the plant cell surface may be associated with induction of hypersensitive responses and limitation of bacterial growth. A potential pathogen interacting with a plant leading to the development of disease is referred to as a *compatible interaction*. The interaction of a potential pathogen with a plant which does not lead to disease is referred to as an *incompatible interaction*. The interaction of a plant with a nonpathogenic bacterium is referred to as a *saprophytic interaction*. Incompatible interactions are often, but not always, associated with a hypersensitive response on the part of the plant (see Chapter 3, Section 4; Chapter 14, Section 4). This response is characterized by restriction of the bacteria to the site of inoculation, limitation of bacterial growth, changes in the plant cell plasmalemma including leakage of electrolytes, and localized necrosis of the plant cells at the site of bacterial inoculation (Klement, 1982).

Many investigators have published electron micrographs showing the association of the hypersensitive response with the attachment of incompatible bacteria to the host cell surface (Goodman *et al.*, 1976; Sequeira *et al.*, 1977; Politis and Goodman, 1978; Roebuck *et al.*, 1978; Cason *et al.*, 1978; Smith and Mansfield, 1982). Compatible bacteria show no such attachment and are free in the intercellular paces (Sequeira *et al.*, 1977; Smith and Mansfield, 1982). Some investigators feel that this adhesion of incompatible bacteria to the host cell surface simply represents the drying of the infiltrated or injected bacterial suspensions (Hildebrand *et al.*, 1980; Daub and Hagedorn, 1980; Al-Issa and Sigee, 1982; Fett and Jones, 1982). Compatible bacteria might not show such drying as the lesions produced by compatible bacteria often have a water-soaked appearance. Extracellular polysaccharide from compatible, but not from incompatible, bacteria injected into bean leaves has been reported to induce formation of water-soaked lesions (El-Banoby and Rudolph, 1979).

In those cases in which incompatible bacteria are observed on the host cell surface, fibrillar or amorphous granular material is often seen covering the microbial cells (Goodman *et al.*, 1976; Sequeira *et al.*, 1977; Politis and Goodman, 1978; Cason *et al.*, 1978). It is unclear whether this material is of host or bacterial cell origin. Attached bacteria may be viable, and in at least one case have been observed to divide and form microcolonies. The growth of incompatible bacteria always ceases, however, at a lower number of cells than does that of compatible bacteria (Essenberg *et al.*, 1979). Living bacteria are not required for bacterial attachment. Heat-killed *P. solanacearum* appear in the electron microscope to be attached to the surface of tobacco mesophyll cells (Sequeira *et al.*, 1977). However, dead bacteria do not induce a hypersensitive response.

The role in the hypersensitive response of attached bacteria and indeed the whole question of whether injected incompatible bacteria are actually attached to the surface of leaf mesophyll cells remain unclear. In one experiment, incompatible bacteria which would oridinarily give rise to a hypersensitive response were injected into leaves in an agar suspension which would prevent the attachment of these bacteria to the mesophyll cells. The bacteria multiplied and gave rise to a lesion characteristic of the susceptible rather than the hypersensitive response in the plant leaf (Stall and Cook, 1979).

A promising approach to this problem would probably be analysis for their adhesion properties of bacterial mutants which fail to induce a hypersensitive response, and analysis for their ability to induce a hypersensitive response of mutants which fail to attach to plant cells. At the present time, a few bacterial mutants are available including one which is temperature-sensitive for the induction of the hypersensitive response. None of these mutants has been characterized sufficiently, however, to allow any conclusions to be drawn (Keen *et al.*, 1981; Anderson and Mills, 1983).

4.2. Role of Plant Proteins Which Agglutinate Bacteria

A number of proteins have been isolated from plants which are capable of agglutinating bacteria (Anderson and Jasalavich, 1979; Fett and Sequeira, 1980; Romeiro *et al.*, 1981a). One class of the most interesting of these is the hydroxyproline-rich glycoproteins (see Chapter 3, Section 2.3). These proteins are secreted by wounded plant tissue and are found on the surface of parenchymal cells, xylem vessels, epidermal cells, and collenchyma cells (Stuart and Varner, 1980; Leach *et al.*, 1982b). They are produced by a variety of plants including potatoes, tobacco, peas, and carrots. They contain about 60% carbohydrate by weight. The protein portion is made up largely of hydroxyproline (about 40%), but is also rich in serine and lysine, giving the molecules a net basic charge (Stuart and Varner, 1980; Leach *et al.*, 1982a). These proteins bind to a wide variety of microorganisms including avirulent nonencapsulated *P. solanacearum* and fungi such as *Phytophthora* (Mellon and Helgeson, 1982; Leach *et al.*, 1982a). They do not bind to virulent encapsulated *P. solanacearum*. Their location and binding properties suggest that they may be involved in the binding of incompatible *P. solanacearum* to the host cell surface (Leach *et al.*, 1982b). The binding of these proteins to bacteria is apparently a charge binding; it is inhibited by high ionic strength (Mellon and Helgeson, 1982). The role of these interesting proteins in the possible adhesion of incompatible bacteria to the host cell surface and in the response of plants to potential pathogens remains to be determined.

A factor which agglutinates avirulent *Erwinia amylovora* lacking extracellular polysaccharide has been isolated from apple tissue (Romeiro *et al.*, 1981a). This material apparently reacts with the O-antigen portion of the LPS of bacterial mutants which have the outer membrane exposed due to a lack of extracellular polysaccharide (Romeiro *et al.*, 1981b). This agglutination reaction apparently prevents the spread of the bacteria in the xylem and the resulting development of fire blight disease (Huang *et al.*, 1975).

Little is known of the possible role in bacterial adhesion of any of the other bacterial agglutinins isolated from plants. This is an area worthy of further investigation.

4.3. Adhesion of Bacteria to the Outer Surface of the Plant

Bacteria are known to adhere to the outer surfaces of plants (Foster, 1981; Blakeman, 1982; Chapter 14). In some cases, these adherent bacteria can cause considerable damage to the plant host. A good example of this phenomenon is the production of frost injury on leaves or fruits by ice-nucleation active bacteria usually belonging to the species *P. syringae* or *E. herbicola* (Lindow *et al.*, 1982). However, the mechanism by which these bacteria attach to the plant surface remains largely unknown although obviously of considerable significance from an agricultural point of view. There is also little information available on the mechanisms of adhesion of the microflora to root surfaces.

5. CONCLUSIONS

Little information is available on the mechanism of adhesion to the plant cell surface for the majority of phytopathogenic and symbiotic bacteria. The two genera which have been examined in detail, *Rhizobium* and *Agrobacterium,* may not be typical. These bacteria are closely related to each other. Their symbiotic and pathogenic mechanisms require the attachment of the bacteria to the plant cell surface. It is not clear whether bacterial adhesion is a component of bacterial pathogenesis or host resistance for other phytopathogens. This question is an important one which needs further study.

The mechanism of adhesion of *Rhizobium* to the root hair surface involves a polysaccharide on the surface of the bacterium (either exopolysaccharide, capsular polysaccharide, or LPS depending on the species) and a protein on the surface of the root hair. In the case of *R. trifolii* this protein is a lectin, trifoliin. In the case of other *Rhizobium* species the nature of the plant receptor site is less clear. Soybean cultivars lacking the major seed lectin are, nonethelss, nodulated by *R. japonicum*. This may reflect the involvment of a nonlectin receptor for *R. japonicum,* of a minor lectin, or of an alternative binding pathway used when the major lectin is absent. Neither the plant receptor nor the bacterial binding site is constitutive; binding of *Rhizobium* to root hairs is dependent on the growth conditions of both the plant and the bacteria. Both the plant and the bacteria are capable of synthesizing enzymes to modify the surface of the opposite partner. The plant produces enzymes which degrade bacterial exopolysaccharide and the bacteria produce enzymes which degrade the host cell wall. Attached bacteria are viable and may divide forming microcolonies. The bacteria may also synthesize cellulose fibrils which could aid in their firm attachment to the host cell surface.

In considering the confusion surrounding the plant receptor for the binding of rhizobia, it must be remembered that closely related bacteria, e.g., *A. tumefaciens* and *A. rhizogenes,* may show significant differences in their attachment mechanisms. Thus, the attempt to find one mechanism involving plant lectins which accounts for the adhesion of all rhizobia may be of limited use.

Unlike rhizobia which typically have a rather narrow host range, agrobacteria generally have a very broad host range. Both the plant receptor site and the bacterial binding site are constitutive. The plant receptor is composed of protein and possibly pectin; the bacterial binding site is composed of protein and LPS. Once the bacteria have attached to the host cell surface, they synthesize cellulose fibrils which anchor the bacteria to the plant surface and prevent their removal by such factors as rainstorms. The bacteria produce enzymes which may degrade the host cell wall and thus aid in DNA transfer across the host plasmalemma.

Attempts to elucidate the mechanisms of attachment of these bacteria to their plant hosts have made clear that for both rhizobia and agrobacteria, adhesion to the plant host cell surface is not a simple one-step process. Instead, bacterial adhesion appears to be a multistep reaction involving considerable "cross-talk" between the bacterium and the plant cell.

REFERENCES

Al-Issa, A. N., and Sigee, D. C., 1982, The hypersensitive reaction in tobacco leaf tissue infiltrated with *Pseudomonas pisi*. 1. Active growth and division in bacteria entrapped at the surface of mesophyll cells, *Phytopathol. Z.* **104**:104–114.

Anand, V. K., and Heberlein, G. T., 1977, Crown gall tumorigenesis in potato tuber tissue, *Am. J. Bot.* **64**:153–158.

Anderson, A. B., and Moore, L. W., 1979, Host specificity in the genus *Agrobacterium*, *Phytopathology* **69**:320–323.

Anderson, A. J., and Jasalavich, C., 1979, Agglutination of pseudomonad cells by plant products, *Physiol. Plant Pathol.* **15**:149–159.

Anderson, D., and Mills, D., 1983, Tn 5 transposon mutagenesis of two phytopathogenic *Pseudomonas syringae* pathovars, *Phytopathology* **73**:824.

Atkinson, M. M., Huang, J. S., and Van Dyke, C. G., 1981, Adsorption of pseudomonads to tobacco cell walls and its significance to bacterium–host interactions, *Physiol. Plant Pathol.* **18**:1–5.

Ausubel, F. M., F. M., 1982, Molecular genetics of symbiotic nitrogen fixation, *Cell* **29**:1–2.

Bal, A. K., Shantharam, S., and Ratnam, S., 1978, Ultrastructure of Rhizobium japonicum in relation to its attachment to root hairs, *J. Bacteriol.* **133**:1393–1400.

Banfalvi, Z., Sakanyan, V., Koncz, C., Kiss, A., Dusha, I., and Kondorosi, A., 1981, Location of nodulation and nitrogen fixation genes on a high molecular weight plasmid of *Rhizobium meliloti*, *Mol. Gen. Genet.* **184**:318–325.

Benyon, J. L., Beringer, J. E., and Johnston, A. W. B., 1980, Plasmids and host-range in *Rhizobium leguminosarum* and *R. phaseoli*, *J. Gen. Microbiol.* **120**:413–420.

Bhuvaneswari, T. V., 1981, Recognition mechanisms and infection process in legumes, *Econ. Bot.* **35**:204–223.

Bhuvaneswari, T. V., and Bauer, W. D., 1978, Role of lectins in plant–microorganism interactions. III. Influence of rhizosphere/rhizoplane culture conditions on the soybean lectin-binding properties of rhizobia, *Plant Physiol.* **62**:71–74.

Bhuvaneswari, T. V., Pueppke, S. G., and Bauer, W. D., 1977, Role of lectins in plant–microorganism interactions. I. Binding of soybean lectin to rhizobia, *Plant Physiol.* **60**:486–491.

Bhuvaneswari, T., Turgeon, B. G., and Bauer, W. D., 1980, Early events in the infection of soybean (*Glycine max* L. Merr) by *Rhizobium japonicum*. I. Localization of infectible root cells, *Plant Physiol.* **66**:1027–1031.

Binns, A. N., Sciaky, D., and Wood, H. N., 1982, Variation in hormone autonomy and regenerative potential of cells transformed by strain A66 of *Agrobacterium tumefaciens*, *Cell* **31**:605–612.

Bishop, P. E., Dazzo, F. B., Applebaum, E. R., Maier, R. J., and Brill, W. J., 1977, Intergeneric transfer of genes involved in the *Rhizobium*–legume symbiosis, *Science* **198**:938–940.

Blakeman, J. P., 1982, Phylloplane interactions, in: *Phytopathogenic Prokaryotes*, Volume 1 (M. S. Mount and G. H. Lacy, eds.), Academic Press, New York, pp. 308–334.

Bohlool, B. B., and Schmidt, E. L., 1974, Lectins: A possible basis for specificity in the *Rhizobium*–legume root nodule symbiosis, *Science* **185**:269–271.

Brewin, N. J., Beringer, J. E., and Johnston, A. W. B., 1980, Plasmid-mediated transfer of host-range specificity between two strains of *Rhizobium leguminosarum*, *J. Gen. Microbiol.* **120**:413–420.

Callaham, D. A., and Torrey, J. G., 1981, The structural basis for infection of root hairs of *Trifolium repens* by *Rhizobium*, *Can. J. Bot.* **59**:1647–1664.

Cason, E. T., Jr., Richardson, P. E., Essenberg, M. K., Brinkerhoff, L. A., Johnson, W. M., and Venere, R. J., 1978, Ultrastructural cell wall alterations in immune cotton leaves inoculated with *Xanthomonas malvacearum*, *Phytopathology* **68**:1015–1021.

Chen, A. P., and Phillips, D. A., 1976, Attachment of *Rhizobium* to legume roots as a basis for specific interactions, *Physiol. Plant.* **38**:83–88.

Chilton, M. D., Drummond, M. H., Merlo, D. J., Sciaky, D., Montoya, A. L., Gordon, M. P., and Nester, E. W., 1977, Stable incorporation of plasmid DNA into higher plants cells: The molecular basis of crown gall tumorigenesis, *Cell* **3**:263–271.

Chilton, M. D., Tepfer, D. A., Petit, A., David, C., Casse Delbart, F., and Tempé, J., 1982, *Agrobacterium rhizogenes* inserts T-DNA into the genomes of host plant root cells, *Nature (London)* **295**:432–434.

Council, O. P., 1983, Interaction of *Agrobacterium rhizogenes* with carrot suspension culture cells, Ph.D. thesis, University of North Carolina, Chapel Hill.

Dart, P. J., 1971, Scanning electron microscopy of plant roots, *J. Exp. Bot.* **22**:163–168.

Daub, M. E., and Hagedorn, D. J., 1980, Growth kinetics and interactions of *Pseudomonas syringae* with susceptible and resistant bean tissues, *Phytopathology* **70**:429–436.

Dazzo, F. B., 1981, Bacterial attachment as related to cellular recognition in the *Rhizobium*–legume symbiosis, *J. Supramol. Struct. Cell. Biochem.* **16:**29–41.

Dazzo, F. B., and Brill, W. J., 1978, Regulation by fixed nitrogen of host–symbiont recognition in the *Rhizobium*–clover symbiosis, *Plant Physiol.* **62:**18–21.

Dazzo, F. B., and Brill, W. J., 1979, Bacterial polysaccharide which binds *Rhizobium trifolii* to clover root hairs, *J. Bacteriol.* **137:**1362–1373.

Dazzo, F. B., and Hubbell, D. H., 1975, Cross-reactive antigens and lectin as determinants of symbiotic specificity in the *Rhizobium*–clover association, *Appl. Microbiol.* **30:**1017–1033.

Dazzo, F. B., and Hubbell, D. H., 1981, Control of root hair infection, in: *Ecology of Nitrogen Fixation,* Volume 2 (W. J. Broughton, ed.), Oxford University Press, London, pp. 274–310.

Dazzo, F. B., Napoli, C. A., and Hubbell, D. H., 1976, Adsorption of bacteria to roots as related to host specificity in the *Rhizobium*–clover symbiosis, *Appl. Environ. Microbiol.* **32:**166–171.

Dazzo, F. B., Yanke, W. E., and Brill, W. J., 1978, Trifoliin: A *Rhizobium* recognition protein from white clover, *Biochim. Biophys. Acta* **539:**276–286.

Dazzo, F. B., Truchet, G. L., Sherwood, J. E., Hrabek, E. M., and Gardiol, A. E., 1982, Alteration of the trifoliin A-binding capsule of *Rhizobium trifolii* 0403 by enzymes released from clover roots, *Appl. Environ. Microbiol.* **44:**478–490.

Dazzo, F. B., Truchet, G. L., and Hooykaas, P. J., 1983, Clover root-recognition of *Agrobacterium tumefaciens* carrying the sym-plasmid of *Rhizobium trifolii*, *Abstr. Annu. Meet. Am. Soc. Microbiol.* K9.

Deasey, M. C., and Matthysse, A. G., 1983, Attachment of wild-type and cellulose-minus *Agrobacterium tumefaciens* to tobacco mesophyll and tissue culture cells, *Phytopathology* **73:**807.

Douglas, C. J., Halperin, W., and Nester, E. W., 1982, *Agrobacterium tumefaciens* mutants affected in attachment to plant cells, *J. Bacteriol.* **152:**1265–1275.

Downie, J. A., Hombrecher, G., Ma, Q. S., Knight, C. D., Wells, B., and Johnston, A. W. B., 1983, Cloned nodulation genes of *Rhizobium leguminosarum* determine specificity, *Mol. Gen. Genet.* **190:**359–365.

El-Banoby, F. E., and Rudolph, K., 1979, Induction of water-soaking in plant leaves by extracellular polysaccharides from phytopathogenic pseudomonads and xanthomonads, *Physiol. Plant Pathol.* **15:**341–349.

Essenberg, M., Cason, E. T., Jr., Hamilton, B., Brinkerhoff, L. A., Gholson, R. K., and Richardson, P. E., 1979, Single cell colonies of *Xanthomonas malvacearum* in susceptible and immune cotton leaves and the local resistant response to colonies in immune leaves, *Physiol. Plant Pathol.* **15:**53–68.

Fåhraeus, G., 1957, The infection of clover root hairs by nodule bacteria studied by a simple glass slide technique, *J. Gen. Microbiol.* **16:**374–381.

Fett, W. F., and Jones, S. B., 1982, Role of bacterial immobilization in race-specific resistance of soybean to *Pseudomonas syringae* pv. *glycinea*, *Phytopathology* **72:**488–492.

Fett, W. F., and Sequeira, L., 1980, New bacterial agglutinin from soybean: Evidence against a role in determining pathogen specificity, *Plant Physiol.* **66:**853–858.

Foster, R. C., 1981, The ultrastructure and histochemistry of the rhizosphere, *New Phytol.* **89:**263–273.

Gade, W., Jack, M. A., Dahl, J. B., Schmidt, E. L., and Wold, F., 1981, The isolation and characterization of a root lectin from soybean (*Glycine max* (L), cultivar chippewa), *J. Biol. Chem.* **257:**12905–12910.

Glogowski, W., and Galsky, A. G., 1978, *Agrobacterium tumefaciens* site attachment as a necessary prerequisite for crown gall tumor formation on potato discs, *Plant Physiol.* **61:**1031–1033.

Goldberg, R. B., Hoschek, G., and Vodkin, L. O., 1983, An insertion sequence blocks the expression of a soybean lectin gene, *Cell* **33:**465–475.

Goldstein, I. J., and Hayes, C. W., 1978, The lectins: Carbohydrate-binding proteins of plants and animals, *Adv. Carbohydr. Chem. Biochem.* **35:**127–340.

Goodman, R. N., Huang, P. Y., and White, J. A., 1976, Ultrastructural evidence for immobilization of an incompatible bacterium, *Pseudomonas pisi*, in tobacco leaf tissue, *Phytopathology* **66:**754–764.

Gurley, W. B., Kemp, J. D., Albert, M. J., Sutton, D. W., and Callis, J., 1979, Transcription of Ti plasmid-derived sequences in three octopine-type crown gall tumor lines, *Proc. Natl. Acad. Sci, USA* **76:**2828–2832.

Gurlitz, R. H. G., Lamb, P. W., and Matthysse, A. G., 1984, Involvement of carrot cell surface proteins in attachment of *Agrobacterium tumefaciens*, submitted for publication.

Hahn, M. G., Darvill, A. G., and Albersheim, P., 1981, Host–pathogen interactions. XIX. The endogeneous elicitor, a fragment of a plant cell wall polysaccharide that elicits phytoalexin accumulation in soybeans, *Plant Physiol.* **68:**1161–1169.

Hildebrand, D. C., Alosi, M. C., and Schroth, M. N., 1980, Physical entrapment of pseudomonads in bean leaves by films formed at air–water interfaces, *Phytopathology* **70:**98–109.

Hirsch, A. M., Long, S. R., Bang, M., Haskins, N., and Ausubel, F. M., 1982, Structural studies of alfalfa roots infected with nodulation mutants of *Rhizobium meliloti*, *J. Bacteriol.* **151:**411–419.

Hirsch, P. R., Van Montagu, M., Johnston, A. W. B., Brewin, N. J., and Schell, J., 1980, Physical identification of bacteriocinogenic nodulation and other plasmids in strains of *Rhizobium leguminosarum, J. Gen. Microbiol.* **120:**403–412.

Hooykaas, P. J. J., Van Brussell, A. A. N., Den Dulk-Ras, H., Van Slogteren, G. M. S., and Shilperoort, R. A., 1981, Sym plasmids of *Rhizobium trifolii* expressed in different *Rhizobium* species and *Agrobacterium tumefaciens, Nature (London)* **291:**351–354.

Hrabek, E. M., Urbano, M. R., and Dazzo, F. B., 1981, Growth-phase-dependent immunodeterminants of *Rhizobium trifolii* lipopolysaccharide which bind trifoliin A, a white clover lectin, *J. Bacteriol.* **148:**697–711.

Huang, P. Y., Huang, J. S., and Goodman, R. N., 1975, Resistance mechanisms of apple shoots to an avirulent strain of *Erwinia amylovora, Physiol. Plant Pathol.* **6:**283–287.

Hubbell, D. H., Morales, V. M., and Umali-Garcia, M., 1978, Pectolytic enzymes in *Rhizobium, Appl. Environ. Microbiol.* **35:**210–213.

Jansen van Rensburg, H., and Strijdom, B. W., 1982, Root surface association in relation to nodulation of *Medicago sativa, Appl. Environ. Microbiol.* **44:**93–97.

Johnston, A. W. B., Benyon, J. L., Buchanan-Wollaston, A. V., Setchell, S. M., Hirsch, P. R., and Beringer, J. L., 1978, High frequency transfer of nodulation ability between strains and species of *Rhizobium, Nature (London)* **276:**634–636.

Kao, J. C., Perry, K. L., and Kado, C. I., 1982, Indoleacetic acid complementation and its relation to host range specifying genes on the Ti plasmid of *Agrobacterium tumefaciens, Mol. Gen. Genet.* **188:**425–432.

Kato, G., Maruyama, Y., and Nakamura, M., 1981, Involvement of lectins in *Rhizobium*–pea recognition, *Plant Cell Physiol.* **22:**759–771.

Keen, N. T., Ersek, T., Long, M., Bruegger, B., and Holliday, M., 1981, Inhibition of the hypersensitive reaction of soybean leaves to incompatible *Pseudomonas* spp. by blasticidin S, streptomycin, or elevated temperature, *Physiol. Plant Pathol.* **18:**325–337.

Kijne, J. W., Vander Schaol, I. A. M., and De Vries, G. E., 1980, Pea lectins and the recognition of *Rhizobium leguminosarum, Plant Sci. Lett.* **18:**65–74.

Klement, Z., 1982, Hypersensitivity, in: *Phytopathogenic Prokaryotes,* Volume 2 (M. S. Mount and G. H. Lacy, eds.), Academic Press, New York, pp. 149–177.

Law, I. J., Yamamoto, Y., Mort, A. J., and Bauer, W. D., 1982, Nodulation of soybean by *Rhizobium japonicum* mutants with altered capsule synthesis, *Planta* **154:**100–109.

Leach, J. E., Cantrell, M. A., and Sequeira, L., 1982a, Hydroxyproline-rich bacterial agglutinin from potato: Extraction, purification, and characterization, *Plant Physiol.* **70:**1353–1358.

Leach, J. E., Cantrell, M. A., and Sequeira, L., 1982b, A hydroxyproline-rich bacterial agglutinin from potato: Its locatization by immunofluorescence, *Physiol. Plant Pathol.* **21:**319–325.

Li, D., and Hubbell, D. H., 1969, Infection thread formation as a basis of nodulation specificity in Rhizobium-strawberry clover associations, *Can. J. Microbiol.* **15:**1133–1136.

Lindow, S. E., Arny, D. C., and Upper, C. D., 1982, Bacterial ice nucleation: A factor in frost injury in plants, *Plant Physiol.* **70:**1084–1089.

Lippincott, B. B., and Lippincott, J. A., 1969, Bacterial attachment to a specific wound site as an essential stage in tumor initiation by *Agrobacterium tumefaciens, J. Bacteriol.* **97:**620–628.

Lippincott, B. B., Whatley, M. H., and Lippincott, J. A., 1977, Tumor induction by *Agrobacterium* involves attachment of the bacterium to a site on the host plant cell wall, *Plant Physiol.* **59:**388–390.

Lippincott, J. A., and Heberlein, G. T., 1965, The quantitative determination of the infectivity of *Agrobacterium tumefaciens, Am. J. Bot.* **52:**856–863.

Lippincott, J. A., and Lippincott, B. B., 1975, The genus *Agrobacterium* and plant tumorigenesis, *Annu. Rev. Microbiol.* **29:**377–407.

Lippincott, J. A., and Lippincott, B. B., 1976, Nature and specificity of the bacterium–host attachment in *Agrobacterium* infection, in: *Cell Wall Biochemistry Related to Specificity in Host–Plant Pathogen Interactions* (B. Solheim and J. Raa, eds.), Universitets-forlaget, Tromsø, Norway, pp. 439–451.

Lippincott, J. A., and Lippincott, B. B., 1978, Cell walls of crown-gall tumors and embryonic plant tissues lack *Agrobacterium* adherence sites, *Science* **199:**1075–1078.

Lippincott, J. A., and Lippincott, B. B., 1980, Microbial adherence in plants, in: *Bacterial Adherence* (E. H. Beachey, ed.), Chapman & Hall, London, pp. 375–398.

Loper, J. E., and Kado, C. I., 1979, Host range conferred by the virulence-specifying plasmid of *Agrobacterium tumefaciens, J. Bacteriol.* **139:**591–596.

Martinez-Molina, E., Morales, V. M., and Hubbell, D. H., 1979, Hydrolytic enzyme production by *Rhizobium, Appl. Environ. Microbiol.* **38:**1186–1188.

Matthysse, A. G., 1983a, The role of bacterial cellulose fibrils in infections by *Agrobacterium tumefaciens, J. Bacteriol.* **154:**906–915.

Matthysse, A. G., 1983b, The use of tissue culture in the study of crown gall and other bacterial diseases, in: *Use of*

Tissue Culture and Protoplasts in Plant Pathology (J. P. Helgeson and B. J. Deverall, eds.), Academic Press, Australia, pp. 39–68.

Matthysse, A. G., and Gurlitz, R. H. G., 1982, Plant cell range for attachment of *Agrobacterium tumefaciens* to tissue culture cells, *Physiol. Plant Pathol.* **21:**381–387.

Matthysse, A. G., and Lamb, P. W., 1981, Soluble factor produced by *Agrobacterium* which promotes attachment to carrot cells, *Abstr. Annu. Meet. Am. Soc. Microbiol.* B95.

Matthysse, A. G., and Stump, A. J., 1976, The presence of *Agrobacterium tumefaciens* plasmid DNA in crown gall tumor cells, *J. Gen. Microbiol.* **95:**9–16.

Matthysse, A. G., Wyman, P. M., and Holmes, K. V., 1978, Plasmid-dependent attachment of *Agrobacterium tumefaciens* to plant tissue culture cells, *Infect. Immun.* **22:**516–522.

Matthysse, A. G., Holmes, K. V., and Gurlitz, R. H. G., 1981, Elaboration of cellulose fibrils by *Agrobacterium tumefaciens* during attachment to carrot cells, *J. Bacteriol.* **145:**583–595.

Matthysse, A. G., Holmes, K. V., and Gurlitz, R. H. G., 1982, Binding of *Agrobacterium tumefaciens* to carrot protoplasts, *Physiol. Plant Pathol.* **20:**27–33.

Matthysse, A. G., Gurlitz, R. H. G., Lamb, P. W., and Van Stee, K., 1983, Binding of *Agrobacterium tumefaciens* and of *Agrobacterium* proteins to carrot suspension cultures, *Abstr. Annu. Meet. Am. Soc. Microbiol.* B12.

Mellon, J. E., and Helgeson, J. P., 1982, Interaction of a hydroxyproline-rich glycoprotein from tobacco callus with potential pathogens, *Plant Physiol.* **70:**401–405.

Moore, L., Warren, G., and Strobel, G., 1979, Involvement of a plasmid in the hairy root disease of plants caused by *Agrobacterium rhizogenes, Plasmid* **2:**619–626.

Mort, A. J., and Bauer, W. D., 1980, Composition of the capsular and extracellular polysaccharides of *Rhizobium japonicum:* Changes with culture age and correlations with binding of soybean seed lectin to the bacteria, *Plant Physiol.* **66:**158–163.

Napoli, C. A., and Hubbell, D. H., 1975, Ultrastructure of *Rhizobium*-induced infection threads in clover root hairs, *Appl. Microbiol.* **30:**1003–1009.

Napoli, C. A., Dazzo, F. B., and Hubbell, D. H., 1975, Production of cellulose microbibrils by *Rhizobium, Appl. Microbiol.* **30:**128–131.

Ohyama, K., Pelcher, L. E., Schaefer, A., and Fowke, L. C., 1979, *In vitro* binding of *Agrobacterium tumefaciens* to plant cells from suspension culture, *Plant Physiol.* **63:**382–387.

Orf, J. H., Hymowitz, T., Pull, S. P., and Pueppke, S. G., 1978, Inheritance of a soybean seed lectin, *Crop Sci.* **18** 899–900.

Paau, A. S., Leps, W. T., and Brill, W. J., 1981, Agglutinin from alfalfa necessary for binding and nodulation by *Rhizobium meliloti, Science* **213:**1513–1515.

Politis, D. J., and Goodman, R. N., 1978, Localized cell wall appositions: Incompatibility response of tobacco leaf cells to *Pseudomonas pisi, Phytopathology* **68:**309–316.

Pueppke, S. G., Freund, T. G., Schulz, B. C., and Freidman, H. P., 1980, Interaction of lectins from soybean and peanut with rhizobia that nodulate soybean, peanut, or both plants, *Can. J. Microbiol.* **26:**1489–1497.

Pueppke, S. G., Friedman, H. P., and Su, L. C., 1981, Examination of Le and lele genotypes of *Glycine max* (L.) Merr. for membrane-bound and buffer soluble soybean lectin, *Plant Physiol.* **68:**905–909.

Pull, S. P., Pueppke, S. G., Hymowitz, T., and Orf, J. H., 1978, Soybean lines lacking the 120,000-dalton seed lectin, *Science* **200:**1277–1279.

Rao, S. S., Lippincott, B. B., and Lippincott, J. A., 1982, *Agrobacterium* adherence involves the pectin portion of the host cell wall and is sensitive to the degree of pectin methylation, *Physiol. Plant* **56:**374–380.

Roebuck, P., Sexton, R., and Mansfield, J. W., 1978, Ultrastructural observations on the development of the hypersensitive reaction in leaves of *Phaeseolus vulgaris* cv. Red Mexican inoculated with *Pseudomonas phaseolicola* (racel), *Physiol. Plant Pathol.* **12:**151–157.

Romeiro, R., Karr, A., and Goodman, R., 1981a, Isolation of a factor from apple that agglutinates *Erwinia amylovora, Plant Physiol.* **68:**772–777.

Romeiro, R., Karr, A. L., and Goodman, R. N., 1981b, *Erwinia amylovora* cell wall receptor for apple agglutinin, *Physiol. Plant Pathol.* **19:**383–390.

Rosenberg, C., Boistard, P., Denarie, J., and Casse-Delbart, F., 1981, Genes controlling early and late functions in symbiosis are located on a megaplasmid in *Rhizobium meliloti, Mol. Gen. Genet.* **184:**326–333.

Rougé, P., and Labroue, L., 1977, Sur le rôle des phytohémagglutinines dans la fixation spécifique des souches compatibles de *Rhizobium leguminosarum* sur le pois, *C.R. Acad. Sci. Ser. D* **284:**2423–2426.

Sadowsky, M. J., and Bohlool, B. B., 1983, Lack of expression of the *Rhizobium leguminosarum* "host range" plasmid, pJB5JI, in a fast-growing isolate of *Rhizobium japonicum, Abstr. Annu. Meet. Am. Soc. Microbiol.* K10.

Sanders, R., Raleigh, E., and Signer, E., 1981, Lack of correlation between extracellular polysaccharide and nodulation ability in *Rhizobium, Nature (London)* **292:**148–149.

Sequeira, L., Gaard, G., and De Zoeten, G. A., 1977, Interaction of bacteria and host cell walls: Its relation to mechanisms of induced resistance, *Physiol. Plant Pathol.* **10**:43–50.

Smith, J. J., and Mansfield, J. W., 1982, Ultrastructure of interactions between pseudomonads and oat leaves, *Physiol. Plant Pathol.* **21**:259–266.

Stacey, G., Paau, A. S., and Brill, W. J., 1980, Host recognition in the *Rhizobium*–soybean symbiosis, *Plant Physiol.* **66**:609–614.

Stacey, G., Paau, A. S., Noel, D., Maier, R. J., Silver, L. E., and Brill, W. J., 1982, Mutants of *Rhizobium japonicum* defective in nodulation, *Arch. Microbiol.* **132**:219–224.

Stall, R. E., and Cook, A. A., 1979, Evidence that bacterial contact with the plant cell is necessary for the hypersensitive reaction but not the susceptible reaction, *Physiol. Plant Pathol.* **14**:77–84.

Stuart, D. A., and Varner, J. E., 1980, Purification and characterization of a salt-extractable hydroxyproline-rich glycoprotein from aerated carrot discs, *Plant Physiol.* **66**:787–792.

Suhayda, C. G., and Goodman, R. N., 1981, Infection counts and systemic movement of ^{32}P-labeled *Erwinia amylovora* in apple petioles and stems, *Phytopathology* **71**:656–660.

Thomashow, M. F., Nutter, R., Montoya, A. L., Gordon, M. P., and Nester, E. W., 1980a, Integration and organization of Ti plasmid sequences in crown gall tumors, *Cell* **19**:729–739.

Thomashow, M. F., Pangopoulos, C. G., Gordon, M. P., and Nester, E. W., 1980b, Host range of *Agrobacterium tumefaciens* is determined by the Ti plasmid, *Nature (London)* **283**:794–796.

Turgeon, B. G., and Bauer, W. D., 1982, Early events in the infection of soybean by *Rhizobium japonicum:* Time course and cytology of the initial infection process, *Can. J. Bot.* **60**:152–161.

Whatley, M. H., Bodwin, J. S., Lippincott, B. B., and Lippincott, J. A., 1976, Role for *Agrobacterium* cell envelope lipopolysaccharide in infection site attachment, *Infect. Immun.* **13**:1080–1083.

Whatley, M. H., Margot, J. B., Schell, J., Lippincott, B. B., and Lippincott, J. A., 1978, Plasmid and chromosomal determination of *Agrobacterium* adherence specificity, *J. Gen Microbiol.* **107**:395–398.

10

Adhesion of Bacteria to Animal Tissues
Complex Mechanisms

GORDON D. CHRISTENSEN, W. ANDREW SIMPSON, and
EDWIN H. BEACHEY

1. INTRODUCTION

Bacteria suspended in body fluids near animal tissue surfaces behave in many respects as waterborne particles near a submerged object. Both bacteria and particles are drawn to the nearby surface by the various physiochemical forces of attraction and repulsion described in other chapters of this volume (see Chapter 5, Sections 1 and 2; Chapter 6, Section 3). The precise degree of attraction or "adsorption" depends upon the balance of the forces acting upon the two surfaces.

Two fundamental differences between the animate and the inanimate systems, however, carry the prokaryotic–eukaryotic cell interaction beyond simple adsorption and give it direction. First of all, both prokaryotic and eukaryotic organisms have complex cell surfaces composed of mosaics of molecular and supramolecular structures (see Chapters 2–4). These structures determine on a regional and cellular scale the intensity of the physiochemical forces that lead to adsorption. More important to this discussion, the cell surfaces often contain specialized adhesive structures which interlock in a stereospecific manner with complementary structures on the opposing surface. These structures mediate attachment of bacteria to tissue, providing that the bacterial and animal cells are close enough to each other for their matching molecular structures to interact. Genetic control of the physiochemical properties, molecular arrangement, and superstructural display of cell surface organelles accounts for the second fundamental difference between the animate and the inanimate systems. Living organisms control their ability to adhere to various surfaces by genetically

Gordon D. Christensen • Veterans Administration Medical Center and Department of Medicine, University of Tennessee Center for the Health Sciences, Memphis, Tennessee 38104.   W. Andrew Simpson • Department of Microbiology and Immunology, University of Tennessee Center for the Health Sciences, Memphis, Tennessee 38104.   Edwin H. Beachey • Veterans Administration Medical Center and Departments of Medicine and of Microbiology and Immunology, University of Tennessee Center for the Health Sciences, Memphis, Tennessee 38104.

Table IA. Examples Where Host Specificity Relates to *in Vitro* Bacterial Adhesion to Host Tissues[a]

Organism	Infection (host)	Comment
N. gonorrhoeae	Pelvic inflammatory disease (human)	Adheres to human oviduct, not to rabbit, pig, or cow oviducts (Johnson *et al.*, 1977)
E. coli (enterotoxigenic isolates)	Diarrhea (human)	Human isolates adhere to human ileal cells; enterotoxigenic isolates from pigs, calves, and rabbits do not adhere (Deneke *et al.*, 1983)
E. coli (RDEC-strain)	Diarrhea (rabbit)	Rabbit strains adhere only to rabbit ileal cells, not to rat, guinea pig, or human ileal cells (Cheney *et al.*, 1980)
Streptococci and diptheroids	Oral microflora (humans and rats)	Isolates of human and rat origin only adhere to tongue scrapings of the homologous (not heterologous) host (Gibbons *et al.*, 1976)
Shigella flexneri	Diarrhea (guinea pig)	Shigella produces diarrhea in humans and guinea pigs. A human isolate adheres to colonic epithelial cells of guinea pig origin > rat, rabbit, hamster (Izhar *et al.*, 1982)
Bordetella pertussis, B. bronchiseptica	Lower respiratory tract infection (humans, mammals)	The human pathogen *B. pertussis* adheres to human lower erspiratory mucosal cells > cells from other mammals. The animal pathogen *B. bronchiseptica* adheres to cells of animal origin > cells of human origin (Tuomanen *et al.*, 1983)

[a]Modified from Christensen *et al.* (1984).

Table IB. Examples Where Host Susceptibility Relates to *in Vitro* Bacterial Adhesion to Host Tissues[a]

Organism	Susceptible host	In comparison to normal patients:
P. aeruginosa	Cystic fibrosis patients	Adhesion to buccal epithelial cells > for cystic fibrosis patients (Woods *et al.*, 1980)
S. aureus	*Staphylococcus aureus* nasal carriers	Adhesion to nasal mucosa cells > for *S. aureus* carrier patients (Aly *et al.*, 1977)
S. pyogenes	Rheumatic fever patients	Adhesion to pharyngeal cells > for rheumatic fever patients (Selinger *et al.*, 1978)
E. coli	Urinary tract infection patients	Adhesion to vaginal, buccal, and urinary epithelial > for patients with frequent urinary tract infections (Kallenius and Winberg, 1978; Svanborg-Eden and Jodal, 1979; Schaeffer *et al.*, 1981)
Streptococcus agalactiae	Neonatal sepsis	Adhesion of *S. agalactiae* to buccal epithelial cells from neonates with sepsis > healthy neonates > adults (Broughton and Baker, 1983)
N. gonorrhoeae	Gonorrhea in women	Adhesion of *N. gonorrhoeae* to vaginal epithelial cells > for postmenopausal women on estrogen (equivalent to young women) than for normal postmenopausal women (Forslin *et al.*, 1980)
E. coli K88	Colibacillosis in piglets	Adhesion to enterocytes and susceptibility to infection segregate together as autosomal dominant trait (Sellwood *et al.*, 1975)

[a]Modified from Christensen *et al.* (1984).

Table IC. Examples Where Tissue Tropism Relates to *in Vitro* Bacterial Adhesion to Host Tissues[a]

Organism	Tissue (host)	Comment
Neisseria meningitidis	Nasopharynx (human)	Adheres to nasopharyngeal cells > buccal, urethral, bladder, anterior nasal epithelial cells (Stephens and McGee, 1981)
Pasteurella multocida	Nasopharynx (rabbit)	Adheres to squamous nasopharyngeal cells > ciliated nasopharyngeal cells or a variety of tissue culture cells (Glorioso *et al.*, 1982)
S. flexneri	Colon (guinea pigs)	Adheres to descending colon > transverse > ascending >> ileum, jejunum, duodenum (Izhar *et al.*, 1982)
Staphylococcus saprophyticus	Urinary bladder (human)	Adheres to urogenital cells > skin and buccal epithelial cells (Colleen *et al.*, 1979)
S. pyogenes	Skin and oropharynx (human)	Strains of skin and oropharyngeal origin adhere better to homologous tissue than to heterologous tissue (Alkan *et al.*, 1977)
Pseudomonas aeruginosa	Trachea and bronchi (human)	Adheres to nasal and tracheal cells > buccal epithelial cells (Niederman *et al.*, 1983)
Corynebacterium renale	Urinary epithelium (bovine)	Adheres to urinary epithelium superficial cells > basal cells (Sato *et al.*, 1982)
Bordetella pertussis	Trachea (hamster)	Adheres only to ciliated epithelium (Muse *et al.*, 1977)

[a]Modified from Christensen *et al.* (1984).

regulating their presentation of specialized adhesive structures, thus bringing the phenomenon of adhesion into the realm of biologic adaptation and evolution.

Specific adhesion mechanisms can be advantageous to the participants of the host–pathogen and host–commensal relationship. To the host, these mechanisms offer the advantage of promoting tissue colonization with commensal or saprophytic organisms that prevent the attachment of pathogenic organisms. To the bacterium, adhesion offers the advantage of a firm attachment to tissue surfaces that can withstand the cleansing action of mucous flow, ciliary movement, cough, urination, peristalsis, swallowing, etc. Perhaps of greatest advantage, adhesion mechanisms allow the bacterium to attach to targeted tissues.

Host and tissue selectivity (tropism), more than any other feature, distinguishes animate from inanimate bacterial adhesion. The adhesion of a particular bacterial species may vary considerably depending on the host species (Table IA), physiology (Table IB), phenotype (Table IB), and tissue (Table IC). It is difficult to imagine how the relatively nonselective process of adsorption can result in the specific interactions that characterize the host–pathogen and host–commensal relationship. Adsorption probably potentiates such interactions by bringing the two surfaces into juxtaposition, but it is doubtful that adsorption by itself directs the process.

Selective adhesion depends primarily upon the union of bacterial surface structures, called *adhesins,* with complementary animal cell structures, called *receptors* (see Chapter 4, Section 4; Chapter 11, Section 5). The specificity of this interaction depends upon two factors. First, the adhesin may interact with a broad variety of molecular structures or may bind only to very specific molecular conformations. Thus, the "fit" of the adhesin–receptor interaction directly determines the range of suitable substrates. The second route to specificity relies upon the utilization of multiple adhesin–receptor systems. It appears that in virtually every system, bacteria employ more than one mechanism of binding to tissues (Table II). Multiple adhesive mechanisms require the targeted tissue to display more than one type of receptor before optimal adhesion can take place; these stringent requirements greatly limit

Table II. Examples of Bacteria Using Multiple Mechanisms to Bind to Animal Tissues

Bacteria	Adhesin	Functional importance
Escherichia coli	Type 1, K88, K99, CFA/, CFA/2, 987P, F41, E8775, P fimbriae	Each fimbria binds to particular hosts and tissues; each fimbrial type is associated with a particular infection, gastroenteritis or urinary tract, in a particular host (reviewed by Christensen *et al.*, 1984)
Neisseria gonorrhoeae	Type II outermembrane protein, fimbriae	Each adhesin binds to different genital tissues or conditions, enabling the organism to persist and infect a variety of microhabitats (reviewed by Swanson, 1983; Christensen *et al.*, 1984)
Streptococcus agalactiae	LTA, lipid end LTA, charged backbone	One mechanism may mediate binding of *S. agalactiae* to immature tissues; the other mechanism may mediate binding to mature tissues (see Section 2.2)
Streptococcus mutans and other oral microorganisms	LTA, glucosyltransferase, cell surface lectins, fimbriae, fibrillae	The oral microorganisms simultaneously employ a variety of adhesins to bind themselves to each other and to the oral surfaces (see Section 4.1)

the chances of coincidental attachment to nontargeted tissues. Multiple systems of adhesion enable bacteria to adhere to a range of host species and tissues.

Most, but not all, of the adhesins reported so far in the literature are proteins that fall in the category of lectins as defined by the capacity to bind to specific sugar residues (see Chapter 9, Section 2.3). This category of adhesins includes fimbriae which are proteinaceous filaments that bind bacteria to carbohydrate residues on animal cell surfaces. The mechanisms of fimbria-mediated adhesion are presented in Chapter 11.

While the specificity of lectin–carbohydrate interactions has been relatively easy to demonstrate, the specificity of the nonlectin–receptor interactions has been more difficult. In general, bacteria utilizing these nonlectin mechanisms bind to tissues through the agency of a third substance which binds the bacterium to the target cell forming a bacterium–cell "complex." These "bridging" ligands may be bacterially produced or host-derived molecules. On the other hand, the bacterium may bind to a tissue surface by binding first to an intermediary cell which in turn binds to the tissue surface. Once again, these "bridging" cells may be other bacteria or host-derived cells. In many cases bacterial adhesion to tissues proceeds simultaneously through several of these mechanisms resulting in a complicated tangle of molecules, bacteria, and host cells on the tissue surface. Despite the seeming chaos of this microscopic jungle, the adherent mass has a predictable composition and structure based upon the specificities of each binding mechanism. These complex mechanisms are the focus of this chapter.

2. "BRIDGING" LIGANDS PRODUCED BY BACTERIA

2.1. The Role of Lipoteichoic Acid (LTA) as an Adhesin for Group A Streptococci

2.1.1. The Biochemistry of LTA

One well-studied example of a complex adhesion mechanism is the binding of LTA, the group A streptococcal adhesin, to fibronectin, a cell surface receptor for group A streptococci on human buccal epithelial cells. LTA is a linear polyglycerolphosphate polymer containing a glycerophosphoryl/diglucosyl/diglyceride moiety at its nonpolar end (McCarty, 1959 and Chapter 2, Section 2.4.1). The polyglycerolphosphate backbone of LTA is substituted to varying degrees with ester-linked alanine residues (McCarty, 1959). The highly charged linear backbone combined with the hydrophobic lipid moiety confers amphipathicity to the LTA molecule (Wicken and Know, 1977). LTA substituted to varying degrees with alanine and carbohydrate residues is common to most, if not all, gram-positive bacteria. It is found at the bacterial cytoplasmic membrane–cell wall interface with the lipid moiety embedded in the membrane and the charged backbone in the cell wall.

In theory, LTA should make a poor candidate for a specific bacterial adhesin considering its ubiquitous nature and the likelihood that the lipid moiety will nonspecifically bind to the bilipid plasma membranes of eukaryotic cells (Jones, 1977). Nevertheless, work in our laboratory indicates that LTA does indeed function as a specific streptococcal adhesin.

2.1.2. The Influence of LTA on Streptococcal Adhesion

Attempts to identify the adhesin of *Streptococcus pyogenes* cells began with tests of the effects on adhesion of a wide variety of known surface components of the microorganisms. Purified M protein, C carbohydrate, peptidoglycan, and LTA were examined for their ability to inhibit the adhesion of streptococci to oral mucosal (buccal epithelial) cells; of these compounds, only LTA exhibited significant inhibitory activity (Beachey, 1975; Ofek et al., 1975; Beachey and Ofek, 1976). Furthermore, only antibodies directed against LTA and not against other surface components inhibited the adhesion (Beachey and Ofek, 1976). Fragmentation of the LTA molecule into the charged backbone and the lipid moiety demonstrated that the inhibitory activity of LTA required an intact ester-linked lipid (Ofek et al., 1975). Reesterification of the inactive polyglycerolphosphate backbone with palmitoylchloride restored the ability of the molecule to block streptococcal attachment (Beachey and Ofek, 1976).

2.1.3. The Binding of LTA to Cell Membranes

Streptococcal LTA binds to all eukaryotic cells tested. Platelets (Beachey et al., 1977), erythrocytes (Beachey et al., 1979b), purified erythrocyte membranes (Chiang et al., 1979), lymphocytes (Beachey et al., 1979a), polymorphonuclear leukocytes (Courtney et al., 1981), oral epithelial cells (Simpson et al., 1980c), and various tissue culture cells (Simpson et al., 1982c) have saturable numbers of LTA-binding sites with dissociation constants ranging from 4.5 μM to 89 μM. In each case, the binding of radiolabeled LTA to the cells was abolished by removing the lipid moiety of LTA with mild ammonia hydrolysis (deacylation).

The biologic activity of LTA is also related to the integrity of the lipid moiety and the ability of the molecule to bind to tissue cells. For instance, stimulation of mitotic activity of lymphocytes by LTA required intact LTA and was maximal when 12% of the available binding sites were occupied (Beachey et al., 1979a). In another example, when half of the available binding sites on human polymorphonuclear leukocytes were occupied with intact LTA, the number of streptococci bound to the cells decreased by 50% (Courtney et al., 1981). Further evidence that the lipid moiety of LTA interacts with specific membrane receptors was reported by Chiang et al. (1979) who found that red cell ghosts bound 10 times more LTA when sealed right-side out (i.e., with their outer surface exposed) than when sealed inside out.

These studies suggest that the binding of LTA to cell membranes involves more than just the simple intercalation of the fatty acid chains of LTA into the lipid bilayer of the plasma membrane. The variation in binding affinities observed with different cell types, the saturable nature of the binding, and the polarity of binding between the inner and the outer surfaces of the erythrocyte membrane all indicate that specific LTA receptors exist on most, if not all, eukaryotic cells. The existence of these receptors does not necessarily mean that the whole streptococcus will also bind to the same cell. For example, streptococci bind poorly to erythrocytes, platelets, and lymphocytes, even though LTA-binding sites are present on these cells. Why this dissociation exists between LTA binding and streptococcal binding remains unclear. It may be that the LTA-binding sites are buried deep within the cell membrane and cannot interact with streptococcal cell surface LTA, or it may be that other binding forces or sites are required for the binding of streptococci to certain tissues.

2.1.4. The Orientation of LTA on the Streptococcal Surface

In order to function, the binding site of a bacterial adhesin must be displayed on the bacterial surface in a configuration that will leave it free to interact with complementary structures on the eukaryotic cell surface. As the lipid moiety of LTA anchors the molecule to the bacterial cytoplasmic cell membrane deep inside the cell wall (Fig. 1), a question arises as to how the lipid moiety can also interact with other substrates (Jones, 1977). This is particularly problematical as the interaction takes place in an aqueous environment that is inimical to the presence of lipid. In further investigations, we sought to answer this question by examining the orientation of LTA on the streptococcal surface.

Electron microscopic studies demonstrated that the first structures of group A streptococci to make contact with the surfaces of oral mucosal cells were fibrillae (Fig. 1). Fibrillae are thin, ill-defined surface appendages, of variable length and size. Collectively, they form a fuzzy coat over the streptococcal surface that is rich in M protein (Beachey and Ofek, 1976), reacts with antibodies specific for LTA (Beachey and Ofek, 1976), and contains other substances such as streptococcal T protein (Johnson et al., 1980). Although these observations confirmed the presence of LTA on the bacterial surface, they did not reveal the configuration LTA takes within the fibrillar network.

Previous work by Doyle et al. (1975) had shown that LTA forms complexes with a variety of proteins and carbohydrates in hydrophobic solutions (83% ethanol), suggesting the complexes were formed by ionic interactions of the charged backbone of LTA. Studies in our laboratory demonstrated that both LTA and deacylated LTA formed complexes with M protein in hydrophobic solutions (83% ethanol), as well as in hydrophilic solutions (phosphate-buffered saline). The maleylation of M protein blocked complex formation with LTA, but activity was restored by demaleylation with pyridine acetate, indicating that complex formation required positively charged residues on M protein (Ofek et al., 1982). The interaction between these two molecules probably occurs because the clusters of positively charged residues on the coiled-coil M protein structure postulated by Phillips et al. (1981) spatially match the negatively charged residues on the polyglycerolphosphate backbone of LTA (Fig. 2). If the two molecules form complexes in this manner, the lipid moiety should remain free to interact with the hydrophobic domains of other substances. To test this hypothesis, we mixed M protein with acylated or deacylated LTA and then precipitated the complexes by reducing the pH to 3.7. The resulting insoluble material was exposed to fatty acid-free bovine serum albumin because previous studies (Simpson et al., 1980a) had established that the lipid moiety of LTA binds specifically to the fatty acid-binding sites of bovine serum albumin. The insoluble LTA–M protein complexes were able to bind 10-fold more bovine serum albumin than the deacylated LTA–M protein complexes, indicating that the fatty acid chains of LTA were, in fact, free to bind with at least one other molecule (Simpson et al., 1980b).

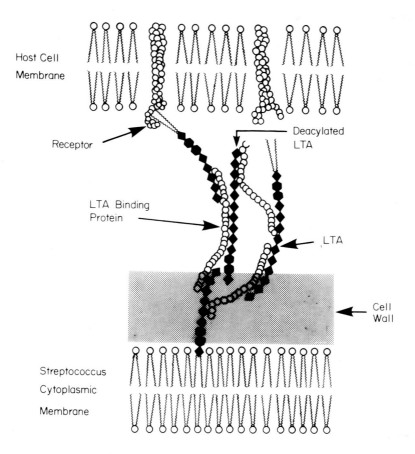

Figure 1. The possible mechanism whereby LTA functions as a "bridging" ligand. LTA is located at the bacterial cell wall–cell membrane interface (bottom) with the lipid moiety in the cell membrane. LTA and deacylated LTA constantly leave the bacterial cell and move across the cell wall. Cross-linking of LTA and deacylated LTA with surface LTA-binding proteins such as M protein would result in a fibrillar network of LTA and protein. This network would permit exposure of the lipid ends of the firmly anchored LTA to interact with receptors, such as fibronectin, on the cell surface (top). Reprinted with permission from Ofek *et al.* (1982).

It has been suggested that LTA is constantly excreted through the streptococcal cell wall (Wicken and Knox, 1977; and see Chapter 2, Section 2.4.1a). Studies conducted in our laboratory (Alkan and Beachey, 1978) found that LTA leaks from group A streptococci into the surrounding medium. The ability of LTA to complex with various molecular components of the streptococcal surface by the charged backbone suggested to us that the LTA might move through the streptococcal surface by alternately binding to and being released from clusters of positively charged residues of other molecules in the cell wall. The final stage of this journey would be an ionic interaction with M protein (or similar molecules) on the streptococcal fibrillae permitting an external display of the lipid moiety that can then interact with other cell membranes (Fig. 1). Consistent with this model is the finding that streptococcal cell surfaces are hydrophobic (Tylewska *et al.*, 1979; Ofek *et al.*, 1983). Miorner *et al.*

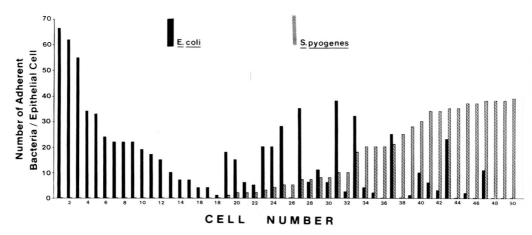

Figure 2. The distribution of gram-positive (*S. pyogenes*) and gram-negative (*E. coli*) bacteria adherent to buccal epithelial cells. One population of buccal epithelial cells has large numbers of adherent *S. pyogenes* (on the right). Further work demonstrated that the gram-positive cocci adhered to cells with surface fibronectin and the gram-negative cocci adhered to cells without surface fibronectin. Reprinted with permission from Abraham *et al.* (1983).

(1983) demonstrated that the surface hydrophobicity of group A streptococci, along with surface LTA, could be removed by proteolytic digestion. They also noted that the degree of surface hydrophobicity in individual strains correlated with the amount of surface LTA (Miorner *et al.*, 1983). Taken together, these studies lend credence to the idea that LTA is anchored by ionic interaction with other bacterial surface molecules and that such anchorage permits external exposure of the lipid end of the molecule.

2.1.5. The Role of Albumin as a Receptor Analog for the Binding of Group A Streptococci

Albumin is the major protein in blood and most other body fluids. It contains a thoroughly characterized binding site for fatty acids composed of a hydrophobic cleft in the tertiary folds of the molecule. Specificities for both the chain length and the head group of the fatty acid have been reported (reviewed by Peters, 1975). We examined the binding of LTA by albumin. LTA, but not deacylated LTA, bound to albumin immobilized on Sepharose and could be eluted with hydrophobic solvents (Simpson *et al.*, 1980a). LTA exhibited an affinity for the fatty acid-binding site of albumin that was midway between that reported for palmitic and octanoic acids (Simpson *et al.*, 1980a). The interaction of LTA with albumin indicates that LTA can engage in specific interactions with other molecular substances through its lipid moiety. It also suggests that cell-bound, albuminlike molecules containing fatty acid-binding sites could act as receptors for LTA-mediated streptococcal adhesion (Simpson *et al.*, 1980b). If epithelial cells contain such molecules, albumin should act as a receptor analog and compete with the human tissue-bound molecules for LTA binding. Indeed, albumin competitively inhibited the binding of LTA to human erythrocytes (Simpson *et al.*, 1980b) and albumin blocked the binding of group A streptococci to buccal epithelial cells in a dose-dependent manner (Beachey *et al.*, 1980).

These studies allowed us to formulate a set of criteria for identifying the mucosal cell surface receptor for group A streptococcal LTA: (1) the receptor must be a molecule present on the cell surface; (2) streptococci should react with the isolated receptor material and this reaction should be blocked by LTA, the isolated streptococcal adhesin; (3) LTA should bind to the isolated receptor in a lipid-dependent reaction; and (4) intact streptococci should interact with the receptor immobilized on a solid surface.

2.1.6. The Role of Fibronectin as a Receptor for Group A Streptococci

Fibronectin is a large glycoprotein ($M_r \sim 450,000$) found in soluble form in body fluids, particularly plasma, and in insoluble form on cell surfaces including buccal epithelial cells (criterion 1) (Zetter et al., 1979) (reviewed by Akiyama et al., 1981; Hormann, 1982). Fibronectin binds to a variety of substances including *Staphylococcus aureus* (Kuusela, 1978) and group A streptococci (Simpson and Beachey, 1983) (criterion 2) (see Section 3.1 for further discussion of the biologic role of fibronectin, and Chapter 4, Sections 3.2 and 4). Binding studies employing radiolabeled LTA and fibronectin indicated that fibronectin contains at least one LTA-binding site with characteristics similar to those reported for albumin (Courtney et al., 1983) (criterion 3). The affinity of LTA for fibronectin appears, however, to be several hundredfold greater than the affinity of LTA for albumin (albumin $nK_a = 0.63$ μM^{-1}, fibronectin $nK_a = 250$ μM^{+1}), where n = number of receptor sites and K_a = affinity constant (Simpson et al., 1980a; Courtney et al., 1983). A recent report by Morgenthaler (1982) supports the observation that fibronectin contains a fatty acid-binding site.

The ability of fibronectin to function as a cell surface receptor for streptococci was evident from the following studies. When fibronectin was immobilized on the surfaces of latex beads and added to suspensions of group A streptococci, the streptococci agglutinated to each other. The agglutination was inhibited by LTA, but not deacylated LTA (Simpson and Beachey, 1983) (criterion 4). Removal of fibronectin from oral epithelial cells decreased the adhesion of streptococci to the cells; when soluble fibronectin was added to the test system, the adhesion of streptococci to the epithelial cells was blocked in a dose dependent manner (Simpson and Beachey, 1983).

Fibronectin is present in human saliva (Simpson et al., 1982a; Babu et al., 1983) as well as on the surface of buccal epithelial cells (Zetter et al., 1979). Immunofluorescence studies indicate that the fibronectin is concentrated on the luminal side of epithelial cells and not on those surfaces which interface with other cells (Zetter et al., 1979). Thus, fibronectin fulfills the four criteria of a cell surface receptor for group A streptococci.

In summary, group A streptococci bind to buccal epithelial cells by forming a molecular "complex" that uses LTA as a "bridging" ligand. Molecules such as M protein anchor LTA to the bacteria by complexing with the charged polyglycerolphosphate backbone of LTA. The cell surface protein, fibronectin, anchors LTA to the buccal mucosal cells by binding the lipid moiety of LTA (Fig. 1). The specificity of this interaction is a function of: (1) the specificity of the fatty acid-binding site of fibronectin, (2) the availability of fibronectin on the eukaryotic cell surface, (3) the production of LTA by the bacterium, and (4) the tenacity of the LTA–bacterial surface protein bond.

2.2. The Role of LTA as an Adhesin for Group B Streptococci

2.2.1. Evidence against LTA as an Adhesin

A concensus has yet to emerge regarding the molecular mechanisms by which *S. agalactiae* (group B streptococci) adheres to animal cells. Nevertheless, LTA appears to play a role in the binding of group B streptococci to certain tissues, particularly immature tissues.

Group B streptococci possess surface fibrillae morphologically similar to the fibrillae of group A streptococci. As with group A streptococci, the fibrillae of group B streptococci can be removed by proteolytic treatment (Bramley and Hogben, 1982). Proteolysis also inhibits

S. agalactiae adhesion to mammary gland epithelial cells (Bramley and Hogben, 1982). In a somewhat conflicting study, Bagg *et al.* (1982) noted that proteolytic treatment inhibited the adhesion of group B streptococci to human buccal epithelial cells. Although both groups of investigators concluded that the *S. agalactiae* adhesin was a protein, the possibility of a protein–LTA complex was not rigorously ruled out.

While studying the adhesion of group B streptococci to human vaginal epithelial cells, Zawaneh *et al.* (1979) noted time, temperature, and pH optima. They also noted that penicillin did not influence this adhesion (Zawaneh *et al.*, 1979). As prior studies conducted in our laboratories had found that exposure of group A streptococci to penicillin diminished their capacity to adhere to human buccal epithelial cells by causing them to lose LTA (Alkan and Beachey, 1978), Zawaneh *et al.* interpreted their findings as evidence against LTA functioning as an *S. agalactiae* adhesin. In a more recent study, Goldschmidt and Panos (1984) noted that group A streptococci readily adhered to HeLa cells and amnion cells in tissue culture. Group B streptococci exhibited a different adhesion pattern; they did not bind to HeLa cells (< 10% of group adhesion) and bound to amnion cells only half as well as the group A streptococci. The adhesion of *S. agalactiae* to human amnion cells was not inhibited by LTA or anti-LTA serum but was inhibited by pretreating the bacterial cells with trypsin and Pronase. These findings indicated that group A and B streptococci adhered to HeLa and amnion cells by different mechanisms. The evidence also suggested that in one case, *S. agalactiae* binding to amnion cells, the adhesin was a protein (Goldschmidt and Panos, 1984).

2.2.2. Evidence for LTA as an Adhesin

The group B streptococcal strain studied by Goldschmidt and Panos (1984) produced very little LTA (< 0.1% of the cell dry weight). In contrast, Nealon and Mattingly (1983) reported that strains of group B streptococci isolated from infected infants were rich in LTA, containing roughly five times more of the macromolecule than strains isolated from asymptomatic infants. [Unfortunately, the methods used by Goldschmidt and Panos and Nealon and Mattingly do not allow direct comparisons of the absolute LTA content of their respective study organisms. Nevertheless, Goldschmidt and Panos stated that their strain had less LTA than the Nealon and Mattingly strains (Goldschmidt and Panos, 1984)]. In a follow-up study, Nealon and Mattingly (1984) reported that LTA markedly inhibited the adhesion of several strains of group B streptococci to human embryonic cells, human fetal brain cells, human fetal lung cells, and human buccal epithelial cells. Deacylated LTA also inhibited the adhesion of the group B streptococci to fetal cells but not to buccal epithelial cells. They interpreted these findings as evidence for two LTA-mediated mechanisms, one relying upon the lipid end which binds the organisms to human buccal epithelial cells and one relying upon the charged backbone which binds the organisms to fetal tissues.

Nealon and Mattingly (1984) further postulated that the mucosal cell receptor for the LTA backbone is found only on immature tissues and is lost as the tissue matures. This hypothesis dovetails nicely with the earlier observations of Broughton and Baker (1983). These investigators noted that as opposed to group A streptococci which selectively adhere to adult buccal epithelial cells versus neonatal buccal epithelial cells (Ofek *et al.*, 1977), group B streptococci selectively adhere to neonatal epithelial cells versus adult epithelial cells (Broughton and Baker, 1983). They also noted that group B streptococci selectively adhere to cells from neonates with *S. agalactiae* infection versus cells from healthy neonates (Broughton and Baker, 1983). These findings suggest that certain neonates have greater concentrations of bacterial receptors on their surfaces, predisposing the infant to group B streptococcal infection.

Earlier studies conducted by Cox (1982) found that pretreatment of human buccal epithelial cells with LTA inhibited group B streptococcal adhesion. Cox (1982) applied this observation by using LTA to prevent colonization of infant mice with group B streptococci. Three-day-old mice were treated with topical applications of LTA to the oral cavity, perineum, and nape; the mother was painted with a slurry of group B streptococci over her oral cavity, vagina, and nipples. After 3 days, 47% of the control pups were culture-positive for group B streptococci while none of the treated pups were culture-positive. Cox (1982) further noted that the oral application of LTA prevented colonization of the oropharynx with *S. agalactiae* in a dose-dependent manner. Topical LTA probably interferes with colonization of neonatal mice by flooding their tissues with excess bacterial adhesin and thereby covering receptor sites and inhibiting group B streptococcal adhesion.

In summary, LTA seems to mediate the binding of *S. agalactiae* to tissues by at least two separate mechanisms, one dependent upon the hydrophobic moiety and the other dependent upon the charged backbone. The formation of molecular complexes with the different populations of cellular receptors that characterize neonatal and adult tissues presumably explains the selectivity of *S. agalactiae* tissue adhesion. Further work will have to be performed to identify these receptors and to determine just how *S. agalactiae* holds the LTA molecule on its surface so that both the hydrophobic moiety and the charged backbone can bind to their respective receptors.

2.3. The Role of LTA as an Adhesin for Other Organisms

Along with group A and B streptococci, LTA also plays a role in the adhesion of *Staphylococcus aureus* to mucosal surfaces. Carruthers and Kabat (1983) studied the adhesion of that bacterium to human buccal epithelial cells. For the sake of comparison, additional parallel studies were conducted with group A streptococci. LTA from both organisms inhibited the adhesion of staphylococci and streptococci to buccal epithelial cells, while deacylated LTA from both organisms had no effect. Of great interest was the finding of specificity; staphylococcal LTA inhibited staphyloccal adhesion to a greater extent than did streptococcal LTA and vice versa (Carruthers and Kabat, 1983).

LTA may mediate the adhesion of *Streptococcus mutans* to dental surfaces (see also Chapter 6, Section 4.7.1). Several adhesion mechanisms have been proposed. For instance, hydroxyapatite, the mineral crystal of teeth, strongly adsorbs LTA *in vitro,* probably by forming ionic bonds between the negatively charged backbone of LTA and positively charged calcium ions in the crystal lattice (Ciardi *et al.,* 1977). These bonds could directly bind the bacterium to the surface of the tooth *in vivo.* On the other hand, LTA has also been found in high concentrations in sucrose-dependent dental plaque, where it probably binds directly to glucan (Rolla *et al.,* 1980). While strengthening the bonds that hold the bacterial plaque cells together, simultaneous adsorption of LTA by hydroxyapatite could serve to anchor the plaque to the dental surface (Rolla *et al.,* 1980).

LTA also interacts with the *S. mutans* surface enzyme, glucosyltransferase. Glucosyltransferase both synthesizes and binds glucan which in turn binds the *S. mutans* plaque bacteria together (see Section 4.1 for further discussion). Kuramitsu *et al.* (1980) found high-molecular-weight aggregates of glucosyltransferase in culture fluids of *S. mutans.* These aggregates were rich in polyglycerolphosphate, indicating the presence of LTA. Purified LTA, but not deacylated LTA, inhibited glucan synthesis and blocked sucrose-dependent colonization of smooth surfaces, indicating the hydrophobic moiety interfered with glucosyltransferase activity. Although deacylated LTA did not inhibit enzyme activity, it nevertheless interacted with glucosyltransferase, as it caused soluble glucosyltransferase to form high-molecular-weight aggregates *in vitro.* These findings suggested to Kuramitsu *et al.*

(1980) that LTA might bind glucosyltransferase by the charged backbone while the lipid end anchored the complex to the surface of the bacteria. In agreement with this hypothesis, they noted that *Lactobacillus casei* and *L. fermentum,* which both contained polyglycerolphosphate teichoic acids, bound glucosyltransferase activity from exogenous *S. mutans* enzyme. Although it has not been examined, the same mechanism may explain the ability of certain organisms, such as *S. sanguis* (Hamada *et al.,* 1981), to bind exogenous *S. mutans* glucosyltransferase. Hamada *et al.* (1979) also noted the formation of LTA–glucosyltransferase complexes in *S. mutans* culture supernatants, and that hydroxyapatite could selectively adsorb these complexes. These findings led them to propose that the dental adsorption of extracellular glucosyltransferase through the agency of LTA could promote the adhesion of *S. mutans* to the dental surface and stabilize the adhesive plaque community (Hamada *et al.,* 1979). In summary, it appears that several mechanisms exist by which LTA can mediate the binding of bacterial plaque communities to dental surfaces.

2.4. LTA Summary

The amphipathic nature of LTA allows the molecule simultaneously to interact with charged (hydrophilic) substances and noncharged (hydrophobic) substances. Thus, LTA can form a molecular bridge between substances which otherwise would never come into contact and thereby bind them together. Gram-positive bacteria utilize amphipathic LTA as a bridging ligand for the binding of bacteria to animal tissues. Exactly how this binding results in specific adhesion is still not quite clear. In the earlier discussion of *S. pyogenes* (Section 2.1.6), we pointed out that specificity was probably a function of the molecular "fit" between LTA and both (1) the anchoring molecule on the surface of the bacteria (M protein) and (2) the anchoring molecule on the surface of the animal cell (fibronectin). Streptococcal adhesion was further dependent upon the availability and display of these anchoring molecules (M protein and fibronectin). This of course is an incomplete explanation. It does not fully explain the differences in specificity between LTA obtained from different species of bacteria; such as the differences between streptococcal and staphylococcal LTA observed by Carruthers and Kabat (1983). Furthermore, the molecular basis for the interaction of LTA in other bacteria–animal cell systems has yet to be examined, although it is presumably analogous to group A streptococcal adhesion. Nevertheless, LTA seems to be a common mechanism by which gram-positive bacteria adhere to cellular substrates.

Gram-negative bacteria do not produce LTA, but they do produce an equivalent amphipathic molecule, lipopolysaccharide (LPS). As with LTA, LPS binds to a discrete population of receptors on many eukaryotic cells, and this binding is dependent upon the lipid portion of the molecule (Springer *et al.,* 1970, 1973; Springer and Adye, 1975). Nevertheless, as far as we are aware, LPS has yet to be identified as an adhesin for gram-negative bacteria. This does not rule out such a biologic role for LPS; the absence of such evidence may simply reflect our limited understanding of bacterial adhesion.

3. HOST-DERIVED "BRIDGING" LIGANDS

3.1. The Role of Fibronectin in the Adhesion of Bacteria to Animal Tissues

Fibronectin is composed of a dimer of two nearly identical polypeptide chains hinged at the COOH-terminal end by a single disulfide bond (for recent reviews, see Mosher and Furcht, 1981; Akiyama *et al.,* 1981; Hormann, 1982; Doran, 1983; Yamada, 1983; Ruoslahti, 1981). Each chain is divided into five domains, each of which binds to charac-

teristic materials, including gelatin, fibrin, factor XIII, actin, DNA, *S. aureus,* and eukaryotic cells. The NH$_2$-terminal fragment in particular accepts covalent cross-linking by
enzyme-catalyzed transamidation with a variety of substances, including collagen, fibrin,
other fibronectin molecules, and staphylococci.

As fibronectin is known to promote the phagocytosis of certain particles (Saba *et al.,*
1983; reviewed by Doran, 1983), its ability to bind bacteria suggested that it may play a role
in host defense. Initial investigations pursued the possibility that fibronectin functions as a
nonimmune opsonin for staphylococci. In *in vitro* situations, many reports suggest that
fibronectin promotes the attachment of staphylococci to the phagocyte cell surface but not
necessarily phagocytosis (Lanser and Saba, 1981; Verbrugh *et al.,* 1981; Proctor *et al.,*
1982b). Fibronectin was also found to promote the attachment of group A streptococci to
phagocytic cells (Simpson *et al.,* 1982b). The physiologic importance of this binding,
however, remains controversial (Lanser and Saba, 1981, 1982; Verbrugh *et al.,* 1981;
Proctor *et al.,* 1982b; Simpson *et al.,* 1982b; Saba *et al.,* 1983); recent reports even question
whether it occurs (van de Water *et al.,* 1983). Nevertheless, these studies do suggest that
soluble fibronectin may function as a ligand to bind bacteria to animal cell surfaces.

In addition to staphylococci and streptococci, fibronectin binds many, but not all, gram-
positive coccal bacteria (Table III). For some species, such as *Staphylococcus epidermidis,*
the binding varies considerably from strain to strain; for other species, such as *Streptococcus
pneumoniae,* the binding takes place at a low, but predictable, level. Little is known about
how each species of bacteria binds to fibronectin, but at least for two organisms, *S. aureus*
and *S. pyogenes,* the mechanism appears to be different. *S. pyogenes* binds fibronectin by
complexing with LTA at a fatty acid-binding site (see Section 2.1.6) [it should be noted,
however, that Speziale *et al.* (1984) reported that fibronectin binds an *S. pyogenes* surface
protein]. In contrast, *S. aureus* binds fibronectin by a surface fibronectin-binding protein
(Verbrugh *et al.,* 1981; Espersen and Clemmensen, 1982; Ryden *et al.,* 1983). This binding
of fibronectin to *S. aureus* is inhibited by lysine and carbamide (Kuusela, 1978; Espersen and
Clemmensen, 1982) and the NH$_2$-terminal fibronectin peptide (Mosher and Proctor, 1980).
In striking contrast to gram-positive cocci, fibronectin binds poorly to gram-negative bacilli
(Kuusela, 1978; Simpson *et al.,* 1982b). This suggests that one of the biologic functions of
fibronectin may be specifically to bind certain gram-positive cocci to cell surfaces.

Woods and Johanson first called attention to the possible role of fibronectin as a
modulator of human oral microecology (Woods *et al.,* 1981a,b). They noted that trypsin
digestion of cell-bound fibronectin increased the adhesion of *Pseudomonas aeruginosa* to the
exposed epithelial cell surface (Woods *et al.,* 1981a,b). Recent studies conducted in our
laboratories found that the buccal mucosal cells could be divided into two distinct populations. One population of cells was coated with fibronectin and probably represented superficial mucosal cells; a second population lacked surface fibronectin and probably represented
deeper, unexposed cells. When adhesion experiments were conducted with a mixture of
gram-positive (*S. pyogenes*) and gram-negative (*P. aeruginosa* or *Escherichia coli*) bacteria,
the epithelial cells that displayed fibronectin bound the streptococci, whereas the cells
without fibronectin bound the gram-negative organisms (Fig. 2) (Abraham *et al.,* 1983).
These observations support the possibility that exogenous fibronectin in salivary secretions
binds to superficial mucosal cells and functions as a "bridging" ligand by specifically
binding gram-positive bacteria to the cell surface (Simpson *et al.,* 1982a).

The hypothesis that fibronectin selectively promotes the colonization of the oropharynx
with gram-positive bacteria may explain a well-recognized clinical observation. Certain
patients (hospitalized, ill, aged) have predominantly gram-negative bacilli colonizing their
oropharynx rather than the usual gram-positive microflora seen in healthy people (Valenti *et*

Table III. Fibronectin Binding by Bacteria[a]

Bacteria which consistently bind fibronectin
   Group A streptococci
   Group C streptococci
   Group G streptococci
   Staphylococcus aureus
   Staphylococcus haemolyticus
Bacteria with variable binding of fibronectin, or bacteria demonstrated to bind fibronectin on the basis of
   selected strains
   Streptococcus mutans
   Streptococcus salivarius
   Streptococcus sanguis
   Streptococcus pneumoniae
   Staphylococcus saprophyticus
   Staphylococcus warneri
   Staphylococcus hyicus
   Staphylococcus simulans
   Staphylococcus hominis
Bacteria with low-level or infrequent binding of fibronectin, or bacteria that do not bind fibronectin on the basis
   of selected strains
   Group B streptococci
   Group D streptococci
   Group M streptococci
   Group N streptococci
   Group P streptococci
   Group U streptococci
   Staphlococcus epidermidis
   Staphylococcus capitis
   Staphylococcus xylosus
   Staphylococcus cohnii
   Streptococcus mitis
   Pseudomonas aeruginosa
   Klebsiella pneumoniae
   Escherichia coli
   Neisseria meningitidis
   Neisseria gonorrhoeae
   Hemophilus influenzae

[a]Summarized from Doran and Raynor (1981), Simpson *et al.* (1982b), Switalski *et al.* (1982, 1983), Babu *et al.* (1983), and Myhre and Kuusela (1983).

al., 1978). As diminished plasma fibronectin levels are found in similar patient populations (burn, trauma, and malnourished patients; reviewed by Saba *et al.*, 1983), these patients may also have decreased fibronectin in their oral secretions. This could result in lower levels of mucosal cell fibronectin and, consequently, less binding of gram-positive bacteria and greater binding of gram-negative bacteria to the mucosal cell surface. On the other hand, Woods *et al.* (1981a,b) have proposed that increased salivary protease digestion of surface fibronectin is responsible for the increased adhesion of gram-negative bacilli in certain patient populations. Either way, it appears likely that fibronectin functions as a modulator of oral microecology, by selectively promoting the adhesion of gram-positive bacteria versus gram-negative bacteria.

    The role of fibronectin as a ''bridging'' ligand for the adhesion of bacteria to animal cell surfaces is not necessarily limited to the oral mucosa. Mosher and Furcht (1981), for example, suggested that fibronectin plays a role in the binding of bacteria to injured tissues.

Although as far as we know this possibility has not yet been investigated, fibronectin may be important in the pathogenesis of certain endothelial infections such as endocarditis (see Section 4.2).

3.2. The Role of Other Serum Proteins in the Adhesion of Bacteria to Animal Tissues

3.2.1. Miscellaneous Serum Proteins

Bacteria, particularly gram-positive cocci, bind many serum proteins in addition to soluble fibronectin. For example, group C and G streptococci bind serum albumin (Myhre and Kronvall, 1980); group A, C, G and various oral streptococci bind aggregated β_2 microglobulin (Kronvall et al., 1978; Ericson et al., 1980), and some strains of streptococci bind haptoglobin (Kohler and Prokop, 1978). At this point, however, the physiologic significance of these interactions is unknown.

Certain other serum proteins, such as immunoglobulin, C-reactive protein, and the complement cascade, have well-described abilities to bind to bacterial surfaces. These proteins, however, function as opsonins to bind bacteria to professional phagocytes for the purposes of phagocytosis. Although these opsonic proteins certainly qualify as ligands, their biologic role is in the arena of host defense and not bacterial colonization.

3.2.2. Immunoglobulin

Under certain circumstances, it is possible that IgG could function as a ligand for bacterial adhesion to animal tissues. It has long been recognized that pathogenic staphylococci have protein A on their surfaces which specifically binds to the Fc receptor of certain classes of IgG (Forsgren and Sjoquist, 1966; Kronvall and Williams, 1969). Immunoglobulin-binding proteins have also been demonstrated on the surfaces of group A, C, and G streptococci (Kronvall, 1973; Shea and Ferretti, 1981; Reis et al., 1983). For S. aureus, protein A may block the alternative pathway of complement activation and subsequent opsonization and phagocytosis (Spika et al., 1981); the functional importance of immunoglobulin binding for streptococci is unknown.

In a recent report, Mogensen and Dishon (1983) demonstrated the in vitro ability of the immunoglobulin–protein A complex to function as a ligand and bind staphylococci to infected tissues. Protein A-rich staphylococci were first incubated in antisera to herpes simplex or varicella–zoster. The staphylococci were then exposed to clinical tissue specimens from patients with viral infections. The immunoglobulin-coated staphylococci bound to corresponding viral antigens expressed on the infected cell surface (Mogensen and Dishon, 1983). In an earlier report, Austin and Daniels (1978) found a fivefold increase in the adhesion of S. aureus to influenza A-infected rabbit kidney cells when the cells were first treated with immune serum to the influenza virus. Adhesion was not influenced by pretreating the cells with nonimmune serum. Although these studies took place under artificial conditions, they suggest the strong possibility that staphylococci and streptococci can exploit IgG in vivo as a "bridging" ligand for targeted adhesion to compromised tissues.

3.2.3. Fibrinogen

A suspension of S. aureus cells clumps together when mixed with plasma. This clumping reaction depends upon a surface protein (clumping factor) that binds the terminal 15-residue amino acid sequence of the γ chain of fibrinogen (reviewed by Hawiger et al., 1983).

A similar staphylococcal binding sequence is present on the opposite side of the fibrinogen molecule making the molecule divalent and allowing it to agglutinate staphylococci. This same 15-amino-acid sequence also binds fibrinogen to platelets (reviewed by Hawiger *et al.*, 1983). In addition to clumping factor, staphylococci also possess a surface polysaccharide (compact colony-forming active substance) which gels fibrinogen (Yoshida *et al.*, 1977). The biologic importance of these fibrinogen-reactive substances is unknown.

Similar to *S. aureus*, *S. pyogenes* also form clumps when mixed with plasma. The fibrinogen-binding protein in this case is surface M protein (Kantor, 1965). The ability to bind fibrinogen explains the antiphagocytic activity of M protein; by binding fibrinogen, M protein coats the streptococcus with a layer of protein that blocks complement activation and subsequent opsonization and phagocytosis (Whitnack and Beachey, 1982).

Fibrinogen very likely binds *S. aureus* and *S. pyogenes* to the platelet–fibrin matrix of injured tissues, but except for endocarditis, we are unaware of any studies exploring this possibility (see Section 4.2 for further discussion). Nevertheless, the functional ability of fibrinogen to bind staphylococci and streptococci to diseased tissues *in vitro* has been demonstrated by the work of Davison and Sanford (1981) with influenza A-infected tissue culture cells.

Influenza is well known to predispose patients to secondary bacterial pneumonia, particularly to pneumonia from staphylococci, pneumococci, and group A streptococci. A number of authors have investigated the possibility that influenza A infection increases the adhesion of bacteria to host tissues. Under *in vitro* conditions, *S. aureus* and *S. agalactiae* adhered in greater numbers to tissues infected with influenza A than to uninfected tissues (Pan *et al.*, 1979; Fainstein *et al.*, 1980; Jones and Menna, 1982). The adhesion of *S. agalactiae* to canine kidney cells infected with influenza A was inhibited by neuraminidase and antibody to viral hemagglutinin. Adhesion was also inhibited by tunicamycin, which inhibites glycosylation of animal cell proteins. These findings indicated that the organism adhered specifically to sialic acid residues on viral hemagglutinin expressed by infected cells (Pan *et al.*, 1979; Sanford *et al.*, 1979).

S. aureus blocked adhesion of *S. agalactiae,* but the *S. aureus* cell receptor was not inhibited by neuraminidase, antibody to hemagglutinin, or tunicamycin, indicating that the cell receptor for *S. aureus* was neither viral hemagglutinin nor sialic acid (Davison and Sanford, 1981). The cell receptor for *S. aureus* was protease sensitive, but could not otherwise be identified (Davison and Sanford, 1982). Davison and Sanford did find, however, that pretreatment of the bacteria with fibrinogen dramatically increased their tissue adhesion and also enabled the staphylococci to adsorb free virus out of solution. Surprisingly, staphylococcal adhesion was not inhibited by partially purified clumping factor added to the system or by protein A or teichoic acid, suggesting that these surface structures did not function as bacterial adhesins by binding to the host fibrinogen. Davison and Sanford (1982) suggested that fibrinogen forms a molecular bridge between *S. aureus* and cells infected with influenza A virus, thereby binding the bacteria to the cell surface.

As opposed to *S. agalactiae, S. pyogenes* does not seem to bind directly to influenza-infected cells. Nevertheless, group A streptococcal pneumonia is common after influenza infections. Sanford and Davison (1982) extended their studies by investigating whether the fibrinogen-binding ability of *S. pyogenes* allows the microorganism to adhere to cells infected with influenza A virus. Fibrinogen treatment did increase group A streptococcal adhesion to influenza A-infected tissue culture cells. The absolute amount of adhesion varied with the test strain of streptococcus or influenza A. Adhesion also varied with the source of fibrinogen; only human fibrinogen, not canine or bovine, promoted the attachment of streptococci to the infected cells. In contrast to their earlier work with *S. aureus,* Davison and Sanford (1982) found that tunicamycin greatly diminished the adhesion of *S. pyogenes* to the infected cells. This presents a problem. While it is reasonable to assume that fibrinogen binds to staphylococci and streptococci by different mechanisms, it is not at all clear why fibrinogen should bind to different eukaryotic cell receptors when linking the eukaryotic cell to staphylococci and streptococci. Nevertheless, the proposal that fibrinogen can bind selected bacteria to neoreceptors on cells infected with influenza A is an intriguing model for the pathogenesis of postinfluenzal pneumonia.

4. BACTERIAL ADHESION TO TISSUES BY "BRIDGING" CELLS

4.1. "Bridging" Cells of Bacterial Origin—The Formation of Dental Plaque

The best example in human microbiology of bacteria employing intermediary bacteria to link themselves to tissue surfaces is in the formation of dental plaque. The genesis of dental plaque also provides further examples of "bridging" ligands derived from host and bacteria.

Dental plaque, the provoking agent of dental disease, includes the bacterial accumulations in the crevices between tooth and gum (subgingival or supragingival plaque) and over the enamel surface of the tooth (coronal plaque). By fixing bacteria onto oral surfaces, the dental plaque allows microbial digestive and toxic products to cause dental demineralization (caries) and mucosal inflammation (periodontitis and gingivitis). The primary microbial constituents of dental plaque are streptococci (including *S. mutans, S. mitis, S. miteor, S. sanguis, S. salivarius*), actinomyces (*Actinomyces viscosus, A. naeslundii*) and Veillonellae, but the microbial membership also includes *Lactobacillus, Neisseria, Bacteroides, Fusobacterium, Capnocytophaga, Actinobacillus, Selenomas,* and *Campylobacter,* to name only a few (reviewed by Gibbons and van Houte, 1980; van Houte, 1982). The bacterial membership of dental plaque is not just a simple random collection of organisms; rather, dental plaque represents microbial climax communities whose precise composition and organization depend upon specific molecular interactions between the plaque components. These interactions include: bacteria to dental surface or oral mucosa, bacteria to homologous bacteria, bacteria to heterologous bacteria, and bacteria to extracellular substances (as summarized by van Houte, 1982).

Extracellular substances are either host-derived, such as salivary agglutinins, fibronectin, and secretory IgA; or bacteria-derived, such as the glucans and mutans synthesized by *S. mutans* glucosyltransferase. These materials constitute the plaque matrix. Bacteria bind by specific interactions to the extracellular substances, resulting in a tightly woven mass. A portion of the extracellular substances adsorb to the dental enamel, coating it with a biofilm or "pellicle" (reviewed by Gibbons and van Houte, 1980; van Houte, 1982). The adsorbed pellicle components then function as bacterial receptors for bacteria adhering to the dental surface and thereby anchor the dental plaque to the tooth.

The multiplicity of bacterial and molecular elements which enter into plaque construction creates a bewildering array of interactions that are far too complicated and controversial for these pages. Instead, for simplicity, the following discussion will dwell only on the proposed mechanisms by which three prominent members of the plaque community, *S. mutans, S. sanguis,* and *A. viscosus,* form coronal plaque.

S. mutans causes dental caries in experimental animals only when the animals are fed a sucrose-rich diet. When grown *in vitro,* sucrose, but not other carbohydrates, added to culture media enables *S. mutans* to coat the culture container with a thick layer of adhesive bacteria. In sucrose media, *S. mutans* produces several types of polysaccharides synthesized by extracellular enzymes, including water-soluble glucans, insoluble mutans, and soluble

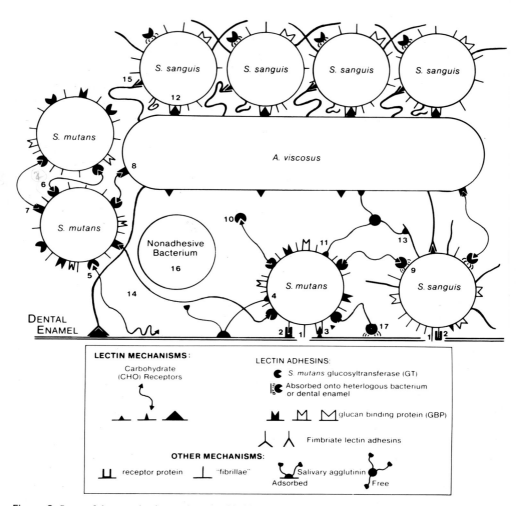

Figure 3. Some of the complex interactions that bind bacteria to dental enamel in the formation of dental plaque. The plaque has a foundation of organic substances adsorbed by the hydroxyapatite mineral crystal. This dental pellicle consists of materials derived from both bacteria and the host. Bacteria bind to the elements of this pellicle which anchors the bacteria to the dental surface. Subsequent layers of organisms are built upon this first layer by interactions with the plaque matrix molecules and through them interactions with other bacteria. On the other hand, bacteria may bind directly to other bacteria. These intermediary binding interactions bind the bacteria to the dental surface by a chain of linking molecules and organisms.

The specific interactions depicted here include: (1) Direct binding by lipoteichoic acid and teichoic acid to the mineral crystal hydroxyapatite. Binding by pellicle proteins such as fibronectin or (2) lectin–carbohydrate interactions such as *S. mutans* binding pellicle galactose residues (3). Pellicle agglutinins (4) may also bind the bacteria to the dental surface. *S. mutans* produces extracellular glucosyltransferase (GT) which synthesizes and binds glucan. *S. mutans* may bind pellicle glucan by the GT (5). Probably more often, *S. mutans* binds other *S. mutans* by both binding glucan through GT (6). This mechanism is accentuated by a surface glucan-binding protein (GBP) on *S. mutans* which also binds glucan (7). The oral bacterium *A. viscosus* also has a surface GBP (8). Organisms such as *S. sanguis* that do not generate GT may adsorb extracellular free *S. mutans* GT to their surfaces and bind to the glucan matrix (9). Free GT continues to synthesize further glucan, increasing the matrix (10). Free salivary agglutinins may also bind bacteria together in the plaque matrix by binding their surfaces (11) or their fimbriae (13). Bacteria may bind directly together by lectin interactions (12, 15) such as in the case of *S. sanguis* and *A. viscosus*. *A. viscosus* further anchors the aggregate to the dental surface by a fimbria-binding pellicle carbohydrate (14). Some nonbinding bacteria may become innocently trapped in the matrix "jungle" (16). And the glucan matrix may be further anchored to the surface of the tooth by adsorption of an LTA–GT complex (17).

fructans. *In vivo*, the polysaccharides form part of the extracellular matrix. They function primarily to bind *S. mutans* cells to each other through surface glucosyltransferases (GT) (Staat *et al.*, 1980) (Fig. 3). *S. mutans* cells are further tied together by cell surface lectins that also bind glucan [glucan-binding protein (GBP)] (Germaine and Schachtele, 1976; McCabe *et al.*, 1977) (Fig. 3). The GT is constantly released into the surrounding fluids where it remains free, or binds to the dental surface through LTA. Either way, the GT remains active, spinning out more glucan, and perhaps further enmeshing *S. mutans* and other bacteria (as suggested by Gibbons and van Houte, 1980) (Fig. 3). Glucan may also become adsorbed to the dental surface. The major function of glucan appears, however, to be the binding of *S. mutans* cells to each through GT and GBP. This creates a coat of bacteria many cells thick on the dental surface with only the lowest level in direct contact with the enamel. On the other hand, *S. mutans* binds to the dental enamel by a variety of mechanisms. Some of these mechanisms are LTA dependent (Section 2.3). They include the direct adsorption by hydroxyapatite of LTA and cell wall teichoic acid (Ciardi *et al.*, 1977) and the adsorption by hydroxyapatite of LTA complexed with GT (Hamada *et al.*, 1981) and complexed with glucan (Rolla *et al.*, 1980). *S. mutans* may, in addition, bind by GT and GBP to glucan that has already been directly incorporated into the dental pellicle (Liljemark and Schauer, 1977). Furthermore, *S. mutans* can bind to host proteins, such as fibronectin (Babu *et al.*, 1983), salivary agglutinins (Magnusson and Ericson, 1976; Magnusson *et al.*, 1976), and aggregated β_2-microglobulin (Ericson *et al.*, 1980), which may have become incorporated into the dental pellicle. Finally, *S. mutans* has a cell surface galactose-specific lectin which further links the organism to the pellicle and the plaque matrix by binding to extracellular substances containing carbohydrate (Gibbons and Qureshi, 1979) (Fig. 3).

The adhesion of *S. mutans* to dental surfaces does not take place independently of other bacteria. Organisms such as *A. viscosus* can also bind to glucan generated by *S. mutans* via their own cell surface GBP (Borgegu and McBride, 1976) (Fig. 3). If the microbe lacks GBP, such as in the case of most *S. sanguis* strains, the organism may still bind to glucan by adsorbing to its surface extracellular GT from *S. mutans* (Hamada *et al.*, 1981) (Fig. 3). *S. sanguis* and *A. viscosus* coaggregate, which further illustrates the complexity of the heterologous interactions. *A. viscosus* binds to *S. sanguis* surface carbohydrates by a fimbrial lactose-specific lectin (Heeb *et al.*, 1982; McIntire *et al.*, 1982; Revis *et al.*, 1982) and by lactose-resistant fimbriae (Kolenbrander, 1982) (Fig. 3). The lactose-resistant fimbriae also appear to directly bind *A. viscosus* to the dental pellicle (Clark *et al.*, 1981). *S. sanguis* seems to bind directly to the dental surface by mechanisms similar to those of *S. mutans* (Rolla *et al.*, 1977; Liljemark and Bloomquist, 1981). *S. sanguis* also possesses fibrillae, the LTA of which bind salivary agglutinins fixed on the dental pellicle (Hogg *et al.*, 1981; Hogg and Embery, 1982).

In summary, because of its bulk, development of the plaque microbial community depends upon the linking of bacteria together in addition to the adhesion of bacteria to the dental surface. As a result, the great majority of plaque bacteria are held to the dental surface by a chain of intermediary cells. The microbial chain, in turn, is anchored to the dental surface by a second set of mechanisms, mechanisms which generally involve the simultaneous linking of bacteria to hydroxyapatite by a ligand that is a component of the dental pellicle.

4.2. "Bridging" Cells of Host Origin—The Pathogenesis of Bacterial Endocarditis

In a manner entirely analogous to dental plaque formation, bacteria may adhere to specific tissues by first binding to host cells which in turn bind to the targeted tissue. This is best exemplified by the pathogenesis of bacterial endocarditis.

Gram-positive bacteria, particularly *S. aureus, S. pyogenes,* and oral streptococci such as *S. mutans* and *S. sanguis,* are the most common causes of bacterial endocarditis (reviewed by Thompson *et al.,* 1982). The predominance of these bacteria is probably due to many factors, but of particular interest is their ability to selectively adhere to damaged endothelial tissue.

Histologic studies conducted on animals with intracardiac plastic catheters have found that experimental bacterial endocarditis develops at the site of valvular injury. Circulating bacteria appear to selectively adhere to the platelet–fibrin network overlying the traumatized valvular leaflets (Garrison and Freedman, 1970; Perlman and Freedman, 1971; Durack and Beeson, 1972). Similar platelet–fibrin "vegetations" or thrombi are pathognomonic findings of subacute bacterial endocarditis. After experimental infection, the vegetation increases in size with bacterial recruitment of additional platelets and fibrin; these blood elements further enmesh the bacteria in the growing thrombus.

The evolution of this infection indicates that specific adhesion reactions take place between the bacteria and the thrombus components. Studies pursuing this possibility have primarily focused upon the glucan-mediated adhesion of oral streptococci to the valvular endothelium (Gould *et al.,* 1975; Scheld *et al.,* 1978; Ramirez-Ronda, 1980) or on the influence of antibiotics on bacterial adhesion (Scheld *et al.,* 1981; Bernard *et al.,* 1981). However, the mechanism by which glucan mediates bacterial adhesion to the endocardium or by which antibiotics inhibit that adhesion is not clear.

Several mechanisms do exist whereby certain gram-positive bacteria could bind to damaged tissues. In the prior discussion of fibronectin and fibrinogen, we have noted that streptococci and staphylococci bind to these proteins. Furthermore, this binding is likely to take place *in vivo* where it could function to bind the bacteria to injured tissues. Fibrinogen (reviewed by Hawiger *et al.,* 1983) and fibronectin (reviewed by Akiyama *et al.,* 1981) also bind to platelets and could mediate bacterial adhesion to these cells. In a recent report, Kurpiewski *et al.* (1983) reported the aggregation of human platelets by *S. pyogenes.* Aggregation depended upon the presence of fibrinogen and was specific for *S. pyogenes;* bacteria like *E. coli* and *S. agalactiae* did not cause aggregation. Gram-positive cocci also directly adhere to platelets and by using platelets as an intermediary bridge, these bacteria can bind to damaged tissues. For example, Clawson and White (1971) studied individual strains of bacteria and found that *S. aureus* and *S. pyogenes* strongly bound and aggregated platelets, while *Streptococcus faecalis* did so to a lesser extent. *S. pneumoniae* and *E. coli,* which do not frequently cause endocarditis, exhibited very weak aggregating ability. Herzberg *et al.* (1983) conducted similar studies examining the adhesion of *S. sanguis* and *S. mutans* to platelet ghosts. Five of seven strains of *S. sanguis* demonstrated appreciable binding while only two of six *S. mutans* strains were bound. In addition to strain variability, Herzberg *et al.* (1983) noted that variations may exist in host susceptibility. The bacterial adhesins were not identified, but *S. sanguis* fibrillae appeared to mediate the attachment of bacteria to platelets (Herzberg *et al.,* 1983). Lowy *et al.* (1983) examined the adhesion of *S. sanguis* to traumatized animal heart valves and to fibrin–platelet clots *in vitro.* Adhesion of the organism was greatly diminished by treatment with penicillin, suggesting a loss of LTA with a loss of adhesiveness. LTA was further implicated as the adhesin, for LTA antiserum and LTA pretreatment inhibited adhesion of *S. sanguis* to the fibrin–platelet clots *in vitro* (Lowy *et al.,* 1983).

Even though the evidence is limited, it seems likely that the tendency for certain gram-positive cocci to cause bacterial endocarditis is in part due to their ability to selectively adhere to heart tissue. The necessary prerequisite for this infection is the initial endocardial injury, as it is the tissue response to the injury which allows the bacteria to adhere to the damaged tissues. The adhesion takes place through intermediary substances, entirely analogous to the formation of dental plaque. The thrombus is anchored onto the tissue surface by the specific reactions of platelets and serum proteins with the damaged tissues. The bacteria, in turn, are bound to each other and the host tissues by interacting with the ''matrix'' of platelets and serum proteins.

5. CONCLUSION

The complex mechanisms by which certain gram-positive bacteria adhere to animal tissues do not have the simple elegance of lectin–carbohydrate interactions. Nevertheless, these complex mechanisms accomplish the same goal of specific tissue adhesion by utilizing intermediary ligands or cells. These intermediary ''bridging'' materials share the common property of multivalent binding by different mechanisms to different substrates. The overall specificity of the interaction depends upon: (1) the availability of the intermediary ''bridging'' ligand or cell, (2) the availability and nature of the bacterial structure that binds the ''bridging'' ligand or cell, and (3) the availability and nature of the animal cell structure that binds the ''bridging'' ligand or cell. The observation that gram-positive cocci, rather than gram-negative bacilli, utilize these mechanisms may only reflect the level of interest of investigators in both fields; or it may point to fundamental biologic differences between these classes of bacteria.

ACKNOWLEDGMENTS. The authors' research studies are supported by research funds from the U.S. Veterans Administration and by Research Grants AI-10085 and AI-13550 from the National Institutes of Allergy and Infectious Diseases. We thank Connie Carrier and Johnnie Smith for excellent secretarial assistance.

REFERENCES

Abraham, S. N., Beachey, E. H., and Simpson, W. A., 1983, Adherence of *Streptococcus pyogenes, Escherichia coli* and *Pseudomonas aeruginosa* to fibronectin-coated and uncoated epithelial cells, *Infect. Immun.* **41**:1261.

Akiyama, S. K., Yamada, K. M., and Hayashi, M., 1981, The structure of fibronectin and its role in cellular adhesion, *J. Supramol. Struct. Cell. Biochem.* **16**:345.

Alkan, M. L., and Beachey, E. H., 1978, Excretion of lipoteichoic acid by group A streptococci: Influence of penicillin on excretion and loss of ability to adhere to human oral epithelial cells, *J. Clin. Invest.* **61**:671.

Alkan, M., Ofek, I., and Beachey, E. H., 1977, Adherence of pharyngeal and skin strain of group A streptococci to human skin and oral epithelial cells, *Infect. Immun.* **18**:555.

Aly, R., Shinefield, H. I., Strauss, W. G., and Maibach, H. I., 1977, Bacterial adherence to nasal mucosal cells, *Infect. Immun.* **17**:546.

Austin, R. M. J., and Daniels, C. A., 1978, The role of protein A in the attachment of staphylococci to influenza-infected cells, *Lab. Invest.* **39**:128.

Babu, J., Simpson, W. A., Courtney, H. S., and Beachey, E. H., 1983, Interaction of human plasma fibronectin with cariogenic and non-cariogenic oral streptococci, *Infect. Immun.* **41**:162.

Bagg, J., Poxton, R., Weir, D. M., and Ross, P. W., 1982, Binding of type III group B streptococci to buccal epithelial cells, *J. Med. Microbiol.* **15**:363.

Beachey, E. H., 1975, Binding of group A streptococci to human oral mucosal cells by lipotheichoic acid, *Trans. Assoc. Am. Physicians* **88**:285.

Beachey, E. H., Ofek, I., 1976, Epithelial cell binding of group A streptococci by lipoteichoic acid on fimbriae denuded of M protein, *J. Exp. Med.* **143**:759.

Beachey, E. H., Chiang, T. M., Ofek, I., and Kang, A. H., 1977, Interaction of lipoteichoic acid of group A streptococci with human platelets, *Infect. Immun.* **16**:694.

Beachey, E. H., Dale, J., Grebe, S., Ahmed, A., Simpson, W. A., Ofek, I., 1979a, Lymphocyte and T-cell mitogenic properties of group A streptococcal lipoteichoic acid, *J. Immunol.* **122**:189.

Beachey, E. H., Dale, J. B., Simpson, W. A., Evans, J. D., Knox, K. W., Ofek, I., and Wicken, A. J., 1979b, Erythrocyte binding properties of streptococci lipoteichoic acids, *Infect. Immun.* **23**:618.

Beachey, E. H., Simpson, W. A., and Ofek, I., 1980, Interaction of surface polymers of *Streptococcus pyogenes* with animal cells, in: *Microbial Adhesion to Surfaces* (R. C. W. Berkeley, J. M. Lynch, J. Melling, P. R. Rutter, and B. Vincent, eds.), Horwood, Chichester, p. 389.

Bernard, J.-P., Francioli, P., and Glauser, M. P., 1981, Vancomycin prophylaxis of experimental *Streptococcus sanguis:* Inhibition of bacterial adherence rather than bacterial killing, *J. Clin. Invest.* **68**:1113.

Bourgegu, G., and McBride, B. C., 1976, Dextran mediates interbacterial aggregation between dextran-synthesizing streptococci and *Actinomyces viscosus, Infect. Immun.* **13**:1228.

Bramley, A. J., and Hogben, E. M., 1982, The adhesion of human and bovine isolates of *Streptococcus agalactiae* to bovine mammary gland epithelial cells, *J. Comp. Pathol.* **92**:131.

Broughton, R. A., and Baker, C. J., 1983, Role of adherence in the pathogenesis of neonatal group B streptococcal infection, *Infect. Immun.* **39**:837.

Carruthers, M. M., and Kabat, W. G., 1983, Mediation of staphylococcal adherence to mucosal cells by lipotheichoic acid, *Infect. Immun.* **40**:444.

Cheney, C. P., Schad, P. A., Formal, S. B., and Boedeker, E. C., 1980, Species specificity of *in vitro Escherichia coli* adherence to host intestinal cell membranes and its correlation with *in vivo* colonization and infectivity, *Infect. Immun.* **28**:1019.

Chiang, T. M., Alkan, M. L., and Beachey, E. H., 1979, Binding of lipoteichoic acid of group A streptococci to isolated human erythrocyte membranes, *Infect. Immun.* **26**:316.

Christensen, G. D., Simpson, W. A., and Beachey, E. H., 1985, Bacterial adherence in infection, in: *Principles and Practice of Infectious Diseases,* 2nd ed. (G. L. Mandell, R. G. Douglas, and J. E. Bennett, eds.), John Wiley and Sons, New York, pp. 6–23.

Ciardi, J. E., Rolla, G., Bowen, W. H., and Reilly, J. A., 1977, Adsorption of *Streptococcus mutans* lipoteichoic acid to hydroxyapatite, *Scand. J. Dent. Res.* **85**:387.

Clark, W. B., Webb, E. L., Wheeler, T. T., Fischlschweiger, W., Birdsell, D. C., and Mansheim, B. J., 1981, Role of surface fimbriae (fibrils) in the adsorption of Actinomyces species to saliva-treated hydroxyapatite surfaces, *Infect. Immun.* **33**:908.

Clawson, C. C., and White, J. G., 1971, Platelet interaction with bacteria. I. Reaction phases and effects of inhibitors, *Am. J. Pathol.* **65**:367.

Colleen, S., Hovelius, B., Wieslander, A., and March, P. A., 1979, Surface properties of *Staphylococcus saphrophyticus* and *Staphylococcus epidermidis* as studied by adherence tests and two-polymer, aqueous phase systems, *Acta Pathol. Microbiol. Scand. Sect. B* **87**:321.

Courtney, H. S., Ofek, I., Simpson, W. A., and Beachey, E. H., 1981, Characterization of lipoteichoic acid binding to polymorphonuclear leukocytes of human blood, *Infect. Immun.* **32**:625.

Courtney, H. S., Simpson, W. A., and Beachey, E. H., 1983, Binding of streptococcal lipoteichoic acid to fatty-binding sites on human plasma fibronectin, *J. Bacteriol.* **153**:763.

Cox, F., 1982, Prevention of group B streptococcal colonization with topically applied lipoteichoic acid in a maternal–newborn mouse model, *Pediat: Res.* **16**:816.

Davison, V. E., and Sanford, B. A., 1981, Adherence of *Staphylococcus aureus* to influenza A virus infected Madin-Darby canine kidney cell cultures, *Infect. Immun.* **32**:118.

Davison, V. E., and Sanford, B. A., 1982, Factors influencing adherence of *Staphylococcus aureus* to influenza A virus-infected cell cultures, *Infect. Immun.* **37**:946.

Deneke, C. F., McGowan, K., Thorne, G. M., Gorbach, S. L., 1983, Attachment of enterotoxigenic *Escherichia coli* to human intestinal cells, *Infect. Immun.* **39**:1102.

Doran, J. E., 1983, A critical assessment of fibronectin's opsonic role for bacteria and micro-aggregates, *Vox Sang* **45**:337.

Doran, J. E., and Raynor, R. H., 1981, Fibronectin binding to protein A containing staphylococci, *Infect. Immun.* **33**:683.

Doyle, R. J., Chattergee, A. N., Streips, U. N., and Young, F. E., 1975, Soluble macromolecular complexes involving bacterial teichoic acids, *J. Bacteriol.* **124**:341.

Durack, D. T., and Beeson, P. B., 1972, Experimental bacterial endocarditis. 1. Colonization of a sterile vegetation, *Br. J. Exp. Pathol.* **53**:44.

Ericson, D., Bjorck, L., and Kronvall, G., 1980, Further characteristics of B_2-microglobulin binding to oral streptococci, *Infect. Immun.* **30**:117.

Espersen, F., and Clemmensen, I., 1982, Isolation of a fibronectin-binding protein from *Staphylococcus aureus, Infect. Immun.* **37**:526.

Fainstein, V., Musher, D. M., and Cate, T. R., 1980, Bacterial adherence to pharyngeal cells during viral infection, *J. Infect. Dis.* **141:**172.

Forsgren, A., and Sjoquist, J., 1966, Protein A from *S. aureus.* 1. Pseudoimmune reaction with human gammaglobulin, *J. Immunol.* **97:**822.

Forshin, L., Danielsson, D., and Falk, V., 1980, Adherence in vitro of *Neisieria-gonorrhea, Escherichia coli,* and group-B streptococci to vaginal epithelial cells of post-menopausal women, *Gynecol. Obstet.* **11:**341.

Garrison, P. G., and Freedman, L. R., 1970, Experimental endocarditis, 1. Staphylococcal endocarditis in rabbits resulting from placement of a polyethylene catheter in the right side of the heart, *Yale J. Biol. Med.* **42:**394.

Germaine, G. R., and Schachtele, C. F., 1976, *Streptococcus mutans* dextran-sucrase: Mode of interaction with high-molecular-weight dextran and role in cellular aggregation, *Infect. Immun.* **13:**365.

Gibbons, R. J., and Qureshi, J. V., 1979, Inhibition of adsorption of *Streptococcus mutans* strains to saliva-treated hydroxyapatite by galactose and certain amines, *Infect. Immun.* **26:**1214.

Gibbons, R. J., and van Houte, J., 1980, Bacterial adherence and the formation of dental plaques, in: *Bacterial Adherence* (E. H. Beachey, ed.), Chapman & Hall, London, p. 60.

Gibbons, R. J., Spinell, D. M., and Skobe, Z., 1976, Selective adherence as a determinant of the host tropisms of certain indigenous and pathogenic bacteria, *Infect. Immun.* **13:**238.

Glorioso, J. C., Jones, G. W., Rush, H. G., Pentler, L. J., Darif, C. A., and Coward, J. E., 1982, Adhesion of type A *Pasteurella multocida* to rabbit pharyngeal cells and its possible role in rabbit respiratory tract infections, *Infect. Immun.* **35:**1103.

Goldschmidt, J. C., and Panos, C., 1984, Teichoic acids of *Streptococcus agalactiae:* Chemistry, cytotoxicity, and effect on bacterial adherence to human cells in tissue culture, *Infect. Immun.* **43:**670.

Gould, K., Ramirez-Ronda, C. H., Holmes, R. K., and Sanford, J. P., 1975, Adherence of bacteria to heart valves in vitro, *J. Clin. Invest.* **56:**1364.

Hamada, S., Mizuno, J., and Kotani, S., 1979, Serological properties of cellular and extracellular glycerol teichoic acid antigens of *Streptococcus mutans, Microbios* **25:**155.

Hamada, S., Torii, M., Kotani, S., and Tsuchitani, Y., 1981, Adherence of *Streptococcus sanguis* clinical isolates to smooth surfaces and interaction of the isolates with *Streptococcus mutans* glucosyltransferase, *Infect. Immun.* **32:**364.

Hawiger, J., Kloczewiak, M., and Timmons, S., 1983, Interaction of fibrinogen with staphylococcal clumping factor and with platelets, *Ann. N.Y. Acad. Sci.* **408:**521.

Heeb, M. J., Costello, A. H., and Gabriel, O., 1982, Characterization of a galactose-specific lectin from *Actinomyces viscosus* by a model aggregation system, *Infect. Immun.* **38:**993.

Herzberg, M. C., Brintzenhofe, K. L., and Clawson, C. C., 1983, Aggregation of human platelets and adhesion of *Streptococcus sanguis, Infect. Immun.* **39:**1457.

Hogg, S. D., and Embery, G., 1982, Blood-group reactive glycoprotein from human saliva interacts with lipoteichoic acid on the surface of *Streptococcus sanguis* cells, *Arch. Oral Biol.* **27:**261.

Hogg, S. D., Handley, P. S., and Embery, G., 1981, Surface fibrils may be responsible for the salivary glycoprotein-mediated aggregation of the oral bacterium *Streptococcus sanguis, Arch. Oral Biol.* **26:**945.

Hormann, H., 1982, Fibronectin—Mediator between cells and connective tissue, *Klin. Wochenschr.* **60:**1265.

Izhar, M., Nuchamowitz, Y., and Mirelman, D., 1982, Adherence of *Shigella flexneri* to guinea pig intestinal cells is mediated by a mucosal adhesion, *Infect. Immun.* **35:**1110.

Johnson, A. P., Taylor-Robinson, D., and McGee, Z. A., 1977, Species specificity of attachment and damage to oviduct mucosa by *Neisseria gonorrhoeae, Infect. Immun.* **18:**833.

Johnson, R. H., Simpson, W. A., Dale, J. B., Ofek, I., and Beachey, E. H., 1980, Lipoteichoic acid-binding and biological properties of T protein of group A streptococcus, *Infect. Immun.* **29:**791.

Jones, G. W., 1977, The attachment of bacteria to the surface of animal cells, in: *Microbial Interactions* (J. L. Reissig, ed.), Chapman & Hall, London, p. 139.

Jones, W. T., and Menna, J. H., 1982, Influenza type A virus-mediated adherence of type 1a group B streptococci to mouse tracheal tissue *in vivo, Infect. Immun.* **38:**791.

Kallenius, G., and Winberg, J., 1978, Bacterial adherence to periurethral epithelial cells in girls prone to urinary-tract infections, *Lancet* **2:**540.

Kanton, F. F. 1965, Fibrinogen precipitation by streptococcal M protein. I. Identity of the reactants and stoichiometry of the reaction, *J. Exp. Med.* **121:**849.

Kohler, W., and Prokop, O., 1978, Relationship between haptoglobin and *Streptococcus pyogenes* T$_4$ antigens, *Nature (London)* **271:**373.

Kolenbrander, P. E., 1982, Isolation and characterization of coaggregate-defective mutants of *Actinomyces viscosus, Actinomyces naelundii,* and *Streptococcus sanguis, Infect. Immun.* **37:**1200.

Kronvall, G., 1973, A surface component of group A, C, and G streptococci with non-immune reactivity for immunoglobulin G, *J. Immunol.* **111:**1401.

Kronvall, G., and Williams, R. C., 1969, Differences in anti-protein A activity among IgG subgroups, *J. Immunol.* **103**:828.

Kronvall, G., Myhre, E. B., Bjorck, L., and Berggard, I., 1978, Binding of aggregated human β_2-microglobulin to surface protein structure in groups A, C, and G streptococcus, *Infect. Immun.* **22**:136.

Kuragmitsu, H. K., Wondrack, L., and McGuinness, M., 1980, Interaction of *Streptococcus mutans* glucosyltransferases with teichoic acids, *Infect. Immun.* **29**:376.

Kurpiewski, G., Forrester, L. J., Campbell, B. J., and Barrett, J. T., 1983, Platelet aggregation by *Streptococcus pyogenes, Infect. Immun.* **39**:704.

Kuusela, P., 1978, Fibronectin binds to *Staphylococcus aureus, Nature (London)* **276**:718.

Lanser, M. E., and Saba, T. M., 1981, Fibronectin as a cofactor necessary for optimal granulocyte phagocytosis of *Staphylococcus aureus, J. Reticoloendothel. Soc.* **30**:415.

Lanser, M. E., and Saba, T. M., 1982, Opsonic fibronectin deficiency and sepsis, cause or effect?, *Ann. Surg.* **195**:340.

Liljemark, W. F., and Bloomquist, C. G., 1981, Isolation of a protein-containing cell surface component from *Streptococcus sanguis* which affects its adherence to saliva-coated hydroxyapatite, *Infect. Immun.* **34**:428.

Liljemark, W. F., and Schauer, S. V., 1977, Competitive binding among oral streptococci to hydroxyapatite, *J. Dent. Res.* **56**:157.

Lowy, F. D., Chang, D. S., Neuhaus, E. G., Horne, D. S., Tomasz, A., and Steigbigel, N. J., 1983, Effect of penicillin on the adherence of *Streptococcus sanguis in vitro* in the rabbit model of endocarditis, *J. Clin. Invest.* **71**:668.

McCabe, M. M., Hamelik, R. M., and Smith, E. E., 1977, Purification of dextran-binding protein on cariogenic *Streptococcus mutans, Biochem. Biophys. Res. Commun.* **78**:273.

McCarty, M., 1959, The occurrence of polyglycerophosphate as the antigenic component of various gram-positive bacterial species, *J. Exp. Med.* **109**:361.

McIntire, F. C., Crosby, L. K., and Vatter, A. E., 1982, Inhibitors of coaggregation between *Actinomyces viscosus* T14V and *Streptococcus sanguis* 34: β-Galactosides, related sugars, and anionic amphipathic compounds, *Infect. Immun.* **36**:371.

Magnusson, I., and Ericson, T., 1976, Effect of salivary agglutinins on reactions between hydroxyapatite and a serotype C strain of *Streptococcus mutans, Caries Res.* **10**:273.

Magnusson, I., Ericson, T., and Pruitt, K., 1976, Effect of salivary agglutinins on bacterial colonization of tooth surfaces, *Caries Res.* **10**:113.

Miorner, H., Johansson, G., and Kronvall, G., 1983, Lipoteichoic acid is the major cell wall component responsible for surface hydrophobicity of group A streptococci, *Infect. Immun.* **39**:336.

Mogensen, S. C., and Dishon, T., 1983, Rapid detection of herpes simplex virus and varicella–zoster virus in clinical specimens by the use of *Staphylococcus aureus* rich in protein A, *Acta Pathol. Microbiol. Scand. Sect. B* **91**:83.

Morgenthaler, J. J., 1982, Hydrophobic chromatography of fibronectin, *FEBS Lett.* **150**:81.

Mosher, D. F., and Furcht, L. T., 1981, Fibronectin: Review of its structure and possible functions, *J. Invest. Dermatol.* **77**:175.

Mosher, D. F., and Proctor, R. A., 1980, Binding and factor XIIIA mediation cross linking of a 27 kilodalton fragment of fibronectin to *Staphylococcus aureus, Science* **209**:927.

Muse, K. E., Collier, A. M., and Baseman, J. B., 1977, Scanning electron microscopic study of hamster tracheal organ cultures infected with *Bordetella pertussis, J. Infect. Dis.* **136**:768.

Myhre, E., and Kussela, P., 1983, Binding of human fibronectin to group A, C, and G streptococci, *Infect. Immun.* **40**:29.

Myhre, E. B., and Kronvall, G., 1980, Demonstration of specific binding sites of a human serum albumin in group C and G streptococci, *Infect. Immun.* **27**:6.

Nealon, T. J., and Mattingly, S. J., 1983, Association of elevated levels of cellular lipoteichoic acids of group B streptococci with human neonatal disease, *Infect. Immun.* **39**:1243.

Nealon, T. J., and Mattingly, S. J., 1984, Role of cellular lipoteichoic acids in mediating adherence of serotype III strains of group B streptococci to human embryonic, fetal, and adult epithelial cells, *Infect. Immun.* **43**:523.

Niederman, M. S., Rafferty, T. D., Sasaki, C. T., Merrill, W. W., Matthay, R. A., and Reynolds, H. Y., 1983, Comparison of bacterial adherence to ciliated and squamous epithelial cells obtained from the human respiratory tract, *Am. Rev. Respir. Dis.* **127**:85.

Ofek, I., Beachey, E. H., Jefferson, W., and Campbell, G. L., 1975, Cell membrane-binding properties of group A streptococcal lipoteichoic acid, *J. Exp. Med.* **141**:990.

Ofek, I., Beachey, E. H., Eyal, F., and Morrison, J. C., 1977, Postnatal development of binding of streptococci and lipoteichoic acid by oral mucosal cells of humans, *J. Infect. Dis.* **135**:267.

Ofek, I., Simpson, W. A., and Beachey, E. H., 1982, Formation of molecular complexes between a structurally defined M protein and acylated or deacylated lipoteichoic acid of *Streptococcus pyogenes, J. Bacteriol.* **149:**426.

Ofek, I., Whitnack, E., and Beachey, E. H., 1983, Hydrophobic interactions of group A streptococci with hexadecane droplets, *J. Bacteriol.* **154:**139.

Pan, Y. T., Schmitt, J. W., and Sanford, B. A., 1979, Adherence of bacteria to mammalian cells: Inhibition of tunicamycin and streptovirudin, *J. Bacteriol.* **139:**507.

Perlman, B. B., and Freedman, L. R., 1971, Experimental endocarditis. II. Staphylococcal infection of the aortic valve following placement of a polyethylene catheter in the left side of the heart, *Yale J. Biol. Med.* **44:** 206.

Peters, T., 1975, Serum albumin, in: *The Plasma Proteins* (F. W. Putnam, ed.), Academic Press, New York, P. 133.

Phillips, G. N., Filcker, P. F., Cohen, C., Manjula, B. N., and Fischetti, V., 1981, Streptococcal M protein: α-Helical coiled coil structure and arrangement on the cell surface, *Proc. Natl. Acad. Sci. USA* **78:**4689.

Proctor, R. A., Mosher, D. F., and Olbrantz, P. J., 1982a, Fibronectin binding to *Staphylococcus aureus, J. Biol. Chem.* **257:**14788.

Proctor, R. A., Pendergrast, E., and Mosher, D. F., 1982b, Fibronectin mediates attachment of *Staphylococcus aureus* to human neutrophils, *Blood* **59:**681.

Ramirez-Ronda, C. H., 1980, Effects of molecular weight of dextran on the adherence of *Streptococcus sanguis* to damaged heart valves, *Infect. Immun.* **29:**1.

Reis, K. J., Ayoub, E. M., and Boyle, M. D. P., 1983, Detection of receptors for the Fc region of IgG on streptococci, *J. Immunol. Methods* **59:**83.

Revis, G. J., Vatter, A. E., Crowle, A. J., and Cisar, J. O., 1982, Antibodies against the Ag2 fimbriae of *Actinomyces viscosus* T14V inhibit lactose-sensitive bacterial adherence, *Infect. Immun.* **36:**1217.

Rolla, G., Robrish, S. A., and Bowen, W. H., 1977, Interaction of hydroxyapatite and protein-coated hydroxyapatite with *Streptococcus mutans* and *Streptococcus sanguis, Acta Pathol. Microbiol Scand. Sect. B* **85:**341.

Rolla, G., Oppermann, R. V., Bowen, W. H., Ciardi, J. E., and Knox, K. W., 1980, High amounts of lipoteichoic acid in sucrose-induced plaque *in vivo, Caries Res.* **14:**235.

Ruoslahti, E., 1981, Fibronectin, *J. Oral Pathol.* **10:**3.

Rutter, J. M., and Jones, G. W., 1973, Protection against enteric disease caused by *Escherichia coli:* A model for vaccination with a virulence determinant, *Nature (London)* **242:**531.

Ryden, C., Rubin, K., Speziale, P., Hook, M., Lindberg, M., and Wadstrom, T., 1983, Fibronectin receptors from *Staphylococcus aureus, J. Biol. Chem.* **258:**3396.

Saba, T. M., Dillon, B. C., and Lanser, M. E., 1983, Fibronectin and phagocytic host defense: Relationships to nutritional support, *J. Parenteral Ent. Nutr.* **7:**62.

Sanford, B. A., Smith, N., Shelokov, A., and Ramsay, M. A., 1979, Adherence of group B streptococci and human erythrocytes to influenza A virus-infected MDCK cells (40424), *Proc. Soc. Exp. Biol. Med.* **160:**226.

Sanford, B. A., Davison, V. E., and Ramsay, M. A., 1982, Fibrinogen-mediated adherence of group A streptococcus to influenza A virus-infected cell cultures, *Infect. Immun.* **38:**513.

Sato, H., Yanagawa, R., and Fukuyama, H., 1982, Adhesion of *Corynebacterium renale, Corynebacterium pilosum,* and *Corynebacterium cystitidis* to bovine urinary bladder epithelial cells of various ages and levels of differentiation, *Infect. Immun.* **36:**1242.

Schaeffer, A. J., Jones, J. M., and Donn, J. K., 1981, Association of *in vitro Escherichia coli* adherence to vaginal and buccal epithelial cells with susceptibility of women to recurrent urinary tract infection, *N. Engl. J. Med.* **304:**1062.

Scheld, W. M., Valone, J. A., and Sande, M. A., 1978, Bacterial adhesion in the pathogenesis of endocarditis, *J. Clin. Invest.* **61:**1394.

Scheld, W. M., Zak, O., Vosbeck, K., and Sande, M. A., 1981, Bacterial adhesion in the pathogenesis of infective endocarditis, *J. Clin. Invest.* **68:**1381.

Selinger, D. S., Julie, N., Reed, W. P., and Williams, R. C., Jr., 1978, Adherence of group A streptococci to pharyngeal cells: A role in the pathogenesis of rheumatic fever, *Science* **201:**455.

Sellwood, R., Gibbons, R. A., Jones, G. W., 1975, Adhesion of enteropathogenic *Escherichia coli* to pig intestinal brush border: The existence of two pig phenotypes, *J. Med. Microbiol.* **8:**405.

Shea, C., and Ferretti, J. J., 1981, Examination of *Streptococcus mutans* for immunoglobulin by Fc reactivity, *Infect. Immun.* **34:**851.

Simpson, W. A., and Beachey, E. H., 1983, Adherence of group A streptococci to fibronectin on oral epithelial cells, *Infect. Immun.* **39:**275.

Simpson, W. A., Ofek, I., and Beachey, E. H., 1980a, Binding of streptococcal lipoteichoic acid to the fatty acid binding sites on serum albumin, *J. Biol. Chem.* **255:**6092.

Simpson, W. A., Ofek, I., and Beachey, E. H., 1980b, Fatty acid binding sites of serum albumin as membrane receptor analogs for streptococcal lipoteichoic acid, *Infect. Immun.* **29:**119.

Simpson, W. A., Ofek, I., Sarasohn, C., Morrison, J. C., and Beachey, E. H., 1980. Characteristics of the binding of streptococcal lipoteichoic acid to human oral epithelial cells, *J. Infect. Dis.* **141:**457.

Simpson, W. A., Courtney, H., and Beachey, E. H., 1982a, Fibronectin—A modulator of the oropharyngeal bacterial flora, *Microbiology 1982* (D. Schlessinger, ed.), American Society for Microbiology, Washington, D.C.

Simpson, W. A., Hasty, D. L., Mason, J. M., and Beachey, E. H., 1982b, Fibronectin-medicated binding of group A streptococci to human polymorphonuclear leukocytes, *Infect. Immun.* **37:**805.

Simpson, W. A., Dale, J. B., and Beachey, E. H., 1982c, Cytotoxicity of the glycolipid region of streptococcal lipoteichoic acid for cultures of human heart cells, *J. Lab. Clin. Med.* **99:**118.

Speziale, P., Hook, M., Switalski, L. M., and Wadstrom, T., 1984, Fibronectin binding to a *Streptococcus pyogenes* strain, *J. Bacteriol.* **57:**420.

Spika, J. S., Vergrugh, H. A., and Verhoef, J., 1981, Protein A effect on alternative pathway complement activation and opsonization of *Staphylococcus aureus, Infect. Immun.* **34:**455.

Springer, G. F., Adye, J. C., 1975, Endoxtoxin-binding substances from human leukocytes and platelets, *Infect. Immun.* **12:**978.

Springer, G. F., Huprikar, S. V., and Neter, E., 1970, Specific inhibition of endotoxin coating of red cells by a human erythrocyte membrane component, *Infect. Immun.* **1:**98.

Springer, G. F., Adye, J. C., Bezkorovainy, A., and Murthy, J. A., 1973, Functional aspects and nature of the lipopolysaccharide-receptor of human erythrocytes, *J. Infect. Dis.* **128:**5202.

Staat, R. H., Langley, S. D., and Doyle, R. J., 1980, *Streptococcus mutans* adherence: Presumptive evidence for protein-mediated attachment followed by glucan-dependent cellular accumulation, *Infect. Immun.* **27:**675.

Stephens, D. S., and McGee, Z. A., 1981, Attachment of *Neisseria meningitidis* to human mucosal surfaces: Influence of pili and type of receptor cell, *J. Infect. Dis.* **143:**525.

Svanborg-Eden, C., and Jodal, U., 1979, Attachment of *Escherichia coli* to urinary sediment epithelial cells from urinary tract infection-prone and healthy children, *Infect. Immun.* **26:**837.

Swanson, J., 1983, Gonococcal adherence: Selected topics, *Rev. Infect. Dis.* **5:**678.

Switalski, L. M., Ljungh, A., Ryden, C., Rubin, K., Hook, M., and Wadstrom, T., 1982, Binding of fibronectin to the surface of group A, C, and G streptococci isolated from human infections, *Eur. J. Clin. Microbiol.* **1:**381.

Switalski, L. M., Ryden, C., Rubin, K., Ljungh, A., Hood, M., and Wadstrom, T., 1983, Binding of fibronectin to *Staphylococcus* strains, *Infect. Immun.* **42:**628.

Thompson, J., Meddens, M. J. M., Thorig, L., and van Furth, R., 1982, The role of bacterial adherence in the pathogenesis of infective endocarditis, *Infection* **10:**196.

Tuomanen, F. I., Nedelman, T., Hendley, J. O., and Hewlet, E. I., 1983, Species specificity of bordetella adherence to human and animal ciliated respiratory epithelial cells, *Infect. Immun.* **42:**692.

Tylewska, S., Hjerten, S., and Wadstrom, T., 1979, Contribution of M protein to the hydrophobic properties of *Streptococcus pyogenes, FEMS Microbiol. Lett.* **6:**249.

Valenti, W. M., Trudell, R. G., and Bentley, D. W., 1978, Factors predisposing to oropharyngeal colonization with gram negative bacilli in the aged, *N. Engl. J. Med.* **298:**1108.

van de Water, L., Destree, A. T., and Hynes, R. O., 1983, Fibronectin binds to some bacteria but does not promote their uptake by phagocytic cells, *Science* **220:**201.

van Houte, J., 1982, Bacterial adherence and dental plaque formation, *Infection* **10:**252.

Verbrugh, H. A., Peterson, P. K., Smith, D. E., Nguyen, B.-Y.T., Hoidal, J. R., Wilkinson, B. J., Verhoef, J., and Furcht, L. T., 1981, Human fibronectin binding to staphylococcal surface proteins and its relative inefficiency in promoting phagocytosis by human polymorphonuclear leukocytes, monocytes, and alveolar macrophages, *Infect. Immun.* **33:**811.

Whitnack, E., and Beachey, E. H., 1982, Antiopsonic activity of fibrinogen bound to M protein on the surface of group A streptococci, *J. Clin. Invest.* **69:**1042.

Wicken, A. J., and Knox, K. W., 1977, Biological properties of lipoteichoic acids, in: *Microbiology* (D. Schlessinger, ed.), American Society for Microbiology, Washington, D.C., p. 360.

Woods, D. E., Bass, J. A., Johanson, W. G., Jr., and Straus, D. C., 1980, Role of adherence in the pathogenesis of *Pseudomonas aeruginosa* lung infection in cystic fibrosis patients, *Infect. Immun.* **30:**694.

Woods, D. E., Straus, D. C., Johanson, W. G., and Bass, J. A., 1981a, Role of salivary protease activity in adherence of gram-negative bacilli to mammalian buccal epithelial cells *in vivo, J. Clin. Invest.* **68:**1435.

Woods, D. E., Straus, D. C., Johanson, W. G., and Bass, J. A., 1981b, Role of fibronectin in the prevention of adherence of *Pseudomonas aeruginosa* to buccal cells, *J. Infect. Dis.* **143:**784.

Yamada, K. M., 1983, Cell surface interactions with extracellular materials, *Annu. Rev. Biochem.* **52:**761.

Yoshida, K., Ohtomo, T., and Minegishi, Y., 1977, Mechanism of compact-colony formation by strains of *Staphylococcus aureus* in serum soft agar, *J. Gen. Microbiol.* **98**:67.

Zawaneh, S. M., Ayoub, E. M., Baer, H., Cruz, A. C. C.,and Spellcy, W. N., 1979, Factors influencing adherence of group B streptococci to human vaginal epithelial cells, *Infect. Immun.* **26**:441.

Zetter, B. R., Daniels, T. E., Quadra-White, C., and Greenspan, J. G., 1979, LETS protein in normal and pathological human oral epithelium, *J. Dent. Res.* **58**:484.

11

Pilus Adhesins

RICHARD E. ISAACSON

1. INTRODUCTION

The adhesion of bacteria to mucosal surfaces has been the focus of study in several laboratories around the world during the past decade and the results have led to a better understanding of the interactions between animal hosts and microbes. The study of bacterial adhesion to animal cells has been stimulated, in part, by the recognition of the importance of these processes in the pathogenesis of diseases caused by bacteria. As such, most of the ensuing discussion will be based on the data obtained from studies of disease pathogenesis. In particular, I shall use as primary examples, results from studies on the pathogenesis of enterotoxigenic *Escherichia coli* diarrheal disease and *E. coli*-induced urinary tract infections as these provide the most extensively studied examples of bacterial adhesion to animal cells. The initiation of most bacterial infections of mucosal surfaces requires the bacterial species in question to colonize and maintain a stable population at the mucosal site. In the cases of pathogenic *E. coli*, colonization is facilitated by adhesion to the mucosa, the adhesion being mediated by bacterial structures called pili. The following discussion will describe in detail the molecular basis of pilus-mediated bacterial adhesion to mucosal surfaces (animal cells). Although the structure and function of pili will be emphasized, the recognition and specificity of receptors on the animal cells is critical and will also be described (see Chapter 4, Section 4).

1.1. Definitions

I shall restrict the usage of the term *adhesion* to denote relatively stable, essentially irreversible attachment of bacteria to the surface of an animal cell. Any structure that mediates adhesion to an animal cell is an adhesin. Adhesins that attach to red blood cells are hemagglutinins. Although hemagglutinins recognize red blood cells as suitable substrata with which to interact, it is likely that the red cells are not recognized in natural infections of animals. In some instances, red blood cells possess receptors that are similar, if not identical, to the receptors on the normal target cells and, therefore, these reactions have been useful

---

Richard E. Isaacson • Department of Epidemiology, School of Public Health, University of Michigan, Ann Arbor, Michigan 48109. *Present address:* Pfizer Central Research, Groton, Connecticut 06340.

during the initial screening of bacterial strains for adhesins. The terms *fimbriae* and *pili* denote identical bacterial structures with a defined morphology—nonflagellar, filamentous structures on the surface of bacterial cells. The two terms have been used by some investigators as synonyms while other investigators have attempted to distinguish between the two. The dichotomy in terminology has led to much confusion. The term *fimbria* was coined by Duguid *et al.* (1955) from the Latin word meaning filament; the term *pilus,* which means hair, was suggested by Brinton (1959). In this chapter, the term *pilus* will be used to denote such structures and *fimbria* will be considered its synonym. The terms *pilus* and *fimbria* define structures with a specific morphology and do not specify function. Pili that are adhesins are therefore called *pilus adhesins.* Sex pili are a unique class of pilus adhesins that cause bacterial cell aggregation and thus promote gene transfer from one cell to another.

Several schemes have been developed to classify pili based upon the diameter, morphology, and hemagglutination properties of the pili (Ottow, 1975; Pearce and Buchanan, 1980). These schemes have been of little use, however, due to several weaknesses. First, morphology as a defining criterion is only useful for those possessing and proficient in the use of the electron microscope, and then sample preparation and interpretation of the results can lead to a variety of different conclusions. For example, the reported diameters of *E. coli* K99 pili have ranged from 2 to 8.4 nm (Isaacson, 1977; Wadstrom *et al.,* 1978; deGraaf *et al.,* 1981). *Myxococcus xanthus* has been reported to produce two types of pili: flacid and rigid (Dobson and McCurdy, 1979). Although these pili differ in morphology, they are composed of the same protein subunits and likely represent slightly different conformations. Hemagglutination has been even less useful, for pili have been divided into only two categories. The division is based upon the effect of D-mannose on the hemagglutination reactions. Thus, the terms *mannose-sensitive* or *mannose-resistant hemagglutinins* were coined. Mannose-sensitivity is a particularly useful trait in classifying pili as it denotes a reaction with a specific sensitivity. The sensitivity to mannose derives from the fact that these pili use D-mannose, D-mannose derivatives, or D-mannose-containing molecules as receptors (Old, 1972; Duguid and Old, 1980). The term *mannose-resistant hemagglutinin* is used for all other pili. This term lacks specificity and assumes a relationship among the pili so classified. This assumption is not correct. Furthermore, the use of the term *resistant* implies an active process, which also is not correct. Pili in this group, simply stated, are not sensitive to mannose. Rather than being dependent upon rigid schemes for classification, it seems more reasonable to put all pili into one group until they are defined functionally. For pili that are adhesins, function could include species of animal cells to which they attach and the specific receptors involved.

1.2. Physicochemical Considerations of Adhesion

A major problem that must be overcome for bacterial cells to adhere to animal cells results from the charge repulsion between bacterium and animal cell. Biological membranes, including bacterial membranes, have an overall negative charge and thus should not come into direct physical contact. Of the several descriptions of the physicochemical interactions between like-charged particles, the DLVO theory and its derivations have provided the most useful descriptions of the interactions between bodies of like charge, and have been used to develop hypotheses about how a bacterium may adhere to the surface of animal cells (Weiss and Harlos, 1972; Pethica, 1980; Tadros, 1980; Rutter and Vincent, 1980; Chapter 5, Section 3.1; Chapter 6, Section 3.1). The DLVO theory states that as rigid bodies of like charge approach each other, they are subject to attractive and repulsive forces that are additive, but vary independently with the distance of separation between the bodies. The overall charge and shape of the bodies are important and contribute significantly to the forces of attraction and repulsion.

Initially, we will consider planar bodies at relatively long distances (> 10 nm). The attractive forces are greater than the repulsive forces, resulting in attraction between the bodies. The forces of attraction at this distance (secondary minimum of potential energy) are weak and easily reversed by fluid shear. The attractive forces at the secondary minimum have been referred to as reversible adhesion. At shorter distances, the repulsive component increases relative to the attractive component. At the potential energy maximum, there is a strong repulsion between the two bodies. If the forces of repulsion can be overcome, at very short distances (1 nm), then there is a mutual attraction (primary minimum of potential energy). At this distance, the attraction between the two bodies is strong and not easily reversed. This has been referred to as irreversible adhesion. Unfortunately, bacteria do not possess sufficient kinetic energy to overcome the forces at the potential energy maximum and therefore should never approach the primary minimum (see Chapter 6, Sections 3.2.1 and 3.2.2).

The forces of attraction and repulsion can be decreased by changing the shape(s) of the bodies. With increasing curvature (decreasing radius), there is a decrease in the forces of attraction and repulsion. The forces of repulsion, however, decrease more rapidly than those of attraction. Therefore, curved bodies come closer together at the secondary minimum and require less kinetic energy to get to the primary minimum. Adhesion of bacteria (ellipsoid structures) to cells with microvilli is energetically more favorable compared to a tissue of planar configuration, for the microvilli also are curved. The effect of curvature may not decrease the repulsive forces sufficiently to permit bacteria to reach the primary minimum. In an organ, such as the small intestine, the effect of fluid shear due to peristalsis, would dislodge the bacteria, resulting in their removal from the tissue surface. Structures on bacterial cells that can bridge the distance between bacteria held at the secondary minimum and the tissue may result in irreversible adhesion. Such structures would, by necessity, have very small radii relative to the bacterial cell and would come in contact with the tissue. Bacterial pili, with radii of 1 to 5 nm fulfill this requirement and should make ideal adhesins. Because of the small radius of pili, molecular interaction between pilus and tissue need not be dependent on the effect of charge although such forces may be important. At the molecular level, the distribution of charge is at least as important as the overall charge of the structures. The interaction between pilus and tissue is probably analogous to other receptor–ligand interactions and may be facilitated by hydrophobic interactions as well as charge interactions (Chapter 6, Sections 3 and 4).

1.3. General Mechanisms of Adhesion

The initial contact between bacterium and tissue is most likely a random event. The rate of contact may also be increased by bacterial motility. Bacteria that are chemotactic (i.e., that migrate toward a chemical attractant) are more likely to contact a tissue surface if the tissue secretes a specific chemical attractant (Freter et al., 1981; Uhlman and Jones, 1982) than nonchemotactic bacteria. The colonization of the small intestine by Vibrio cholerae has been shown to be dependent on a chemotactic response (Freter et al., 1981; Chapter 15, Section 2.2.1). The mucosa secretes an attractant that results in the active migration of the vibrios to the mucosal surface. Chemotactic bacteria may be maintained at the surface without direct contact or adhesion as long as the attractant is secreted resulting in the maintenance of a chemical gradient. This type of adhesion would appear to be reversible and would not require the bacterial cell to come any closer to the tissue than the distance of the secondary minimum. Reversible adhesion at the secondary minimum, whether or not facilitated by motility and chemotaxis, serves to stabilize the bacterium at that site and permits a more permanent interaction(s) between adhesin and tissue to occur.

The precise site of colonization or adhesion may also be dependent upon bacterial motility and/or chemotaxis. *E. coli* that produce the pilus K88 uniformly colonize pig small intestines from anterior to posterior (Arbuckle, 1970; Jones and Rutter, 1972). Such strains also are motile. On the other hand, *E. coli* that produce the pilus 987P colonize only the posterior small intestine (Nagy *et al.*, 1976) although specific receptors are distributed equally throughout the small intestine (Isaacson *et al.*, 1978b). These strains are not motile. This phenomenon may be related to the differences in bacterial motility. Because motility increases the rate of contact between bacterium and surface, the probability of motile *E. coli* contacting the surface of the anterior small intestines is greater than the frequency for nonmotile strains. In the posterior small intestine, the probability of contact by nonmotile strains increases due to the higher numbers of bacterial cells present due to cell replication, and a decreased rate of fluid flow at that end of the small intestine which increases exposure time.

After the bacterial cell attaches to the surface, replication results in the colonization of that site. Attached bacteria may remain attached, be drawn closer to the tissue surface, invade the tissue, or detach (see Chapter 15, Section 2.2.1).

2. PILI AS ADHESINS

2.1. *In Vivo*

The only examples of pilus-mediated bacterial adhesion to animal cells *in vivo* pertain to *E. coli*. Enterotoxigenic *E. coli* (ETEC) is a major cause of diarrhea in humans (Gorbach *et al.*, 1975) and neonatal animals (Barnum *et al.*, 1967; Chapter 15, Section 2.2.1). The disease in animals is characterized by two phenomena: the secretion of large amounts of water in response to enterotoxins which frequently leads to the death of the animal, and the colonization of the animal's small intestine. In normal animals the number of *E. coli* in the small intestine is quite low. Typically, organisms are consumed orally and as they descend the small intestine their concentration increases due to replication. Fluid flow in the small intestine due to secretion and peristalsis acts to keep the *E. coli* population low. In the large intestine where the flow of bacteria becomes quite slow, the *E. coli* population increases. In the adult large intestine, there may be 10^7 to 10^8 *E. coli* cells/g tissue.

Neonatal pigs challenged orally with 10^9 virulent ETEC cells may contain 10^9 to 10^{10} virulent bacterial cells per 10-cm segment of small intestine (Smith and Linggood, 1971b; Jones and Rutter, 1972; Nagy *et al.*, 1976). Pigs challenged with a nonvirulent *E. coli* strain contain less than 10^8 bacterial cells per 10-cm segment of small intestine and usually less than 10^6. The mucosal surfaces of tissues obtained from the small intestines of pigs challenged with ETEC are covered with *E. coli* cells (Arbuckle, 1970; Bertschinger *et al.*, 1972; Nagy *et al.*, 1976). The bacteria appear to adhere to the tissue surface based on direct observation after Giemsa staining (Nagy *et al.*, 1977) or by fluorescent antibody staining (Arbuckle, 1970; Bertschinger *et al.*, 1972) of bacterial cells. By electron microscopy, the layer of adherent bacteria is characteristically several bacterial cell lengths in thickness (Fig. 1) (Bertschinger *et al.*, 1972; Moon *et al.*, 1977). Few, if any, bacterial cells come in direct contact with the microvilli. It is presumed that the distance between tissue and bacteria is occupied in part by adhesins. The putative adhesin may be seen as a filamentous structure

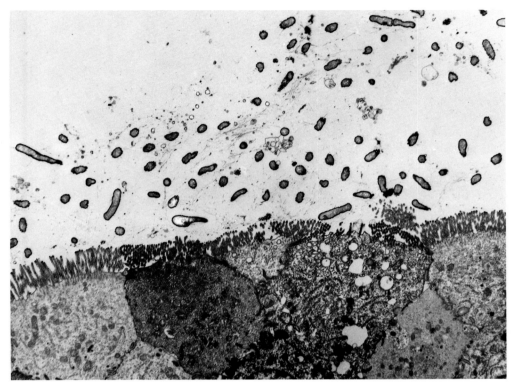

Figure 1. Transmission electron micrograph of pig ileum from an animal challenged 18 hr prior to sample collection with a virulent enterotoxigenic strain of *E. coli*. A layer of bacteria can be seen on the luminal side of the tissue, the layer extending several bacterial lengths from the microvillus surface. Occasionally, an electron lucent region can be observed around the *E. coli* cells, the region presumably containing the relevant adhesin.

radiating from the bacterial cell and contacting microvilli (Fig. 2). Although these filaments may indeed be pili, no direct evidence supports this conclusion. Bacteria at greater distances from the mucosal surface may adhere to other bacteria in a matrix or may adhere to a receptor-containing glycocalyx. Some of the bacteria (Fig. 1) may also have been dislodged from the surface in processing the sample for electron microscopy.

The strength of the interaction between bacteria and tissue as well as a demonstration that the bacteria adhere to the tissue rather than soluble or loosely associated cellular components was shown in experiments employing vigorous washing of the tissue (Nagy *et al.*, 1977). Washing of tissue with dithiothreitol results in the removal of loosely associated material including the mucous gel. Tissues colonized by ETEC and washed with dithiothreitol contain as many adherent bacteria as unwashed tissues, whereas control samples from animals challenged with nonvirulent *E. coli* are devoid of bacteria subsequent to washing. The interpretation of these experiments is consistent with the hypothesis that colonization of the small intestine by ETEC is facilitated by adhesion of bacteria to the mucosal surfaces.

The conclusion that the *E. coli* adhesins are pili is based, in part, on the results of genetic manipulations. Some ETEC that cause diarrheal disease in pigs possess the K88 pilus which is plasmid encoded. Smith and Linggood (1971b) selected K88-negative strains after treating a wild-type, K88-positive ETEC with the plasmid-curing agent acridine orange. The

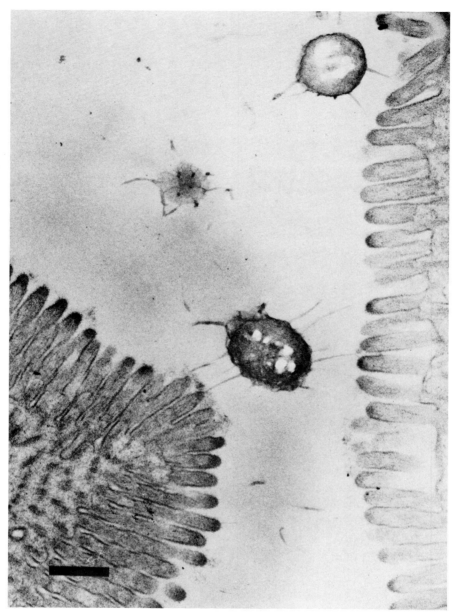

Figure 2. Transmission electron micrograph of pig ileum from an animal challenged 18 hr prior to sample collection with a virulent enterotoxigenic strain of *E. coli*. Surface appendages that may be adhesins can be seen on the bacterial cell surface and contacting microvilli.

cured strain used in further experiments remained enterotoxigenic. The two strains (K88-positive and K88-negative) were then used to challenge separate pigs. The K88-positive strain was virulent, colonizing the small intestine and causing diarrheal disease, while the K88-negative strain did not colonize the small intestine or cause diarrheal diease. The K88-specifying plasmid was then reintroduced into the K88-negative strain by conjugal mating.

K88-positive transconjugants regained their virulence and capacity to colonize. These experiments demonstrated that something, presumably K88, encoded on the K88-specifying plasmid was required for virulence. Similar experiments were performed using ETEC containing the plasmid-encoded pilus K99 (Smith and Linggood, 1971a) or by selection of nonpiliated mutants from 987P-positive ETEC (Isaacson et al., 1977). The results obtained from these experiments demonstrated that K99 and 987P also promoted colonization of the small intestine.

If the assumption that K88, K99, and 987P are adhesins is correct, then they must be expressed in vivo. Segments of small intestines colonized by piliated ETEC have a characteristic fluorescent layer covering the mucosal surface when stained with pilus-specific fluorescent antibodies and observed by fluorescence microscopy, thus demonstrating their synthesis in vivo (Isaacson et al., 1978a; Moon et al., 1980).

Not only can pilus-specific antibodies be used to demonstrate pili on bacteria (in vivo or in vitro) but as pili must be expressed in vivo to promote colonization, the antibodies also can be used to prevent colonization of small intestines and therefore diarrheal disease. Vaccination of pregnant pigs (Jones and Rutter, 1974; Morgan et al., 1978; Nagy et al., 1978) or calves (Acres et al., 1979) with purified E. coli pili stimulates production of antibodies against the pili (Nagy et al., 1978; Isaacson et al., 1980). Piglets that suckle vaccinated dams receive pilus-specific antibody in colostrum and milk and consequently are protected against diarrheal disease caused by ETEC. In an experimental challenge, pigs from dams vaccinated with 987P or K99 were challenged with either a 987P- or a K99-producing ETEC strain. Protection resulted when the pilus in the vaccine and in the challenge organism were homologous (Morgan et al., 1978), and was conferred by preventing or reversing small intestinal colonization.

Although the results of in vivo experiments support a role for pili in colonization of the small intestine, they may not be used to draw conclusions about the specific role(s) of pili. The observations that ETEC colonize by adhesion to the mucosal surfaces of the small intestine and that occasionally filamentous projectiles are observed by electron microscopy on the surface of bacteria that also contact microvilli suggest that the pili are the relevant adhesins.

2.2. In Vitro

The best evidence supporting the hypothesis that some pili are adhesins was derived from in vitro adhesion experiments. Several in vitro assays have been developed to measure the adhesive capacity of bacteria. The assays can be divided into two groups: assays using the natural target cells and assays using substitute target cells. Examples of cells used in the former assays include small intestinal enterocytes and urinary tract epithelium; cells used in the latter assays include red blood cells, buccal cells, and cells grown in tissue culture. In general, all in vitro adhesion assays are performed by mixing bacteria with the target cells, incubating the mixtures, and determining the number of adherent bacteria per target cell.

Piliated E. coli producing K88, K99, 987P, or type 1 pili adhere to slices of small intestine (Jones and Rutter, 1974), to intact enterocytes (Wilson and Hohmann, 1974; Nagy et al., 1977; Isaacson et al., 1978b), and to brush borders prepared from the enterocytes (Sellwood et al., 1975; Rutter et al., 1975), whereas isogenic but nonpiliated strains do not. Nonagglutinating antibody (Fab fragments) prepared against purified pili inhibits attachment of pilated bacteria to enterocytes (Isaacson et al., 1978b). However, Fab fragments against surface antigens other than pili do not affect adhesion in vitro.

The results of competitive inhibition experiments using purified pili as competitor also demonstrate the adhesive properties of *E. coli* pili (Isaacson *et al.*, 1978b). Purified pili compete with *E. coli* cells producing the same pili for sites on enterocytes but do not compete with bacteria producing different pili. It can be concluded, therefore, that pili bind to enterocytes, that the pili on *E. coli* cells mediate the adhesion of the cells to the enterocyte, and that the binding is specific. Results from these experiments and others have shown that specific receptors exist on pig enterocytes for K88, K99, 987P, and type 1 pili.

E. coli strains that cause upper urinary tract infections adhere *in vitro* to urinary tract epithelial cells (Svanborg-Eden, 1978). Experiments similar to those used for ETEC strains (e.g., antibody inhibition and pilus competition) demonstrated that pili were relevant adhesins on these strains.

The adhesion of bacterial species other than *E. coli* to mucosal surfaces is predicted because the virulence of bacteria, with few exceptions, is dependent upon the capacity to colonize mucosal surfaces. Most evidence supporting this prediction has been derived from *in vitro* adhesion experiments. A variety of different pathogenic and nonpathogenic bacteria have been shown to adhere *in vitro* to one or more types of animal cells. In some cases, there is a good correlation between adhesion and piliation. Such results, however, must be interpreted with caution. As described above, porcine enterocytes possess a receptor for the type 1 pili of *E. coli* and yet there is no evidence that type 1 pili confer *in vivo* adhesiveness to *E. coli* in the small intestine. The demonstration of adhesiveness *in vitro* is only an indicator of adhesive capacity but cannot be used to judge the relevance of adhesion *in vivo*. Unfortunately, not all organisms are as exquisitely suited to study as *E. coli* and the chances of performing appropriate *in vivo* experiments in many cases are almost nonexistent. Therefore, the selection of appropriate target cells is critical in interpreting the results of *in vitro* adhesion experiments.

One example where *in vitro* adhesion experiments established the potential to adhere *in vivo* is *Neisseria gonorrhoeae* (see Chapter 15, Section 2.2.2). *N. gonorrhoeae* normally colonizes the mucosa of the urogenital tract. Adhesion to the mucosa is predicted because the organism would presumably be washed out due to fluid (urine and mucus) flow. *N. gonorrhoeae* has been shown to adhere to several types of target cells including mucosal cells from human fallopian tubes. There is a predilection for nonciliated cells and adhesion is specific to the human fallopian tube and not to that of the rabbit, pig, or cow. Piliated *N. gonorrhoeae* cells adhere better than nonpiliated cells although nonpiliated cells adhere better than avirulent cells (McGee *et al.*, 1982). This suggests that something in addition to pili may also be important in the adhesive process. Although piliation is important in determining the degree of adhesiveness to mucosal cells, adhesion to leukocytes, for example, is independent of piliation (Swanson *et al.*, 1975; Heckels, 1982).

Outer membrane protein II also appears to contribute to the adhesiveness of *N. gonorrhoeae* cells (Heckels, 1982). This protein may exist in several forms. One form of protein II promotes leukocyte-associating activity, while a second variant of protein II enhances mucosal cell adhesion. The adhesive capacity of *N. gonorrhoeae* therefore may be altered with respect to degree of adhesiveness and tissue specificity by altering the form or amount of specific surface components (i.e., protein II and pili).

3. STRUCTURE OF PILI

3.1. Morphology

The filamentous nature of pili is often demonstrated by electron microscopy. On most gram-negative bacteria, the pili are distributed peritrichously, the number of pili per cell

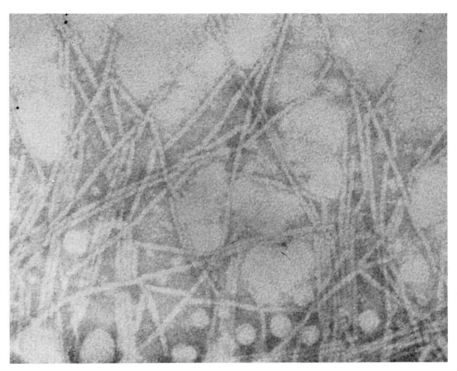

Figure 3. Purified 987P pili.

being dependent upon the organism and conditions of growth. On the other hand, the pili of myxobacteria are usually observed to be tufts at the poles of the cells (MacRae *et al.*, 1977).

In general, pili that are readily observed by electron microscopy appear to be morphologically similar, rigid structures. Exemplary of this group are *E. coli* type 1 pili and 987P pili (Fig. 3) which also have axial holes (Brinton, 1965; Isaacson and Richter, 1981). These structures have a uniform cross-sectional diameter of approximately 7 nm although other pili in this group are as narrow as 2 nm or wider than 10 nm (Brinton, 1965; Ottow, 1975). The length of these pili is variable and may be as long as 4 μm. The rigid pili have a strong tendency to aggregate into large parallel bundles. Treatment with salt promotes the formation of angle-layered crystals (Fig. 4), the angle between the planes of the two pilus layers being characteristic of the specific type of pilus (Brinton, 1965). Studies on the structure of *E. coli* type 1 pili, in which aggregates of parallel bundles and angle-layered crystals were examined by X-ray diffraction, provided evidence that type 1 pili are helical structures (Brinton, 1965, 1971). The assembly of repeating subunits results in a pilus structure that has a right-hand configuration with $3\frac{1}{8}$ subunits per revolution and a subunit pitch of 23.2 Å. Pili with different subunit pitch produce angle-layered crystals with different angles of pilus orientation.

The *E. coli* K99 pilus is exemplary of a group of pili that morphologically appear to be flexible (Fig. 5). K99 on bacterial cells is very difficult to visualize by electron microscopy due to its morphology. Often, K99-positive cells appear to have a fuzzy surface (Moon *et al.*, 1977; Isaacson, 1980b). As with pili in the rigid pilus group, K99 also forms aggregates (Isaacson, 1977). Aggregated, purified K99 can readily be seen by electron microscopy (Fig. 5) (Isaacson, 1977; Wadstrom *et al.*, 1978; deGraaf *et al.*, 1981). The measured diameters for K99 vary between 2 and 8.4 nm. Two nanometers probably is a good approximation for

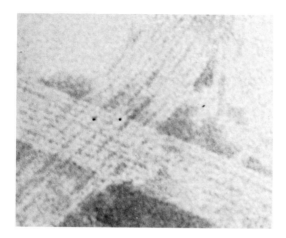

Figure 4. An angle-layered crystal of 987P pili. Note that the angle between the pili in the two planes is characteristic of pilus types.

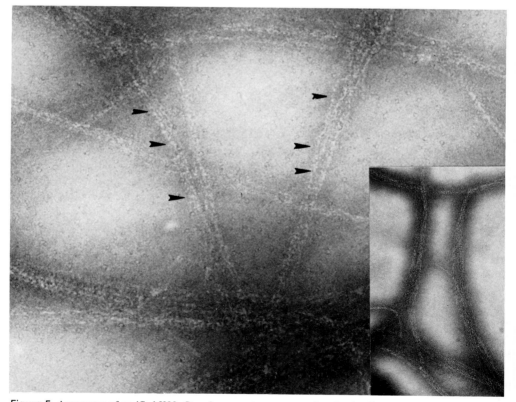

Figure 5. Aggregates of purified K99. Strands appear to wrap around each other in a helical orientation. Arrows point to sites where the helix is apparent. The insert is a lower magnification.

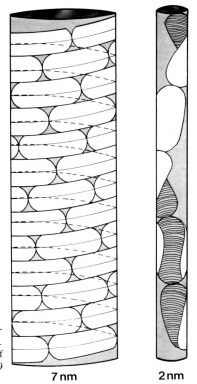

Figure 6. Schematic diagram of pili. The left structure is a representation of type 1 pili with $3\frac{1}{8}$ subunits (shown in white) per revolution. The diameter is 7 nm. The right structure is a possible configuration of K99, generated by stretching lengthwise the type 1 pilus structure. K99 as depicted would be more flexible than the type 1 pilus.

7 nm   **2 nm**

the diameter of the pilus. The larger structures probably are aggregates; at 8.4 nm the aggregate is composed of four strands. The aggregates of K99 have several different characteristics compared to the aggregates of rigid pili. Most obvious of these is that K99 aggregates do not have an ordered parallel alignment, but appear to wrap around each other to form a loose, helical structure.

The demonstrations that rigid pili are helical in structure and that flexible pili such as K99 may also form helices suggest that the physical orientation of subunits to form pilus polymers is similar among different types of pili. The observed morphologic differences are probably a result of the number of subunits per revolution of the helix. The schematic representation of type 1 pili and K99 shown in Fig. 6 was constructed to illustrate the possible similarities and differences between the two structures. Type 1 pili have few subunits per revolution ($3\frac{1}{8}$) which results in a tight, rigid structure. The structure of K99 was obtained by stretching lengthwise the type 1 pilus structure. K99 pili therefore have more subunits per revolution which would be expected to produce an open, flexible structure.

Physicochemical variability among pilus structures makes difficult the categorization of pili by morphology. As already described, the pili of *M. xanthus* can take two morphologies: rigid and flacid (Dobson and McCurdy, 1979). Other pili that appear ''wavy'' may also pose a problem in defining a morphologic group. Fortunately, very little emphasis has been placed on the morphologic differences of pili and, as may be concluded, does not provide an adequate means of classifying pili.

3.2. Biochemistry

Bacterial pili are polymers, composed of identical protein subunits called *pilin* (for a recent review, see Jones and Isaacson, 1983). The number of subunits defines the molecular weight and length of an intact pilus. The forces holding pilin subunits together are unknown but do not include covalent bonds. Pili readily dissociate into free pilin in the presence of SDS, guanidine, or dilute acid. Free type 1 pilin obtained by treatment with guanidine will assemble into pili upon the removal of the guanidine by dialysis although the assembly process requires from several hours to several days (Eshdat *et al.*, 1981). Based upon the long time required to assemble *in vitro*, it is assumed that assembly is catalyzed by other cellular components. It is reasonable to believe that hydrogen and hydrophobic bonds are important in pilus formation and maintenance. The NH_2-terminal end of *N. gonorrhoeae* pilin contains an abundance of hydrophobic amino acids which has been suggested to be a recognition site and/or to provide stabilizing forces in pilus assembly (Schoolnik, 1982).

The molecular weights of some pilin molecules are shown in Table I. Several nonamino acid substitutions have been detected on certain pili and include phosphate (Brinton, 1965; Stirm *et al.*, 1967), simple sugars (Brinton, 1965; Robertson *et al.*, 1977; Isaacson and Richter, 1981), or phospholipid (Isaacson, unpublished) (Table I). Whether all of these substituents are integral parts of the pili or simply remnants of cellular attachment sites remains to be determined. The sugars associated with *E. coli* 987P (Isaacson and Richter, 1981) and *N. gonorrhoeae* pili (Gubish *et al.*, 1982) have been shown to be involved in binding of the pili to receptors. Periodate cleavage of these pilus-bound sugars abolishes the capacity of the pilus to subsequently bind to its cognate receptor. Furthermore, galactosidase treatment of gonococcal pili also abolishes receptor-binding activity.

The amino acid compositions of several pilin proteins are shown in Table II. Each contains about 50% nonpolar amino acids (43.7 to 64.9%). Frequently, such figures have been cited in assigning hydrophobic properties to pili. The determination of hydrophobicity based on nonpolar amino acid content leads to rather erroneous results as the amount of hydrophilic amino acids and the degree of hydrophobicity and hydrophilicity assigned to each amino acid are not taken into account. The relative hydrophobicity of the pilin molecules have been calculated using the values of amino acid hydrophobicity derived by Nozaki and Tanford (1971) and are listed in Table II. Values range from -459 to -894 cal/mole per amino acid. For comparison, human serum albumin has a calculated hydrophobicity of -769 cal/mole per amino acid and is composed of 55% nonpolar amino acids. Human serum albumin is therefore more hydrophobic than most of the pili listed.

The calculation of protein hydrophobicity based upon amino acid composition results in a value that assumes a uniform distribution of hydrophobicity which is unrealistic. Other methods have been developed and used to measure the relative hydrophobicity of piliated and nonpiliated bacteria. Type 1 pili reduce the electrophoretic mobility of *E. coli* cells and alter the phase partitioning of cells in polyethylene glycol–dextran (Brinton, 1965). Pili also promote the adsorption of cells to hydrophobic gels such as phenyl or octyl Sepharose (Wadstrom *et al.*, 1978; Smyth *et al.*, 1978). Surface hydrophobicity has also been measured in a salting-out procedure involving ammonium sulfate (Lindahl *et al.*, 1981). The greater the amount of ammonium sulfate required to salt the bacteria out, the lower its hydrophobicity. By this procedure, piliated *E. coli* have greater surface hydrophobicity than nonpiliated *E. coli*. Not all *E. coli* pili equally increase surface hydrophobicity. Therefore, it is possible to rank each pilus type with respect to hydrophobicity. Although this procedure has been given limited use, the results are in close agreement with the calculated values (CFA/I > CFA/II > K88 ~ K99 > type 1).

Table I. Properties of Various Pilin Molecules

Organism	Pilus	Subunit molecular weight (SDS–PAGE)	pI	Nonamino acid substitutions	Reference
Escherichia coli	Type 1	17,100	3.9		Brinton (1965)
	F7	22,000	—		Klemm *et al.* (1982)
	987P	20,000	3.7	Hexosamine	Isaacson and Richter (1981)
	K88ab	26,200	4.2		Mooi and deGraaf (1979)
	K99	19,500	10.1	Phosphatidyl-ethanolamine	Isaacson (1977), Isaacson *et al.* (1981)
	CFA/I	14,500	—		Klemm (1979)
	018ac	21,000	5.1	Phosphate	Weaver *et al.* (1980)
	3048	17,000	4.9		Korhonen *et al.* (1980a)
	F	11,800	—	Glucose, phosphate	Brinton (1965)
Klebsiella pneumoniae	Type 1	21,500	—		Fader *et al.* (1982)
Salmonella typhimurium	Type 1	21,000	4.1		Korhonen *et al.* (1980b)
Neisseria gonorrhoeae	P9	19,500	4.9	Phosphate	Robertson *et al.* (1977)
	201	19,600	4.9	Phosphate	Robertson *et al.* (1977)
Pseudomonas aeruginosa	PAK	17,800	3.9		Paranchych *et al.* (1979)
Moraxella nonliquifaciens		17,000	—		Froholm and Stetten (1977)
Corynebacterium renale		19,000	4.5		Kumazawa and Yanogawa (1972)
Myxococcus xanthus		8,000	—		Dobson and McCurdy (1979)
Actinomyces viscosus	WVU627	64,000	—	Reducing sugar	Cisar and Vatter (1979)

Hydrophobic domains at the receptor-recognition site may promote the interaction with the receptor. These domains may, however, represent minor regions of the pilus with respect to overall surface area and hydrophobic content and thus although critically important for the biologic activity of the pilus may not contribute significantly to the overall hydrophobicity of the protein. Consequently, the results obtained by measuring pilus hydrophobicity may not be useful in defining the specific interactions between pilus and receptor or for that matter the interactions between pilin molecules.

Charge interactions between bacteria and target cell are, as described above, particularly significant in the adhesive process. The charge of a pilus may therefore contribute to the

Table II. Amino Acid Compositions of Various Pilin Molecules

Pilus	Escherichia coli									Klebsiella pneumoniae type I	Salmonella typhimurium type I	Neisseria gonorrhoeae						Actinomyces viscosus	
Amino acid	Type I	F7	987P	K88ab	K99	CFA/I	018ac	3048	F			P9	201	Pseudomonas aeruginosa PAK	Moraxella nonliquefaciens	Corynebacterium renale	Myxococcus xanthus	T14$_a$	WVU627
Lys	3	13	10	12	9	8	14	6	9	8	9	16	16	15	14	12	2	95	30
Arg	3	3	2	7	4	3	0	4	0	5	4	8	4	4	4	6	4	20	16
His	2	2	0	0	0	1	3	2	0	2	3	3	3	0	0	3	1	8	8
Asp	20	24	34	28	25	18	22	22	7	27	22	31	30	15	20	23	6	139	46
Thr	20	22	28	27	20	13	16	22	7	25	25	12	14	15	15	17	4	136	42
Ser	10	15	23	16	12	11	16	12	11	14	23	16	18	10	8	8	4	35	30

Glu	13	19	16	16	7	12	15	14	4	17	19	21	20	15	13	23	6	102	46
Pro	2	6	8	6	3	4	7	4	0	5	11	1	7	10	5	9	4	62	34
Gly	17	22	26	35	15	11	23	18	13	18	23	17	18	18	14	16	6	83	48
Ala	34	21	26	27	46	16	25	32	14	30	34	24	27	24	20	18	8	95	44
Cys	2	2	2	0	2	0	2	ND	0	4	ND	5	4	4	1–2	1	0	ND	2
Val	13	22	15	18	14	13	17	16	19	18	16	18	18	9	7	15	4	62	32
Met	0	1	1	2	3	2	0	0	8	2	—	4	2	3	1	1	0	1	4
Ile	4	10	12	10	13	6	12	6	4	8	7	6	6	12	16	6	4	35	18
Leu	10	11	17	18	7	11	10	12	8	13	12	12	13	14	14	10	4	75	26
Tyr	0	3	1	8	0	4	3	0	2	6	4	5	6	2	8	6	2	33	10
Phe	8	11	3	10	0	3	8	8	6	6	9	3	2	2	6	4	4	19	12
Trp	0	ND	1	ND	3	ND	0	ND	2	1	ND	6	4	2	ND	1	ND	ND	ND
Hyp					3														
MePhe											1	1	1	1	1				
Orn																			3
Total	161	207	225	240	186	136	193	178	114	209	221	208	212	175	166–167	179	63	—	443
Percent nonpolar	54.6	50.2	48.4	52.5	54.7	48.5	52.8	53.9	64.9	48.3	50.6	43.7	45.7	53.7	50.0	44.6	53.9	43.2	48.3
hydrophobicity (−cal/mole/amino acid)	511	614	459	609	577	595	592	533	894	583	500	623	572	635	735	550	649	567	537

overall attractive and repulsive forces between bacterium and target cell. Pili that reduce or reverse the overall surface charge on a bacterial cell contribute to the adhesive process, in part, by reducing the repulsive forces between bacterium and target. Most pili, however, are negatively charged structures with isoelectric points in the range of pH 3.7 to 5.6 (see Table I). The *E. coli* K99 pilus is an exception with an isoelectric point of pH 10 (Isaacson, 1977). This property makes K99 a particularly useful adhesin. K99-positive *E. coli* may adhere to gut epithelial cells totally on the basis of charge interaction. Although it is presumed that K99-specific receptors exist on the epithelial cell surface, the charge interaction between K99 and the target cell may well increase the stability of the bacterial layer, thereby increasing the probability of permanent, irreversible attachment.

Although the pilus structure does not contain interpilin covalent bonds, it is likely that in some pili intrapilin covalent bonds are essential. For example, pilin molecules that contain disulfide bonds may take on three different conformations depending upon the integrity of the disulfide bond (McMichael and Ou, 1979; Isaacson, 1980b; Isaacson and Richter, 1981; Isaacson *et al.*, 1981; Jann *et al.*, 1981). Native molecules undergo a conformational change upon cleavage of the disulfide bond. Molecules treated with the reducing agent 2-mercaptoethanol have a slower electrophoretic mobility in SDS–polyacrylamide gels than nontreated pilin. If the disulfide bonds are permitted to re-form, then pilin molecules may regain their original conformation or assume a third conformation. These differences again can be demonstrated by the differences in electrophoretic mobility. Pilin in the third confirmation migrates slower than native molecules and faster than molecules with disrupted disulfide bonds.

The facts that many, although not all, pilin molecules contain disulfide bonds and that each of them subsequently may assume three different conformations suggest the existence of conserved amino acid sequences in pili. The intrachain disulfide bonds affect the shape of pilin which may be necessary for subunit assembly, adhesive function, or antigenicity.

The notion of amino acid sequence homology among pili is partially supported by the results of NH_2-terminal amino acid sequence analysis. The remarkable sequence similarity of the NH_2-termini of *N. gonorrhoeae* (Hermodson *et al.*, 1978), *Moraxella nonliquifaciens* (Hermodson *et al.*, 1978), and *Pseudomonas aeruginosa* (Paranchych *et al.*, 1978) has been described (Pearce and Buchanan, 1980). The NH_2-terminal amino acid for each is an unusual amino acid, methylphenyl alanine, and only three amino acid substitutions exist in the following 20, three isoleucines substituted for valine (Table III). Each of the amino acid substitutions could result from a single nucleic acid base change. This sequence homology is particularly interesting as the bacterial genera are phylogenetically unrelated and the pili are antigenically unrelated.

The NH_2-terminal amino acid sequences of several other pili show remarkable relatedness. The sequences of the first 20 amino acids of different type 1 pili of *E. coli* (Klemm *et al.*, 1982), type 1 pili of *Klebsiella pneumoniae* (Fader *et al.*, 1982), and *E. coli* F7 pilus (Klemm *et al.*, 1982) are nearly identical and yet the five pili are antigenically unrelated. The hemagglutination of erythrocytes by the F7 pilus is not inhibited by D-mannose whereas the four other pili are.

Although less remarkable, NH_2-terminal amino acid sequence homology between *E. coli* K88 pili (Gaastra *et al.*, 1979; Klemm, 1981) and K99 pili (deGraaf *et al.*, 1981) and to a lesser extent F41 pili (deGraaf and Roorda, 1982) have been observed. Six of the first 20 amino acids are identical between K88 and K99 and another 6 differ by a change of a single base pair per codon. The physical properties of the amino acids that differ by a single base pair in K88 and K99 are, for the most part, so similar that the overall properties of the structures would not be expected to differ greatly. Analysis of pilus sequence by DNA–DNA

Table III. NH$_2$-Terminal Amino Acid Sequence of Several Pili

	1	5	10	15	20
E. coli		Gln			
F7	Ala Ala Thr Ile Pro Gln Gly Gly Glu Val Ala Phe Lys Gly Thr Val Val Asx Ala				
Type 1a	Ala Ala Thr Thr Val Asn Gly Gly Thr Val His Phe Lys Gly Glu Val Val Asn Ala				
Type 1b	Ala Thr Thr Val Asn Gly Gly Thr Val His Phe Lys Glu Glu Val Val				
Type 1c	Val Thr Thr Val Asn Gly Gly Thr Val His Phe Lys Gly Glu Val Val				
K. pneumoniae	Asn Thr Thr Thr Val Asn Gly Gly Thr Val Ala Phe Lys Gly Glu Val Val Asp Ala				
E. Coli					
K88	Trp Met Thr Gly Asp Phe Asn Gly Ser Val Asp Ile Gly Gly Ser Ile Thr Ala Asp				
K99	Asn Thr Gly Thr Ile Asn Phe Asn Gly Lys Ile Glu Thr Ala Thr Ser X Ile Glu Pro				
F41	Ala Asp Trp Thr Glu Gly Gln Pro Gly Asp Ile Leu Gly Gly Glu Ile Thr X Pro				
N. gonorrhoeae	MePhe Thr Leu Ile Glu Leu Met Ile Val Ile Ala Ile Val Gly Ile Leu Ala Ala Val Ala				
P. aeruginosa PAK	MePhe Thr Leu Ile Glu Leu Met Ile Val Val Ala Ile Ile Gly Ile Leu Ala Ala Ile				
M. nonliquifaciens	MePhe Thr Leu Ile Glu Leu Met Ile Val Ile Ala Ile Ile Gly Ile Leu Ala Ala Ile				

hybridization has also demonstrated limited homology between K88 and K99 (Isaacson, unpublished).

3.3. Antigenicity

The proteinaceous nature of pili makes them good antigens. Pili have been detected on bacteria with a variety of serologic tests including serum agglutination and enzyme-linked immunosorbent assays (Jones and Isaacson, 1983). Some pili readily diffuse in agar and can be detected by double-diffusion precipitation assays (Jones and Isaacson, 1983).

Antigenic cross-reactions among pili are limited; antigenic diversity is the rule even within a single species such as *E. coli*. The F7 pilus of *E. coli* is nearly identical in NH$_2$-terminal amino acid sequence with *E. coli* type 1 pili, but is antigenically unrelated (Ørskov *et al.*, 1980; Klemm *et al.*, 1982). Antigenic diversity also appears to occur among the group of *E. coli* pili classified as type 1 (i.e., D-mannose-sensitive hemagglutinins).

Three antigenic variations of K88 pili have also been recognized and analyzed. Each variant possesses a common *a* antigen and a *b, c,* or *d* antigen (Guinée and Jansen, 1979; Guinée *et al.*, 1980).

Even more impressive is the diversity of *N. gonorrhoeae* pili. There are at least 50 different antigenic types (Buchanan, 1975; Brinton *et al.*, 1978). Between 1 and 10% of these pili have limited cross-reactions. The P9 strain of *N. gonorrhoeae* has been shown to produce individually four different pili (α, β, γ, δ) (Lambden *et al.*, 1981; Heckels, 1982). The α and β pili have subunits with identical molecular weights and nearly identical amino acid compositions and tryptic peptide fingerprints, but do not cross-react in serum agglutination tests. The shift from one antigenic variant to another is quite rapid and can be demonstrated after several serial passages *in vitro*.

Analyses of cyanogen bromide fragments of the P9 pili by immunologic and biochemical techniques have provided some insight into the mechanism of antigenic variation (Schoolnik, 1982). At the NH$_2$ terminus is a highly conserved sequence that may be involved in pilin assembly. A middle conserved sequence specifies receptor-binding activity. At the

COOH-terminus is a highly variable type-specific antigen changes in which do not cause functional or structural alterations of the pilus. The COOH-terminal region is immunodominant. Polyclonal sera prepared against *N. gonorrhoeae* pili are almost exclusively specific for this region. Thus, closely related pili, based on sequence homology, may appear to be antigenically unrelated.

4. GENETICS

4.1. Gene Loci

A variety of genetic techniques have been used to determine the genetic loci of several pili. The most intensively studied example which will only be briefly mentioned is the *E. coli* F pilus. F pili are plasmid-encoded, the pilin structural gene residing in a region called *tra* (Willetts and Skurray, 1980). The *tra* region includes different cistrons that specify, in the aggregate, plasmid transfer via conjugal mating.

The gene locus of *E. coli* type 1 pili was determined by a classical genetic approach (Brinton *et al.*, 1961). Various Hfr strains were used in mating experiments with a stably nonpiliated recipient. Initial matings showed a linkage with *ara*. Using interrupted matings, the map site was shown to correspond to 98 min. The F' plasmid F101 which contains the markers *thr, leu,* and *ara* also specifies production of type 1 pili.

Complementation analysis (Swaney *et al.*, 1977) was performed by construction of double mutants, one mutation residing on the chromosome and the other on F101. Three complementation groups were identified and designated *pil* A, *pil* B, *pil* C. A fourth complementation was identified that appeard to indicate mutants defective in both *pil* A and *pil* B.

Recent experiments have confirmed a chromosomal location for type 1 pili and demonstrated that another pilus produced by the same *E. coli* strain and associated with adhesion to urinary tract epithelium is also encoded on the chromosome (Hull *et al.*, 1981). The genes specifying mannose-sensitive (type 1) and mannose-resistant (the other pilus) hemagglutinins were individually isolated from a urinary tract pathogen by DNA cloning. The *E. coli* strain used as the DNA source in the cloning experiment contained a single plasmid known to specify colicin V. Restriction endonuclease digests of the Col V plasmid and the two recombinant DNA molecules were compared by separation of the digests by agarose gel electrophoresis. The three plasmids (Col V and the two recombinant plasmids) had different digestion patterns, demonstrating that the cloned DNA in the recombinant DNA molecules was not from the Col V plasmid and that the two cloned DNA segments were different. Thus, it was concluded that type 1 pili and the mannose-resistant pili were encoded at different sites on the chromosome.

Analysis of several *N. gonorrhoeae* strains has demonstrated a lack of relationship between piliation and plasmids, suggesting a chromosomal locus (Mayer *et al.*, 1974).

The genes encoding several other *E. coli* pilus adhesins have been associated with plasmids. Bak *et al.* (1972) demonstrated that K88, a heat-labile antigen on the surface of *E. coli* pathogens of porcine origin, was plasmid-encoded. Smith and Linggood (1971b) examined pig enteropathogens and showed that K88 was frequently transmissible to K88-negative strains. In some cases, K88 appeared to be self-transmissible. Subsequently, Shipley *et al.* (1978) showed that K88 was associated with a 50-megadalton nonconjugated plasmid or larger conjugative plasmids. The genes encoding K99 have been associated with a 58-megadalton plasmid (Isaacson *et al.*, 1984) and the genes for CFA/I have been associated with a 60-megadalton plasmid which also encodes heat-stable enterotoxin (McConnell *et al.*,

1981; Chapter 15, Section 2.2.1). Confirmation of plasmid loci for K88 and K99 has been established by DNA cloning (Mooi *et al.*, 1979; van Embden *et al.*, 1980; Shipley *et al.*, 1981; Isaacson *et al.*, 1984).

The medical significance of plasmid-mediated pilus production, especially with regard to pili that serve as adhesins, is enormous. Adhesins are an important attribute of microbial virulence. Their capacity to be spread epidemically within species or closely related species results in the potential formation of new pathogens or pathogens of increased virulence. Several other attributes related to microbial virulence also are plasmid-encoded including toxins (see Chapter 15, Section 2.2.1), hemolysins, iron chelation, and resistance to various antibiotics. Most *E. coli* pathogens contain several plasmids which is often interpreted as an indication of their potential virulence. Irrespective of the number of plasmids present in an organism, it is clear that the potential to produce strains of increased virulence may be facilitated by the acquisition of a single plasmid. It may be of greater importance to recognize that totally innocuous organisms (e.g., normal flora) may become pathogens by acquisition of virulence-associated plasmids.

Several virulence genes have also been shown to be on transposons (So and McCarthy, 1980). The rapid evolution of plasmids containing multiple virulence attributes is thus potentiated, making the *in vivo* construction of new pathogens easier. Interestingly, although the potential to create pathogens by acquisition of plasmids appears to be virutally limitless, distribution and diversity of *E. coli* serotypes associated with disease are limited. Of the hundreds of different *E. coli* serotypes known, relatively few have been recognized as pathogens. It is quite possible that other bacterial traits not plasmid-mediated also contribute to virulence. For example, some ETEC strains require a capsule for virulence (Isaacson *et al.*, 1977).

4.2. Expression

The expression of pili can be divided into two general categories. The first is a qualitative process called phase variation (Brinton, 1965; Ottow, 1975; Duguid and Old, 1980; Isaacson and Richter, 1981). This is an all-or-none reversible process with the rates of conversion from one phase to the other being greater that mutation rates and being characteristic of the organism, adhesin, and cultural conditions. Not all pili are subject to phase variation. A second process is quantitative variation. As the name implies, this process governs the actual amount of pili produced.

4.2.1. Phase Variation

E. coli type 1 and 987P pili undergo phase variation. In the laboratory, depending upon the method of culture, cultures can be grown containing cells in either phase. In general, aerobic growth selects for the nonpiliated phase, whereas growth at reduced levels of oxygen selects for the piliated phase (Brinton, 1965; Old *et al.*, 1968; Old and Duguid, 1970). Growth of 987P-producing *E. coli* in pigs (*in vivo*) results in the selection of cells in the piliated phase (Nagy *et al.*, 1977), suggesting that piliation *in vivo* is advantageous compared to the nonpiliated phase. Accompanying the change from one phase to the other is a corresponding change of the morphology of colonies growing on semisolid media (Brinton, 1965). Colonies containing piliated cells tend to be translucent, small, and cohesive whereas colonies containing nonpiliated cells are the opposite.

The precise mechanism of phase variation is not known. However, several lines of investigation are consistent with a mechanism similar to flagellar phase variation of *Salmonella* (Silverman and Simon, 1980). Flagellar phase variation in *Salmonella* results in the switching of flagellar phenotype by regulating which of the two flagellar genes is expressed. This process is accomplished by the inversion of a DNA sequence containing the promoter for one of the flagellar genes (H2) and a repressor of the other flagellar gene. (H1) When the invertible sequence is in one orientation, the H2 gene and H1 repressor are expressed, resulting in the H2 phenotype. When the invertible sequence flips to the other orientation, the H2 gene and H1 repressor are not made and H1 flagella are produced.

One line of evidence supporting a similar mechanism for pilus phase variation comes from the results obtained with Mudlac bacteriophage fusions to *pil* genes (Eisenstein, 1981). The mu phage contains the gene for β-galactosidase (*lac* Z); however, this gene is not expressed because it lacks a promoter. When the phage integrates into a gene near a promoter, the *lac* Z gene may be expressed using the host promoter to initiate transcription. If the bacteria used for the fusions are *lac*-negative, such fusions will result in a *lac*-positive phenotype. Fusions of mu phage to *pil* genes were constructed and expression of *lac* Z was shown to be regulated by the *pil* gene promoter. *lac*-positive and *lac*-negative clones were isolated; each was shown to undergo a transition to the other phenotype. As promoters control transcription, it was concluded that phase variation was controlled at the transcriptional level. The mu–*pil* fusions were cloned in a lambda phage and used to construct merodiploids in *pil*-positive cells. The switching of phase was found to be *cis*-active.

A second line of evidence comes from studies on expression of *N. gonorrhoeae* pili (Meyer *et al.*, 1982). The pilus genes from *N. gonorrhoeae* were cloned into *E. coli*. Recombinant plasmid DNA was isolated from one such recombinant and used as a probe for DNA–DNA hybridization. Chromosomal DNA purified from piliated and nonpiliated phase variants was digested with a restriction endonuclease and separated by agarose gel electrophoresis. The fragments were blotted to filters and hybridized with the probe from the recombinant plasmid. The results showed that the conversion from the piliated phase to the nonpiliated phase resulted in the appearance of probe-specific fragments of different sizes. This result indicates that phase variation was accompanied by a chromosomal rearrangement. Such results are consistent with an invertible sequence analogous to *Salmonella* flagella.

4.2.2. Quantitative Variation

Pili such as K88, K99, and CFA/I do not appear to be subject to phase variation. Most, if not all, pili are subject, however, to quantitative variation. The most extensively studied example of quantitative variation pertains to K99.

Environmental factors known to influence K99 production include aeration, the presence of glucose, L-alanine, cAMP, and temperature of incubation (Brinton, 1965; Guinée *et al.*, 1977; Wadstrom *et al.*, 1978; Isaacson, 1980a, 1983; deGraaf *et al.*, 1980; Isaacson and Richter, 1981).

K99-positive *E. coli* incubated at 37°C produce pili on the cell surface (i.e., assembled K99) exclusively during the exponential phase of growth (Isaacson, 1980a, 1983). The degree of aeration affects the amount of pili on the cell surface. Vigorous shaking increases aeration which results in increased piliation; poor aeration results in the opposite effect. Glucose added to the growth medium suppresses piliation, the degree of suppression being correlated with the basal medium. Glucose in a rich medium such as trypticase–soy broth results in a high degree of K99 suppression. cAMP alleviates the glucose-mediated repression. Glucose in minimal medium does not significantly reduce piliation.

A study has been made of whether the K99 genes are subject to glucose-mediated catabolite repression. The K99 plasmid was transferred to a mutant in which the enzyme adenylate cyclase (*cya*) was not expressed (Isaacson, 1983). This strain produced normal quantities of cell surface-associated K99, suggesting that the glucose effect, observed using rich medium, was not related to K99 genes. However, when the actual synthesis of K99 subunits was measured, rather than the degree of piliation (i.e., the amount of assembled pilus on the cell surface), the *cya* mutation effected a dramatic reduction in subunit synthesis ($\simeq$ 33-fold). The synthesis of subunits was also determined during the various phases of bacterial growth and compared to the amount of K99 on the cell surface. While K99 appeared to be assembled on the cell surface exclusively during the exponential phase, subunits were synthesized continuously during all phases of growth. It was concluded, therefore, that K99 expression could be divided into two interactive but independent steps: subunit synthesis which is subject to catabolite repression and pilus assembly.

In a study to test that hypothesis, the effect of inhibiting protein synthesis on K99 expression was determined. Chloramphenicol added to actively growing K99-positive cultures stopped subunit synthesis while assembly proceeded normally.Therefore, assembly does not require active protein synthesis and probably draws on a pool of subunits prexisting in the outer membrane of the cell. The results from the *cya* mutant experiments indicate that a 33-fold reduction of the subunit pool does not affect assembly. Thus, a substantial pool must normally exist which even when subunit synthesis is inhibited is still ample for pilus assembly. In wild-type strains, K99 subunits are synthesized in great excess and stored in the outer membrane. A small fraction is assembled into functional pili.

The facts that K99 is a membrane protein and that an outer membrane subunit pool exists suggest that the membrane(s) plays an integral role in expression. Cells grown at 18°C fail to synthesize and assemble K99. Likewise, cells grown at 37°C in the presence of excess L-alanine also fail to synthesize and assemble the pilus. Both conditions result in altered cell membranes which are therefore presumed to be necessary for K99 expression. Consistent with these observations is the fact that K99 subunits are synthesized with a 3-kilodalton leader sequence (Gaastra and deGraaf, 1982, Isaacson *et al.*, 1984). Interestingly, most other *E. coli* pili also are not produced when cells are incubated at 18°C.

4.2.3. Pilus Gene Organization

The genetic organization of several *E. coli* pilus adhesins has been determined with recombinant DNA technology.

4.2.3a. K88.

Both K88ab and K88ac have been individually cloned from naturally occurring plasmids (Mooi *et al.*, 1979; Shipley *et al.*, 1981). Each is encoded on a 6.5-kb HindIII–EcoRI restriction fragment with a second EcoRI site being located in the K88 structural genes (Fig. 7). Minicell-producing *E. coli* containing either of the K88 recombinant plasmids were used to determine the number and size of polypeptides encoded by the cloned K88 fragments. Each encodes five polypeptides of 81, 27.5, 27.0, 26.0, and 17 kilodaltons for K88ab and 70, 29, 26, 23.5, and 17 kilodaltons for K88ac in addition to the vector-encoded polypeptides. The 26 and 23.5 polypeptides are from the K88ab and the K88ac structural genes, respectively. The organization and locations of the specific cistrons on the cloned fragments were determined by deletion mapping (for K88ab) (Mooi *et al.*, 1981) and insertional mutagenesis (for K88ac) (Kehoe *et al.*, 1981) and are shown in Fig. 7. The organization of both are nearly identical. A major difference between the two is that K88ab appears to be encoded by a single transcriptional unit that utilizes a promoter on the vector (Mooi *et al.*, 1983), while K88ac is encoded by two transcriptional units (Kehoe *et al.*, 1981).

The functions of the various polypeptides, other than the K88 structural genes, have

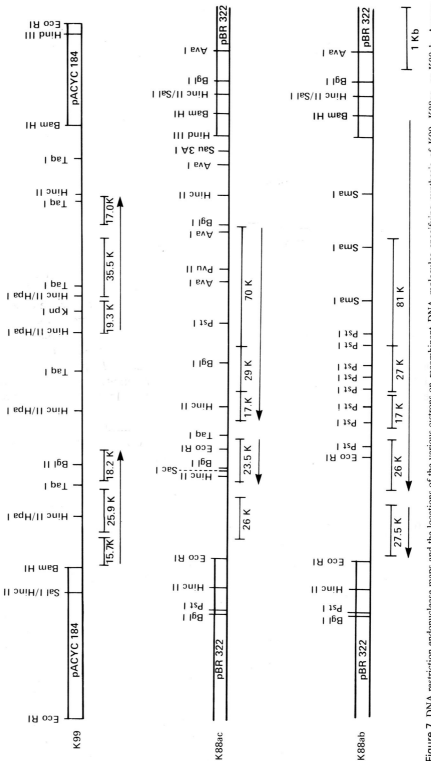

Figure 7. DNA restriction endonuclease maps and the locations of the various cistrons on recombinant DNA molecules specifying synthesis of K99, K88ac, or K88ab. Arrows indicate the direction of transcription.

been deduced by a variety of techniques. For K88ab (Mooi *et al.*, 1983) the 81K polypeptide is probably involved in subunit translocation across the outer membrane. The 27K polypeptide is possibly involved in stabilizing a specific subunit confirmation required for translocation from the periplasm to the outer membrane; the 17K polypeptide modifies the subunit such that it is functional when assembled. A role for the 27.5K polypeptide in K88ab expression, if any, has not been determined.

The K88ac polypeptides are assumed to function in the following manner (Dougan *et al.*, 1983). The 70K polypeptide is assumed to be an anchor for K88 in the outer membrane. The 29 and 17K polypeptides are probably involved in polymerization, although the 17K polypeptide was previously believed to be a positive regulator of the K88ac subunit cistron (Kehoe *et al.*, 1981). The role of the 26K polypeptide in K88ac expression, if any, is not known.

4.2.3b. K99.

4.2.3b. K99. A 7.15-kb BamHI restriction fragment from a naturally occurring plasmid was cloned and cells containing the recombinant plasmid produced K99 (van Embden *et al.*, 1980; Isaacson *et al.*, 1984). Recombinants produce 4- to 40-fold greater amounts of assembled K99. The recombinant plasmid specifies six polypeptides in addition to those encoded by the cloning vector with molecular weights of 35.5, 25.9, 19.3, 18.2, 17.0, and 15.7K (Isaacson *et al.*, 1984). Deletions and insertional mutants were created and used to map the six cistrons (Fig. 7). The cistrons are organized into two clusters of three. The left cluster appeared to be a single transcriptional unit, while the right cluster appeared to contain three separately transcribed cistrons. Based upon the effect of mutations on expression of the six polypeptides, the following functions were deduced. The 18.2K polypeptide is the K99 subunit, the 19.3K polypeptide is required for subunit assembly, and the 35.5 and 17K polypeptides probably are gene regulators; the functions of the 15.7 and 25.9K polypeptides are unknown.

The two gene clusters of K99 represent separate complementation groups (Isaacson *et al.*, 1984). Two complementation groups representing the two transcriptional units of K88ac were also detected (Kehoe *et al.*, 1981). Genes for a pilus from an *E. coli* urinary tract pathogen have been cloned and appear to contain at least two complementation groups (Pierce and Clegg, 1983).

4.3. The Role of Pilus (Adhesin) Expression in the Establishment of Populations

The role of pilus–adhesin expression in the establishment of bacterial populations is implicit. Organisms that colonize habitats, such as the gastrointestinal tract where washout due to specific host secretion and fluid removal may occur, have evolved mechanisms to overcome host-mediate clearance. Some such bacteria may adhere to mucosal surfaces via adhesins. The adhesive process is dependent upon the expression of the adhesin. Geneotypically adhesive but phenotypically nonadhesive organisms will not colonize the same habitats as those with an adhesive phenotype. Regulation of adhesin expression therefore is considered an attribute of virulence. Although we view adhesin expression as a process involved in habitat colonization, adhesin expression is also important in transmission of bacterial pathogens and/or invasion. Cells adherent to a mucosal surface are, by virtue of their interaction with the target cell, maintained at that site. For cells to be transmitted to another individual, they must leave the first individual. Under these circumstances, pilus production is no longer advantageous and to the contrary becomes a liability. Transmission may be accomplished by sloughing of epithelial cells or by reversal of adhesin expression followed by excretion of the bacterial cells by the host. It is predicted that should expression be important for bacterial transmission, then the bacteria should be capable of growing in piliated or nonpiliated phases.

Remaining piliated also becomes a liability to bacteria that are invasive. The initial invasive process is an endocyticlike process that probably occurs independent of attachment and piliation. Piliation becomes a liability for bacteria that enter the circulatory system or any site accessed by macrophages or antibodies. The highly antigenic nature of pili may lead to synthesis of pilus-specific antibodies and possibly enhanced phagocytosis although in some instances pili appear to inhibit phagocytosis.

It should also be recognized that many bacteria have the capability of existing in more than one habitat. It may ultimately be concluded that the need for adhesins is also dependent upon the habitat. For example, *E. coli* strains, whether pathogenic or not, colonize the gastrointestinal tracts of mammals and can be found in other environments such as water and soil. The need for pilus adhesins in these habitats, particularly in soil, is questionable. Related to this is the observation that cells grown at 18°C do not produce pili (Brinton, 1965; Ottow, 1975). This phenomenon may well have evolved as a mechanism of preventing pilus production in a habitat where they are not needed.

5. ADHESIN RECEPTORS

The interaction between adhesin and receptor must be specific and result in selective adhesion of the bacterium to a surface (see Chapter 4, Section 4). Direct evidence for the existence of specific receptors may, in part, be derived from studies on the pathogenesis of K88-positive ETEC. These strains of *E. coli* extensively colonize the small intestines of neonatal pigs by adhesion and also adhere *in vitro* to purified brush borders prepared from porcine enterocytes. Some pigs have been shown to be resistant to disease induced by K88-positive ETEC (Rutter *et al.*, 1975; Sellwood *et al.*, 1975). K88-positive ETEC do not colonize the small intestines or adhere to brush borders from resistant pigs. These results are interpreted as evidence for K88-specific receptors. In another study, a relationship was observed between resistance to *E. coli* urinary tract infections and attachment *in vitro* of uropathogenic *E. coli* to urinary tract epithelium (Lomberg *et al.*, 1981). The results from experiments on the competitive inhibition by purified pili of adhesion *in vitro* described above also provide strong evidence for adhesin-specific receptors (Isaacson *et al.*, 1978b).

Receptors on the target cells provide the apparent selectivity or specificity to the adhesive process. The distribution of receptors on an animal cell or among different types of animal cells may be diverse. Thus, some adhesins specific for the gastrointestinal tract also recognize receptors on certain erythrocytes. Most receptors appear to be glycoconjugates with the sugar residue providing an essential recognition site. Interactions of adhesins with substrates containing the correct carbohydrate, although not the natural support, may occur and appear to be selective. Type 1 pili of *E. coli* recognize D-mannose as a receptor component (Old, 1972; Duguid and Old, 1980). They recognize free-D-mannose as well as D-mannose on erythrocytes or the yeast cell polymer mannan. The relevance of any receptor is therefore dependent upon the demonstration of biological activity.

The identification of receptors has been based on several experimental approaches:

1. Inhibition of adhesion by soluble substances
2. Inhibition of adhesion by lectins
3. Inhibition of adhesion by enzymatic destruction of the receptor
4. Attachment to inert surfaces coated with the putative receptor
5. Comparison of genetically resistant and sensitive animal cells.

5.1. P-Pili (Fimbriae)-Specific Receptors

Perhaps the most elegant study of bacterial receptors for bacterial pili is that on the P-pili or fimbriae of a uropathogenic *E. coli* strain (Leffler and Svanborg-Eden, 1980). The observations that a group of individuals are resistant to colonization with this organism and that the urinary tract epithelium of these individuals does not contain the pilus-specific receptor (based on *in vitro* adhesion experiments) led to recognition of a relationship between adhesive phenotype and P-blood group (hence the name P-pili). Subsequently, globo-series glycolipids were extracted from epithelial cells of an adhesive phenotype and shown to inhibit *in vitro* adhesion of the uropathogenic *E. coli* strain to urinary tract epithelial cells. Purified globo-series glycolipids with the minimal polysaccharide sequence α-Gal1→4β-Gal had receptor activity. Adhesion of uropathogenic *E. coli* to nonreactive (receptorless) cells is facilitated when the cells are coated with the globo-series glycolipid.

5.2. Other Pilus-Specific Receptors

The hemagglutinating activity of K88 can be inhibited by glycoproteins obtained from porcine colostrum (Gibbons *et al.*, 1975). Treatment of the glycoproteins to remove terminal β-D-galactose residues results in the loss of inhibitor activity. Thus, β-D-galactose is part of the K88 receptor. Furthermore, β-D-galactose prevents attachment of purified K88 to the brush borders of neonatal porcine enterocytes (Sellwood, 1980).

The rabbit receptor for the RDEC strain (see Chapter 15, Section 2.2.1) has been partially purified and appears to be associated with the isomaltase enzyme complex (Cheney and Boedeker, 1984).

The adult rabbit small intestine has been used as a model for studying 987P pilus–receptor interaction as well as 987P receptor structure (Dean and Isaacson, 1982). Although 987P-positive ETEC are not pathogens of adult rabbits, several properties of the rabbit make it well suited to such studies. First, there is age specificity. Infant rabbits lack the receptor which means that they may be used, in essence, as mutants. Second, the interaction is specific in that purified 987P pili prevent *in vitro* attachment of piliated cells to the epithelium of the small intestine. Third, other piliated *E. coli* do not adhere to adult rabbit small intestine, except for the rabbit pathogen RDEC and to a small extent K99-positive strains. (The K99-positive strains adhere best to the small intestines of infant rabbits and considerably less to those of adults.) The rabbit receptor has been purified from brush-border membranes by methanol–chloroform extraction and gel filtration (Dean and Isaacson, 1985). The receptor, which is homogeneous, as assessed by SDS–polyacrylamide gel electrophoresis and silver staining, also can be stained with periodic acid–Schiff stain and agglutinates piliated 987P-positive cells. Treatment with periodate abolishes receptor activity. Receptor activity in crude preparations is also abolished by digestion with Pronase and proteinase K. Several lipases, phospholipases, and boiling have no effect on receptor activity. The lectin soybean agglutinin which recognizes D-galactose and *N*-acetylgalactosamine also prevents pilus–receptor interaction, suggesting that one of these sugars is important for receptor activity.

6. CONCLUDING REMARKS

The preceding discussion has presented evidence of the importance of pilus adhesins for the colonization of mucosal surfaces, and the results of studies on the molecular structure of some pilus adhesins and the cognate receptors. Analyses of the predicted and measured physicochemical properties of pili in particular and of bacterial cells and mucosal surfaces in general led to hypotheses defining the potential interactions between pilus and receptor and between bacterium and host.

The relevance of the adhesive process, in some instances, is obvious. Adhesion provides an ecological advantage over nonadhesive organisms in a given habitat. For example, *E. coli* normally colonizes the mammalian colon. In this habitat, *E. coli* need not adhere to the mucosa, for the rate of flow through the colon is slower than the rate of replication of the bacterial cells. However, this habitat is also conducive to the proliferation of other bacterial species which can achieve very large populations (10^{10} to 10^{11} per gram of tissue and content). Thus, in the colon *E. coli* must compete against other organisms for nutrients. *E. coli* strains that have the capacity to colonize the small intestine by adhering to mucosa therefore can be said to have a competitive edge over strains in the colon of a particular host.

In general, adhesins have evolved as a means for bacterial cells to colonize specific habitats and in so doing contributing to the overall economy of the ecosystem. Occasionally, organisms that have evolved an aggressive means of surviving cause diseases when they colonize specific habitats. In most instances, however, the relationship between host and microbe in a given habitat is mutually beneficial (see Chapter 15, Section 2.1).

REFERENCES

Acres, S. D., Isaacson, R. E., Babiuk, L. A., and Kapitany, R. A., 1979, Immunization of calves against enterotoxigenic colibacillosis by vaccinating dams with purified K99 antigen and whole cell bacterins, *Infect. Immun.* **25**:121–126.

Arbuckle, J. B. R., 1970, The location of *Escherichia coli* in the pig intestine, *J. Med. Microbiol.* **3**:333–340.

Bak, A. L., Christiansen, G., Christiansen, C., Stenderup, A., Ørskov, I., and Ørskov, F., 1972, Circular DNA molecules controlling synthesis and transfer of the surface antigen (K88) in *Escherichia coli*, *J. Gen. Microbiol.* **73**:373–385.

Barnum, D. A., Glantz, P. J., and Moon, H. W., 1967, Colibacillosis, Ciba Veterinary Monograph Series 2, Ciba Pharmaceutical Company, Summit, N.J.

Bertschinger, H. V., Moon, H. U., and Whipp, S. C., 1972, Association of *Escherichia coli* with the small intestinal epithelium. I. Comparison of enteropathogenic and nonenteropathogenic porcine strains in pigs, *Infect. Immun.* **5**:595–605.

Brinton, C. C., 1959, Non-flagellar appendages of bacteria, *Nature (London)* **183**:782–786.

Brinton, C. C., 1965, The structure, function, synthesis and genetic control of bacterial pili and a molecular model for DNA and RNA transport in gram negative bacteria, *Trans. N.Y. Acad. Sci.* **27**:1003–1054.

Brinton, C. C., 1971, The properties of sex pili, the viral nature of "conjugal" genetic transfer systems, and some possible approaches to the control of bacterial drug resistance, *Crit. Rev. Microbiol.* **1**:105–160.

Brinton, C. C., Gemski, P., Falkow, S., and Baron, L. S., 1961, Location of the piliation factor on the chromosome of *Escherichia coli*, *Biochem. Biophys. Res. Commun.* **5**:293–298.

Brinton, C. C., Bryan, J., Dillon, J. A., Guerina, N., and Jacobsen, J., 1978, Uses of pili in gonorrhoea control: Role of bacterial pili in disease, purification and properties of gonococcal pili, and progress in the development of a gonococcal pilus vaccine for gonorrhoea, in: *Immunobiology of Neisseria gonorrhoeae* (G. F. Brooks, E. C. Gotschlich, K. K. Holmes, W. D. Sawyer, and F. E. Young, eds.), American Society for Microbiology, Washington, D.C., pp. 155–178.

Buchanan, T. M., 1975, Antigenic heterogeneity of gonococcal pili, *J. Exp. Med.* **141**:1470–1475.

Cheney, C. P., and Boedeker, E. C., 1984, Appearance of host intestinal receptors for pathogenic *E. coli* with age, in: *Attachment of Organisms to the Gut Mucosa*, Vol. II (E. C. Boedecker, ed.), CRC Press, Boca Raton, pp. 157–166.

Cisar, J. O., and Vatter, A. E., 1979, Surface fibrils (fimbriae) of *Actinomyces viscosus* T14V, *Infect. Immun.* **24**:523–531.

Dean, E. A., and Isaacson, R. E., 1982, *In vitro* adhesion of piliated *Escherichia coli* to small intestinal villous epithelial cells from rabbits and the identification of a soluble 987P pilus receptor-containing fraction, *Infect. Immun.* **36**:1192–1198.

Dean, E. A., and Isaacson, R. E., 1985, Purification and characterizaton of a receptor for the 987P pilus of *Escherichia coli*, *Infect. Immun.* **47**:98–105.

deGraaf, F. K., and Roorda, I., 1982, Production, purification, and characterization of the fimbrial adhesive antigen F41 isolated from calf enteropathogenic *Escherichia coli* strain B41M, *Infect. Immun.* **36**:751–753.

deGraaf, F. K., Boor, P. K., and vanHess, J. E., 1980, Biosynthesis of the K99 surface antigen is repressed by alanine, *Infect. Immun.* **30:**125–128.

deGraaf, F. K., Klemm, P., and Gaastra, W., 1981, Purification, characterization, and partial covalent structure of *Escherichia coli* adhesive antigen K99, *Infect. Immun.* **33:**877–883.

Dobson, W. J., and McCurdy, H. D., 1979, The function of fimbriae in *Myxococcus xanthus*. I. Purification and properties of *M. xanthus* fimbriae, *Can. J. Microbiol.* **25:**1152–1160.

Dougan, G., David, G., and Kehoe, M., 1983, Organization of K88ac encoded polypeptides in the *Escherichia coli* cell envelope: Use of minicells and outer membrane protein mutants for studying assembly of pili, *J. Bacteriol.* **153:**364–370.

Duguid, J. P., and Old, D. C., 1980, Adhesive properties of Enterobacteriaceae, in: *Bacterial Adherence* (E. H. Beachey, ed.), Chapman & Hall, London, pp. 185–217.

Duguid, J. P., Smith, I. W., Dempster, G., and Edmunds, P. N., 1955, Non-flagellar filamentous appendages ("fimbriae") and haemagglutinating activity in *Bacterium coli*, *J. Pathol. Bacteriol.* **70:**335–348.

Eisenstein, B. I., 1981, Phase variation of type 1 fimbriae in *Escherichia coli* is under transcriptional control, *Science* **214:**337–339.

Eshdat, Y., Silverblatt, F. J., and Sharon, N., 1981, Dissociation and reassembly of *Escherichia coli* type 1 pili, *J. Bacteriol.* **148:**308–314.

Fader, R. C., Duffy, L. K., Davis, C. P., and Kurosky, A., 1982, Purification and chemical characterization of type 1 pili isolated from *Klebsiella pneumoniae*, *J. Biol. Chem.* **257:**3301–3305.

Freter, R., Allweiss, B., O'Brien, P. C. M., Halstead, S. A., and Macsai, M. S., 1981, Role of chemotaxis in the association of mobile bacteria with intestinal mucosae: *In vitro* studies, *Infect. Immun.* **34:**241–249.

Froholm, L. O., and Sletten, K., 1977, Purification and N-terminal sequence of fimbrial protein from *Moraxella nonliquifaciens*, *FEBS Lett.* **73:**29–32.

Gaastra, W., and deGraaf, F. K., 1982, Host-specific fimbrial adhesins of noninvasive enterotoxigenic *Escherichia coli* strains, *Microbiol. Rev.* **46:**129–161.

Gaastra, W., Klemm, P., Walker, J. M., and deGraaf, F. K., 1979, K88 fimbrial proteins: Amino and carboxyl terminal sequences of intact proteins and cyanogen bromide fragments, *FEMS Microbiol. Lett.* **6:**15–18.

Gibbons, R. A., Jones, G. W., and Sellwood, R., 1975, An attempt to identify the intestinal receptor for the K88 adhesin by means of a haemagglutination inhibition test using glycoproteins and fractions from sow colostrum, *J. Gen. Microbiol.* **86:**228–240.

Gorbach, S. L., Kean, B. H., Evans, S. G., Evans, D. J., and Bessudo, D., 1975, Travelers' diarrhea and toxigenic *Escherichia coli*, *N. Engl. J. Med.* **292:**933–936.

Gubish, E. R., Chen, K. D. S., and Buchanan, T. M., 1982, Attachment of gonococcal lectin-resistant clones of Chinese hamster ovary cells, *Infect. Immun.* **37P:**189–194.

Guinée, P. A. M., and Jansen, W. H., 1979, Behavior of *Escherichia coli* antigens K88ab, K88ac, and K88ad in immunoelectrophoresis, double diffusion and hemagglutination, *Infect. Immun.* **23:**700–705.

Guinée, P. A. M., Veldkamp, J., and Hansen, W. H., 1977, Improved Minca medium for the dilution of K99 antigen in calf enterotoxigenic strains of *Escherichia coli*, *Infect. Immun.* **15:**676–678.

Guinée, P. A. M., Mooi, F. R., and Jansen, W. H., 1980, Preparation of specific *Escherichia coli* K88 antisera by means of purified K88ab and K88ad antigens, *Zentralbl. Bakteriol. Parasitenkd. Infektionskr. Hyg. Abt. 1 Orig. Reihe A* **248:**182–189.

Heckels, J. E., 1982, Role of surface proteins in pathogenic adhesion, in: *Microbiology 1982* (D. Schlessinger, ed.), American Society for Microbiology, Washington, D.C., pp. 301–304.

Hermodson, M. H., Chen, K. C. S., and Buchanan, T. M., 1978, Neisseria pili proteins: Amino-terminal amino acid sequences and identification of an unusual amino acid, *Biochemistry* **17:**442–445.

Hull, R. A., Gill, R. E., Hsu, P., Minshew, B., and Falkow, S., 1981, Construction and expression of recombinant plasmids encoding type 1 or D-mannose-resistant pili from a urinary tract infection *Escherichia coli* isolate, *Infect. Immun.* **33:**933–938.

Isaacson, R. E., 1977, K99 surface antigen of *Escherichia coli:* Purification and partial characterization, *Infect. Immun.* **15:**272–279.

Isaacson, R. E., 1980a, Pili of enterotoxigenic *Escherichia coli*, In: *Third International Symposium on Neonatal Diarrhea* (S. Acres, ed.), Veterinary Infectious Disease Organization, Saskatoon, pp. 213–236.

Isaacson, R. E., 1980b, Factors affecting expression of the *Escherichia coli* pilus K99, *Infect. Immun.* **28:**190–194.

Isaacson, R. E., 1983, Regulation of expression of the *Escherichia coli* pilus K99, *Infect. Immun.* **40:**633–639.

Isaacson, R. E., and Richter, P., 1981, *Escherichia coli* 987P pilus: Purification and partial characterization, *J. Bacteriol.* **146:**784–789.

Isaacson, R. E., Nagy, B., and Moon, H. W., 1977, Colonization of porcine small intestines by *Escherichia coli:* Colonization and adhesion factors of pig enteropathogens that lack K88, *J. Infect. Dis.* **135:**531–539.

Isaacson, R. E., Moon, H. W., and Schneider, R. A., 1978a, Distribution and virulence of *Escherichia coli* in the small intestines of calves with and without diarrhea, *Am. J. Vet. Res.* **39:**1750–1755.

Isaacson, R. E., Fusco, P. C., Brinton, C. C., and Moon, H. W., 1978b, *In vitro* adhesion of *Escherichia coli* to porcine small intestinal epithelial cells: Pili as adhesive factors, *Infect. Immun.* **21**:392–397.

Isaacson, R. E., Dean, E. A., Morgan, R. L., and Moon, H. W., 1980, Immunization of suckling pigs against enterotoxigenic *Escherichia coli*-induced diarrheal disease by vaccinating dams with purified K99 or 987P pili: Antibody production in response to vaccination, *Infect. Immun.* **29**:824–826.

Isaacson, R. E., Colmenero, J., and Richter, P., 1981, *Escherichia coli* K99 pili are composed of one subunit species, *FEMS Microbiol. Lett.* **12**:229–232.

Isaacson, R. E., Betzold, J. A., and Petre, J. O., 1984, Cloning, genetic organization, and expression of the *Escherichia coli* pilus K99, *J. Bacteriol.*, Submitted.

Jann, K., Jann, B., and Schmidt, G., 1981, SDS polyacrylamide gel electrophoresis and serological analysis of pili from *Escherichia coli* of different pathogenic origin, *FEMS Microbiol. Lett.* **11**:21–25.

Jones, G. W., and Isaacson, R. E., 1983, Proteinaceous bacterial adhesins and their receptors, *CRC Crit. Rev. Microbiol.* **10**:229–260.

Jones, G. W., and Rutter, J. M., 1972, Contribution of the K88 antigen of *Escherichia coli* to enteropathogenicity: In the pathogenesis of neonatal diarrhea caused by *Escherichia coli* in piglets, *Am. J. Clin. Nutr.* **278**:1414–1419.

Jones, G. W., and Rutter, J. M., 1974, Role of the K88 antigen. Protection against disease by neutralizing the adhesive properties of K88 antigen, *Infect. Immun.* **6**:918–927.

Kehoe, M., Sellwood, R., Shipley, P. L., and Dougan, G., 1981, Genetic analysis of K88-mediated adhesion of enterotoxigenic *Escherichia coli*, *Nature (London)* **291**:122–126.

Klemm, P., 1979, Fimbrial colonization factor CFA/1 protein from human enteropathogenic *Escherichia coli* strains, *FEBS Lett.* **108**:107–110.

Klemm, P., 1981, The complete amino acid sequence of the K88 antigen—A fimbrial protein from *Escherichia coli*, *Eur. J. Biochem.* **117**:617–627.

Klemm, P., Ørskov, I., and Ørskov, F., 1982, F7 and type 1-like fimbriae from three *Escherichia coli* strains isolated from urinary tract infections: Protein, chemical and immunological aspects, *Infect. Immun.* **36**:462–468.

Korhonen, T., Nurmiaho, E.-L., Ranta, H., and Svanborg-Eden, C., 1980a, New method for isolation of immunologically pure pili from *Escherichia coli*, *Infect. Immun.* **27**:569–575.

Korhonen, T. K., Lownatma, K., Ranta, H., and Kwasi, N., 1980b, Characterization of the type 1 pili of *Salmonella typhimurium* LT2, *J. Bacteriol.* **144**:800–805.

Kumazawa, N., and Yanogawa, R., 1972, Chemical properties of the pili of *Corynebacterium renale*, *Infect. Immun.* **5**:27–30.

Lambden, P. R., Robertson, J. N., and Watt, P. J., 1981, The preparation and properties of α and β pili from variants of *Neisseria gonorrhoeae* P9, *J. Gen. Microbiol.* **124**:109–116.

Leffler, H., and Svanborg-Eden, C., 1980, Chemical identification of a glycosphingolipid receptor of *Escherichia coli* attaching to human urinary tract epithelial cells and agglutinating human erythrocytes, *FEMS Microbiol. Lett.* **8**:127–134.

Lindahl, M., Faris, A., Waström, T., and Hjerton, S., 1981, A new test based on "salting out" to measure relative surface hydrophobicity of bacterial cells, *Biochim. Biophys. Acta* **677**:471–476.

Lomberg, H., Jodal, U., Svanborg-Eden, C., Leffler, H., and Samuelsson, B., 1981, P₁ blood group and urinary tract infection, *Lancet* **1**:551–552.

McConnell, M. M., Smith, H. R., Willshaw, G. A., Field, A. M., and Rowe, B., 1981, Plasmids coding for colonization factor antigen I and heat-stable enterotoxin production isolated from enterotoxigenic *Escherichia coli*: Comparison of their properties, *Infect. Immun.* **32**:927–936.

McGee, Z. A., Stephens, D. S., Melly, M. A., Gregg, C. R., Schlech, W. F., and Hoffman, L. H., 1982, Role of attachment in the pathogenesis of disease caused by *Neisseria gonorrhoeae* and *Neisseria meningitis*, in: *Microbiology 1982* (D. Schlessinger, ed.), American Society for Microbiology, Washington, D.C., pp. 292–295.

McMichael, J. C., and Ou, J. T., 1979, Structure of common pili from *Escherichia coli*, *J. Bacteriol.* **138**:969–975.

MacRae, T. H., Dobson, W. J., and McCurdy, H. D., 1977, Fimbriation in gliding bacteria, *Can. J. Microbiol.* **23**:1096–1108.

Mayer, L. W., Holmes, K. K., and Falkow, S., 1974, Characterization of plasmid deoxyribonucleic acid from *Neisseria gonorrhoeae*, *Infect. Immun.* **10**:712–717.

Meyer, T. F., Mlawer, N., and So, M., 1982, Pilus expression in *Neisseria gonorrhoeae* involves chromosomal rearrangement, *Cell* **30**:45–52.

Mooi, F. R., and deGraaf, F. K., 1979, Isolation and characterization of K88 antigens, *FEMS Microbiol. Lett.* **5**:17–20.

Mooi, F. R., deGraaf, F. K., and van Embden, J. D. A., 1979, Cloning, mapping and expression of the genetic determinant that encodes for the K88ab antigen, *Nucleic Acids Res.* **6**:849–865.

Mooi, F. R., Harms, N., Bakker, D., and deGraaf, F. K., 1981, Organization and expression of genes involved in the production of the K88ab antigen, *Infect. Immun.* **32**:1155–1163.

Mooi, F. R., Wÿfjes, A., and deGraaf, F. K., 1983, Identification and characterization of precursors in the biosynthesis of the K88ab fimbria of *Escherichia coli*, *J. Bacteriol.* **154**:41–49.

Moon, H. W., Nagy, B., and Isaacson, R. E., 1977, Intestinal colonization and adhesion by enterotoxigenic *Escherichia coli*: Some ultrastructural observations on adherence to ileal epithelium of the pig, *J. Infect. Dis.* **136**:S124–S129.

Moon, H. W., Kohler, E. M., Schneider, R. A., and Whipp, S. C., 1980, Prevalence of pilus antigens, enterotoxin types and enteropathogenicity among K88-negative enterotoxigenic *Escherichia coli* from neonatal pigs, *Infect. Immun.* **27**:222–230.

Morgan, R. L., Isaacson, R. E., Moon, H. W., Brinton, C. C., and To, C. C., 1978, Immunization of suckling pigs against enterotoxigenic *Escherichia coli*-induced diarrheal disease by vaccinating dams with purified 987 or K99 pili: Protection correlates with pilus homology of vaccine and challenge, *Infect. Immun.* **22**:771–777.

Nagy, B., Moon, H. W., and Isaacson, R. E., 1976, Colonization of porcine small intestines by *Escherichia coli*: Ileal colonization and adhesion by pig enteropathogens that lack K88 antigen and by some acapsular mutants, *Infect. Immun.* **13**:1214–1220.

Nagy, B., Moon, H. W., and Isaacson, R. E., 1977, Colonization of porcine intestine by enterotoxigenic *Escherichia coli*: Selection of piliated forms *in vivo*, adhesion of piliated forms to epithelial cells *in vitro*, and incidence of a pilus antigen among porcine enteropathogenic *E. coli*, *Infect. Immun.* **16**:344–352.

Nagy, B., Moon, H. W., Isaacson, R. E., To, C. C., and Brinton, C. C., 1978, Immunization of suckling pigs against enterotoxigenic *Escherichia coli* infection by vaccinating dams with purified pili, *Infect. Immun.* **21**:269–274.

Nozaki, L., and Tanford, C., 1971, The solubility of amino acid and two glycine peptides in aqueous ethanol and dioxane solutions, *J. Biol. Chem.* **246**:2211–2217.

Old, D. C., 1972, Inhibition of the interaction between fimbrial hemagglutinins and erythrocytes by D-mannose and other carbohydrates, *J. Gen. Microbiol.* **71**:149–157.

Old, D. C., and Duguid, J. P., 1970, Selective outgrowth of fimbriate bacteria in static medium, *J. Bacteriol.* **103**:447–456.

Old, D. C., Cornell, I., Gibson, L. F., Thomson, A. D., and Duguid, J. P., 1968, Fimbriation pellicle formation and the amount of growth of *Salmonella* in broth, *J. Gen. Microbiol.* **51**:1–16.

Ørskov, I., Ørskov, F., and Birch-Andersen, A., 1980, Comparison of *Escherichia coli* fimbrial antigen F7 with type 1 fimbriae, *Infect. Immun.* **27**:657–666.

Ottow, J. C. G., 1975, Ecology, physiology and genetics of fimbriae and pili, *Annu. Rev. Microbiol.* **29**:79–108.

Paranchych, W., Frost, L. S., and Carpenter, M., 1978, N-terminal amino acid sequence of pilin isolated from *Pseudomonas aeruginosa*, *J. Bacteriol.* **134**:1179–1180.

Paranchych, W., Sastry, P. A., Frost, L. S., Carpenter, M., Armstrong, G. D., and Watts, T. H., 1979, Biochemical studies on pili isolated from *Pseudomonas aeruginosa* strain PAO, *Can. J. Microbiol.* **25**:1175–1181.

Pearce, W. A., and Buchanan, T. M., 1980, Structure and cell membrane-binding properties of bacterial fimbriae, in: *Bacterial Adherence* (E. H. Beachey, ed.), Chapman & Hall, London, pp. 289–343.

Pethica, B. A., 1980, Microbial and cell adhesion, in: *Microbial Adhesion to surfaces* (R. C. W. Berkeley, J. M. Lynch, J. Melling, P. R. Rutter, and B. Vincent, eds.), Horwood, Chichester, pp. 19–45.

Pierce, J., and Clegg, S., 1983, Identification of the genetic determinants coding for expression of mannose-resistant fimbriae of *Escherichia coli*, *Abstr. Annu. Meet. Am. Soc. Microbiol.* B46.

Robertson, J. N., Vincent, P., and Ward, M. E., 1977, The preparation and properties of gonococcal pili, *J. Gen. Microbiol.* **102**:169–177.

Rutter, J. M., Burrows, M. R., Sellwood, R., and Gibbons, R. A., 1975, A genetic basis for resistance to enteric disease caused by *E. coli*, *Nature (London)* **257**:135–136.

Rutter, P. R., and Vincent, B., 1980, The adhesion of microorganisms to surfaces: Physico-chemical aspects, in: *Microbial Adhesion to Surfaces* (R. C. W. Berkeley, J. M. Lynch, J. Melling, P. R. Rutter, and B. Vincent, eds.), Horwood, Chichester, pp. 79–92.

Schoolnik, G. K., 1982, Mechanism of binding of gonococcal pili, in: *Microbiology 1982* (D. Schlessinger, ed.), American Society for Microbiology, Washington, D.C., pp. 312–316.

Sellwood, R., 1980, The interaction of the K88 antigen with porcine intestinal epithelial cell brush-borders, *Biochim. Biophys. Acta* **632**:326–335.

Sellwood, R., Gibbons, R. A., Jones, G. W., and Rutter, J. M., 1975, Adhesion of enteropathogenic *Escherichia coli* to pig intestinal brush-borders: The existence of two pig phenotypes, *J. Med. Microbiol.* **8**:405–411.

Shipley, P. L., Gyles, C. L., and Falkow, S., 1978, Characterization of plasmids that encode for the K88 colonization antigen, *Infect. Immun.* **20:**559–566.

Shipley, P. L., Dougan, G., and Falkow, S., 1981, Identification and cloning of the genetic determinant that encodes for the K88ac adherence antigen, *J. Bacteriol.* **145:**920–925.

Silverman, M., and Simon, M., 1980, Phase variation: Genetic analysis of switching mutants, *Cell* **19:**845–854.

Smith, H. W., and Linggood, M. A., 1971a, Further observations on *Escherichia coli* enterotoxins with particular regard to those produced by atypical piglet strains and by calf and lamb strains: The transmissible nature of these enterotoxins and of a K antigen possessed by calf and lamb strains, *J. Med. Microbiol.* **5:**243–250.

Smith, H. W., and Linggood, M. A., 1971b, Observations on the pathogenic properties of the *K88, Hly*, and *Ent* plasmids of *Escherichia coli* with particular reference to porcine diarrhea, *J. Med. Microbiol.* **4:**467–485.

Smyth, C., Jonsson, P., Olsson, E., Söderlünd, O., Rosengren, J., Hjesten, S., and Wadstrom, T., 1978, Differences in hydrophobic surface characteristics of porcine enteropathogenic *Escherichia coli* with and without K88 antigen as revealed by hydrophobic interaction chromatography, *Infect. Immun.* **22:**462–472.

So, M., and McCarthy, B. J., 1980, Nucleotide sequence of the bacterial transposon TN 1681 encoding a heat-stable (ST) toxin and its identification in enterotoxigenic *Escherichia coli* strains, *Proc. Natl. Acad. Sci. USA* **77:**4011–4015.

Stirm, S., Ørskov, F., Ørskov, I., and Mansa, B., 1967, Episome-carried surface antigen K88 of *Escherichia coli*. II. Isolation and chemical analysis, *J. Bacteriol.* **93:**731–739.

Svanborg-Eden, C., 1978, Attachment of *Escherichia coli* to human urinary tract epithelial cells, *Scand. J. Infect. Dis. Suppl.* **15:**1–54.

Swaney, L. M., Liu, Y.-P., Ippen-Ihler, K., and Brinton, C. C., 1977, Genetic complementation analysis of *Escherichia coli* type 1 somatic pilus mutants, *J. Bacteriol.* **130:**506–511.

Swanson, J., Sparks, E., Young, D., and King, G., 1975, Studies on gonococcus infection. X. Pili and leukocyte association factor as mediators of interactions between gonococci and eukaryotic cells *in vitro, Infect. Immun.* **11:**1352–1361.

Tadros, T. F., 1980, Particle–surface adhesion, in: *Microbial Adhesion to Surfaces* (R. C. W. Berkeley, J. M. Lynch, J. Melling, P. R. Rutter, and B. Vincent, eds.), Horwood, Chichester, pp. 93–116.

Uhlman, D., and Jones, G. W., 1982, Chemotaxis as a factor in interactions between HeLa cells and *Salmonella typhimurium, J. Gen. Microbiol.* **128:**415–417.

van Embden, J. D. A., deGraaf, F. K., Schouls, L. M., and Teppema, J. S., 1980, Cloning and expression of a deoxyribonucleic acid fragment that encodes for the adhesive antigen K99, *Infect. Immun.* **29:**1125–1133.

Wadstrom, T., Smyth, C. J., Faris, A., Jonsson, P., and Freer, J. H., 1978, Hydrophobic adsorptive and hemagglutinating properties of enterotoxigenic *Escherichia coli* with different colonizing factors: K88, K99, and colonization factor antigens and adherence factor, in: *Second International Symposium on Neonatal Diarrhea* (S. Acres, ed.), Veterinary Infectious Disease Organization, Saskatoon, pp. 29–45.

Weaver, P., Pikken, R., Schmidt, G., Jann, B., and Jann, K., 1980, Characterization of pili associated with *Escherichia coli* 018ac, *Infect. Immun.* **29:**685–691.

Weiss, L., and Harlos, J. P., 1972, Short-term interactions between cell surfaces, *Prog. Surf. Sci.* **1:**355–405.

Willetts, N. S., and Skurray, R., 1980, The conjugation system of F-like plasmids, *Annu. Rev. Genet.* **14:**41–76.

Wilson, M. R., and Hohmann, A. W., 1974, Immunity to *Escherichia coli* in pigs: Adhesion of enteropathogenic *Escherichia coli* to isolated intestinal epithelial cells, *Infect. Immun.* **10:**776–786.

III

Consequences of Adhesion

12

Effect of Solid Surfaces on the Activity of Attached Bacteria

MADILYN FLETCHER

1. INTRODUCTION

It has been realized for some time that solid surfaces, such as clay particles or glass beads, added to a bacterial suspension may produce an increase in bacterial activity and that bacterial attachment to these surfaces may be involved. As early as 1917, Douglas *et al.* reported that the addition of ''inert'' materials, such as glass, asbestos, charcoal, or chalk, allowed the growth of anaerobic bacilli in broth culture. In the 1930s and 1940s, a much more extensive study of the effects of surfaces on marine bacterial activity was conducted by ZoBell and co-workers (ZoBell and Anderson, 1936; ZoBell, 1943; ZoBell and Grant, 1943). However, it was not until recently that workers refocused attention on the effects of solid surfaces on bacterial physiological processes and renewed attempts to determine the underlying mechanisms for this surface effect.

In many of the early studies, it was not always clear whether activity was stimulated by actual attachment of the bacteria to the surfaces or whether the bacterium–surface interaction was only temporary or superficial. Moreover, in some cases, the activity of suspended cells may have been promoted by the addition of surfaces through removal of inhibitors from the medium by surface adsorption, in which case attachment would have been a disadvantage to the cell.

The influence of solid surfaces on bacterial activity is complex and, particularly in natural environments, may be extremely difficult to evaluate for a number of reasons. The first difficulty is deciding what is meant by ''activity'' or selecting a relevant parameter to measure, which will provide a valid indication of metabolic processes. Such parameters have included: changes in the number of suspended (Harwood and Pirt, 1972) or attached (Hattori, 1972) cells; change in cell size (Kjelleberg *et al.*, 1982); respiration rate, measured as CO_2 production (Bright and Fletcher, 1983b; Kefford *et al.*, 1983), oxygen uptake (Jannasch and Pritchard, 1972) or intensity of electron transport system activity (Bright and Fletcher,

Madilyn Fletcher • Department of Environmental Sciences, University of Warwick, Coventry CV4 7AL, England.

1983a); substrate uptake (Fletcher, 1979; Bright and Fletcher, 1983a,b) or breakdown (Estermann *et al.*, 1959); product formation (Hattori and Hattori, 1963) and heat production (Gordon *et al.*, 1983). On the one hand, some of these processes may be directly related to one another, e.g., growth and respiration rates where nutrient supply is not low. On the other hand, changes in one type of activity may not necessarily be accompanied by corresponding changes in another; thus, care should be taken whenever possible to specify the type of activity being considered. For the purposes of this chapter, *activity* will be used as a general term, which includes any of the processes mentioned above. Where given processes are considered, they are specified.

A second reason why the effect of surfaces on bacterial activity is difficult to determine is that in many cases it is impossible to characterize and quantify the attached and suspended populations. In natural environments or mixed cultures, the attached population will almost certainly differ in both size (Harvey and Young, 1980; Mills and Maubrey, 1981) and/or species composition (Marsh *et al.*, 1983) from the free-living one. Thus, activity of attached versus free-living cells may be due to differences between the two communities, as well as to differences in specific activities.

Because of these difficulties, it is an even greater problem to try to explain the mechanisms which underlie observed differences in attached and suspended bacterial activity. Indeed, it is still only possible to speculate about such mechanisms, except in somewhat specialized cases where the surface has specific properties, such as electrostatic charge, or provides a microenvironment which is buffered from perturbations in the bulk phase (see Section 4.1).

The purpose of this chapter is twofold. First, evidence dealing with the effect of solid surfaces on the activity of attached or surface-associated bacteria is examined, and, where possible, the specific conditions which promote or reduce bacterial activity are identified. Second, an attempt is made to propose mechanisms by which surfaces could affect activity, by considering possible relationships between bacterial metabolic processes and properties of the solid–liquid interface. The emphasis is placed on interactions between solid surfaces and heterotrophic or chemolithotrophic bacteria, as complex assemblages in thick biofilms are dealt with in Chapter 13.

2. COMPARATIVE ACTIVITIES OF ATTACHED AND FREE-LIVING BACTERIA IN THE NATURAL ENVIRONMENT

The data collected from communities in the natural environment are frequently complex and difficult to interpret, primarily because the "experimental system" is an open one, and it is often impossible to define or quantify all the relevant variables. Nevertheless, information is available from studies on at least two basic types of aquatic, attached bacterial communities, i.e., bacteria attached to suspended particles and to submerged surfaces.

2.1. Bacteria Associated with Suspended Particles in Aquatic Environments

In aquatic environments, the number of bacteria found attached to suspended particles can vary considerably. However, there appears to be a general relationship between the proportion of the total flora which is particle-bound and the amount of suspended particulate material in the system (Goulder, 1977; Bell and Albright, 1982; Kirchman and Mitchell, 1982). In a salt marsh, where the water was heavily laden with particles, 93% of the bacteria sampled from the surface water were particle-bound, whereas only approximately 22% were bound in subsurface (0.2 m) samples (Harvey and Young, 1980). By contrast, in a survey of five coastal ponds and marshes, bacterial colonization of inorganic particles was low, at less than 10 cells per particle (Kirchman and Mitchell, 1982).

Different types of measurements have been used to compare the activities of particle-bound and free-living bacteria, and, in a number of cases, attached bacteria have been found to be more active. In the subsurface waters of a salt marsh, measurement of respiratory activity using a tetrazolium dye found that about 63% of the respiring bacteria were particle-bound, whereas only about 22% of the total were attached to particles (Harvey and Young, 1980). Similarly, substrate uptake by attached bacteria has been shown to be greater than that of suspended cells (Goulder, 1977; Kirchman and Mitchell, 1982). However, results are sometimes not so clear-cut when systems are examined in more detail. For example, comparisons of substrate uptake by attached and free-living cells may depend upon the substrate, as was found in a study of 44 marine, estuarine, and freshwater environments (Bell and Albright, 1982). Particle-bound bacteria were generally more active than suspended cells with respect to amino acid uptake, whereas the reverse was true for glucose assimilation. Results may also present a different interpretation when activities are compared on some basis other than cell number. In a study of uptake of dissolved ATP by marine bacteria, the rate of assimilation by attached cells was about 66 times that of suspended cells, on the basis of cell number. However, a comparison of uptake rates on the basis of cell volume gave similar results for the two populations, as the free-living bacteria were smaller, with about 1/40th the volume of attached cells (Hodson et al., 1981). Higher assimilation rates per attached cell were compensated for by a corresponding increase in cell size, so that assimilation rate per biovolume remained the same.

Thus, the relative activities of particle-bound and suspended bacteria appear to differ in a number of cases, with particle-bound bacteria often showing higher levels of activity. Nevertheless, relative activities of the two populations also depend on a number of environmental factors, such as size of the load of suspended solids (Kirchman and Mitchell, 1982), the substrates available (Bell and Albright, 1982), salinity (Bell and Albright, 1982), and cell size (Hodson et al., 1981).

2.2. Bacterial Biofilms on Submerged Solid Surfaces

The second type of attached bacterial community in aquatic environments occurs on submerged surfaces which are stationary (e.g. rocks) or do not move with the dominant current flow (e.g., ships, platforms) (see Chapters 6 and 13). It is quite difficult to evaluate the effect of surfaces on bacterial activity in such communities, because of the complexity of the biofilm and the extensive nutrient recycling interactions which may take place between different organisms and taxonomic groups (Haack and McFeters, 1982a,b). One approach has been to evaluate the potential activity of attached cells by removing them from the surface and measuring their activity after suspending the cells in medium. Such experiments have shown substrate uptake by biofilm cells from oligotrophic waters to be about 26 times that of free-living cells (Ladd et al., 1979), whereas assimilation by biofilm bacteria from a polluted river was of the same magnitude (although somewhat higher) as that of suspended cells (Straškrabová et al., 1978). Such measurements of suspended biofilm cells cannot be expected to be a true reflection of their activity in situ. Ladd et al. (1979) found that glutamic acid uptake (measured as a V_{max} specific activity index) was about 4 times greater for dispersed biofilm cells than for those in situ. However, in situ values were still about 8 times greater than previously measured values for free-living cells, and, thus, in these oligotrophic waters, biofilm bacteria assimilated substrate at higher rates than free-living cells. Apparently, *in situ* uptake by biofilm bacteria was being restricted, probably by the surrounding slime matrix (see Chapter 1, Section 1), and potential uptake was considerably higher. Such studies illustrate the differences in activity which can occur between biofilm and free-living bacteria; much more work is needed to evaluate the comparative influence of substratum and polymer matrix properties, community composition, and substrate availability on biofilm activity.

3. COMPARATIVE ACTIVITIES OF ATTACHED AND FREE-LIVING BACTERIA IN THE LABORATORY

A large amount of laboratory data are now available which illustrate that solid surfaces can affect bacterial activity (see Marshall, 1976; Fletcher and Marshall, 1982b). In many cases, surfaces appear to promote activity, although they may alternatively have no effect, be inhibitory, or produce various results, depending on experimental conditions. Moreover, it has not always been clear whether actual attachment to the surfaces is a prerequisite for the observed effect.

3.1. Enhancement by Solid Surfaces of Suspended and Surface-Associated Bacterial Activity

There is clear evidence that inert (nonsubstrate) solid surfaces added to a bacterial culture may result in an increase in measured total activity (i.e., combined activity of both attached and free-living cells). However, some workers have observed this effect only at low nutrient concentrations (Heukelekian and Heller, 1940); there may be no such stimulation by solid surfaces at higher nutrient levels. For example, respiration has been stimulated by suspended chitin particles at peptone concentrations of 0.5 mg/liter (Jannasch and Pritchard, 1972) (Fig. 1) and by surfaces such as glass at organic nutrient concentrations of 5 to 10 mg/liter (ZoBell, 1943), with a disappearance of these effects at higher substrate concentrations. Moreover, such stimulation was not obtained with all nutrients tested, leading ZoBell (1943) to suggest that the solubility, dispersion, or molecular size of the nutrient was significant. Certainly, the evidence is strong that, in many cases, when surfaces enhance activity, they do so by promoting nutrient accessibility to the organisms. This could be accompanied by the adsorption and concentration of scarce nutrients on the surfaces. There

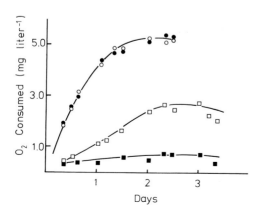

Figure 1. The stimulation of respiration by chitin particles. An *Achromobacter* sp. was grown in batch culture in medium containing 5.0 or 0.5 mg peptone/liter and in the absence or presence of protein-free chitin particles. Symbols: 5.0 mg peptone/liter, with chitin (○); 0.5 mg peptone/liter, with chitin (□); 5.0 mg peptone/liter, without chitin (●); 0.5 mg peptone/liter, without chitin (■). Note that at the lower peptone concentration, chitin stimulates respiration, whereas at the higher peptone concentration, chitin has no effect. Adapted from Jannasch and Pritchard (1972).

are other possible mechanisms, however, which will be considered in detail in Sections 4.4.1 and 4.2.2.

The influence of solid surfaces on activity also depends on the type of surface added to the culture (ZoBell, 1943). One type of surface which has been studied in some detail is that of particles with ion-exchange properties, such as clays (Filip, 1973) and ion-exchange resins. Clays have been found to promote bacterial growth (Conn and Conn, 1940), respiration of *Agrobacterium radiobacter* and other species (Stotzky, 1966a,b; Stotzky and Rem, 1966), and amino acid uptake (Zvyagintsev and Velikanov, 1968). However, these effects usually depended on the type of clay; for example, montmorillonite stimulated bacterial respiration, whereas kaolinite did not (Stotzky and Rem, 1966). One mechanism by which clays apparently stimulated respiration was maintenance of the pH of the environment at a level suitable for the organisms; this was related to the relative basicity of the exchangeable cations on the clays (Stotzky, 1966a). Other ion exchangers, such as glass, have also been found to affect pH, as well as stimulate an increase in growth of *Bacillus megaterium* (Ou and Alexander, 1974) (see also Section 3.2). The adsorbed minerals and water on ion-exchange surfaces may also affect activity by providing a source of inorganic nutrients (Stotzky and Rem, 1966) and protecting against cell desiccation (Bushby and Marshall, 1977), respectively.

When surfaces added to a bacterial culture result in a change in activity, it is frequently assumed that the bacteria are interacting with, and perhaps attaching to, the surfaces. For example, an increase in suspended cell numbers occasionally has been thought to indicate increased growth and subsequent desorption of bacteria associated with the surfaces. However, the same observed effect could be achieved by a completely different mechanism, i.e., the adsorption of inhibitors on the surfaces and their consequent removal from the medium, thus promoting growth of suspended cells (Sutherland and Wilkinson, 1961; Harwood and Pirt, 1972). Therefore, whenever possible, some attempt should be made to determine whether any modification of activity is associated with the free-living and/or attached populations.

Another important cell-surface system, where direct association does not always occur between the bacteria and the "solid" surface, is the immobilized cell fermenters utilized in industrial production processes (Vieth and Venkatsubramanian, 1979; Fukui and Tanaka, 1982). Product formation in immobilized cell reactors generally exceeds that in stirred reactors. To some extent, this is due to the higher cell concentrations and reduced product- and/or substrate-inhibition maintained in immobilized systems (Fukui and Tanaka, 1982); however, enhanced specific activity is also sometimes involved. The cells may be immobilized on solid supports (Sitton and Gaddy, 1980) or in gel matrices (Slowinski and Charm, 1973), both of which have been found to promote certain activities, as compared with freely suspended cells. For example, the yeast *Saccharomyces cerevisiae,* when immobilized on ceramic supports, produced ethanol at a rate 4.2 times greater than that achieved with suspended cells in a stirred reactor (Sitton and Gaddy, 1980). This is an area of intensive research and rapid development, which should help to clarify many questions on bacterium–surface interactions and their effects on bacterial activity.

3.2. Enhancement of Bacterial Activity through Attachment to Solid Surfaces

The importance of determining whether bacteria affected by surfaces are actually attached to them has been stressed (Section 3.1). This is not always easy to do, because of the exchange which may occur between attached and free-living populations, as additional cells become attached and as attached cells desorb or are sloughed off. Nevertheless, it is possible to make comparative evaluations of activity of attached and free-living populations by using short incubation times, so that exchange between the two populations is minimized. Also, microscopic techniques, e.g., microautoradiography and histochemical staining, can be used for direct assessment of cell activity. Accordingly, a wide variety of experimental systems have demonstrated that the activity of bacteria on surfaces may exceed (albeit sometimes marginally) that of free-living cells in the same system. Both continuous culture (Hendricks, 1974; Brown *et al.*, 1977; Ellwood *et al.*, 1982) and batch systems have been used. Bacterial activity has been evaluated by measuring respiration (Hendricks, 1974; Bright and Fletcher, 1983a,b; Kjelleberg *et al.*, 1983), increase in cell number (growth) (Hattori, 1972; Hattori and Hattori, 1981; Ellwood *et al.*, 1982), increase in cell size (Kjelleberg *et al.*, 1982), and increase in nutrient assimilation (Fletcher, 1979; Bright and Fletcher, 1983a,b). The effects have been observed with a range of organisms, including various aquatic bacteria (Hendricks, 1974; Ellwood *et al.*, 1982; Kjelleberg *et al.*, 1982, 1983; Bright and Fletcher, 1983a,b), *Escherichia coli* (Hattori, 1972; Hattori and Hattori, 1981), and yeasts, e.g., *Saccharomyces* spp. (Velikanov and Zvyagintsev, 1967; Navarro and Durand, 1977), as well as with a number of different surfaces, such as glass (Hendricks, 1974; Navarro and Durand, 1977; Ellwood *et al.*, 1982; Kjelleberg *et al.*, 1983), dialysis membrane (Kjelleberg *et al.*, 1982), ion-exchange resins (Kuwajima *et al.*, 1957; Hattori, 1972; Hattori and Hattori, 1981), minerals (Velikanov and Zvyagintsev, 1967), and plastics (Fletcher, 1979; Bright and Fletcher, 1983a,b). Data such as these and those described in Sections 2.1, 2.2, 3.1, and 3.2 have led to a general belief that surfaces tend to enhance bacterial activity.

3.3. Lack of Effect or Inhibition of Activity by Solid Surfaces

Although some studies clearly illustrate that attachment is accompanied by increased activity, other investigations have produced different results. For example, Hattori and Hattori (1981) found that the growth rate and glucose-molar growth yield of *E. coli* adsorbed on an anion-exchange resin were much higher than those of free-living cells. In other studies (Hattori and Furusaka, 1959b, 1960), however, *E. coli* adsorbed to an ion-exchange resin demonstrated lower oxidation rates of glucose, as well as lactose, xylose, fumarate, adenine (Hattori and Furusaka, 1959b), succinate, and asparagine (Hattori and Furusaka, 1960), as compared with suspended cells. Similar oxidative rate differences were obtained with *Micrococcus luteus* (Hattori and Hattori, 1963). As the pH optimum in the medium of oxidation by adsorbed cells was approximately 1.2 to 1.5 units higher than that for free-living cells, the differences in oxidation rate were probably due to different hydrogen ion concentrations on the anion-exchange resin and in the medium (Fig. 2).

Only recently, microcalorimetry has been applied to measure heat production, which is largely determined by catabolic activities (Gordon, 1982), by free-living marine bacteria (*Vibrio alginolyticus*) and those attached to hydroxyapatite (Gordon *et al.*, 1983) or polyacrylamide microcarriers (A. S. Gordon and F. J. Millero, personal communication). With low (5 μM) glucose concentration, no difference was found between heat production by free-living cells and those adsorbed to hydroxyapatite (Gordon, 1982). At a higher glucose concentration (6.5 mM), cells attached to polyacrylamide microcarriers showed somewhat lower heat production than did free-living cells (A. S. Gordon and F. J. Millero, personal communication). Thus, these workers observed no increase in attached bacterial activity with glucose, or with glutamic acid (Gordon *et al.*, 1983). In some cases, activity of adsorbed cells was lower than that of free-living cells. Moreover, in complementary experiments in which respiration and assimilation of ^{14}C-labeled glucose or glutamic acid by suspended cells and those attached to hydroxyapatite (Gordon *et al.*, 1983) were measured, activity of attached cells was lower (Fig. 3).

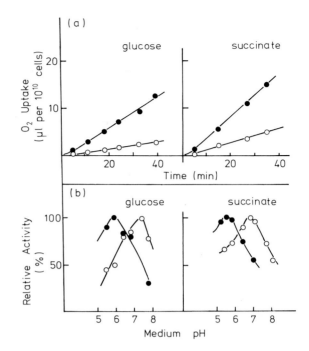

Figure 2. (a) The effect of adsorption of *M. luteus* to an anion-exchange resin on oxidation of glucose and succinate at optimum pH. Symbols: Adsorbed cells (○), suspended cells (●). (b) The oxidation of glucose and succinate by adsorbed (○) and suspended (●) cells as a function of medium pH. Adapted from Hattori and Hattori (1963).

Particulates, such as flowers of sulfur, fluorapatite, glass beads, and pyrite, considerably inhibited the activity of *Thiobacillus ferrooxidans* (Dispirito *et al.*, 1981). Direct association between the bacteria and surfaces was not demonstrated; indeed, in one case (pyrite), inhibition was due to leaching of a soluble toxic component. Nevertheless, it was suggested that inhibition by the other substances was due to microenvironmental differences at the liquid–solid interfaces, particularly differences in proton concentration.

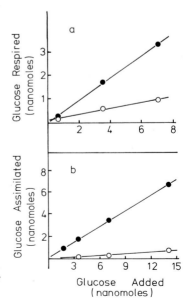

Figure 3. The respiration and assimilation of glucose (5 μM) by *V. alginolyticus* which were attached to hydroxyapatite particles (○) or free-living (●). Adapted from Gordon *et al.* (1983).

3.4. Complex Interactions between Bacterium, Substrate, and Substratum

In some experimental systems, it has been possible to obtain either stimulation, inhibition, or no effect by adding solid surfaces, with the results depending upon the organism used, the substrate and its concentration, and the nature of the solid surface.

3.4.1. The Organism

The bacterium used for an investigation will have a considerable influence on the results obtained, not only because of metabolic differences between strains, but also because of different attachment capabilities (Kefford *et al.*, 1983). For example, the ability to attach to surfaces will confer an advantage on colonizing organisms in systems with a large amount of particulate material. This was clearly illustrated by enrichment of seawater bacteria in continuous culture, in which kaolinite added to the culture selected for organisms which were able to attach to the clay particles (Jannasch and Pritchard, 1972). Moreover, the transformation of valeric acid by a strain of attaching organisms was also affected by clay, whereas transformation by a nonattaching strain remained unaffected (Jannasch and Pritchard, 1972). Such differences in microorganisms could account for many of the inconsistencies between reported results.

3.4.2. The Substrate

The effect of solid surfaces on bacterial activity can depend heavily upon the substrate and its concentration. In this respect, perhaps one of the most important distinctions between substrate types is whether they are low-molecular-weight compounds, which can be transported directly into the cell, or macromolecular nutrients, which must be broken down by extracellular enzymes before being assimilated. ZoBell (1943) did not find surface enhancement of activity with low-molecular-weight substrates, e.g., glucose, glycerol, or lactate, but did with sodium caseinate, lignoprotein, or emulsified chitin. However, other workers (Hattori and Furusaka, 1959a) observed surface effects with low-molecular-weight substrates, as the oxidation of succinate (but not asparagine) differed for free-living cells and those attached to an anion-exchange resin. In those situations, the role of electrostatic interactions in adsorption of low-molecular-weight substrates may become important (see also Section 4.4.1). A result obtained with the mineralization of peptone by soil bacteria in the presence of kaolinite clearly emphasizes how complex substrate–surface–bacterial interactions can be (Nováková, 1970). During the initial stages of peptone mineralization, kaolinite retarded the process, whereas at later stages, mineralization was accelerated in the presence of clay.

3.4.3. The Substratum

The composition of the solid surface can affect attached bacterial activity in two principal ways. First, the number of attached cells and their distribution on the surface will depend on surface composition (Murray and van den Berg, 1981; Shimp and Pfaender, 1982; see also Chapter 6, Sections 4 and 6), which, for example, has been found to influence the growth and conversion of acetic acid to methane by methanogenic bacteria (Murray and van den Berg, 1981). Second, the substratum can apparently modify attached bacterial activity, because of the special physicochemical conditions and forces at the solid–liquid interface. This was illustrated by a microautoradiographic study which demonstrated variations in amino acid uptake by a marine *Pseudomonas* sp. attached to a range of surfaces (Bright and Fletcher, 1983a). The surfaces (tissue culture–polyethylene terephthalate, glass, poly-

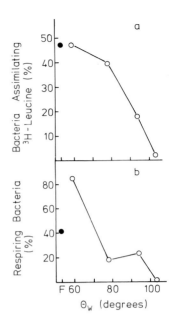

Figure 4. (a) Microautoradiographic determination of the percentage of cells of a marine *Pseudomonas* sp. which have assimilated detectable levels of [³H]leucine. The horizontal axis represents the contact angle of water (θ_w) on the attachment substrata, which were tissue culture–polyethylene terephthalate (59.4°), polyvinylidene fluoride (77.3°), glass (78.3°), polyethylene (93.6°), and polytetrafluoroethylene (103.5°). Symbols: Attached (○) and free-living (●) bacteria after a 2-hr incubation with 1.4 μM leucine. (b) Comparison of respiratory activity of attached (○) and free-living (●) marine pseudomonads by tetrazolium dye binding. The substrata are the same as in (a), and the leucine concentration was 14 μM. The vertical axis represents the percentage of bacteria with respiration levels above the minimum which can be detected by the method. Adapted from Bright and Fletcher (1983a).

vinylidene fluoride, polyethylene, and polytetrafluoroethylene) comprised a range of water-wettabilities (giving water contact angles of 59.4, 78.3, 77.3, 93.6, and 103.5°, respectively). This decrease in water-wettability with the range of surfaces also reflects a decrease in surface free energy (Fletcher and Marshall, 1982a,b). The proportion of cells assimilating [³H]leucine clearly decreased with an increase in substratum hydrophobicity at leucine concentrations of 50 or 100 μg carbon/liter (Fig. 4a). Somewhat similar results were obtained in corresponding experiments in which the proportion of respiring cells was measured with a tetrazolium dye [2-(p-iodophenyl)-3-(p-nitrophenyl)-5-phenyl tetrazolium chloride] stain, although results were different for cells attached to glass (water contact angle = 78.3°) (Fig. 4b). In some cases, differences in activity on various surfaces appear to be linked with strong electrostatic charges or pH effects (Hattori and Hattori, 1963; Stotzky, 1966a); such effects are not likely, however, with covalent organic polymers, such as plastics. The possible influence of substratum physicochemical properties on bacterial metabolic processes is discussed in more detail in Section 4.

It is not surprising that different bacterial strains are not affected in the same way by various solid surfaces and that the results obtained can also vary with substrate and its concentration. It is somewhat disquieting, however, that with a given bacterium, various types of activity apparently can be influenced by surfaces in quite different ways, and that the nature of the results obtained depends heavily on the method used for evaluating activity. For example, the apparently inconsistent results obtained with *E. coli* adsorbed on an anion-exchange resin have been mentioned (Section 3.3); with attached cells, growth was stimulated (Hattori and Hattori, 1981), but oxidative activity was reduced (Hattori and Furusaka, 1959b; 1960). Similarly, when microautoradiography was used to assess the uptake of [³H]leucine (Bright and Fletcher, 1983a), the proportion of assimilating cells decreased relative to increasing substratum hydrophobicity, so that on polytetrafluoroethylene, less than 10% of the attached cells assimilated leucine (Fig. 4a). This contrasted with scintillation counts of [¹⁴C]leucine uptake under similar conditions (Bright and Fletcher, 1983b), where

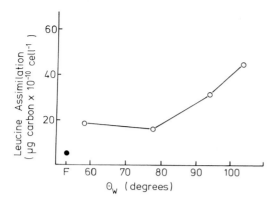

Figure 5. The assimilation of [^{14}C]leucine (14 μM) after a 2-hr incubation by free-living (●) marine pseudomonads and those attached (○) to tissue culture–polyethylene terephthalate (59.4°), polyvinylidene fluoride (77.3°), polyethylene (93.6°), and polytetrafluoroethylene (103.5°). Adapted from Bright and Fletcher (1983b).

there was little difference between assimilation on the various surfaces, except for polyethylene and, particularly, polytetrafluoroethylene, which, in this case, had the highest assimilation rate (Fig. 5).

Clearly, it is impossible to generalize about the effects of surfaces on activity. The influence of attachment on measured activity depends upon a number of factors; it is important, therefore, to bear in mind the experimental conditions when considering results. No doubt, certain factors will be found to dominate in particular systems. It may then be possible to make generalizations which apply to these specific situations. For example, there is some strong evidence that nutrients accumulating at particular interfaces facilitate their assimilation by adsorbed bacteria (Sims and Little, 1973; Kjelleberg et al., 1982).

3.5. Effects of Surfaces on Bacterial Morphology

Bacterial attachment to surfaces has been found at times to stimulate certain morphological transformations, e.g., shape or size of cells, spore formation, or flagellation. It is possible that some sort of ''surface-recognition'' mechanism is involved, as frequently occurs with developmental processes in eukaryotic cells. However, nutritional factors also appear to play a role. For example, when B. megaterium was grown in the presence of glass microbeads, many cells became filamentous (up to lengths of 600 μm), a morphological change which was inhibited by Mg^{2+}, but not other cations (Ou and Alexander, 1974).

An interesting aspect of the influence of surfaces on cell form is the effect of adsorption of bacteria on their subsequent division and miniaturization. When nutrient levels are low, bacteria may continue to divide with no increase in biomass, resulting in small cells, which have been called dwarfs, ultramicrobacteria, ultramicrocells, or minibacteria (Morita, 1982). Such miniaturization appears, at least in some cases, to be a response to starvation conditions which increases the surface area–volume ratio of the cell and thus its ability to scavenge dissolved substrates. It is still not clear whether all observed dwarfs are the result of a starvation–survival response. However, certainly with some heterotrophic strains adapted to high nutrient concentrations, miniaturization is a response to starvation (Novitsky and Morita, 1977). The cells grow and return to normal size when nutrients become available (Torrella and Morita, 1981).

This miniaturization and regrowth has been found to be influenced by both air–liquid (Kjelleberg et al., 1982) and solid–liquid (Humphrey et al., 1983; Kjelleberg et al., 1983) interfaces. Out of 15 marine bacteria examined, 12 decreased in size more rapidly when associated with a dialysis membrane surface than when free-living, whereas 3 decreased in size more rapidly when freely suspended (Humphrey et al., 1983) (Fig. 6). These differences

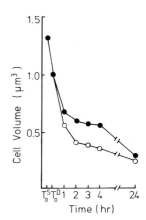

Figure 6. The spherical volume of a hydrophobic marine pseudomonad measured at various times after the onset of starvation in the liquid culture (○) and at a cellulose dialysis membrane surface (●). T_O^D is time at beginning of dwarfing, and T_O^S is time of contact with the surface. Adapted from Humphrey *et al.* (1983).

in miniaturization response to surfaces were related to cell surface hydrophobicity, as the first group of 12 strains had overall hydrophilic surfaces and the remaining 3 strains were hydrophobic (determined by hydrophobic interaction chromatography) (see also Chapter 6, Section 2.2.4). Similarly, starved cells of a marine *Vibrio* sp. at an air–water interface were smaller (peak cell volume of 0.32 μm³) than free-living cells (peak cell volume of 0.42 μm³) (Kjelleberg *et al.*, 1982). The role of substrate limitation and surface interactions in dwarf formation is considered further in Section 4.2.1.

Other surface-mediated changes in morphological development were the promotion of spore development by *Bacillus subtilus* when adsorbed to an anion-exchange resin (Hattori, 1976) and the production of peritrichous flagella by vibrios, as compared with the polar flagellation characteristic of suspended cells (de Boer *et al.*, 1975). Thus, the proximity of surfaces can affect a wide range of metabolic and developmental features. The underlying mechanisms of these effects are still largely unknown.

3.6. Promotion of Bacterial Survival at Surfaces

In some situations, particularly certain natural environments, the growth of attached populations may only appear to be greater than that of free-living populations because their numbers are greater. The same effect would be achieved, however, if attachment to a surface offered some sort of protection or survival advantage not available to free-living cells. The promotion of survival is probably an important factor in large surface communities of bacteria, which are embedded in polymeric matrices, or "slimes." These highly hydrated, and frequently charged matrix polymers, which are largely polysaccharides, may protect cells from potential toxic effects by binding heavy metals, by retarding diffusion of biocides or lytic agents (e.g., bacteriophages, *Bdellovibrio* spp.), or by resisting desiccation (Geesey, 1982) (see Section 4.1.2). The specific activity of cells in such bacterial slime layers may, in fact, be comparatively low; cell growth may be significant only at the surfaces of the layer.

4. POSSIBLE MECHANISMS FOR THE EFFECTS OF SURFACES ON ACTIVITY

In most cases, the specific causes of observed surface effects are not known. With a few experiments, it has been possible to pinpoint a likely mechanism, e.g., influence on pH

(Hattori and Hattori, 1963; Stotzky, 1966a) or nutrient availability (Kjelleberg *et al.*, 1982). Nevertheless, the basis for many observations is still a matter of conjecture. This section will examine the characteristics of surfaces and of bacterial physiology which may be important in solid surface–cell interactions and in associated increases or reductions in attached bacterial activity.

4.1. The Solid Surface as a Microenvironment

4.1.1. Influence of the Surface on Substrate Availability

4.1.1a. Substrate Adsorption.   Enhancement of bacterial activity at surfaces has frequently been observed only at low nutrient concentrations (see Section 3.1). Thus, at least in some cases, activity may be promoted through the adsorption (concentration) of potential nutrients on surfaces making them more accessible to attached bacteria (Marshall,1976). Potential nutrients include both low-molecular-weight and macromolecular compounds. The adsorption patterns of these two types of nutrients are quite different. Low-molecular-weight substances usually enter into an adsorption equilibrium, so that, although the concentration of the adsorbed species remains higher at the interface, adsorption and desorption of individual molecules occur continuously; the time spent at the surface by any individual molecule is finite.

This observation contrasts with adsorption of macromolecules, which have a large number of adsorption sites. Although the individual sites follow an adsorption equilibrium, at any given time some sites will be adsorbed. The net result is irreversible adsorption of the macromolecule. For example, protein adsorption tends to be irreversible. However, the configuration (Graham and Phillips, 1979) and concentration (Norde, 1980) of the adsorbed species will depend on properties of the adsorbate and the surface (Norde, 1980), as well as on possible competitive interactions between two or more species (Roe, 1980).

The prerequisites for bacterial utilization of the two types of adsorbed substrates differ. Low-molecular-weight compounds are usually just transported into the cells, whereas macromolecules must first be broken down by extracellular or membrane-bound (usually hydrolytic) enzymes before products can be assimilated. It is not always easy to predict how adsorption of molecules might affect their susceptibility to substrate recognition and subsequent transport or to enzymatic attack; observations may be difficult to interpret. For example, when arginine was bound to montomorillonite homoionic to aluminum, microorganisms could use it as either a carbon source or a nitrogen source (with the addition of complementary nitrogen or carbon sources, respectively), but not as both at the same time (Stotzky and Burns, 1982). It is unclear how organisms can selectively use either carbon or nitrogen moieties of an extracellular amino acid, where substrate breakdown prior to transport would not be expected. Such results emphasize the complexities of microbial utilization of adsorbed substrates.

The adsorption of potential substrates has been well documented in natural and laboratory environments. Dissolved organic substances are quickly adsorbed on solid surfaces exposed to seawater (ZoBell, 1943; Neihof and Loeb, 1972; Loeb and Neihof, 1975; Kristoffersen *et al.*, 1982) or brackish water (Hunter and Liss, 1982). These organic substances, possibly along with metallic oxides (Hunter and Liss, 1982), confer a net negative charge on the surfaces, irrespective of their original net charge (Neihof and Loeb, 1972; Loeb and Neihof, 1975), although there can be variations in elemental composition of adsorbed material on different surfaces (Kristoffersen *et al.*, 1982). The adsorption of specific low-molecular-weight substances dissolved in seawater is less predictable. For example, glutamic acid, but not glucose, was found to adsorb to hydroxyapatite surfaces (Gordon *et al.*, 1983).

As potential nutrients may be adsorbed on surfaces, the question then arises whether attached microorganisms can assimilate such substrates or whether they may be too tightly bound. Generally, adsorbed low-molecular-weight species appear to be utilized, as illustrated by the uptake of various amino acids adsorbed to clays (Zvyagintsev and Velikanov, 1968). The adsorption of substrates, such as amino acids (Velikanov and Zvyagintsev, 1968), benzylamine (Subba-Rao and Alexander, 1982), or ammonium ions (Sims and Little, 1973), may even facilitate their assimilation. Difficulty in uptake of low-molecular-weight substances would not generally be expected, because of the continuous desorption of species in the adsorption equilibrium. Nevertheless, the inhibition of amino acid utilization by apparent firm binding has been observed, as neither aspartic acid nor cysteine adsorbed to clays could be assimilated by bacteria (Stotzky and Burns, 1982).

By comparison, the utilization of macromolecules is more likely to be prevented by adsorption. Extracellular enzymatic breakdown may be inhibited when (1) the terminal residues involved in the initial cleavage are either bound or masked by bound groups, or (2) adsorption has altered the configuration of the macromolecule, making it no longer susceptible to enzymatic attack (Stotzky and Burns, 1982). Accordingly, the enzymatic breakdown of substances, such as certain proteins (Estermann *et al.*, 1959; Stotzky and Burns, 1982), dextran (Olness and Clapp, 1972), and diquat (Weber and Coble, 1968), was retarded when they were adsorbed to clays. The availability of clay-bound proteins has been found to depend not only on the properties of the protein, but also on the amount bound, its location on the clay, and the cation saturating the clay (Stotzky and Burns, 1982).

4.1.1b. Substrate Transport to the Surface.

In flowing systems, nutrients may be more accessible to bacteria attached to stationary surfaces, as compared with freely suspended cells, because of increased substrate transport to the surface. This has been shown to occur with laboratory biofilms, in which substrate uptake rates were directly related to the flow rate of the aqueous phase over the biofilm (up to a flow rate maximum, above which substrate transport to the surface was not limiting) (La Motta, 1976). By comparison, a planktonic bacterium or one attached to a surface in still water may, by nutrient uptake, deplete nutrients from their surrounding microzones at a rate which cannot be replenished by diffusion (Pasciak and Gavis, 1974). Moreover, motility is not usually sufficient to offset such depletion through a "grazing" effect (Pasciak and Gavis, 1974; Purcell, 1977), although it may allow the organism to reach a site of richer nutrient concentration (Purcell, 1977). Thus, for microorganisms in a biofilm in flowing water, the problem of such diffusion limitation through water may be eliminated, although diffusion limitation through the polymer matrix may then be significant (Ladd *et al.*, 1979).

4.1.1c. Attachment to the Substrate.

A situation where nutrient availability is clearly greater at the solid–liquid interface than in the liquid phase is when the substratum is also the substrate. One example is the transformation of minerals, such as by chemolithotrophic bacteria. The oxidation of inorganic substrates, such as sulfur or reduced iron compounds, probably requires contact between the bacterial and mineral surfaces. The adhesion of *Thiobacillus thiooxidans* was inhibited to sulfur by certain metal chelators and inhibitors, an observation tending to support a relationship between adhesion and sulfur oxidation (Takakuwa *et al.*, 1979). Also, the separation of iron-reducing bacteria from ferric oxides by a semipermeable membrane almost totally prevented the production of ferrous iron (Munch and Ottow, 1982). When substrate utilization involves extracellular enzyme activity, such as with collagen (Monboisse *et al.*, 1979) or cellulose (Minato and Suto, 1976), the enzyme may even be involved in the adhesion mechanism. Other examples demonstrating the necessity of adsorption to the substrate for its utilization are the dissolution of cortisone and cortisone acetate crystals by *Mycobacterium globiforme* (Zvyagintseva and Zvyagintsev, 1969) and the utilization of hydrocarbons by *Acinetobacter calcoaceticus* (Rosenberg and Rosenberg, 1981; Rosenberg *et al.*, 1982).

4.1.2. Influence of the Surface on Physicochemical Conditions

Bacterial metabolism is to a large extent influenced by local physicochemical factors, and these conditions may differ, favorably or unfavorably, at the interface as compared with the bulk phase. One of the clearest examples of the significance of local surface conditions is found with strongly charged surfaces, at which pH or ion concentrations are modified. This has been discussed with respect to clays and ion-exchange resins in Section 3.1. Water activity may also be reduced at the surface, through the adsorption of water by either the surface or adsorbed hydrophilic substances. Such water would be more highly structured and may not be "free" to act as a solvent for substrates and thus limit nutrient diffusion. Redox potential and oxygen tension can also be affected when oxidizing or reducing compounds are adsorbed. Moreover, such effects may be enhanced through diffusion limitation if the cells are embedded in a polymer matrix.

The production and accumulation of bacterial extracellular polymers is an extremely important factor in determining the nature of the microenvironment at the solid–liquid interface (Geesey, 1982). The polymers which have been isolated from periphytic bacteria and analyzed are largely composed of polysaccharides (Sutherland, 1980), which may be acidic (Corpe, 1970; Sutherland, 1980; Fazio *et al.*, 1982); proteins may also be present (Danielsson *et al.*, 1977; Corpe, 1980; Fletcher and Marshall, 1982a). The associated bacteria are thus enmeshed in a highly hydrated ($\sim 98\%$ water), gel matrix, which may promote or retard diffusion of various substrates and/or metabolic products. Such polymer matrices can also protect the cells from outside perturbations or toxins (Exner *et al.*, 1982; Geesey, 1982), maintain conditions suitable for growth, e.g., anoxic conditions supporting growth of anaerobes (Douglas *et al.*, 1917), or stabilize extracellular enzyme activity (Burchard, 1981).

The production of extracellular polysaccharides involves several stages including: (1) uptake of substrate for polysaccharide synthesis, (2) transfer of the appropriate monosaccharides from sugar nucleotides within the cell to carrier lipid–isoprenoid alcohol phosphate, (3) accumulation of oligosaccharide repeating units on carrier lipids, and (4) release from the lipids (Sutherland, 1977). It is feasible that the proximity of the surface could indirectly affect any of these stages of polymer synthesis, by influencing energy generation or cell membrane or outer membrane structure (see Section 4.2).

The physiological status of the bacteria affects both quantity and composition of synthesized extracellular polymers. For example, with *Pseudomonas atlantica*, both the total amount of polymer produced and the proportion of the uronic acid component increased with culture age (Uhlinger and White, 1983). Moreover, the genotypic control of mucoid or rough variants may be affected by nutrient conditions (Govan, 1975). Alternatively, mucoid or rough strains may be enriched by selection pressures, e.g., flow rate (Pringle *et al.*, 1983). Thus, not only do accumulated polymer matrices affect the activity of attached bacteria by conditioning the microenvironment, but also the activity of bacteria and their physiological status may in turn influence the properties of the polymers.

4.2. Direct Influence of the Surface on Bacterial Physiological Processes

Surfaces may affect bacterial activity directly by influencing physiological processes, such as substrate transport or energy conservation. Such effects might be negligible in nutrient-rich environments but could become significant in oligotrophic habitats, where carbon and energy sources may be severely limited. This, as well as the surface concentration of nutrients (see Section 4.1.1a), could explain why the enhancement of activity by surfaces has been observed most often in low-nutrient environments.

4.2.1. Bacterial Responses to Low-Nutrient Environments

It is possible to divide bacteria into two nutritional types on the basis of their nutrient level requirements. Bacteria which are adapted to growth in low-nutrient environments (1 to 15 mg organic carbon/liter) are oligotrophs, whereas those suited for high-nutrient environments are copiotrophs (Poindexter, 1981; Chapter 7, Section 3.1.2). Although they are adapted to nutrient abundance, at least some copiotrophs are able to survive periods of famine and thus must have a mechanism for shifting into a relatively inactive state. There is evidence that natural populations may respond to nutrient starvation by becoming "dormant" and that this reversible physiological state is an adaptation to nutrient availability (Wright, 1978). The physiological shift as a response to starvation conditions, at least in some cases, appears to be accompanied by rapid division and the resultant formation of dwarfs (Morita, 1982) (see also Section 3.5) and by an increase in substrate affinity (Reichardt and Morita, 1982).

As the formation of dwarfs is thus to some extent an indication of nutrient availability, this miniaturization response of copiotrophs under starvation conditions has been used to gain some understanding of the effects of surfaces on nutrient assimilation. Cells at air–water and solid–water interfaces (Kjelleberg et al., 1982) were shown to respond (by increase in cell size) more quickly than suspended bacteria to the addition of nutrient. Moreover, dwarfs may even be better able to attach to surfaces than normal-size cells (Dawson et al., 1981); this increase in adhesiveness with miniaturization may be due to a concomitant increase in cell surface hydrophobicity (Kjelleberg et al., 1983; Chapter 7, Section 3.1.1) or production of bridging polymer (Dawson et al., 1981). It has been suggested that this increase in attachment ability of dwarfs may be an adaptation promoting their survival by increasing their opportunity for attachment to surfaces, where nutrient levels may be higher than in the bulk phase (Dawson et al., 1981). However, one of the difficulties in interpreting such data is that the proximate (or only) stimulus for miniaturization may not always be low nutrient concentration. Thus, with bacteria on surfaces, physicochemical factors may combine with nutrient conditions to induce rapid growth and dwarf formation. The different responses of hydrophobic and hydrophilic bacteria at an interface to starvation (see Section 3.5) could be an example of the significance of physicochemical interactions, as opposed to nutritional factors. Similarly, in a microscopic study of bacteria attached to different types of surfaces (M. Fletcher, unpublished observations), dwarf formation occurred on two of the substrata, tissue culture–polystyrene and –polyethylene terephthalate (LUX, Naperville, Ill.) but not on glass, even when nutrients were apparently abundant (Fig. 7). Such observations suggest that surfaces may be able to affect bacterial physiological processes directly, and that effects on nutrient availability are not the only explanation for surface enhancement of attached bacterial activity.

4.2.2. Bacterial Transport of Substrates

Attachment or close approach to a solid surface can affect bacterial envelope structure in two basic ways. First, approach toward a charged surface and adsorbed electrolytes (Fletcher et al., 1980) will affect charge densities in the bacterial envelope and thus may modify cell membrane processes which are influenced by ion concentrations or pH, e.g., enzyme reactions (Pethica, 1980). Second, the attractive forces between two surfaces result in elastic

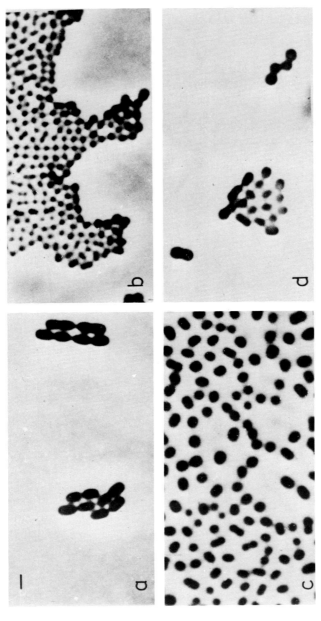

Figure 7. An *Acinetobacter* sp. which was first allowed to attach to coverslips, and, after rinsing, the coverslips were placed on glass slides coated with nutrient agar. The attached bacteria were thus "sandwiched" between the attachment substratum and nutrient agar and were incubated at 15°C for 29 hr. The coverslips were made of (a) glass, (b,c) tissue culture–polystyrene, and (d) tissue culture–polyethylene terephthalate. Note that miniaturization has not occurred on glass, but rapid division has resulted in the formation of dwarfs on the other two surfaces. Bar = 2 μm.

flattening at their interface (Dahneke, 1972; Tadros, 1980), the degree of deformation increasing with the flexibility of the surface. The flattening of "soft," spheroidal bodies is readily observed with attached animal cells. Bacteria are more rigid, however, and apparently cannot be observed microscopically to be deformed (Rutter and Vincent, 1980). Nevertheless, because of the attractive forces between the solid and bacterial surfaces, it is quite possible that some deformation of the components of the cell membrane and outer membrane of gram-negative bacteria occurs on a submicroscopic scale and that associated processes are thus affected. The strength of the adhesive forces between the two surfaces, and thus the degree of deformation, is to some extent related to the surface free energies of the surfaces (Fletcher and Marshall, 1982b). Thus, membrane deformation could be involved in the differences in activity found with bacteria attached to surfaces with different surface free energies, and thus hydrophilicities (Figs. 4 and 5).

The first step in substrate transport in gram-negative bacteria is passage by diffusion pathways through the outer membrane. There are a number of routes, including at least three nonspecific and several specific pathways (Nikaido and Nakae, 1980). Nonspecific diffusion pathways include (1) protein porin channels for diffusion of hydrophilic compounds, (2) hydrophobic pathways through the interior of the membrane for transport of hydrophobic compounds, and, probably, (3) rupture and resealing (Nikaido and Nakae, 1980). There are a large number of porins per cell ($\sim 10^5$). This concentration of porins is necessary, however, so that transport is not diffusion-limited for less permeable substrates or at low substrate concentrations. Thus, a decrease in outer membrane permeability would not be detectable except at low substrate concentrations, and perhaps not then if the V_{max} of the cell membrane transport system was low. Evidence for such a decrease in permeability could be a change in the overall transport rate, with an increase in the measured K_m value and no change in V_{max} (Nikaido and Nakae, 1980). Possible causes would be configurational changes in the diffusion channels or substrate diffusion limitation where the outer membrane was closely associated with a solid surface. The outer membrane in Enterobacteriaceae is thermodynamically unstable (Nikaido and Nakae, 1980). Structural organization may be modified by perturbations, such as changes in local ion concentration or deformation forces, which could thus affect substrate diffusion.

Transport across the cell membrane is more complicated. There are at least four mechanisms for transport of carbohydrates, amino acids, ions, and vitamins (Wilson, 1978; Dills et al., 1980); these are (1) facilitated diffusion, i.e., the rapid equilibrium of a substrate, e.g., glycerol, across the membrane, which does not require metabolic energy; (2) binding protein transport systems, in which the proteins are soluble, and in gram-negative bacteria are located in the periplasmic space; (3) proton-linked active transport in which proton extrusion is coupled to uptake, i.e., by protonmotive force as described by chemiosmotic theory (Mitchell, 1963), and in which the permease proteins involved are firmly bound to the cell membrane, and (4) group translocation systems, in which the transported solute is altered chemically, e.g., the phosphoenolpyruvate phosphotransferase system.

Soluble binding protein systems (2) appear to utilize energy derived from ATP or other high-energy compounds, whereas group translocation (4) uses chemical energy of the modification reaction (Wilson, 1978). Transport via membrane-bound proteins (3) is driven by protonmotive force (see also Section 4.2.3), and uptake may be linked directly by proton symport (both substances transported in the same direction on the same carrier), or indirectly by symport with some other ion or by antiport (two substances move in opposite directions). The most common indirect method is sodium symport (Wilson, 1978). In addition to these principal transport mechanisms, transport may also be facilitated by (1) the excretion of substances which bind extracellularly and specifically with substrates to promote their uptake, e.g., the excretion of iron-chelating compounds, siderochromes, (2) the presence of binding sites on the cell surface for substrate concentration prior to transport, e.g., magnesium binding on teichoic acids in gram-positive bacteria, and (3) the secretion of enzymes (e.g., phosphatases, proteinases, amylases) for mobilization of otherwise inaccessible substrates (Tempest and Neijssel, 1978). Transport systems may also be under the influence of various control mechanisms; e.g., permease systems may be influenced either positively or negatively by changes in membrane potential and/or intracellular pH (Dills *et al.*, 1980). All of these transport phenomena are dependent upon membrane integrity. Thus, any perturbation of membrane structure by attachment to a surface could affect substrate transport.

The situation is complicated even further, however, as most substrates are transported by several systems, as indicated by more than one measured K_m value. Multiple K_m values for single substrates can be due to several independent systems or to a single system with multiple binding sites (Wilson, 1978). Thus, it may also be feasible that proximity to a surface could affect substrate transport by effecting a shift in the substrate uptake mechanism.

It is clear that there are a number of ways in which substrate transport could be affected by attachment. It is likely, therefore, that proximity to a surface can induce perturbations which facilitate or disrupt certain, if not all, transport mechanisms. The difficulty lies in identifying the relevant transport processes in a given experimental situation, in monitoring the changes in substrate uptake that occur with attachment, and in determining the physiological reasons for these transport modifications. One type of transport which may lend itself most easily to this type of study is that related to cation concentration or movement, e.g., sodium symport. For example, amino acid uptake by marine bacteria was found to be reduced by exposure of the bacteria to hypotonic salt solutions (Geesey and Morita, 1981). Although this may well have been due to structural disruption of the cell membrane, it is also possible, where salt concentration was only slightly reduced, that direct or indirect coupling to protonmotive force was a factor. In another study investigating the effect of salinity on activity of attached bacteria, amino acid turnover time was increased with attached cells. It was suggested that salt concentration may have been affecting chemiosmosis (Bell and Albright, 1982). Certainly, modification of chemiosmotic processes is a likely mechanism for modifications in activity of attached cells, as chemiosmosis is not only linked to certain transport mechanisms, but also to energy conservation via protonmotive force.

4.2.3. Bacterial Energy Metabolism

Bacteria utilize two quite different mechanisms for obtaining energy for metabolic processes. First, ATP and other energy-rich compounds are generated by substrate-level phosphorylation, e.g., glycolysis, arginine fermentation. Such energy-rich compounds are active in intermediary metabolism and some substrate transport mechanisms (Section 4.2.2). Second, as described by chemiosmotic theory (Mitchell, 1963), certain energy pathways, e.g., aerobic and anaerobic redox chains, translocate protons across the cell membrane to outside the cell, and the return flow of the protons across the membrane can drive energetically unfavorable processes, such as ATP synthesis or substrate transport. In bacteria, the initial translocation of protons out of the cell may be coupled to the respiratory chain (Harold, 1972), to light-sensitive bacteriorhodopsin, or to electron transfer in photosynthetic bacteria or the membrane-bound Ca^{2+}/Mg^{2+}-ATPase. Thus, energy generation by the cell is intimately associated with membrane structure and configuration, either by (1) being dependent upon efficient functioning of enzymes (e.g., the ATP phosphohydrolase complex, which transforms the redox energy of the respiratory chain to hydrolysis energy of ATP, or the various dehydrogenases, oxidases, and reductases of the respiratory chain) or (2) being conditioned by local proton or ion concentrations, which would influence transmembrane potential, and thus chemiosmosis. It is not possible, at this stage, to predict whether, or in what way, membrane function would be affected by the proximity of a surface. The membrane should be protected to some extent from direct effects by peptidoglycan, extracellular polymers, and, in gram-negative bacteria, the outer membrane. Nevertheless, bacterial membrane structure and function are susceptible to changes in the surrounding environment; e.g., the concentration of cytochrome oxidase is typically increased in response to low-oxygen conditions (Jones,1981). Thus, forces and conditions at the solid–liquid interface could be expected to exert some influence on energy-generating membrane processes.

4.2.4. Metabolic Pathways

Proximity to a surface may also affect attached bacterial activity by indirectly influencing metabolic pathways. Mathiasson and Hahn-Hägerdal (1982) proposed a model which accounted for observed changes in activity of immobilized cells by implicating alterations in water activity or low oxygen tension. For example, low water activity has been shown to stimulate the formation of certain products, such as diacetyl by lactobacilli, glycerol accumulation by *Dunaliella,* and amino acid accumulation by various nonhalophilic bacteria. It is possible that other "increases in activity," such as more efficient fermentation of glucose by immobilized yeasts (Sitton and Gaddy, 1980), could also be associated with decreased water activity, due to increased binding of water by the surface (Mathiasson and Hahn-Hägerdal, 1982) or by accumulated polymers (see Section 4.1.2). It was also suggested that immobilized cells could have a high maintenance energy and thus produce a greater proportion of products associated with maintenance metabolism, at the expense of cell growth, and that such reactions could also have a smaller demand for oxygen. Possibly, this could account for increased production of α-amylase by *Bacillus amyloliquefaciens* at low oxygen tension or increased pigment production by immobilized plant cells (Mathiasson and Hahn-Hägerdal, 1982).

It is tempting to speculate about whether similar modifications in metabolism are involved in the production and accumulation of extracellular polymers at interfaces and the formation of biofilms. It is not yet clear whether these polymers are normally produced to the same extent by both suspended and attached cells, and that they are observed at surfaces because they are adsorbed and thus accumulated, or whether proximity to the surface in some way triggers an increase in polymer production. Exopolymer production may be stimulated by added solid particles, e.g., in sediments (Uhlinger and White, 1983), although, in many cases, this is possibly a nutritional effect due to the presence of adsorbed nutrients, rather than a purely surface-induced effect.

5. CONCLUSIONS

The influence of solid surfaces on attached bacterial activity is a complex phenomenon. The various components of bacterial metabolism, such as substrate uptake, respiration, or product formation, are likely to be influenced by different mechanisms and to different extents, depending on nutritional and physicochemical conditions. Moreover, surfaces may exert their influence both indirectly, by modifying nutrient and ion concentrations or microenvironmental conditions, and directly, through electrostatic interactions or possible deformation of the cell envelope. Analysis of these various factors and their significance for attached organisms has proved to be extremely difficult, because it is usually impossible to maintain separate attached and free-living populations, with no exchange between the two, and at the same time measure their separate activities and the environmental variables in the two microenvironments. The evidence suggests, however, that some features of the solid–liquid interface have a significant influence on attached bacterial metabolism. These include nutrient–surface–bacterium interactions and ion (including H^+) concentrations at charged surfaces. Other, as yet unexplored, mechanisms, such as influence on substrate transport and energy-generating systems, may be equally important. Their possible role in the modification of bacterial physiological processes on solid surfaces deserves detailed examination.

REFERENCES

Bell, C. R., and Albright, L. J., 1982, Attached and free-floating bacteria in a diverse selection of water bodies, *Appl. Environ. Microbiol.* **43:**1227–1237.

Bright, J. J., and Fletcher, M., 1983a, Amino acid assimilation and electron transport system activity in attached and free-living marine bacteria, *Appl. Environ. Microbiol.* **45:**818–825.

Bright, J. J., and Fletcher, M., 1983b, Amino acid assimilation and respiration by attached and free-living populations of a marine *Pseudomonas* sp., *Microbial Ecol.* **9:**215–226.

Brown, C. M., Ellwood, D. C., and Hunter, J. R., 1977, Growth of bacteria at surfaces: Influence of nutrient limitation, *FEMS Microbiol. Lett.* **1:**163–166.

Burchard, R. P., 1981, Gliding motility of prokaryotes: Ultrastructure, physiology, and genetics, *Annu. Rev Microbiol.* **35:**497–529.

Bushby, H. V. A., and Marshall, K. C., 1977, Water status of rhizobia in relation to their susceptibility to desiccation and to their protection by montmorillonite, *J. Gen. Microbiol.* **99:**19–27.

Conn, H. J., and Conn, J. E., 1940, The stimulating effect of colloids upon the growth of certain bacteria, *J Bacteriol.* **39:**99–100.

Corpe, W. A., 1970, An acid polysaccharide produced by a primary film-forming marine bacterium, *Dev. Ind Microbiol.* **11:**402–412.

Corpe, W. A., 1980, Microbial surface components involved in adsorption of microorganisms to surfaces, in *Adsorption of Microorganisms to Surfaces* (G. Bitton and K. C. Marshall, eds.), Wiley, New York, pp. 105–144.

Dahneke, B., 1972, The influence of flattening on the adhesion of particles, *J. Colloid Interface Sci.* **40:**1–13.

Danielsson, A., Norkrans, B., and Björnsson, A., 1977, On bacterial adhesion—The effect of certain enzymes on adhered cells of a marine *Pseudomonas* sp., *Bot. Mar.* **20:**13–17.

Dawson, M. P., Humphrey, B. A., and Marshall, K. C., 1981, Adhesion: A tactic in the survival strategy of a marine vibrio during starvation, *Curr. Microbiol.* **6:**195–199.

De Boer, W. E., Golten, C., and Scheffers, W. A., 1975, Effects of some physical factors on flagellation and swarming of *Vibrio alginolyticus, Neth. J. Sea Res.* **9:**197–213.

Dills, S. S., Apperson, A., Schmidt, M. R., and Saier, M. H., Jr., 1980, Carbohydrate transport in bacteria, *Microbiol. Rev.* **44:**385–418.

Dispirito, A. A., Dugan, P. R., and Tuovinen, O. H., 1981, Inhibitory effects of particulate materials in growing cultures of *Thiobacillus ferrooxidans, Biotechnol. Bioeng.* **23:**2761–2769.

Douglas, S. R., Fleming, A., and Colebrook, M. B., 1917, On the growth of anaerobic bacilli in fluid media under apparently aerobic conditions, *Lancet* **2:**530–532.

Ellwood, D. C., Keevil, C. W., Marsh, P. D., Brown, C. M., and Wardell, J. N., 1982, Surface-associated growth, *Philos. Trans. R. Soc. London Ser. B* **297:**517–532.

Estermann, E. F., Peterson, G. H., and McLaren, A. D., 1959, Digestion of clay–protein, lignin–protein and silica–protein complexes by enzymes and bacteria, *Soil Sci. Soc. Am. Proc.* **23:**31–36.

Exner, M., Tuschewitzki, G.-J., and Haun, F., 1982, Rasterelektronenoptische Darstellung der Wandbesiedlung wasserführender Kunststoffschläuche, *Zentralbl. Bakteriol. Parasitenkd. Infektionskr. Hyg. Abt. 1 Orig. Reihe B* **176:**425–434.

Fazio, S. A., Uhlinger, D. J., Parker, J. H., and White, D. C., 1982, Estimations of uronic acids as quantitative measures of extracellular and cell wall polysaccharide polymers from environmental samples, *Appl. Environ. Microbiol.* **43:**1151–1159.

Filip, Z., 1973, Clay minerals as factors influencing biochemical activity of soil microorganisms, *Folia Microbiol. (Prague)* **18:**56–74.

Fletcher, M., 1979, A microautoradiographic study of the activity of attached and free-living bacteria, *Arch. Microbiol.* **122:**271–274.

Fletcher, M., and Marshall, K. C., 1982a, Bubble contact angle method for evaluating substratum interfacial characteristics and its relevance to bacterial attachment, *Appl. Environ. Microbiol.* **44:**184–192.

Fletcher, M., and Marshall, K. C., 1982b, Are solid surfaces of ecological significance to aquatic bacteria?, in: *Advances in Microbial Ecology,* Volume 6 (K. C. Marshall, ed.), Plenum Press, New York, pp. 199–236.

Fletcher, M., Latham, M. J., Lynch, J. M., and Rutter, P. R., 1980, The characteristics of interfaces and their role in microbial attachment, in: *Microbial Adhesion to Surfaces* (R. C. W. Berkeley, J. M. Lynch, J. Melling, P. R. Rutter, and B. Vincent, eds.), Horwood, Chichester, pp. 67–78.

Fukui, S., and Tanaka, A., 1982, Immobilized microbial cells, *Annu. Rev. Microbiol.* **36:**145–172.

Geesey, G. G., 1982, Microbial exopolymers: Ecological and economic considerations, *ASM News* **48:**9–14.

Geesey, G. G., and Morita, R. Y., 1981, Relationship of cell envelope stability to substrate capture in a marine psychrophilic bacterium, *Appl. Environ. Microbiol.* **42:**533–540.

Gordon, A. S., 1982, The effect of inorganic particles on metabolism by marine bacteria, Ph.D. thesis, University of Miami.

Gordon, A. S., Gerchakov, S. M., and Millero, F. J., 1983, Effects of inorganic particles on metabolism by a periphytic marine bacterium, *Appl. Environ. Microbiol.* **45:**411–417,

Goulder, R., 1977, Attached and free bacteria in an estuary with abundant suspended solids, *J. Appl. Bacteriol.* **43:**399–405.

Govan, J. R. W., 1975, Mucoid strains of *Pseudomonas aeruginosa:* The influence of culture medium on the stability of mucus production, *J. Med. Microbiol.* **8:**513–522.

Graham, D. E., and Phillips, M. C., 1979, Proteins at liquid interfaces. III. Molecular structures of adsorbed films, *J. Colloid Interface Sci.* **70:**427–439.

Haack, T. K., and McFeters, G. A., 1982a, Microbial dynamics of an epilithic mat community in a high alpine stream, *Appl. Environ. Microbiol.* **43:**702–707.

Haack, T. K., and McFeters, G. A., 1982b, Nutritional relationships among microorganisms in an epilithic biofilm community, *Microb. Ecol.* **8:**115–126.

Harold, F. M., 1972, Conservation and transformation of energy by bacterial membranes, *Bacteriol. Rev.* **36:**172–230.

Harvey, R. W., and Young, L. Y., 1980, Enumeration of particle-bound and unattached respiring bacteria in the salt marsh environment, *Appl. Environ. Microbiol.* **40:**156–160.

Harwood, J. H., and Pirt, S. J., 1972, Quantitative aspects of growth of the methane oxidizing bacterium *Methylococcus capsulatus* on methane in shake flask and continuous chemostat culture, *J. Appl. Bacteriol.* **35:**597–607.

Hattori, R., 1972, Growth of *Escherichia coli* on the surface of an anion-exchange resin in continuous flow system, *J. Gen. Appl. Microbiol.* **18:**319–330.

Hattori, R., 1976, Growth and spore formation of *Bacillus subtilis* adsorbed on an anion-exchange resin, *J. Gen. Appl. Microbiol.* **22:**215–226.

Hattori, R., and Hattori, T., 1963, Effect of a liquid–solid interface on the life of microorganisms, *Ecol. Rev.* **16:**64–70.

Hattori, R., and Hattori, T., 1981, Growth rate and molar growth yield of *Escherichia coli* adsorbed on an anion-exchange resin, *J. Gen. Appl. Microbiol.* **27:**287–298.

Hattori, T., and Furusaka, C., 1959a, Chemical activities of *Escherichia coli* adsorbed on a resin, Dowex-1, *Nature (London)* (Suppl. 20) **184:**1566–1567.

Hattori, T., and Furusaka, C., 1959b, Chemical activities of *Escherichia coli* adsorbed on a resin, *Biochim. Biophys. Acta* **31:**581–582.

Hattori, T., and Furusaka, C., 1960, Chemical activities of *Escherichia coli* adsorbed on a resin, *J. Biochem. (Tokyo)* **48:**831–837.

Henricks, C. W., 1974, Sorption of heterotrophic and enteric bacteria to glass surfaces in the continuous culture of river water, *Appl. Microbiol.* **28:**572–578.

Heukelekian, H., and Heller, A., 1940, Relation between food concentration and surface for bacterial growth, *J. Bacteriol.* **40:**547–558.

Hodson, R. E., Maccubbin, A. E., and Pomeroy, L. R., 1981, Dissolved adenosine triphosphate utilization by free-living and attached bacterioplankton, *Mar. Biol.* **64:**43–51.

Humphrey, B., Kjelleberg, S., and Marshall, K. C., 1983, Responses of marine bacteria under starvation conditions at a solid–water interface, *Appl. Environ. Microbiol.* **45:**43–47.

Hunter, K. A., and Liss, P. S., 1982, Organic matter and the surface charge of suspended particles in estuarine waters, *Limnol. Oceanogr.* **27:**322–335.

Jannasch, H. W., and Pritchard, P. H., 1972, The role of inert particulate matter in the activity of aquatic microorganisms, *Mem. Ist. Ital. Idrobiol.* **29**(Suppl.):289–308.

Jones, C. W., 1981, *Biological Energy Conversation,* Chapman & Hall, London.

Kefford, B., Kjelleberg, S., and Marshall, K. C., 1983, Bacterial scavenging: Utilization of fatty acids localized at a solid/liquid interface, *Arch. Microbiol.* **133:**257–260.

Kirchman, D., and Mitchell, R., 1982, Contribution of particle-bound bacteria to total microheterotrophic activity in five ponds and two marshes, *Appl. Environ. Microbiol.* **43:**200–209.

Kjelleberg, S., Humphrey, B. A., and Marshall, K. C., 1982, Effect of interfaces on small, starved marine bacteria, *Appl. Environ. Microbiol.* **43:**1166–1172.

Kjelleberg, S., Humphrey, B. A., and Marshall, K. C., 1983, Initial phases of starvation and activity of bacteria at surfaces, *Appl. Environ. Microbiol.* **46:**978–984.

Kristoffersen, A., Rölla G., Skjörland, K., Glantz, P. O., and Ivarsson, B., 1982, Evidence for the formation of organic films on metal surfaces in seawater, *J. Colloid Interface Sci.* **86:**196–203.

Kuwajima, Y., Matsui, T., and Kishigami, M., 1957, The growth-supporting effect of some anion exchange resins for phase I *Haemophilus pertussis, Jpn. J. Microbiol.* **1:**375–381.

Ladd, T. I., Costerton, J. W., and Geesey, G. G., 1979, Determination of the heterotrophic activity of epilithic microbial populations, in: *Native Aquatic Bacteria: Enumeration, Activity and Ecology* (J. W. Costerton and R. R. Colwell, eds.), ASTM, Philadelphia, pp. 180–195.

La Motta, E. J., 1976, Kinetics of growth and substrate uptake in a biological film system, *Appl. Environ. Microbiol.* **31:**286–293.

Loeb, G. I., and Neihof, R. A., 1975, Marine conditioning films, *Adv. Chem. Ser.* **145:**319–335.

Marsh, P. D., Hunter, J. R., Bowden, G. H., Hamilton, I. R., McKee, A. S., Hardie, J. M., and Ellwood, D. C., 1983, The influence of growth rate and nutrient limitation on the microbial composition and biochemical properties of a mixed culture of oral bacteria grown in a chemostat, *J. Gen. Microbiol.* **129:**755–770.

Marshall, K. C., 1976, *Interfaces in Microbial Ecology,* Harvard University Press, Cambridge, Mass.

Mathiasson, B., and Hahn-Hägerdal, B., 1982, Microenvironmental effects on metabolic behaviour of immobilized cells: A hypothesis, *Eur. J. Appl. Microbiol. Biotechnol.* **16:**52–55.

Mills, A. L., and Maubrey, R., 1981, Effect of mineral composition on bacterial attachment to submerged rock surfaces, *Microb. Ecol.* **7:**315–322.

Minato, H., and Suto, T., 1976, Technique for fractionation of bacteria in rumen microbial ecosystem. I. Attachment of rumen bacteria to starch granules and elution of bacteria attached to them, *J. Gen. Appl. Microbiol.* **22:**259–276.

Mitchell, P., 1963, Molecule, group and electron translocation through natural membranes, *Biochem. Soc. Symp.* **22:**142–168.

Monboisse, J. C., Labadie, J., and Gouet, P., 1979, Attachment d'une bactérie collagénolytique à son substrat, *Ann. Microbiol. (Inst. Pasteur)* **130A:**435–440.

Morita, R. Y., 1982, Starvation-survival of heterotrophs in the marine environment, in: *Advances in Microbial Ecology,* Volume 6 (K. C. Marshall, ed.), Plenum Press, New York, pp. 171–198.

Munch, J. C., and Ottow, J. C. G., 1982, Einfluss von Zellkontakt und Eisen (III)—Oxidform auf die Bacterielle Eisenreduktion, *Z. Pflanzenernaehr. Bodenkd.* **145:**66–77.

Murray, W. D., and van den Berg, L., 1981, Effect of support material on the development of microbial fixed films converting acetic acid to methane, *J. Appl. Bacteriol.* **51:**257–265.

Navarro, J. M., and Durand, G., 1977, Modification of yeast metabolism by immobilization onto porous glass, *Eur. J. Appl. Microbiol.* **4:**243–254.

Neihof, R. A., and Loeb, G. I., 1972, The surface charge of particulate matter in seawater, *Limnol. Oceanogr.* **17:**7–16.

Nikaido, H., and Nakae, T., 1980, The outer membrane of gram-negative bacteria, in: *Microbial Cell Walls and Membranes* (H. J. Rogers, H. R. Perkins, and J. B. Ward, eds.), Chapman & Hall, London, pp. 163–250.

Norde, W., 1980, Adsorption of proteins at solid surfaces, in: *Adhesion and Adsorption of Polymers* (L.-H. Lee, ed.), Plenum Press, New York, pp. 801–825.

Nováková, J., 1970, Effect of clay minerals on the mineralisation of peptone in liquid medium, *Folia Microbiol. (Prague)* **15:**217.

Novitsky, J. A., and Morita, R. Y., 1977, Survival of a psychrophilic marine vibrio under long-term nutrient starvation, *Appl. Environ. Microbiol.* **33:**635–641.

Olness, A., and Clapp, C. E., 1972, Microbial degradation of a montmorillonite–dextran complex, *Soil Sci. Soc. Am. Proc.* **36:**179–181.

Ou, L.-T., and Alexander, M., 1974, Influence of glass microbeads on growth, activity and morphological changes of *Bacillus megaterium, Arch. Microbiol.* **101:**35–44.

Pasciak, W. J., and Gavis, J., 1974, Transport limitation of nutrient uptake in phytoplankton, *Limnol. Oceanogr.* **19:**881–888.

Pethica, B. A., 1980, Microbial and cell adhesion, in: *Microbial Adhesion to Surfaces* (R. C. W. Berkeley, J. M. Lynch, J. Melling, P. R. Rutter, and B. Vincent, eds.), Horwood, Chichester, pp. 19–45.

Poindexter, J. S., 1981, Oligotrophy: Fast and famine existence, in: *Advances in Microbial Ecology*, Volume 5 (M. Alexander, ed.), Plenum Press, New York, pp. 63–89.

Pringle, J. H., Fletcher, M., and Ellwood, D. C., 1983, Selection of attachment mutants during the continuous culture of *Pseudomonas fluorescens* and relationship between attachment ability and surface composition, *J. Gen. Microbiol.* **129**:2557–2569.

Purcell, E. M., 1977, Life at low Reynolds number, *Am. J. Phys.* **45**:3–11.

Reichardt, W., and Morita, R. Y., 1982, Survival stages of a psychrotrophic *Cytophaga johnsonae* strain, *Can. J. Microbiol.* **28**:841–850.

Roe, R.-J., 1980, Selective adsorption of polymers from solution, in: *Adhesion and Adsorption of Polymers* (L.-H. Lee, ed.), Plenum Press, New York, pp. 629–641.

Rosenberg, M., and Rosenberg, E., 1981, Role of adherence in growth of *Acinetobacter calcoaceticus* RAG-1 on hexadecane, *J. Bacteriol.* **148**:51–57.

Rosenberg, M., Rottem, S., and Rosenberg, E., 1982, Cell surface hydrophobicity of smooth and rough *Proteus mirabilis* strains as determined by adherence to hydrocarbons, *FEMS Microbiol. Lett.* **13**:167–169.

Rutter, P., and Vincent, B., 1980, The adhesion of micro-organisms to surfaces: Physico-chemical aspects, in: *Microbial Adhesion to Surfaces* (R. C. W. Berkeley, J. M. Lynch, J. Melling, P. R. Rutter, and B. Vincent, eds.), Horwood, Chichester, pp. 79–92.

Shimp, R. J., and Pfaender, F. K., 1982, Effects of surface area and flow rate on marine bacterial growth in activated carbon columns, *Appl. Environ. Microbiol.* **44**:471–477.

Sims, R. C., and Little, L., 1973, Enhanced nitrification by addition of clinoptilolite to tertiary activated sludge units, *Environ. Lett.* **4**:27–34.

Sitton, O. C., and Gaddy, J. L., 1980, Ethanol production in an immobilized-cell reactor, *Biotechnol. Bioeng.* **22**:1735–1748.

Slowinski, W., and Charm, S. E., 1973, Glutamic acid production with gel-entrapped *Corynebacterium glutamicum*, *Biotechnol. Bioeng.* **15**:973–979.

Stotzky, G., 1966a, Influence of clay minerals on microorganisms. II. Effect of various clay species, homoionic clays, and other particles on bacteria, *Can. J. Microbiol.* **12**:831–848.

Stotzky, G, 1966b, Influence of clay minerals on microorganisms. III. Effect of particle size, cation exchange capacity, and surface area on bacteria, *Can. J. Microbiol.* **12**:1235–1246.

Stotzky, G., and Burns, R. G., 1982, The soil environment: Clay–humus–microbe interactions, in: *Experimental Microbial Ecology* (R. G. Burns and J. H. Slater, eds.), Blackwell, Oxford, pp. 105–133.

Stotzky, G., and Rem, L. T., 1966, Influence of clay minerals on microorganisms. I. Montmorillonite and kaolinite on bacteria, *Can. J. Microbiol.* **12**:547–563.

Straškrabová, V., Punčochář, P., Bojanovski, B. B., and Fuksa, J., 1978, Suspended and attached microorganisms in a polluted river, *Verh. Int. Ver. Limnol.* **20**:2278–2283.

Subba-Rao, R. V., and Alexander, M., 1982, Effect of sorption on mineralization of low concentrations of aromatic compounds in lake water samples, *Appl. Environ. Microbiol.* **44**:659–668.

Sutherland, I. W., 1977, Microbial exopolysaccharide synthesis, in: *Extracellular Microbial Polysaccharides* (P. A. Sandford and A. Laskin, eds.), American Chemical Society, Washington, D.C., pp. 40–57.

Sutherland, I. W., 1980, Polysaccharides in the adhesion of marine and freshwater bacteria, in: *Microbial Adhesion to Surfaces* (R. C. W. Berkeley, J. M. Lynch, J. Melling, P. R. Rutter, and B. Vincent, eds.), Horwood, Chichester, pp. 329–338.

Sutherland, I. W., and Wilkinson, J. R., 1961, A new growth medium for virulent *Bordetella pertussis*, *J. Pathol. Bacteriol.* **82**:431–438.

Tadros, T. F., 1980, Particle–surface adhesion, in: *Microbial Adhesion to Surfaces* (R. C. W. Berkeley, J. M. Lynch, J. Melling, P. R. Rutter, and B. Vincent, eds.), Horwood, Chichester, pp. 93–116.

Takakuwa, S., Fujmori, T., and Iwasaki, H., 1979, Some properties of cell–sulfur adhesion in *Thiobacillus thiooxidans*, *J. Gen. Appl. Microbiol.* **25**:21–29.

Tempest, D. W., and Neijssel, O. M., 1978, Eco-physiological aspects of microbial growth in aerobic nutrient-limited environments, in: *Advances in Microbial Ecology*, Volume 2 (M. Alexander, ed.), Plenum Press, New York, pp. 105–153.

Torrella, F., and Morita, R. Y., 1981, Microcultural study of bacterial size changes and microcolony and ultramicrocolony formation by heterotrophic bacteria in seawater, *Appl. Environ. Microbiol.* **41**:518–527.

Uhlinger, D. J., and White, D. C., 1983, Relationship between physiological status and formation of extracellular polysaccharide glycocalyx in *Pseudomonas atlantica*, *Appl. Environ. Microbiol.* **45**:64–70.

Velikanov, L. L., and Zvyagintsev, D. G., 1967, Vliyanie adsorbentov na zhizned eyatel 'nost' drozhzhei, *Biol. Nauki (Moscow)* **10**:99–103.

Velikanov, L. L., and Zvyagintsev, D. G., 1968, Dustup nost' adsorbirovannykh aminokislot dlya defitsitnykh po nim mutantov *Escherichia coli*, *Biol. Nauki (Moscow)* **11**:86–89.

Vieth, W. R., and Venkatsubramanian, K., 1979, Immobilized microbial cells in complex biocatalysis, in: *Immobilized Microbial Cells* (K. Venkatsubramanian, ed.), American Chemical Society, Washington, D.C., pp. 1–11.

Weber, J. B., and Coble, H. D., 1968, Microbial decomposition of diquat adsorbed on montmorillonite and kaolinite clays, *J. Agric. Food Chem.* **16**:475–478.

Wilson, D. B., 1978, Cellular transport mechanisms, *Annu. Rev. Biochem.* **47**:933–965.

Wright, R. T., 1978, Measurement and significance of specific activity in the heterotrophic bacteria of natural waters, *Appl. Environ. Microbiol.* **36**:297–305.

ZoBell, C. E., 1943, The effect of solid surfaces upon bacterial activity, *J. Bacteriol.* **46**:39–56.

ZoBell, C. E., and Anderson, D. Q., 1936, Observations on the multiplication of bacteria in different volumes of stored seawater and the influence of oxygen tension and solid surfaces, *Biol. Bull.* **71**:324–342.

ZoBell, C. E., and Grant, C. W., 1943, Bacterial utilization of low concentrations of organic matter, *J. Bacteriol.* **45**:555–564.

Zvyagintsev, D. G., and Velikanov, L. L., 1968, Influence of adsorbents on the activity of bacteria growing on media with amino acids, *Microbiology (USSR)* **37**:861–866.

Zvyagintseva, I. S., and Zvyagintsev, D. G., 1969, Effect of microbial cell adsorption onto steroid crystals on the transformation of the steroid, *Microbiology (USSR)* **38**:691–694.

13

Influence of Attachment on Microbial Metabolism and Growth in Aquatic Ecosystems

HANS W. PAERL

1. INTRODUCTION—THE AQUATIC ENVIRONMENT: SURFACE VERSUS PLANKTONIC ENVIRONMENTS

Among diverse microbial habitats, the aquatic environment offers readily available energy sources to both autotrophic and heterotrophic organisms. Photoreductive energy in the form of photosynthetically active radiation (PAR) can be transmitted to at least a depth of 100 m of much of the world's pelagic oceanic and freshwater systems (Hutchinson, 1957, Raymont, 1980). Products of photosynthesis are either directly released by photoautotrophs (Fogg, 1966) or are introduced into the water column through cellular lysis and decomposition processes (Golterman, 1964; Mankaeva, 1966; Daft and Stewart, 1973). Such products, as well as dissolved and particulate inorganic nutrients, are effectively dispersed either through molecular diffusion or turbulent mixing, the latter being particularly important in radiant-energy-rich, near-surface layers (epilimnia). Despite the abundance of PAR and positive effects of convective, diffusive, and turbulent mixing on microbial growth, vast regions, approximating 70 to 80% of our oceans and at least 50% of our freshwater systems, exhibit exceedingly low rates of photosynthetic and chemosynthetic primary production (Parsons and Takahashi, 1973; Wetzel, 1983). As a result, biomass production among heterotrophic microorganisms and higher-ranked animal consumers of organic matter is also restricted. Waters where such production processes are severely limited in magnitude are termed *oligotrophic* (Hutchinson, 1957). Most often, inorganic nutrient deficiencies have been identified as the chief factors controlling, and hence limiting, photosynthetic production in illuminated (euphotic) surface waters (Ryther, 1963; Likens, 1972). Among these nutrients, both nitrogen and phosphorus are the most commonly encountered limiting factors, followed at times by iron, silicon, inorganic carbon, and trace metals (such as Co, Mo, B, Cu, Zn, and Mn), roughly in this order of frequency. Hence, as is the case with terrestrial (agricultural) environments, restricted availability of inorganic nutrients rather than energy is the most common form of limitation to micro- and macrophytic production. Among aquatic micro-

Hans W. Paerl • Institute of Marine Sciences, University of North Carolina, Morehead City, North Carolina 28557.

organisms, nutrient limitation can to some degree be compensated for in two ways. First, specialized biochemical processes such as nitrogen fixation, alkaline and acid phosphatase (phosphorus cleaving) activities, intracellular phosphorus and nitrogen storage, metal chelation, and carbonic anhydrase activity (in order to utilize HCO_3^- under CO_2-limited conditions) can supplement available nutrients. Second, as Hutchinson (1961) initially suggested and Lehman and Scavia (1982) have recently experimentally confirmed, inorganic (and organic) nutrients are not homogeneously dispersed in aquatic environments, but are present in patches (Shanks and Trent, 1979). Patches can range from micrometers to kilometers in size. Microorganisms able to orient and maintain themselves in such patches often demonstrate greater growth rates, cell sizes, and population densities, when compared to similar populations residing in nutrient-depleted patches (Hutchinson, 1961; Richerson *et al.,* 1970).

As a general rule, limitation of primary production among micro- and macrophytes leads to decreased dissolved and particulate organic matter (DOM and POM, respectively) availability to heterotrophic micro- and macroorganisms. Patchy distributions of primary producers therefore lead to a similar patchiness of DOM and POM, and ultimately heterotrophic consumers (Richerson *et al.,* 1970). Spatially, nonhomogeneous limitation of autotrophic and/or heterotrophic production appears to be commonplace in natural waters, despite the presence, in varying degrees, of molecular and turbulent dispersion processes (Hutchinson, 1957). Given the ubiquity of nutrient limitation in the dilute media constituting aquatic environments, localized nutrient-rich micro- or macrohabitats would appear to be attractive, if not essential, zones for optimal growth of primary and heterotrophic producers.

Collectively, research conducted during the past three decades has identified submersed surfaces as relatively nutrient-rich habitats in natural waters, particularly in oligotrophic pelagic (open water) habitats (Marshall *et al.,* 1971a,b; Jannasch and Pritchard, 1972; Paerl and Goldman, 1972; Shanks and Trent, 1979). Not only do surfaces provide an interface for the concentration of charged particles and molecules (Hull and Kitchener, 1969; Bitton and Marshall, 1980; Kefford *et al.,* 1982), but numerous surfaces, particularly those of biotic or geological origin, also represent direct nutrient sources, either due to mineral or organic matter content. In waters commonly supporting part-per-billion (or lower) levels of dissolved nutrients, surfaces are thought to harbor orders of magnitude higher nutrient content than planktonic, counterpart environments (ZoBell and Anderson, 1936; Stark *et al.,* 1938; Marshall, 1976; Paerl and Merkel, 1982). A range of both organic and inorganic molecules are readily concentrated on surfaces. In particular, studies have illustrated the ability of particle surfaces to concentrate phosphate (Paerl and Merkel, 1982) iron (Stumm and Morgan, 1981), calcium, copper (Wallace, 1982), other trace metals (Wallace, 1982), and a variety of organic molecules (Neihof and Loeb, 1972; Loeb and Neihof, 1975), most notably amino acids, peptides, lipids, organic acids, and a variety of polymers, under natural conditions. Other studies have shown microbial metabolism to be altered (most often stimulated) in the presence of particles (Jannasch and Pritchard, 1972; Paerl and Goldman, 1972; Hendricks, 1974; Fletcher, 1979a,b; Kirchman and Mitchell, 1982). Recently, we (Paerl and Merkel, 1982) have shown cellular uptake rates for specific nutrients (organics and PO_4^{3-}) to be higher among particle-associated, as opposed to free-floating, bacteria. In general, bacteria associated with particles are larger than free-floating counterparts in both marine and freshwater habitats (Marshall, 1979; Paerl, 1980). Also, diverse studies have indicated that, at times, particles concentrate metabolically active bacterial populations, when compared to free-floating densities (Jannasch and Pritchard, 1972; Paerl and Goldman, 1972; Wangersky, 1977), implying that particles themselves act as growth- and production-enhancing micropatches in otherwise dilute aquatic environments.

The evidence presented is by no means unequivocal. A great deal of disagreement has existed as to the trophic roles that surfaces play in diverse habitats thus far examined (Melchiorri-Santolini and Hopton, 1972). Some investigators have found surfaces to be microbe concentrating sites (Jannasch and Pritchard, 1972; Paerl and Goldman, 1972; Madsen, 1972), while others have found no significant degrees of patchiness, in terms of population distributions, when comparing attached and free-floating communities (Wiebe and Pomeroy, 1972). Both temporal (seasonal) and spatial (horizontal) patterns of aquatic habitats play roles in determining the relative distributions of attached and free-floating bacteria. As many of the above-mentioned studies have not considered the complications of seasonality and physical structure of the water column or benthic environments, it is not surprising to encounter open disagreement on the occurrences and ecological roles of microbe–surface interactions in natural waters. Another set of factors worth considering are the trophic states of systems under investigation. It is thought that in systems having exceedingly low dissolved nutrient levels, such as the oligotrophic Sargasso Sea, North Pacific Gyre, and oligotrophic lakes such as Lake Tahoe, Crater Lake, Lake Superior (U.S.A.), and numerous alpine streams and lakes, surfaces may play a relatively more important role in both attracting and supporting microbial populations than in highly nutrient-enriched waters, such as polluted lakes, streams, and rivers, marine upwelling habitats, and estuaries receiving terrigenous dissolved nutrient inputs. It would appear that a great deal of relativity exists in both defining and understanding the trophic roles that surfaces play in the world's aquatic habitats (Paerl, 1980). The mere fact that divergent results and opinions exist as to the roles that surfaces play in regulating microbial metabolism and growth is testimony to the potentially diverse way in which surfaces contribute to the overall productivity of aquatic environments.

In recognizing surfaces as sites for both chemical and microbial interactions, we cannot ignore the fact that both potentially stimulatory and inhibitory substances may interact with surfaces. Toxic wastes from industry, mining, petroleum exploration, agriculture, and municipalities are known to reside in many of the world's freshwater, coastal, and, at times, truly pelagic (offshore) habitats (Wellman, 1973; Higgins and Burns, 1975; Hutzinger et al., 1977). Recent work has shown that toxic organic and inorganic substances readily interact with and are concentrated by surfaces, whether they be suspended biotic or abiotic particles or submersed metallic, concrete, or wooden surfaces (i.e., vessel hulls, pilings, dams, seawalls, casons, etc.) (Young and Mitchell, 1973; Chet et al., 1975; F. K. Pfaender and A. V. Palumbo, personal communication). Such substances are known to repel and at times prove lethal to microorganisms (Young and Mitchell, 1973; Chet et al., 1975). Accordingly, their mode of action at the liquid–solid interface may well play a role in determining the types, frequencies, and magnitudes of microbe–surface interactions (Chet and Mitchell, 1976). It follows that conflicting observations on the presence or lack of microbe–surface interactions may result from a variety of interacting growth-promoting, attracting, growth-retarding, and repelling substances which have either accumulated on, or are integral components of, submerged surfaces.

Aside from nutritional advantages to attachment, particularly in nutrient-deficient waters, there are likely to be additional factors regulating microbial colonization. The metabolic alterations (either enhancement or reduction) associated with microbial growth at the liquid–solid interface themselves exert an effect on the physical, chemical, and ultimately biotic characteristics of this interface (Hendricks, 1974; Fletcher, 1979b; Harvey and Young, 1980; Kirchman and Mitchell 1982; Paerl and Merkel, 1982). First, surfaces can provide localized, high-density habitats for specific microorganisms which are absent or present in very low concentrations in free-floating states (Marshall, 1976; Brown et al., 1977; Paerl, 1980). Second, basic metabolic processes associated with microorganisms found in interface habitats have been shown to alter the physicochemical microenvironment (microzone) present in such habitats (Caldwell and Caldwell, 1978; Paerl, 1978; Paerl and Kellar, 1978). Oxygen levels alone can be radically altered in such microzones, as can be illustrated among marine microbial mats anchored on sand substrates (see Fig. 3). Depending on the composition of communities and proportions of photosynthetic vs. heterotrophic microorganisms associated with the liquid–solid interface, oxygen supersaturation or total depletion may result within micrometer distances (Revsbech and Jorgensen, 1981; Jorgensen and Revsbech, 1983; Bautista, 1983). Oxygen levels can vary not only spatially within a few micrometers of the liquid–solid interface, but extreme temporal variations have also been observed (Bautista, 1983). Finally, the activity of higher-ranked organisms (starting with protozoans on up to grazing invertebrates and fish) further affects the distribution as well as periodicity of oxygen levels in microzones. One could potentially study the dynamics of inorganic and organic carbon, various oxidation states of sulfur, metals, silicon, phosphorus, nitrogen along similar micrometer orders of magnitude at the liquid–solid interface. This chapter will tend to focus largely on consequences of microbe–surface interactions with respect to oxygen regimes because (1) data on oxygen characteristics and dynamics of surface microzones have, to a limited extent, been obtained; (2) oxygen evolution and utilization characteristics in microenvironments can be monitored by recently developed techniques (Paerl, 1980; Revsbech and Jorgensen, 1981); and (3) oxygenation characteristics of surface microzones are important determinants in dictating potentials for, as well as quantitative importances of, specific biochemical nutrient-transforming processes regulating overall aquatic production.

2. MICROBIAL ATTACHMENT IN AQUATIC ENVIRONMENTS

2.1. Factors Initiating Attachment

Although the physicochemical conditions typifying the liquid–solid interface in natural environments are discussed in detail elsewhere (Chapters 3 and 6), general features relevant to initial adhesion and colonization are reviewed below.

Immersed solids are normally accompanied by surface changes which can be attributed to ionic characteristics of the material making up the surface, as well as ions adsorbed from the aqueous media (Neihof and Loeb, 1972; Loeb and Neihof, 1975; Matjevic, 1977). Diverse aqueous media lead to distinctly different ionic accumulations at the surface, accompanied by contrasting surface charges (Arends and Jongebloed, 1977). In particular, the presence of hydrated metal ions, calcium, phosphate, and organic carboxyl and amino groups can be instrumental in determining surface charge characteristics of submersed objects (Matjevic, 1977; Arends and Jongebloed, 1977).

Charged surfaces will attract oppositely charged ions from surrounding waters. This form of attraction results in the accumulation of oppositely charged ions in excess of similar concentrations of such ions remaining in solution (Clark, 1974; Chapter 5, Section 3). This concentrating region, the Gouy/Chapman diffuse electric double layer, is attributable to the combined additive effects of electrostatic attraction and thermal motion of oppositely charged ions (Haydon, 1964). Most submersed surfaces are negatively charged, and as such, will accumulate excess positive ions in the diffuse double layer, which typically extends approximately 1 μm or less into the aqueous medium (Kitchener, 1973). Bacteria similarly exhibit negatively charged surfaces with positively charged diffuse double layers (Harden and Harris, 1953; Fletcher, 1979b). Therefore, repulsion will occur when both diffuse double layers overlap, whereas attraction will result at a slightly more distant point of separation (Fletcher, 1979b). In this manner, negatively charged surfaces can electrostatically either repel or attract bacteria. Interestingly, electron microscopic views of bacteria attached to submerged solids seldom reveal close contact between cell and solid surfaces, but more often demonstrate a 1- to 2-μm space between associated objects (Fletcher and Floodgate, 1973; Corpe, 1980; Costerton, 1980). This space may be indicative of the nonoverlapping double diffuse layer which would allow for successful attraction to take place. Surface wettability or hydrophobicity (or hydrophilicity) is another determinant regulating microbial adhesion (Fletcher and Loeb, 1979; Chapter 6, Section 4).

The events leading to the attachment of an individual bacterium to a surface have been described by Marshall (1976). Surface attachment by a cell is a two-stage process consisting of a reversible stage, in which the bacterium can be easily dislodged by changing the ionic strength of the bulk solution, and an irreversible stage, in which the bacterium becomes firmly attached after producing a polymeric material which helps bind the cell to the surface (Chapter 6, Sections 3 and 4).

According to Marshall (1976; Chapter 6) reversible adsorption can be modeled after the DVLO theory of colloid interaction (Derjaguin and Landau, 1941; Verwey and Overbeck, 1948): the attraction between two similarly charged particles is the result of a balance between electrostatic repulsion and van der Waals attraction.

After a bacterium is attracted to a surface by van der Waals forces, it may become permanently attached to the surface. The types of permanent attachment are varied. Some bacteria extrude thin pili or fibrils which are able to bind to the surface after bridging the gap formed by the interacting double layers (Rogers, 1979). Cells may be surrounded by primary polysaccharides which are chemically "sticky" and permit initial permanent attachment, while the cell secretes a secondary polysaccharide that can form a strong fibrous connection with the surface (Fletcher and Floodgate, 1973). Motile bacteria may possibly attach permanently to a surface by using kinetic energy to penetrate the double layer and get close enough to the surface to permit polymeric binding (Fletcher, 1979b). In fact, marine pseudomonads have been observed to spin on the axis of their flagella in an effort to attach to glass (Marshall *et al.*, 1971a).

After permanent attachment occurs, the cell may utilize adsorbed substrates and grow. Attached cells increase their number exponentially and form small microcolonies, (10 to 100 individual cells) on the surface (Meadows, 1966).

These physicochemical characteristics play key roles in determining whether submersed surfaces are amenable to attachment and colonization. However, whether or not colonization in fact results may well be regulated by an independent set of variables more relevant to chemical characteristics, and particularly nutritional status, of surfaces. Some proportion of the microbial community, either as dormant or actively growing cells, randomly encounters submersed surfaces and, depending on ionic and wettability characteristics, may adhere to such surfaces (Fletcher, 1979b). Increasingly, recent evidence from natural environments has pointed to nonrandom events and processes initiating specific microbe–surface associations (Chet and Mitchell, 1976). Most notable are the highly specific algae–bacteria associations which at times actively proliferate, while at other times are mysteriously absent even though individual partners in the association can be simultaneously isolated from free-floating populations (Herbst and Overbeck, 1978; Caldwell and Caldwell, 1978; Paerl and Kellar, 1978; Gallucci, 1981). Often, associations between dominant diatom, chlorophycean, and cyanobacterial hosts and specific bacterial epiphytes can be observed in nature (Paerl, 1980). When such associations proliferate, algal hosts and specific bacterial epiphytes can be shown to be metabolically active (Paerl, 1978; Gallucci, 1981); hence, the associations are not necessarily due to the senescence or active decomposition of host algae, but involve actively growing populations. It is unlikely that altered ionic surface characteristics or wettability would induce or disperse such associations. In the case of N_2-fixing filamentous cyanobacteria, the associations have been shown to be specific among the N_2-fixing cells (heterocysts) and bacterial epiphytes, and appear operative only during N_2-fixing periods (Paerl and Kellar, 1978; Lupton and Marshall, 1981). What induces such specific types of associations?

An ecologically important process mediating encounters, and ultimately associations, between surfaces (including plants and animals) and aquatic microorganisms is chemotaxis. As early as 1881, Engelmann discovered the ability of aquatic microorganisms to orient themselves along physical and chemical gradients. Orientation can occur in either negative or positive manners with respect to chemicals which diffuse into the aqueous medium (Adler, 1969; Koshland, 1980). Positive chemotactic substances have been identified in both freshwater and marine environments (Fogel et al., 1971; Bell and Mitchell, 1972). They include numerous organic constituents of both the particulate and dissolved organic matter pools, including simple sugars, organic acids, amino acids, and peptides (Chet and Mitchell, 1976). Attraction by inorganic substances, including nitrogen- and phosphorus-containing salts, has yet to be investigated in detail (Koshland, 1980). Chemically diverse substances, such as crude oil components, organic polymers, and algal hydrolysate, are known to chemotactically attract microorganisms (Bell and Mitchell, 1972; Wellman and Paerl, 1980; Gallucci, 1981). Likewise, negative chemotaxis has been shown to play a role in the repulsion of microorganisms from surfaces containing heavy metals (Young and Mitchell, 1973), certain organic compounds (Chet et al., 1975), and humic substances (K. Gallucci and H. W. Paerl, unpublished data).

Clearly, positive and negative chemotaxis play roles in mediating initial attraction and encounters with submersed surfaces. Taxonomically diverse, motile bacterial genera, including Bacillus, Pseudomonas, Vibrio, Escherichia, and Salmonella, have the ability to move along chemical (and in some cases physical) gradients. Commonly, such gradients originate from surfaces either through leaching and molecular diffusion of parent material or the loss, by desorption, of organic and inorganic substances accumulating at the liquid–solid interface. Hence, from an ecological point of view, chemotaxis must be considered as a factor mediating microbial encounters at this interface (Young and Mitchell, 1973; Chet and Mitchell, 1976).

Numerous bacterial and fungal genera are capable of gliding in order to disperse and to colonize surfaces (Chapter 6, Section 5). Gliding is mediated along chemical as well as physical gradients (Koshland, 1980). It is therefore conceivable that some microorganisms can seek nutritionally or physically favorable microenvironments at liquid–solid interfaces by gliding along an attractant or repellent gradient. Such instances have been documented for the filamentous, colorless sulfur bacterium, Beggiatoa (Jorgensen and Revsbech, 1983).

In summary, both the physical characteristics of surfaces as well as their chemical constituents regulate microbial attachment to submersed surfaces. Because of diversity in attachment-promoting mechanisms, it is not surprising to find diversity of microbial attachment to either newly submersed, previously submersed, or permanently submersed surfaces.

2.2. Postadhesive Events and Their Effects on Modification of the Liquid–Solid Interface

Structural modifications attributable to the production of extracellular slimes, appendages (pili, fimbriae), fibrils, and polysaccharide pads generally and quickly follow attachment at the liquid–solid interface (Fletcher and Floodgate, 1973; Corpe, 1980; Costerton,

1980; Chapter 1). Details on the specificity, chemical composition, morphology, and physical appearance of extracellular appendages and excretions instrumental in permanent (nonreversible) adhesion are addressed in depth elsewhere (Chapters 6, 9, 10, and 11). In this chapter, consideration will be given to the ecological–physiological ramifications of attachment. That is, how do surfaces function as microenvironments supporting metabolic activities of microbes and associated biotic communities following microbial colonization? Furthermore, what are the potentially modifying impacts of attachment on microbial metabolism, growth, and resultant nutrient cycling in aquatic habitats?

In addressing these two general questions, we must focus on parameters relevant to the influence of attachment on microbial metabolism and growth. Some of these parameters affected by, and in turn affecting, metabolism and growth include (1) redox potentials, attributable to the metabolism of microorganisms implicated in attachment, (2) nutrient release and uptake, which to a large extent are controlled by metabolic requirements for growth and maintenance of the microbial community, (3) fluctuations and alterations in oxygen conditions at the liquid–solid interface (respiration and photosynthesis are important offsetting or balancing processes with regard to the establishment of oxygen regimes and gradients), (4) metabolic coupling, which is the coupling of the products of one (or several) metabolic reactions to the use or requirements by other processes, or in the case of diverse interface communities, other populations, and (5) consumption of one organism by another, including phagocytosis and grazing by protozoans, invertebrates, and larval and mature vertebrates.

3. AQUATIC MICROZONES AS SITES OF POSTATTACHMENT ALTERATION (MODIFICATION) OF METABOLISM

Microzones can be operationally defined as regions bordering the liquid–solid interface where concentrated (by association and attachment) microbial populations lead to and function under environmental conditions distinct from the surrounding aqueous environment (Nikitin, 1973). Conditions which distinguish microzones from surrounding waters are, to a large extent, products of the integrated metabolic processes outlined in Section 2.2. Unique chemical and structural characteristics often separate microzones at the liquid–solid interface from truly planktonic conditions in the liquid medium. Some constituents which characterize microzones are derived from the following:

1. Deposition of some particulate organic carbon, nitrogen, phosphorus, etc. may be due to both the biomass and excretory products of attached microorganisms (Paerl, 1973a,b; Marshall, 1976). Included are mucilaginous, slime, and capsular materials, and specific appendages such as flagellae, pili, and microfibrils (Corpe, 1980)

2. Conversely, metabolic breakdown products of organic and inorganic structural components of the parent surface may at times accumulate and form significant proportions of the chemical environment constituting the microzone (Paerl, 1978; Trulear and Characklis, 1979; Eighmy et al., 1983). Breakdown products emanating from biotic detrital (nonliving organic matter) surfaces may be particularly abundant in microzones

3. Microzones serve as sites of both adsorption and absorption of dissolved substances, including organic and inorganic nutrients (MacRitchie and Alexander, 1963; Marshall, 1976; Kefford et al., 1982). Adsorption can, in part, be attributed to alteration and modification of attached cell surface charges and morphology result-

ing from microbial growth. Absorption also can be directly attributable to nutrient or metabolite uptake by microorganisms situated in microzones (Paerl, 1978; Paerl and Merkel, 1982). It follows that the establishment of microzones due to microbial attachment does not always signal metabolic breakdown of that surface, but may at times be a prelude to organic and/or inorganic enrichment at the interface due to microbial assimilation of dissolved substances, hence "fixing" them in the microzone (Paerl, 1978; Robinson *et al.*, 1982; Haack and McFeters, 1982)

4. Both diffusion and nutrient solubilization (from the particulate to dissolved states) can be dramatically altered in microzones supporting actively metabolizing microbial populations (Marshall, 1976). Accordingly, resultant nutrient release and uptake characteristics of surfaces can differ substantially between poorly colonized and heavily colonized microzones, or between microzones harboring inactive and active microorganisms. Diffusion of nutrients, dissolved gases, and metabolites can be altered by the formation of mucilaginous coatings, sheaths, and slimes in the microzone (Matson and Characklis, 1976; Caldwell and Caldwell, 1978). Resultant alterations of chemical and gaseous microenvironments can lead to the establishment of microzone environment conditions in stark contrast to the adjacent aqueous environment (Marshall, 1976; Ellwood *et al.*, 1979; Bitton and Marshall, 1980)

5. Alterations of redox potentials due to the effects of respiration and/or photosynthesis of microorganisms, as well as the reduction or oxidation of substrates required for growth, can also lead to surface microzones distinctly different from the adjacent environment (Jorgensen *et al.*, 1979). Oxygen concentration alone is often decisive in promoting, reducing, or eliminating specific nutrient-consuming or -recycling processes, as well as in establishing specific aerobic, anaerobic, or microaerophilic communities (Cappenberg and Jongejan, 1977; Indrebo *et al.*, 1977; Jorgensen and Revsbech, 1983).

4. DIVERSIFICATION OF MICROZONES

During microzone development at the liquid–solid interface, specific groups of microorganisms may sequentially dominate to the extent of preventing other members of the planktonic community from attaching. Such events are particularly common if the surface being inhabited is abundant in specific organic or inorganic constituents amenable to breakdown or solubilization by highly specialized microbial populations. Diverse plant tissues (Dazzo, 1980; see also Chapter 9), chitinous exoskeletons of invertebrates (Kaneko and Colwell, 1975), mucilaginous sheaths surrounding colonial algae (Caldwell and Caldwell, 1978; Paerl and Kellar, 1978), seaweeds (Laycock, 1974) and higher plants, wood structures (Austin *et al.*, 1978), special metals, and organic polymers (Dempsey, 1981) are all prone to colonization by specific bacterial species. Normally, continued dominance by single-species populations is short-lived, largely because (1) nutrients other than the major surface constituents fall in short supply, (2) metabolites generated by the breakdown of surface constituents accumulate and must either be sloughed off or metabolized by other microorganisms, and (3) the possibility exists that extremely active metabolic breakdown of surface constituents may drive the resident microbial community from aerobic to anaerobic conditions, thereby paving the way for microorganisms tolerant to or preferring such conditions. Due to these potentially rapid-changing environmental conditions, successional patterns are often observed in microzones bordering newly introduced organic and inorganic surfaces (Marshall *et al.*, 1971b; Mack *et al.*, 1975). Following settlement of pioneer species, complementary species often aggregate at the same site or in well-defined layers radiating away from the parent surface (Paerl, 1975; Geesey *et al.*, 1978; Eighmy *et al.*, 1983). In this manner, a microbial community made up of complementary taxa is evolved for the dual purpose of (1) establishing each species in nutrient-rich microzones (Dawson *et al.*, 1981; Kjelleberg *et al.*, 1982) and (2) taking advantage of the exchange of metabolites and nutrients either liberated or utilized by other species (Haack and McFeters, 1982). The accumulation of biomass in microzones leads to enhancement of organic deposition (Paerl, 1973a,b, 1978). A proportion of these deposits can be degraded by metabolically diverse microorganisms, and as a result localized portions of microzones can become oxygen-depleted under the influence of oxygen-consuming degradation processes (Paerl, 1980; Bautista, 1983). Anaerobic microorganisms can achieve a foothold in such patchy microzone environments, while aerobic producers and consumers of organic matter reside in oxygenated regions bordering the aqueous medium (Bautista, 1983). Active exchange of metabolites can continue between anaerobic, microaerophilic, and aerobic components of the microzone community, largely through the diffusion of nutrients and metabolites through successional layers superimposed on the liquid–solid interface (Jorgensen *et al.*, 1979; Bautista, 1983).

With the evolution of the microzone environment from one harboring unispecific populations to a diverse community where patchiness and layering along oxygen–nutrient gradients occur, the physical (structural) microenvironment also becomes more complex. Periphytic bacteria and algae generally become established as "climax communities," bordering mineral (rocks, soils, sands), plant, animal, and detrital substrates (Round, 1972; Geesey *et al.*, 1978; Wetzel, 1979). Such communities often impart a feltlike, slimy, or spongelike texture to the liquid–solid interface. The complex textures due to stalk formation, excretions, appendages, and microbial mucous webs offer an excellent substrate as well as protective environment for protozoan as well as invertebrate consumers of microbial biomass (Marszelak *et al.*, 1979; Wetzel, 1983). The resulting interface community grows more complex both in terms of its species composition and its metabolite and nutrient exchange processes. Evolution from single-species bacterial communities to bacterial, algal, protozoan, invertebrate surface communities can occur on both a microscopic and a macroscopic scale. An example of succession on a microscopic scale can be obtained from observations on sand grains and intact algal colonies (Meadows and Anderson, 1968; Stotzky, 1972; Paerl, 1975; Marshall, 1980). Initial establishment of single-species bacterial communities rapidly commences after the introduction of either sand grains or axenic algal colonies in natural waters. In the case of sand grains, pioneer bacterial epilithic populations lay down copious amounts of extracellular polymeric material following attachment (Fig. 1). This material effectively binds grains together to form a thin layer (100 to 200 μm) of aggregated sand in marine tidal flats where erosional wave and tidal surge actions are not strong enough to overcome aggregation processes (Monty, 1967; Polimeni, 1976; Hirsch, 1978; Potts, 1980). Despite some sensitivity to the potential physical perturbations of seawater, turbulent nutrient and gaseous exchange between aggregated sand grain surfaces and seawater is necessary for the enhancement of biomass and resultant establishment of a more diverse microbial community (Polimeni, 1976). Both periphytic bacteria (particularly stalked genera such as *Caulobacter*) and filamentous algae become established at the sand–water interface. Both groups of microorganisms add further structural integrity to the sand aggregates, intertwining between sand grains and thereby forming intertidal and subtidal microbial mats which are more resistant to disturbances by wave and wind action (Monty, 1967; Neumann *et al.*, 1970; Polimeni, 1976). With the increased production of biomass (largely due to photosynthesis by stalked diatoms and filamentous cyanobacteria), organic deposition is greatly enhanced in the mat community (Fig. 2), leading to anaerobic conditions in the lower mat

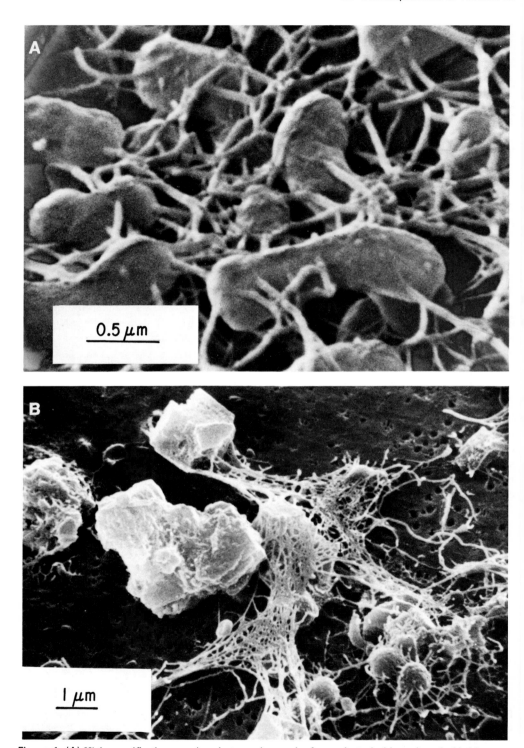

Figure 1. (A) High-magnification scanning electron micrograph of several attached bacteria embedded in extra-cellular polymeric fibrillar "webbing." Marine sand grains provided the surface substratum for these bacteria. (B) Lower-magnification view of bacterially produced fibrillar "webbing" aggregating sand grains in the benthic region of Lake Tahoe, California.

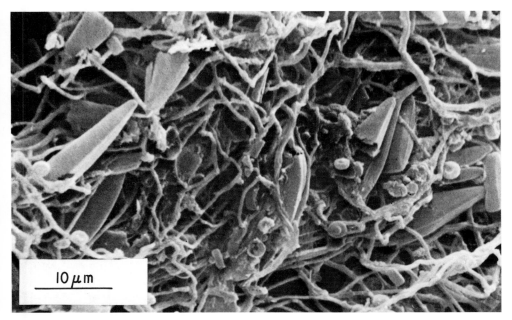

Figure 2. Surface morphology of a marine microbial mat freshly sampled from Shackleford Banks, North Carolina. A sample was cut out of the mat, fixed in buffered (pH 8.0) 2% (v/v) glutaraldehyde, prepared by critical point drying, and examined by scanning electron microscopy. Note stalked diatoms, filamentous cyanobacteria, and attached bacteria constituting the bulk of the mat's surface layer.

layers. Such conditions are favorable for the establishment of both photosynthetic and heterotrophic anaerobic bacteria (Bautista, 1983). Meanwhile, the upper portion of the mat community, being composed of an active algal photosynthetic community, provides supersaturated oxygen conditions (Bautista, 1983) (Fig. 3). Fully developed intertidal microbial mats which initially were established through microzone formation around sand particles are complex microbial communities often only a few millimeters thick (Fig. 3). Within this mat,

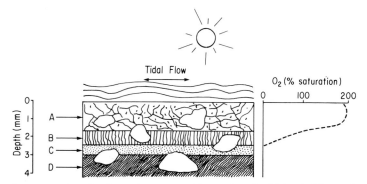

Figure 3. Schematic cross-section of various microzones making up the microbial mat at the sand–water interface on Shackleford Island, North Carolina. The upper layer (A) is largely composed of cyanobacteria, diatoms, and aerobic bacteria; this layer commonly has supersaturated O_2 concentrations. The next layer (B) contains the N_2-fixing cyanobacterium *Microcoleus* in a sub-O_2-saturated habitat. Underneath this region is the aerobic–anaerobic interface (C), where the anaerobic, photosynthetic bacterium *Chromatium* resides in a dense, purple layer. The bottom layer (D) is composed of anoxic bacteria, reduced organic and inorganic compounds, and anoxic detritus interspersed with sand grains. Both representative oxygen regimes and mat thickness are indicated.

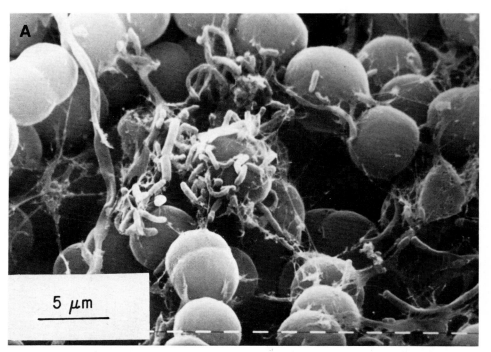

Figure 4. (A) Scanning electron micrograph of bacteria embedded in mucilaginous slime excreted by the host cyanobacterium *Anabaena*. This sample was obtained from eutrophic Heart Lake, Ontario, Canada. (B) Transmission electron micrograph illustrating the extensive network of mucilaginous fibrils in which bacteria can be found in association with host cyanobacteria (*Anabaena*). (B) courtesy of Dr. A. M. Massalski.

oxygen concentrations can range from 200 to 300% saturation at the surface to anaerobic conditions at the bottom. Such large oxygen gradients serve to accommodate highly specialized anaerobic or aerobic microorganisms in distinct layers. These layers are not chemically isolated. Active and passive exchange of reduced and oxidized substrates maintains each component of the mat (Jorgensen *et al.*, 1979; Haack and McFeters, 1982). In this manner one microorganism's waste product serves as a potential energy source, electron acceptor or donor, or growth substrate for another component of the community. Structural complexity of microbial mats forms specialized habitats for protozoan and invertebrate grazers, which, in turn, utilize primary-produced organic compounds and liberate inorganic nutrient, such as phosphate and ammonia. Fish grazing can add to the trophic complexity embodied in such microzones.

Algal cells and colonies and suspended nonliving organic and inorganic particles (detritus) can also serve as sites of microzone formation, leading to the establishment of an actively metabolizing and nutrient-exchanging surface community (Paerl, 1978; Haack and McFeters, 1982). Colonial cyanobacteria offer an attractive surface for microzone formation because (1) organic nutrients are produced and excreted by cyanobacterial hosts, and (2) fibrillar, mucilaginous cyanobacterial excretions offer excellent attachment and colonization sites for associated microorganisms. Naturally occurring, actively growing *Microcystis* and *Anabaena* colonies are typically observed coated with a layer of bacteria, often embedded in mucilaginous slime (Fig. 4). A specific proportion of these bacteria are known to be chemotactically attracted to the cyanobacterial host (Gallucci, 1981). Following the establishment of bacterial epiphytes, algal epiphytes (including diatoms and flagellated chlorophyceans) as

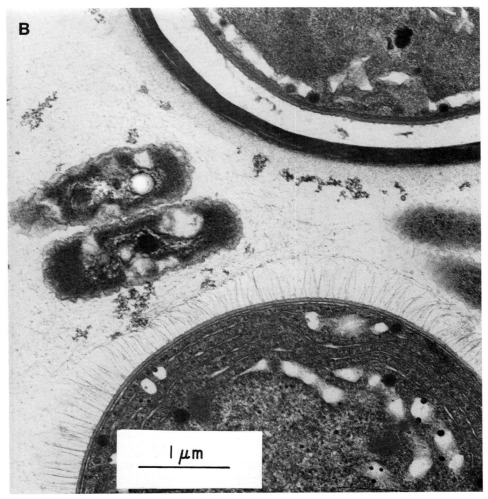

Figure 4. (*Continued*)

well as a grazing protozoan- and invertebrate- (worms, crustaceans) epizoic community becomes established (Fig. 5). Protozoans appear to be particularly active members of microzones coating host algal colonies. Both sessile (attached) protozoans (*Stentor, Vorticella*) as well as gliding *Amoeba* cells, free-swimming ciliates, and larval invertebrates form a bulk of the epizoic biomass (Wetzel, 1983) (Fig. 5). Microzone-associated protozoans have been observed consuming attached bacteria and, at times, algal cells (H. W. Paerl, unpublished results). These observations indicate that the complexity of microzones not only promotes active nutrient exchange, but also leads to the establishment of microbial food chains, starting with primary-producer hosts whose photosynthetic products are utilized by a heterotrophic, secondary-producer community (epiphytic bacteria), which in turn is consumed by a grazer community (Menge, 1976; Nicotri, 1977; Harrison and Harrison, 1980). Ultimately, microzone-enriched particles, microbial mats, and other epiphytized surfaces can be consumed by larger fish, invertebrates, and benthic molluscan and crustacean grazers.

    The formation of microzones leads to the promotion of specific biochemical processes which would otherwise be unfavorable in planktonic or surface-free environments. With

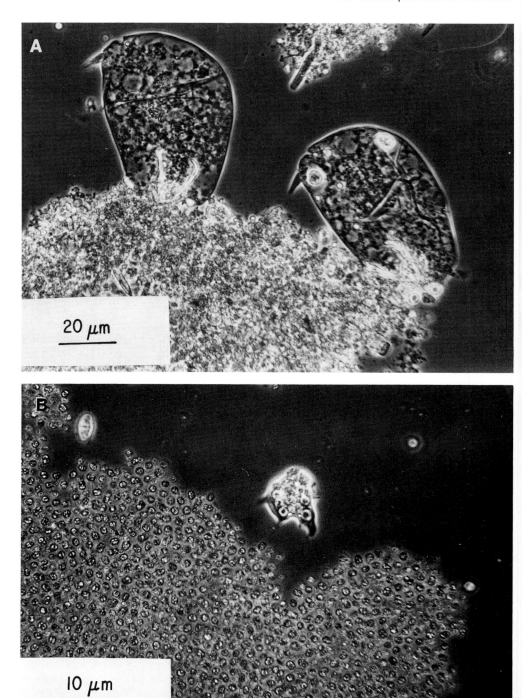

Figure 5. (A) Light (phase contrast) micrograph of unidentified rotifers grazing on epiphytic bacteria and diatoms associated with a large cyanobacterial (*Microcystis*) colony residing in the Neuse River, North Carolina. Protozoan, rotifer, and crustacean epizoic populations rapidly assume important grazing and nutrient cycling roles following the establishment of both freshwater and marine microbial surface communities. (B) *Amoeba* actively grazing microorganisms associated with the surface of a *Microcystis* colony, also obtained from the Neuse River.

respect to nitrogen cycling, for example, both denitrification and N_2-fixation require reduced microenvironments to proceed (Alexander, 1971). Some aerobic, filamentous cyanobacteria have circumvented this constraint by confining N_2-fixation to specialized oxygen-free cells termed *heterocysts* (Fogg et al., 1973). However, single-celled or nonheterocystous filamentous prokaryotes require reduced conditions as well as an energy source (either photosynthetically, chemosynthetically, or heterotrophically mediated) in order to fix N_2 (Gallon et al., 1975; Carpenter and Price, 1976). Microzones such as microbial mats, as well as submersed, colonized biotic and abiotic surfaces, offer both a potential energy source (organic carbon adsorption and deposition, close proximity to primary producers) and reduced microenvironments (subsurface layers of microzones) allowing N_2-fixation to occur (Paerl and Kellar, 1978; Paerl and Bland, 1982; Bautista, 1983). In nitrogen-depleted waters, which characterize much of the world's oceanic and nearshore marine environments, N_2-fixation constitutes the only "new" source of nitrogen input and is mediated through microzone communities (Paerl et al., 1981). One can therefore visualize mutually useful processes (autotrophic CO_2-fixation and N_2-fixation), whose end products and metabolites can serve as exchange products essential for both the development and the maintenance of microzone communities.

Denitrification is a strictly anaerobic process, regulated by (1) the presence of energy sources, (2) absence of O_2, and (3) availability of NO_3^-, which serves as the main electron acceptor. It is therefore an interface-oriented process, for NO_3^- normally only exists in oxygenated waters (NO_3^- is formed by obligate aerobic, nitrifying bacteria), and yet its reduction by denitrifiers must occur in the absence of O_2 (Alexander, 1971). Both benthic and intertidal microbial mats, aquatic macrophytes, decomposing macrophyte and leaf litter, animal and suspended inorganic particles, and rocky bottoms serve as potentially ideal locations for denitrification to proceed. Active microbial metabolism can potentially reduce O_2 in subsurface layers of the microzone, while proximity to adjacent aerobic waters is essential for NO_3^- supplies (Sugahara et al., 1972).

Several investigators have independently observed the promotion of both denitrification and nitrification in aerobic waters and soils through the addition of either nutrient-rich or inert particulate matter (Carlucci and Strickland, 1968; Paerl et al., 1975; Kunc and Stotzky, 1980). In the case of denitrification, it has to be postulated (as discussed above) that ready access to reduced microenvironments in microzones bordering particles and surfaces, as well as the presence of NO_3^-, are the reasons for observed promotion of this process in aerobic waters. Sewage lagoons having extensive bottom regions exposed to aerated NO_3^--rich waters reveal similar stimulation of denitrification, particularly if organic energy sources (such as sugars and alcohols) are provided (R. Gersberg and C. R. Goldman, personal communication).

Nitrification has also been promoted by the addition of particles, especially in nutrient-poor oligotrophic waters. Particle surfaces offer a potential source of ammonia (which is the reduced substrate in nitrification), both through surface adsorption of dissolved ammonia as well as ammonia release from particles themselves, especially particles of organic origin (Paerl et al., 1975). Decomposition, followed by deamination of amino groups from proteinaceous constituents, would be the most likely source of ammonia for particle- or surface-associated nitrifiers. In oligotrophic environments, known to contain undetectable levels of dissolved ammonia, a surface microzone existence provides a favorable habitat for nitrifiers. Such favorable conditions appear to explain the stimulatory effects of surface on nitrification.

Diverse electron-transferring, oxidation–reduction-dependent processes are also reliant on the presence of surface microzones. Included are the sulfur-transforming processes, starting with SO_4^{2-} reduction (Jorgensen, 1977) and ending with H_2S oxidation (Alexander, 1971; Jorgensen and Revsbech, 1983). The filamentous gliding bacterium *Beggiatoa*, although not firmly embedded in a microzone, shows strong reliance on aerobic–anaerobic interfaces, such as water–sediment interfaces and surfaces of decaying organic particles (Strohl and Larkin, 1978). This microorganism has a need for microaerophilic conditions (O_2 is needed as a terminal electron acceptor), organic nutrition, and H_2S (which is oxidized to elemental sulfur, often seen deposited in filaments). *Beggiatoa* commonly proliferates on marine and freshwater sediment surfaces shortly after overturn or complete oxygenation of the water column. Such events assure dissolved O_2 supplies in the water as well as H_2S diffusing out of anaerobic sediments (Jorgensen and Revsbech, 1983). Once the surface sediments have become thoroughly oxygenated (following continued oxygenation of the overlying water column), *Beggiatoa* populations rapidly decline; their preferred microzone has been disrupted through continued inward diffusion of O_2. Destruction of such microzones commonly occurs several weeks after complete oxygenation of the water column. *Beggiatoa* populations are not observed again until deoxygenation of surface sediments and oxygenation of the water column reestablish the proper surface–interface conditions.

Other interface-dependent processes worth mentioning include methanogenesis (Lovley *et al.*, 1982), methane oxidation (Reeburgh and Heggie, 1977), iron and manganese reduction (Mortimer, 1971) and oxidation (Nealson and Ford, 1980), phosphorous solubilization (Mortimer, 1971), and calcite and carbonate deposition (Wetzel, 1983).

5. EXAMINATION OF MICROZONE GRADIENTS AND PATCHINESS

The identification and monitoring of microzone habitats which are the potential sites for the microbially mediated, nutrient-transforming and oxidation–reduction reactions (Section 4) have remained difficult. Most microzones range from no more than a few micrometers to several millimeters in thickness. Hence, macroanalyses, such as pore water extraction followed by chemical determinations, and even microprobe analyses, often fail to detect microzones chemically distinct from adjacent waters, as well as microniches or ''patches'' of specialized biochemical processes occurring within microzones. Microscopic observations often preclude being able to examine microenvironments in an intact fashion, except for suspended, and generally less than 1-mm-diameter, particles. Therefore, techniques which can be readily employed on undisturbed microzones, yet yield sufficient sensitivity in detection, are highly desirable. Among these, two novel approaches appear attractive. They include the use of tetrazolium redox indicator salts (Kuhn and Jerchel, 1941; Altman, 1972; Zimmermann *et al.*, 1978; Paerl and Bland, 1982) and autoradiographic detection of incorporated and exchanged radioisotopes (Watt, 1971; Stull, 1972; Paerl, 1978; Fletcher, 1979a). Both techniques can be applied to undisturbed particulates, sediments, and submersed surfaces. Tetrazolium salts are low-molecular-weight (a general range is from 200 to 600), soluble, organic compounds in oxidized form. Upon reduction they form brightly colored, insoluble (in aqueous media) formazan crystals which are readily detected by both brightfield and phase-contrast microscopy (Altman, 1972). Each tetrazolium salt has its own specific reduction potential. Hence, the observed reduction of individual salts yields direct evidence of specific redox regimes present both inside microorganisms and their neighboring microenvironments. Approximately 20 to 30 tetrazolium salts, varying in color as well as specific reduction potentials, are commercially available (Altman, 1972).

Following the addition of dilute (0.01 to 0.05% w/v) tetrazolium salts to waters of interest, exposure (reduction) is allowed to proceed from 15 min to several hours (depending on the diffusion characteristic of microzone layers and density of microorganisms) (Paerl and Bland, 1982). Tetrazolium reduction can be terminated by the addition of biological fixatives; most commonly, 2 to 3% (v/v) glutaraldehyde or formaldehyde solutions are used. Following fixation, microzones on small particles can be directly observed by microscopy. Larger particles, surfaces, and microbial mats can be partially dissected, followed by microscopic observations of sections.

Microautoradiography has also been employed, to a somewhat more limited extent, in the differentiation and detection of specifically metabolically active regions of microzones (Paerl, 1978). This technique is more limited than tetrazolium salt applications because samples must be incubated with tracer, rinsed (to remove excess tracer), freeze-dried, and flattened on a microscopic slide prior to exposure. Thus, more manipulation is required for this technique. Nevertheless, it is useful for (1) identifying specific biochemical processes occurring in microzones, (2) relatively intact observations on small (< 100-μm width) surface microzones, and (3) separation of symbiotic components of microbial communities and heterotrophic populations. Furthermore, coupled with freeze-drying, the technique offers excellent preservation of cells, associated structures, and appendages of interest (Fig. 6). The incorporation of biologically significant elements (in various forms), such as carbon, sulfur, phosphorus, hydrogen, calcium, and iron, can be readily followed by autoradiography of microzone communities. Using pulse-labeling experiments, exchange of nutrients from one segment of the community to another can be observed (Paerl, 1978). In particular, the use of the soft β-emitter ^3H offers the best resolution available to this technique. Accordingly, in examining the fate of organic substrates, ^3H-labeling of relevant compounds is recommended (Paerl, 1978).

Oxygen, pH, and redox microelectrodes have recently been introduced and applied to diverse microenvironmental studies. Most notable is the work of Jorgensen et al. (1979) and Jorgensen and Revsbech (1983) on benthic microbial mats located in saline lakes in Israel

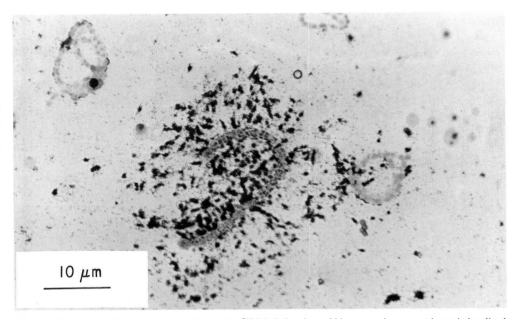

10 μm

Figure 6. Microautoradiograph demonstrating active ^3H-labeled amino acid incorporation among bacteria localized in the mucilaginous sheath coating the N_2-fixing cyanobacterium *Anabaena*. Bacterial ^3H-labeled amino acid incorporation can be seen as the accumulation of clusters of optically dense, reduced (by radioactive decay) silver grains in the mucilaginous sheath.

and in marine sediments in Denmark. Good resolution of O_2, pH, and E_h gradients was obtained in mats only 4 to 5 mm thick. Microelectrode technology is expected to advance rapidly over the next few years. This will perhaps allow investigators to distinguish physicochemical gradients within a 100-μm-thick microzone. Such a technological breakthrough would certainly facilitate rapid detection of gross oxidation–reduction (and perhaps nutrient) features of microzones established at the liquid–solid interface.

6. DYNAMICS OF NATURAL PLANKTONIC AND BENTHIC AQUATIC MICROZONE HABITATS

An expanding number of microzone habitats have recently received attention because (1) such habitats can be responsible for enhancement and acceleration of specific nutrient transformation rates and growth rates (see Section 7) and (2) microbial fouling and conditioning of surfaces for the settlement of higher-ranked organisms occur in such microzones, which in itself may give an indication of microzone surface properties. Both planktonic and benthic environments host microzones. Planktonic environments invariably contain suspended sediments and detrital particles as well as suspended or motile biota. All represent surfaces which have been shown, at various stages and frequencies, to be inhabited by microbial (bacterial, fungal, protozoan) communities (Paerl, 1973a, 1980; Sieburth, 1975). The frequencies and magnitudes of microbial associations with suspended marine and freshwater sediments have been subjects of dispute for at least the past two decades (Melchiorri-Santolini and Hopton, 1972). Largely due to the application of electron microscopy, improved microbial staining techniques, the use of cellular redox indicators such as tetrazolium salts, and advancements in the application of tracer techniques, investigators have become more aware of the diversity, and at times, subtleness of microbe–particle and microbe–surface interactions in aquatic environments. At our present state of knowledge it is an undeniable fact that microbe–surface interactions are a universal feature of planktonic environments (Marshall, 1976, 1980). Frequencies and magnitudes of association are disputable, and likely to depend on the specific interacting ecological variables, including (1) the chemical and physical nature of surfaces, (2) nutritional values of particles and surfaces, (3) dissolved nutrient regimes, (4) water temperatures, ionic strength, pH, and hydrostatic pressure, and (5) overall trophic states of ecosystems under investigation. Surface-associated microzones have been reported on (1) detrital surfaces (Paerl, 1973a), (2) mineral (erosion) particles (Paerl and Goldman, 1972), (3) live algae, particularly colonial taxa (Caldwell and Caldwell, 1978; Paerl and Kellar, 1978), (4) zooplankton and their fecal pellets (Turner, 1979), (5) suspended invertebrates, including tunicates, larval mollusks, ctenophores, ascidian larvae, and round and flatworms (Sieburth, 1975), (6) fish, and (7) anthropogenically derived, suspended sediments, including industrial and municipal wastes (Paerl, 1980).

In studies investigating potential attachment habitats in planktonic environments, virtually all types of nontoxic particulates have been shown to become colonized at some stage. Often, only a few bacterial cells can be detected on surfaces, falling far short of the populations needed to establish a microzone having physicochemical (and biotic) features distinct from the planktonic environment. However, the application of both isotope techniques and redox indicator salts has shown that microzone establishment is not necessarily dependent on the presence of vast numbers of associated bacteria (Paerl and Kellar, 1978; Paerl and Bland, 1982). Bacterial (and presumably fungal) activities, including turnover rates of organic and inorganic nutrients, O_2 consumption, and phytosynthetic O_2 evolution rates are often better indicators of the potential extent to which microzones are formed than are bacterial numbers alone. Evidence points to a vast proportion of enumerable bacteria as being metabolically inactive (Stevenson, 1978; Zimmermann et al., 1978), and hence contributing little to community metabolism and resultant microzone formation.

Planktonic microzone formation has been shown to lead to the enhancement of physical particle breakdown, as well as the building and diversification of small particles into detrital aggregates through the formation of mucilaginous adhesives (Paerl, 1973a), pads (Fletcher and Floodgate, 1973), fibrils (Massalski and Leppard, 1979), and other anchoring appendages (Massalski and Leppard, 1979), which act to cement particles together. This leads to larger detrital aggregates containing a relatively large proportion of bacterial and fungal organic matter in the form of cellular excretion, structures, or partially decomposed biomass attributable to prior colonizing populations (Seiburth, 1975; Paerl, 1980). Hence, microzone formation, leading to both POM losses and gains, is largely regulated through the offsetting processes of decomposition of surface or particle material and the uptake of dissolved constituents, with some portion of these constituents being "fixed" into particulate organic matter (Paerl, 1975; Robinson et al., 1982).

Active transfer of photosynthate from algal hosts to microbial (bacterial and fungal) epiphytes has been shown to occur among a variety of phytoplankton, including cyanobacteria (Paerl, 1978), chlorophyceans (Nalewajko et al., 1976), diatoms (Larsson and Hagstrom, 1979), dinoflagellates (Cole et al., 1982), and macroalgae (Robinson et al., 1982), as well as higher-rooted plants residing in the water column. Among the planktonic algae, filamentous and colonial cyanobacteria have perhaps been most thoroughly documented. Both freshwater and marine cyanobacteria (including N_2- and non-N_2-fixing genera) are virtually always found colonized by bacteria in nature (Paerl, 1980). This is particularly true among the bloom-forming genera, which often reach explosive population densities during spring and summer growth periods in temperate regions, and are ubiquitous in tropical regions (Fogg et al., 1973). These include Anabaena, Aphanizomenon, Gloeotrichia, Microcystis and Oscillatoria in freshwater habitats and Trichodesmium (Oscillatoria) in marine habitats. Benthic marine cyanobacteria (see Section 7) also serve as hosts to bacterial epiphytes. Bacteria appear embedded in mucilaginous sheaths (Fig. 7) or are firmly attached to the fibrillar surfaces of hosts (Fig. 8). Such epiphytic bacteria have been shown to be attracted to cyanobacterial hosts through positive chemotactic responses (Gallucci, 1981). A high degree of metabolic activity is virtually always observed among these epiphytic populations, including enhanced uptake of dissolved organic substrates (sugars, amino acids, organic acids), phosphorous, and thymidine (an indicator of DNA replication or growth rates) (Paerl, 1978, 1980). In general, cellular assimilation of nutrients, particularly organic nutrients, is enhanced in populations associated in this manner, as compared with free-living bacteria (Caldwell and Caldwell, 1978; Paerl and Kellar, 1978). Marked reduction of tetrazolium salts has also been observed in microzones with bacterial epiphytes, while similar reduction among free-living bacteria was often absent (Paerl and Kellar, 1978; Paerl, 1980). Considering the fact that waters supporting cyanobacterial blooms are often O_2-supersaturated, it can be concluded that such microzones harbor O_2 regimes truly distinct from the bordering aqueous environment. Accordingly, highly reductive (having a demand for reduced environmental conditions) processes, such as denitrification, N_2-fixation, and microaerophilic decomposition processes (such as cellulose degradation), as well as numerous low-O_2 nutrient solubilization processes (such as ammonification and phosphorus solubilization), are more likely to occur in such microzones than in surrounding waters (Paerl and Kellar, 1978; Lupton and Marshall, 1981).

Bacterial epiphytes stand to gain (in terms of metabolic activities and growth potentials)

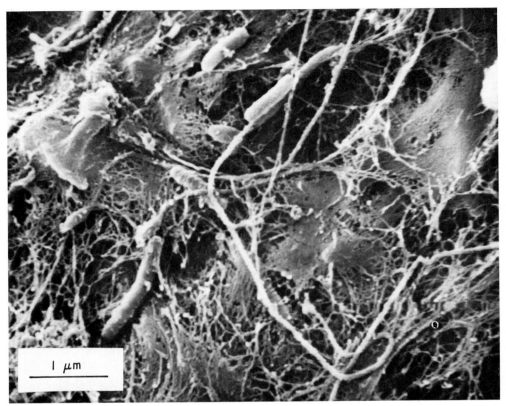

Figure 7. Bacteria embedded in mucilaginous sheaths produced by the benthic marine cyanobacterium *Micro-coleus*. Use of tetrazolium redox indicator salts has shown reduced (in terms of O_2) microzones to be present in sheaths hosting these bacteria; such results illustrate active metabolism attributable to epiphytic bacteria.

a great deal by residing in association with host cyanobacteria and other phytoplankton. It has been shown that a certain proportion of organic compounds photosynthetically produced by host phytoplankton is either actively or passively excreted through the cell wall (Fogg, 1966; Nalewajko *et al.*, 1976). Some of the excreted material can actually be physically observed in the form of mucilaginous slime and fibrillar extracellular appendages which are actively sloughed off during growth (Paerl, 1978, 1980; Massalski and Leppard, 1979). Such substances, as well as dissolved excreta diffusing outward through mucilaginous sheaths, serve as growth factors and substrates for associated bacteria. Through their metabolic activities, the bacteria consume O_2 and generate CO_2, both of which are known to benefit host algae faced with potential localized CO_2-limitation (Lange, 1973; Paerl, 1983), photorespiration (Burris, 1977), and photooxidation (Eloff *et al.*, 1976) during aquatic blooms. Studies on algal–bacterial exchange and its impact on growth have concluded that microzones bordering colonial algae, as well as microalgae and higher aquatic plants, serve an important function in maintaining optimal primary productivity of host organisms (Lange, 1973; Paerl and Kellar, 1978). Without the presence of a microzone recycling nutrients, CO_2, and O_2, lower productivity and less intense bloom conditions could result.

While benefiting host algae, epiphytic microzones also lead to the establishment of complex microbial communities, which, among themselves, promote active exchange of nutrients and metabolites. In certain cases microzones have a marked impact on rates of

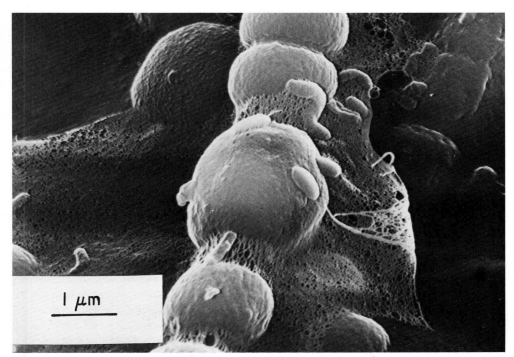

Figure 8. Scanning electron micrograph of bacteria firmly attached to benthic cyanobacteria. The host cyanobacterium is *Anabaena*.

extracellular production of photosynthate from host plants. During a study of epiphytic cyanobacterial and bacterial communities inhabiting leaves of the rooted macrophyte *Myriophyllum spicatum* (Eurasian Millfoil) in a mesotrophic Canadian lake, it was shown that as much as 60% of the photosynthate excreted by *Myriophyllum* was assimilated by microbial epiphytes in a 20- to 30-μm-thick layer bordering the leaves (Paerl, 1979). It was therefore concluded, as has been similarly found among marine macroalgae (seaweeds) (Robinson *et al.*, 1982), that surface microzones have a highly significant impact on both the fate and the recycling of excreted organic matter.

Microzones are often semipermanent (annual growth), long-lasting features of intertidal and subtidal marine, as well as freshwater submersed benthic habitats (Monty, 1967; Potts and Whitton, 1979). As described in Section 4, benthic microzones, such as the microbial mats located in intertidal regions of North Carolina's barrier islands, are initially established through microbial (algal–bacterial) attachment to sand grains followed by deposition of excreted organic matter (Polimeni, 1976). This chain of events is similar to microbial colonization of submersed rocks in freshwater streams (epilithic communities) (Round, 1964; Geesey *et al.*, 1978). Following "conditioning" of the substrate by deposition of organic matter (Chapter 6, Section 4.8), diversification of a microzone community takes place. Both a food chain and a chemical recycling community are then evolved. Marine microzones appear ubiquitous in regions supporting attachment substrata, e.g., sand, rocks, attached and rooted plants, wood, metal, concrete, and plastic (polymer) submersed structures. Microzones are particularly well established in regions supporting active flow of nutrient-rich waters, such as intertidal marshes, inlets, intertidal sand spits, mangrove regions, coral reef habitats, canals, and harbors. There appear to be distinct ecological advantages to the presence of microorganisms in such regions. First, virtually all of these regions exist in segments of the marine environment supporting nutrient (both organic and inorganic) levels exceeding those found in oligotrophic open ocean, and particularly oceanic gyre environments. Second, location in a benthic region experiencing tidal, or in the case of estuaries freshwater, flushing, offers a distinct advantage over stagnant habitats, for nutrient exchange and resultant continual nutrient availability would be optimal in such high-turbulence environments (Hynes, 1970). However, there are distinctions between desirable and undesirable degrees of turbulence. For example, although turbulence due to tidal flow is highly desirable in sustaining microbial mats located in sand flats, storm surges, hurricanes, and even man-made sources of turbulent disturbances (e.g., motorboat wakes, uprooting of submersed vegetation, dredging) can provide disruptive energy in excess of stored energy accumulated in the structural integrity of microbial mats and surface microzones (Monty, 1967).

Fouling communities are somewhat more resistant to disturbance by a high degree of turbulence for several reasons. First, and most important, microorganisms are attached to surfaces whose physical features are not easily altered (e.g., ship hulls, pier pilings, concrete caissons), as opposed to movable mud, sand, and plant materials in benthic regions. Second, fouling communities are typified by microzones rich in polymeric substances laid down by bacterial populations (Chapter 1, Section 1). Such glycocalyx polymeric bridging structures are effective in forming strong bonds between attachment surface molecules and microbial adhesives (Corpe, 1980; McCoy et al., 1981). Finally, fouling communities form generally thin (10 to 20 μm thick), high-tensile-strength microzones, highly resistant to surrounding turbulent regimes (Dempsey, 1981). There exists some degree of structural similarity between fouling communities and epilithic communities in high-velocity streams, where both are well anchored and abundant in adhesive polymers (Geesey et al., 1978).

Reef habitats, being relatively high-productivity microenvironments in oligotrophic marine systems, support diverse microzones bordering mineral, plant, and animal constituents. Virtually all surface segments of coral reefs support epimicrobial growth (Whitton and Potts, 1982). Normally, low suspended-sediment concentrations can be found in waters bordering reefs. Microbial microzones are therefore largely confined to benthic and intertidal surface regions. Little is known of the metabolism and functional roles that microzones play in the productivity and nutrient cycling characteristics of reef communities.

7. MICROBIAL METABOLISM AND GROWTH CHARACTERISTICS IN SURFACE MICROZONES

Pioneering studies by Stark et al. (1938), Heukelekian and Heller (1940), and ZoBell (1943) demonstrated growth stimulation of bacteria by surfaces of containers in which low-nutrient seawater was stored (Chapter 12, Section 3.1). This "surface effect" has since been repeatedly confirmed, both in freshwater and in seawater. As a general rule, the smaller the storage container, i.e., the greater the surface area/volume ratio, the greater the relative surface effect. Surface stimulation effects have been attributed to two processes beneficial to microbial growth (Chapter 12, Section 4.1). The first is the set of adsorptive characteristics that diverse surfaces exhibit. Both organic and inorganic ions are readily adsorbed by glass, plastic, metal, and wooden surfaces (Neihof and Loeb, 1972). In particular, charged compounds can be concentrated on submersed surfaces to the extent that surface concentrations are significantly higher than dissolved concentrations of identical compounds (Marshall, 1976). Charged organic compounds, such as amino acids, fatty acids, humic acids, and proteins, can be shown to be readily adsorbed to and concentrated on surfaces (MacRitchie and Alexander, 1963). Likewise, inorganic nutrient ions, such as PO_4^{3-}, Fe^{2+}, Fe^{3+}, as well as a range of other metals, are effectively accumulated on surfaces (Clark, 1974). Therefore, in oligotrophic environments where dissolved nutrient concentrations are low enough to limit microbial growth, concentration by surface adsorption constitutes a mechanism for accumulating nutrients in concentrations high enough to circumvent nutrient-limited microbial growth (Jannasch and Pritchard, 1972; Paerl and Goldman, 1972; Smith and Oliver, 1981). The second way in which surfaces may stimulate growth is that surfaces themselves can constitute a concentrated nutrient source. Particularly in oligotrophic systems, such surfaces may represent a concentrated nutrient patch (Paerl and Goldman, 1972). Clearly, the nutritive value of surfaces and particles depends on the proportion of mineral and organic constituents.

It has been shown that microorganisms settle onto such surfaces in a variety of ways, including chemotactic attraction (Adler, 1969). Hence, the ability of microorganisms to place themselves in close proximity to a surface with elevated nutrient concentrations gives them a distinct advantage over those microorganisms existing in nutrient-limited, truly planktonic environments. Close contact at the liquid–solid interface assures proximity to nutrients. Effective use of adsorbed or leached nutrients by surface microorganisms therefore supports enhanced growth rates, particularly if such nutrients are growth-limiting.

Through the use of isotope uptake studies it has been shown that a variety of inorganic and organic nutrients are concentrated on surfaces (see Paerl and Goldman, 1972; Paerl, 1978; Paerl and Merkel, 1982). Furthermore, it was shown by Jannasch and Pritchard (1972), as well as Paerl and Goldman (1972), that particles do not necessarily need to contain nutrients in order to prove biostimulatory. In both above studies, biostimulation by inert surfaces was attributed to such surfaces acting as concentration sites for dissolved nutrients and in this manner exposing attached bacteria to nutrient concentrations in excess of those in the nutrient-poor dissolved state. Several lines of evidence support this thesis. First, when isotopic forms of known growth-limiting nutrients (e.g., PO_4^{3-}) are administered to water samples containing inert surfaces, both nonbiological adsorption and microbial uptake of the isotope are notably higher in surface microzones than in solution (free-floating bacteria) (Paerl and Merkel, 1982). Second, within a single community, surface-associated bacteria are generally larger in size than free-floating individuals, even when inert surfaces serve as attachment sites (Fig. 9). Finally, these larger bacteria reveal differences in uptake kinetics, in terms of substrate saturation levels and affinities, than planktonic counterparts. In general, larger bacteria (among a variety of species) exhibit lower affinities and higher saturation levels than smaller bacteria with regard to uptake rates of isotopically labeled nutrients administered (Fig. 10). These results indicate that greater nutrient availability may exist for larger bacteria and that generally higher concentrations of nutrients are likely to be available in habitats supporting large as opposed to small bacteria. These results might also be interpreted to mean that large bacteria simply grow slower than small bacteria and that size may be attributable to factors not related to nutrients.

Evidence compiled by a variety of workers indicates that in most instances attached bacteria demonstrate enhanced metabolic activities in comparison to free-floating bacteria (ZoBell, 1943; Estermann and McLaren, 1959; Marshall, 1976; Kholdebarin and Oertli, 1977; Steinberg, 1978; Paerl and Merkel, 1982). Results indicate that both growth rates and concentrations of limiting nutrients are generally higher in regions supporting attached, as opposed to free-floating, populations of the same species. It is therefore concluded that attached bacteria are often larger in size than corresponding free-floating bacteria because higher nutrient concentrations present in surface microzones lead to enhanced growth and resultant size differentials.

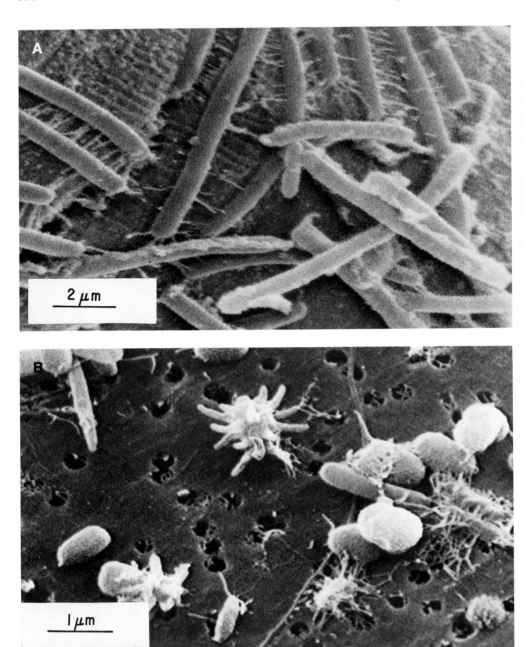

Figure 9. A size comparison of representative attached (to an empty diatom frustule) (A) and free-floating (B) bacteria sampled from oligotrophic Lake Taupo, New Zealand. Observations were made on the same water sample. Note the fibrillar appendages tightly anchoring the attached bacteria.

Enhanced uptake rates of attached versus free-floating microorganisms as observed by Fletcher (1979a) and Paerl and Merkel (1982) may largely be due to elevated concentrations of those nutrients in microzones bordering particles. Uptake rates reflect both community metabolism (and resultant growth) as well as concentrations of substrates available. According to Michaelis–Menten kinetics generally applicable to aquatic microbial populations,

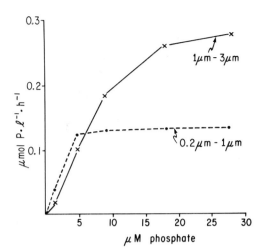

Figure 10. Uptake rates versus substrate concentrations, using $^{32}PO_4^{3-}$ as an isotopic tracer. Uptake characteristics of small ($<$ 1 μm linear dimension) and large (1 to 3 μm) bacteria were compared in freshly sampled Chowan River (N.C.) waters obtained during June 1981. Samples were size-fractionated using sterile Nuclepore membrane filters.

biomass-specific uptake rates are directly related to substrate availability, up to a certain point, where the rate-limiting reaction responsible for either uptake, transport, incorporation, or metabolism of the substrate is operating at maximum velocity. At subsaturating substrate levels, assimilation of a growth-limiting nutrient is related (in a positive manner) to the concentration of that nutrient. Uptake characteristics of aquatic microorganisms are thought to fall into two categories (Chapter 7, Section 3.1; Chapter 12, Section 4.2). The first category comprises microorganisms able to utilize and grow on exceedingly low nutrient concentrations. These microorganisms have been named *oligophiles, oligotrophs,* or more anthropomorphically *sippers.* Typically, sippers reveal high affinities for nutrients at relatively low nutrient concentrations, while uptake saturation also occurs at relatively low nutrient concentrations. Generally, free-floating microorganisms in nutrient-depleted (oligotrophic) waters conform to this type of kinetic relationship between nutrient availability and growth. Such microorganisms are usually slow-growing and conservative, in terms of cell size and in the amounts of excreted (extracellular) organic matter, either as appendages, slimes, or capsular material outside the cell wall. On the other end of the spectrum are the eutrophic microorganisms, copiotrophs, or "gulpers," which are capable of greatly enhanced uptake rates, growth, and extracellular production in high-nutrient environments. This group of microorganisms generally possesses lower affinities and higher saturation levels for growth-limiting nutrients. Because of their abilities to grow luxuriantly at high saturating nutrient concentrations, the production of cellular biomass (and resultant cell sizes) is often greatly enhanced over those of oligotrophic microorganisms. Microbial genera can reveal some plasticity, growing as small cells in oligotrophic environments while proliferating as large cells in nutrient-rich environments. From a generalized kinetic perspective, cell-specific growth versus substrate concentrations in oligotrophic and in eutrophic bacteria are shown in Fig. 11. At low substrate concentrations, oligotrophs have a distinct advantage in supporting high growth rates; however, as nutrient supplies become more plentiful, the eutrophic bacteria can attain higher maximum growth rates by growing at nutrient saturation concentrations in excess of those for oligotrophs. As stated above, oligotrophs generally exhibit small cell sizes, as witnessed among an array of both bacterial and algal cells examined from oligotrophic freshwater and marine habitats, while eutrophic bacteria and algae, present in such nutrient-enriched environments as estuarine waters, eutrophic lakes, and marine upwelling regions, attain cell volumes generally from 2 to 10 times those of oligotrophs.

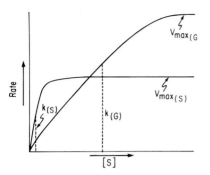

Figure 11. Idealized kinetic plots relating uptake and growth rates to substrate saturation characteristics among small "sipper" and large "gulper" microbial populations. In general, sippers reveal high substrate affinities at low substrate concentrations while gulpers reveal lower affinities at low substrate concentrations. Conversely, gulpers appear to greatly benefit under high concentrations, as saturation of uptake and growth rates are much higher at these concentrations than corresponding rates for sipper populations. Both half-saturation constants, $k_{(S)}$ and $k_{(G)}$ for sippers and gulpers, respectively, and respective V_{max} regimes are indicated.

Within a single system, ranges in cell sizes can commonly be observed between attached and free-floating cells, with attached cells generally being larger than planktonic counterparts. On a seasonal basis, similar variations in size can be observed. Wintertime, low-primary-productivity periods in temperate regions generally yield small bacterial sizes, particularly in ice-covered lakes and deep-water layers devoid of fresh photosynthate input. Spring and summer, high-primary-productivity periods lead to increased cell sizes among a variety of bacterial cells. When small cells of the common heterotrophic species *Pseudomonas aeruginosa* previously grown in low-nutrient planktonic suspensions were transferred to a medium containing nutritive particles (suspended silts or algal colonies), both cell sizes and specific uptake rates per cell significantly increased (Table I). The results were presumably due to enhancement of growth and resultant increases in cellular biomass brought on by the concentrated nutrient supplies represented by such particles. Scanning electron microscope observations of natural microbial communities in ultraoligotrophic lake water (Lake Tahoe, California) enriched with either biotic (algal) or abiotic (eroded mineral sediments) particles showed distinct size differences between planktonic and attached bacteria (Paerl, 1975). Volumetric biomass estimates indicated the average biomass of the dominant rod-shaped bacterial cells to be 2 to 4 times greater in attached as compared with free-floating cells (Table II). Once particles were filtered from these waters, the smaller cells again became dominant.

Combined with the findings that submersed surfaces (1) act as concentrating sites for dissolved nutrients, (2) are often of nutritive value themselves, and (3) act as anchoring substrates in fluvial systems, where attached microorganisms can potentially come into contact with a large volume of dissolved nutrients passing by the attachment surface, it is not surprising to report both enhanced growth and enlarged cell sizes among surface-associated microorganisms. In microautoradiographic examinations of orthophosphate ($^{33}PO_4^{3-}$) uptake partitioning between particle-associated and free-floating bacteria in freshwater and estuarine habitats, Paerl and Merkel found both (1) larger bacterial cells associated with

Table I. *Pseudomonas aeruginosa* Cell Sizes and Cellular Uptake Rates of an ^{3}H-Labeled Amino Acid Mixture Determined by Autoradiography[a]

Growth condition	$\bar{X}$ cell size (μm^3)	$\bar{X}$ cellular uptake rate[b]
Free-floating	0.14 ± 0.05	3.1 ± 0.24
Attached[c]	0.31 ± 0.08	5.2 ± 0.47

[a]*P. aeruginosa* was initially isolated from the Chowan River, North Carolina (Gallucci, 1981).
[b]Silver grains accumulated per cell per hour.
[c]Attached to detrital particles.

Table II. Mean Cell Sizes of Attached and Free-Floating Rod-Shaped Bacteria, Collected in a Region of Lake Tahoe, California, Receiving the Sediment-Laden Waters of the Upper Truckee River[a]

Date[a]	$\bar{X}$ cell size of attached bacteria (μm^3)	$\bar{X}$ cell size of free-floating bacteria (μm^3)
28 May, 1971	0.28 ± 0.05	0.09 ± 0.02
3 June, 1971	0.35 ± 0.07	0.13 ± 0.04
12 June, 1971	0.37 ± 0.06	0.15 ± 0.04

[a]Samples were collected during the late spring runoff season.

particles as opposed to those in planktonic habitats, and (2) consistently higher $^{33}PO_4^{3-}$ uptake rates per cell with attached as compared with free-floating cells. The enhancement of cellular $^{33}PO_4^{3-}$ uptake in attached cells (versus free-floating cells) was most marked in oligotrophic waters. Similar results have been obtained with autoradiographic ^3H glucose and amino acid mixture uptake, on a per cell basis, by attached and free-floating bacteria in the Newport River Estuary, the Neuse Estuary, and the lower Chowan River, all located in eastern North Carolina (H. W. Paerl, unpublished results). Among bacteria associated with bloom-forming cyanobacteria in the Chowan River, the greatest enhancement of amino acid uptake was noted in cells situated in mucilaginous sheaths. In this instance, attached cells exhibited at least twice the ^3H-labeled amino acid assimilation rates of free-floating forms (Table III). A dominant member of the river's microbial community, *P. aeruginosa*, demonstrated both enhancement of amino acid uptake when associated with the cyanobacteria *Anabaena oscillarioides* and *Aphanizomenon flos-aquae* and consistent positive chemotaxis toward a wide array of individual amino acids (Gallucci, 1981). In this system it could be shown that organic substrates which yielded elevated growth rates also served to attract bacteria to potential sources or concentration sites of these substrates.

As with any generalized scheme depicting growth relationships with nutrient availability, there are exceptions. First, nutrients essential for the growth of particular bacterial species may not be readily concentrated on surfaces. As a result, no particular advantage may be provided with respect to the attachment of that microorganism to a surface. However, during the establishment of a microzone community, it is conceivable that other microbial members of the epilithic, epibenthic, or epiphytic community may be able to provide these growth-limiting nutrients or growth factors. In this case, the initially growth-limited microorganisms may later benefit from the establishment of a microzone community, rather than from the initial concentration of a growth-limiting nutrient at the liquid–solid interface.

Table III. Cellular ^3H-Labeled Amino Acid (Added at 0.02 μM) Uptake in Free-Floating and Attached Bacteria during a Bloom of the N_2-Fixing Cyanobacteria *Anabaena* and *Aphanizomenon* in the Chowan River, North Carolina[a]

Growth condition	$\bar{X}$ cellular uptake rates
Free-floating	4.7 ± 0.25
Attached	10.5 ± 0.78

[a]Bacteria were exclusively attached to cyanobacterial hosts. Both cyanobacteria accounted for 92% of the river's phytoplankton biomass. Uptake rates are given as mean numbers of silver grains accumulated per cell per hour, as determined from autoradiographic examinations.

Second, although it could be demonstrated that microorganisms display positive chemotaxis toward known growth-limited nutrients (Section 2.1), it can also be shown among other microbial populations that positive chemotaxis exists toward compounds which are not readily metabolized by the same population (Adler, 1969; Koshland, 1980). Hence, in some cases chemotaxis can be linked to meeting metabolic requirements, while in other instances no clear ecological relationship or motive for chemotaxis can be hypothesized.

Obviously, a great deal is still to be learned concerning the roles that surfaces and associated microzones play in regulating microbial growth, ultimately leading to complex communities which actively assimilate and exchange nutrients, metabolites, and growth factors at rates greatly exceeding those detectable in planktonic free-living environments. Collectively, evidence points to the fact that surfaces can, at times, greatly enhance metabolic rates, resultant growth, and overall microbial production.

8. ECOLOGICAL AND TROPHIC IMPLICATIONS OF MICROZONE DEVELOPMENT AND PROLIFERATION

The establishment of microzones at the liquid–solid interface expands the number of habitats available to microorganisms. In effect, both microbial biomass and nutrient generation and cycling processes associated with specific microflora are enhanced by microzone formation. Nitrogen-fixation, ammonification, and soluble phosphorus, iron, and trace metal release are only a few of the ways in which nutrient availability can be enhanced in the photosynthetically productive zones of benthic and planktonic communities. The nutrient cycling processes embodied in microzones therefore exert a measurable degree of control over primary productivity, and ultimately trophic states, in freshwater and marine habitats whose plant growth is regulated by nutrient availability. Microzones such as benthic algal mats, periphytized stream bottoms, microbial marsh and epimacrophytic communities can themselves be regulated by nutrient cycling processes located within such microzones (Whitton and Potts, 1982; Wetzel, 1983; Jorgensen and Revsbech, 1983). Hence, strong positive relationships normally exist between the development and the proliferation of microzones and trophic states of systems hosting such microzones.

Examples of concomitant expansion and increase in complexity of microzones and acceleration of primary productivity, leading to alteration of trophic states, can be found in the eutrophication of formerly pristine freshwater and marine habitats. As a representative ultraoligotrophic freshwater habitat, Lake Tahoe, California–Nevada, has undergone recent dramatic trophic changes, particularly in near-shore benthic and planktonic regions (Goldman and De Amezaga, 1974, 1975). Much of the blame for this rapid increase lies with man-made causes: (1) accelerated soluble nutrient inputs, due to discharge of tertiary treated sewage and septic tank leaching; (2) land disturbance, bringing with it erosion products, which can act as both nutrient sources and attachment sites for microorganisms, and (3) physical alteration, including stagnation, of specific inflow regimes, due to construction of marinas, harbors, and other man-made impoundments. Controlled by very low yearly inputs and resultant concentrations of growth-limiting nutrients (particularly nitrogen, and to a lesser extent phosphorus and iron), Lake Tahoe had been, since recorded time, an ultraoligotrophic lake, enjoying remarkable clarity, full oxygenation, and the absence of nuisance plant and animal species. With rapid post-World War II development of the Tahoe Basin has come an alarming acceleration of both soluble and particulate nutrient loading, to the extent that algal primary productivity has increased by 200 to 300% in some regions of the lake (Goldman and De Amezaga, 1975). Increased algal production has led to increased bacterial and fungal production, due to a greater increase in organic constituents and products attributable to primary production (Paerl, 1973b). Increased sedimentation (erosion) has led to the dispersion of particulate matter in both the water column and benthic regions (Paerl and Goldman, 1972). The combination of organic and inorganic nutrient loading with increased sedimentation has greatly diversified and increased potential microbial growth at the liquid–solid interface (Paerl and Goldman, 1972). The resultant alarming rate of eutrophication, although initially fueled by nutrient enrichment, has become a permanent feature of some near-shore regions, partially but largely because expanding epiphytic and epilithic microbial communities have made the littoral regions of this lake a more efficient nutrient cycling locale. Whereas barren rocks were a common feature of the lake's shoreline 20 years ago, a heavily epiphytized lake bottom is now commonplace. Analyses of epiphytic and epilithic communities reveal a complex community of aerobic, microaerophilic, and even anaerobic microorganisms, and protozoan and invertebrate grazers (Loeb and Reuter, 1981). Increased primary productivity in overlying waters has led to increased sedimentation. Both detrital sediments, constituting decaying phytoplankton biomass, and receiving benthic sediments have, in turn, become enriched with organic matter resulting from increased primary productivity (Goldman and De Amezaga, 1974). Increased deposition of organic (and inorganic) matter has provided more luxuriant growth conditions for microorganisms associated with these nutrient sources. It follows that microbial metabolism in liquid–solid microzones has dramatically increased. As a result, benthic oxygen depletion, soluble nutrient release, and highly reduced biochemical processes such as N_2-fixation are now more commonplace, particularly in littoral benthic regions (Reuter *et al.*, 1985). The expansion of Lake Tahoe's trophic status is now, in part, sustained by internal nutrient cycling processes, which previously were not able to become established due to nutrient and substrate limitation (Loeb and Reuter, 1981).

In large lakes, as well as oceanic environments, we must face the grim reality that undesirable alterations of trophic status due to enhanced nutrient loading will not be easily reversed. Part of this skeptical outlook is based on the fact that water retention or flushing times are often on the order of hundreds to thousands of years in such systems (Hutchinson, 1957). Hence, effects of indiscriminant and excessive nutrient loading will need to be reckoned with well beyond our lifetimes. Accelerated internal cycling of nutrients, occurring to a large extent in microenvironments on epilithic, periphytic, detrital, and intact biotic surfaces, is also a factor to be considered in stemming undesirable eutrophication trends (Wetzel, 1983). Formerly pristine aquatic systems, which previously hosted highly restricted (in terms of available environmental conditions) and sparsely populated attached microbial communities, are faced with more complex (in terms of biochemical processes) capabilities and a greater number of alternative ways of generating nutrients required for growth. For example, nitrogen inputs which previously may have originated from atmospheric and terrigenous sources can now be provided through N_2-fixation within a microbial surface habitat provided with carbon, phosphorus, and other essential nutrients by algal and invertebrate cohabitants. This form of "internal" nutrient cycling is thus capable of sustaining microbial productivity for extensive periods of time (in some cases on the order of decades), because nutrients generated and cycled are not likely to be flushed out in the immediate future. Even increased flushing through hydrological means may be of limited value if metabolic processes, such as N_2-fixation, algal primary productivity, and animal grazing, followed by nutrient release and recycling into the plant community, are tightly coupled, as they appear to be in lotic and lentic interface habitats. Microzones, once established as a result of enhanced fertility, appear to maintain a certain degree of homeostasis (largely through close coupling of such processes), which is capable of resisting, to a degree, subsequent changes in nutrient and hydrological status of surrounding waters. This means of organization among surface-dwelling microorganisms has led to persistence of surface growth in marine and freshwater systems which are virtually devoid of planktonic growth. Alpine streams, featuring heavily colonized rocks but running waters devoid of flora and fauna, are one example (Geesey *et al.*, 1978; Haack and McFeters, 1982). Coral reef communities in ultraoligotrophic tropical seas are another (Paerl *et al.*, 1981), as are hulls of vessels and submersed structures residing in low-productivity waters (Sieburth, 1975). There exists little doubt that the liquid–solid interfaces discussed in this chapter are of considerable trophic importance, from both nutrient cycling and food production standpoints (Marshall, 1976).

The importance of microzones as food production sites, having integral roles in aquatic food webs, has received little attention. Only recently have microbial ecologists considered surface-associated microbial growth from a production perspective. This is despite the fact that earlier investigations by ZoBell (1943) revealed that glass surfaces hosting bacteria were capable of adsorbing 2 to 27% of the DOM in seawater and the finding of Heukelekian and Heller (1940) that bacteria grew best in low-nutrient solutions if the surface area of the incubation flask was increased by the addition of surfaces in the form of glass beads. They concluded that bacteria can reproduce in nutrient regimes which are otherwise too dilute for growth, by associations with surfaces to which vital nutrients were adsorbed.

A proportion, usually no greater than 50% of the nutrients assimilated, are directly respired (in the case of carbon) (Hobbie and Crawford, 1969) or excreted (in the case of inorganic nutrients) by aquatic bacteria (Paerl, 1978), whether they be surface-associated or free-floating. What happens to the remaining 50% is a question addressed by recent research efforts aimed at understanding the productive capacity of heterotrophic aquatic bacteria (Cole *et al.*, 1982; Robinson *et al.*, 1982). Furthermore, although it is known that photosynthetically active primary producers, as well as decaying and lysing flora and fauna, are sources of DOM in aquatic environments, concentrations of DOM seldom appear enhanced and deviate little from a mean value for individual systems (Wetzel, 1983). These findings would indicate that constant and effective removal of DOM mediated by heterotrophic organisms must be a universal feature of aquatic systems. It would also mean that a large portion of organic matter initially formed by photosynthetic organisms may be more readily available as a grazeable food source to heterotrophs than previously assumed (Calow, 1977; Paerl, 1978; Robinson *et al.*, 1982). Although bacterial biomass, as determined by cell counts, may only constitute 2 to 15% of planktonic and benthic biomass in natural environments, bacterial productivity, being the rate of change of biomass over time, may substantially exceed static biomass (standing crop) figures. Combined, these results picture the heterotrophic microbial community as one experiencing not only rapid growth, but also rapid biomass turnover. In other words, the potential contributions of heterotrophic communities to food chains may ecologically overshadow the enumerable standing stocks involved in production and nutrient cycling. In work using natural bacteria maintained in dialysis (diffusive) bag cultures, it could be shown that bacteria readily assimilated from 4 to 10 times their observable biomass from DOM sources (Paerl, 1975). Assuming a high mineralization rate of 50%, how can we account for assimilation of DOM which was not detected as biomass? The vast excess of assimilated carbon apparently went into the production of extracellular capsular materials, appendages, fibrils, and slime which, in the case of attached bacteria, served to anchor the organisms (Fig. 1). In addition, much of this excreted material became structural components of attached surfaces, whether they be detritus, mineral, or biotic in origin (Paerl, 1978). At least 2 to 5 times the attached bacterial production went into structural modification of the microzones which respective organisms inhabited. Similar findings have become available in studies of bacteria inhabiting seaweeds, freshwater macrophytes, as well as streamborne epilithic and epiphytic bacteria (Laycock, 1974; Robinson *et al.*, 1982). It can therefore be concluded that direct counts converted to volumetric biomass seriously underestimate true bacterial production in natural waters.

In surface habitats where successive populations of bacteria often coat each other in luxurious mucilaginous slime and fibrillar layers, a vast proportion of bacterial "production" resides in such excreta, in a manner somewhat similar to extracellular plaque formation due to bacterial growth on teeth (Chapter 10, Section 4.1). In such a manner, nutrition values of particles originally devoid of organic matter may be greatly altered and enhanced by microbial conditioning. Organic enhancement of mineral particles, silts, nonreactive organic matter, lignins, and tannins by bacterial conditioning is likely to render such particles as more palatable food sources for grazing invertebrates of both trophic (zooplankton) and commercial (e.g., shrimp) importance (Finenko and Zaika, 1970). The adsorption and ultimate incorporation of dissolved organic carbon, as well as nitrogenous, phosphorous, and metallic nutrients required for attached microbial growth, represents a means by which such soluble nutrients are incorporated into grazing components of food chains. Surface-associated bacteria "fix" such nutrients onto particles and surfaces. Grazing or filter-feeding consumers account for transfer of these fixed structural and nutritional constituents through the aquatic food chain.

Surface-associated microorganisms therefore serve dual ecological functions: (1) they are instrumental in assimilating and mineralizing components of both the attachment substratum as well as surrounding dissolved substances, thereby enhancing nutrient cycling in aquatic ecosystems, and (2) microbially assimilated organic and inorganic particulate and dissolved substances not respired or solubilized represent particulate matter fixed into cells and onto surfaces and particles, to be potentially available for biota residing in or near microzones. In this regard, microzones can be regarded as nutrient scavenging sites, instrumental in recovery of dissolved nutrients and incorporating them into particulate matter which can then be available to animal consumers. Soluble nutrient scavenging is likely to be instrumental in the function of consumable particulate matter in diverse microzone habitats, including suspended particulate matter, detritus, microbial mats, and fouling layers on submersed biotic and abiotic surfaces. Fouling, being the conditioning of submersed surfaces by microbial layers, normally precedes establishment of a more diverse community, including microbial grazers and predators, filter-feeders, and predators on invertebrate grazers and herbivores.

9. FUTURE OUTLOOKS AND RESEARCH NEEDS

With our current knowledge of the biochemical and ecological significance of microzone formation at the liquid–solid interface, both basic research on, and applied use of, microzone development and proliferation are likely to expand in the near future. Our understanding of nutrient processing through both adsorptive and assimilatory capacities of aquatic microzones (biofilms) is of central importance to the treatment of wastewater, i.e., water purification using surface-active ion-exchange and activated carbon columns, as well as desalination treatments (Trulear and Characklis, 1979). In this regard, the formulation of nutrient and ion-exchange surfaces compatible with and making use of biofilms will be a challenge to both microbiologists and engineers. In desalinization alone, microbial growth may need to be minimized because of the need for particle-free feed water to come in contact with reverse osmosis membranes, commonly employed in this technology. By examining the effects of grain size (exposed surface area) and flow rates on colonization, as well as the relationship between adsorption and colonization, knowledge can be gathered which may help expedite the control of microbial growth on activated carbon and polymeric ion-exchange columns. A working knowledge of factors regulating and ultimately controlling microbial colonization on such surfaces will be essential.

As it has been shown that enhancement and regulation of specific biochemical processes (N_2-fixation, denitrification, sulfate reduction, methane formation, phosphorus and metal solubilization, aerobic and anaerobic carbon decomposition) is dependent on types and magnitudes of microzone formation, controlling the means of attraction of microorganisms to liquid–solid interfaces, the rates of microzone formation, and the factors regulating qualitative and quantitative degrees of microzone formation will be of crucial importance in controlling nutrient cycling in aquatic ecosystems. Potentially, man can manipulate microzones in a beneficial manner, either by minimizing their development in systems faced with water-quality degradation, or by optimizing their development in systems where specific biochemical transformations are desirable. As examples, stimulation of microzones enhancing denitrification is desirable in waste treatment facilities, while stimulation of microzone N_2-fixation is useful in agricultural projects dependent on nitrogenous fertilizer inputs (e.g., rice paddies). Hence, there exists a need for agricultural, sanitary engineering, and water-quality experts to work in cooperation with microbial ecologists in exploring and resolving a diverse set of environmental problems focused on microbial nutrient cycling dynamics at the liquid–solid interface. We have only begun to recognize the value of interdisciplinary approaches to solving water-quality management-oriented problems.

A related research area with a vast potential is the use of microzones as biological purification "filters," i.e., employing surface-dwelling microorganisms as immobilizers and potential degraders of dissolved and particulate toxic materials, including pesticides, petroleum derivatives, industrial wastes, and heavy metals.

As microzone development has been shown to be instrumental in improving both the quality and the quantity of grazeable food items, pond and impoundment aquaculture systems relying on optimal grazeable food production should rely heavily on both basic research and technological breakthroughs in the development of nutritionally attractive particulate materials and surfaces supporting this form of food production. This facet of research requirements is especially evident in tropical countries where needs for aquaculture based on the production of grazer organisms (shrimp and fish cultures) are steadily increasing as sources of food supplementation as well as economic income. Coordinated efforts aimed at biomanipulation and structural designs of such ponds and impounds will call for the expertise of both basic research and process-oriented microbiologists in resolving a wide variety of production-related questions and management options.

Clearly, technology useful for identifying physical and chemical processes and their rates in microzones is urgently needed among the variety of applications listed above. The development of biochemical indicators, microprobe systems, and autoradiographic and immunofluorescent assays instrumental in identifying relevant microorganisms, processes, and production potentials in microzone habitats is a rewarding area of activity for chemical, microbial, and aquatic ecologists.

Methodological benefits can be obtained by specifically examining chemotactic behavior of microorganisms at the liquid–solid interface. Positive chemotaxis occurs in response to increases in attractant concentration (Koshland, 1980). If the attractants are nutrients, which appear to be the case in the Chowan River (Gallucci, 1981), then specific microorganisms revealing a positive chemotactic response to these nutrients can be used as "assay" organisms to detect microenvironments rich (having elevated levels) in such nutrients. Similarly, localized nutrient depletion in such microenvironments can also be detected with this technique.

Formulation of effective antifouling agents depends on a better understanding of factors regulating the establishment of submersed microzones, as well as interdisciplinary information on the physicochemical alteration of diffusion and leaching properties through biofilms colonizing microzones on surfaces treated with antifouling substances. The effective chelation of metals by microbial films is likely to be another factor influencing the effectiveness of antifouling substances. These are just a few of the microbe mediated processes requiring future examination by microbial ecologists in conjunction with chemical and structural engineers in our quests for more effective biorepellents and antifouling substances. A more thorough understanding of microbial degradation of submersed wood, metal, and polymeric substances will create additional coordinated interdisciplinary research efforts.

In summary, we have just begun to initiate research avenues in improving knowledge of the dynamics of nutrient cycling, regulation of microbial production, trophic implications, and community organization in microzones at the liquid–solid interface. The promise of much work lies ahead for microbiologists, chemists, and ecologists able to apply their basic knowledge and expertise to aquatic ecosystems.

ACKNOWLEDGMENTS. The joint efforts of field and laboratory co-workers were instrumental in obtaining relevant data and results discussed. Particular thanks go to P. Bland, P. Kellar, K. Gallucci, H. and V. Page for their time and expertise. I appreciate the typing and proofreading efforts of J. Garner and B. Bright. Research activities discussed were supported in part by the National Science Foundation (BSR 8314702), North Carolina Sea Grant and the Water Resources Research Institute of the University of North Carolina (Grant B 127 NCB and A-122NC).

REFERENCES

Adler, J., 1969, Chemoreceptors in bacteria, *Science* **166:**1588–1597.

Alexander, M., 1971, *Microbial Ecology,* Wiley, New York.

Altman, F. P., 1972, *An Introduction to the Use of Tetrazolium Salts in Quantitative Enzyme Cytochemistry,* Koch–Light Laboratories, Colnbrook, England.

Arends, J., and Jongebloed, W. L., 1977, The enamel substrate characteristics of the enamel surface, *Swed. Dent. J.* **1:**215–221.

Austin, B., Allen, D. A., Zachary, A., Belas, M. R., and Colwell, R. R., 1978, Ecology and taxonomy of bacteria attaching to wood surfaces in a tropical harbor. *Can. J. Microbiol.* **25:**447–461.

Bautista, M. F., 1983, Photosynthesis and N_2 fixation in microbial intertidal mat (Shackleford Island), North Carolina, M.Sc. thesis, University of North Carolina, Chapel Hill.

Bell, W., and Mitchell, R., 1972, Chemotactic and growth responses of marine bacteria to algal extracellular products, *Biol. Bull.* **143:**265–277.

Bitton, G., and Marshall, K. C. (eds.), 1980, *Adsorption of Microorganisms to Surfaces,* Wiley, New York.

Brown, C. M., Ellwood, D. C., and Hunter, J. R., 1977, Growth of bacteria on surfaces: Influence of nutrient limitation, *FEMS Microbiol. Lett.* **1:**163–166.

Burris, J. E., 1977, Photosynthesis, photorespiration, and dark respiration in eight species of algae, *Mar. Biol.* **39:**371–379.

Caldwell, D. E., and Caldwell, S. J., 1978, A *Zoogloea* sp. associated with blooms of *Anabaena flos-aquae, Can. J. Microbiol.* **24:**922–931.

Calow, P., 1977, Conversion efficiencies in heterotrophic organisms, *Biol. Rev. Cambridge Philos. Soc.* **52:**385–409.

Cappenberg, T. E., and Jongejan, E., 1977, Microenvironments for sulfate reduction and methane production in freshwater sediments, in: *Environmental Biogeochemistry and Geomicrobiology*, Volume 1 (W. E. Krumbein, ed.), Ann Arbor Science, Michigan, pp. 129–138.

Carlucci, A. F., and Strickland, J. D. H., 1968, The isolation, purification and some kinetic studies of marine nitrifying bacteria, *J. Exp. Mar. Biol. Ecol.* **2:**156–166.

Carpenter, E. J., and Price, C. C., IV, 1976, Marine *Oscillatoria (Trichodesmium):* Explanation for aerobic nitrogen fixation without heterocysts, *Science* **191:**1278–1279.

Chet, I., and Mitchell, R., 1976, Ecological aspects of microbial chemotactic behavior, *Annu. Rev. Microbiol.* **30:**221–239.

Chet, I., Asketh, P., and Mitchell, R., 1975, Repulsion of bacteria from marine surfaces, *Appl. Microbiol.* **30:**1043–1045.

Clark, A., 1974, *The Chemisorptive Bond,* Academic Press, New York.

Cole, J. J., Likens, G. E., and Strayer, D., 1982, Photosynthetically produced dissolved organic carbon: An important carbon source for planktonic bacteria, *Limnol. Oceanogr.* **27:**1080–1090.

Corpe, W. A., 1980, Microbial surface components involved in adsorption of microorganisms onto surfaces, *Adsorption of Microorganisms to Surfaces* (G. Bitton and K. C. Marshall, eds.), Wiley, New York, pp. 105–144.

Costerton, J. W., 1980, Some techniques involved in study of adsorption of microorganisms to surfaces, in: *Adsorption of Microorganisms to Surfaces* (G. Bitton and K. C. Marshall, eds.), Wiley, New York. pp. 403–424.

Daft, M. J., and Stewart, W. D. P., 1973, Light and electron microscope observations on algal lysis by bacterium CP-1, *New Phytol.* **72:**799–808.

Dawson, M. P., Humphrey, B., and Marshall, K. C., 1981, Adhesion: A Tactic in the survival strategy of a marine vibrio during starvation, *Curr. Microbiol.* **6:**195–198.

Dazzo, F., 1980, Adsorption of microorganisms to roots and other plant surfaces, in: *Adsorption of Microorganisms to Surfaces* (G. Bitton and K. C. Marshall, eds.), Wiley, New York, pp. 253–316.

Dempsey, M. J., 1981, Colonisation of antifouling paints by marine bacteria, *Bot. Mar.* **24:**185–191.

Derjaguin, B. V., and Landau, L., 1941, Theory of stability of strongly charged lyophobic soils and the adhesion of strongly charged particles in solutions of electrolytes, *Acta Physiochim. URSS* **14:**633–651.

Eighmy, T. T., Maratea, D., and Bishop, P. L., 1983, Electron microscopic examination of wastewater biofilm formation and structural components, *Appl. Environ. Microbiol.* **45:**1921–1931.

Ellwood, D. C., Melling, J., and Rutter, P. (eds.), 1979, *Adhesion of Microorganisms to Surfaces,* Academic Press, New York.

Eloff, J. N., Steinitz, Y., and Shilo, M., 1976, Photooxidation of cyanobacteria in natural conditions, *Appl. Environ. Microbiol.* **31:**119–126.

Engelmann, T. W., 1881, Neue Methode zur Untersuchung der Sauerstoffauscheidung Pflanzicher und Thierischer Organismen, *Pfluegers Arch. Gesamte Physiol. Menschen Tiere* **25:**285–292.

Estermann, E. F., and McLaren, A. D., 1959, Stimulation of bacterial proteolysis by adsorbents, *J. Soil Sci.* **10:**64–69.

Finenko, Z. Z., and Zaika, V. E., 1970, Particulate organic matter and its role in the productivity of the sea, in: *Marine Food Chains* (J. H. Steele, ed.), University of California Press, Berkeley, pp. 32–44.

Fletcher, M., 1979a, A microautoradiographic study of the activity of attached and free-living bacteria, *Arch. Microbiol.* **122:**271–274.

Fletcher, M., 1979b, The attachment of bacteria to surfaces in aquatic environments, in: *Adhesion of Microorganisms to Surfaces* (D. C.Ellwood, J. Melling, and P. Rutter, eds.), Academic Press, New York, pp. 87–108.

Fletcher, M., and Floodgate, G. D., 1973, An electron microcopic demonstration of an acidic polysaccharide involved in the adhesion of a marine bacterium to solid surfaces, *J. Gen. Microbiol.* **74:**325–334.

Fletcher, M., and Loeb, G., 1979, The influence of substratum characteristics on the attachment of a marine pseudomonad to solid surfaces, *Appl. Environ. Microbiol.* **37:**67–72.

Fogel, S., Chet, I., and Mitchell, R., 1971, The ecological significance of bacterial chemotaxis, *Bacteriol. Proc.* **28:**G31 (abstract).

Fogg, G. E., 1966, The extracellular products of algae, *Oceanogr. Mar. Biol. Annu. Rev.* **4:**195–212.

Fogg, G. E., Stewart, W. D. P., Fay, P., and Walsby, A. E., 1973, *The Blue-Green Algae, Academic Press,* New York.

Gallon, J. R., Kurz, W. G. W., and LaRue, T. A., 1975, The physiology of nitrogen fixation by a *Gloeocapsa* sp., in: *Nitrogen Fixation by Free-Living Microorganisms* (W. D. P. Stewart, ed.), Cambridge University Press, London, pp. 159–173.

Gallucci, K. K., 1981, Algal–bacterial symbiosis in the Chowan River, N. C., M.Sc. thesis, University of North Carolina, Chapel Hill.

Geesey, G. G., Mutch, R., Costerton, J. W., and Green, R. B., 1978, Sessile bacteria: An important component of the microbial population in a small mountain stream, *Limnol. Oceanogr.* **23:**1214–1223.

Goldman, C. R., and De Amezaga, E., 1974, Primary productivity of the littoral zone of Lake Tahoe, California–Nevada, *Symp. Biol. Hung.* **15:**49–62.

Goldman, C. R., and De Amezaga, E., 1975, Spatial and temporal changes in the primary productivity of Lake Tahoe, California–Nevada between 1959 and 1971, *Verh.-Int. Ver. Theor. Limnol. Angew.* **19:**812–819.

Golterman, H. L., 1964, Mineralization of algae under sterile conditions or by bacterial breakdown, *Verh.-Int. Ver. Theor. Angew. Limnol.* **15:**544–548.

Haack, T. K., and McFeters, G. A., 1982, Nutritional relationships among microorganisms in an epilithic biofilm community, *Microb. Ecol.* **8:**115–126.

Harden, F. P., and Harris, J. O., 1953, The isoelectric point of bacterial cells, *J. Bacteriol.* **65:**198–202.

Harrison, P. G., and Harrison, B. J., 1980, Interactions of bacteria, microalgae and copepods in a detritus microcosm through a flask darkly, in: *Marine Benthic Dynamics* (K. R. Tenore and B. C. Coull, eds.), University of South Carolina Press, Columbia, pp. 373–386.

Harvey, R. W., and Young, L. Y., 1980, Enumeration of particle-bound and unattached respiring bacteria in the salt marsh environment, *Appl. Environ. Microbiol.* **40:**156–160.

Haydon, D. A., 1964, The electric double layer and electrokinetic phenomena, *Recent Prog. Surf. Sci.* **1:**94–158.

Hendricks, C. W., 1974, Sorption of heterotrophic and enteric bacteria to glass surfaces in the continuous culture of river water, *Appl. Microbiol.* **28:**572–578.

Herbst, V., and Overbeck, J., 1978, Metabolic coupling between the alga *Oscillatoria redekei* and accompanying bacteria, *Naturwissenschaften* **65:**598–599.

Heukelekian, H., and Heller, A., 1940, Relation between food concentration and surface for bacterial growth, *J. Bacteriol.* **40:**547–558.

Higgins, I. V., and Burns, R. G., 1975, *The Chemistry and Microbiology of Pollution*, Academic Press, New York.

Hirsch, P., 1978, Microbial mats in a hypersaline solar lake: Types, composition, in: *Abstr. Third Int. Symp. Environ. Biogeochem.* (W. E. Krumbein, ed.) Occasional Publication No. 1, University of Oldenburg, West Germany, pp. 56–57.

Hobbie, J. E., and Crawford, C. C., 1969, Bacterial uptake of organic substrate: New Methods of study and application to eutrophication, *Verh.-Int. Ver. Theor. Angew. Limnol.* **17:**725–730.

Hull, M., and Kitchener, J. A., 1969, Interactions of spherical colloidal particles with planar surfaces, *Faraday Soc. Trans.* **65:**304–310.

Hutchinson, G. E., 1957, *A Treatise on Limnology*, Volume 2, Wiley, New York.

Hutchinson, G. E., 1961, The paradox of the plankton, *Am. Nat.* **115:**137–143.

Hutzinger, O., Van Lelyveld, I. H., and Zoeteman, B. C. J., 1977, *Aquatic Pollutants: Transformation and Biological Effects*, Pergammon Press, Elmsford, N.Y.

Hynes, H. B. N., 1970, *The Ecology of Running Waters*, University of Toronto Press, Toronto.

Indrebo, B., Pengerud, B., and Dundas, I., 1977, Microbial activities in a permanently stratified estuary. II. Microbial activities at the oxic–anoxic interface, *Mar. Biol.* **51:**305–309.

Jannasch, H. W., and Pritchard, P. H., 1972, The role of inert particulate matter in the activity of aquatic microorganisms, *Mem. Ist. Ital. Idrobiol. Suppl.* **29:**289–308.

Jorgensen, B. B., 1977, Bacterial sulfate reduction within reduced microniches of oxidised marine sediments, *Mar. Biol.* **41:**7–17.

Jorgensen, B. B., and Revsbech, N. P., 1983, Colorless sulfur bacteria, *Beggiatoa* spp. and *Thiovulum* spp. in O_2 and H_2S microgradients, *Appl. Environ. Microbiol.* **45:**1261–1270.

Jorgensen, B. B., Revsbech, N. P., Blackburn, T. H., and Cohen, Y., 1979, Diurnal cycle of oxygen and sulfide microgradients and microbial photosynthesis in cyanobacterial mat sediments, *Appl. Environ. Microbiol.* **38:**46–58.

Kaneko, T., and Colwell, R. R., 1975, Adsorption of *Vibrio parahemolyticus* onto chitin and copepods, *Appl. Microbiol.* **29:**269–274.

Kefford, B., Kjelleberg, S., and Marshall, K. C., 1982, Bacterial scavenging: Utilization of fatty acids located at a solid–liquid interface, *Arch. Microbiol.* **133:**257–260.

Kholdebarin, B., and Oertli, J. J., 1977, Effect of suspended particles and their sizes on nitrification in surface water, *J. Water Pollut. Control Fed.* **49:**1693–1697.

Kirchman, D., and Mitchell, R., 1982, Contribution of particle bound bacteria to total microheterotrophic activity in five ponds and two marshes, *Appl. Environ. Microbiol.* **43:**200–209.

Kitchener, J. A., 1973, Surface forces in the deposition of small particles, *J. Soc. Cosmet. Chem.* **24:**709–713.

Kjelleberg, S., Humphrey, B. A., and Marshall, K. C., 1982, Effect of interfaces on small, starved marine bacteria, *Appl. Environ. Microbiol.* **43**:1166–1172.

Koshland, D. E., 1980, *Bacterial Chemotaxis as a Model Behavioral System,* Raven Press, New York.

Kuhn, R., and Jerchel, D., 1941, Über Investseifen. VIII. Mitteil. Reduktion von Tetrazoliumsalzen durch Bacterien, garende Hefe und Keimende Samen, *Ber. Dtsch. Chem. Ges.* **74**:949–952.

Kunc, F., and Stotzky, G., 1980, Acceleration by montmorillonite of nitrification in soil, *Folia Microbiol. (Prague)* **25**:106–125.

Lange, W., 1973, Bacteria assimilable organic compounds, phosphate and enhanced growth of bacteria associated blue-green algae, *J. Phycol.* **9**:507–511.

Larsson, V., and Hagstrom, A., 1979, Phytoplankton extracellular release as an energy source for bacterial growth in a pelagic ecosystem, *Mar. Biol.* **52**: 199–206.

Laycock, A., 1974, The detrital food chain based on seaweed. I. Bacteria associated with the surface of *Laminaria* fronds, *Mar. Biol.* **25**:223–231.

Lehman, J. T., and Scavia, D., 1982, Microscale patchiness of nutrients in plankton communities, *Science* **216**:729–730.

Likens, G. E. (ed.), 1972, *Nutrients and Eutrophication: The Limiting Nutrient Controversy,* Am. Soc. Limnol. Oceanogr. Spec. Symp. Volume 1.

Loeb, G. I., and Neihof, R. A., 1975, Marine conditioning films, *Adv. Chem. Serv.* **145**:319–335.

Loeb, S. L., and Reuter, J. E., 1981, The epilithic periphyton community: A five lake comparative study of community productivity, nitrogen metabolism and depth-distribution of standing crops, *Verh.-Int. Ver. Theor. Limnol. Angew.* **21**:346–352.

Lovley, D. R., Dwyer, D. F., and Klug, M. J., 1982, Kinetic analysis of competition between sulfate reducers and methanogens for hydrogen in sediments, *Appl. Environ. Microbiol.* **43**:1373–1379.

Lupton, F. S., and Marshall, K. C., 1981, Specific adhesion of bacteria to heterocysts of *Anabaena* spp. and its ecological significance, *Appl. Environ. Microbiol.* **42**:1085–1092.

Mack, W. N., Mack, J. P., and Ackerson, A. O., 1975, Microbial film development in a trickling filter, *Microb. Ecol.* **2**:215–221.

McCoy, W. F., Bryers, J. D., Robbins, J., and Costerton, J. W., 1981, Observations of fouling biofilm formation, *Can. J. Microbiol.* **27**:910.

MacRitchie, F., and Alexander, A. E., 1963, Kinetics of adsorption of proteins at interfaces. III. The role of electrical barriers in adsorption, *J. Colloid Sci.* **18**:464–469.

Madsen, B. L., 1972, Detritus on stones in small streams, *Mem. Ist. Ital. Idrobiol. Suppl.* **29**:385–403.

Mankaeva, K. A., 1966, Studies of lysis in culture of *Chlorella, Microbiology* **35**:724–728.

Marshall, K. C., 1976, *Interfaces in Microbial Ecology,* Harvard University Press, Cambridge, Mass.

Marshall, K. C., 1979, Growth at interfaces, in: *Proceedings of the Dahlem Workshop on Strategy of Life in Extreme Environments* (M. Shilo, ed.), Dahlem Konferenzen–Chemie Verlag, Berlin, pp. 281–290.

Marshall, K. C., 1980, Adsorption of microorganisms to soils and sediments, in: *Adsorption of Microorganisms to Surfaces* (G. Bitton and K. C. Marshall, eds.), Wiley, New York, pp. 317–330.

Marshall, K. C., Stout, R., and Mitchell, R., 1971a, Mechanism of the initial events in the sorption of marine bacteria to surfaces, *J. Gen. Microbiol.* **68**:337–348.

Marshall, K. C., Stout, R., and Mitchell, R., 1971b, Selective sorption of bacteria from seawater, *Can. J. Microbiol.* **17**:1413–1416.

Marszalek, D. S., Gerchako, S. M., and Utey, L. R., 1979, Influence of substrate composition on marine microfouling, *Appl. Environ. Microbiol.* **38**:987–995.

Massalski, A., and Leppard, G. G., 1979, Morphological examination of fibrillar colloids associated with algae and bacteria in lakes, *J. Fish. Res. Board Can.* **36**:922–938.

Matjevic, E., 1977, The role of chemical complexing in the formation and stability of colloidal dispersions, *J. Colloid Interface Sci.* **58**:374–389.

Matson, J. V., and Characklis, W. G., 1976, Diffusion into microbial aggregates, *Water Res.* **10**:877–885.

Meadows, P. S., 1966, Microorganisms attached to marine and freshwater sand grains, *Nature (London)* **212**:1059.

Meadows, P. S., and Anderson, J. G., 1968, Microorganisms attached to marine sand grains, *J. Mar. Biol. Assoc. U.K.* **48**:161.

Melchiorri-Santolini, U., and Hopton, J. W. (eds.), 1972, *Detritus and Its Role in Aquatic Ecosystems, Mem. Ist. Ital. Idrobiol. Suppl.* **29.**

Menge, B. A., 1976, Organization of the New England rocky intertidal community: Role of predation, competition and environmental heterogeneity, *Ecol. Monogr.* **46**:355–381.

Monty, L. V., 1967, Distribution and structure of recent stromatolitic algal mats, eastern Andros Islands, Bahamas, *Ann. Soc. Geol. Belg.* **90**:55–68.

Mortimer, C. H., 1971, Chemical exchanges between sediments and water in the Great Lakes—Speculations on probable regulatory mechanisms, *Limnol. Oceanogr.* **16**:387–404.

Nalewajko, C., Dunstall, T. G., and Shear, H., 1976, Kinetics of extracellular release in axenic algae and in mixed algal bacterial cultures: Significance in estimation of total (gross) phytoplankton excretion rates, *J. Phycol.* **12**:1–5.

Nealson, K. M., and Ford, J., 1980, Surface enhancement of bacterial manganese oxidation: Implications for aquatic environments, *Geomicrobiol. J.* **2**:21–38.

Neihof, R. A., and Loeb, G., 1972, The surface charge of particulate matter in seawater, *Limnol. Oceanogr.* **17**:7–16.

Neumann, A. C., Gebelein, C., and Scoffin, T. P., 1970, The composition, structure and erodibility of subtidal mats, Abaco, Bahamas, *J. Sediment, Petrol.* **40**:274–297.

Nicotri, M. E., 1977, Grazing effects of four marine littoral herbivores on the microflora, *Ecology* **58**:1020–1028.

Nikitin, D. I., 1973, Electron microscopic studies of attached microorganisms, in: *Modern Methods in the Study of Microbial Ecology* (T. Rosswall, ed.), *Bull. Swed. Natl. Sci. Res. Counc.* **17**:85–91.

Paerl, H. W. 1973a, Detritus in Lake Tahoe: Structural Modification by attached microflora, *Science* **180**:496–498.

Paerl, H. W., 1973b, The regulation of heterotrophic activity by environmental factors at Lake Tahoe, California–Nevada, Ph.D. thesis, University of California, Davis.

Paerl, H. W., 1975, Microbial attachment to particles in marine and freshwater ecosystems, *Microb. Ecol.* **2**:73–83.

Paerl, H. W., 1978, Microbial organic carbon recovery in aquatic ecosystems, *Limnol. Oceanogr.* **23**:927–935.

Paerl, H. W., 1979, Study of the importance of nitrogen fixation to the growth of *Myriophyllum spicatum* (Eurasian Millfoil), Report: National Water Research Institute, Burlington, Ontario, Canada.

Paerl, H. W., 1980, Attachment of micro-organisms to living and detrital surfaces in freshwater systems, in: *Adsorption of Microorganisms to Surfaces* (G. Bitton and K. C. Marshall, eds.), Wiley–Interscience, New York, pp. 375–402.

Paerl, H. W., 1983, Partitioning of CO_2 fixation in the colonial cyanobacterium *Microcystis aeruginosa:* Mechanism promoting formation of surface scums, *Appl. Environ. Microbiol.* **46**:252–259.

Paerl, H. W., and Bland, P. T., 1982, Localized tetrazolium reduction in relation to N_2 fixation, CO_2 fixation and H_2 uptake in aquatic filamentous cyanobacteria, *Appl. Environ. Microbiol.* **43**:218–226.

Paerl, H. W., and Goldman, C. R., 1972, Stimulation of heterotrophic and autotrophic activities of a planktonic microbial community by siltation at Lake Tahoe, California, *Mem. Ist. Ital. Idrobiol. Suppl.* **29**:129–147.

Paerl, H. W., and Kellar, P. E., 1978, Significance of bacterial *Anabaena* (Cyanophyceae) associations with respect to N_2 fixation in aquatic habitats, *J. Phycol.* **14**:254–260.

Paerl, H. W., and Merkel, S., 1982, The effects of particles on phosphorus assimilation in attached vs. free floating microorganisms, *Arch. Hydrobiol.* **93**:125–134.

Paerl, H. W., Richards, R. C., Leonard, R. C., and Goldman, C. R., 1975, Seasonal nitrate cycling as evidence for complete vertical mixing in Lake Tahoe, California–Nevada, *Limnol. Oceanogr.* **20**:1–8.

Paerl, H. W., Webb, K. L., Baker, J., and Wiebe, W. J., 1981, Nitrogen fixation in waters, in: *Nitrogen Fixation,* Volume I (W. J. Broughton, ed.), Oxford Science, Oxford, pp. 193–241.

Parsons, T. R., and Takahashi, K., 1973, *Biological Oceanographic Processes,* Pergamon Press, Elmsford, N.Y.

Polimeni, C., 1976, Seasonality and life history of a blue-green algal mat on Shackleford Banks, North Carolina, M.Sc. thesis, University of North Carolina, Chapel Hill.

Potts, M., 1980, Blue-green algae (Cyanophyta) in marine coastal environments of the Sinai Peninsula: Distribution, zonation, stratification and taxonomic diversity, *Phycologia* **19**:60–73.

Potts, M., and Whitton, B. A., 1979, pH and E_h on Aldabra Atoll. II. Intertidal photosynthetic microbial communities showing zonation, *Hydrobiologia* **67**:99–105.

Raymont, J. E. G., 1980, *Plankton and Productivity in the Oceans,* Volume 1, Pergamon Press, Elmsford, N.Y.

Reeburgh, W. S., and Heggie, D. T., 1977, Microbial methane consumption reactions and their effect on methane distributions on freshwater and marine environments, *Limnol. Oceanogr.* **22**:1–9.

Reuter, J. E., Loeb, S. L., and Goldman, C. R., 1985, Nitrogen fixation in oligotrophic Lake Tahoe, in: *Periphyton in Freshwater Ecosystems* (R. G. Wetzel, ed.), in press.

Revsbech, N. P., and Jorgensen, B. B., 1981, Primary production of microalgae in sediments measured by oxygen microprofile, $H^{14}CO_3{}^-$ fixation, and oxygen exchange methods, *Limnol. Oceanogr.* **26**:717–730.

Richerson, P. J., Armstrong, R., and Goldman, C. R., 1970, Contemporaneous disequilibrium, a new hypothesis to explain the "paradox of the plankton," *Proc. Natl. Acad. Sci. USA* **67**:1710–1714.

Robinson, J. D., Mann, K. H., and Novitsky, J. A., 1982, Conversion of the particulate fraction of seaweed detritus to bacterial biomass, *Limnol. Oceanogr.* **27**:1072–1079.

Rogers, H. J., 1979, Adhesion of microorganisms to surfaces: Some general considerations of the role of the envelope, in: *Adhesion of Microorganisms to Surfaces* (D. C. Ellwood, J. Melling, and P. Rutter, eds.), Academic Press, New York, pp. 30–55.

Round, F. E., 1964, The ecology of benthic algae, in: *Algae and Man* (D. F. Jackson, ed.), Plenum Press, New York, pp. 138–184.

Round, F. E., 1972, Patterns of seasonal succession of freshwater epipelic algae, *Br. Phycol. J.* **7**:213–221.

Ryther, J. H., 1963, Geographic variations in productivity, in: *The Sea,* Volume 2 (M. N. Hill, ed.), Wiley–Interscience, New York, pp. 347–380.

Shanks, A. L., and Trent, J. D., 1979, Marine snow: Microscale nutrient patches, *Limnol. Oceanogr.* **24**:850–854.

Sieburth, J. M., 1975, *Microbial Seascapes,* University Park Press, Baltimore.

Smith, J. E., and Oliver, J. D., 1981, The significance of bacterial attachment in the metabolic activity of bacteria in the deep sea, *Abstr. Annu. Meet. Am. Soc. Microbiol.*

Stark, W. H., Stadler, J., and McCoy, E., 1938, Some factors affecting the bacterial population of freshwater lakes, *J. Bacteriol.* **36**:653–654.

Steinberg, C., 1978, Bacteria and their activity in the surfaces of profundal sediments of Lake Walchensea, Upper Bavaria, West Germany, *Arch. Hydrobiol.* **84**:29–41.

Stevenson, L. M., 1978, A case for bacterial dormancy in aquatic ecosystems, *Microb. Ecol.* **4**:127–132.

Stotzky, G., 1972, Activity, ecology, and population dynamics of microbes in soil, *Crit. Rev. Microbiol.* **2**:59–137.

Strohl, W. R., and Larkin, J. M., 1978, Enumeration, isolation and characterization of *Beggiatoa* from freshwater sediments, *Appl. Environ. Microbiol.* **36**:755–770.

Stull, E. A., 1972, Autoradiographic measurement of the primary productivity of individual species of algae from Castle Lake, California, Ph.D. thesis, University of California, Davis.

Stumm, W., and Morgan, J. J., 1981, *Aquatic Chemistry,* 2nd ed., Wiley, New York.

Sugahara, I., Sugiyama, M., and Kawai, J., 1972, Distribution and activity of nitrogen-cycle bacteria in water-sediments with different concentrations of oxygen, in: *Effect of the Open Ocean Environment on Microbial Activities* (R. R. Colwell and R. Y. Morita, eds.), University Park Press, Baltimore, pp. 327–340.

Trulear, M. G., and Characklis, W. G., 1979, Dynamics of biofilm processes, in: *34th Annual Purdue Industrial Wastewater Conference,* Ann Arbor Science, Ann Arbor, Mich., pp. 838–853.

Turner, J. T., 1979, Microbial attachment to copepod fecal pellets and its possible ecological significance, *Trans. Am. Microsc. Soc.* **98**:131–137.

Verwey, E. J. W., and Overbeck, J. T. G., 1948, *Theory of Stability of Lyophobic Colloids,* Elsevier, Amsterdam.

Wallace, G. T., 1982, The association of copper, mercury and lead with surface-active organic matter in coastal seawater, *Mar. Chem.* **11**:379–385.

Wangersky, P. J., 1977, The role of particulate matter in the productivity of surface waters, *Helgol. Wiss. Meeresunters.* **30**:546–564.

Watt, W. D., 1971, Measuring the primary production rates of individual phytoplankton species in natural mixed populations, *Deep-Sea Res.* **18**:329–339.

Wellman, A. M., 1973, Oil floating in the North Atlantic, *Mar. Pollut. Bull.* **4**:190–191.

Wellman, A., and Paerl, H. W., 1980, Rapid chemotaxis assay using radioactively labeled bacterial cells, *Appl. Environ. Microbiol.* **42**:216–221.

Wetzel, R. G., 1979, The role of the littoral zone and detritus in lake metabolism, *Arch. Hydrobiol. Beih. Ergebn. Limnol.* **13**: 145–153.

Wetzel, R. G., 1983, *Limnology,* 2nd ed., Saunders, Philadelphia.

Whitton, B. A., and Potts, N., 1982, Marine littoral, in: *The Biology of Cyanobacteria* (N. G. Carr and B. A. Whitton, eds.), Blackwell, Oxford, pp. 515–542.

Wiebe, W. J., and Pomeroy, L. R., 1972, Microorganisms and their association with aggregates and detritus in the sea: A microscopic study, *Mem. Ist. Ital. Idrobiol. Suppl.* **29**:325–351.

Young, L. Y., and Mitchell, R., 1973, Negative chemotaxis of marine bacteria to toxic chemicals, *Appl. Microbiol.* **25**:972–975.

Zimmerman, R., Iturriaga, R., and Becker-Birk, J., 1978, Simultaneous determination of the total number of aquatic bacteria and the number thereof involved in respiration, *Appl. Environ. Microbiol.* **36**:926–935.

ZoBell, C. E., 1943, The effect of solid surfaces upon bacterial activity, *J. Bacteriol.* **46**:39–56.

ZoBell, C. E., and Anderson, D. Q., 1936, Observations on the multiplication of bacteria in different volumes of sea water and the influence of folid surfaces, *Biol. Bull.* **71**:324–342.

14

Responses of Plant Cells to Adsorbed Bacteria

STEVEN G. PUEPPKE and DANIEL A. KLUEPFEL

1. INTRODUCTION

To our knowledge this is the first review devoted entirely to the responses of plant cells to bacteria adhering to their surfaces. The choice of the term *response* is deliberate. It is intended to emphasize the fact that plant cells react to microorganisms, and that they are more than just passive substrates for bacterial colonization. Research on quantitative and qualitative aspects of bacterial adhesion to plant cells has gained momentum in recent years (see Dazzo, 1980; Sequeira, 1983; Pueppke, 1984; and Chapter 9). There also is a large body of data on the general responses of plants to challenge by pathogenic bacteria. Unfortunately, we often do not know how adherent bacteria influence plant responses. Distinctions between bacteria that are indeed bound and those that simply repose on plant cells also are difficult to make. As a consequence, we offer a substantial amount of speculative material in this section.

The plant–bacterium interactions of interest here are divided into three broad categories. The first group of organisms includes species of *Rhizobium, Frankia,* and *Agrobacterium.* Both *Rhizobium* and *Frankia* infect plant roots and produce nitrogen-fixing nodules. *Agrobacterium,* although closely related to *Rhizobium,* is a parasite. It infects dicotyledonous plants and a few monocots and induces the formation of tumors that may under appropriate conditions weigh as much as 44 kg (Galloway, 1919). The adhesion of each of these microorganisms to plant cells precedes and is associated with infection (see Chapter 9).

The second group of organisms, sometimes termed the bacterial invaders, includes parasites that normally cause necrotic plant diseases. Included are members of the genera *Pseudomonas* and *Xanthomonas.* The interactions between these bacteria and plants are classified as either compatible or incompatible (Klement and Goodman, 1967a). Compatibility describes the combination of a virulent bacterial strain with a susceptible host. The bacterial population increases and symptoms of disease soon appear. Incompatibility de-

Steven G. Pueppke and Daniel A. Kluepfel • Department of Plant Pathology, University of Florida, Gainesville, Florida 32611. *Present address of S. G. P.:* Department of Plant Pathology, University of Missouri, Columbia, Missouri 65211. *Present address of D. A. K.:* Laboratory for Microbiology, Agricultural University, Wageningen, The Netherlands.

scribes the interaction of virulent bacteria with a resistant host, or alternatively, the interaction of mutant, avirulent bacteria with an otherwise susceptible host. In many incompatible interactions, bacterial growth is curtailed and active resistance by the plant is evident. Bacterial adhesion often is associated with incompatibility and precedes the appearance of hypersensitive resistance responses. For bacterial invaders, therefore, we must consider the significance of adhesion to the induction of resistance. This relationship is precisely the opposite of that observed in interactions between plants and organisms such as *Rhizobium*, *Frankia*, and *Agrobacterium*, where adhesion is thought to condition compatibility.

The final group of bacteria includes nonparasitic organisms that colonize the phylloplane and rhizoplane. The associations of these organisms with plants are loose in the sense that plant tissues usually are not invaded. As a consequence, surface colonizers often fail to elicit marked plant responses. Bacterial inhabitants of the phylloplane and rhizoplane nevertheless warrant attention, because they serve as buffers between the plant and the environment. These bacteria also may exert direct and indirect effects on the underlying plant tissues.

2. NITROGEN-FIXING BACTERIA THAT INFECT ROOT HAIRS

2.1. Rhizobia

2.1.1. Root Hair Deformation

The roots of many species of legumes contain nodules in which atmospheric nitrogen is reduced to ammonia. Although several workers in the last century found that plant nodules contain soil organisms, Beijerinck in 1888 was the first to fulfill Koch's postulates and prove that nodules are the result of infection by the bacteria now known as rhizobia. Because of their obvious agricultural significance, legume–*Rhizobium* symbioses have been studied intensively over the course of the past 100 years. In many but not all legume–*Rhizobium* combinations, plant root hairs participate in the initiation of the symbiosis. Rhizobia from the rhizosphere contact and eventually bind to the root hair surface (Fig. 1) (Dazzo, 1980; Bauer, 1981; Pueppke, 1984; and Chapter 9, Section 2). One early root hair response to the rhizobia is deformation, which often causes the root hair to curl back upon itself (Fig. 1). Side branches are produced by root hairs of some plants, and the root hair tips may become contorted into knotlike structures (Haack, 1964; Yao and Vincent, 1969, 1976; Dart, 1974, 1977; Ranga Rao, 1977). Root hairs are infected by rhizobia that are enveloped within the folds of the hair or pocketed between two adhering hairs (Ward, 1887; Bieberdorf, 1938; Fåhraeus, 1957; Nutman, 1959; Sahlman and Fåhraeus, 1963; Napoli and Hubbell, 1975; Newcomb *et al.*, 1979; Higashi and Abe, 1980; Callaham and Torrey, 1981; Turgeon and Bauer, 1982). Recent microscopic evidence indicates that in soybean, infection threads are initiated by rhizobia entrapped between emerging root hairs and the walls of adjacent epidermal cells (Turgeon and Bauer, 1984). Infection threads also may originate from microcolonies of rhizobia on exposed root hair surfaces (Ljunggren, 1969; Napoli and Hubbell, 1975; Callaham and Torrey, 1981). Although the precise mechanism of infection is unclear, there is apparent dissolution of the root hair cell wall beneath the invading microorganisms. *Rhizobium* enzymes may function in this process (Callaham and Torrey, 1981; Hubbell, 1981). The plant replaces the degraded wall with a layer of newly synthesized wall material that gives rise to the infection thread—a cellulosic tube encasing the invading rhizobia (Fig. 1). At a rate estimated to be from 8 to 30 μm/hr, the infection thread carries the rhizobia to

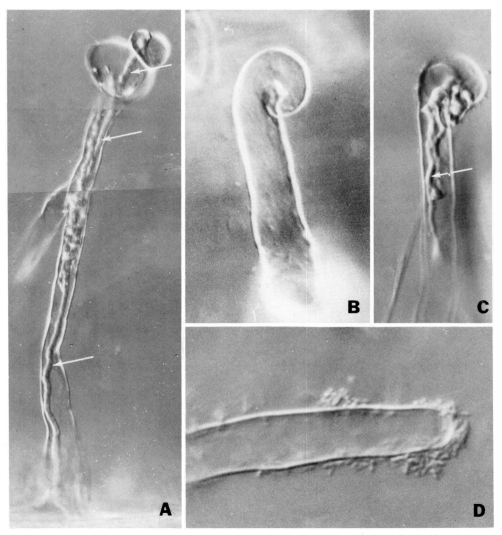

Figure 1. Interaction of rhizobia with legume root hairs. (A) Composite photograph of an infection thread (arrows) in an elongated alfalfa root hair. The plant was inoculated with *Rhizobium meliloti* strain 102F51 1 week before the micrograph was taken. The infective rhizobia have deformed the root hair tip into a knotlike structure. (B) Tip-curled root hair of cowpea, 2 days after inoculation with *Rhizobium* sp. strain 229. Infection threads are not visible. (C) Tip-curled root hair of alfalfa containing infection threads of *R. meliloti* 102F51 (arrow). (D) Adsorption of cells of infective *Rhizobium* sp. strain 3G4b16 to the tip of a cowpea root hair. Bacteria (5 × 10⁸ cells/ml) were incubated with intact roots for 1 hr, the roots were rinsed briefly, and freehand longitudinal sections were made. All micrographs are with Nomarski interference-contrast optics.

the interior of the root (Dart, 1975, 1977; Robertson *et al.*, 1981). The bacteria then are released into the developing nodule. Many agriculturally important legumes—alfalfa (*Medicago sativa*), clover (*Trifolium* spp.), soybean (*Glycine max*), pea (*Pisum sativum*), bean (*Phaseolus vulgaris*), cowpea (*Vigna unguiculata*), and vetch (*Vicia* spp.)—are infected via root hairs (Dart, 1974, 1977). Other legumes, notably peanut (*Arachis hypogaea*) and *Stylosanthes* spp. (Chandler, 1978; Chandler *et al.*, 1982), apparently are invaded by means of ruptures in the root surface. Little is known of this mode of infection, and it will not be discussed here.

Host selectivity is a hallmark of the legume—*Rhizobium* symbiosis. Six species of the genus *Rhizobium* are recognized in part on the basis of host affinities (Holt, 1977). Thus, rhizobia isolated from soybeans are *R. japonicum,* and rhizobia from alfalfa, sweet clover (*Melilotus* spp.), and fenugreek (*Trigonella foenum-graecum*) are *R. meliloti* (Vincent, 1977; Trinick, 1982). A seventh large group of rhizobia, termed *Rhizobium* spp., serves as an ill-defined repository for strains that do not fit into any of the six defined species. The bounds of *Rhizobium* host selectivity may be very strict or rather loose (Vincent, 1974, 1977, 1980). Alfalfa is nodulated only by *R. meliloti.* Soybean, in contrast, is nodulated not only by *R. japonicum,* but also by some strains of *R. lupini* and *Rhizobium* spp.

Rhizobia adsorb to legume root hairs and many other surfaces. The subject of binding mechanisms is beyond the scope of the present discussion, but it has been considered in detail elsewhere (Pueppke, 1984; Chapter 9, Section 2). Here we shall consider how the root hairs respond to bacterial adsorption, how these responses facilitate infection, and whether such responses are selective, i.e., occur only in the presence of nodulating rhizobia. Those who have examined the *Rhizobium*—root hair interface will realize that systematic observation of the initiation of the symbiosis is an arduous task. The distribution of deformed root hairs usually is not uniform (Nutman, 1959; Callaham and Torrey, 1981; Turgeon and Bauer, 1982). A small and highly variable fraction of the deformed hairs are infected, and only a few of the infected hairs yield nodules (McCoy, 1932; Thornton, 1936; Purchase, 1958; Nutman, 1959, 1962). Progress is further complicated by the very nature of infection—a process that normally occurs within the confines of a curled or otherwise deformed root hair.

McCoy (1932) was one of the first to examine legume roots inoculated with nodulating and nonnodulating rhizobia. She found that 75, 24.5, and 0.5% of noninoculated alfalfa root hairs were straight, bent, and curled, respectively. The corresponding values for plants that had been inoculated 3 weeks earlier with *R. meliloti* were 40% straight, 45% bent, and 2.5% curled. Deformation does not seem to be restricted to nodulating pea–*R. leguminosarum* combinations. Although only 9% of noninoculated pea root hairs were assigned to the combined bent and curled categories, 77% were bent and curled after inoculation with nonnodulating *R. meliloti.* McCoy (1932) also observed that sterile filtrates of *Rhizobium* culture fluids elicit deformation and concluded that "the curling which normally precedes infection of the root hairs is produced by a substance secreted by the bacteria before they enter." Subsequent studies confirmed that substrates from cell-free *Rhizobium* culture filtrates cause various sorts of deformation (Thornton, 1936; Li and Hubbell, 1970; Hubbell, 1970; Solheim and Raa, 1973; Dazzo and Hubbell, 1975).

Haack (1964) and Yao and Vincent (1969, 1976) reinvestigated the phenomenon of deformation and observed that both the number of deformed hairs and the magnitude of the curling response are most evident in the presence of nodulating rhizobia. Although this response was detected with serradella (*Ornithopus sativus*), alfalfa, and siratro (*Macroptilium atropurpureum*), cluster clover (*Trifolium glomeratum*) was examined in greatest detail (Yao and Vincent, 1969, 1976). Inoculation of *T. glomeratum* with nodulating strains of *R. trifolii* induced root hair branching, moderate curling (angle of rotation of the tip at least 90°, but less than 360°), and marked curling (angle of tip rotation greater than 360°). Culture filtrates contained branching and moderate curling activity, but little marked curling activity (Yao and Vincent,1969). Nonnodulating strains of *R. trifolii,* rhizobia of other species, and bacteria of other genera caused some branching and moderate curling, but little or no marked curling. The most noteworthy exceptions are two strains of *R. leguminosarum* that stimulated levels of marked curling comparable to those induced by *R. trifolii*. Although these strains did not nodulate *T. glomeratum, R. leguminosarum* is closely related to *R. trifolii*, and certain strains nodulate clover (Kleczkowska *et al.*, 1944; Vincent, 1977; Hepper, 1978; Hepper and Lee, 1979).

In later experiments bacteria were immobilized within the agar medium supporting the roots, and in some cases a dialysis membrane was inserted between bacteria and plant roots (Yao and Vincent, 1976). Although rhizobia induced root hair branching both from a distance and across a dialysis membrane, marked curling required that the bacteria and their high-molecular-weight metabolites have unrestricted access to the roots. Because infection threads ordinarily form in markedly curled hairs, Yao and Vincent (1969) argue that such extreme curling is a prerequisite for infection. Moreover, marked curling is viewed as a relatively host-selective response—more selective than branching and moderate curling, but less so than infection per se. It should be kept in mind that infection threads form in branched, moderately curled, slightly bent, and straight root hairs (Bieberdorf, 1938; Fåhraeus, 1957; Purchase, 1958; Nutman, 1959; Ranga Rao, 1977; Higashi and Abe, 1980; Callaham and Torrey, 1981). Even epidermal cells can be infected infrequently (Bieberdorf, 1938; Nutman, 1959), which indicates that marked curling clearly is not a prerequisite for infection. The available data nevertheless suggest, but do not prove, that *Rhizobium* binding induces marked root hair curling, which traps the rhizobia and in some way facilitates infection. One possibility is that the chamber containing entrapped rhizobia is a kind of reaction vessel, serving to sequester *Rhizobium* enzymes and other factors necessary for invasion (Hubbell, 1981).

At least two models to explain *Rhizobium*-induced root hair curling have been proposed. Both are based on the assumption that *Rhizobium* adsorption precedes curling. Bauer (1981) speculates that a *Rhizobium* cell attached to the tip of an elongating root hair inhibits the localized deposition of the root hair cell wall. The continued deposition of cell wall material (Chapter 3, Section 2) at sites not influenced by the bacterium causes unequal tip elongation. The bacterium consequently is displaced to the side and tucked in a cleft formed as the root hair curls back upon itself. Hubbell's (1981) model is similar, except that enzymatic damage to the root hair cell wall is invoked as the cause of localized cessation of cell wall synthesis. The model of Hubbell (1981) thus predicts that the initiation of infection precedes or occurs coincident with the initiation of curling. In fact, microcolonies of adsorbed rhizobia on undeformed root hairs sometimes cover a refractile spot or swelling that may mark the point of incipient infection (Thornton, 1936; Fåhraeus, 1957 and older references cited therein; Dart and Mercer, 1964; Dart, 1975, 1977). Although this can be taken as evidence in support of Hubbell's view, it unfortunately is difficult to predict the extent to which such root hairs will deform or indeed if they are being infected at all. Several authors have pointed out that rhizobia are conspicuously absent on the surfaces of curled root hairs (Fåhraeus, 1957; Callaham and Torrey, 1981). This observation may indicate that only a few very active, but hard-to-detect rhizobia can induce pronounced changes in root hair cells.

2.1.2. Mechanism and Regulation of Deformation

Virtually nothing is known of the actual mechanism of deformation or of the signals that pass from adsorbed rhizobia to the underlying plant cell. Migration of the root cell nucleus to the vicinity of incipient infection often is accompanied by intense cytoplasmic streaming (Fåhraeus, 1957; Nutman, 1959; Fåhraeus and Ljunggren, 1968; Dart, 1977; Callaham and Torrey, 1981). Both responses are associated with the tip of the elongating infection thread. The position of the nucleus is thought to specify the direction of thread elongation and to be indispensible for thread growth (Dart, 1977). One important unanswered question is why some root hairs appear to curl in response to adsorbed rhizobia while other nearby hairs bind rhizobia, but do not curl (Dazzo *et al.*, 1976; Stacey *et al.*, 1980; Zurkowski, 1980; Callaham and Torrey, 1981). Information on the regulation of the plant's response also is lacking. For example, why does the relative number of branched versus markedly curled hairs depend on the *Rhizobium* strain, the host variety, and the region of the root under examination (Nutman, 1959, 1967; Yao and Vincent, 1976; Bhuvaneswari and Solheim, 1982)?

Nitrate has a well-known inhibitory influence on nodulation (Dart, 1974, Dart, 1975; Dazzo and Brill, 1978; Truchet and Dazzo, 1982). It appears to prevent nodulation by suppressing the adsorption of rhizobia to white clover root hairs (Dazzo and Brill, 1978) and by inhibiting deformation of alfalfa root hairs (Thornton, 1936; Munns, 1968b; Truchet and Dazzo, 1982). In the latter case the activities of the root hair, and not those of the microsymbiont, seem to be sensitive to nitrate (Munns, 1968b). Acid and salinity also interfere with alfalfa root hair deformation, but the physiological basis of the effect is unknown (Munns, 1968a; Tu, 1981).

It has long been observed that growing root hairs are maximally responsive to rhizobia and that "mature" parts of roots do not nodulate (Nutman, 1951, 1956, 1959; Dart and Pate, 1959; Fåhraeus and Ljunggren, 1968; Munns, 1968a; Skrdleta, 1970). Bhuvaneswari *et al.* (1980, 1981) provided quantitative evidence that infections giving rise to nodules on the primary root of alfalfa, cowpea, and soybean (but not white clover) are restricted to zones having no or immature root hairs at the time of inoculation. It later was demonstrated that infection thread formation is restricted to these zones in cowpea, wild soybean (*Glycine soja*), and soybean (Pueppke, 1983). White clover is unique in that nodules form routinely both in the zones of no or immature root hairs and in portions of the roots having mature, fully elongated root hairs at the time of inoculation (Bhuvaneswari *et al.*, 1981). Many of the nodules in the mature root hair zone, however, are late in appearing. Bhuvaneswari *et al.* (1981) speculated that mature white clover root hairs, unlike immature hairs, cannot respond immediately to rhizobia, but are induced to do so by exposure to rhizobia. Recent preliminary evidence suggests that most infections in the zone of fully elongated white clover root hairs originate in branches, but that infections in the zone of no or immature root hairs usually arise in root hair tips (Bhuvaneswari and Solheim, 1982). A similar relationship between the mode of infection and root hair development in *T. glomeratum* can be inferred from the data of Nutman (1965). These observations emphasize that deformation responses of root hairs to rhizobia are finely tuned, vary among plant species, and may be temporally and spatially distinct.

2.1.3. Genetics

Although a number of legume genes that regulate the *Rhizobium* symbiosis have been identified (Nutman, 1969, 1981; Caldwell and Vest, 1977; Vincent, 1980), only two appear to influence events prior to infection thread biogenesis. In red clover, the recessive gene *r* acts in concert with a maternally inherited factor to preclude infection thread formation (Nutman, 1949). In soybean, the recessive gene rj_1 prevents nodulation by many strains of *R. japonicum* (Williams and Lynch, 1954; Clark, 1957), and infection threads apparently are not produced (J. H. Payne and S. G. Pueppke, preliminary observations). Root hairs of the red clover mutants are markedly curled by *R. trifolii* (Nutman, 1949); thus, the genetic lesion in these plants appears to influence some aspect of infection subsequent to the curling response.

Soybeans of the rj_1 genotype are especially interesting, because they are not totally resistant to nodulation. Certain strains of *R. japonicum* induce foliar chlorosis in soybean, and successfully infect and produce a small number of nodules on roots of rj_1 soybean varieties (Devine and Weber, 1977). If cells of rj_1-nonnodulating strain I-110 are included in an inoculum containing cells of chlorosis-inducing strain 61, strain I-110 surprisingly seems to be the sole occupant of about one-third of the nodules (Devine *et al.*, 1980). Devine *et al.* (1981) sought but could not detect a diffusible metabolite from strain 61 that endows strain I-110 with the ability to nodulate rj_1 soybean. They concluded that contact between cells of strain 61 and rj_1 roots renders plant cells competent to respond to the otherwise nonnodulating strain I-110 (Devine *et al.*, 1981). Although we know of no evidence that this phenomenon depends upon cells of strain 61 adhering to the root, the possibility warrants investigation.

Mutant bacteria that have lost their ability to markedly curl host root hairs have been described (Vincent, 1974; Yao and Vincent, 1976; Napoli and Albersheim, 1980; Hooykaas *et al.*, 1981; Badenoch-Jones *et al.*, 1982; Hirsch *et al.*, 1982; Scott and Ronson, 1982; Forrai *et al.*, 1983; Morrison *et al.*, 1983; Kondorosi *et al.*, 1984). Such mutants sometimes occur spontaneously, or they can be obtained by treatments that induce deletions in or loss of large *Rhizobium* (*sym*)biosis plasmids. *Sym* plasmids are readily transferable to some nonnodulating bacteria. Avirulent *Agrobacterium* and *Rhizobium* strains that apparently do not deform legume root hairs become competent to do so coincident with the acquisition of *sym* plasmids (Hooykaas *et al.*, 1981, 1982; Djordjevic *et al.*, 1983; Truchet *et al.*, 1984). With one exception (Truchet *et al.*, 1984), the deformed hairs are infected. Although systematic analysis of root hair curling by means of genetically manipulated bacteria is still in its infancy, it is likely to become a powerful tool for use in unraveling the mechanism of deformation and infection.

2.2. *Frankia* spp.

Between 140 and 160 species of nonleguminous plants are nodulated by actinomycetes (Bond, 1976; Becking, 1977; Torrey, 1978; Akkermans and van Dijk, 1981). All are dicotyledonous, and most are woody shrubs or trees that grow in nitrogen-deficient soils. The actinomycete symbionts are considered to be members of the filamentous bacterial genus, *Frankia* (Holt, 1977). These organisms are poorly understood; taxonomic relationships at the species level are not yet settled. The nitrogen-fixing root nodules produced by *Frankia* spp. are known as actinorhizae.

Although nitrogen-fixing nonlegume symbioses have been studied for decades, research has lagged behind that on the legume symbiosis. Historically, the most serious handicap was the inability of researchers to isolate an organism from surface-sterilized nodules that would, upon reinoculation, produce nodules. Researchers thus were forced to rely on the largely unsatisfactory use of crushed nodule preparations as inocula. Although Pommer (1959) may have been the first to obtain pure cultures of infective *Frankia* spp., his cultures were lost (Quispel and Burggraaf, 1981). The inoculum problem remained unsolved until 1978, when Callaham *et al.* succeeded in isolating and cultivating an infective *Frankia* isolate from *Comptonia peregrina* (sweet fern). The "*Comptonia* isolate" nodulates *C. peregrina,* the closely related *Myrica gale* (sweet gale) and *M. cerifera* (wax myrtle), and several more distantly related actinorhizal plants (Callaham *et al.*, 1979; Lalonde and Calvert, 1979; Torrey and Callaham, 1979; Torrey *et al.*, 1980). More recently, infective *Frankia* spp. have been isolated routinely from nodules of additional actinorhizal plant species (Berry and Torrey, 1979; Lechevalier and Lechevalier, 1979; Baker and Torrey, 1979, 1980; Lalonde *et al.*, 1981; Quispel and Burggraaf, 1981; Diem *et al.*, 1982b).

Table I. Observations of Infection of Deformed Root Hairs of Actinorhizal Plants
by *Frankia* spp.

Plant species	Inoculum	Reference
Alnus glutinosa	Crushed nodules	Pommer (1956)
	Crushed nodules	Taubert (1956)
	Crushed nodules	Angulo Carmona (1974)
	Crushed nodules	Becking (1975)
	Crushed nodules	Lalonde (1977)
	Crushed nodules	Lalonde and Quispel (1977)
Casuarina cunninghamiana	Crushed nodules	Callaham *et al.* (1979)
C. equisetifolia	Cultured organism	Diem *et al.* (1982a)
Comptonia peregrina	Crushed nodules	Callaham *et al.* (1979)
	Cultured organism	Callaham *et al.* (1979)
Myrica gale	Cultured organism or crushed nodules	Callaham *et al.* (1979) Torrey and Callaham (1979)
M. cerifera	Crushed nodules	Callaham *et al.* (1979)

Actinorhizal plants inoculated with *Frankia* develop root hair deformation that is strikingly reminiscent of the legume root hair response to rhizobia. Only young, immature root hairs appear to be deformed by *Frankia* (Pizelle, 1972; Angulo Carmona *et al.*, 1976; Torrey, 1976; Callaham and Torrey, 1977; Lalonde, 1977; Lalonde and Quispel, 1977; Callaham *et al.*, 1979; Torrey *et al.*, 1980; Diem *et al.*, 1982a); infection invariably occurs via such deformed hairs (Table I). Preliminary evidence exists that the deformation process occurs rapidly and involves changes in the root hair cell walls and nuclear volume (Berry and Torrey, 1983). Deformed root hairs sometimes have marked tip curling, but most often they twist, branch repeatedly, form candelabralike protuberances, or adhere to one another (Becking, 1975, 1977; Torrey, 1976; Callaham and Torrey, 1977; Callaham *et al.*, 1979; Diem *et al.*, 1982a; Berry and Torrey, 1983). The events that give rise to deformation remain speculative. Becking (1975, 1977) believes that root hairs are deformed in response to cell-free, auxinlike substances elaborated by the *Frankia* inoculum. Although cell-free substances seem to elicit deformation (Pizelle, 1972), hyphal filaments of *Frankia* adsorb to both deformed and undeformed root hairs as the rhizoplane is colonized (Becking, 1975, 1977; Callaham *et al.*, 1979; Knowlton *et al.*, 1980; Lalonde *et al.*, 1981; Diem *et al.*, 1982a). Because infections appear to originate from filaments trapped at sites where deformed hairs fold back on themselves (Becking, 1970; Callaham and Torrey, 1977; Callaham *et al.*, 1979), it is likely that the deformations giving rise to infection are responses of the plant to contact with the microsymbiont.

Some cultured *Frankia* isolates infect seedlings under axenic conditions. *Alnus glutinosa* (black alder) is a case in point (Lalonde *et al.*, 1981; Périnet and Lalonde, 1983). Other isolates readily infect plant seedlings under nonsterile conditions, but fail to do so if the plants and *Frankia* inoculum are grown aseptically. Knowlton *et al.* (1980) found that 63 to 100% of *A. rubra* (red alder) seedlings inoculated under nonsterile conditions with a *Frankia* isolate from *Comptonia peregrina* were nodulated. Aseptically grown plants usually failed to nodulate, and root hairs were not deformed (Fig. 2). Four soil bacteria added to the inoculum facilitated deformation (Fig. 2) and permitted 70% of the seedlings to nodulate. By themselves, the "helper" bacteria did not nodulate, but did cause significant root hair deformation. Although a second *Frankia* isolate occasionally nodulated red alder by itself, root hair deformation and nodulation were stimulated markedly by either the above-men-

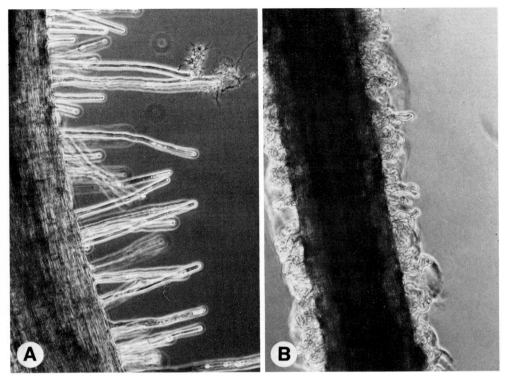

Figure 2. Root hairs of *Alnus rubra*. (A) Aseptically grown roots inoculated with *Frankia* sp. CpI1. A *Frankia* colony clings to the root hairs, but there is no root hair deformation. (B) Aseptically grown roots inoculated with *Frankia* sp. CpI1 and four bacterial helpers. The helper bacteria induce root hair deformation and facilitate nodulation by *Frankia*. Courtesy of Susan Knowlton and the National Research Council of Canada.

tioned helpers or laboratory cultures of *Pseudomonas cepacia. P. cepacia* bound to root hairs and stimulated the walls of deformed root hairs to fluoresce orange-yellow in the presence of acridine orange (Berry and Torrey, 1983). Berry and Torrey believe that this fluorescence is due to a pectic polymer formed in response to the bacteria.

The artificiality of the plant growth system of Knowlton *et al.* (1980) has received some criticism (Lalonde *et al.*, 1981); there is recent evidence that *P. cepacia* alters the rhizosphere pH such that nodulation is promoted (Knowlton and Dawson, 1983). The observed multiple interaction of the symbiotic bacteria and other soil microorganisms nevertheless is likely to occur in the soil and therefore is worthy of additional investigation. One obvious point of interest is whether helper bacteria must be adsorbed to induce deformation and other changes in root hair cell walls.

3. TUMOR-FORMING *AGROBACTERIUM* SPP.

3.1. Events Leading to Tumorigenesis

Four species of *Agrobacterium* are known; all inhabit the soil. Three of these organisms are plant parasites: *A. tumefaciens* incites crown-gall on many plants; the closely related *A. rubi* incites cane-gall on raspberries (*Rubus* spp.); *A. rhizogenes* incites hairy-root on a number of plant species. The fourth bacterium, *A. radiobacter,* is a saprophyte. Although the parasitic *Agrobacterium* spp. are not the only bacteria that produce tumors in plants (Nester and Kosuge, 1981; Kahl and Schell, 1982), the ontogeny of *Agrobacterium* tumors is unique. During tumorigenesis the bacteria transfer plasmid DNA (the T-DNA) to living host cells, where it eventually is integrated into the nuclear genome. Plant cells containing T-DNA are induced to proliferate autonomously, and give rise to tumors. Such genetically transformed tumor cells continue to divide indefinitely, even in the absence of the causal bacteria (Braun, 1978; Gordon, 1981; Nester and Kosuge, 1981).

Agrobacterium cells are not taken up by living plant cells (Coulomb, 1971; Braun, 1982). Thus, T-DNA must pass between bacteria and plant cells, and cannot be simply donated by an intracellular parasite. The DNA transfer process must be complex, because both *Agrobacterium* and plant cells are surrounded by membranes and rather formidable cell walls. Because *Agrobacterium* cells cannot invade intact plant tissues, host tissue damage is a prerequisite for tumorigenesis (Riker, 1923b; Lippincott and Lippincott, 1976). Such damage is likely to wound plant cell walls that otherwise act as barriers to the passage of DNA. Although the precise mechanism of DNA transfer is one of nature's best-kept secrets, it seems unlikely that damaged plant cells merely absorb DNA that has been released into the environment by the bacteria. An alternate hypothesis—that *Agrobacterium* cells are adsorbed by host cells and literally deliver T-DNA—is more appealing (Lippincott and Lippincott, 1976; Braun, 1978; Pueppke, 1984). The binding hypothesis is supported indirectly by repeated observations of the attachment of *Agrobacterium* cells to wounded host cells and tissues (for reviews, see Lippincott and Lippincott, 1980; Pueppke, 1984; and Chapter 9, Section 3), and requires that the plant cells be responsive to adsorbed *Agrobacterium* cells containing T-DNA.

In addition to an extensive literature on *Agrobacterium* attachment to plant cells, a voluminous literature exists on the histology and cytology of *Agrobacterium* tumorigenesis (Gee *et al.,* 1967; Lipetz, 1970; Manocha, 1970; for summaries of older data, see Braun and Stonier, 1958; Braun, 1982). One thus might expect that the responses of host cells to adsorbed *Agrobacterium* cells are thoroughly documented. This unfortunately is not the case. In fact, these responses for the most part are inadequately described, much less understood. The difficulty is largely one of pinpointing the location of *Agrobacterium* cells within the inoculated plant tissue (Riker, 1923b; Lipetz, 1970; Manocha, 1970).

Alterations in the cell wall are one generalized response of plant cells to adsorbed *Agrobacterium* cells. Tomato (*Lycopersicum esculentum*) and castor bean (*Ricinus communis*) cell walls thicken and turn yellow in response to *Agrobacterium* (Riker, 1923a,b, 1927). Walls also lose birefringence, which is evidence of cellulose degradation. These changes were easily seen and serve as an aid in tracing the distribution of bacterial cells during tumor formation (Riker, 1923a). Similar effects are observed in thin sections of pinto bean (*Phaseolus vulgaris*) leaf tumors, but bacteria are not detected (Lippincott and Heberlein, 1965). Plant cell walls beneath adsorbed avirulent *A. radiobacter* cells also thicken noticeably. The bacteria may become enveloped by an electron-dense layer that is considered to be of host cell wall origin (Bogers, 1972). Plant cells apparently respond in the same way to adsorbed, virulent *A. tumefaciens* cells (Bogers, 1972). Analogous structures sometimes envelop virulent *Agrobacterium* cells attached to pinto bean leaf cells (Kino *et al.,* 1979).

Meristematic activity leading to cell division is the second host response to adsorbed *Agrobacterium* cells (Riker, 1923a; Lippincott and Heberlein, 1965; Kino *et al.,* 1979). Riker (1923a) examined the division of tomato pith cells in detail and found that each cell characteristically divides such that the new cell wall is deposited parallel to the location of the bacteria. A large and a small daughter cell are produced. The smaller cell is adjacent to the bacteria (Fig. 3); its nucleus divides frequently. Riker (1923a) considered these divisions to be associated with tumor formation. Enlargement and changes in the staining properties of nuclei also occur in response to *Agrobacterium* cells. These changes are observed in *Chrysanthemum frutescens* (Robinson and Walkden, 1923), *Vicia faba* (Therman, 1956), and tissue-cultured tobacco (*Nicotiana tabacum*) cells (Sigee *et al.*, 1982). In the latter case, however, the bacteria were not visibly adsorbed to plant cells.

Transmission electron microscopy has been used to examine transformed cells of sunflower (Gee *et al.*, 1967; Manocha, 1970), *Kalanchoë diagremontiana* (Lipetz, 1970), and pea (Coulomb, 1971). Although thin sections reveal a myriad of fine structural changes accompanying transformation, these have not yet been related to bacterial adsorption. The

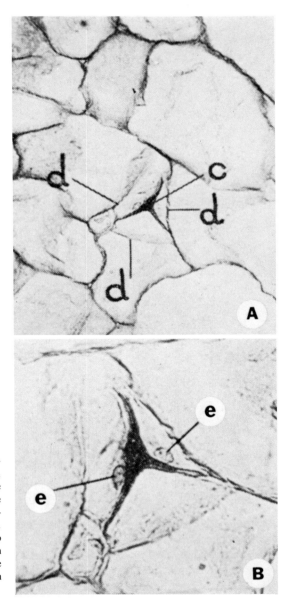

Figure 3. Response of tomato stems to *Agrobacterium tumefaciens*. (A) Cross-section of the pith, 8 days after inoculation. The intercellular space (c) contains *Agrobacterium* cells (not visible here), and three surrounding plant cells have divided. The new plant cell walls are marked (d). Note that the smaller daughter cells are adjacent to the bacteria. (B) Enlargement of the tissues shown at left. Plant cell nuclei (e) are adjacent to the intercellular spaces occupied by bacteria. From Riker (1923a).

scanning electron microscope also has been used to monitor the attachment of *Agrobacterium* to tissue-cultured plant cells and protoplasts (Matthysse *et al.*, 1978, 1981, 1982; Douglas *et al.*, 1982; Matthysse, 1983, and Chapter 9, Section 3) and to inoculated plant tissues (Kluepfel and Pueppke, 1983), but plant cell responses have not been cataloged. Sophisticated methods for the localization of plant–bacterium interaction sites at both light and electron microscope levels are now available (Aist, 1976; Zeyen and Bushnell, 1981). In the near future, these techniques should provide insight into the relationship between adsorbed *Agrobacterium* cells and the plant cell responses that lead to transformation.

3.2. Nontumorigenic Responses to *Agrobacterium* spp.

Perhaps the most dramatic example of plant cell responses to adsorbed *Agrobacterium* cells involves several genera of mosses. *Agrobacterium* cells do not form tumors on these plants, but they do cause protonema to bud and differentiate into gametophores (Spiess *et al.*, 1971). Similar developmental changes are induced by cytokinins. It originally was thought that hormones released by *Agrobacterium* cells resulted in gametophore induction (Spiess *et al.*, 1972, 1973). *A. tumefaciens* cells, however, are active only when permitted direct physical access to the moss (Spiess *et al.*, 1976); they attach to moss cell surfaces in large numbers (Spiess *et al.*, 1977b). There is no distinct relationship between the number of *A. tumefaciens*, *A. rhizogenes*, *A. rubi*, and *A. radiobacter* cells that attach to *Pylaisiella selwynii* and *Funaria hygrometrica* and budding of protonema (Spiess *et al.*, 1977b). Indeed, *P. selwynii* also forms buds in response to several strains of rhizobia (Spiess *et al.*, 1977a). Although it is clear that only certain moss–bacteria combinations are effective, the nature of the developmental stimulus is unknown. Spiess *et al.* (1977a) speculate that the adsorbed bacteria transfer hormones to the underlying plant cells.

4. RESISTANCE TO BACTERIAL INVADERS

In nature, plant–bacterium interactions leading to necrotic plant disease responses are the exception rather than the rule. When plant cells are disturbed by incompatible bacteria, normal host physiological functions often are altered and affected cells rapidly collapse and die. These responses, which collectively are termed the hypersensitive reaction (HR) (see Chapter 9, Section 4.1), are associated with two general resistance mechanisms that often work in concert. One type of resistance is biochemical (see Chapter 3, Section 4), the other is morphological. Biochemical resistance takes place when the plant cell produces a specific compound(s) or an environment, that is toxic to the pathogen. Morphological resistance includes the deposition of physical barriers to halt the spread of the pathogen. Both types of resistance are believed to contribute to the inhibition and ultimate cessation of bacterial proliferation. Bacteria adsorbed to the plant cell surface are associated with the initiation of the HR, and thus may trigger this plant response.

4.1. Events Associated with the Hypersensitive Reaction

The term *hypersensitivity* was borrowed from medical terminology, where it describes an organism that is abnormally sensitive to a pathogenic agent (Müller, 1959). The HR in plants has been known for many years (Ward, 1902). Although plants exhibit the HR to a host of fungi (Király, 1980) and viruses (Holmes, 1929), bacterial elicitation of the HR was not documented until the 1960s (Klement and Lovrekovich, 1961, 1962; Klement *et al.*, 1964). The basic test for the HR involves injecting or vacuum infiltrating a bacterial suspension (10^6 to 10^{10} cells/ml) into the intercellular spaces of a leaf. Injection is accomplished by placing the beveled end of a hypodermic syringe containing the bacterial suspension against the leaf surface and applying pressure (Klement, 1963). Vacuum infiltration is accomplished by submerging a leaf into a bacterial suspension, drawing a vacuum to remove the air, and then releasing the vacuum to allow the bacterial suspension to enter the leaf through the stomata. The intercellular spaces usually dry within 30 min and hypersensitive flecks appear within 8 to 24 hr.

The sequence of events comprising the HR has been divided into three stages (Klement, 1971, 1972) (Fig. 4). The first stage, termed the induction period, defines the time required for the irreversible activation of the HR. During this period viable bacteria are thought to adsorb to host cell walls to initiate the HR (Goodman et al., 1976, 1977). Viable bacteria are required throughout the induction period to trigger the HR (Klement and Goodman, 1967b). Evidence exists that initiation of the HR requires plant cell–bacterial cell contact (Cook and Stall, 1977; Klement, 1977; Stall and Cook, 1979), which probably leads to bacterial adsorption.

The induction period is defined with antibiotics injected into leaves at various time intervals subsequent to bacterial inoculation. The interval between inoculation and the time at which antibiotic injection no longer prevents the HR is by definition the induction period. Reported induction periods vary from investigator to investigator and depend on the plant–bacterium combination under study. For example, in tobacco the induction period for either *Erwinia amylovora* or *Pseudomonas syringae* pv. *syringae* is 20 min (Klement and Goodman, 1967a). In pepper (*Capsicum annuum*), the induction period is 3 to 4 hr with *Xanthomonas campestris* pv. *vesicatoria* (Meadows and Stall, 1981), and in soybean leaves incompatible *Pseudomonas* spp. require a 2-hr induction period (Keen et al., 1981). Induction times also seem to depend on the antibiotic used. When pepper is inoculated with antibiotic-sensitive, incompatible *X. campestris* pv. *vesicatoria,* the induction period defined by streptomycin is less than 1 hr. But with chloramphenicol or rifampicin, the period is 3 to 4 hr (Meadows and Stall, 1981). The notion of an induction period thus may be somewhat arbitrary. The important concept, however, is that at some point the host resistance mechanism is set into motion and viable bacteria are no longer needed.

The second stage, termed the latent period (Fig. 4), is defined as the time between the onset of measurable plant reactions and the first signs of turgor loss by host cells. During this period no symptoms are obvious to the naked eye, but a number of detectable physiological and ultrastructural host responses occur. Alterations in host cell respiration, electrolyte leakage, and nucleic acid synthesis all are evident (Fig. 4). Each of these will be discussed below with respect to bacterial adsorption.

The third stage of the HR, termed tissue collapse (Fig. 4), is defined as that period encompassing the onset of turgor loss to the ultimate disruption of the affected cells. This stage is marked by maximal electrolyte leakage, alterations in chloroplast proteins, phytoalexin accumulation, and general membrane deterioration. Some authors have suggested a fourth stage, necrotization of host tissue, which results in the bacteriostatic action of the HR (Roebuck et al., 1978; Klement, 1982). This final stage in the HR will not be dealt with in this chapter, because it seems to be the result of generalized cell disruption and not an immediate and direct response to bacterial adsorption.

4.2. Immobilization and Associated Morphological Responses

The mechanisms by which plants differentiate compatible from incompatible bacteria and initiate the HR against the latter are matters of great interest. One attractive hypothesis is

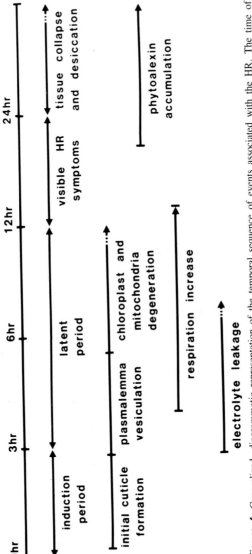

Figure 4. Generalized, diagrammatic representation of the temporal sequence of events associated with the HR. The time of infiltration is designated "0 hr." The relative lengths of the time intervals vary substantially as a function of bacteria–plant combination, inoculum level, and environmental conditions.

that compatible bacteria multiply because they are allowed to remain free in the intercellular spaces of the leaf tissues. Incompatible bacteria, on the other hand, are thought to be adsorbed onto mesophyll cell wall surfaces and then enveloped by wall materials. These events are thought to immobilize the bacteria and to initiate the HR through an unknown series of biochemical events that culminate in bacteriostasis (Király, 1980; Sequeira, 1980; Klement, 1982).

The initial stages of the HR have been studied in some detail at the ultrastructural level (Goodman and Plurad, 1971; Goodman et al., 1976; Sequeira et al., 1977; Politis and Goodman, 1978; Hildebrand et al., 1980). Plant cells beneath and adjacent to adsorbed bacteria respond in a localized manner. The plasmalemma separates from the wall and becomes highly vesiculated. Microfibril aggregates occupy the space between plasmalemma and wall, forming a complex structure by 6 hr after inoculation. This structure is termed the cell wall apposition (Goodman et al., 1976; Politis and Goodman, 1978) (Fig. 5A). In several other systems degradation of the outer cell wall occurs at the site of bacterial adsorption; localized accumulation of electron-dense material (apposition) between the withdrawn plasmalemma and cell wall also is seen (Sequeira et al., 1977; Roebuck et al., 1978; Smith and Mansfield, 1982). Roebuck et al. (1978) observed an aggregation of host cytoplasm, which often contained membrane-bound vesicles, adjacent to adsorbed bacteria.

The host tissue response beginning in the latent period and continuing into the tissue collapse period is dramatic and has been documented well at the ultrastructural level in several different hosts (Goodman and Plurad, 1971; Goodman, 1972; Goodman et al., 1976; Sequeira et al., 1977; Cason et al., 1978; Roebuck et al., 1978; Sigee and Al-Issa, 1983). Seven hours after inoculation of tobacco with 10^8 cells/ml, widespread cellular disruption can be noted (Goodman and Plurad, 1971). Mitochondrial cristae and outer membranes lose their integrity. The lamellar structure of chloroplasts is lost in a manner that may be unique to the HR (Goodman and Plurad, 1971). Chloroplast degeneration also is visible in scanning electron micrographs of spongy mesophyll cells (Sigee and Al-Issa, 1983) (Fig. 6). In addition, both the plasmalemma and the tonoplast are damaged beyond recognition. These responses coincide with significant increases in electrolyte leakage (Goodman, 1968), which are one of the earliest measurable consequences of the HR. Loss of electrolytes occurs early in the latent period and progresses well into the tissue collapse period. This is in contrast to compatible interactions, in which electrolyte leakage is delayed well into the latent period and progresses at a much slower rate (Goodman, 1972; Cook and Stall, 1977; Stall and Cook, 1979). Disruption of the integrity of the plasmalemma and the subsequent loss of selective permeability are consistent with increased electrolyte leakage.

Early adsorption and immobilization events initially were described in tobacco leaves infiltrated with incompatible Pseudomonas syringae pv. pisi (Goodman and Plurad, 1971; Goodman et al., 1976). During the induction period, a portion of the outer wall (termed the cuticle) separates from the mesophyll cells in response to incompatible bacterial cells. This sheetlike material then folds away from the host cell surface into the intercellular space. In time, the cuticle appears to increase in complexity, and it eventually encompasses the bacteria, immobilizing them on mesophyll cell surfaces (Fig. 5B). Saprophytic P. fluorescens cells also are enveloped by an extremely fine cuticle that differs from that found in incompatible cells. Compatible P. syringae pv. tabaci cells are not enveloped, but they may be associated with microfibrils of host cell wall origin. On the basis of this ultrastructural evidence, Goodman et al. (1976) suggests that the sequence of envelopment events in the incompatible reaction is unique.

A similar response was observed when compatible and incompatible strains of P. solanacearum are injected into tobacco leaves (Sequeira et al., 1977). In response to the

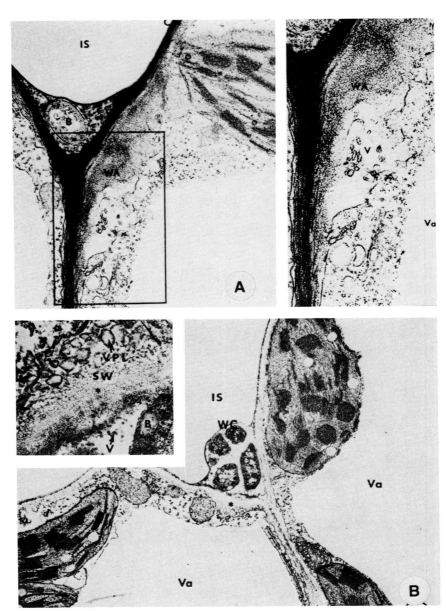

Figure 5. Responses of tobacco mesophyll cells to incompatible *Pseudomonas syringae* pv. *pisi*, 4 hr postinocula-
tion. (A) Wall apposition (WA) is adjacent to a bacterial cell (B) immobilized at the junction of two plant cells.
Inset: enlargement of the wall apposition, showing the associated vesicles (V). CW: cell wall. (B) Bacterial cells are
enveloped by wall cuticle (WC) at the junction of two plant cells. The inset shows the vesiculated plasmalemma
(VPL) and swollen plant cell wall (SW). IS: intercellular space; Va: vacuole. Courtesy of R. N. Goodman and the
American Phytopathological Society.

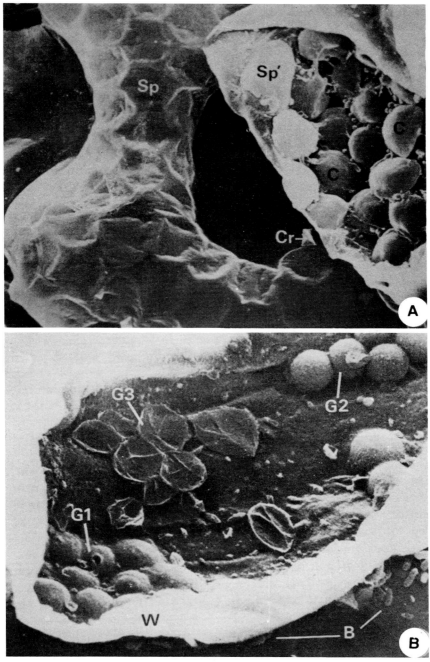

Figure 6. Scanning electron microscopy of the response of tobacco leaf cells to incompatible *Pseudomonas syringae* pv. *pisi*. (A) Outer surface of a spongy mesophyll cell (SP) and the exposed contents of a ruptured cell (SP'). The cells were fixed immediately after infiltration. Intact chloroplasts (C) are in the ruptured cell. CR: surface crystal. (B) Fractured spongy mesophyll cell, 7 hr postinoculation. Chloroplasts occur in three irregular groups (G1, G2, G3). Chloroplasts in G1 are covered by the tonoplast, but those in G2 are not. Chloroplasts in G3 are ruptured and flattened. B: bacteria on the cell surface. Courtesy of D. C. Sigee and Verlag Paul Parey.

adsorbed bacteria, a cuticlelike structure termed the pellicle originates from the host cell wall and envelops incompatible bacterial cells. Nonphytopathogens such as *Escherichia coli* and *Bacillus subtilis* are also enveloped as are heat-killed incompatible *P. solanacearum*. As none of these organisms induces the HR (Sequeira *et al.*, 1977), it is clear that adsorption is not sufficient for the HR.

Data from other systems do not support the hypothesis that adsorption and envelopment of incompatible bacteria are required to initiate the HR. For example, the saprophyte *P. putida* is enveloped in bean, but not an incompatible, HR-inducing strain of *P. syringae* pv. *tomato* (Sing and Schroth, 1977). In addition, although compatible *P. syringae* pv. *phaseolicola* cells are not enveloped by bean, numerous bacterial cells are adsorbed onto the plant cell walls (Sing and Schroth, 1977). In this host, envelopment does not appear to be required for the initiation of the HR. Cason *et al.* (1978) also suggested that the envelopment of *Xanthomonas campestris* pv. *malvacearum* in cotton (*Gossypium hirsutum*) is not necessarily related to the initiation of the HR. In the cotton–*Xanthomonas* system no bacteria were observed to be adsorbed in compatible interactions, although fibril production is evident (Al-Mousawi *et al.*, 1982b). Bacterial adsorption and loose fibrillar envelopment structures occur in the incompatible interaction (Al-Mousawi *et al.*, 1982a). The structures appear to be fragile, easily ruptured, and morphologically dissimilar, however, to those viewed by Goodman *et al.* (1976). It was concluded that the fibrils enveloping *X. campestris* pv. *malvacearum* in cotton may be of little significance. Al-Mousawi *et al.* (1983) also reported that starch grains, killed *X. campestris* pv. *malvacearum,* and the gram-positive saprophyte *Micrococcus lysodeikticus* are enveloped. As none of these particles induces the HR, it seems clear that in cotton, as in tobacco, envelopment per se is not sufficient for induction of the HR.

Several authors have quantified bacterial adsorption and envelopment during induction of the HR. Most were unable to relate the numbers of attached or enveloped bacteria to the development of incompatibility (Daub and Hagedorn, 1980; Atkinson *et al.*, 1981; Fett and Jones, 1982; Smith and Mansfield, 1982). *P. syringae* pv. *pisi* and *P. syringae* pv. *tabaci* are adsorbed differentially by tobacco leaf tissue. Centrifugation of leaves at time 0 and at 2 hr after inoculation removes 61 and 8.4%, respectively, of the incompatible, HR-inducing pv. *pisi* cells from intercellular spaces (see Fig. 6). The analogous values for compatible pv. *tabaci* cells are 17.5% at time 0 and 1.84% at 2 hr (Atkinson *et al.*, 1981). Fett and Jones (1982) did not observe adsorption of incompatible *P. syringae* pv. *glycinea* cells to leaf cells of two resistant soybean cultivars, each of which reacts hypersensitively. Although statistical analysis was not included, envelopment of bacteria does not seem to relate to the response of oats (*Avena* spp.) to compatible, incompatible, or saprophytic bacteria (Smith and Mansfield, 1982) (Table II).

Hildebrand *et al.* (1980) have offered an alternative interpretation of the cuticle (or pellicle) layer originally thought to be a host response to bacteria. When water infiltrated into bean leaves was allowed to evaporate or to be absorbed by the surrounding cells, amorphous fibrillar structures similar to the cuticle or pellicle were observed (Fig. 7A). Hildebrand *et al.* (1980) believe that the water dissolves cell wall surface material, which Slusarenko and Wood (1983) suggest may consist of pectic polysaccharides. Upon drying, a film of the fibrillar material is deposited along the air–water interface. These fibrillar films occasionally are observed even in uninfiltrated plants (Fig. 7B). As in the case of plants infiltrated with water, these films most often span the space between juxtaposed cells. Hildebrand *et al.* (1980) believe that bacteria are merely trapped by such retreating water films. This interpretation also is suggested by Al-Issa and Sigee (1982a). Interestingly, Al-Mouswai *et al.* (1983) showed that hydrophobic polystyrene spheres are not enveloped by host cells in

Table II. Localization of Pseudomonads Infiltrated into Oat Leaves[a]

Time after inoculation (hr)	Reaction type[c]	% of observed bacteria[b]	
		Near or attached to plant cell walls	Free in intercellular spaces
2.5	Compatible	49	51
2.5	Incompatible (HR)	66	34
2.5	Saprophyte	80	20
12.0	Compatible	47	53
12.0	Incompatible (HR)	61	39
12.0	Saprophyte	86	14
24.0	Compatible	31	69
24.0	Incompatible (HR)	56	44
24.0	Saprophyte	83	17

[a]Data from Smith and Mansfield (1982).
[b]For each treatment, a total of 100 bacterial cells in thin sections of infiltrated tissues were examined.
[c]The compatible strain was *Pseudomonas coronafaciens*, the incompatible HR-inducing strain was *P. coronafaciens* pv. *atropurpurea*, and the saprophyte was *P. fluorescens*.

cotton cotyledons. They suggested that the water meniscus is repelled from the hydrophobic surfaces of the spheres, thus preventing a film from being deposited.

If adsorption and envelopment are active processes unrelated to drying but involved in the HR, maintaining a fluid-filled intercellular space should not inhibit envelopment and the HR. Although envelopment under these conditions has not been examined, water congestion in fact inhibits the HR (Stall and Cook, 1979). It nevertheless is difficult to explain how physical drying can account for the strain-dependent differences in immobilization observed by Sequeira *et al.* (1977) and Goodman *et al.* (1976).

Are plants able to distinguish compatible from incompatible bacterial strains and actively embed the latter beneath a layer of cutin (Goodman *et al.,* 1976; Chapter 3, Section 3.2)? Is this active host response then responsible for the marked bacteriostasis that occurs in some incompatible interactions? These questions may be answered in part by histochemical staining of envelopment structures and immobilized bacterial cells. In a recent study, Obukowicz and Kennedy (1981) illustrated cytochemically that polyphenyloxidase is activated in cells responding hypersensitively. This enzyme is thought to be responsible for the phenolic compounds that appear within the vesicles that form between the cell wall and the plasmalemma (Obukowicz and Kennedy, 1981). The enzyme also may contribute to the accumulation of tannins, which Obukowicz and Kennedy (1981) suggest may be responsible for the encapsulation structures. A histochemical technique that may be useful for staining cutin has been reported by Berg (1983). After digestion of tissue with chromic acid and enbloc staining with $KMnO_4$ (a stain for polar molecules), the suberinlike nature of host walls of actinorhizal plants is evident. Suberin and cutin are similar polymers (see Chapter 3, Sections 3.2 and 3.3); both are insoluble in chromic acid and neither is stained by $KMnO_4$ (Johansen, 1940; Scott, 1948, 1950). If applied to the envelopment problem in the HR, these and similar techniques may yield useful information about the composition and origin of the cuticle (pellicle).

Adsorption and envelopment are still matters of debate; nevertheless, the above data are consistent with several provisional conclusions: (1) Metabolically active incompatible cells are required to trigger the HR (Klement and Goodman, 1967a); (2) Close contact and possibly bacterial adsorption onto the plant cell surface are prerequisites for induction of the

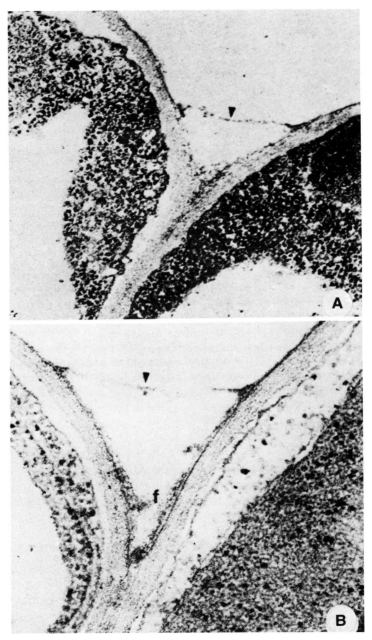

Figure 7. Cuticlelike films (arrowheads) spanning intercellular junctions in uninoculated bean leaves. (A) Untreated leaf, (B) 3 hr postinfiltration with water. Note the large amount of fibrillar material (f) on the wall beneath the film. Courtesy of D. C. Hildebrand and the American Phytopathological Society.

HR (Klement, 1977; Stall and Cook, 1979); (3) As envelopment is seen in some incompatible interactions and not in others, this postulated defense mechanism does not appear to be a generalized plant response to incompatible invaders.

4.3. Biochemical Host Responses

As might be expected, plants challenged with HR-inducing bacterial strains undergo changes in their respiration rates (Németh and Klement, 1967; P.-Y. Huang and R. N. Goodman, personal communication). When intact tobacco leaves are injected with either compatible *P. syringae* pv. *tabaci,* incompatible *P. syringae* pv. *syringae,* or the saprophyte *P. fluorescens,* markedly different respiration rates are observed (Németh and Klement, 1967). Incompatible interactions result in a twofold increase in respiration in 5.5 hr, and the respiration rate drops precipitously after symptoms of the HR appear. Compatible interactions exhibit a much slower, constant rate that takes nearly 10 hr to double. The saprophytic combination shows an even smaller increase in respiration. In each case the host tissue is postulated to be responsible for the observed respiration changes. Enzyme-separated leaf cells appear to behave as intact tissues in response to compatible, incompatible, and saprophytic bacteria (P.-Y. Huang and R. N. Goodman, personal communication). Oxygen consumption occurs at a much higher rate in the incompatible interaction than in either of the other interactions. However, increased respiration of separated cells (which is sensitive to antimycin A and therefore probably of host origin) occurs within 2 min, rather than in hours, as with intact tissues. The significance of altered respiration rates remains unclear. The rapidity of these events suggests, however, that they may be responses to adsorbed bacteria.

Membrane disruption most probably involves alteration of phospholipids, membrane-bound enzymes, and structural proteins. Huang and Goodman (1970) showed that phospholipid-altering enzymes are not causally related to the HR. On the other hand, there is evidence for HR-associated alterations in the nature and properties of chloroplast structural proteins (Huang and Goodman, 1972; Huang *et al.,* 1973, 1974). More recently, changes in the membrane potential (Em) of plant cell membranes incubated with HR-inducing bacterial strains were measured over time (Pavlovkin and Novacky, 1984). Electrolyte leakage data and ultrastructural observations of membrane disruption suggest alterations in the lipid matrix or embedded proteins.

Pavlovkin and Novacky (1984) were able to differentiate the Em of the plasmalemma into its component parts: lipid-mediated membrane diffusion and energy-requiring pumps. Response of the membrane pumps to inhibitors late in the latent period led these authors to conclude that electrolyte leakage in the HR is due to alterations in the lipid matrix and not the protein comprising the pumps (Pavlovkin and Novacky, 1984).

4.3.1. Host Nucleic Acid Synthesis

Changes in host nucleic acid synthesis associated with the HR also have been examined. To study the synthesis of DNA, *in situ* uptake of [^{3}H]thymidine was investigated by autoradiography at the light microscope level (Al-Issa and Sigee, 1982b). DNA synthesis in mesophyll cells occurred in two stages. During the induction period and into the early latent period, host cell DNA synthesis was not significantly different from that in controls. By the second half of the latent period, however, a significant drop in DNA synthesis was observed in both nuclei and cytoplasmic organelles (Al-Issa and Sigee, 1982b). Because changes in DNA synthesis occur late in the latent period, i.e., after the onset of membrane disruption, they are not thought to be involved in the triggering of the HR or to be a direct consequence of adsorption (Al-Issa and Sigee, 1982b).

The effects of bacteria on RNA synthesis are somewhat different. Tobacco leaves infiltrated with incompatible *P. syringae* pv. *pisi* undergo rapid decrease in RNA synthesis within the 2-hr induction period, as determined by autoradiographic examination of [^{3}H]uridine uptake (Al-Issa and Sigee, 1983). RNA synthesis dropped more rapidly in the cytoplasm than in the nucleus. This decrease, hours before ultrastructural alterations are apparent, may be in response to bacterial adsorption culminating in the initiation of the HR.

4.3.2. Phytoalexin Accumulation

Phytoalexins are antimicrobial plant metabolites that undergo enhanced or *de novo* synthesis in response to microbial challenge (VanEtten and Pueppke, 1976). These compounds accumulate in tissues responding hypersensitively to incompatible bacteria, and may function to decrease or prevent bacterial multiplication. Phytoalexins also are produced in tissues inoculated with compatible or saprophytic bacteria, although the rate and extent of accumulation are relatively low (for a recent review, see Keen and Holliday, 1982). The phytoalexin response has been studied most extensively in bean leaves challenged with incompatible *P. syringae* pv. *phaseolicola* and in soybean leaves challenged with incompatible *P. syringae* pv. *glycinea* (Stholasuta *et al.*, 1971; O'Brien and Wood, 1973; Keen and Kennedy, 1974; Lyon and Wood, 1975; Patil and Gnanamanickam, 1976; Gnanamanickam and Patil, 1977a,b; Bruegger and Keen, 1979; Keen *et al.*, 1981; Holliday and Keen, 1982). Unfortunately, phytoalexin accumulation and bacterial adsorption have not been examined concurrently.

Markedly elevated levels of bean and soybean phytoalexins occur within 24 hr after inoculation with incompatible bacteria (Keen and Kennedy, 1974; Lyon and Wood, 1975; Gnanamanickam and Patil, 1977a; Keen *et al.*, 1981). Envelopment of such bacteria is seen within 4 to 8 hr after inoculation (Sigee and Epton, 1976; Roebuck *et al.*, 1978; Daub and Hagedorn, 1980). Moreover, cell envelope preparations (Bruegger and Keen, 1979) and lipopolysaccharides (Keen and Holliday, 1982) from incompatible *P. syringae* pv. *glycinea* function as phytoalexin elicitors. These findings are consistent with the hypothesis that phytoalexin accumulation is a response, at least in part, to adsorbed bacteria. Nevertheless, this relationship remains unclear. The rate and extent of phytoalexin accumulation are sharply distinct in bean and soybean leaves challenged with compatible and incompatible bacterial strains (Keen and Kennedy, 1974; Lyon and Wood, 1975; Gnanamanickam and Patil, 1977a; Keen *et al.*, 1981). In contrast the extent to which compatible and incompatible bacteria are adsorbed by leaves of the same plant species are indistinct (see Section 4.2). Several soluble compounds are efficient elicitors of phytoalexins in soybean leaves (Keen *et al.*, 1981). In addition, pseudomonads cause suspension-cultured soybean cells to elaborate phytoalexins but do not visibly bind to and agglutinate such cells (Fett and Zacharius, 1983). Final resolution of the possible relationship between the phytoalexin response and bacterial adsorption awaits systematic examination of both phenomena in a single plant–bacterium system.

5. NONPARASITIC RHIZOPLANE AND PHYLLOPLANE BACTERIA

Both the rhizoplane and the phylloplane harbor large and metabolically active populations of bacteria and fungi (Last and Warren, 1972; Bowen and Rovira, 1976; Blakeman, 1981; Lynch, 1982). These organisms serve as a buffer between the plant and the environment, exerting direct and indirect effects on the underlying plant tissues. Most researchers studying rhizoplane and phylloplane microflora have not concerned themselves directly with the subject of attachment. As a consequence distinctions between microorganisms that literally stick to plant surfaces and those that merely rest on such surfaces are fuzzy. The rhizosphere presents an additional challenge, because in nature the root surface is intimately associated with the soil. Soil particles adsorb tenaciously to the root, and complicate both the analysis and the interpretation of root–microbe events. Although plants can be grown in artificial media lacking soil, systems free of soil and those containing soil may not behave similarly. Much of the literature dealing with bacteria on unwashed plant surfaces is excluded from the following discussion, because experimental procedures used do not separate adsorbed and unadsorbed microbes. Thus, the ability of a leaf or root surface-associated bacterium to withstand rinsing has been taken as the empirical test for adsorption.

5.1. Responses to Associative Nitrogen-Fixing Bacteria

A diverse number of nitrogen-fixing bacteria are found in association with the external surfaces of various plant parts. The microorganisms include species of *Azospirillum, Azotobacter, Bacillus, Beijerinckia, Campylobacter, Clostridium,* and *Derxia,* as well as *Enterobacter cloacae, Erwinia herbicola,* and *Klebsiella pneumoniae* (Rennie, 1981). Yield increases have been reported for plants inoculated with such bacteria (Smith *et al.,* 1978). It is believed that the plants derive benefit from bacterially supplied nitrogen and growth hormones (vanBerkum and Bohlool, 1980).

Azospirillum brasilense colonizes the root mucigel of several grass species (Schank *et al.,* 1979). The bacteria seem to prefer grooves between surface cells, where their distribution is patchy. They also sometimes occupy the internal spaces between cortical cells. *Azotobacter paspali* becomes embedded in the mucigel (Chapter 3, Section 3.1) of *Paspalum notatum* (Döbereiner *et al.,* 1972), as do *Beijerinckia* and *Enterobacter cloacae* in the mucigel near the root tips of rice (*Oryza sativa*) seedlings grown on agar (Diem *et al.,* 1978). *Klebsiella pneumoniae* cells adsorb to roots of *Poa pratensis* by means of type 3 fimbriae (Korhonen *et al.,* 1983). Nitrogen-fixing bacteria also are localized in the cortex and stele and within root hairs of several grass species (Diem *et al.,* 1978; Kavimandan *et al.,* 1978; Patriquin and Döbereiner, 1978; Watanabe and Barraquio, 1979). A portion of these bacteria seem to be adsorbed tightly to the plant and cannot be dislodged by vigorous washing. Estimates of the populations of nitrogen-fixing bacteria adsorbed to rice roots range from 1.2 $\times$ 10^5 to 1.1 $\times$ 10^8 per g (Diem *et al.,* 1978; Watanabe and Barraquio, 1979).

Three morphological responses of guinea grass (*Panicum maximum*) and pearl millet (*Pennisetum americanum*) roots can be observed to *Azospirillum brasilense:* Levels of root mucigel are elevated, the number of lateral roots is increased, and root hair production is enhanced (Fig. 8) (Umali-Garcia *et al.,* 1978, 1979, 1980; Tien *et al.,* 1979). However, root hairs of wheat branch in response to *A. brasilense* (Patriquin *et al.,* 1983). Although cells of *A. brasilense* adsorb to both root hair and epidermal cells of these grasses, the contribution of the adsorbed bacteria to the altered root morphology is uncertain. It is conceivable that some of the "mucigel" on inoculated roots is in fact bacterial extracellular polysaccharide (Umali-Garcia *et al.,* 1979; see also Chapter 3, Section 3.1). *A. brasilense* also produces several compounds with hormonal activity; certain combinations of hormones mimic the bacterial effects on lateral root growth and root hair development (Tien *et al.,* 1979). It remains to be seen whether such substances account for the plant responses to the bacteria and whether bacterial adsorption plays a role in delivery of hormones to particular regions of the roots.

Bacteria also are found in association with leaves, where they may contribute nitrogen and hormones to the plant, and even form nodules (Lange, 1966). *Psychotria bacteriophila,*

Figure 8. Response of pearl millet roots to *Azospirillum brasilense*. (A) Uninoculated roots, (B) inoculated roots. Luxurient root hairs have formed in response to the bacteria. Courtesy of T. M. Tien, D. H. Hubbell, and the American Society for Microbiology.

which is obligately symbiotic with bacteria of uncertain taxonomy (Horner and Lersten, 1972), is the best understood example of the leaf-nodulating plant species (Lersten and Horner, 1967; Horner and Lersten, 1968; Whitmoyer and Horner, 1970). The bacteria survive in mucilage surrounding the shoot apex and migrate into substomatal chambers of leaf primordia. The first apparent plant response is the enlargement and intrusion of mesophyll cells into the substomatal chambers occupied by bacteria. The intrusive mesophyll cells become branched, and their cell walls thicken. The chamber enlarges to form a nodule and soon is filled with a reticulum of plant cells and bacteria embedded in mucilage. Movement of vesicles from the Golgi to the thickened plant cell walls is marked. The external surfaces of the walls become ill-defined and seem to merge with the mucilaginous material surrounding the bacteria. Although Lersten and Horner (1967) hypothesized that the mesophyll cell walls serve as a carbon source for the bacteria, there is no physiological evidence to support their contention. Moreover, there is no clear-cut relationship between nodule initiation and the physical adsorption of bacteria to plant cell surfaces within the substomatal cavities.

5.2. General Responses to Nonparasitic Bacteria

Although a large number of plant responses probably are influenced by adsorbed bacteria (Dazzo, 1980), it often is difficult to disentangle responses to adsorbed bacteria from those to unadsorbed bacteria and fungi or to soluble products of bacterial metabolism. Table III lists the effects of some nonparasitic rhizoplane and phylloplane organisms on plants. Some of the plant responses are nonspecific. Heightened production of mucigel is elicited by many bacteria including rhizobia and associative nitrogen-fixing organisms. Resistance to parasites is induced by both nonparasites and avirulent parasites. Altered uptake of ions appears to be a general phenomenon that may be mediated in part by soluble bacterial metabolites (Barber and Lee, 1974). The nature of the substrate seems to be unimportant, for bacteria increase the adsorption of phosphate by both living plant roots and inert strands of cotton (Barber and Frankenberg, 1971).

Table III. Some Effects of Nonparasitic Rhizoplane and Phylloplane Bacteria on Plants

Response	Plant/bacterium combination	Reference
Altered ion uptake	Barley/gram-negative bacteria	Nissen (1971)
	Barley/soil bacteria (and fungi)	Barber and Frankenburg (1971), Barber and Lee (1974) Rovira and Campbell (1974) Barber et al. (1976)
Increased mucigel	Various crop plants/Pseudomonas sp., Cytophaga johnsonii, soil bacteria	Greaves and Darbyshire (1972)
	Wheat/soil bacteria (and fungi)	Foster (1981)
Induction of proteoid roots	Banksia grandis/soil bacteria (and fungi)	Malajczuk and Bowen (1974)
Shoot chlorosis	Wheat, barley/Pseudomonas sp.	Bennett and Lynch (1981)
Induced resistance to parasites	Cherry/Erwinia sp.	Crosse (1965)
	Apple, pear/Erwinia herbicola	Goodman (1967) McIntyre et al. (1973) Wrather et al. (1973)
	Rice/E. herbicola	Hsieh and Buddenhagen (1974)

The interaction of several cereals with nonparasitic *Pseudomonas* sp., *Mycoplana* sp., or *Curtobacterium* sp. is more specific (Bennett and Lynch, 1981). The pseudomonad adsorbs to wheat (*Triticum vulgare*) and barley (*Hordeum vulgare*) roots. Such adsorption leads to pronounced shoot chlorosis. The remaining organisms apparently do not attach and have no observable deleterious effects on shoots. There also is good reason to believe that uptake of choline sulfate is influenced directly by bacteria adsorbed to roots (Nissen, 1971). The involvement of nonparasitic bacteria in other plant responses, however, awaits further study.

6. RESEARCH PRIORITIES FOR THE FUTURE

Because our knowledge of the responses of plant cells to adsorbed bacteria is at best imperfect, it is not difficult to suggest directions for future research. The plant–microbe interactions described here can be divided arbitrarily into two categories: those about which we know a little and those about which we know virtually nothing. The responses of root hairs to rhizobia and *Frankia* fit into the first category. In these interactions the responses of the plant are reasonably well documented, and in the case of the *Rhizobium* symbiosis, there is good reason to believe that adsorbed bacteria elicit the responses. The next logical steps are to identify the signal(s) that passes from adsorbed bacteria to underlying host cells and then to discover and characterize the mechanism by which the plant responds to the signal(s).

The remainder of the plant–microbe interactions examined here unfortunately must be assigned to the second category. The interfaces between these bacteria and plant cells are inadequately described; in many cases the relationship between adsorbed bacteria and ob-served plant responses is tenuous or nonexistent. There is urgent need, for example, for systematic descriptions of the plant cell–*Agrobacterium* cell interaction during tumorigenesis and the precise relationship between adsorbed bacteria relative to the HR also are needed. In the past, attachment of nonparasitic bacteria to external plant surfaces rarely has been separated either conceptually or experimentally from the process of colonization. It also is apparent, and surprising, that the subject of the consequences of bacterial adsorption to leaf surfaces has been largely overlooked. Both of these oversights should be corrected in the future.

A final comment deals primarily with the rhizoplane but is applicable to all plant–microbe interactions. Great strides have been made in unraveling how parasitic and non-parasitic bacteria interact with roots. Many of the experiments rely upon plants grown either in artificial media lacking soil or in soil devoid of its normal microflora. Although the use of such conditions often simplifies experimental design, there is some danger in removing experiments too far from the real world (Costerton, 1980; and Chapter 1). There is evidence, for example, that soil bacteria may play a normal role in the curling of root hairs, which primes the plant to respond to *Frankia* (Knowlton *et al.*, 1980; Berry and Torrey, 1982). The interaction of associative nitrogen-fixing bacteria with rice roots is profoundly influenced by soil (Diem *et al.*, 1978), and the types of root hairs that are deformed and invaded by rhizobia also may be modified by the root environment (Bieberdorf, 1938; Ranga Rao and Keister, 1978; Turgeon and Bauer, 1982; Pueppke, 1983). These observations underscore an obvious, but often overlooked point: plants in nature respond to a diversity of bacteria within the arena of a complex environment. Insofar as is possible, laboratory experiments should be relevant to this situation.

ACKNOWLEDGMENTS. This work is supported by NSF Grant PCM82-00110 and by funds from the Florida Agricultural Experiment Station. We thank Frank Dazzo and our colleagues at the University of Florida for review of the manuscript.

REFERENCES

Aist, J. R., 1976, Cytology of penetration and infection—Fungi, in: *Physiological Plant Pathology* (R. Heitefuss and P. H. Williams, eds.), Springer-Verlag, Berlin, pp. 197–221.

Akkermans, A. D. L., and van Dijk, C., 1981, Non-leguminous root–nodule symbioses with actinomycetes and *Rhizobium*, in: *Nitrogen Fixation*, Volume 1 (W. J. Broughton, ed.), Oxford University Press (Clarendon), London, pp. 57–103.

Al-Issa, A. N., and Sigee, D. C., 1982a, The hypersensitive reaction in tobacco leaf tissue infiltrated with *Pseudomonas pisi*. 1. Active growth and division in bacteria entrapped at the surface of mesophyll cells, *Phytopathol. Z.* **104**:104–114.

Al-Issa, A. N., and Sigee, D. C., 1982b, The hypersensitive reaction in tobacco leaf tissue infiltrated with *Pseudomonas pisi*. 3. Changes in the synthesis of DNA in bacteria and mesophyll cells, *Phytopathol. Z.* **105**:198–213.

Al-Issa, A. N., and Sigee, D. C., 1983, The hypersensitive reaction in tobacco leaf tissue infiltrated with *Pseudomonas pisi*. 5. Inhibition of RNA synthesis in mesophyll cells, *Phytopathol. Z.* **106**:23–34.

Al-Mousawi, A. H., Richardson, P. E., Essenberg, M., and Johnson, W. M., 1982a, Cotyledon and leaf ultrastructure of a bacterial blight-immune cotton line inoculated with a low level of *Xanthomonas campestris* pv. *malvacearum*, *Phytopathology* **72**:1230–1234.

Al-Mousawi, A. H., Richardson, P. E., Essenberg, M., and Johnson, W. M., 1982b, Ultrastructural studies of a compatible interaction between *Xanthomonas campestris* pv. *malvacearum* and cotton, *Phytopathology* **72**:1222–1230.

Al-Mousawi, A. H., Richardson, P. E., Essenberg, M., and Johnson, W. M., 1983, Specificity of the envelopment of bacteria and other particles in cotton cotyledons, *Phytopathology* **73**:484–489.

Angulo Carmona, A. F., 1974, La formation des nodules fixateurs d'azote chez *Alnus glutinosa* (L.) Vill, *Acta Bot. Neerl.* **23**:257–303.

Angulo Carmona, A. F., von Dijk, C., and Quispel, A., 1976, Symbiotic interactions in non-leguminous root nodules, in: *Symbiotic Nitrogen Fixation in Plants* (P. S. Nutman, ed.), Cambridge University Press, London, pp. 475–484.

Atkinson, M. M., Huang J.-S., and vanDyke, C. G., 1981, Adsorption of pseudomonads to tobacco cell walls and its significance to bacterium–host interactions, *Physiol. Plant Pathol.* **18**:1–5.

Badenoch-Jones, J., Summons, R. E., Djordjevic, M. A., Shine, J., Letham, D. S., and Rolfe, B. G., 1982, Mass spectrometric quantification of indole-3-acetic acid in *Rhizobium* culture supernatants: Relation to root hair curling and nodule initiation, *Appl. Environ. Microbiol.* **44:**275–280.

Baker, D., and Torrey, J. G., 1979, The isolation and cultivation of actinomycetous root nodule endophytes, in: *Symbiotic Nitrogen Fixation in the Management of Temperate Forests* (J. C. Gordon, C. T. Wheeler, and D. A. Perry, eds.), Oregon State University Press, Corvallis, pp. 38–56.

Baker, D., and Torrey, J. G., 1980, Characterization of an effective actinorhizal microsymbiont, *Frankia* sp. AvcI1 (Actinomycetales), *Can. J. Microbiol.* **26:**1066–1071.

Barber, D. A., and Frankenburg, U. C., 1971, The contribution of micro-organisms to the apparent adsorption of ions by roots grown under non-sterile conditions, *New Phytol.* **70:**1027–1034.

Barber, D. A., and Lee, R. B., 1974, The effect of micro-organisms on the absorption of manganese by plants, *New Phytol.* **73:**97–106.

Barber, D. A., Bowen, G. D., and Rovira, A. D., 1976, Effects of microorganisms on absorption and distribution of phosphate in barley, *Aust. J. Plant Physiol.* **3:**801–808.

Bauer, W. D., 1981, Infection of legumes by rhizobia, *Annu. Rev. Plant Physiol.* **32:**407–449.

Becking, J. H., 1970, Plant–endophyte symbiosis in non-leguminous plants, *Plant Soil* **32:**611–654.

Becking, J. H., 1975, Root nodules in non-legumes, in: *The Development and Function of Roots* (J. G. Torrey and D. T. Clarkson, eds.), Academic Press, New York, pp. 507–566.

Becking, J. H., 1977, Endophyte and association establishment in nonleguminous nitrogen-fixing plants, in: *Recent Developments in Nitrogen Fixation* (W. E. Newton, J. R. Postgate, and C. Rodriguez-Barrueco, eds.), Academic Press, New York, pp. 551–567.

Bennett, R. A., and Lynch, J. M., 1981, Bacterial growth and development in the rhizosphere of gnotobiotic cereal plants, *J. Gen. Microbiol.* **125:**95–102.

Berg, R. H., 1983, Preliminary evidence for the involvement of suberization in infection of *Casuarina, Can. J. Bot.* **61:**2910–2918.

Berry, A., and Torrey, J. G., 1979, Isolation and characterization in vivo and in vitro of an actinomycetous endophyte from *Alnus rubra* Bong, in: *Symbiotic Nitrogen Fixation in the Management of Temperate Forests* (J. C. Gordon, C. T. Wheeler, and D. A. Perry, eds.), Oregon State University Press, Corvallis, pp. 69–83.

Berry, A., and Torrey, J. G., 1983, Root hair deformation in the infection process of *Alnus rubra, Can. J. Bot.* **61:**2863–2876.

Bhuvaneswari, T. V., and Solheim, B., 1982, Induction of root hair branching in white clover, *Plant Physiol.* **69:** Suppl. 22 (abstract).

Bhuvaneswari, T. V., Turgeon, B. G., and Bauer, W. D., 1980, Early events in the infection of soybean (*Glycine max* L. Merr) by *Rhizobium japonicum*. I. Localization of infectible root cells, *Plant Physiol.* **66:**1027–1031.

Bhuvaneswari, T. V., Bhagwat, A. A., and Bauer, W. D., 1981, Transient susceptibility of root cells in four common legumes to nodulation by rhizobia, *Plant Physiol.* **68:**1144–1149.

Bieberdorf, F. W., 1938, The cytology and histology of the root nodules of some Leguminosae, *J. Am. Soc. Agron.* **30:**375–389.

Blakeman, J. P. (ed.), 1981, *Microbial Ecology of the Phylloplane,* Academic Press, New York.

Bogers, R. J., 1972, On the interaction of *Agrobacterium tumefaciens* with cells of *Kalanchoe diagremontiana*, in: *Proc. Third Int. Congr. Plant Pathogenic Bacteria* (H. P. Maas Geesteranus, ed.), Cent. Agric. Publ. Doc., Wageningen, pp. 239–250.

Bond, G., 1976, The results of the IBP survey of root-nodule formation in nonleguminous angiosperms, in: *Symbiotic Nitrogen Fixation in Plants* (P. S. Nutman, ed.), Cambridge University Press, London, pp. 443–474.

Bowen, G. D., and Rovira, A. D., 1976, Microbial colonization of plant roots, *Annu. Rev. Phytopathol.* **14:**121–144.

Braun, A. C., 1978, Plant tumors, *Biochim. Biophys. Acta* **516:**167–191.

Braun, A. C., 1982, A history of the crown gall problem, in: *Molecular Biology of Plant Tumors* (G. Kahl and J. S. Schell, eds.), Academic Press, New York, pp. 155–210.

Braun, A. C., and Stonier, T., 1958, Morphology and physiology of plant tumors, *Protoplasmatologia* **10:**1–93.

Bruegger, B. B., and Keen, N. T., 1979, Specific elicitors of glyceollin in the *Pseudomonas glycinea*–soybean host–parasite system, *Physiol. Plant Pathol.* **15:**43–51.

Caldwell, B. E., and Vest, H. G., 1977, Genetic aspects of nodulation and dinitrogen fixation by legumes: The macrosymbiont, in: *A Treatise on Dinitrogen Fixation,* Section III (R. W. F. Hardy and W. S. Silver, eds.), Wiley–Interscience, New York, pp. 557–576.

Callaham, D. A., and Torrey, J. G., 1977, Prenodule formation and primary nodule development in roots of *Comptonia* (Myricaceae), *Can. J. Bot.* **55:**2306–2318.

Callaham, D. A., and Torrey, J. G., 1981, The structural basis for infection of root hairs of *Trifolium repens* by *Rhizobium, Can. J. Bot.* **59:**1647–1664.

Callaham, D. A., Del Tredici, P., and Torrey, J. G., 1978, Isolation and cultivation in vitro of the actinomycete causing root nodulation in *Comptonia, Science* **199**:899–902.

Callaham, D. A., Newcomb, W., Torrey, J. G., and Peterson, R. L., 1979, Root hair infection in actinomycete-induced root nodule initiation in *Casuarina, Myrica,* and *Comptonia, Bot. Gaz.* **140**(Suppl.):S1–S9.

Cason, E. T., Richardson, P. E., Essenberg, M. K., Brinkerhoff, L. A., Johnson, W. M., and Venere, R. J., 1978, Ultrastructural cell wall alterations in immune cotton leaves inoculated with *Xanthomonas malvacearum, Phytopathology* **68**:1015–1021.

Chandler, M. R., 1978, Some observations on infection of *Arachis hypogaea* L. by *Rhizobium, J. Exp. Bot.* **29**:749–755.

Chandler, M. R., Date, R. A., and Roughley, R. J., 1982, Infection and root nodule development in *Stylosanthes* species by *Rhizobium, J. Exp. Bot.* **33**:47–57.

Clark, F. E., 1957, Nodulation of two near-isogenic lines of the soybean, *Can. J. Microbiol.* **3**:113–123.

Cook, A. A., and Stall, R. E., 1977, Effect of watersoaking on response to *Xanthomonas vesicatoria* in pepper leaves, *Phytopathology* **67**:1101–1103.

Costerton, J. W., 1980, Some techniques involved in study of adsorption of microorganisms to surfaces, in: *Adsorption of Microorganisms to Surfaces* (G. Bitton and K. C. Marshall, eds.), Wiley, New York, pp. 403–423.

Coulomb, P., 1971, Sur la présence de phytolysosomes dans les cellules de tumeurs de la plantule de Pois (*Pisum sativum* L.) induites par l'*Agrobacterium tumefaciens, C. R. Acad. Sci. Ser. D* **272**:1229–1231.

Crosse, J.E., 1965, Bacterial canker of stone-fruits. VI. Inhibition of leaf-scar infection of cherry by a saprophytic bacterium from the leaf surfaces, *Ann. Appl. Biol.* **56**:149–160.

Dart, P. J., 1974, The infection process, in: *Biology of Nitrogen Fixation* (A. Quispel, ed.), North-Holland, Amsterdam, pp. 381–429.

Dart, P. J., 1975, Legume root nodule initiation and development, in: *The Development and Function of Roots* (J. G. Torrey and D. T. Clarkson, eds.), Academic Press, New York, pp. 467–506.

Dart, P. J., 1977, Infection and development of leguminous nodules, in: *A Treatise on Dinitrogen Fixation,* Section III (R. W. F. Hardy and W. S. Silver, eds.), Wiley–Interscience, New York, pp. 367–472.

Dart, P. J., and Mercer, F. V., 1964, The legume rhizosphere, *Arch. Mikrobiol.* **47**:344–378.

Dart, P. J., and Pate, J. S., 1959, Nodulation studies in legumes. III. The effects of delaying the inoculation on the seedling symbiosis of barrel medic (*Medicago tribuloides* Desr.), *Aust. J. Biol. Sci.* **12**:427–444.

Daub, M. E., and Hagedorn, D. J., 1980, Growth kinetics and interactions of *Pseudomonas syringae* with susceptible and resistant bean tissues, *Phytopathology* **70**:429–436.

Dazzo, F. B., 1980, Adsorption of microorganisms to roots and other plant surfaces, in: *Adsorption of Microorganisms to Surfaces* (G. Bitton and K. C. Marshall, eds.), Wiley, New York, pp. 253–316.

Dazzo, F. B., and Brill, W. J., 1978, Regulation by fixed nitrogen of host–symbiont recognition in the *Rhizobium–*clover symbiosis, *Plant Physiol.* **62**:18–21.

Dazzo, F. B., and Hubbell, D. H., 1975, Cross-reactive antigens and lectin as determinants of symbiotic specificity in the *Rhizobium–*clover association, *Appl. Microbiol.* **30**:1017–1033.

Dazzo, F. B., Napoli, C. A., and Hubbell, D. H., 1976, Adsorption of bacteria to roots as related to host specificity in the *Rhizobium–*clover symbiosis, *Appl. Environ. Microbiol.* **32**:166–171.

Devine, T. E., and Weber, D. F., 1977, Genetic specificity of nodulation, *Euphytica* **26**:527–535.

Devine, T. E., Kuykendall, L. D., and Breithaupt, B. H., 1980, Nodulation of soybeans carrying the nodulation-restrictive gene, rj_1, by an incompatible *Rhizobium japonicum* strain upon mixed inoculation with a compatible strain, *Can. J. Microbiol.* **26**:179–182.

Devine, T. E., Breithaupt, B. H., and Kuykendall, L. D., 1981, Tests for a diffusable compound endowing rj_1-incompatible strains of *Rhizobium japonicum* with the ability to nodulate the rj_1rj_1 soybean genotype, *Crop Sci.* **21**:696–699.

Diem, H. G., Rougier, M., Hamad-Fares, I., Balandreau, J. P., and Dommergues, Y. R., 1978, Colonization of rice roots by diazotroph bacteria, *Ecol. Bull. (Stockholm)* **26**:305–311.

Diem, H. G., Gauthier, D., and Dommergues, Y. R., 1982a, Extranodular growth of *Frankia* on *Casuarina equisetifolia, FEMS Microbiol. Lett.* **15**:181–184.

Diem, H. G., Gauthier, D., and Dommergues, Y. R., 1982b, Isolement et culture in vitro d'une souche infective et effective de *Frankia* isolée de nodules de *Casuarina* sp., *C.R. Acad. Sci. Ser. C* **295**:759–763.

Djordjevic, M. S., Zurkowski, W., Shine, J., and Rolfe, B. G., 1983, Sym plasmid transfer to various symbiotic mutants of *Rhizobium trifolii, R. leguminosarum,* and *R. meliloti, J. Bacteriol.* **156**:1035–1045.

Döbereiner, J., Day, J. M., and Dart, P. J., 1972, Nitrogenase activity and oxygen sensitivity of the *Paspalum notatum–Azotobacter paspali* association, *J. Gen. Microbiol.* **71**:103–116.

Douglas, C. J., Halperin, W., and Nester, E. W., 1982, *Agrobacterium tumefaciens* mutants affected in attachment to plant cells, *J. Bacteriol.* **152**:1265–1275.

Fåhraeus, G., 1957, The infection of clover root hairs by nodule bacteria studied by a simple glass slide technique, *J. Gen. Microbiol.* **16**:374–381.

Fåhraeus, G., and Ljunggren, H., 1968, Pre-infection phases of the legume symbiosis, in: *The Ecology of Soil Bacteria* (T. R. G. Gray and D. Parkinson, eds.), University of Toronto Press, Toronto, pp. 396–421.

Fett, W. F., and Jones, S. B., 1982, Role of bacterial immobilization in race-specific resistance of soybean to *Pseudomonas syringae* pv. *glycinea, Phytopathology* **72**:488–492.

Fett, W. F., and Zacharius, R. M., 1983, Bacterial growth and phytoalexin elicitation on soybean cell suspension cultures inoculated with *Pseudomonas syringae* pathovars, *Physiol. Plant Pathol.* **22**:151–172.

Forrai, T., Vincze, E., Banfalvi, Z., Kiss, G. B., Randhawa, G. S., and Kondorosi, A., 1983, Localization of symbiotic mutations in *Rhizobium meliloti, J. Bacteriol.* **153**:635–643.

Foster, R. C., 1981, The ultrastructure and histochemistry of the rhizosphere, *New Phytol.* **89**:263–273.

Galloway, B. T., 1919, Giant crowngalls from the Florida Everglades, *Phytopathology* **9**:207–208.

Gee, M. M., Sun, C. N., and Dwyer, J. D., 1967, An electron microscope study of sunflower crown gall tumor, *Protoplasma* **64**:195–200.

Gnanamanickam, S. S., and Patil, S. S., 1977a, Accumulation of antibacterial isoflavonoids in hypersensitively responding bean leaf tissues inoculated with *Pseudomonas phaseolicola, Physiol. Plant Pathol.* **10**:159–168.

Gnanamanickam, S. S., and Patil, S. S., 1977b, Phaseotoxin suppresses bacterially induced hypersensitive reaction and phytoalexin synthesis in bean cultivars, *Physiol. Plant Pathol.* **10**:169–179.

Goodman, R. N., 1967, Protection of apple stem tissue against *Erwinia amylovora* infection by avirulent strains and three other bacterial species, *Phytopathology* **57**:22–24.

Goodman, R. N., 1968, The hypersensitive reaction in tobacco a reflection of changes in host cell permeability, *Phytopathology* **58**:872–873.

Goodman, R. N., 1972, Electrolyte leakage and membrane damage in relation to bacterial population, pH, and ammonia production in tobacco leaf tissue inoculated with *Pseudomonas pisi, Phytopathology* **62**:1327–1331.

Goodman, R. N., and Plurad, S. B., 1971, Ultrastructural changes in tobacco undergoing the hypersensitive reaction caused by plant pathogenic bacteria, *Physiol. Plant Pathol.* **1**:11–15.

Goodman, R. N., Huang, P.-Y., and White, J. A., 1976, Ultrastructural evidence for immobilization of an incompatible bacterium, *Pseudomonas pisi,* in tobacco leaf tissue, *Phytopathology* **66**:754–764.

Goodman, R. N., Politis, D. J., and White, J. A., 1977, Ultrastructural evidence of an ''active'' immobilization process of incompatible bacteria in tobacco leaf tissue: A resistance reaction, in: *Cell Wall Biochemistry Related to Specificity in Host–Plant Pathogen Interactions* (B. Solheim and J. Raa, eds.), Universitetsforlaget, Oslo, pp. 423–437.

Gordon, M. P., 1981, Tumor formation in plants, in: *The Biochemistry of Plants,* Volume 6 (P. K. Stumpf and E. E. Conn, eds.), Academic Press, New York, pp. 531–570.

Greaves, M. P., and Darbyshire, J. F., 1972, The ultrastructure of the mucilaginous layer on plant roots, *Soil Biol. Biochem.* **4**:443–449.

Haack, A., 1964, Uber den Einfluss der Knöllchenbakterien auf die Wurzelhaare von Leguminosen und Nichtleguminosen, *Zentralbl. Bakteriol. Parasitenkd. Infektionskr. Abt. 2 Hyg.* **117**:343–366.

Hepper, C. M., 1978, Physiological studies on nodule formation: The characteristics and inheritance of abnormal nodulation in *Trifolium pratense* by *Rhizobium leguminosarum, Ann. Bot.* **42**:109–115.

Hepper, C. M., and Lee, L., 1979, Nodulation of *Trifolium subterraneum* by *Rhizobium leguminosarum, Plant Soil* **51**:441–445.

Higashi, S., and Abe, M., 1980, Scanning electron microscopy of *Rhizobium trifolii* infection sites on root hairs of white clover, *Appl. Environ. Microbiol.* **40**:1094–1099.

Hildebrand, D. C., Alosi, M. C., and Schroth, M. N., 1980, Physical entrapment of pseudomonads in bean leaves by films formed at air–water interfaces, *Phytopathology* **70**:98–109.

Hirsch, A. M., Long, S. R., Bang, M., Haskins, N., and Ausubel, F. M., 1982, Structural studies of alfalfa roots infected with nodulation mutants of *Rhizobium meliloti, J. Bacteriol.* **151**:411–419.

Holliday, M. J., and Keen, N. T., 1982, The role of phytoalexins in the resistance of soybean leaves to bacteria: Effect of glyphosate on glyceollin accumulation, *Phytopathology* **72**:1470–1474.

Holmes, F. O., 1929, Local lesions in tobacco mosaic, *Bot. Gaz.* **87**:39–55.

Holt, J. G. (ed), 1977, *Bergey's Manual of Determinative Bacteriology,* 8th ed., Williams & Wilkins, Baltimore.

Hooykaas, P. J. J., vanBrussel, A. A. N., den Dulk-Ras, H., vanSlogteren, G. M. S., and Schilperoort, R. A., 1981, Sym plasmid of *Rhizobium trifolii* expressed in different rhizobial species and *Agrobacterium tumefaciens, Nature (London)* **291**:351–353.

Hooykaas, P. J. J., Snijdewint, F. G. M., and Schilperoort, R. A., 1982, Identification of the sym plasmid of *Rhizobium leguminosarum* strain 1001 and its transfer to and expression in other rhizobia and *Agrobacterium tumefaciens, Plasmid* **8**:73–82.

Horner, H. T., Jr., and Lersten, N. R., 1968, Development, structure and function of secretory trichomes in *Psychotria bacteriophila* (Rubiaceae), *Am. J. Bot.* **55:**1089–1099.

Horner, H. T., Jr., and Lersten, N. R., 1972, Nomenclature of bacteria in leaf nodules of the families Myrsinaceae and Rubiaceae, *Int. J. Syst. Bacteriol.* **22:**117–122.

Hsieh, S. P. Y., and Buddenhagen, I. W., 1974, Suppressing effects of *Erwinia herbicola* on infection by *Xanthomonas oryzae* and on symptom development in rice, *Phytopathology* **64:**1182–1185.

Huang, J.-S., and Goodman, R. N., 1970, The relationship of phosphatidase activity to the hypersensitive reaction in tobacco induced by bacteria, *Phytopathology* **60:**1020–1021.

Huang, J.-S., and Goodman, R. N., 1972, Alterations in structural proteins from chloroplast membranes of bacterially induced hypersensitive tobacco leaves, *Phytopathology* **62:**1428–1434.

Huang, J.-S., Huang, P.-Y., and Goodman, R. N., 1973, Reconstitution of a membrane-like structure with structural proteins and lipids isolated from tobacco thylakoid membranes, *Am. J. Bot.* **60:**80–85.

Huang, J.-S., Huang, P.-Y., and Goodman, R. M., 1974, Ultrastructural changes in tobacco thylakoid membrane protein caused by a bacterially induced hypersensitive reaction, *Physiol. Plant Pathol.* **4:**93–97.

Hubbell, D. H., 1970, Studies on the root hair "curling factor" of *Rhizobium, Bot. Gaz.* **131:**337–342.

Hubbell, D. H., 1981, Legume infection by *Rhizobium:* A conceptual approach, *BioScience* **31:**832–837.

Johansen, D. A., 1940, *Plant Microtechnique,* McGraw–Hill, New York.

Kahl, G., and Schell, J. S. (eds.), 1982, *Molecular Biology of Plant Tumors,* Academic Press, New York.

Kavimandan, S. K., Subba Rao, N. S., and Mohrir, A. V., 1978, Isolation of *Spirillum lipoferum* from the stems of wheat and nitrogen fixation in enrichment cultures, *Curr. Sci. (Bangalore)* **47:**96–98.

Keen, N. T., and Holliday, M. J., 1982, Recognition of bacterial pathogens by plants, in: *Phytopathogenic Prokaryotes,* Volume 2 (M. S. Mount and G. H. Lacy, eds.), Academic Press, New York, pp. 179–217.

Keen, N. T., and Kennedy, B. W., 1974, Hydroxyphaseollin and related isoflavonoids in the hypersensitive resistant response of soybeans against *Pseudomonas glycinea, Physiol. Plant Pathol.* **4:**173–185.

Keen, N. T., Ersek, T., Long, M., Bruegger, B. B., and Holliday, M., 1981, Inhibition of the hypersensitive reaction of soybean leaves to incompatible *Pseudomonas* spp. by blasticidin S, streptomycin or elevated temperature, *Physiol. Plant Pathol.* **18:**325–337.

Kino, Y., Nozu, M., and Itoi, S., 1979, Interaction between French bean primary leaf and *Agrobacterium tumefaciens* (Smith et Townsend) Conn. in earlier stages of infection, *Ann. Phytopathol. Soc. Jpn.* **45:**275–278.

Király, Z., 1980, Defenses triggered by the invader: Hypersensitivity, in: *Plant Disease: An Advanced Treatise,* Volume 5 (J. G. Horsfall and E. B. Cowling, eds.), Academic Press, New York, pp. 201–224.

Kleczkowska, J., Nutman, P. S., and Bond, G., 1944, Note on the ability of certain strains of *Rhizogium* from peas and clover to infect each other's host plants, *J. Bacteriol.* **48:**673–675.

Klement, Z., 1963, Rapid detection of the pathogenicity of phytopathogenic pseudomonads, *Nature (London)* **199:**299–300.

Klement, Z., 1971, The hypersensitive reaction of plants to bacterial infection, *Acta Phytopathol. Acad. Sci. Hung.* **6:**115–118.

Klement, Z., 1972, Development of the hypersensitive reaction induced by phytopathogenic bacteria, in: *Proc. Third Int. Congr. Plant Pathogenic Bacteria* (H. P. Maas Geesteranus, ed.), Cent. Agric. Publ. Doc., Wageningen, pp. 157–164.

Klement, Z., 1977, Cell contact recognition versus toxin action in induction of bacterial hypersensitive reaction, *Acta Phytopathol. Acad. Sci. Hung.* **12:**257–261.

Klement, Z., 1982, Hypersensitivity in: *The Phytopathogenic Prokaryotes,* Volume 2 (M. S. Mount and G. H. Lacy, eds.), Academic Press, New York, pp. 149–177.

Klement, Z., and Goodman, R. N., 1967a, The hypersensitive reaction to infection by bacterial plant pathogens, *Annu. Rev. Phytopathol.* **5:**17–44.

Klement, Z., and Goodman, R. N., 1967b, The role of the living bacterial cell and induction time in the hypersensitive reaction of the tobacco plant, *Phytopathology* **57:**322–323.

Klement, Z., and Lovrekovich, L., 1961, Defense reactions induced by phytopathogenic bacteria in bean pods, *Phytopathol. Z.* **41:**217–227.

Klement, Z., and Lovrekovich, L., 1962, Studies on host–parasite relations in bean pods infected with bacteria, *Phytopathol. Z.* **45:**81–88.

Klement, Z., Farkas, G. L., and Lovrekovich, L., 1964, Hypersensitive reaction induced by phytopathogenic bacteria in the tobacco leaf, *Phytopathology* **54:**474–477.

Kluepfel, D. A., and Pueppke, S. G., 1983, Surface ultrastructure of the adherence of *Agrobacterium tumefaciens* to potato tuber cells, *Abstr. Annu. Meet. Am. Soc. Microbiol.* **J20.**

Knowlton, S., and Dawson, J. O., 1983, Effects of *Pseudomonas cepacia* and cultural factors on the nodulation of *Alnus rubra* roots by *Frankia, Can. J. Bot.* **61:**2877–2882.

Knowlton, S., Berry, A., and Torrey, J. G., 1980, Evidence that associated soil bacteria may influence root hair infection of actinorhizal plants by *Frankia, Can. J. Microbiol.* **26:**971–977.

Kondorosi, E., Banfalvi, Z., and Kondorosi, A., 1984, Physical and genetic analysis of a symbiotic region of *Rhizobium meliloti:* Identification of nodulation genes, *Mol. Gen. Genet.* **193:**445–452.

Korhonen, T. K., Tarkka, E., Ranta, H., and Haahtela, K., 1983, Type 3 fimbriae of *Klebsiella* sp.: Molecular characterization and role in bacterial adhesion to plant roots, *J. Bacteriol.* **155:** 860–865.

Lalonde, M., 1977, Infection process of *Alnus* root nodule symbiosis, in: *Recent Developments in Nitrogen Fixation* (W. E. Newton, J. R. Postgate, and C. Rodriguez-Barrueco, eds.), Academic Press, New York, pp. 569–589.

Lalonde, M., and Calvert, H. E., 1979, Production of *Frankia* hyphae and spores as an infective inoculant for *Alnus* species in: *Symbiotic Nitrogen Fixation in the Management of Temperate Forests* (J. C. Gordon, C. T. Wheeler, and D. A. Perry, eds.), Orgeon State University Press, Corvallis, pp. 95–110.

Lalonde, M., and Quispel, A., 1977, Ultrastructural and immunological demonstration of the nodulation of the European *Alnus glutinosa (L.) Gaertn.* host plant by the North-American *Alnus crispa* var. *Mollis* Fern. root nodule endophyte, *Can. J. Microbiol.* **23:**1529–1547.

Lalonde, M., Calvert, H. E., and Pine, S., 1981, Isolation and use of *Frankia* strains in actinorhizae formation, in: *Current Perspectives in Nitrogen Fixation* (A. H. Gibson and W. E. Newton, eds.), Australian Academy of Science, Canberra, pp. 296–299.

Lange, R. T., 1966, Bacterial symbiosis with plants, in: *Symbiosis,* Volume 1 (S. M. Henry. ed.), Academic Press, New York, pp. 99–170.

Last, F. L., and Warren, R. C., 1972, Non-parasitic microbes colonizing green leaves: Their form and functions, *Endeavour* **31:**143–150.

Lechevalier, M. P., and Lechevalier, H. A., 1979, The taxonomic position of the actinomycetic endophytes, in: *Symbiotic Nitrogen Fixation in the Management of Temperate Forests* (J. C. Gordon, C. T. Wheeler, and D. A. Perry, eds.), Oregon State University Press, Corvallis, pp. 111–122.

Lersten, N. T., and Horner, H. T., Jr., 1967, Development and structure of bacterial leaf nodules in *Psychotria bacteriophila* Val. (Rubiaceae), *J. Bacteriol.* **94:**2027–2036.

Li, D., and Hubbell, D. H., 1970, Infection thread formation as a basis of nodulation specificity in *Rhizobium*–strawberry clover associations, *Can. J. Microbiol.* **15:**1133–1136.

Lipetz, J., 1970, The fine structure of plant tumors. I. Comparison of crown gall and hyperplastic cells, *Protoplasma* **70:**207–216.

Lippincott, J. A., and Heberlein, G. T., 1965, The induction of leaf tumors by *Agrobacterium tumefaciens, Am. J. Bot.* **52:**396–403.

Lippincott, J. A., and Lippincott, B. B., 1976, Morphogenic determinants as exemplified by the crown-gall disease, in: *Physiological Plant Pathology* (R. Heitefuss and P. H. Williams, eds.), Springer-Verlag, Berlin, pp. 356–388.

Lippincott, J. A., and Lippincott, B. B., 1980, Microbial adherence in plants, in: *Bacterial Adherence* (E. H. Beachey, ed), Chapman & Hall, London, pp. 375–398.

Ljunggren, H., 1969, Mechanism and pattern of *Rhizobium* invasion into leguminous root hairs, *Physiol. Plant. Suppl.* **V:**1–82.

Lynch, J. M., 1982, Interactions between bacteria and plants in the root environment, in: *Bacteria and Plants* (M. Rhodes-Roberts and F. A. Skinner, eds.), Academic Press, New York, pp. 1–23.

Lyon, F. M., and Wood, R. K. S., 1975, Production of phaseollin, coumestrol and related compounds in bean leaves inoculated with *Pseudomonas* spp., *Physiol. Plant Pathol.* **6:**117–124.

McCoy, E., 1932, Infection by *Bact. radicicola* in relation to the microchemistry of the host's cell walls, *Proc. R. Soc. London Ser. B* **110:**514–533.

McIntyre, J. L., Kuć, J., and Williams, E. B., 1973, Protection of pear against fire blight by bacteria and bacterial sonicates, *Phytopathology* **63:**872–877.

Malajczuk, N., and Bowen, G. D., 1974, Proteoid roots are microbially induced, *Nature (London)* **251:**316–317.

Manocha, M. A., 1970, Fine structure of crown gall tissue, *Can. J. Bot.* **48:**1455–1458.

Matthysse, A. G., 1983, Role of bacterial cellulose fibrils in *Agrobacterium tumefaciens* infection, *J. Bacteriol.* **154:**906–915.

Matthysse, A. G., Wyman, P. M., and Holmes, K. V., 1978, Plasmid-dependent attachment of *Agrobacterium tumefaciens* to plant tissue culture cells, *Infect. Immun.* **22:**516–522.

Matthysse, A. G., Holmes, K. V., and Gurlitz, R. H. G., 1981, Elaboration of cellulose fibrils by *Agrobacterium tumefaciens* during attachment to carrot cells, *J. Bacteriol.* **145:**583–595.

Matthysse, A. G., Holmes, K. V., and Gurlitz, R. H. G., 1982, Binding of *Agrobacterium tumefaciens* to carrot protoplasts, *Physiol. Plant Pathol.* **20:**27–33.

Meadows, M. E., and Stall, R. E., 1981, Different induction periods for hypersensitivity in pepper to *Xanthomonas vesicatoria* determined with antimicrobial agents, *Phytopathology* **71:**1024–1027.

Morrison, N. A., Hau, C. Y., Trinick, M. J., Shine, J., and Rolfe, B. G., 1983, Heat curing a sym plasmid in a fast-growing *Rhizobium* sp. that is able to nodulate legumes and the nonlegume *Parasponia* sp., *J. Bacteriol.* **153:**527–531.

Müller, K. O., 1959, Hypersensitivity in: *Plant Pathology: An Advanced Treatise* (J. G. Horsfall and A. E. Dimond, eds.), Academic Press, New York, pp. 469–519.

Munns, D. N., 1968a, Nodulation of *Medicago sativa* in solution culture. I. Acid-sensitive steps, *Plant Soil* **28:**129–146.

Munns, D. N., 1968b, Nodulation of *Medicago sativa* in solution culture. III. Effects of nitrate on root hairs and infection, *Plant Soil* **29:**33–47.

Napoli, C., and Albersheim, P., 1980, *Rhizobium leguminosarum* mutants incapable of normal extracellular polysaccharide production, *J. Bacteriol.* **141:**1454–1456.

Napoli, C. A., and Hubbell, D. H., 1975, Ultrastructure of *Rhizobium*-induced infection threads in clover root hairs, *Appl. Microbiol.* **30:**1003–1009.

Németh, J., and Klement, Z., 1967, Changes in respiration rate of tobacco leaves infected with bacteria in relation to the hypersensitive reaction, *Acta Phytopathol. Acad. Sci. Hung.* **2:**303–308.

Nester, E. W., and Kosuge, T., 1981, Plasmids specifying plant hyperplasias, *Annu. Rev. Microbiol.* **35:**531–565.

Newcomb, W., Sippel, D., and Peterson, R. L., 1979, The early morphogenesis of *Glycine max* and *Pisum sativum* root nodules, *Can. J. Bot.* **57:**2603–2616.

Nissen, P., 1971, Choline sulfate permease: Transfer of information from bacteria to higher plants? II. Induction processes, in: *Information Molecules in Biological Systems* (L. Ledoux, ed.), North-Holland, Amsterdam, pp. 201–212.

Nutman, P. S., 1949, Nuclear and cytoplasmic inheritance of resistance to infection by nodule bacteria in red clover, *Heredity* **3:**263–291.

Nutman, P. S., 1951, Host factors influencing infection and nodule development in leguminous plants, *Proc. R. Soc. London Ser. B* **139:**176–185.

Nutman, P. S., 1956, The influence of the legume in root–nodule symbiosis: A comparative study of host determinants and functions, *Biol. Rev. Cambridge Philos. Soc.* **31:**109–151.

Nutman, P. S., 1959, Some observations on root-hair infection by nodule bacteria, *J. Exp. Bot.* **10:**250–263.

Nutman, P. S., 1962, The relation between root hair infection by *Rhizobium* and nodulation in *Trifolium* and *Vicia, Proc. R. Soc. London Ser. B* **156:**122–137.

Nutman, P. S., 1965, The relation between nodule bacteria and the legume host in the rhizosphere and in the process of infection, in: *Ecology of Soil-borne Plant Pathogens* (K. F. Baker and W. C. Snyder, eds.), University of California Press, Berkeley, pp. 231–246.

Nutman, P. S., 1967, Varietal differences in the nodulation of subterranean clover, *Aust. J. Agric. Res.* **18:**381–425.

Nutman, P. S., 1969, Genetics of symbiosis and nitrogen fixation in legumes, *Proc. R. Soc. London Ser. B* **172:**417–437.

Nutman, P. S., 1981, Hereditary host factors affecting nodulation and nitrogen fixation, in: *Current Perspectives in Nitrogen Fixation* (A. H. Gibson and W. E. Newton, eds.), Australian Academy of Science, Canberra, pp. 194–204.

O'Brien, F., and Wood, R. K. S., 1973, Antibacterial substances in hypersensitive responses induced by bacteria, *Nature (London)* **242:**532–533.

Obukowicz, M., and Kennedy, G. S., 1981, Phenolic ultracytochemistry of tobacco cells undergoing the hypersensitive reaction to *Pseudomonas solanacearum, Physiol. Plant Pathol.* **18:**339–344.

Patil, S. S., and Gnanamanickam, S. S., 1976, Suppression of bacterially-induced hypersensitive reaction and phytoalexin accumulation in bean by phaseotoxin, *Nature (London)* **259:**486–487.

Patriquin, D. G., and Döbereiner, J., 1978, Light microscopy observations of tetrazolium-reducing bacteria in the endorhizosphere of maize and other grasses in Brazil, *Can. J. Microbiol.* **24:**734–742.

Patriquin, D. G., Döbereiner, J., and Jain, D. K., 1983, Sites and processes of association between diazotrophs and grasses, *Can. J. Microbiol.* **29:**900–915.

Pavlovkin, J., and Novacky, A., 1984, Bacterial hypersensitivity-related alterations in cell membranes, in: *Membrane Transport in Plants* (W. J. Cram, K. Janacek, R. Rybova, and K. Sigler, eds.), Academia, Prague, p. 423.

Périnet, P., and Lalonde, M., 1983, In vitro propagation and nodulation of the actinorhizal host plant *Alnus glutinosa* (L.) Gaertn., *Plant Sci. Lett.* **29:**9–17.

Pizelle, G., 1972, Observations sur les racines de plantules d'Aune glutineux (*Alnus glutinosa* Gaertn.), *Bull. Soc. Bot. Fr.* **119:**571–580.

Politis, D. J., and Goodman, R. N., 1978, Localized cell wall appositions: Incompatibility response of tobacco leaf cells to *Pseudomonas pisi, Phytopathology* **68:** 309–316.

Pommer, E. H., 1956, Beitrage zur Anatomie und Biologie der Wurzelknollchen von *Alnus glutinosa* Gaertn., *Flora (Jena)* **143**:603–634.

Pommer, E. H., 1959, Uber die Isolierung des Endophyten aus den Wurzelknollchen *Alnus glutinosa* Gaertn. und uber erfolgreiche Re-Infektionsversuche, *Ber. Dtsch. Bot. Ges.* **72**:138–150.

Pueppke, S. G., 1983, *Rhizobium* infection threads in root hairs of *Glycine max* (L.) Merr., *Glycine soja* Sieb. & Zucc., and *Vigna unguiculata* (L.) Walp., *Can. J. Microbiol.* **29**:69–76.

Pueppke, S. G., 1984, Adsorption of bacteria to plant surfaces, in: *Plant–Microbe Interactions, Molecular and Genetic Perspectives* (T. Kosuge and E. W. Nester, eds.), Macmillan Co., New York, pp. 215–261.

Purchase, H. F., 1958, Restriction of infection threads in nodulation of clover and lucerne, *Aust. J. Biol. Sci.* **11**:155–161.

Quispel, A., and Burggraaf, A. J. P., 1981, *Frankia,* the diazotrophic endophyte from actinorhiza's, in: *Current Perspectives in Nitrogen Fixation* (A. H. Gibson and W. E. Newton, eds.), Australian Academy of Science, Canberra, pp. 229–236.

Ranga Rao, V., 1977, Effect of root temperature on the infection processes and nodulation in *Lotus* and *Stylosanthes, J. Exp. Bot.* **28**:241–259.

Ranga Rao, V., and Keister, D. L., 1978, Infection threads in the root hairs of soybean (*Glycine max*) plants inoculated with *Rhizobium japonicum, Protoplasma* **97**:311–316.

Rennie, R. J., 1981, Diazotrophic biocoenosis—The workshop consensus paper, in: *Associative N₂-Fixation,* Volume II (P. B. Vose and A. P. Ruschel, eds.), CRC Press, Boca Raton, Fla., pp. 253–258.

Riker, A. J., 1923a, Some morphological responses of the host tissue to the crowngall organism, *J. Agric. Res.* **26**:425–435.

Riker, A. J., 1923b, Some relations of the crowngall organism to its host tissue, *J. Agric. Res.* **25**:119–132.

Riker, A. J., 1927, Cytological studies of crowngall tissue, *Am. J. Bot.* **14**:25–37.

Robertson, J. G., Lyttleton, P., and Pankhurst, C. E., 1981, Preinfection and infection processes in the legume–*Rhizobium* symbiosis, in: *Current Perspectives in Nitrogen Fixation* (A. H. Gibson and W. E. Newton, eds.), Australian Academy of Science, Canberra, pp. 280–291.

Robinson, W., and Walkden, H., 1923, A critical study of crown gall, *Ann. Bot.* **37**:299–324.

Roebuck, P., Sexton, R., and Mansfield, J. W., 1978, Ultrastructural observations on the development of the hypersensitive reaction in leaves of *Phaseolus vulgaris* cv. Red Mexican inoculated with *Pseudomonas phaseolicola* (race 1), *Physiol. Plant Pathol.* **12**:151–157.

Rovira, A. D., and Campbell, R., 1974, Scanning electron microscopy of microorganisms on the roots of wheat, *Microb. Ecol.* **1**:15–23.

Sahlman, K., and Fåhraeus, G., 1963, An electron microscope study of root-hair infection by *Rhizobium, J. Gen. Microbiol.* **33**:425–427.

Schank, S. C., Smith, R. L., Weiser, G. C., Zuberer, D. A., Bouton, J. H., Quesenberry, K. H., Tyler, M. E., Milam, J. R., and Littell, R. C., 1979, Fluorescent antibody technique to identify *Azospirillum brasilense* associated with roots of grasses, *Soil Biol. Biochem.* **11**:287–295.

Scott, D. B., and Ronson, C. W., 1982, Identification and mobilization by cointegrate formation of a nodulation plasmid in *Rhizobium trifolii, J. Bacteriol.* **151**:36–43.

Scott, F. M., 1948, Internal suberization of plant tissues, *Science* **108**:654–655.

Scott, F. M., 1950, Internal suberization of tissues, *Bot. Gaz.* **111**:378–394.

Sequeira, L., 1980, Defense triggered by the invader: Recognition and compatibility phenomena, in: *Plant Disease: An Advanced Treatise,* Volume 5 (J. G. Horsfall and E. B. Cowling, eds.), Academic Press, New York, pp. 179–200.

Sequeira, L., 1983, Recognition and specificity between plants and pathogens, in: *Challenging Problems in Plant Health* (T. Kommedahl and P. H. Williams, eds.), American Phytopathological Society, St. Paul, pp. 301–310.

Sequeira, L., Gaard, G., and DeZoeten, G. A., 1977, Attachment of bacteria to host cells walls: Its relation to mechanisms of induced resistance, *Physiol. Plant Pathol.* **10**:43–50.

Sigee, D. C., and Al-Issa, A. N., 1983, The hypersensitive reaction in tobacco leaf tissue infiltrated with *Pseudomonas pisi.* 4. Scanning electron microscope studies on fractured leaf tissue, *Phytopathol. Z.* **106**:1–15.

Sigee, D. C., and Epton, H. A. S., 1976, Ultrastructural changes in resistant and susceptible varieties of *Phaseolus vulgaris* following artificial inoculation with *Pseudomonas phaseolicola, Physiol. Plant Pathol.* **9**:1–8.

Sigee, D. C., Smith, V. A., and Hindley, J., 1982, Passage of bacterial DNA into host cells during in vitro transformation of *Nicotiana tabacum* by *Agrobacterium tumefaciens, Microbios* **34**:113–132.

Sing, V. O., and Schroth, M. N., 1977, Bacteria–plant cell surface interaction: Active immobilization of saprophytic bacteria in plant leaves, *Science* **197**:759–761.

Skrdleta, V., 1970, Competition for nodule sites between two inoculum strains of *Rhizobium japonicum* as affected by delayed inoculation, *Soil Biol. Biochem.* **2**:167–171.

Slusarenko, A. J., and Wood, R. K. S., 1983, Agglutination of *Pseudomonas phaseolicola* by pectic polysaccharide from leaves of *Phaseolus vulgaris, Physiol. Plant Pathol.* **23:**217–227.

Smith, J. J., and Mansfield, J. W., 1982, Ultrastructure of interactions between pesudomonads and oat leaves, *Physiol. Plant Pathol.* **21:**259–266.

Smith, R. L., Schank, S. C., Bouton, J. H., and Quesenberry, K. H., 1978, Yield increases of tropical grasses after inoculation with *Spirillum lipoferum, Ecol. Bull. (Stockholm)* **26:**380–385.

Solheim, B., and Raa, J., 1973, Characterization of the substances causing deformation of root hairs of *Trifolium repens* when inoculated with *Rhizobium trifolii, J. Gen. Microbiol.* **77:**241–247.

Spiess, L. D., Lippincott, B. B., and Lippincott, J. A., 1971, Development and gametophore induction in the moss *Pylaisiella selwynii* as influenced by *Agrobacterium tumefaciens, Am. J. Bot.* **58:**726–731.

Spiess, L. D., Lippincott, B. B., and Lippincott, J. A., 1972, Influence of certain plant growth regulators and crown-gall related substances on bud formation and gametophore development of the moss *Pylaisiella selwynii, Am. J. Bot.* **59:**233–241.

Spiess, L. D., Lippincott, B. B., and Lippincott, J. A., 1973, Effect of hormones and vitamin B_{12} on gametophore development in the moss *Pylaisiella selwynii, Am. J. Bot.* **60:**708–716.

Spiess, L. D., Lippincott, B. B., and Lippincott, J. A., 1976, The requirement of physical contact for moss gametophore induction by *Agrobacterium tumefaciens, Am. J. Bot.* **63:**324–328.

Spiess, L. D., Lippincott, B. B., and Lippincott, J. A., 1977a, Comparative response of *Pylaisiella selwynii* to *Agrobacterium* and *Rhizobium* species, *Bot. Gaz.* **138:**35–40.

Spiess, L. D., Turner, J. C., Mahlberg, P. G., Lippincott, B. B., and Lippincott, J. A., 1977b, Adherence of agrobacteria to moss protonema and gametophores viewed by scanning electron microscopy, *Am. J. Bot.* **64:**1200–1208.

Stacey, G., Paau, A. S., and Brill, W. J., 1980, Host recognition in the *Rhizobium*–soybean symbiosis, *Plant Physiol.* **66:**609–614.

Stall, R. E., and Cook, A. A., 1979, Evidence that bacterial contact with the plant cell is necessary for the hypersensitive reaction but not the susceptible reaction, *Physiol. Plant Pathol.* **14:**77–84.

Stholasuta, P., Bailey, J. A., Severin, V., and Deverall, B. J., 1971, Effect of bacterial inoculation of bean and pea leaves on the accumulation of phaseollin and pisatin, *Physiol. Plant Pathol.* **1:**177–183.

Taubert, H., 1956, Über den Infektionsvorgang und die Entwicklung der Knöllchen bei *Alnus glutinosa* Gaertn., *Planta* **48:**135–156.

Therman, E., 1956, Dedifferentiation and differentiation of cells in crown gall of *Vicia faba, Caryologia* **8:**325–348.

Thornton, H. G., 1936, The action of sodium nitrate upon the infection of lucerne root-hairs by nodule bacteria, *Proc. R. Soc. London Ser. B* **119:**474–492.

Tien, T. M., Gaskins, M. H., and Hubbell, D. H., 1979, Plant growth substances produced by *Azospirillum brasilense* and their effect on the growth of pearl millet (*Pennisetum americanum* L.), *Appl. Environ. Microbiol.* **37:**1016–1024.

Torrey, J. G., 1976, Initiation and development of root nodules of *Casuarina* (Casuarinaceae), *Am. J. Bot.* **63:**335–344.

Torrey, J. G., 1978, Nitrogen fixation by Actinomycete-nodulated angiosperms, *BioScience* **28:**586–592.

Torrey, J. G., and Callaham, D. A., 1979, Early nodule development in *Myrica gale, Bot. Gaz.* **140**(Suppl.):10–14.

Torrey, J. G., Baker, D., Callaham, D., Del Tredici, P., Newcomb, W., Peterson, R. L., and Tjepkema, J. D., 1980, On the nature of the endophyte causing root nodulation in *Comptonia*, in: *Nitrogen Fixation*, Volume II (W. E. Newton and W. H. Orme-Johnson, eds.), University Park Press, Baltimore, pp. 217–227.

Trinick, M. J., 1982, Host–*Rhizobium* association, in: *Nitrogen Fixation in Legumes* (J. M. Vincent, ed.), Academic Press, New York, pp. 111–122.

Truchet, G. L., and Dazzo, F. B., 1982, Morphogenesis of lucerne root nodules incited by *Rhizobium meliloti* in the presence of combined nitrogen, *Planta* **154:**352–360.

Truchet, G. L., Rosenberg, C., Vasse, J., Julliot, J.-S., Camut, S., and Denarie, J., 1984, Transfer of *Rhizobium meliloti* pSym genes into *Agrobacterium tumefaciens*: Host-specific nodulation by atypical infection, *J. Bacteriol.* **157:**134–142.

Tu, J. C., 1981, Effect of salinity on *Rhizobium*–root hair interaction, nodulation and growth of soybean, *Can. J. Plant Sci.* **61:**231–239.

Turgeon, B. G., and Bauer, W. D., 1982, Early events in the infection of soybean by *Rhizobium japonicum:* Time course and cytology of the initial infection process, *Can. J. Bot.* **60:**152–161.

Turgeon, B. G., and Bauer, W. D., 1984, Ultrastructure of infection–thread development during the infection of soybean by *Rhizobium japonicum, Planta*, in press.

Umali-Garcia, M., Hubbell, D. H., and Gaskins, M. H., 1978, Process of infection of *Panicum maximum* by *Spirillum lipoferum, Ecol. Bull. (Stockholm)* **26:**373–379.

Umali-Garcia, M., Hubbell, D. H., Gaskins, M. H., and Dazzo, F. B., 1979, Adsorption and mode of entry of *Azospirillum brasilense* to grass roots, in: *Associative N₂-fixation*, Volume I (P. B. Vose and A. P. Ruschel, eds.), CRC Press, Boca Raton, Fla., pp. 49–62.

Umali-Garcia, M., Hubbell, D. H., Gaskins, M. H., and Dazzo, F. B., 1980, Association of *Azospirillum* with grass roots, *Appl. Environ. Microbiol.* **39:**219–226.

vanBerkum, P., and Bohlool, B. B., 1980, Evaluation of nitrogen fixation by bacteria in association with roots of tropical grasses, *Microbiol. Rev.* **44:**491–517.

VanEtten, H. D., and Pueppke, S. G., 1976, Isoflavonoid phytoalexins, in: *Biochemical Aspects of Plant–Parasite Relationships* (J. Friend and D. R. Threlfall, eds.), Academic Press, New York, pp. 239–289.

Vincent, J. M., 1974, Root–nodule symbioses with *Rhizobium*, in: *Biological Nitrogen Fixation* (A. Quispel, ed.), North-Holland, Amsterdam, pp. 265–341.

Vincent, J. M., 1977, *Rhizobium:* General Microbiology, in: *A Treatise on Dinitrogen Fixation*, Section III (R. W. F. Hardy and W. S. Silver, eds.), Wiley–Interscience, New York, pp. 277–366.

Vincent, J. M., 1980, Factors controlling the legume–*Rhizobium* symbiosis, in: *Nitrogen Fixation*, Volume II (W.E. Newton and W. H. Orme-Johnson, eds.), University Park Press, Baltimore, pp. 103–129.

Ward, H. M., 1887, On the tubercular swellings on the roots of *Vicia faba*, *Philos. Trans. R. Soc. London Ser. B* **178:**539–562.

Ward, H. M., 1902, On the relations between host and parasite in the Bromes and their brown rust, *Puccinia dispersa* (Erikss.), *Ann. Bot.* **16:**233–315.

Watanabe, I., and Barraquio, W. L., 1979, Low levels of fixed nitrogen required for isolation of free-living N₂-fixing organisms from rice roots, *Nature (London)* **277:**565–566.

Whitmoyer, R. E., and Horner, H. T., Jr., 1970, Developmental aspects of bacterial leaf nodules in *Psychotria bacteriophila* Val. (Rubiaceae), *Bot. Gaz.* **131:**193–200.

Williams, L. F., and Lynch, D. L., 1954, Inheritance of a non-nodulating character in the soybean, *Agron. J.* **46:**28–29.

Wrather, J. A., Kuć, J., and Williams, E. B., 1973, Protection of apple and pear fruit against fireblight with nonpathogenic bacteria, *Phytopathology* **63:**1075–1076.

Yao, P. Y., and Vincent, J. M., 1969, Host specificity in the root hair "curling factor" of *Rhizobium* spp., *Aust. J. Biol. Sci.* **22:**413–423.

Yao, P. Y., and Vincent, J. M., 1976, Factors responsible for the curling and branching of clover root hairs by *Rhizobium*, *Plant Soil* **45:**1–16.

Zeyen, R. J., and Bushnell, W. R., 1981, An in-block, light microscope viewing procedure for botanical materials in plastic embedments; with emphasis on location and selection of host cell–microbe encounter sites, *Can. J. Bot.* **59:**397–402.

Zurkowski, W., 1980, Specific adsorption of bacteria to clover root hairs, related to the presence of the plasmid pWZ2 in cells of *Rhizobium trifolii*, *Microbios* **27:**27–32.

15

Effects on Host Animals of Bacteria Adhering to Epithelial Surfaces

DWAYNE C. SAVAGE

1. INTRODUCTION

1.1. Association and Adhesion

Bacterial strains of many genera and species, both pathogenic and nonpathogenic, are known to associate with epithelial surfaces in animals of several taxonomic orders (Savage, 1972, 1980a,b, 1983; Jones, 1977; Lee, 1980). As documented in early chapters in this volume, in some cases the bacterial cells adhere to the epithelial cells with which they associate. As discussed elsewhere (Savage, 1980a,b, 1983), however, in other cases the microbial cells may not adhere to the epithelial cells, but rather may colonize secretions, principally mucins, overlying the cells with which they associate. In some, perhaps most, instances the mechanisms by which bacterial cells associate with particular living surfaces are unknown. Methods used for detecting microbial cells associating with animal cell surfaces do not always yield unequivocal evidence that microorganisms seen actually adhere to the epithelial cell membranes (Savage, 1983). This problem may be of little importance, however, for a discussion of the mechanisms by which microorganisms associated with a surface affect their animal host.

With only a few exceptions, whether microorganisms adhere to a surface or colonize secretions over it, they may have much the same overall impact on the functions of the epithelial surface or other tissues of an animal. [Exceptions are microorganisms that alter the architecture of or invade into or through the epithelium after adhering to it; see below (Neutra, 1980; Savage, 1980a).] In the paragraphs to follow, therefore, examples are drawn from cases where bacteria are known to adhere to epithelial cells, and where information, usually from microscopy, is not yet conclusive that the microbial cells adhere to substratum epithelial cells.

1.2. Effects on Hosts of Bacteria Associating with Epithelia

Bacterial pathogens of several species induce disease in their animal hosts while adhering to epithelial surfaces (Table I). The mechanisms by which some of these microorganisms

---

Dwayne C. Savage • Department of Microbiology, University of Illinois, Urbana, Illinois 61801.

Table I. Mechanisms by Which Some Bacterial Pathogens Associated with Mucosal Membranes Induce Disease in Man[a]

Mode	Diseases (representative)	Etiological agent	Area of body infected[b]	Action of protein toxins	References
Epicellular pathogens Bacteria may adhere to mucosal epithelium; rarely if ever penetrate into epithelium or submucosa; induce disease by multiplying on epithelium; may produce extracellular protein toxin	Primary atypical pneumonia	*Mycoplasma pneumoniae*	Upper respiratory	No known extracellular toxin	Hu et al. (1982)
	Diphtheria	*Corynebacterium diphtheriae*	Upper respiratory	Adenosine diphosphoribosyl transferase; inhibits protein synthesis in susceptible cells	Moynihan and Pappenheimer (1981)
	Pertussis	*Bordetella pertussis*	Upper respiratory	*Pertussis toxin: Adenosine diphosphoribosyl transferase*; increase intracellular levels of cAMP in susceptible cells *Adenylate cyclase:* Increases intracellular levels of cAMP in susceptible cells	Confer and Eaton (1982), Tuomanen and Hendley (1983), Burns et al. (1983).
	Cholera	*Vibrio cholerae*	Small intestine	Adenosine diphosphoribosyl transferase; increases intracellular levels of cAMP in susceptible cells	Mekalanos et al. (1979), Finkelstein et al. (1983)
	E. coli enteritis	*Escherichia coli*	Small intestine	*Heat-labile toxin:* Same as cholera toxin *Heat-stable toxin:* Increases intracellular levels of cGMP in susceptible cells *Shigella-like toxin:* Inhibits protein synthesis in susceptible cells	Gill et al. (1981) Frantz et al. (1984), Guerrant et al. (1980) O'Brien and LaVeck (1983)

Category	Disease	Organism	Area of body[b]	Toxin	References
Intraepithelial pathogens Bacteria may adhere to and penetrate into mucosal epithelia; rarely penetrate into subepithelial tissues; induce disease by multiplying within and killing epithelial cells; some species may produce extracellular protein toxins	Cystitis and pyelonephritis	*Escherichia coli* (other species of gram-negative bacteria)	Bladder, ureters, kidney tubules	No known extracellular toxin	Svanborg-Eden and Jodal (1979)
	Gonorrhea	*Neisseria gonorrhoeae*	Urogenital tract	No known protein toxin	Johnson (1981)
	Bacillary dysentery	*Shigella dysenteriae*	Colon	Inhibits protein synthesis in susceptible cells	O'Brien et al. (1980), Sansonetti et al. (1982)
	E. coli dysentery	*Escherichia coli*	Colon	No toxin yet reported	O'Brien et al. (1980)
Subepithelial pathogens Bacteria adhere to and penetrate through mucosal epithelia into subepithelial tissues; induce disease by multiplying in submucosal tissues; some species grow intracellularly in macrophages; some may produce extracellular protein toxins	Bacterial meningitis	*Neisseria meningitidis*	Pharynx[c]	No known toxin	Salit and Morton (1981)
	Streptococcal pharyngitis	*Streptococcus pyogenes*	Pharynx	*Scarlatinal toxin*: Induces fever and rash by unknown mechanisms, disrupts action of cell membranes	Duncan (1975)
	Salmonella enteritis	*Salmonella* sp.	Small intestine	Similar to cholera toxin	Molina and Peterson (1980)
	E. coli enteritis	*Escherichia coli*	Small intestine	No toxin yet reported	Clancy and Savage (1981)
	Enterocolitis	*Yersinia enterocolitica*	Small intestine	No toxin yet reported	Gemski et al. (1980)
	"Staph" infection	*Staphylococcus aureus*	Skin, nasal mucosa	Numerous toxins	Rogolsky (1979)

[a] Adapted from Savage (1972).
[b] Area of body commonly infected.
[c] In carriers.

adhere to the epithelial cells are described in Chapters 10 and 11. The mechanisms by which the bacteria impact on the host tissues are known in some molecular detail in a few of these cases, but are not well understood in most instances. In most cases, however, by comparison with indigenous microorganisms ["normal flora" (Savage, 1977b)], pathogenic bacteria associate with tissue for a relatively short time during a disease process.

Some pathogens adhere to epithelial cells only transiently as a step in a process in which they invade into or through the cells (Table I). Others remain adherent to the epithelium throughout the disease process. Even in these cases, however, the cell–cell interaction is usually of short duration (a few days or weeks). Either the host eliminates the cells of the pathogen from its body surface and recovers, or fails to do so and dies. Whatever the result, an interaction of a host and a microbial pathogen leading to disease is usually a transient one of short duration by comparison with an interaction between a host and an indigenous bacterium.

Indigenous bacteria associate with epithelial surfaces in all areas of the body exposed to the external environment of animals of most species studied, including man (Lee, 1980; Savage, 1980a, 1983). Populations of such organisms are believed to occupy their epithelial habitats during much of the lifetime of their animal hosts. Thus, in contrast to bacterial pathogens, indigenous microorganisms can impact on host function throughout the lives of the animals. Largely because of the findings from extensive research with germfree animals [animals free of all microorganisms except a few viruses (Sasaki *et al.,* 1981)], indigenous microorganisms, especially those that colonize habitats in the gastrointestinal canal, are well known to alter many physiological properties and some anatomical characteristics of mammals and birds. In sharp contrast with the case with some bacterial pathogens, however, little is known about the precise mechanisms by which indigenous microorganisms affect host functions, even though such influences may have profound consequences for the health and welfare of the animals involved. These issues are amplified in succeeding paragraphs.

1.3. Limitations and Goals

Many examples can be cited of bacterial pathogens or members of the "normal flora" associating with epithelial surfaces in animals of numerous species. However, most detailed research has been accomplished with domestic fowl and mammals of a few species, including humans. Therefore, the discussion is limited to findings with those animals. Even in those animals, pathogenic and nonpathogenic bacteria may influence many physiological properties. As a consequence, detailed discussion is limited to cases which in my opinion best achieve the primary goal, which is to describe the state-of-the-art in research on interactions between animals and bacteria adherent to their epithelial surfaces. One of the best studied cases in which bacteria are known to induce disease in an animal is dental caries. The "tissue" affected is not an epithelium, as such, however, but is the hardened surface of teeth. Moreover, the issue is developed well in Chapter 10 (Section 4.1), and is discussed briefly in Chapters 6 (Section 4.1) and 13 (Section 8). Therefore, it is not covered in this chapter.

2. EFFECTS ON HOSTS LEADING TO DISEASE

2.1. Introduction: Bacterial Pathogens That Associate with Epithelial Surfaces

Microbial disease results when microbial cells or their products and host tissues interact in such ways that the animal feels ill-at-ease or otherwise manifests some detectable signs that cells or tissues are malfunctioning in some way. In the total spectrum of intimate interactions between animals and microbial cells, however, very few such interacttions ever damage the host; some, perhaps most, are even favorable to or protective of host function (Savage, 1972). Nevertheless, because of their genetic makeup, some bacterial strains may virtually always induce disease when they contact ("infect") an epithelial surface in an animal body. Such strains might be referred to as "frank" pathogens (Table I). Some of these strains affect host function while only associating superficially with epithelial cells ("epicellular pathogens"). Others penetrate into and grow inside of epithelial cells ("interaepithelial pathogens"). Still others penetrate into and through the epithelium into submucosal tissues and may travel throughout the body where they grow in organs sometimes far removed from the primary site of contact ("subepithelial pathogens").

Indigenous inhabitants of various surfaces of the animal body normally live in harmony with their hosts, in equilibrium with the animal's resistance mechanisms. If the resistance mechanisms are disturbed in some way, however, then even some strains of these organisms may induce an interaction that leads to disease (Freter, 1971, 1983; Savage, 1977a). Such species are regarded as "opportunistic" pathogens. Likewise, even when host resistance mechanisms may seem to be controlling microorganisms colonizing body surfaces, some indigenous bacteria may be involved in inducing disease, requiring long periods to develop, e.g., colonic cancer (Hill, 1983). Such microorganisms might be called "subterranean" pathogens. These concepts are amplified below.

2.2. Mechanisms by Which Bacterial Pathogens Associated with Epithelial Surfaces Induce Disease

2.2.1. Epicellular Pathogens

Pathogenic strains of certain bacterial species associate with epithelial surfaces, but rarely if ever penetrate into or through the epithelium. Rather, they remain associated with the epithelial membranes and synthesize and secrete complex proteins which either enter the substratum epithelial cells and disturb their function, or pass through blood and lymph to parts of the body remote from the surface where they enter host cells and alter their functions. Such proteins bind to receptors in the cell membranes of the host (Chapter 4; Eidels *et al.*, 1983), enter or pass through the membranes, and alter the function of the cells in elaborate and complex ways (Table I).

Certain pathogens of this type have been studied extensively. They are *Vibrio cholerae,* the etiologic agent of cholera, and certain strains of *Escherichia coli* that cause "traveler's" or "neonatal" diarrhea ["enterotoxigenic strains of *E. coli*" (ETEC)]. These bacterial pathogens enter the gastrointestinal tract via food or water, and localize predominantly in the small intestine. In that area of the tract, *V. cholerae,* a motile bacterium, is atttracted by chemotactic influences into the mucous gel overlying the mucosal epithelium (Freter *et al.,* 1981), where it probably colonizes (i.e., multiplies) in the gel and may adhere to the membranes of the substratum epithelial cells (Freter and Jones, 1976; Finkelstein *et al.,* 1983). ETEC strains also presumably enter the mucous gel, because their cells somehow make their way to the epithelial cells and adhere via specific receptors to the microvillous membranes of the epithelium. (The mechanisms mediating that adhesion are detailed in Chapter 11.)

Both *V. cholerae* and *E. coli* synthesize and secrete proteins ("enterotoxins"), which bind to specific receptors in the substratum and disturb the physiology of the epithelial cells, causing them to hypersecrete fluids and electrolytes (Field, 1979). Enteropathogenic strains of *E. coli* secrete one or both of two classes of enterotoxic proteins. One of these classes is the so-called heat-stable ("ST") toxins. [ST_a, which has biological activity in humans, suckling mice, rats, rabbits, and piglets, and ST_b, with biological activity only in piglets (So and McCarthy, 1980; Dreyfus *et al.*, 1983; Mosely *et al.*, 1983) have been described so far.] These toxins are small polypeptides of molecular weights from about 2000 to 5000. The other class, the heat-labile toxin ("LT"), is a large, oligomeric protein of molecular weight of about 84,000 (Gill *et al.*, 1981). The toxin ("CT") secreted by *V. cholerae* is similar molecularly and functionally to *E. coli* LT. It too is a complex protein of molecular weight of about 84,000 (Field, 1979). All of these toxins induce intestinal enterocytes to hypersecrete fluids. That physiological activity is induced, however, by at least two different mechanisms.

   E. coli STs function by one mechanism, whereas *E. coli* LT and cholera toxin function by another mechanism that is similar if not identical for both toxins (Field, 1979; Gill and Richardson, 1980). *E. coli* STs bind to specific high-affinity receptors in the membranes of enterocytes (Frantz *et al.*, 1984), and from there enter or pass through the membranes, setting in motion a complex series of events (Dreyfus *et al.*, 1984), the result of which is an increase over the levels in unaffected cells of the activity of guanylate cyclase and of intracellular levels of cGMP. The high levels of cGMP function in some way to cause the cells to secrete more fluids and electrolytes than normal. The net effect for the animal host is watery diarrhea (Field, 1979; Guerrant *et al.*, 1980).

   E. coli LT and *V. cholerae* toxins also induce watery diarrhea in man and animals of other species. (In nature, *V. cholerae* causes disease only in humans.) In these cases, however, the large proteins involved have at least two functional regions (A and B) that can be distinguished by peptidase activity and sulfhydryl reduction. The B region of the molecules consists of five polypeptides each of about 11,500 daltons, and serves to bind the entire protein to specific macromolecular receptors in the substratum membrane. The receptor for cholera toxin is known to be GM_1 ganglioside (Field, 1979), The receptor for *E. coli* LT has not yet been defined with certainty but may be a macromolecule other than GM_1 (Holmgren *et al.*, 1982). The B part of the toxin functions as well, perhaps in some cooperative interaction with the membrane receptor, to "translocate" the A polypeptide (molecular weight of about 30,000) into or through the membrane. Once inside the cell, the A polypeptide functions as an enzyme to catalyze the overall reaction:

$$\text{NAD + GTP-binding protein} \overset{\text{CT or LT}}{\rightleftharpoons} \text{ADP-ribosyl–GTP-binding protein + nicotinamide + H}$$

where NAD is nicotinamide adenine dinucleotide, GTP-binding protein is a regulatory element for the enzyme adenylate cyclase (see below), and "ADP-ribosyl–" is the adenosine diphosphoribosyl moiety of NAD.

   This reaction serves to "activate" adenylate cyclase and thus to induce in the epithelial cells increased levels of cAMP (Mekalanos *et al.*, 1979). The reactions that may function in the "activation" of the enzyme are as follows:

Normal Function Regulating Secretion in Enterocytes

$$\frac{\text{GTP-binding protein} \quad \text{GTP} \rightleftharpoons \text{GDP}}{\underset{\text{ATP} \rightleftharpoons \text{cAMP}}{\text{adenylate cyclase}}}$$

GTP-binding protein, GTP, and adenylate cyclase form a so-called "ternary catalytic complex" regulated by the GDP$\rightleftharpoons$GTP reaction which is under hormonal control

Abnormal Function of Regulation of Secretion

ADP-ribosyl–

GTP-binding protein   GTP

――――――――――――――――――――

adenylate cyclase

ATP⇌cAMP

Adenylate cyclase is "activated," when ternary complex behaves as if GTP is in unlimited supply, when GTP-binding protein is ADP-ribosylated.
More cAMP, more secretion, electrolyte loss, watery diarrhea

As with the high activities of cGMP induced by the STs, cAMP at high activity leads to hypersecretion of fluids and electrolytes in the enterocytes. Again, the net effect for the host is watery diarrhea. None of these toxins apparently alters the absorptive function of the enterocytes. The enterotoxins of *E. coli* (as well as the adhesins; see Chapter 11) are encoded by genetic sequences on plasmids (Field, 1979; Mosley *et al.*, 1983). The genetic sequence encoding ST_a is on a transposon (Mosley *et al.*, 1983). That of *V. cholerae* is encoded by chromosomal genes (Saunders *et al.*, 1983; Sporecke *et al.*, 1984).

In spite of the sophistication of our knowledge about the functions of the enterotoxins, surprisingly little is known about how the strains of *V. cholerae* and *E. coli* maintain themselves and increase their populations in the environment of the intestinal epithelium. As noted earlier, *V. cholerae* is motile, and is now known to be attracted by chemotaxis into the mucous gel overlying the intestinal epithelium (Freter *et al.*, 1981). The chemotactic factors have yet to be described chemically, but have been shown to derive from the intestinal mucosa. *V. cholerae* is known as well to synthesize and secrete into its environment an enzyme which hydrolyzes mucinous glycoproteins (Schneider and Parker, 1982; Finkelstein *et al.*, 1983). The products of the reaction catalyzed by the enzyme could serve as carbon, energy, and nitrogen sources for the organism. ETEC strains may have similar capacities for colonizing the intestinal epithelium (Savage, 1980a).

Some bacterial pathogens of taxonomic classes other than *V. cholerae* and *E. coli* also bind to epithelial surfaces, rarely penetrate into or through them, and produce and secrete protein toxins (Table I). In these cases, however, the toxins affect the metabolism of the epithelial cells in such a way that the epithelium is damaged and eroded. With the epithelium destroyed, the bacterial cells may enter the exposed submucosa. In addition, an inflammatory response may ensue (which is not the case with the *V. cholerae* and *E. coli* strains just discussed). Examples in these cases are *Corynebacterium diphtheriae* and *Bordetella pertussis* in the upper respiratory tract.

The protein toxins of these bacterial pathogens, apparently unlike those of *V. cholerae* and the diarrheagenic *E. coli* strains, affect not only the epithelial cells in the substratum colonized by the bacteria, but also other tissues in the body remote from the primary site. Diphtheria toxin is encoded by a structural gene located on the chromosomes of some lysogenic bacteriophages and is a protein of molecular weight of about 80,000 (Tweten and Collier, 1983). It kills eukaryotic cells it can enter by blocking protein synthesis in them. To be active, the protein must be separated into two peptides (A and B) by peptidase "nicking" and sulfhydryl reduction. Peptide B binds to membranes of cells of any type containing GM_1 in their membranes. If attached to peptide B via a sulfhydryl group, peptide A (and B?) then passes into the cell by receptor-mediated endocytosis and other processes (Morris and Saelinger, 1983). Once inside the cell, peptide A functions as an enzyme and catalyzes the reaction (Collier, 1975; Moynihan and Pappenheimer, 1981):

$$EF2 + NAD \rightleftharpoons ADPR-EF2 + nicotinamide$$

where EF2 is a protein, elongation factor 2, which is involved in translocating a growing polypeptide chain from the donor to the receptor site on the ribosome, NAD is nicotinamide adenine dinucleotide, and ADPR–EF2 is elongation factor 2 with an adenosine diphosphoribosyl moiety from NAD bound covalently to it; in this form the EF2 is inactive in protein synthesis. The equilibrium of the reaction is strongly in the direction of formation of ADPR–EF2 ($F° = 5.2$ Kcal per mole at pH 7.0 and 25°C).

With their capacity to synthesize proteins blocked, cells of many tissues collapse, leading to hemorrhage (e.g., in the adrenal cortex), and inflammation, leading to deposition in tissues of abnormal amounts of certain macromolecules (e.g., "fatty degeneration" of the heart muscle).

B. pertussis produces at least two extracellular proteins that can enter (by as yet undefined mechanisms) and alter the function of animal cells. One of the proteins is an enzyme that activates cellular adenylate cyclase, increasing intracellular concentrations of cAMP, and thus deranging any cellular functions in which cAMP acts in a regulatory role. This enzyme ("pertussis toxin") catalyzes in animal cells the adenosine diphosphoribosylation (i.e., is an adenosyl diphosphoribosyl transferase) of a protein that regulates adenylate cyclase (Burns *et al.*, 1983), and in that respect has a function identical to that of cholera and *E. coli* LT toxins. However, the GTP-binding protein affected by pertussis toxin may differ from that which is ADP-ribosylated by the two enterotoxins (Codina *et al.*, 1983).

The second protein produced by *B. pertussis* is itself an adenylate cyclase, i.e., has such catalytic activity. It can enter and act catalytically inside animal cells, increasing their intracellular cAMP concentrations. As with pertussis toxin, the increased cAMP levels alter and derange cellular function (Confer and Eaton, 1982).

Because these pertussis proteins travel throughout the body, they have a potential for deranging the functions of many tissues. The adenylate cyclase is known to inhibit functions of human phagocytes (Confer and Eaton, 1982). The adenosyl diphosphoribosyl transferase also depresses host resistance mechanisms. Cells involved in immunological responses are particularly affected by this toxin. It is responsible, for example, for the lymphocytosis and heightened sensitivity to the effects of certain inflammatory mediators (e.g., histamine) of animals infected by *B. pertussis* (Munoz *et al.*, 1981).

Some strains of *E. coli* function as epicellular pathogen of the intestine, apparently without producing enterotoxins. In this case, however, the strains alter the ultrastructure and undoubtedly the absorptive as well as the secretive functions of the microvillous membranes of enterocytes. The best described of such a strain, known as RDEC-1, adheres to the microvillous membranes of the small bowel enterocytes in rabbits (Cheney *et al.*, 1983). In adhering to the membranes, the bacteria destroy the microvillous architecture, but do not penetrate through the membrane into the epithelial cells. The bacteria may secrete a labile substance, or have in their outer membrane a macromolecule that alters the architecture of the epithelial cell membranes. As will be discussed later, certain members of the indigenous intestinal microflora of animals of some species, including man, may affect enterocyte membranes in apparently much the same way as do these particular *E. coli* strains (Neutra, 1980; Savage, 1983). In these cases, however, little evidence supports a hypothesis that the indigenous bacteria cause the animal host any obvious harm.

2.2.2. Intraepithelial Pathogens

Some frank bacterial pathogens adhere to epithelial cell membranes only as a first step in entering the cells (Table I). The best studied examples of such pathogens are the *Shigella* species that invade into, multiply inside of, and destroy colonic epithelial cells in humans and certain subhuman primates (Sansonetti *et al.*, 1982), and *Neisseria gonorrhoeae*, which has the capacity to invade into the epithelial cells of fallopian tubes and other sites in the human genitourinary tract (Johnson, 1981). The *Shigella* species [and also *N. gonorrhoeae* (Johnson, 1981)] probably enter into epithelial cells via induced endocytosis (Michl, 1980). Intestinal enterocytes (and the epithelial cells of the fallopian tube) are regarded as "nonprofessional" phagocytes, i.e., they are capable of endocytosis, but must be induced to engulf particles by some exogenous mechanism, and are not equipped well with lysosomal systems for killing ingested bacteria (Michl, 1980). The mechanisms by which the bacterial pathogens induce the endocytic mechanism in the epithelial cells are unknown, but must involve some intimate interaction between the bacterial outer membranes and the epithelial cell membranes, perhaps during the process of adhesion (Jones *et al.*, 1982). In some *Shigella* species, the capacity of the bacterial cells to invade into animal cells (and adhere to them?) is encoded by genes on large plasmids (Sansonetti *et al.*, 1982). Some strains of *N. gonorrhoeae* are known to contain plasmids (Johnson *et al.*, 1983). Whether genes on plasmids encode the capacity of those bacteria to induce endocytosis in epithelial cells is unknown.

2.2.3. Subepithelial Pathogens

Some bacterial pathogens adhere to epithelial cells as a step in penetrating through or around them into subepithelial tissues (Table I). Once having penetrated into such tissues, the pathogens may be carried by blood or lymph to organs remote from the initial site of invasion, and sometimes throughout the body. Some such bacterial cells may be ingested by, and multiply intracellularly in, macrophages in various tissues, particularly the liver and spleen. The best studied of such "facultative intracellular bacterial pathogens" known to adhere to epithelial cells is the etiological agent of typhoid fever (*Salmonella typhi*). Relatives of *S. typhi* (e.g., *S. typhimurium*, *S. enteritidis*) have been demonstrated to adhere to the membranes of enterocytes in the small intestines of infected animals (Tannock *et al.*, 1975), and human cells of certain types maintained in culture (Jones *et al.*, 1982; Uhlman and Jones, 1982). Little is known about the mechanisms by which the organisms adhere to the cells, or how they invade into subepithelial tissues. The properties are probably encoded, however, by genes on transmissible plasmids (Jones *et al.*, 1982).

Some years ago, the mechanism by which *S. typhimurium* invades into intestinal epithelial cells was studied by transmission electron microscopy (Takeuchi, 1967). In that study, *S. typhimurium* cells induced degeneration of the microvilli of intestinal epithelial cells when about 3500 nm from the membrane of those structures in experimentally infected guinea pigs. Subsequently, the bacterial cells induced the epithelial cells to engulf them by a process that appears to be endocytic and to disgorge them into tissues below the epithelium by a process that appears to be exocytic. The bacteria also passed between epithelial cells, penetrating through tight junctions (so-called desmosomes) between cells and the basement membrane into the subepithelial tissues.

This study can be criticized as the animals involved (guinea pigs) had been treated with opium to retard peristaltic and villous motility in their intestines, and were given by gavage a quite large dosage of bacterial cells. Such treatment of the animals could have permitted the bacteria to interact with the intestinal epithelium in ways not characteristic of natural infections. Indeed, *S. typhimurium* probably invades into subepithelial tissues in the small intestines more through the endocytic "M" cells in the epithelium covering Peyer's patches (Owen and Jones, 1974), than through the absorptive epithelium of the intestine per se (Carter and Collins, 1974). Nevertheless, the study of the invasion of guinea pig intestine by *S. typhimurium* (Takeuchi, 1967) was seminal, by showing that processes resembling endocytosis–exocytosis were involved in that invasion.

2.3. Effects of Bacterial Pathogens on Host Resistance Functions

2.3.1. Induction of Immunologic Responses

Bacterial pathogens associated with epithelial surfaces may induce in their animal hosts immunological responses that may eventually function to dislodge the pathogen from the surface and force it from the body. In mammals of several species, immunoglobulins in secretions on epithelial surfaces are principally of the IgA class (McNabb and Tomasi, 1981). In humans, this class of antibodies includes two complex proteins of somewhat differing structure, IgA1 and IgA2. They are produced principally by lymphocytes located in the submucosa underlying epithelial surfaces and glands, and are organized into the so-called "secretory" molecular configuration by epithelial cells during their transport to the surface. Secretory IgA antibodies induced by surface antigens of epicellular bacterial pathogens such as *V. cholerae* undoubtedly can function to clear the bowel of the bacteria (Freter, 1971).

Just as importantly, IgA antibodies and antibodies of other classes that are induced by toxic proteins of epicellular pathogens can combine with and neutralize the function of those proteins. Secretory immunoglobulins in mother's milk are known to function in preventing diarrheal diseases caused by epicellular bacteria (principally *E. coli* strains) in neonatal mammals. Undoubtedly, therefore, secreted antibodies are an important specific host defense against such bacterial pathogens (McNabb and Tomasi, 1981). Evidence is not available, however, to support a hypothesis that such antibodies function importantly in the resistance of animals and man to invasive bacterial pathogens.

2.3.2. Repression of Resistance Mechanisms

Certain bacterial pathogens associated with epithelial surfaces can repress the immunological responses of their hosts. For example, as noted (Section 2.2.1), the protein toxins of *B. pertussis* repress the functions of phagocytes, and probably also lymphocytes and other immunological cells, depressing host defenses in the process. Likewise, diptheria toxin undoubtedly represses the functions of such cells by interfering with their protein synthesis.

Certain bacterial species that associate with epithelial surfaces in humans and some other animals produce hydrolytic enzymes that catalyze the hydrolysis of immunoglobulins on the surfaces. Such enzymes, known as IgA-proteases, catalyze the hydrolysis of human IgA1, at a specific amino sequence of the immunoglobulin molecule (Kornfeld and Plaut, 1981; Plaut, 1983). Bacterial strains that produce such enzymes presumably have an advantage over strains not producing them in colonizing epithelial surfaces. However, little direct evidence supports such a hypothesis (Plaut, 1983).

2.4. Diseases of Unknown Etiology That May Involve Bacteria Associated with Epithelial Surfaces

Some diseases of unknown etiology, such as cancer of the large intestine in man, may be induced by carcinogens or cocarcinogens produced by bacteria of the "normal" intestinal microflora from compounds synthesized or consumed by the host (Hill, 1983). Some of the microorganisms involved ("subterranean pathogens") could be associated with the epithelial surfaces of the intestine (Savage, 1983). Little information supports such a hypothesis at this time (see below). Indeed, evidence that indigenous bacteria are involved at all in the etiology of malignancies has been controversial (Hill, 1983). The diseases at issue are as yet of uncertain specific etiology. The conditions usually manifest themselves late in life, presumably after many years of contact between the host tissues and the bacterial products. Moreover, the incriminated bacterial strains need never leave their intestinal habitat, i.e., be "invasive," and their incriminated chemical products may be molecular concomitants of biochemical processes regarded as ordinary. As with diseases caused by opportunistic pathogens, the malignancies at issue may form only when host resistance mechanisms are deranged. Thus, while "subterranean" bacterial pathogens may have a role in the etiology of certain cancers, that role is difficult to define at this time (Hill, 1983).

One other interesting possibility for an activity of a "subterranean pathogen" is the discovery that *Streptococcus faecium* represses the growth of chicks (Houghton *et al.*, 1981). This bacterium adheres to and colonizes the epithelium of the upper small intestine in those animals, and when present in that region from a time shortly after hatching, substantially depresses their growth. The mechanism of the effect is unknown at this time.

2.5. Summary: How Bacteria That Associate with Epithelial Surfaces Induce Disease

Depending upon their genetic equipment, bacterial strains that associate with epithelial surfaces may be frank, opportunistic, or subterranean pathogens. Some frankly pathogenic strains induce disease while adherent to membranes of epithelial cells. Such epicellular pathogens produce and secrete protein toxins that pass into or through the epithelium and induce the symptoms of disease. Some strains are modestly invasive, intracellular pathogens that can enter (probably by inducing endocytosis), grow inside of, and destroy epithelial cells. Some strains can invade into and through epithelial cells into submucosal spaces and may spread throughout the body. As do epicellular pathogens, some strains of intracellular and subepithelial pathogens may produce and secrete protein toxins that can enter and alter the function of host cells. Some such pathogens may also release into their environment toxic lipopolysaccharides or lipoteichoic acids present in their cell membranes and walls (Schockman and Wicken, 1981). Thus, symptoms of diseases caused by invasive pathogens result not only from damage to host tissues caused by proliferating bacterial cells, but also from effects on host tissues of a variety of toxic macromolecules. The genes encoding mediators of the invasive processes as well as the protein toxins are often on mobile genetic elements, i.e., transposons and transmissible plasmids. The genetics and biochemistry of many of the protein toxins, and in a few cases the mechanisms by which the pathogens adhere to epithelial cells (Chapters 10 and 11), have received much investigative attention in recent years. Much less attention has been given, however, to the environmental and nutritional conditions influencing growth of the pathogens on epithelial surfaces, and the molecular mechanisms by which the invasive ones penetrate into and through host cells.

3. EFFECTS ON HOSTS NOT NORMALLY LEADING TO DISEASE

3.1. Introduction: Indigenous Bacteria That Associate with Epithelial Surfaces

As noted earlier, bacterial members of the normal microflora are now well known to associate with epithelial surfaces in all areas of the body exposed to the outside world, including the gastrointestinal canal. A description of the taxonomic groups of the bacteria and epithelial surfaces involved is beyond the scope of this chapter (reviewed by Hardie and Bowden, 1974; Lee, 1980; Savage, 1980a, 1983). However, a few general points will be made about the biochemistry of such organisms as a background for discussion of their impact on host tissue.

Most bacterial inhabitants of surfaces of mammals and birds are capable of growth without oxygen. Inhabitants of surfaces exposed to air (skin, nasopharynx, etc.) are often able to use oxygen as a terminal electron acceptor in their energy-generating metabolism (Rosebury, 1962; Noble and Pitcher, 1978). Such bacteria are often also capable of anaerobic growth, however, commonly by fermentation processes. Thus, indigenous bacteria living in even so-called aerobic environments on epithelial surfaces of animals often find it advantageous to be able to derive energy for growth from anaerobic processes. Presumably, therefore, epithelial habitats in mammals and birds, even those on surfaces ostensibly exposed to air, must be relatively free of oxygen.

This phenomenon is most striking for epithelial habitats in the alimentary and urogenitcal tracts. In those areas, the predominating microbial populations are composed of strictly anaerobic bacteria (Savage, 1977b; Larsen et al., 1977). Any strains of strictly aerobic microorganisms found are undoubtedly transients entering the areas with material from outside of the body (Alexander, 1971).

Because most indigenous bacterial species colonizing body surfaces are obligately anaerobic in their energy-generating metabolism, the end products of their metabolism are predominantly organic compounds in a more-or-less reduced state of oxidation–reduction. Anaerobic bacteria of most species indigenous to epithelial habitats ferment saccharides as carbon and energy sources. Some of the organisms, however, can gain energy from oxidation of organic compounds other than sugars, e.g., amino and fatty acids (Prins, 1977). Predominating among the end products of the metabolism are lactic acid, ethanol, and short-chain volatile fatty acids, such as acetic, propionic, butyric, caproic, and caprylic acids (Prins, 1977; Wolin, 1981). Therefore, such compounds must be present in respectable concentrations on epithelial surfaces, especially in the gastrointestinal tract.

Bacterial strains associated with epithelial surfaces in animals may also produce enzymes that catalyze reactions not known to be involved in energy-generating metabolism. Again, however, such reactions are limited to those that can proceed in an environment without O_2 and at low oxidation–reduction potentials. For example, some bacterial strains make enzymes that catalyze deconjugation and transformation of bile (Eyssen et al., 1983) and hormonal steroids (Winter and Bokkenheuser, 1979). Some of the strains may even secrete enzymes that catalyze hydrolysis of polymers of β-glucuronides (Hill, 1983). The products of all such reactions can be found on epithelial surfaces.

Indigenous bacteria adherent to epithelial surfaces may produce as well enzymes that hydrolyze macromolecular polymers produced and secreted by the host. For example, bacteria associated with the epithelium of the intestines may produce glycosidases that hydrolyze mucus (Salyers et al., 1977; Roberton and Stanley, 1982). Bacteria adherent to skin may secrete enzymes that catalyze hydrolysis of fats secreted by sebaceous glands (Noble and Pitcher, 1978). The end products of such reactions may serve the bacteria as carbon, energy, and nitrogen sources, and may also accumulate in the area, altering the local environment on the surface. For example, long-chain fatty acids accumulating as a result of bacterial hydrolysis of lipids on skin (Noble and Pritchard, 1978) and in the vagina (Larsen *et al.*, 1977) are known to lower local pH. Likewise, lactic acid and some other short-chain organic acids produced as end products of fermentative metabolism by lactic acid bacteria lower the pH on the nonsecreting epithelium in the stomachs of rodents (Kunstyr *et al.*, 1976) and in the crops of chickens (Fuller, 1977). Any of these microbial end products could be absorbed into the bloodstream. Thus, the nutritional and biochemical activities of indigenous bacteria have much potential for altering host function, either of the local epithelium or of organs far removed in the body from the bacteria involved. In addition, the bacterial cells themselves (Berg and Owens, 1979) or macromolecules from their cell walls (Schockman and Wicken, 1981) may enter host tissues and alter their function.

3.2. Effects on Host of Adherent Indigenous Bacteria

3.2.1. Nutritional Effects

In almost no case are the influences on their host of indigenous bacteria associated with epithelial surfaces understood in mechanistic detail. By contrast with bacterial pathogens (see Section 2), the specific mechanisms by which indigenous bacteria influence properties of their hosts are only now beginning to receive investigative attention. Thus, much in the following discussion has been largely inferred from what is known about the biochemical capacities of indigenous bacterial species and about how germfree animals and animals with a microflora (so-called "conventional" animals) differ physiologically and anatomically (Sasaki *et al.*, 1981).

Since their development as laboratory artifacts at the turn of the century (Luckey, 1963), germfree animals have been studied extensively. (For recent approaches, see, e.g., Welling *et al.*, 1980; Umesaki *et al.*, 1980; Koopman *et al.*, 1982; Savage and Whitt, 1982; Freter *et al.*, 1983). In some of the most interesting of the studies, physiological and anatomical properties of germfree animals have been compared with those of animals with a microflora ("conventional" animals), often with surprising results. Based upon the experience of investigators worldwide, a general statement about such comparisons can be fairly posited—germfree animals differ from comparable conventional ones in most physiological and even some anatomical parameters. Some such differences are dramatic, such as the enlarged cecum of germfree rodents discovered 80 years ago (Luckey, 1963), and the recent finding that conventional mice contain queuosine-tRNA, whereas germfree mice fed a diet free of queuine do not (Farkas, 1980).

A detailed exposition of all of the physiological properties of germfree animals that differ from those of conventional animals is beyond the scope and focus of this chapter (see, e.g., Luckey, 1963; Sasaki *et al.*, 1981). Moreover, in most cases, too little is known about the precise effects on host animals of specific microorganisms associated with epithelia. In some instances, however, enough is known about the bacterial activities to permit some informed speculation about how such microorganisms may affect their hosts.

For example, bacteria associated with epithelial surfaces undoubtedly contribute at least to some extent to host energy, carbon, and nitrogen nutrition. As noted above, most bacterial species associated with epithelial surfaces produce as products of their energy-yielding metabolism carbon compounds such as lactic acid and short-chain volatile fatty acids. These compounds can be absorbed by epithelia in certain regions of the gastrointestinal tract in animals of some species, e.g., the rumen epithelium in ungulate animals (Wolin, 1981) and the colonic epithelium in monogastric mammals such as rats (Umesaki *et al.*, 1980). In ungulates at least (Wolin, 1981), and undoubtedly other mammals and some birds, the animal tissues can utilize the compounds as carbon and energy sources. Such phenomena have been understood well for many years for ungulate animals, e.g., cows, sheep, etc. (Wolin, 1981), but have only recently been recognized for monogastric mammals, e.g., humans, swine, rodents, etc. (Byrne and Dankert, 1979; Umesaki *et al.*, 1980; Wrong *et al.*, 1981). Such compounds may also influence electrical activity in the colon (Yajima *et al.*, 1981).

For some time, microorganisms in the rumens of ungulates have been known to function importantly in the nitrogen metabolism of the animal hosts (Bryant, 1974). Such organisms hydrolyze urea (Okumura *et al.*, 1976) and other nitrogenous compounds and incorporate the released ammonia nitrogen into their own nitrogenous compounds. Because the animals digest the bacterial cells in their intestines, they regain some of the nitrogen excreted earlier as urea. Recently, bacteria adherent to the rumen epithelium in bovines have been found to be active in hydrolyzing urea (Cheng *et al.*, 1981; Chapter 1, Section 3.2). A hypothesis has been advanced that urea is excreted into the rumen through the rumen wall, and thus adherent bacteria are strategically situated to be important in converting urea nitrogen to ammonia, which can then be utilized by ruminant microorganisms as a source of nitrogen. That hypothesis requires further experimental confirmation but stands as an interesting possibility, even for bacteria adherent to epithelia at other sites in the host. Bacteria adherent to epithelia may well be important in conserving nitrogen that would otherwise be excreted by the animal. In addition to nitrogen, microorganisms in ruminant animals may also conserve other tissue molecules by digesting epitheial cells, the components of which are then recycled when the animal digests the bacteria (Dinsdale *et al.*, 1980).

3.2.2. Physiological Effects

Bacteria associated with epithelial surfaces are also situated strategically for impacting on the physiology of the substratum epithelium and underlying host tissues (Table II). As discussed, bacteria associated with a surface contribute numerous compounds to the environment immediately overlying the substratum. Moreover, many such organisms make enzymes that hydrolyze macromolecules in secretions overlying the host cell membranes. The products from such biochemical activity, and even components of the cell walls and membranes of bacteria [e.g., lipopolysaccharides from outer membranes of gram-negative bacteria and lipoteichoic acids from the cytoplasmic membranes and walls of gram-positive bacteria (Bradley, 1979; Schockman and Wicken, 1981)] may absorb through or intercalate into the membranes of the epithelial cells and alter the function of those membranes.

The membranes of the microvilli of epithelial cells of the small intestines contain enzymes that catalyze digestive transport of disaccharides, peptides, and compounds containing phosphate, such as purines and pyrimidines. The activities of such enzymes are three- to fivefold higher in the intestinal mucosa from germfree mice than in that from conventional mice (Table II). Thus, indigenous microorganisms somehow depress the activities of the enzymes.

In both germfree and conventional mice, the enzymatic activities are highest in the duodenal mucosa, i.e., the part of the small intestine just distal to the stomach, and decrease progressively down the intestine to lowest levels in the ileum, the part of the small intestine just proximal to the large intestine and the part containing the largest populations of indigenous microorganisms. (Savage and Whitt, 1982). In rodents, no microorganisms have yet been found to be associated intimately with the duodenal epithelium. In mice and rats from

Table II. Activities of Digestive–Absorptive Enzymes in Extracts of the Small Intestinal Mucosa of Germfree Mice and Mice with a Microflora ("Associated")[a]

Enzyme	Activity[b]			
	Total[c]		Specific[d]	
	Germfree	Associated[e]	Germfree	Associated[e]
Alkaline phosphatase	46.2 ±19.2	15.5 ± 5.1	3.7 ± 1.1	1.8 ± 0.6
Maltase	1.18 ± 0.23	0.50 ± 0.17	0.097 ± 0.016	0.059 ± 0.022
Sucrase	0.238 ± 0.056	0.088 ± 0.042	0.020 ± 0.005	0.011 ± 0.005

[a]See Whitt and Savage (1981).
[b]Each number represents the arithemetic mean ± S.D. of values obtained from 15 mice (three experiments with five mice each).
[c]Micromoles of product formed per minute per 3-cm segment of upper small intestine.
[d]Micromoles of product formed per minute per milligram of protein.
[e]Associated: ex-germfree mice associated for at least 3 weeks with an indigenous microflora.

conventional colonies located in various regions of the world, however, filamentous bacteria have been found adherent in high population levels to the small bowel epithelium beginning about one-third the way down from the stomach and extending (in ever-increasing numbers) to the junction of the small and large bowels (Davis and Savage, 1974; Blumershine and Savage, 1978; Savage, 1980a; Garland et al., 1982). Moreover, proximal to the duodenum, i.e., in the stomach, the epithelial surfaces are colonized by lactic acid bacteria and yeasts in rodents of all colonies of conventional animals studied (Savage, 1980a). The products of such organisms, as well as their cells, pour into the small bowel. Thus, microorganisms associated with epithelial surfaces in rodents are located strategically for maximal direct impact on the microvillous membranes of the small intestine. Attempts to demonstrate that the microorganisms impact directly on the membranes and to study the mechanisms of that impact have proved, however, to be surprisingly difficult.

Intestinal epithelial cells can be isolated from murine small intestines by techniques that remove the cells progressively from the tips of the villi to the crypts of Lieberkuhn (Savage and Whitt, 1982). When enterocytes are obtained in such fashion from germfree mice, only the cells obtained from near the tips of the villi have higher activities of digestive–absorptive enzymes than cells removed by the same procedure from the intestines of comparable mice given a microflora ("ex-germfree") (Savage and Whitt, 1982). In other words, except for cells nearest the villous tips, each enterocyte from a germfree mouse expresses the same activities of enzymes in their microvillous membranes as conventional animals. That finding argues against a hypothesis that microorganisms associated with the epithelial surfaces in the stomach and small intestine impact directly in some way upon the enterocyte membranes. Nevertheless, such microorganisms may still be responsible for the low enzymatic activities in the small bowel mucosae of conventional mice.

Enterocytes in the small intestine divide through mitosis only in the crypts of Lieberkuhn (Savage et al., 1981). The cells then glide in a monolayer on or with a basement membrane from the crypts along the surface of the villi to a zone at the tip where they are extruded (discarded) into the lumen of the intestine. As the cells glide along the villus to the extrusion zone, they no longer undergo mitosis, but do differentiate in several ways. In one such process, the activities of microvillous enzymes increase, probably through processing and insertion into the membrane of proteins produced in the cells while they are still in the crypts of Lieberkuhn. Thus, as the cells glide along the villus, the activities of their digestive–absorptive enzymes increase progressively.

Table III. Time of Transit from the Crypts of Lieberkuhn to the Villous
Tips of Epithelial Cells in the Upper Small Intestines of Germfree Mice
and Mice with Microflora ("Associated")[a]

| Mice | Expt | Transit time | |
		Time (hr)	95% confidence interval[b]
Germfree	1	119.0	105.8–134.5
	2	113.0	101.7–125.6
	3	113.2	98.5–130.6
Associated[c]	1	51.6	31.3–106.8
	2	43.0	23.3–124.8
	3	63.4	53.0– 77.0

[a]See Savage et al. (1981).
[b]In hours; as estimated by segmental regression analysis of kinetic data.
[c]Ex-germfree mice associated for at least 3 weeks with an indigenous microflora.

The time required for small bowel enterocytes to transit from the crypts of Lieberkuhn to the tips of the villi is influenced by the indigenous microbiota. In the duodena of conventional mice, the cells make the transit in about 2 days, whereas in germfree mice the trip requires over 4 days (Table III). Thus, enterocytes remain on the villous surface almost twice as long in the small bowels of germfree mice as in those of conventional animals, and have almost twice as long to develop activities of microvillous enzymes. As a consequence, each enterocyte near the tips of the villi in the intestines of germfree mice should have and, as noted above, do have higher activities of the enzymes than each such cell in conventional mice.

Nevertheless, the greater activity in the cells on the tips of the villi in germfree mice is insufficient to account for the much greater total activity of the enzymes in the mucosae of germfree animals than in those of conventional animals. Indeed, most of the enzymatic activity in excess of the conventional levels in the germfree mice can be explained in terms of the number of enterocytes. Germfree mice have almost twice as many enterocytes in the epithelium of their small bowels as do animals with a microflora (Savage and Whitt, 1982). Thus, turnover of the epithelium requires 4 days in germfree mice, rather than 2 days as in conventional animals, because twice as many cells are making the transit in the former than in the latter animals. The microflora in the conventional animals exerts its main influence on the activities of microvillous enzymes, therefore, by somehow limiting the total number of enterocytes to levels considerably below those in germfree animals. The mechanisms of that process remain obscure, but involve undoubtedly the microorganisms adherent to the wall of the stomach and intestine.

Such findings have important implications for studies of the etiology of the disease called "small bowel overgrowth" syndrome (Donaldson, 1967; Gorbach, 1971; Simon and Gorbach, 1984) and of the consequences of the activities of microbial populations growing in the small intestine due to obstructions, "blind loops," and other similar structures produced by surgical procedures (Nagy and Weipers, 1968; Sykes et al., 1976). Small bowel overgrowth syndrome is a condition in which microorganisms of genera and species similar to those that normally colonize the large bowel colonize the small intestine. Individuals with the condition usually have some underlying disease that has neutralized their mechanisms for preventing the flora from colonizing that area. As example is scleroderma (Mallory et al., 1973), in which several organs of the body alter in function, including smooth muscle

(Julkunen, 1971), interfering with peristalsis and villous contraction; some major immunological deficiencies may also occur (Brown *et al.*, 1972) interfering with immunological mechanisms for clearing microorganisms from the intestine. Microorganisms of species and genera characteristic of the flora in the large intestine are often isolated as well from pouches and loops created in the small intestines by obstructions and surgical manipulation (Nagy and Weipers, 1968; Gorbach, 1971).

When such circumstances exist, the individuals involved may have a variety of symptoms, including malabsorption of fats and vitamin B_{12} and reduced absorption of saccharides, peptides, and other essential compounds (Donaldson, 1967; Gorbach, 1971; Simon and Gorbach, 1984). Microorganisms reduce fat absorption by enzymatically removing taurine or glycine from conjugated bile salts, thus reducing the concentration of the conjugated salts below the critical concentration required for micelles to form (Donaldson, 1967; Gracey *et al.*, 1971). Many species of intestinal bacteria produce such deconjugating enzymes (Shimada *et al.*, 1969). Likewise, many intestinal bacteria compete with the host epithelial cells for vitamin B_{12}, and undoubtedly produce enzymes that alter or destroy intrinsic factor, which is required for vitamin B_{12} to be absorbed by intestinal cells (Giannella *et al.*, 1972).

Sugars and amino acids are malabsorbed in individuals with microbial populations in their small intestines probably because changes occur in the mechanisms for absorbing such compounds (Gracey *et al.*, 1971). In such individuals, the mucosa of the intestine may undergo changes in architecture (Gracey *et al.*, 1971; Wehman *et al.*, 1978). Villi may grow blunt and even disappear altogether. In the latter case, the mucosa of the small intestine may resemble that of the large intestine. In that process, the number of epithelial cells may be reduced significantly from levels found in normal intestines. In addition, the apical membranes of the cells may alter so that microvilli are few in number and malformed in shape. Enzymes involved in digestive absorption of sugars and amino acids are localized in the microvillous membranes (Gracey *et al.*, 1971; Savage and Whitt, 1982). Thus, any process that reduces the area of or damages those membranes undoubtedly alters their absorptive functions. The functions and anatomy of the microvillous membranes may change because of some activities of the microflora that impact directly on the membranes. Loss of epithelial cells, however, may be due to activities of the flora that impact on the mechanisms controlling multiplication and loss (extrusion into the epithelial lumen) of the cells. The activities of the microorganisms involved may be the same as, or similar to, those by which the flora alters the numbers of such cells in the small intestines of conventional rodents (see above).

Such a hypothesis is difficult to test directly at this time. Moreover, little can be said about possible mechanisms that may be involved. Nevertheless, a safe speculation is that microorganisms associated with the epithelial surface will be most active in any direct influences on the architecture and function of the microvillous membranes. Moreover, such organisms and their products may have virtually direct access to the crypts of Lieberkuhn where epithelial cells multiply and to the tips of the villi where the cells extrude into the lumen and are lost. Thus, unless the mechanisms by which the flora alters intestinal cell activities are mediated through the bloodstream, then microorganisms associated with epithelial surfaces are undoubtedly important in such processes.

Bacteria have been found associated with epitheial surfaces of the small bowels of persons ostensibly suffering from "overgrowth" syndrome (Bhat *et al.*, 1972; Peach *et al.*, 1978). Ironically, however, microorganisms, particularly bacteria, were reported to associate with epithelial surfaces in the small bowels of ostensibly normal individuals examined in the same studies (Peach *et al.*, 1978; Bhat *et al.*, 1980; Savage, 1983). The "normal" persons were said to be healthy for their populations. Presumably, therefore, they absorb nutrients as well as anyone else in those populations. Because the normal controls had microorganisms adherent to their small bowel epithelium, just as did the persons with symptoms of "overgrowth," no obvious relationship could be found between the symptoms and the microflora. These important discoveries justify further research on the mechanisms by which intestinal microorganisms, and especially those that associate with epithelia, induce changes in the physiological functions of intestinal epithelial cells.

3.2.3. Effects on Host Resistance Functions

Indigenous bacteria associated with epithelial surfaces are undoubtedly important as well in exerting certain influences by which the animal host resists some microbial diseases. Such organisms have long been known to inhibit nonindigenous bacteria of related and unrelated species from colonizing niches in epithelial habitats. That phenomenon, popularly known as "bacterial interference," has been studied intensively by numerous investigators in recent years (Barrow et al., 1980; Tannock, 1981; Freter, 1983). Evidence now supports a hypothesis that the interference phenomenon has many possible mechanisms, and may be exerted by the indigenous microflora either directly or indirectly (Table IV).

Direct influences that have received particular study are bacteriocins, nutritional competition, metabolic end products operating at low oxidation–reduction potentials, and competition for adhesion ("binding") sites (see Chapters 10 and 11). Bacteriocins are proteins, encoded by genes on plasmids, that are produced and secreted by certain bacteria and have the capacity to bind to the surface of and kill susceptible bacteria of the same or a related genus or species (Konisky, 1982). Such a protein produced by an indigenous bacterium could function in a natural habitat to kill any bacterial strains that find their way into the region and are related to the indigenous producer, but are not immune to the effects of its bacteriocin. Evidence for such a hypothesis is not strong for bacteria associated with epithelial surfaces in higher animals (Konisky, 1978). Arguments can be made, however, that the hypothesis has yet to receive adequate testing. Indigenous bacterial strains known to associate with epithelial surfaces have recently been found to produce in the test tube bacteriocins active against related organisms not indigenous to the same habitats (McCormick and Savage, 1983). Thus, research in the future should be aimed at adequate testing of the actions of bacteriocins in a natural epithelial habitat.

Indigenous bacteria that associate with particular epithelial surfaces are undoubtedly well adapted to the nutritional and environmental conditions prevailing on the surface. As a consequence, once established in a particular epithelial habitat to which it is indigenous, a microbial strain may be well able to compete successfully with any invaders for nutrients available. Such competition functioning in environmental conditions prevailing in a habitat is now believed to be, along with the capacity to associate with a surface, a fundamental property stabilizing indigenous microbial communities and protecting them from invading microorganisms (Guiot, 1982; Freter et al., 1983).

As noted earlier (Section 3.1), the environmental conditions prevailing in an epithelial surface also can be influenced considerably by an indigenous community, principally by the end products of the metabolism of the microbial members of the community. Such end products tend to be acidic and to contribute to low pH values. At low oxidation–reduction potentials, the compounds may be active in inhibiting growth of invading bacteria of some species (Tannock, 1981). Hydrogen sulfide produced by indigenous bacteria may be particularly potent as a killer of such unwelcome nonindigenous trespassers (Freter et al., 1983). Thus, with the capacity to compete nutritionally with invaders and to poison them with toxic metabolites, the indigenous (primary) colonizers of an epithelial surface must have a considerable competitive edge over invader strains in maintaining themselves in the habitat.

Table IV. Direct and Indirect Mechanisms by Which Bacteria Indigenous to Habitats on Epithelial Surfaces May Inhibit Nonindigenous Bacteria ("Invaders") from Colonizing a Niche in the Habitats[a]

Mechanism	Effect on invaders	References[a]
	DIRECT	
Nutritional competition	Inhibits growth	Guiot (1982), Freter et al. (1983)
Bacteriocins	Kills	McCormick and Savage (1983)
Antibiotics other than bacteriocins	Inhibits growth or kills	Goldman (1978)
Volatile organic acids	Inhibits growth, especially at low oxidation–reduction potential; lowers local pH	Byrne and Dankert (1979)
Lactic acid	Same as above	
H_2S	Inhibits growth	Freter et al. (1983)
Motility	None; motile bacterium competes better for access to space and nutrients	Stanton and Savage (1984)
Adhesion–aggregation	Inhibits access to space; adhesion receptors	Savage (1980a)
	INDIRECT	
Stimulate mucosal motility	Prevents localization	Savage (1977b)
Lower local pH	Inhibits growth; kills	Kunstyr et al. (1976)
Lower local oxidation–reduction potential	Prevents growth of aerobes	Koopman et al. (1975)
Enhance epithelial turnover	Limits access to space	Savage (1977b)
Enhance immunological responses	Prevents colonization	Savage (1977b)
Synergism with local immunological responses	Prevents colonization	Freter (1971)

[a]Adapted from Savage (1977b), which may be consulted for additional references.

Competition for adhesion (binding) sites also has been postulated to be a mechanism by which indigenous bacteria may restrict access of invaders to epithelial surfaces (Savage, 1972, 1977b, 1980b). The corollary hypothesis is that the native cells bind more avidly to available sites than do invader cells. That is, the natives occupy sites by secondary binding mechanisms (e.g., cell membranes, capsules) and can exclude invaders which must first bind by primary mechanisms (e.g., fimbriae). Primary binding mechanisms may be less powerful in binding than are secondary forces, their function being mainly to overcome net negative charges repelling apart cell surfaces (see Chapter 1; Chapter 6, Section 3.1). Recently, certain staphylococcal strains have been noted to inhibit other staphylococcal strains from colonizing mucous membranes in nasal passages. A hypothesis has been advanced that the primary colonizers prevent the invaders from binding to receptors in or on the membranes (Bibel et al., 1983). Such findings presage other such discoveries in the future.

Indigenous bacteria may also interfere indirectly with invaders by stimulating functions of the host that serve to limit access of the latter bacteria to habitats (Table IV). Host functions such as intestinal peristalsis, mucous secretion, and immunological defenses are well known to be stimulated by the indigenous microflora. At this time, however, although they undoubtedly do so, little specific evidence supports a suggestion that bacteria associated with epithelia are involved in stimulating such functions.

Evidence does support at this time a hypothesis that bacteria associated with epithelial surfaces can function synergistically with host immunological mechanisms to inhibit access of bacterial invaders to host epithelia. An indigenous microbiota, containing elements that associate with epithelial surfaces, along with IgA in intestinal secretions prevent V. *cholerae* from colonizing the small bowel epithelium in experimental mice (Freter, 1971). Similarly, indigenous bacteria that associate with epithelial surfaces prevent S. *typhimurium* from reducing cecal size in ex-germfree mice vaccinated with killed S. *typhimurium* cells (Tannock and Savage, 1976). Thus, in addition to contributing to enhanced general functioning of the host's immunological mechanisms, indigenous bacteria on epithelial surfaces may also function synergistically (possibly by the "interference" mechanisms discussed above) to enhance immunological responses.

3.3. Summary: How Indigenous Bacteria Associated with Epithelial Surfaces Influence Their Animal Hosts

Indigenous bacterial populations associated with epithelial surfaces can influence their host for better or for worse. For better, the populations feed the host, help conserve energy, and help resist disease. For worse, some members of the populations can act as opportunistic and subterreanean pathogens. The mechanisms by which the bacterial cells effect these functions are most complex and at best poorly understood in detail. Details are emerging, however, that point to some general patterns.

The products that bacteria produce on epithelial surfaces are largely limited to those made in processes that function in environments of little oxygen and low oxidation–reduction potential. Some such products produced in significant concentrations are reduced carbon compounds produced by the energy-generating metabolism of the organisms. Those compounds are often absorbed by the host and serve, in some proportion, as part of the animal's own carbon and energy nutrition. The compounds also function to help the host resist disease by limiting access of microbial pathogens to epithelial surfaces, sometimes in synergy with host immunological mechanisms.

Indigenous bacteria also produce compounds of many molecular classes other than end products of their energy-generating functions. Such products result from enzymatic alteration of exogenous (from outside the host) and endogenous (produced by the host) compounds. The products undoubtedly influence many host properties, such as digestion and absorption, by mechanisms yet to be defined.

4. RESEARCH IN THE FUTURE

4.1. Present State of Research

As has been amplified in the foregoing, with the exception of some bacterial pathogens, the influences on animal hosts of bacteria that associate with epithelial surfaces have not received as much experimental attention as have the mechanisms by which some of the microorganisms adhere to epithelial surfaces (see Chapters 10 and 11). Indeed, even in the cases of the pathogens, the focus of most recent research (that not on the mechanisms of adhesion) has been on the genetics, structure, and molecular activities of protein toxins of a relatively few species. In spite of the immense knowledge about how germfree animals differ physiologically from conventional ones, little specific research is under way on the mechanisms by which indigenous bacteria influence their host's physiology.

Work with pathogens (i.e., on the toxins produced by any given bacterial strain and the mechanisms by which it adheres to a surface) has therapeutic and preventative goals (Levine *et al.*, 1983). Investigators in the areas strive to learn at the molecular level how the toxins affect host function or how the bacteria producing the toxins adhere to epithelial surfaces, because they believe that knowledge generated from such research may lead to drugs for treating the diseases or vaccines for immunizing hosts against the diseases. Such is not the case, however, for investigators interested in how indigenous microorganisms associated with epithelial surfaces influence their host animal. In such research, the goals are more nebulous than are those of students of bacterial diseases. Except when they are behaving as opportunistic pathogens, indigenous microorganisms influence the host animals in subtle, undramatic ways in comparison with pathogenic bacteria. Those influences can be life-long, however, and in terms of impact on longevity and quality of life, may be more influential than the effects of frank pathogens.

Every conventional living animal of most studied species has an indigenous microbiota. Thus, the influences of indigenous bacteria on their animal hosts deserve more experimental attention than they are receiving at this time. Research in this area is well past the "discovery phase"; the ways in which germfree animals differ from conventional ones are well described. [Some surprises may remain; the recent discovery that germfree animals fed diets free of queuine do not contain Q-tRNA (Farkas, 1980) suggests that at the molecular level further differences between germfree and conventional animals may yet be detected.] Intensive effort is needed in this area to provide information at the molecular level on how indigenous bacteria exert their influences on the host.

4.2. Focus for the Future

Recently, research with penetrating pathogens has focused on how such microorganisms induce endocytosis in epithelial cells that are not "professional" phagocytes (Michl, 1980). These pathogens must secrete, or more likely have in their outer membranes, substances that trigger in epithelial cells organization of the cytoplasmic microfilaments and other machinery that function when the cell ingests foreign particles. Tests of such a hypothesis may be facilitated by recent discoveries that invasive properties of several bacterial pathogens are encoded by genes on plasmids (Harris *et al.*, 1982; Jones *et al.*, 1982; Sansonett *et al.*, 1982; Hale *et al.*, 1983).

Bacterial pathogens of one species (certain strains of *E. coli*), which adhere to the rabbit small bowel, some indigenous filamentous bacteria that colonize the epithelium of the murine small bowel, and some spirochetes and helical bacteria that adhere to the epithelium of the colons of monkeys and humans adhere to the membranes of enterocytes by inducing an attachment site in the membranes and underlying cytoplasm (Neutra, 1980). In such sites, the membranes of the enterocytes appear to have undergone changes that mimic sol-to-gel conversions (Snellen and Savage, 1978), while the underlying cytoplasm appears as if it has prepared for endocytosis (Neutra, 1980). In other words, the bacteria appear to have the capacity to induce the epithelial cells to form endocytic mechanisms, but then to arrest those mechanisms before the microbial cells can be drawn fully into the host cell interior. Such adhesion sites may be induced by mechanisms analogous to those by which some bacterial pathogens induce endocytosis in epithelial cells, except that the process is arrested before ingestion takes place. The molecular mechanisms of these phenomena deserve intense, and perhaps comparative, study, especially for their influence on the function of the membranes of the affected epithelial cells.

In the small or large intestines of animals, the membranes altered under such circumstances are those that mediate absorption and secretion. Presumably, any bacterium that alters the structure of the membranes would have some influence on the functions of the epithelium. Mice and rats containing filamentous bacteria in their small bowels and monkeys and humans with helical bacteria and spirochetes adherent to the epithelium of their large bowels appear to be healthy and show no signs of any impaired function, even though in the latter case the absorptive membrane is almost totally altered in architecture. Could it be that the bacteria in these cases function to facilitate in some way the absorptive function of the enterocytes? Such a speculation is difficult to test at this time. In neither case mentioned have the bacteria involved been cultured from their natural habitats into culture media. Presumably, they are highly adapted to their life of intimate contact with the epithelial cells.

These and other indigenous bacteria that associate with epithelial surfaces obviously impact on their animal hosts in profound and subtle ways. Among other things, they provide food for hosts of some species, alter many physiological functions of hosts, and protect hosts against invasion by bacterial pathogens. As discussed, however, in the main these phenomena await study at the molecular level, especially those involved in altering host function and protecting hosts from disease. The nutrition, end products of the energy-yielding metabolism and products other than those made as a result of energy-yielding metabolism should be defined for indigenous bacteria in epithelial habitats. However, some bacterial products that are produced and affect host function in epithelial habitats may not be produced in artificial nutrient media. Care must be taken, therefore, so that important microbial compounds are not overlooked (Goldman, 1978). Moreover, attention should be given to the ways bacteria on epithelial surfaces may function in conserving urea nitrogen for an animal. The effort involved in any such work is great. However, the work must be considered to be worth the time and expense; microorganisms associated with epithelial surfaces are extremely important to the health and welfare of man and other animals.

5. SUMMARY

Frank bacterial pathogens and indigenous bacteria, both nonpathogenic and potentially pathogenic, associate with body surfaces in mammals and other animals. In animals of some species, certain bacteria adhere to membranes of substratum epithelial cells. All bacterial strains involved have the potential to exert influences on their animal hosts. Frank pathogens induce disease in animals by producing protein toxins that enter and alter the function of host cells, or by invading into and destroying host cells or tissues. Indigenous bacteria exert influences on animals throughout their lives by producing compounds that can be used by the host as nutrients, or that alter host physiology and resistance functions. Indigenous bacteria may damage host tissues while functioning as opportunistic or "subterranean" pathogens. The mechanisms of the effects on animals of some frank bacterial pathogens are under active study, and are becoming understood well at the molecular level. The mechanisms of the effects of indigenous bacteria are not being studied as actively as are those of frank pathogens, and are not well understood at the molecular level.

ACKNOWLEDGMENT. The author's research and preparation of the manuscript were supported by Grants AI-11858 and AI-19518 from the National Institute of Allergy and Infectious Diseases.

REFERENCES

Alexander, M., 1971, *Microbial Ecology,* Wiley, New York.
Barrow, P. A., Brooker, B. E., Fuller, R., and Newport, M. J., 1980, The attachment of bacteria to the gastric epithelium of the pig and its importance in the microecology of the intestine, *J. Appl. Bacteriol.* **48**:147–154.

Berg, R. D., and Owens, W. E., 1979, Inhibition of translocation of viable *Escherichia coli* from the gastrointestinal tract of mice by bacterial antagonism, *Infect. Immun.* **25:**820–827.

Bhat, P., Shantakumari, S., Rajan, D., Mathan, V. I., Kapadia, C. R., Swarnabai, C., and Baker, S. J., 1972, Bacterial flora of the gastrointestinal tract in southern Indian control subjects and patients with tropical sprue, *Gastroenterology* **62:**11–21.

Bhat, P., Albert, M. J., Rajan, D., Ponniah, J., Mathan, V. I., and Baker, S. J., 1980, Bacterial flora of the jejunum: A comparison of luminal aspirate and mucosal biopsy, *J. Med. Microbiol.* **13:**247–256.

Bibel, D. J., Aly, R., Bayles, C., Strauss, W. G., Shinefield, H. R., and Maibach, H. I., 1983, Competitive adhesion as a mechanism of bacterial interference, *Can. J. Microbiol.* **29:**700–703.

Blumershine, R. V., and Savage, D. C., 1978, Filamentous microbes indigenous to the murine small bowel: A scanning electron microscopic study of their morphology and attachment to the epithelium, *Microb. Ecol.* **4:**95–103.

Bradley, S. G., 1979, Cellular and molecular mechanisms of action of bacterial endotoxins, *Annu. Rev. Microbiol.* **33:**67–94.

Brown, W. R., Savage, D. C., Dubois, R. S., Alp, M. H., Mallory, A., and Kern, F., Jr., 1972, Intestinal microflora of immunoglobulin-deficient and normal subjects, *Gastroenterology* **62:**1143–1152.

Bryant, M. P., 1974, Nutritional features and ecology of predominant anaerobic bacteria of the intestinal tract, *Am. J. Clin. Nutr.* **27:**1313–1319.

Burns, D. L., Hewlett, E. L., Moss, J., and Vaughn, M., 1983, Pertussis toxin inhibits enkephalin stimulation of GTPase of NG108-15 cells, *J. Biol. Chem.* **258:**1435–1438.

Byrne, B. M., and Dankert, J., 1979, Volatile fatty acids and aerobic flora in the gastrointestinal tract of mice under various conditions, *Infect. Immun.* **23:**559–563.

Carter, P. B., and Collins, F. M., 1974, The route of enteric infection in mice, *J. Exp. Med.* **139:**1189–1203.

Cheney, C. P., Formal, J. B., Schad, P. A., and Boedeker, E. C., 1983, Genetic transfer of a mucosal adherence factor (R1) from an enteropathogenic *Escherichia coli* strain into a *Shigella flexneri* strain and phenotypic suppression of this factor, *J. Infect. Dis.* **147:**711–723.

Cheng, K.-J., Irvin, R. T., and Costerton, J. W., 1981, Autochthonous and pathogenic colonization of animal tissues by bacteria, *Can. J. Microbiol.* **24:**461–490.

Clancy, J., and Savage, D. C., 1981, Another colicin V phenotype: In vitro adhesion of *Escherichia coli* to mouse intestinal epithelium, *Infect. Immun.* **32:**343–352.

Codina, J., Hildebrandt, J., Iyengar, R., Binnbaumer, L., Sekura, R. D., and Mandark, C. R., 1983, Pertussis toxin substrate, the putative Ni component of adenylyl cyclases, is an αβ heterodimer regulated by guanine nucleotide and magnesium, *Proc. Natl. Acad. Sci. USA* **80:**4276–4280.

Collier, J. R., 1975, Diphtheria toxin: Mode of action and structure, *Bacteriol. Rev.* **39:**54–85.

Confer, D. L., and Eaton, J. W., 1982, Phagocyte impotence caused by an invasive bacterial adenylate cyclase, *Science* **217:**948–950.

Davis, C. P., and Savage, D. C., 1974, Habitat, succession, attachment and morphology of segmented, filamentous microbes indigenous to the murine gastrointestinal tract, *Infect. Immun.* **10:**948–956.

Dinsdale, D., Cheng, K.-J., Wallace, R. J., and Goodlad, R. A., 1980, Digestion of epithelial tissue of the rumen wall by adherent bacteria in infused and conventionally fed sheep, *Appl. Environ. Microbiol.* **39:**1059–1066.

Donaldson, R. M., Jr., 1967, Role of enteric microorganisms in malabsorption, *Fed. Proc.* **26:**1426–1431.

Dreyfus, L. A., Franz, J. C., and Robertson, D. C., 1983, Chemical properties of heat-stable enterotoxins produced by enterotoxigenic *Escherichia coli* of different host origins, *Infect. Immun.* **42:**539–548.

Dreyfus, L. A., Jaso-Friedman, L., and Robertson, D. C., 1984, Characterization of the mechanism of action of *Escherichia coli* heat-stable enterotoxin, *Infect. Immun.* **44:**493–501.

Duncan, J. L., 1975, Streptococcal toxins, in: *Microbiology 1975,* (D. Schlessinger, ed.), American Society for Microbiology, Washington, D.C., pp. 257–262.

Eidels, L., Proia, R. L., and Hart, D. A., 1983, Membrane receptors for bacterial toxins, *Microbiol. Rev.* **47:**596–620.

Eyssen, H., DePauw, G., Stragier, J., and Verhulst, A., 1983, Cooperative formation of σ-muricholic acid by intestinal microorganisms, *Appl. Environ. Microbiol.* **45:**141–147.

Farkas, W. R., 1980, Effect of diet on the queuosine family of tRNAs of germ-free mice, *J. Biol. Chem.* **255:**6832–6835.

Field, M., 1979, Modes of action of enterotoxins from *Vibrio cholerae* and *Escherichia coli, Rev. Infect. Dis.* **1:**918–925.

Finkelstein, R. A., Boesman-Finkelstein, M., and Holt, P., 1983, *Vibrio cholerae* hemagglutinin/lectin/protease hydrolyzes fibronectin and ovomucin: F. M. Burnet revisited, *Proc. Natl. Acad. Sci. USA* **80:**1091–1095.

Frantz, J. C., Jaso-Friedman, L., and Robertson, D. C., 1984, Binding of *Escherichia coli* heat-stable enterotoxin to rat intestinal cells and brush-border membranes, *Infect. Immun.* **43:**622–630.

Freter, R., 1971, Host defense mechanisms in the intestinal tract, *Recent Adv. Microbiol.* **7:**333–339.

Freter, R., 1983, Mechanisms that control the microflora in the large intestine, in: *Human Intestinal Microflora in Health and Disease* (D. J. Hentges, ed.), Academic Press, New York, pp. 33–55.

Freter, R., and Jones, G. W., 1976, Adhesive properties of *Vibrio cholerae:* Nature of the interaction with intact mucosal surfaces, *Infect. Immun.* **14:**246–256.

Freter, R., Allweiss, B., O'Brien, P. C. M., Halstead, S. A., and Macsai, M. S., 1981, Role of chemotaxis in the association of motile bacteria with intestinal mucosa: In vitro studies, *Infect. Immun.* **34:**241–249.

Freter, R., Brickner, H., Fekete, J., Vickerman, M. M., and Carey, K. E., 1983, Survival and implantation of *Escherichia coli* in the intestinal tract, *Infect. Immun.* **39:**686–703.

Fuller, F., 1977, The importance of lactobacilli in maintaining normal microbial balance in the crop, *Br. Poult. Sci.* **18:**85–94.

Garland, C. D., Lee, A., and Dickson, M. R., 1982, Segmented filamentous bacteria in the rodent small intestine: Their colonization of growing animals and possible role in host resistance to *Salmonella, Microb. Ecol.* **8:**181–190.

Gemski, P., Lazere, J. R., and Casey, T., 1980, Plasmid associated with pathogenicity and calcium dependency of *Yersinia enterocolitica, Infect. Immun.* **27:**682–685.

Giannella, R. A., Broitman, S. A., and Zamchek, N., 1972, Competition between bacteria and intrinsic factor for vitamin B_{12}: Implications for vitamin B_{12} malabsorption in intestinal bacterial overgrowth, *Gastroenterology* **62:**255–260.

Gill, D. M., and Richardson, S. H., 1980, Adenosine diphosphate-ribosylation of adenylate cyclase catalyzed by heat-labile enterotoxin of *Escherichia coli:* Comparison with cholera toxin, *J. Infect. Dis.* **141:**64–70.

Gill, D. M., Clements, J. D., Robertson, D. C., and Finkelstein, R. A., 1981, Subunit number and arrangement in *Escherichia coli* heat-labile enterotoxin, *Infect. Immun.* **33:**677–682.

Goldman, P., 1978, Biochemical pharmacology of the intestinal flora, *Annu. Rev. Pharmacol. Toxicol.* **18:**523–539.

Gorbach, S. L., 1971, Intestinal microflora, *Gastroenterology* **60:**1110–1129.

Gracey, M., Burke, V., Oshin, A., Barker, J., and Glasgow, E. F., 1971, Bacteria, bile salts, and intestinal monosaccharide malabsorption, *Gut* **12:**683–692.

Guerrant, R. L., Hughes, J. M., Chang, B., Robertson, D. C., and Murad, F., 1980, Activation of intestinal guanylate cyclase by heat-stable enterotoxin of *Escherichia coli:* Studies of tissue specificity, potential receptors, and intermediates, *J. Infect. Dis.* **142:**220–228.

Guiot, H. F. L., 1982, Role of competition for substrate in bacterial antagonism in the gut, *Infect. Immun.* **38:**887–892.

Hale, T. L., Sansonetti, P. J., Schad, P. A., Austin, S., and Formal, S. B., 1983, Characterization of virulence plasmids and plasmid-associated outer membrane proteins in *Shigella flexneri, Shigella sonnei,* and *Escherichia coli, Infect. Immun.* **40:**340–350.

Hardie, J. M., and Bowden, G. H., 1974, The normal microbial flora of the mouth, in: *The Normal Microbial Flora of Man* (F. A. Skinner and J. G. Carr, eds.), Academic Press, New York, pp. 47–84.

Harris, J. R., Wachsmuth, I. K., Davis, B. R., and Cohen, M. L., 1982, High-molecular-weight plasmid correlates with *Escherichia coli* enteroinvasiveness, *Infect. Immun.* **37:**1295–1298.

Hill, M. J., 1983, Bile, bacteria and bowel cancer, *Gut* **24:**871–875.

Holmgren, J., Fredman, P., Lindblad, M., Svennerholm, A.-M., and Svennerholm, L., 1982, Rabbit intestinal glycoprotein receptor for *Escherichia coli* heat-labile enterotoxin lacking affinity for cholera toxin, *Infect. Immun.* **38:**424–433.

Houghton, S. B., Fuller, R., and Coates, M. E., 1981, Correlation of growth depression of chicks with the presence of *Streptococcus faecium* in the gut, *J. Appl. Bacteriol.* **51:**113–120.

Hu, P. C., Cole, R. M., Huang, Y. S., Graham, J. A., Gardner, D. E., Collier, A. M., and Clyde, W. A., Jr., 1982, *Mycoplasma pneumoniae* infection: Role of a surface protein in the attachment organelle, *Science* **216:**313–315.

Johnson, A. P., 1981, The pathogenesis of gonorrhoea, *J. Infect.* **3:**299–308.

Johnson, S. R., Anderson, B. F., Biddle, J. W., Perkins, G. H., and DeWitt, W. E., 1983, Characterization of concatemeric plasmids of *Neisseria gonorrhoeae, Infect. Immun.* **40:**843–846.

Jones, G., 1977, The attachment of bacteria to the surfaces of animal cells, in: *Receptors and Recognition*, Series B, Volume 3 (J. L. Reissig, ed.), Chapman & Hall, London, pp. 141–176.

Jones, G. W., Rabert, D. K., Svinarich, D. M., and Whitfield, H. J., 1982, Association of adhesive, invasive, and virulent phenotypes of *Salmonella typhimurium* with autonomous 60-megadalton plasmids, *Infect. Immun.* **38:**476–486.

Julkunen, H., 1971, Scanning electron microscope study in scleroderma, *Ann. Med. Exp. Biol. Fenn.* **49:**180–185.

Konisky, J., 1978, The bacteriocins, in: *The Bacteria,* Volume VI (L. N. Ornston and J. R. Sokatch, eds.), Academic Press, New York, pp. 71–136.

Konisky, J., 1982, Colicins and other bacteriocins with established modes of action, *Annu. Rev. Microbiol.* **36:**125–144.

Koopman, J. P., Janssen, F. G. J., and van Druten, J. A. M., 1975, Oxidation–reduction potentials in the cecal contents of rats and mice, *Proc. Soc. Exp. Biol. Med.* **149:**995–999.

Koopman, J. P., Kennis, H. M., Lankhorst, A., Prins, R. A., Stadhouders, A. M., and deBoer, H., 1982, The influence of microflora and diet on gastro-intestinal parameters, *Z. Versuchstierkd.* **24:**184–192.

Kornfeld, S. J., and Plaut, A. G., 1981, Secretory immunity and bacterial IgA proteases, *Rev. Infect. Dis.* **3:**521–534.

Kunstyr, I., Peters, K., and Gartner, K., 1976, Investigations on the function of the rat fore-stomach, *Lab. Anim. Sci.* **26:**166–170.

Larsen, B., Markovetz, A. Z., and Galask, R. P., 1977, Role of estrogen in controlling the genital microflora of female rats, *Appl. Environ. Microbiol.* **34:**534–540.

Lee, A., 1980, Normal flora of intestinal surfaces, in: *Adsorption of Microorganisms to Surfaces* (G. Bitton and K. C. Marshall, eds.), Wiley, New York, pp. 145–174.

Levine, M. M., Kaper, J. B., Black, R. E., and Clements, M. L., 1983, New knowledge on pathogenesis of bacterial enteric infections as applied to vaccine development, *Microbiol. Rev.* **47:**510–550.

Luckey, T. D., 1963, *Germfree Life and Gnotobiology,* Academic Press, New York.

McCormick, E. L., and Savage, D. C., 1983, Characterization of *Lactobacillus* sp. strain 100-37 from the murine gastrointestinal tract: Ecology, plasmid content and antagonistic activity toward *Clostridium ramosum* H1, *Appl. Environ. Microbiol.* **46:**1103–1112.

McNabb, P. C., and Tomasi, T. B., 1981, Host defense mechanisms at mucosal surfaces, *Annu. Rev. Microbiol.* **35:**477–496.

Mallory, A., Savage, D., Kern, F., Jr., and Smith, J. G., 1973, Patterns of bile acids and microflora in the human small intestine, *Gastroenterology* **64:**34–42.

Mekalonas, J. J., Collier, R. J., and Romig, W. R., 1979, Enzymatic activity of cholera toxin. I. New method of assay and the mechanism of ADP-ribosyl transfer, *J. Biol. Chem.* **254:**5849–5854.

Michl, J., 1980, Receptor-mediated endocytosis, *Am. J. Clin. Nutr.* **33:**2462–2471.

Molina, N. C., and Peterson, J. W., 1980, Cholera toxin-like toxin released by *Salmonella* species in the presence of mitomycin C, *Infect. Immun.* **30:**224–230.

Morris, R. E., and Saelinger, C. B., 1983, Diptheria toxin does not enter resistant cells by receptor-mediated endocytosis, *Infect. Immun.* **42:**812–817.

Mosely, S. L., Hardy, J. H., Huq, M. I., Echeverria, P., and Falkow, S., 1983, Isolation and nucleotide sequence determination of a gene encoding a heat-stable enterotoxin of *Escherichia coli, Infect. Immun.* **39:**1167–1174.

Moynihan, M. R., and Pappenheimer, A. W., Jr., 1981, Kinetics of adenosinediphosphoribosylation of elongation factor 2 in cells exposed to diphtheria toxin, *Infect. Immun.* **32:**575–582.

Munoz, J. J., Arai, H., Bergman, R. K., and Sakowski, P. L., 1981, Biological activities of crystalline pertussigen from *Bordetella pertussis, Infect. Immun.* **33:**820–826.

Nagy, L., and Weipers, W. L., 1968, A study of bacteria and lethal factors in fluids from experimental intestinal obstruction in dogs, *J. Pathol. Bacteriol.* **95:**199–210.

Neutra, M. R., 1980, Prokaryotic–eukaryotic cell junctions: Attachment of spirochetes and flagellated bacteria to primate large intestinal cells, *J. Ultrastruct. Res.* **70:**186–203.

Noble, W. C., and Pitcher, D. G., 1978, Microbial ecology of the human skin, in: *Advances in Microbial Ecology,* Volume 2 (M. Alexander, ed.), Plenum Press, New York, pp. 245–289.

O'Brien, A. D., and LaVeck, G. D., 1983, Purification and characterization of a *Shigella dysenteriae* 1-like toxin produced by *Escherichia coli, Infect. Immun.* **40:**675–683.

O'Brien, A. D., LaVeck, G. D., Griffin, D. E., and Thompson, M. R., 1980, Characterization of *Shigella dysenteriae* 1 (shiga) toxin purified by anti-shiga toxin affinity chromatography, *Infect. Immun.* **30:**170–179.

Okumura, J., Hewitt, D., Salter, D. N., and Coates, M. E., 1976, The role of the gut microflora in the utilization of dietary urea by the chick, *Br. J. Nutr.* **36:**265–272.

Owen, R. L., and Jones, A. L., 1974, Epithelial cell specialization within human Peyer's patches: An ultrastructural study of intestinal lymphoid follicles, *Gastroenterology,* **66:**189–203.

Peach, S., Lock, M. R., Katz, D., Todd, I. P., and Tabaqchali, S., 1978, Mucosal-associated bacterial flora of the intestine in patients with Crohn's disease and in a control group, *Gut* **19:**1034–1042.

Plaut, A. G., 1983, The IgA1 proteases of pathogenic bacteria, *Annu. Rev. Microbiol.* **37:**603–622.

Prins, R. A., 1977, Biochemical activities of gut microorganisms, in: *Microbial Ecology of the Gut* (R. T. J. Clark and T. B. Bauchop, eds.), Academic Press, New York, pp. 73–184.

Roberton, A. M., and Stanley, R. A., 1982, In vitro utilization of mucin by *Bacteroides fragilis, Appl. Environ. Microbiol.* **43:**325–330.

Rogolsky, M., 1979, Nonenteric toxins of *Staphylococcus aureus, Microbiol. Rev.* **43:**320–360.

Rosebury, T., 1962, *Microorganisms Indigenous to Man,* McGraw–Hill, New York.

Salit, I. E., and Morton, G., 1981, Adherence of *Neisseria meningitidis* to human epithelial cells, *Infect. Immun.* **31:**430–435.

Salyers, A. A., West, S. E. H., Vercellotti, J. R., and Wilkins, T. D., 1977, Fermentation of mucins and plant polysaccharides by anaerobic bacteria from the human colon, *Appl. Environ. Microbiol.* **34:**529–533.

Sansonetti, P. J., Kopecko, D. J., and Formal, S. B., 1982, Involvement of a plasmid in the invasive ability of *Shigella flexneri, Infect. Immun.* **35:**852–860.

Sasaki, S., Ozawa, A., and Hashimoto, K. (eds.), 1981, *Recent Advances in Germfree Research,* Tokai University Press, Tokyo.

Saunders, D. W., Kubala, G. J., Vaidya, A. B., and Bramucci, M. G., 1983, Evidence indicating that the cholera toxin structural genes of *Vibrio cholerae* RJ1 and 3083-2 are between *met* and *trp, Infect. Immun.* **42:**427–430.

Savage, D. C., 1972, Survival on mucosal epithelia, epithelial penetration and growth in tissues of pathogenic bacteria, in: *Microbial Pathogenecity in Man and Animals* (H. Smith and J. H. Pearce, eds.), Cambridge University Press, London, pp. 25–57.

Savage, D. C., 1977a, Interactions between the host and its microbes, in: *Microbial Ecology of the Gut* (R. T. J. Clarke and T. Bauchops, eds.), Academic Press, New York, pp. 277–310.

Savage, D. C., 1977b, Microbial ecology of the gastrointestinal tract, *Annu. Rev. Microbiol.* **31:**107–133.

Savage, D. C., 1980a, Adherence of normal flora to mucosal surfaces, in: *Bacterial Adherence* (E. H. Beachey, ed.), Chapman & Hall, London, pp. 33–59.

Savage, D. C., 1980b, Colonization by and survival of pathogenic bacteria on intestinal mucosal surfaces, in: *Adsorption of Microorganisms to Surfaces* (G. Bitton and K. C. Marshall, eds.), Wiley, New York, pp. 175–206.

Savage, D. C., 1983, Associations of indigenous microrganisms in gastrointestinal epithelial surfaces, in: *Human Intestinal Microflora in Health and Disease* (D. J. Hentges, ed.), Academic Press, New York, pp. 55–78.

Savage, D. C., and Whitt, D. D., 1982, Influence of the indigenous microbiota on amounts of protein, DNA, and alkaline phosphatase activity extractable from epithelial cells of the small intestines of mice, *Infect. Immun.* **37:**539–549.

Savage, D. C., Siegel, J. E., Snellen, J. E., and Whitt, D. D., 1981, Transit time of epithelial cells in the small intestines of germfree mice and ex-germfree mice associated with indigenous microorganisms, *Appl. Environ. Microbiol.* **42:**996–1001.

Schneider, D. R., and Parker, C. D., 1982, Purification and characterization of the mucinase of *Vibrio cholerae, J. Infect. Dis.* **145:**474–482.

Schockman, G. D., and Wicken, A. J. (eds.), 1981, *Chemistry and Biological Activities of Bacterial Surface Amphiphiles,* Academic Press, New York.

Shimada, K., Bricknell, K. S., and Finegold, S. M., 1969, Deconjugation of bile acids by intestinal bacteria: Review of the literature and additional studies, *J. Infect. Dis.* **119:**273–281.

Simon, G. L., and Gorbach, S. L., 1984, Intestinal flora, in health and disease, *Gastroenterology* **86:**174–193.

Snellen, J. E., and Savage, D. C., 1978, Freeze-fracture study of the filamentous, segmented microorganism attached to the murine small bowel, *J. Bacteriol.* **134:**1099–1107.

So, M., and McCarthy, B. J., 1980, Nucleotide sequence of the bacterial transposon Tn1681 encoding a heat-stable (ST) toxin and its identification in enterotoxigenic *Escherichia coli* strains, *Proc. Natl. Acad. Sci. USA* **77:**4011–4015.

Sporecke, I., Castro, D., and Mekalanos, J. J., 1984, Genetic mapping of *Vibrio cholerae* enterotoxin structural genes, *J. Bacteriol.* **157:**253–261.

Stanton, T. B., and Savage, D. C., 1984, Motility as a factor in bowel colonization by *Roseburia cecicola,* an obligately anaerobic bacterium from the mouse cecum, *J. Gen. Microbiol.* **130:**173–183.

Svanborg-Eden, C., and Jodal, U., 1979, Attachment of *Escherichia coli* to urinary sediment epithelial cells from urinary tract infection-prone and healthy children, *Infect. Immun.* **26:**837–840.

Sykes, P. A., Boulter, K. H., and Schofield, P. F., 1976, Alterations in small-bowel microflora in acute intestinal obstruction, *J. Med. Microbiol.* **9:**13–22.

Takeuchi, A., 1967, Electron microscope studies of experimental salmonella infection. 1. Penetration into the intestinal epithelium by *Salmonella typhimurium, Am. J. Pathol.* **50:**109–136.

Tannock, G. W., 1981, Microbial interference in the gastrointestinal tract, *Holan. J. Clin. Sci.* **2:**2–24.

Tannock, G. W., and Savage, D. C., 1976, Indigenous microorganisms prevent reduction in cecal size induced by *Salmonella typhimurium* in vaccinated gnotobiotic mice, *Infect. Immun.* **13:**172–179.

Tannock, G. W., Blumershine, R. V. H., and Savage, D. C., 1975. Association of *Salmonella typhimurium* with, and its invasion of, the ileal mucosa in mice, *Infect. Immun.* **11:**365–370.

Tuomanen, E. I., and Hendley, J. O., 1983, Adherence of *Bordetella pertussis* to human respiratory epithelial cells, *J. Infect. Dis.* **148:**125–130.

Tweten, R. K., and Collier, R. J., 1983, Molecular cloning and expression of gene fragments from cornyebacteriophage encoding enzymatically active peptides of diphtheria toxin, *J. Bacteriol.* **156:**680–685.

Uhlman, D. L., and Jones, G. W., 1982, Chemotaxis as a factor in interactions between HeLa cells and *Salmonella typhimurium*, *J. Gen. Microbiol.* **128:**415–418.

Umesaki, Y., Yajima, T., Tohyama, K., and Mutai, M., 1980, Characterization of acetate uptake by the colonic epithelial cells of the rat, *Pfluegers Arch.* **388:**205–209.

Wehman, H. J., Lifshitz, F., and Teichberg, S., 1978, Effects of enteric microbial growth on small intestinal ultrastructure in the rat, *Am. J. Gastroenterol.* **70:**249–258.

Welling, G. W., Groen, G., Tuinte, J. H. M., Koopman, J. P., and Kennis, H. M., 1980, Biochemical effects on germ-free mice of associations with several strains of anaerobic bacteria, *J. Gen. Microbiol.* **117:**57–63.

Whitt, D. D., and Savage, D. C., 1981, Influence of indigenous microbiota on amount of protein and activities of alkaline phosphatases and disaccharidases in extracts of intestinal mucosa in mice, *Appl. Environ. Microbiol.* **42:**513–520.

Winter, J., and Bokkenheuser, V. D., 1979, Bacterial metabolism of corticoids with particular reference to the 21-dehydroxylation, *J. Biol. Chem.* **254:**2626–2629.

Wolin, M. J., 1981, Fermentation in the rumen and human large intestine, *Science* **213:**1463–1468.

Wrong, O. M., Edmonds, C. J., and Chadwick, V. S., 1981, *The Large Intestine: Its Role in Mammalian Nutrition and Homeostasis,* Wiley, New York.

Yajima, T., Kojima, K., Tohyama, K., and Mutai, M., 1981, Effect of short-chain fatty acids on electrical activity of the small intestinal mucosa of rat, *Life Sci.* **28:**983–989.

Index